A Level
Salters Advanced Chemistry
for OCR

Fourth Edition

B

Project director
Chris Otter

Authors
Frank Harriss
Dave Waistnidge
Lesley Johnston
Mark Gale
Dave Newton
Mutlu Cukurova
David Goodfellow
Adelene Cogill

THE SALTERS' INSTITUTE

UNIVERSITY OF YORK
SCIENCE EDUCATION GROUP

OXFORD
UNIVERSITY PRESS

OXFORD
UNIVERSITY PRESS

Great Clarendon Street, Oxford, OX2 6DP, United Kingdom

Oxford University Press is a department of the University of Oxford. It furthers the University's objective of excellence in research, scholarship, and education by publishing worldwide. Oxford is a registered trade mark of Oxford University Press in the UK and in certain other countries

British Library Cataloguing in Publication Data
Data available

978-0-19-833290-9

10 9 8 7

Paper used in the production of this book is a natural, recyclable product made from wood grown in sustainable forests. The manufacturing process conforms to the environmental regulations of the country of origin.

Printed in Great Britain

This resource is endorsed by OCR for use with specification H033 AS Level GCE Chemistry B (Salters) and H433 A Level GCE Chemistry B (Salters). In order to gain endorsement this resource has undergone an independent quality check. OCR has not paid for the production of this resource, nor does OCR receive any royalties from its sale. For more information about the endorsement process please visit the OCR website www.ocr.org.uk.

Index compiled by INDEXING SPECIALISTS (UK) Ltd., Indexing House, 306A Portland Road, Hove, East Sussex BN3 5LP United Kingdom.

First edition
George Burton, Margaret Ferguson, John Holman, Gwen Pilling, David Waddington, Malcolm Churchill, Derek Denby, Frank Harriss, Miranda Stephenson, Brian Ratcliff, Ashley Wheway

Second edition
John Lazonby, Gwen Pilling, David Waddington, Derek Denby, John Dexter, Margaret Ferguson, Frank Harriss, Gerald Keeling, Dave Newton, Brian Ratcliff, Mike Shipton, Terri Vine

Third edition
Chris Otter, Adelene Cogill, Frank Harriss, Dave Newton, Gill Saville, Kay Stephenson, David Waistnidge, Ashley Wheway,

We are grateful for sponsorship from the Salters' Institute, which has contributed to support the Salters Advanced Chemistry project and has enabled the development of these materials.

Thanks to Derek Denby and Emily Perry who wrote a range of activities that are referenced in the book.

Thanks also to Joanna MacDonald who has worked as the administrator for the Salters' Advanced Chemistry project.

Practical skills: NASA/ESA/STSCI/J. Hester & A. Loll, ASU/Science Photo Library; Chapter Header 1: Vladyslav Danilin/Shutterstock; Chapter Header 2: Twenty20 Inc/Shutterstock; Chapter Header 3: LOOK Die Bildagentur der Fotografen GmbH/Alamy; Chapter Header 4: Peter J. Kovacs/Shutterstock; Chapter Header 5: Tischenko Irina/Shutterstock; Header Photo 6: Nostal6ie/Shutterstock; Header Photo 7: Eye of Science/Science Photo Library; Header Photo 8: Dean Pennala/Shutterstock; Header Photo9: Sarin Kunthong/Shutterstock; Header Photo 10: Nengloveyou/Shutterstock;

COVER: Image Source / Alamy **p2-3:** Tischenko Irina/Shutterstock; **p4-5:** Tischenko Irina/Shutterstock; **p6-7:** NASA/ESA/STSCI/J. Hester & A. Loll, ASU/Science Photo Library; **p8:** NASA/ESA/STSCI/J. Hester & A. Loll, ASU/Science Photo Library; **p9:** NASA/ESA/STSCI/J. Hester & A. Loll, ASU/Science Photo Library; **p17:** Science Photo Library; **p19**(TL): Zhykova/Shutterstock; **p19**(TR): Richard A McMillin/Shutterstock; **p19**(BL): Ponsulak/Shutterstock; **p19**(BR): Joshua Resnick/Shutterstock; **p25:** Ria Novosti/Science Photo Library; **p42:** ggw1962/Shutterstock; **p46**(L): Philippe Psaila/Science Photo Library; **p46**(R): Herve Conge, ISM/Science Photo Library; **p47**(T): Jiri Hera/Shutterstock; **p47**(CT): Joshelerry/iStockphoto; **p47**(C): Hong xia/Shutterstock; **p47**(CB): Martyn F. Chillmaid/Science Photo Library; **p47**(B): Kletr/Shutterstock; **p50:** Charles D. Winters/Science Photo Library; **p55**(L): Thinkstock; **p55**(C): Lucky Business/Shutterstock; **p55**(R): Photodisc; **p56:** Science Photo Library; **p59:** Andrew Lambert Photography/Science Photo Library; **p61:** Geoff Tompkinson/Science Photo Library; **p71**(T): Mariusz S. Jurgielewicz/Shutterstock; **p71**(B): E2dan/Shutterstock; **p72-73:** Vladyslav Danilin/Shutterstock; **p74**(T): Car Culture/Getty Images; **p74**(B): Teddy Leung/Shutterstock; **p87:** Georgeclerk/iStockphoto; **p94**(T): Victor De Schwanberg/Science Photo Library; **p94**(B): brainmaster/iStockphoto; **p95:** nitinut380/Shutterstock; **p102:** Ria Novosti/Science Photo Library; **p105**(L): JPC-PROD/Shutterstock; **p105**(R): CatLane/iStockphoto; **p108**(L): Andrew Lambert Photography/Science Photo Library; **p108**(R): Andrew Lambert Photography/Science Photo Library; **p111**(A): Martin Wierink; **p111**(B): Toa55/Shutterstock; **p111**(C): Difught/Shutterstock; **p111**(D): Dorling Kindersley RF; **p111**(BR): Lucie Lang/Shutterstock; **p111**(TR): Sergio Schnitzler/Shutterstock; **p112**(T): Valeri Potapova/Shutterstock; **p112**(B): US National Archives and Records Administration/Science Photo Library; **p113:** Sebastien_B/iStockphoto; **p122:** Leonid Andronov/Shutterstock; **p130:** Hung Chung Chih/Shutterstock; **p132:** Julof90/iStockphoto; **p134:** Dorling Kindersley/UIG/Science Photo Library; **p136:** Martin Bond/Science Photo Library; **p137:** Roger Job/Reporters/Science Photo Library; **p139**(T): Steve Morgan/Alamy; **p139**(B): Agencja Fotograficzna Caro/Alamy; **p144-145:** Twenty20 Inc/Shutterstock; **p146:** Vicspacewalker/Shutterstock; **p148**(T): Martyn F. Chillmaid/Science Photo Library; **p148**(B): Andrew Lambert Photography/Science Photo Library; **p150**(T): Andrew Lambert Photography/Science Photo Library; **p150**(B): Andrew Lambert Photography/Science Photo Library; **p155:** Martyn F. Chillmaid/Science Photo Library; **p160:** Author's Photo; **p163:** M_Stankov/iStockphoto; **p166:** Science Photo Library; **p173:** Robert Brook/Science Photo Library; **p174:** Greenshoots Communications/Alamy; **p177:** AVS Technology AG; **p182:** Grenville Collins Postcard Collection/Mary Evans; **p183:** Science Photo Library; **p192-193:** LOOK Die Bildagentur der Fotografen GmbH/Alamy; **p194:** IM_photo/Shutterstock; **p199**(T): Pictorial Press Ltd/Alamy; **p199**(B): Barcin/iStockphoto; **p200:** Home Bird/Alamy; **p203:** Dorling Kindersley/UIG/Science Photo Library; **p205:** Karim Agabi/Eurelios/Science Photo Library; **p206:** Jim Wileman/Alamy; **p210:** Charles D. Winters/Science Photo Library; **p212:** Charles D. Winters/Science Photo Library; **p222:** Science Photo Library; **p230:** LOOK Die Bildagentur der Fotografen GmbH/Alamy; **p234:** NASA/Science Photo Library; **p238:** PawelG Photo/Shutterstock; **p246-247:** Peter J. Kovacs/Shutterstock; **p248**(L): Peter J. Kovacs/Shutterstock; **p248**(R): Fenlin/Shutterstock; **p249:** Gines Romero/Shutterstock; **p269:** John Mclean/Science Photo Library; **p273:** Dmitry Kalinovsky/Shutterstock; **p275:** Bildagentur Zoonar GmbH/Shutterstock; **p276:** CTR Photos/Shutterstock; **p282-283:** Nostal6ie/Shutterstock; **p284**(T): Olaf Schulz/Shutterstock; **p284**(B): Anna Omelchenko/Shutterstock; **p290:** Meryll/Shutterstock; **p298:** Nigel Cattlin/Science Photo Library; **p299:** Nigel Cattlin/Science Photo Library; **p300:** Charles D. Winters/Science Photo Library; **p302**(L): JoHo/Shutterstock; **p302**(R):

AS/A Level course structure

This book has been written to support students studying for OCR AS Chemistry B (Salters) and OCR A Level Chemistry B (Salters). The content covered is shown in the contents list, which also shows you the page numbers for the main topics within each chapter. There is also an index at the back to help you find what you are looking for. If you are studying for OCR AS Chemistry B (Salters), you will only need to know the content in the blue box.

AS exam

Year 1 content

1 Elements of life
2 Developing fuels
3 Elements from the sea
4 The ozone story
5 What's in a medicine

Year 2 content

6 The chemical industry
7 Polymers and life
8 Oceans
9 Developing metals
10 Colour by design

A level exam

A Level exams will cover content from Year 1 and Year 2 and will be at a higher demand. You will also carry out practical activities throughout your course.

Science Photo Library; **p312**: Lculig/Shutterstock; **p312**: Grandpa/Shutterstock; **p322**(TR): Davemhuntphotography/Shutterstock; **p322**(BL): Steve Allen/Shutterstock; **p324**: Paul Rapson/Science Photo Library; **p328**: Marjolaine Cady/BSIP/Science Photo Library; **p329**: Chris Hill/Shutterstock; **p335**(Bkgd): Nostal6ie/Shutterstock; **p336-337**(Bkgd): Eye of Science/Science Photo Library; **p338**(L): Heritage Image Partnership Ltd/Alamy; **p338**(R): Yuri Samsonov/Shutterstock; **p341**: D. Pimborough/Shutterstock; **p345**: Ria Novosti/Science Photo Library; **p348**: Charles D. Winters/Science Photo Library; **p350**: Custom Medical Stock Photo/Science Photo Library; **p353**(L): Shawn Hempel/Shutterstock; **p353**(R): Aspen Photo/Shutterstock; **p358**: Alexlukin/Shutterstock; **p361**(TL): Alfred Pasieka/Science Photo Library; **p361**(TR): Alfred Pasieka/Science Photo Library; **p361**(BL): Molekuul.be/Shutterstock; **p361**(BR): Evan Oto/Science Photo Library; **p363**: Laguna Design/Science Photo Library; **p364**(T): Dr. P. Marazzi/Science Photo Library; **p364**(B): Marco Mayer/Shutterstock; **p365**: Div. of Computer Research & Technology, National Institute of Health/Science Photo Library; **p370**: Monkey Business Images/Shutterstock; **p374**(T): Sofiaworld/Shutterstock; **p374**(B): Everett Historical/Shutterstock; **p380**: Ria Novosti/Science Photo Library; **p381**(T): Epstock/Shutterstock; **p381**(B): Kondor83/Shutterstock; **p395**(Bkgd): Eye of Science/Science Photo Library; **p396-397**(Bkgd): Dean Pennala/Shutterstock; **p398**(TR): David Wrobel, Visuals Unlimited/Science Photo Library; **p398**(BL): NOAA/Science Photo Library; **p409**: Lebendkulturen.de/Shutterstock; **p410**: Martyn F. Chillmaid/Science Photo Library; **p416**: Sheila Terry/Science Photo Library; **p417**: Tugodi/iStockphoto; **p421**(T): Evlakhov Valeriy/Shutterstock; **p421**(B): Patrick Wang/Shutterstock; **p426**: Galyna Andrushko/Shutterstock; **p437**(Bkgd): Dean Pennala/Shutterstock; **p438-439**(Bkgd): Sarin Kunthong/Shutterstock; **p440**(BR): The Art Archive/Alamy; **p440**(BL): Renata Sedmakova/Shutterstock; **p441**: www.BibleLandPictures.com/Alamy; **p450**(BL): LiliGraphie/Shutterstock; **p450**(BR): Alphonse Tran/Shutterstock; **p451**: Paul Atkinson/Shutterstock; **p452**: Science Photo Library; **p455**: Andrew Lambert Photography/Science Photo Library; **p457**: Martyn F. Chillmaid/Science Photo Library; **p458**: Royal Institution of Great Britain/Science Photo Library; **p459**(TR): Dr. Kari Lounatmaa/Science Photo Library; **p459**(CR): Stockcam/iStockphoto; **p459**(BR): Science Photo Library; **p474**: Lasse Ansaharju/Shutterstock; **p477**: Terence Walsh/Shutterstock; **p481**: Kallista Images/Custom Medical Stock Photo/Science Photo Library; **p489**(Bkgd): Sarin Kunthong/Shutterstock; **p489**: Asaf Eliason/Shutterstock; **p490-491**(Bkgd): Nengloveyou/Shutterstock; **p492**: Pagina/Shutterstock; **p502**(T): Marilyn barbone/Shutterstock; **p502**(B): Salajean/Shutterstock; **p507**: Trevor Kittelty/Shutterstock; **p513**: Jonathan & Angela Scott/AWL Images/Getty Images; **p518**: SSPL/Getty Images; **p523**: Stocksolutions/Shutterstock; **p526**: Stephen VanHorn/Shutterstock; **p532**: Andrew Lambert Photography/Science Photo Library; **p533**: Science Photo Library; **p551**(Bkgd): Nengloveyou/Shutterstock; **p551**: Fotografixx/iStockphoto; **p536**: Olha Rohulya/Shutterstock; **p576**: Fotohunter/Shutterstock;

Artwork by Q2A Media

Although we have made every effort to trace and contact all copyright holders before publication this has not been possible in all cases. If notified, the publisher will rectify any errors or omissions at the earliest opportunity.

Links to third party websites are provided by Oxford in good faith and for information only. Oxford disclaims any responsibility for the materials contained in any third party website referenced in this work.

Contents by chemical storylines

Chapter 4 The ozone story — 192

Chapter 5 What's in a medicine — 246

Chapter 6 The chemical industry — 282

Contents by chemical ideas

How to use this book

This book contains many different features. Each feature is designed to support and develop the skills you will need for your examinations, as well as foster and stimulate your interest in chemistry.

Each Topic has storylines content which contains engaging and contemporary contexts that are relevant to the concepts you will cover. Storylines content is highlighted by a coloured background.

Chemical ideas:

The chemical ideas sections contain the concepts you need to know for your examinations. Each chemical ideas section starts with a chemical ideas reference. This tells you what concepts you will be covering and links to the Contents page by chemical ideas. You can use the Contents page by chemical ideas to see how the different chapters interlink – an important skill for your examinations.

Study Tips

Study tips contain prompts to help you with your understanding and revision.

Extension features

These features contain material that is beyond the specification. They are designed to stretch and provide you with a broader knowledge and understanding and lead the way into the types of thinking and areas you might study in further education. As such, neither the detail nor the depth of questioning will be required for the examinations. But this book is about more than getting through the examinations.

1 Extension features also contain questions that link the off-specification material back to your course.

Synoptic link

These highlight the key areas where topics relate to each other. As you go through your course, knowing how to link different areas of chemistry together becomes increasingly important. Many exam questions, particularly at A Level, will require you to bring together your knowledge from different areas. Synoptic links that link to the Techniques and procedures chapter have a practical symbol

Summary Questions

1 These are short questions at the end of each topic.

2 They test your understanding of the topic and allow you to apply the knowledge and skills you have acquired.

3 The questions are ramped in order of difficulty. Lower-demand questions have a paler background, with the higher-demand questions having a darker background. Try to attempt every question you can, to help you achieve your best in the exams.

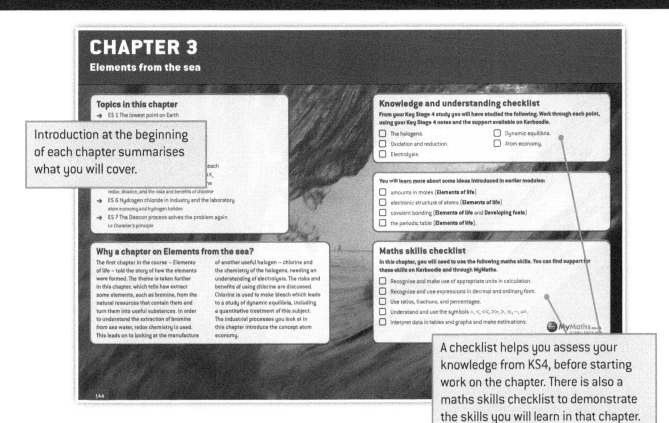

CHAPTER 3
Elements from the sea

Topics in this chapter
→ ES 1 The lowest point on Earth

> Introduction at the beginning of each chapter summarises what you will cover.

...each
K_s...
redox, titration, and the risks and benefits of chlorine
→ ES 6 Hydrogen chloride in industry and the laboratory
atom economy and hydrogen halides
→ ES 7 The Deacon process solves the problem again
Le Chatelier's principle

Why a chapter on Elements from the sea?

The first chapter in the course – Elements of life – told the story of how the elements were formed. The theme is taken further in this chapter, which tells you how we extract some elements, such as bromine, from the natural resources that contain them and turn them into useful substances. In order to understand the extraction of bromine from sea water, redox chemistry is used. This leads on to looking at the manufacture

of another useful halogen – chlorine and the chemistry of the halogens, needing an understanding of electrolysis. The risks and benefits of using chlorine are discussed. Chlorine is used to make bleach which leads to a study of dynamic equilibria, including a quantitative treatment of this subject. The industrial processes you look at in this chapter introduce the concept atom economy.

Knowledge and understanding checklist
From your Key Stage 4 study you will have studied the following. Work through each point, using your Key Stage 4 notes and the support available on Kerboodle.

- [ ] The halogens.
- [ ] Oxidation and reduction.
- [ ] Electrolysis.
- [ ] Dynamic equilibria.
- [ ] Atom economy.

You will learn more about some ideas introduced in earlier modules:

- [ ] amounts in moles (**Elements of life**)
- [ ] electronic structure of atoms (**Elements of life**)
- [ ] covalent bonding (**Elements of life** and **Developing fuels**)
- [ ] the periodic table (**Elements of life**).

> A checklist helps you assess your knowledge from KS4, before starting work on the chapter. There is also a maths skills checklist to demonstrate the skills you will learn in that chapter.

Maths skills checklist
In this chapter, you will need to use the following maths skills. You can find support for these skills on Kerboodle and through MyMaths.

- [ ] Recognise and make use of appropriate units in calculation.
- [ ] Recognise and use expressions in decimal and ordinary form.
- [ ] Use ratios, fractions, and percentages.
- [ ] Understand and use the symbols =, <, <<, >>, >, ∝, ~, ⇌.
- [ ] Interpret data in tables and graphs and make estimations.

MyMaths.co.uk

144

> Visual summaries emphasise how the key concepts of that chapter interlink, and how they interlink with content covered in other chapters.

> Application task brings together some of the key concepts of the chapter in a new context.

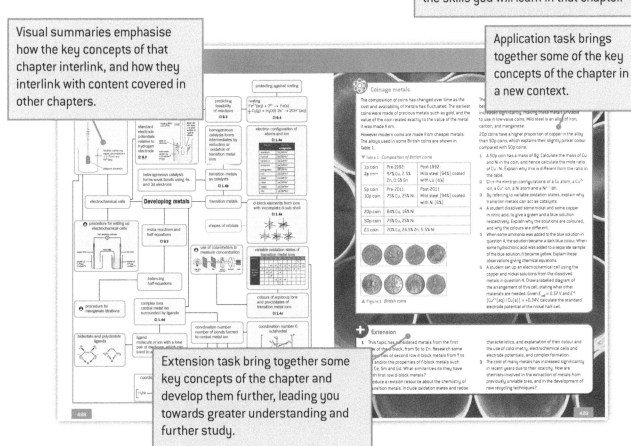

> Extension task bring together some key concepts of the chapter and develop them further, leading you towards greater understanding and further study.

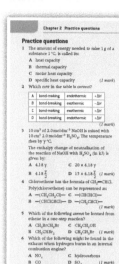

> Practice questions at the end of each chapter including questions that cover practical and maths skills.

> Dedicated Techniques and procedures chapter, detailing all of the practical skills you need to know for your assessment. This chapter is referenced to throughout the book by practical synoptic links.

> Scientific literacy in chemistry chapter to help prepare you for the literacy elements of your assessment.

Kerboodle

This book is supported by next generation Kerboodle, offering unrivalled digital support for independent study, context, differentiation, assessment, and the new practical endorsement.

If your school subscribes to Kerboodle, you will also find a wealth of additional resources to help you with your studies and with revision.

- Study guides
- Activity sheets for hands on application of your knowledge and practicals to try
- Maths skills boosters and calculation worksheets
- Animations and revision podcasts

Apply your knowledge by trying the activities signposted throughout the book.

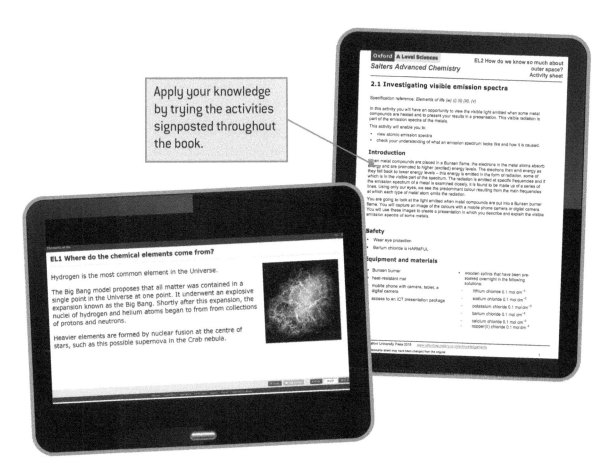

For teachers, Kerboodle also has plenty of further support, including answers to the questions in the book and a digital markbook. There are also full teacher notes for the activities and worksheets, which include suggestions on how to support students and engage them in their own learning. All of the resources are pulled together into teacher guides that suggest a route through each chapter.

Development of practical skills in chemistry

Chemistry is a practical subject and experimental work provides you with important practical skills, as well as enhancing your understanding of chemical theory. You will be developing practical skills by carrying out experiments including investigative work in the laboratory throughout both the AS and the A level Chemistry course. You will be assessed on your practical skills is two different ways:

- written examinations (AS and A level)
- practical endorsement (A level only)

Practical coverage throughout this book

Practical skills are a fundamental part of a complete education in science, and you should keep a record of your practical work from the start of your A level course that you can later use as part of your practical endorsement. You can find more details of the practical endorsement from your teacher or from the specification.

In this book and its supporting materials practical skills are covered in a number of ways. By studying the Techniques and procedures chapter and the Exam-style questions in this student book, and by using the Practical activities and Skills sheets in Kerboodle, you will have many opportunities to learn about the scientific method and carry-out practical activities.

1.1 Practical skills assessed in written examinations

In the written examination papers for AS and A level, at least 15% of the marks will be from questions that assess practical skills. The questions will cover four important skill areas, all based on the practical skills that you will develop by carrying out experimental work during your course.

- Planning – your ability design experiments including solving problems in a practical context.
- Implementing – your ability to use important practical techniques and processes and to present observations and data.
- Analysing – your ability to process, anaylse, and interpret experimental results.
- Evaluating – your ability to evaluate experimental results and to draw conclusions from them including suggesting improvements to procedures and apparatus.

1.1.1 Planning

- Designing experiments
- Identifying variables to be controlled
- Evaluating the experimental method

Skills checklist

- [ ] Selecting apparatus and equipment
- [ ] Selecting appropriate techniques
- [ ] Selecting appropriate quantities of chemicals and scale of working
- [ ] Solving chemical problems in a practical context
- [ ] Applying chemistry concepts to practical problems

1.1.2 Implementing

- Using a range of practical apparatus
- Carrying out a range of techniques
- Using appropriate units for measurements
- Recording data and observations in an appropriate format

Skills checklist

- ☐ Understanding practical techniques and processes
- ☐ Identifying hazards and safe procedures
- ☐ Using appropriate units
- ☐ Recording qualitative observations accurately
- ☐ Recording a range of quantitative measurements
- ☐ Using the appropriate apparatus

1.1.3 Analysis

- Processing, analysing, and interpreting results
- Analysing data using appropriate mathematical skills
- Using significant figures appropriately
- Plotting and interpreting graphs

Skills checklist

- ☐ Interpreting qualitative observations
- ☐ Analysing quantitative experimental data
- ☐ For graphs,
    - selecting and labelling axes with appropriate scales, quantities, and units
    - drawing tangents and measuring gradients and intercepts

1.1.4 Evaluation

- Evaluating results to draw conclusions
- Identify anomalies
- Explain limitations in experimental procedures
- Identifying uncertainties and calculating percentage errors
- Suggesting improvements to procedures and apparatus

Skills checklist

- ☐ Reaching conclusions from qualitative observations
- ☐ Identifying uncertainties and calculating percentage errors
- ☐ Identifying procedural limitations and measurement errors
- ☐ Refining procedures and apparatus to suggest improvements

1.2 Practical skills assessed in practical endorsement

You will also be assessed on how well you carry out a wide range of practical work and how to record the results of this work. These hands-on skills are divided into 12 categories and form the practical endorsement. This is assessed for A level Chemistry qualification only.

The endorsement requires a range of practical skills from both years of your course. If you are taking only AS Chemistry, you will not be assessed through the practical endorsement but the written AS examinations will include questions that relate to the skills that naturally form part of the AS common content to the A level course.

1.2.1 Practical skills

By carrying out experimental work through the course, you will develop your ability to:

- design and use practical techniques to investigate and solve problems
- use a wide range of experimental and practical apparatus, equipment, and materials, including chemicals and solutions
- carry out practical procedures skillfully and safely, recording and presenting results in a scientific way
- research using online and offline tools.

Along with the experimental work, these skills are covered in practical skills questions throughout the book.

1.2.2 Use of apparatus and techniques

To meet the requirements for the practical endorsement, you will be assessed in at least 12 practical experiments to enable you to experience a wide range of apparatus and techniques. These practical experiments are incorporated throughout the book in practical application boxes.

This will help to give you the necessary skills to be a competent and effective practical chemist.

Practical Activity Group (PAG) overview

The PAG labels identify activities that could count towards the practical endorsement. The table below shows where these PAG references are covered throughout this course. You will find further details on each PAG in the Techniques and procedures chapter.

Specification reference	Chapter reference
PAG1 *Moles determination EL(b)(i)*	EL, DF
PAG2 *Acid–base titration EL(c)(i)*	EL, OZ
PAG3 *Enthalpy determination DF(f), O(b)*	DF, O
PAG4 *Qualitative analysis of ions EL(s), ES(k), CI(j)*	EL, ES, CI
PAG5 *Synthesis of an organic liquid WM(f)*	WM
PAG6 *Synthesis of an organic solid WM(e)*	WM
PAG7 *Qualitative analysis of organic functional groups DF(c), WM(c), WM(d), PL(h), PL(j), PL(n), CD(f)*	DF, WM, PL, CD
PAG8 *Electrochemical cells DM(d)*	DM
PAG9 *Rates of reaction – continuous monitoring method CI(c), DM(n)*	CI, DM
PAG10 *Rates of reaction – initial rates method CI(c)*	CI
PAG11 *pH measurement O(k)*	O
PAG12 *Research skills*	EL, DF, OZ, CI, DM, O, CD

Maths skills and How Science Works across Module 1

Maths skills are very useful for scientists and as you study your course you will learn maths techniques and equations that support the development of your science knowledge. Each module opener in this book has an overview of the maths skills that relate to the theory in the chapter. There are also questions using maths skills throughout the book that will help you practice.

How Science Works skills help you to put science in a wider context, and to develop your critical and creative thinking skills and help you solve problems in a variety of contexts. How Science Works is embedded throughout this book, particularly in application boxes and practice questions.

You can find further support for maths and How Science Works on Kerboodle.

CHAPTER 1
Elements of life

Topics in this chapter

Why a chapter on Elements of life

This chapter tells the story of the elements of life – what they are, how they originated, and how they can be detected and measured. It shows how studying the composition of stars can throw light on the formation of the elements that make up our own bodies and considers how these elements combine to form the molecules of life. The chapter also takes the opportunity to look at some aspects of how science works, in particular, developing models and seeing how the scientific community validates work.

The chapter begins with a journey through the Universe. Starting with deep space, the story unfolds through the galaxies, the stars, and our own Sun and solar system. This section looks at the origin of the elements, introducing ideas about the structure of atoms, and briefly considers how elements combine to form compounds and the formation of molecules in the apparently inhospitable dense gas clouds of space, such molecules possibly being the origin of the molecules of life, which make up our bodies.

Later in the chapter you are brought back down to Earth. You learn how to measure amounts of elements (in terms of atoms) and how to calculate chemical formulae. The story then leads into learning about patterns in the properties of elements and the periodic table.

Knowledge and understanding checklist

From your Key Stage 4 study you should have studied the following. Work through each point, using your Key Stage 4 notes and the support available on Kerboodle.

☐ the periodic table

☐ protons, neutrons, and electrons

☐ chemical bonding

☐ writing chemical equations

☐ the wave model of light

☐ the electromagnetic spectrum

☐ relative atomic masses, relative molecular masses, and relative formula masses

☐ chemical formulae

☐ ionic and covalent bonding

☐ acids

☐ precipitation.

Maths skills checklist

In this chapter, you will need to use the following maths skills. You can find support for these skills on Kerboodle and through MyMaths.

☐ Identify and use appropriate units in calculations.

☐ Recognise and use expressions in decimal and ordinary form.

☐ Use ratios, fractions, and percentages.

☐ Use calculators to find and use power functions.

☐ Use appropriate numbers of significant figures.

☐ Find arithmetic means.

☐ Identify uncertainty in measurements.

☐ Change the subject of an equation.

☐ Substitute numerical values into algebraic equations using appropriate units for physical quantities.

☐ Solve algebraic equations.

☐ Translate information between graphical, numerical, and algebraic form.

☐ Use angles and shapes in 2D and 3D structures.

EL 1 Where do the chemical elements come from?

Specification reference: EL(a), EL(g), EL(h), EL(x)

Learning outcomes

Demonstrate and apply knowledge and understanding of:

→ atomic number, mass number, isotope, relative isotopic mass, relative atomic mass A_r

→ how knowledge of the structure of the atom developed in terms of a succession of gradually more sophisticated models – interpretation of these and other examples of such developing models

→ fusion reactions – lighter nuclei join to give heavier nuclei (under conditions of high temperature and pressure), and this is how certain elements are formed

→ use of data from a mass spectrum to determine relative abundance of isotopes and calculate the relative atomic mass of an element.

Various models have been proposed to explain the origin of the Universe, but the Big Bang theory is the one agreed on by the majority of cosmologists.

The Big Bang theory

At some moment, all matter in the Universe was contained in a single point. This matter underwent an explosive expansion known as the Big Bang.

After about three minutes, the nuclei of hydrogen and helium formed from hot collections of tiny particles such as protons and neutrons.

After about 10 000 years, the Universe cooled sufficiently so that electrons moved slowly enough to be captured by oppositely charged protons in nuclei to form atoms. The Universe was made of mainly hydrogen and helium.

As the Universe continued to cool, dust and gas was pulled together by their gravity forming gas clouds. Particles had low kinetic energy and moved around relatively slowly, so gravitational forces were able to keep them together.

Parts of the clouds contracted in on themselves, compressing the gases and forming clumps of denser gas. The densest part of a clump was at the centre, where temperatures were hot enough so that atoms could not retain their electrons. Matter became a **plasma** of ionised atoms and unbound electrons.

Nuclear reactions, such as nuclear fusion, occur releasing vast amounts of energy and causing the dense gas cloud to glow – dense gas cloud becomes a star.

Nuclear reactions generate a hot wind that drives away some of the dust and gas leaving behind the stars. Planets condense out of the remaining dust cloud around these stars.

Nuclear fusion is common at the centre of stars. In nuclear fusion, lighter nuclei are fused together to form heavier nuclei, such as hydrogen atoms joining together to form helium. Other heavier elements are also produced in stars by fusion. Nuclei approach each other at high speed, with a large kinetic energy to overcome repulsion by positive charges on the two nuclei. It requires extreme conditions of temperature and gravitational pressure for the reacions to occur. However, the vast amount of energy released, with no pollution, could make it a useful source of energy if such conditions could be controlled. This is currently being researched by scientists.

▲ Figure 1 *Formation of stars in the Eagle Nebula – the thin 'fingers' at the top of the pillar of gas are embryonic stars*

The life cycle of stars

Hydrogen is still the most common element in the Universe. All stars turn hydrogen into helium by nuclear fusion. The theory of the evolution of the stars shows how heavy elements can be formed from lighter ones, and helps to explain the way elements are distributed throughout the Universe.

Heavyweight stars

The temperatures and pressures at the centre of heavyweight stars mean that, along with hydrogen fusion, other fusion reactions can take place producing heavier elements than helium. Layers of elements form within the star, with the heaviest elements near the centre (Figure 2).

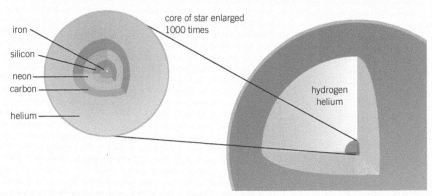

▲ Figure 2 *A model of the core of a typical heavy-weight star after a few million years – long enough for extensive fusion to have taken place*

Eventually, the element at the centre of the core is iron. When iron nuclei fuse together they do not release energy but *absorb* it. When the core of a heavyweight star contains mostly iron it becomes unstable and explodes – a supernova (Figure 3). A supernova causes the elements in the star to be dispersed into the Universe as clouds of dust and gas and the life cycle begins again.

Lightweight stars

The Sun is a lightweight star – it is not as hot as most other stars and will last longer than heavyweight stars. It will keep on shining until all the hydrogen has been used up and the core stops producing energy – there will be no supernova. Once the hydrogen is used up, the Sun will expand into a red giant, swallowing up the planets Mercury and Venus. The oceans on Earth will start to boil and eventually it too will be engulfed by the Sun. The good news for Earth is that the Sun still has an estimated 5000 million years' supply of hydrogen left.

▲ Figure 3 *A possible supernova in the Crab nebula*

As red giants get bigger they become unstable and the outer gases drift off into space, leaving behind a small core called a white dwarf, about $\frac{1}{100}$ of the size of the original star.

In order to understand the nuclear reactions in stars, you need to look at atomic structure and fusion reactions.

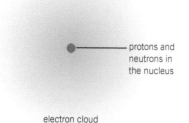

electron cloud

▲ **Figure 4** *A simple model of the atom – not to scale*

protons and neutrons in the nucleus

Chemical Ideas: Atomic structure 2.1

A simple model of the atom

Scientists cannot see inside atoms, but experimental evidence provides a working model of atomic structure. Models are useful as they help to explain observations. Scientists spend lots of time building, testing, and revising scientific models. They should not be considered as the truth.

In a simplified model of atomic structure, atoms can be considered to be made of three types of sub-atomic particle – protons, neutrons, and electrons. Many chemical and nuclear processes can be explained using this simplified model.

Protons and neutrons form the nucleus of atoms. Electrons move around the nucleus. The nucleus is tiny compared with a volume occupied by the electrons. If an atom was the size of Wembley Stadium, the nucleus would be the size of a pea!

Sub-atomic particles

Table 1 summarises some properties of protons, neutrons, and electrons.

▼ Table 1 *Some properties of sub-atomic particles*

Particle	Mass on relative atomic mass scale	Charge (relative to neutron)	Location in atom
proton	1	+1	in nucleus
neutron	1	0	in nucleus
electron	0.000 55	−1	around nucleus

Protons and electrons have equal but opposite electrical charges. Neutrons have no charge. Protons and neutrons have almost equal masses, and are much more massive than electrons. The nucleus accounts for almost all the mass of the atom but hardly any of its volume. Most of the atom is empty space.

It is the electrons in the outer parts of atoms that interact together in chemical reactions.

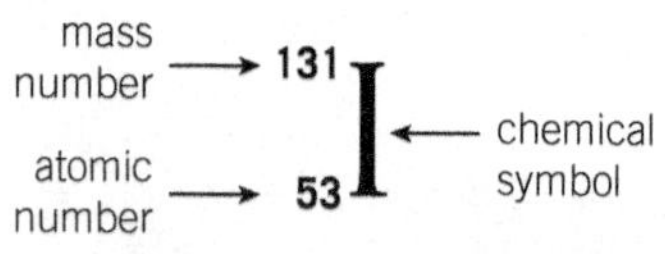

Nuclear symbols

The nucleus can be described by just two numbers – the **atomic number** Z and the **mass number** A. The atomic number is the number of protons in the nucleus. It is numerically equal to the charge on the *nucleus*. The atomic number is the same for every atom of an element, for example, $Z = 6$ for *all* carbon atoms. The mass number is the number of protons *and* neutrons in the nucleus.

mass number A = atomic number Z + number of neutrons N

Nuclear symbols identify the mass number and the atomic number as well as the symbol for the element (Figure 5).

mass number $\longrightarrow$ **131**

atomic number $\longrightarrow$ **53** I

chemical symbol

▲ **Figure 5** *The nuclear symbol for iodine-131*

What are isotopes?

Isotopes are atoms of the *same* element with different *mass* numbers. All atoms of an element have the same number of protons. Differences in mass are caused by different numbers of neutrons.

Most elements exist naturally as a mixture of isotopes (Table 2). The **relative atomic mass** A_r is an average of the relative isotopic masses, taking into account their abundances. A technique called **mass spectrometry** is used to find the atomic mass of elements and compounds.

▼ Table 2 *Isotopes of some elements*

Element	Isotope	Abundance / %
chlorine	^{35}Cl	75.0
	^{37}Cl	25.0
iron	^{54}Fe	5.8
	^{56}Fe	91.7
	^{57}Fe	2.2
	^{58}Fe	0.3
bromine	^{79}Br	50.0
	^{81}Br	50.0
calcium	^{40}Ca	96.9
	^{42}Ca	0.7
	^{43}Ca	0.1
	^{44}Ca	2.1
	^{48}Ca	0.2

Chemical ideas: Radiation and matter 6.5a

Mass spectrometry with elements

Mass spectrometry measures the atomic or molecular mass of different particles (i.e., atoms or molecules) in a sample and the relative abundance of different **isotopes** in an element.

In a mass spectrometer, sample atoms or molecules are ionised to positively charged **cations**. These **ions** are separated according to their mass m to charge z ratios, m/z. The separated ions are detected, together with their relative abundance.

Using mass spectra to calculate relative atomic mass

Figure 6 shows the mass spectrum for naturally occurring iron. If you assume that all the ions are singly charged, $z = 1$, m/z is the same as the mass of the ion detected. The relative abundance of each ion can be calculated from the height of each peak.

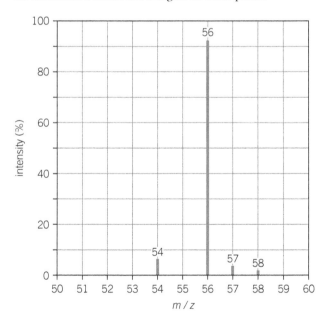

Relative isotopic mass	Relative abundance / %
54	5.8
56	91.7
57	2.2
58	0.3

▲ Figure 6 *The mass spectrum for naturally occurring iron (left) and a table summarising the information from the mass spectrum (right)*

Samples of an element may vary slightly in average A_r depending on their source. This is due to slightly different abundances of individual isotopes between samples.

Synoptic link

You will find out more about mass spectrometry in Chapter 5, What's in a medicine?

Activity EL 1.2

In this activity you use data from a mass spectrometer to calculate an A_r.

 Worked example: Calculating the relative atomic mass of iron

What is the relative atomic mass of iron?

Step 1: Using the table in Figure 6, calculate the average mass of 100 atoms.

(relative isotopic mass × relative abundance)

$(54 \times 5.8) + (56 \times 91.7) + (57 \times 2.2) + (58 \times 0.3)$
$= 313.2 + 5135.2 + 125.4 + 17.4$
$= 5591.2$

Step 2: Calculate the mass of one atom.

$$5591.2 \div 100 = 55.91$$

Average mass of one atom of iron is 55.9 (1 d.p.)

Chemical ideas: Atomic structure 2.2

Nuclear fusion

In a **nuclear fusion** reaction, two light atomic nuclei fuse together forming a single heavier nucleus of a new element, releasing enormous quantities of energy. For two nuclei to fuse, they must come very close together. At the normal temperatures found on the Earth, the positive nuclei repel so strongly that fusion cannot happen. At very high temperatures, such as in a star, the nuclei are moving much more quickly and collide with so much energy that this repulsion is overcome.

Atomic numbers and mass numbers must balance in a nuclear equation. The equation below shows fusion of different isotopes of hydrogen.

$$^{1}_{1}H + ^{2}_{1}H \rightarrow ^{3}_{2}He + \gamma$$

The γ symbol indicates the release of energy when the fusion reaction occurs. This is a common occurrence in fusion reactions.

Summary questions

1 Copy and complete the table. *(5 marks)*

Isotope	Symbol	Atomic number	Mass number	Number of neutrons
carbon-12	$^{12}_{6}C$	6	12.0	6
carbon-13			13.0	7
oxygen-16	$^{16}_{8}O$	8	16.0	
strontium-90	$^{90}_{38}Sr$	38		52
iodine-131	$^{131}_{53}I$			
	$^{121}_{53}I$			

2 State the number of protons, neutrons, and electrons present in each of the following atoms.

a $^{79}_{35}Br$ (*1 mark*)   c $^{35}_{17}Cl$ (*1 mark*)

b $^{81}_{35}Br$ (*1 mark*)   d $^{37}_{17}Cl$ (*1 mark*)

3 Use the data in Table 2 to calculate the relative atomic masses of:

a bromine (*2 marks*)

b calcium. (*2 marks*)

4 The relative atomic mass of the element iridium is 192.2. Iridium occurs naturally as a mixture of iridium-191 and iridium-193. Calculate the percentage of each isotope in naturally occurring iridium using the following steps.

a If the percentage of iridium-193 in naturally occurring iridium is $x\,\%$, there must be x atoms of iridium-193 in every 100 atoms of the element. How many atoms of iridium-191 must there be in every 100 atoms of iridium? (*1 mark*)

b Write an expression for the total relative mass of the atoms of iridium-193 in the 100-atom sample. (*1 mark*)

c Write an expression for the total relative mass of the atoms of iridium-191 in the 100-atom sample. (*1 mark*)

d Write an expression for the total relative mass of all 100 iridium atoms. (*1 mark*)

e Write an expression for the relative atomic mass of one iridium atom. (*1 mark*)

f Use your answer to part e and the value of the relative atomic mass of iridium to calculate the percentage of each isotope in naturally occurring iridium. (*3 marks*)

5 Write a nuclear equation for each of the following.

a A $^{7}_{3}Li$ nucleus absorbs a colliding proton and then disintegrates into two identical fragments. (*2 marks*)

b The production of carbon-14 by collision of a neutron with an atom of nitrogen-14. (*2 marks*)

6 The relative atomic mass of antimony is 121.8. Antimony exists as two isotopes – antimony-121 and antimony-123. Calculate the relative abundances of the two isotopes. (*4 marks*)

EL 2 How do we know so much about outer space?

Specification reference: EL(v), EL(w)

The work of chemists has made a vital contribution to the understanding of the origin, structure, and composition of our Universe. To do this, they have used a method called **spectroscopy**.

Spectroscopy

Many spectroscopic techniques exist but all are based on the same scientific principle – under certain conditions, a substance can absorb or emit electromagnetic radiation in a characteristic way. There are different types of electromagnetic radiation (Figure 1). By analysing electromagnetic radiation you can identify a substance or find information about it, such as its structure and the way atoms are held together.

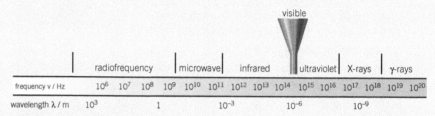

▲ **Figure 1** *The electromagnetic spectrum*

Absorption spectra

Glowing stars emit all of the light frequencies between the ultraviolet and infrared parts of the electromagnetic spectrum. The Sun emits mainly visible light, whereas stars hotter than the Sun emit mainly ultraviolet radiation.

Outside a star's surface – the photosphere – is a region called the chromosphere that contains ions, atoms, and small molecules. These particles *absorb* some of the emitted radiation so the light analysed from stars is missing certain frequencies. The absorption lines appear black as these are the missing frequencies of light – they have been absorbed by the particles in the chromosphere (Figure 2).

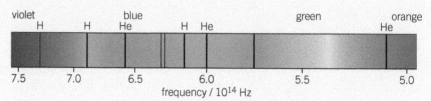

▲ **Figure 2** *Absorption spectrum of a B-type star (e.g., β Centauri). This spectrum allows scientists to detect the presence of hydrogen and helium. Black lines occur where frequencies are missing from the otherwise continuous spectrum. (The frequency increases from right to left, which is a common convention for absorption spectra)*

Emission spectra

When the atoms, molecules, and ions in the chromosphere absorb energy, they are raised from their ground state (their lowest energy state) to higher energy states called excited states. The particles can lose their extra energy by *emitting* electromagnetic radiation. The resulting emission spectra can also be detected. Emission spectra appear as coloured lines on a black background – corresponding to the frequencies emitted (Figure 3).

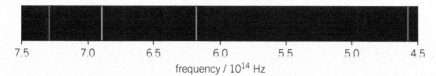

▲ Figure 3 *The hydrogen emission spectrum in the visible region. Coloured lines occur where frequencies are emitted*

Continuous and atomic spectra

White light contains all the visible wavelengths and its spectrum is normally continuous, like a rainbow. However light seen from stars is not continuous. It consists of lines, corresponding to the absorption or emission of specific frequencies of light – atomic spectra.

The atomic spectrum of hydrogen atoms

The Sun's chromosphere consists mainly of hydrogen and helium atoms and so hydrogen atoms dominate the chromosphere's *emission* spectrum, but helium emission lines can also be seen. Comparing the emission spectrum of hydrogen (Figure 3) with the absorption spectrum shown in Figure 2, the emitted frequencies match up exactly with those of the absorbed frequencies for hydrogen.

Hydrogen atoms also have a characteristic emission spectrum in the ultraviolet region of the electromagnetic spectrum (Figure 4). The hydrogen emission spectrum in visible light is known as the Balmer series. The hydrogen emission spectrum in ultraviolet light is known as the Lyman series.

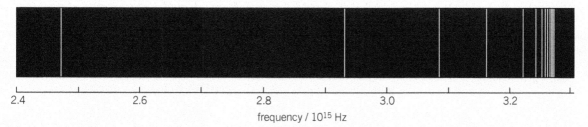

▲ Figure 4 *The Lyman series in the hydrogen atomic spectrum*

These spectra are the result of the interaction of light and matter.

Light and electrons

Spectroscopy is the study of how light and matter interact. In order to understand spectroscopy you need to know about the behaviour of light.

Bohr's theory and wave-particle duality

Chemists use two models to describe the behaviour of light – the wave theory and the particle theory. Neither fully explains all the properties of light, but some are explained by the wave model and others by the particle model.

The wave theory of light

Light is one form of electromagnetic radiation. Like all electromagnetic radiation, it behaves like a wave with a characteristic **wavelength** λ and **frequency** v.

A wave of light travels the distance between two points in a certain time – it doesn't matter what kind of light it is, the time is always the same. The **speed** the wave moves, the speed of light c, is the same for all kinds of light and electromagnetic radiation. It has a value of $3.00 \times 10^{8} \, \text{m s}^{-1}$ when the light is travelling in a vacuum.

Different colours of light have different wavelengths. Waves can also have different frequencies. Both waves in Figure 5 travel at the same speed, so they both travel the same *distance* per second but wave B has twice the frequency of wave A.

▲ Figure 5 *Wavelength measures the distance (in metres) travelled by the wave during one cycle. Wave A has twice the wavelength of wave B*

Frequency and wavelength are very simply related. In Figure 5, wave B has twice the frequency but half the wavelength of wave A. Multiplying the wavelength and frequency together give the speed of light c. A simple equation links together c, λ, and v.

speed of light c (m s^{-1}) = wavelength λ (m) × frequency v (s^{-1})

Though the term light is often meant to refer to just visible light, it is only one small part of the **electromagnetic spectrum** (Figure 1). Other regions include radio waves, ultraviolet, infrared, and gamma rays.

The particle theory of light

In some situations, the behaviour of light is easier to explain by thinking of it not as waves but as particles. This regards light as a stream of tiny packets of energy called **photons**. The energy of the photons is

Synoptic link

You will look further at this in Topic OZ 2, Screening the Sun.

Study tip

As the wavelength increases, the frequency decreases.

Worked example: Calculations using the wave theory

Calculate the wavelength of an ultraviolet wave of frequency $1.4 \times 10^{15} \, \text{s}^{-1}$.

Step 1: Rearrange the equation

$$\text{speed of light } c \, (\text{m s}^{-1}) = \text{wavelength } \lambda \, (\text{m}) \times \text{frequency } v \, (\text{s}^{-1})$$

to make wavelength the subject.

$$\lambda = \frac{c}{v}$$

Step 2: Calculate the wavelength.

$$\frac{3.00 \times 10^{8}}{1.4 \times 10^{15}} = 2.14 \times 10^{-7} \, \text{m}$$

related to the position of the light in the electromagnetic spectrum, for example, photons with energy 3×10^{-19} J correspond to red light.

The wave and the photon models of light are linked.

energy of a photon E (J) = Planck constant h (Js^{-1}) × frequency v (s^{-1})

The value of the Planck constant is 6.63×10^{-34} Js^{-1}.

Bohr's theory

Using the hydrogen emission spectrum (Figure 3) the scientist Niels Bohr came up with a theory to explain why the hydrogen atom only emits a limited number of specific frequencies.

When an atom is **excited**, electrons jump into higher energy levels. Later, they drop back into lower levels, emitting the extra energy as electromagnetic radiation and giving off an **emission spectrum**.

When white light is passed through a relatively cool sample of a gaseous element, black lines appear in the otherwise continuous absorption spectrum. The black lines in the **absorption spectrum** correspond to light that has been *absorbed* by the atoms in the sample. Electrons have been raised to higher levels without then dropping back again. They correspond exactly with the coloured lines in the emission spectrum of that element.

The sequence of lines in an atomic spectrum is characteristic of the atoms of an element. They can be used to identify the element, even when the element is present in a compound or is part of a mixture. The *intensities* of the lines provide a measure of the element's abundance.

Bohr's theory not only explains how you get absorption spectra and emission spectra, it also gave scientists a model for the electronic structure of atoms. It was controversial when first proposed as it relied on the new theory of the *quantisation* of energy. However, Bohr's theory predicted experimental observations accurately and lent support to the new *quantum theory*.

The main points of Bohr's theory were:

- the electron in the hydrogen atom exists only in certain definite energy levels or electron shells
- a photon of light is emitted or absorbed when the electron changes from one energy level to another
- the energy of the photon is equal to the difference between the two energy levels ΔE
- since $E = hv$ it follows that the frequency of the emitted or absorbed light is related to ΔE by $\Delta E = hv$.

Energy levels and quanta

An electron can only possess definite quantities of energy, or **quanta**. The electron's energy cannot change continuously – it is not able to change to any value, only those values that are allowed.

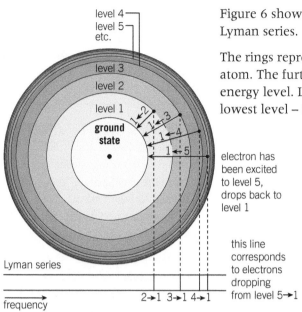

Figure 6 shows how Bohr's ideas explained the emission lines of the Lyman series.

The rings represent the energy levels of the electron in the hydrogen atom. The further away from the nucleus an electron is, the higher the energy level. Levels are labelled with numbers, starting at one for the lowest level – the ground state.

The frequencies of the lines of the Lyman series correspond to changes in electronic energy from various upper levels to one common lower level, level 1. Each line corresponds to a particular energy level change, such as level 4 to level 1.

An alternative way of representing emission spectra is by an energy level diagram. Each arrow represents an electron dropping from a *higher* energy level to a *lower* energy level. It can be seen from Figure 7 that the larger the energy gap ΔE between the two levels, the higher the frequency of electromagnetic radiation emitted.

▲ **Figure 6** *How the Lyman series in the emission spectrum is related to energy levels in the hydrogen atom*

Summary questions

1 Lithium carbonate is used to give the bright red colour to a firework.

   The visible region of the emission spectrum of lithium contains several coloured lines with a particularly intense line at the red end of the spectrum. The emission spectrum is shown below.

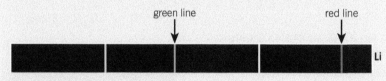

   Which line, green or red, has a greater frequency? (*1 mark*)

2 Flame tests can be used to identify some metal cations.

   Describe the flame colours for the following cations.

   a Li^+ b Na^+ c K^+ d Ca^{2+} e Ba^{2+} f Cu^{2+} (*6 marks*)

3 Describe the similarities and differences between the above emission spectrum of lithium and the absorption spectrum of lithium. (*2 marks*)

4 Copy the diagram that represents electronic energy levels in a lithium atom.
   a Draw an arrow to represent an electron energy level change that might give rise to the red line seen in the emission spectrum. (*1 mark*)
   b Draw an arrow to represent an electron energy level change that might give rise to the green line seen in the emission spectrum. (*1 mark*)

5 Calculate the frequency of electromagnetic radiation with a wavelength of 5.5×10^{-7} m (*1 mark*)

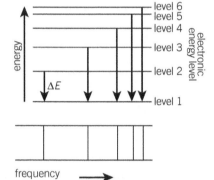

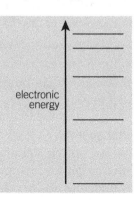

▲ **Figure 7** *An energy level diagram and corresponding emission spectrum for the Lyman series. A similar diagram can be used to represent an absorption spectrum – the difference between the diagrams would be that the arrows would point upwards as the electron is raised from a lower energy level to a higher energy level*

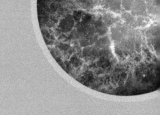

Specification reference: EL(e), EL(f)i, EL(f)iii

Electrons are involved in our lives and the world around us in many ways. Figure 1 to 4 show some examples.

▲ Figure 1 *Laser light shows*

▲ Figure 2 *Bioluminescence in jellyfish*

▲ Figure 3 *Modern technology*

▲ Figure 4 *Chemical reactions like cooking*

Learning outcomes

Demonstrate and apply knowledge and understanding of:

→ conventions for representing the distribution of electrons in atomic orbitals – the shapes of s- and p-orbitals

→ the electronic configuration, using sub-shells and atomic orbitals, of:

- atoms from hydrogen to krypton
- the outer sub-shell structures of s- and p-block elements of other periods.

In order to understand why electrons have such wide-ranging effects, you need to find out a bit more about atomic structure.

Determining atomic structure

Both models of electron structure – negative particles or negative wave form – can be used to understand the structure of atoms. Electrons can be thought of as particles when filling the shells of atoms, but the electron orbitals can be thought of as standing (or stable) waves with a maximum overall negative charge equivalent to two electrons.

Chemical ideas: Atomic structure 2.3

Electronic structure: shells, sub-shells, and orbitals

Shells of electrons

Bohr's theory to describe the emission spectrum of the hydrogen atom needed expanding to describe the structure of atoms with more than one electron. For these atoms, the energy levels 2, 3, 4, and so on have a more complex structure than the single levels that exist in hydrogen. It is more appropriate to talk about the first, second, third electron shells rather than energy level 1, 2, 3. The shells are labelled by giving each

Synoptic link

The filling of shell 3 is not straightforward. This is further explained in Chapter 9, Developing metals.

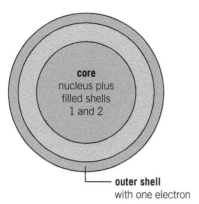

▲ Figure 5 *The core and outer shell model of a sodium atom (2.8.1)*

▼ Table 1 *Maximum number of electrons in the s-, p-, d-, and f-sub-shells*

Sub-shell	Maximum number of electrons
s	2
p	6
d	10
f	14

a principal quantum number n. For the first shell, $n = 1$, for the second shell $n = 2$, and so on. The higher the value of n, the further the shell is from the nucleus and the higher the energy associated with the shell.

Although each shell can hold more than one electron, there is a limit. The maximum numbers of electrons which can be held are:

- first shell   ($n = 1$)   2 electrons
- second and third shells   ($n = 2$ and $n = 3$)   8 electrons
- fourth and fifth shells   ($n = 4$ and $n = 5$)   18 electrons
- sixth and seventh shells   ($n = 6$ and $n = 7$)   32 electrons

A shell that contains its maximum number of electrons is called a filled shell. Electrons are arranged so that the lowest energy shells are filled first. Electron shell configurations can be written, for example, the electron shell configuration of sodium is 2, 8, 1. This means:

- two electrons in first shell
- eight electrons in second shell
- one electron in third shell.

Chemists explain many of the properties of atoms without needing to use a detailed theory of atomic structure. Much chemistry is decided only by the outer shell electrons, and one very useful model treats the atom as being composed of a core of the nucleus and the inner electrons shells, surrounded by an outer shell (Figure 5).

Sub-shells of electrons

Much of the knowledge of electron shells has come from studying the emission spectrum of hydrogen. The hydrogen atom has only one electron and its spectrum is relatively simple to interpret. The spectra of elements other than hydrogen are much more complex – electron shells are not the whole story. The shells are themselves split up into sub-shells.

The sub-shells are labelled s, p, d, and f. The $n = 1$ shell has only an s-sub-shell. The $n = 2$ shell has two sub-shells – s and p. The $n = 3$ shell has three sub-shells – s, p, and d. The $n = 4$ shell has four sub-shells – s, p, d, and f.

The different types of sub-shells can hold different numbers of electrons (Table 1).

- the $n = 1$ shell can hold two electrons in the s-sub-shell
- the $n = 2$ shell can hold two electrons in the s-sub-shell and six electrons in the p-sub-shell – a total of eight electrons
- the $n = 3$ shell can hold two electrons in the s-sub-shell, six electrons in the p-sub-shell, and 10 electrons in the d-sub-shell – a total of 18 electrons
- the $n = 4$ shell can hold two electrons in the s-sub-shell, six electrons in the p-sub-shell, 10 electrons in the d-sub-shell, and 14 electrons in the f-sub-shell – a total of 32 electrons.

In atoms other than hydrogen, the sub-shells within a shell have different energies. The energy of a sub-shell is not fixed, but falls as the charge on the nucleus increases from one element to the next in the periodic table. Figure 6 shows the relative energies of the sub-shell for each of the shells $n = 1$ to $n = 4$ in a typical many-electron atom. The overlap in energy between the $n = 3$ and $n = 4$ shells has important consequences that you will meet later.

Atomic orbitals

The s- p- d- and f-sub-shells are themselves divided further into atomic orbitals. An electron in a given orbital can be found in a particular region of space around the nucleus.

- An s-sub-shell always contains *one* s-orbital
- A p-sub-shell always contains *three* p-orbitals
- A d-sub-shell always contains *five* d-orbitals
- An f-sub-shell always contains *seven* f-orbitals

In an isolated atom, orbitals in the same sub-shell have the same energy (Figure 7).

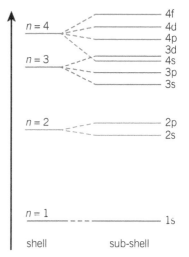

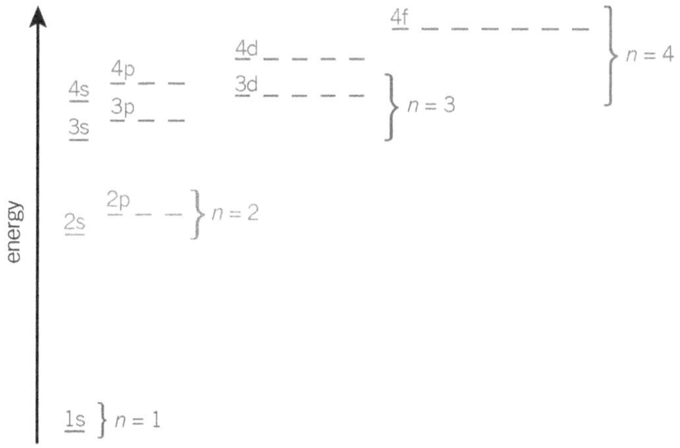

▲ Figure 6 *Energies of electron sub-shells from n = 1 to n = 4 in a typical many-electron atom. The order shown in the diagram is correct for the elements in Period 3 and up to nickel in Period 4 (periods are the horizontal rows on the periodic table). After nickel, the 3d-sub-shell has lower energy than 4s*

▲ Figure 7 *Energy levels of atomic orbitals in the n = 1 to n = 4 shells*

The position of an electron cannot be mapped exactly. For an electron in a given atomic orbital, is is the *probability* of finding the electron in any region that is known (Figure 8).

Each atomic orbital can hold a maximum of two electrons. Electrons in atoms have a *spin*, which is pictured as a spinning motion in one of two directions. Every electron spins at the same rate in either a clockwise ↑ or anticlockwise ↓ direction. Electrons can only occupy the same orbital if they have opposite, or paired, spins. They can be written as ↑↓.

The box represents the atomic orbital and the arrows represent the electrons. Four pieces of information are needed when describing an electron:

- the electron shell it is in
- its sub-shell

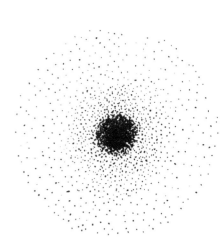

▲ Figure 8 *One way of representing a 1s electronic orbital. The dots represent the probability of finding the electron in that region – the denser the dots, the higher the probability of finding the electron there*

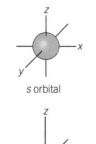

s orbital

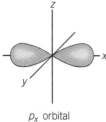

p_x orbital

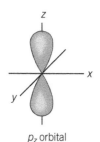

p_y orbital

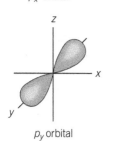

p_z orbital

▲ **Figure 9** *Shapes of s-orbitals and p-orbitals*

● its orbital within the sub-shell

● its spin.

Atomic orbitals have a three dimensional shape that represents the volume of space where there is a high probability of finding up to two electrons (Figure 9).

Filling up atomic orbitals

The arrangement of electrons in shells and orbitals is called the **electronic configuration** of an atom. The orbitals are filled in a specific order to produce the lowest energy arrangement possible.

The orbitals are filled in order of increasing energy. Where there is more than one orbital with the same energy, these orbitals are first occupied singly by electrons. This keeps the electrons in an atom as far apart as possible. Only when every orbital is singly occupied do the electrons pair up in orbitals. For the lowest energy arrangement, electrons in singly occupied orbitals have parallel spins.

Figure 10 shows how the 11 electrons in a sodium atom are arranged in atomic orbitals. The electronic configuration of a sodium atom can also be represented as $1s^2 2s^2 2p^6 3s^1$. The large numbers show the principal quantum number of each shell, the letters show the sub-shells, and the small superscripts indicate the numbers of electrons in each sub-shell.

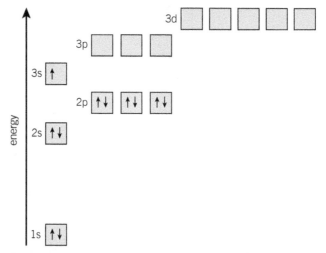

▲ **Figure 10** *Arrangement of electrons in atomic orbitals in a ground state sodium atom*

Writing electronic configurations

Hydrogen is the simplest element, with atomic number $Z = 1$. It has one electron that occupies the s-orbital of the $n = 1$ shell.

1s ⬚ H $1s^1$

The next element, helium ($Z = 2$) has two electrons that both occupy the 1s orbital with paired spins.

1s ⬚ He $1s^2$

Lithium ($Z = 3$) has three electrons. The third electron cannot fit in the $n = 1$ shell, so occupies the next lowest orbital – the 2s orbital.

1s $\uparrow\downarrow$ 2s $\uparrow$ Li $1s^22s^1$

Nitrogen ($Z = 7$) has seven electrons.

1s $\uparrow\downarrow$ 2s $\uparrow\downarrow$ 2p $\uparrow$ $\uparrow$ $\uparrow$ N $1s^22s^22p^3$

The three electrons in the 2p-sub-shell occupy the three separate p-orbitals singly and their spins are parallel.

Oxygen ($Z = 8$) has eight electrons

1s $\uparrow\downarrow$ 2s $\uparrow\downarrow$ 2p $\uparrow\downarrow$ $\uparrow$ $\uparrow$ O $1s^22s^22p^4$

However, to write the electronic configuration of the element, scandium ($Z = 21$), look back to Figure 6. The energy level of the 3d-sub-shell lies just above that of the 4s-sub-shell but just below the 4p-sub-shell. This means that the 4s-orbital fills before the 3d-orbital. Once the 4s level is filled in calcium ($Z = 20$), scandium has the electronic structure $1s^22s^22p^63s^23p^6\mathbf{3d^14s^2}$. The 3d-sub-shell continues to be filled across the period in the elements scandium to zinc. Zinc ($Z = 30$) has the electronic configuration $1s^22s^22p^63s^23p^6\mathbf{3d^{10}4s^2}$. The electronic configurations of elements with atomic numbers 1 to 36 are shown in Table 2.

> **Study tip**
>
> The 3d-sub-shell is written alongside the other $n = 3$ sub-shells even though it is filled after the 4s-sub-shell.

Worked example: Working out electronic configuration of an atom

What is the electronic configuration of vanadium?

Step 1: Identify the number of electrons in a vanadium atom.

From the periodic table, $Z = 23$

Step 2: Electrons go into lowest energy shells and sub-shells, using the appropriate numbers and letters, until they are full, so for vanadium $1s^22s^22p^63s^23p^6$

Step 3: The energy of the 4s-sub-shell is lower than the 3d-sub-shell so this is filled with two electrons first.

Step 4: This leaves the three remaining electrons to go into the 3d-sub-shell

Step 5: Group together the same-shell electrons. The final overall configuration is therefore $1s^22s^22p^63s^23p^63d^34s^2$

Electron configurations show how the periodic table is built up. Period 2 (lithium to neon) corresponds to the filling of the 2s- and 2p-orbitals, Period 3 (sodium to argon) corresponds to the filling of the 3s-and 3p-orbitals, and so on. You can also write electron configurations of ions, which you will do in Topic EL 4.

> **Study tip**
>
> You will have access to a periodic table in examinations.

> **Study tip**
>
> You need to be able to write the electron configuration of an atom from hydrogen to krypton, but not copper and chromium.

> **Synoptic link**
>
> You will find out more about copper and chromium in Topic DM 1, Metals of antiquity.

▼ Table 2 *Ground state electronic configurations of elements with atomic numbers 1–36*

Period	Atomic number / Z	Element	Electronic configuration	Period	Atomic number / Z	Element	Electronic configuration
$n = 1$	1	hydrogen	$1s^1$		19	potassium	$1s^2 2s^2 2p^6 3s^2 3p^6 4s^1$
	2	helium	$1s^2$		20	calcium	$1s^2 2s^2 2p^6 3s^2 3p^6 4s^2$
$n = 2$	3	lithium	$1s^2 2s^1$		21	scandium	$1s^2 2s^2 2p^6 3s^2 3p^6 3d^1 4s^2$
	4	beryllium	$1s^2 2s^2$		22	titanium	$1s^2 2s^2 2p^6 3s^2 3p^6 3d^2 4s^2$
	5	boron	$1s^2 2s^2 2p^1$		23	vanadium	$1s^2 2s^2 2p^6 3s^2 3p^6 3d^3 4s^2$
	6	carbon	$1s^2 2s^2 2p^2$		24	chromium	$1s^2 2s^2 2p^6 3s^2 3p^6 3d^5 4s^1$
	7	nitrogen	$1s^2 2s^2 2p^3$		25	manganese	$1s^2 2s^2 2p^6 3s^2 3p^6 3d^5 4s^2$
	8	oxygen	$1s^2 2s^2 2p^4$		26	iron	$1s^2 2s^2 2p^6 3s^2 3p^6 3d^6 4s^2$
	9	fluorine	$1s^2 2s^2 2p^5$	$n = 4$	27	cobalt	$1s^2 2s^2 2p^6 3s^2 3p^6 3d^7 4s^2$
	10	neon	$1s^2 2s^2 2p^6$		28	nickel	$1s^2 2s^2 2p^6 3s^2 3p^6 3d^8 4s^2$
$n = 3$	11	sodium	$1s^2 2s^2 2p^6 3s^1$		29	copper	$1s^2 2s^2 2p^6 3s^2 3p^6 3d^{10} 4s^1$
	12	magnesium	$1s^2 2s^2 2p^6 3s^2$		30	zinc	$1s^2 2s^2 2p^6 3s^2 3p^6 3d^{10} 4s^2$
	13	aluminium	$1s^2 2s^2 2p^6 3s^2 3p^1$		31	gallium	$1s^2 2s^2 2p^6 3s^2 3p^6 3d^{10} 4s^2 4p^1$
	14	silicon	$1s^2 2s^2 2p^6 3s^2 3p^2$		32	germanium	$1s^2 2s^2 2p^6 3s^2 3p^6 3d^{10} 4s^2 4p^2$
	15	phosphorous	$1s^2 2s^2 2p^6 3s^2 3p^3$		33	arsenic	$1s^2 2s^2 2p^6 3s^2 3p^6 3d^{10} 4s^2 4p^3$
	16	sulfur	$1s^2 2s^2 2p^6 3s^2 3p^4$		34	selenium	$1s^2 2s^2 2p^6 3s^2 3p^6 3d^{10} 4s^2 4p^4$
	17	chlorine	$1s^2 2s^2 2p^6 3s^2 3p^5$		35	bromine	$1s^2 2s^2 2p^6 3s^2 3p^6 3d^{10} 4s^2 4p^5$
	18	argon	$1s^2 2s^2 2p^6 3s^2 3p^6$		36	krypton	$1s^2 2s^2 2p^6 3s^2 3p^6 3d^{10} 4s^2 4p^6$

Summary questions

1 $1s^2$ is an example of a notation for electronic configuration.
  a What does the number 1 refer to?   *(1 mark)*
  b What does the s refer to?   *(1 mark)*
  c What does the superscript 2 refer to?   *(1 mark)*

2 Write out the electronic configurations of the following atoms.
  a boron ($Z = 5$)   *(1 mark)*   b phosphorus ($Z = 15$)   *(1 mark)*

3 The electronic configuration of the outermost shell of an atom of an element X is $3s^2 3p^4$. What is the atomic number and name of the element?   *(2 marks)*

4 Electronic configurations are sometimes abbreviated by labelling the core of filled inner shells as the electronic configuration of the appropriate noble gas. For example, the electronic configuration of neon is $1s^2 2s^2 2p^6$ and that of sodium is $1s^2 2s^2 2p^6 3s^1$ so you can write the electronic configuration of sodium as $[Ne]3s^1$. Name the elements from the electronic configurations.
  a $[Ne]3s^2 3p^5$ *(1 mark)*   c $[Ar]3d^2 4s^2$ *(1 mark)*
  b $[Ar]4s^1$ *(1 mark)*   d $[Kr]4d^{10} 5s^2 5p^2$ *(1 mark)*

EL 4 Organising the elements of life

Specification reference: EL(f), EL(m), EL(n)

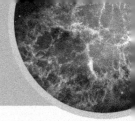

The periodic table

As elements were discovered and more was learnt about their properties, chemists looked for patterns so the elements could be grouped. Much of the work on finding patterns in the elements was done by Johann Döbereiner and Lothar Meyer in Germany, John Newlands in England, and Dmitri Mendeleev in Russia. These chemists looked at similarities in the chemical reactions of the elements they knew about, and also patterns in physical properties such as melting point, boiling point, and density.

Mendeleev arranged elements in order of increasing atomic mass and so that elements with similar properties were in the same vertical group. His grouping was seen as the most credible. Mendeleev's values for atomic masses were not accurate because the existence of isotopes was not known at that time.

▲ Figure 1 Dimitri Mendeleev

Learning outcomes

Demonstrate and apply knowledge and understanding of:

→ the electronic configuration using sub-shells and atomic orbitals of ions of the s- and p-block of Periods 1–4

→ the periodic table as a list of elements in order of atomic (proton) number that groups elements together according to their common properties; using given information, make predictions concerning the properties of an element in a group; the classification of elements into s-, p-, and d-blocks

→ periodic trends in the melting points of elements in Periods 2 and 3, in terms of structure and bonding.

▼ Table 1 A form of Mendeleev's periodic table – the asterisks denote elements that he thought were yet to be discovered

	Group 1	Group 2	Group 3	Group 4	Group 5	Group 6	Group 7	Group 8
Period 1	H							
Period 2	Li	Be	B	C	N	O	F	
Period 3	Na	Mg	Al	Si	P	S	Cl	
Period 4	K Cu	Ca Zn	* *	Ti *	V As	Cr Se	Mn Br	Fe, Co, Ni
Period 5	Rb Ag	Sr Cd	Y In	Zr Sn	Nb Sb	Mo Te	* I	Ru, Rh, Pd

Making predictions using the periodic table

Mendeleev left gaps in his table of elements. These gaps corresponded to his predictions of elements that had not been discovered at the time. He was so confident of his table that he even made predictions about the properties of these undiscovered elements. When the element germanium was discovered in 1886, its properties were found to be in excellent agreement with the predictions Mendeleev had made using his table of elements.

Some elements do not exist naturally and are made synthetically in a laboratory. The first two elements to be made synthetically were neptunium ($Z = 93$) and plutonium ($Z = 94$). They were formed by bombarding uranium atoms with neutrons. The heaviest element synthesized had an atomic number of 118 but only existed for 200 microseconds!

The modern periodic table is based on the one originally drawn up by Mendeleev. It is one of the most amazingly compact stores of information ever produced – with a copy of the periodic table in front of you, and some knowledge of how it was put together, you have thousands of facts at your fingertips!

(1)	(2)											(3)	(4)	(5)	(6)	(7)	(0)
1																	18
1 H 1.0	2			key								13	14	15	16	17	2 He 4.0
3 Li 6.9	4 Be 9.0		atomic number symbol relative atomic mass									5 B 10.8	6 C 12.0	7 N 14.0	8 O 16.0	9 F 19.0	10 Ne 20.2
11 Na 23.0	12 Mg 24.3	3	4	5	6	7	8	9	10	11	12	13 Al 27.0	14 Si 28.1	15 P 31.0	16 S 32.1	17 Cl 35.5	18 Ar 39.9
19 K 39.1	20 Ca 40.1	21 Sc 45.0	22 Ti 47.9	23 V 50.9	24 Cr 52.0	25 Mn 54.9	26 Fe 55.8	27 Co 58.9	28 Ni 58.7	29 Cu 63.5	30 Zn 65.4	31 Ga 69.7	32 Ge 72.6	33 As 74.9	34 Se 79.0	35 Br 79.9	36 Kr 83.8
37 Rb 85.5	38 Sr 87.6	39 Y 88.9	40 Zr 91.2	41 Nb 92.9	42 Mo 95.9	43 Tc 98]	44 Ru 101.1	45 Rh 102.9	46 Pd 106.4	47 Ag 107.9	48 Cd 112.4	49 In 114.8	50 Sn 118.7	51 Sb 121.8	52 Te 127.6	53 I 126.9	54 Xe 131.3
55 Cs 132.9	56 Ba 137.3	57–71	72 Hf 178.5	73 Ta 180.9	74 W 183.8	75 Re 186.2	76 Os 190.2	77 Ir 192.2	78 Pt 195.1	79 Au 197.0	80 Hg 200.6	81 Tl 204.4	82 Pb 207.2	83 Bi 209.0	84 Po [209]	85 At [210]	86 Rn [222]
Fr	Ra		Rf	Db	Sg	Bh	Hs	Mt	Ds	Rg	Cn		Fi		Lv		

57 La 138.9	58 Ce 140.1	59 Pr 140.9	60 Nd 144.2	61 Pm 144.9	62 Sm 150.4	63 Eu 152.0	64 Gd 157.2	65 Tb 158.9	66 Dy 162.5	67 Ho 164.9	68 Er 167.3	69 Tm 168.9	70 Yb 173.0	71 Lu 175
Ac	90 Th 232.0	Pa	92 U 238.1	Np	Pu	Am	Cm	Bk	Cf	Es	Fm	Md	No	Lr

▲ Figure 2 *A modern periodic table*

Superheavy elements and the Island of Stability

The periodic table arranges elements in order by the number of protons residing in an element's nucleus. Neutrons and protons (collectively known as **nucleons**) reside in the nucleus. The protons are positively charged and repel each other, with the repulsion increasing as the protons get closer. However another force – the strong nuclear force – is stronger than this repulsion of charge, and holds the nucleus together. Neutrons don't have an overall charge, but they have some magnetic properties. Neutrons act as a buffer between protons. Usually, the more protons you have, the more neutrons you need. Beyond element 92, uranium, the strong nuclear force is

Activity EL 4.1

In this activity you will use the Internet to look for patterns in the properties of elements.

insufficient to hold the nucleus of the element together and all isotopes are unstable, breaking down by radioactive decay. The Island of Stability follows from theoretical predictions that some superheavy elements with a total number of nucleons around 300 may hang around for a while longer than their lighter cousins, before radioactively decaying into other types of nuclei.

1   Look up the neutron to proton ratio for naturally occurring isotopes of the elements (the so called 'band of stability')
    a   State the neutron/proton ratio for the first 20 elements.
    b   Describe how this ratio changes as the atomic number increases beyond 20.
2   Explain, in terms of neutron/proton ratio and relative stability, why either alpha or beta radioactive decay might be expected to occur for elements above atomic number 83.

Chemical ideas: The periodic table 11.1

Periodicity

The modem periodic table (Figure 2) is based on one proposed by the Russian chemist Mendeleev in 1869. In Mendeleev's periodic table the elements were arranged in order of increasing relative atomic mass. At first glance, the same seems to be true of the modern version – but not quite. Some pairs of elements are out of order, for example, tellurium, Te (A_r = 128) comes before iodine, I (A_r = 127). It is atomic number – the number of protons in the nucleus – that actually determines the place of an element in the periodic table.

With the exception of hydrogen, the elements can be organised into four blocks labelled s, p, d, and f (Figure 3). Elements in the same block show general similarities. For example, all the non-metals are in the p-block – many of the reactive metals (like sodium, potassium, and strontium) are in the s-block.

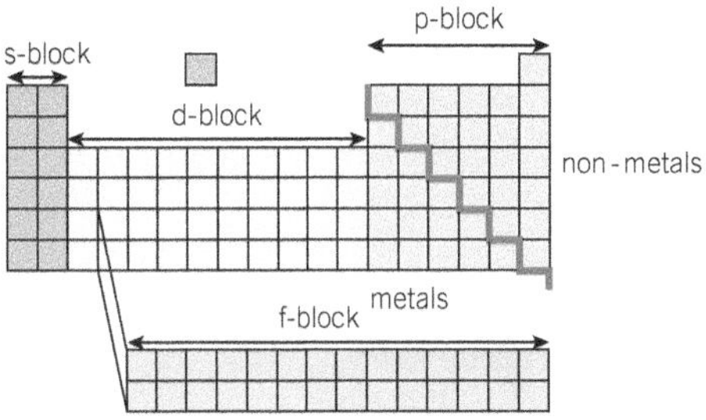

Vertical columns in the periodic table are called **groups**. Horizontal rows in the periodic table are called **periods**.

▲ Figure 3   *Blocks in the periodic table*

Physical properties

The elements in a group show patterns in their physical properties. There are *trends* in the properties as you go down a group. Because they cut across the groups there are fewer common features among the elements of a period but many properties vary in a fairly regular way as you move across a period from left to right. The pattern is then repeated as you go across the next period. The occurrence of periodic patterns is called **periodicity**.

One of the most obvious periodic patterns is the change from metals to non-metals as you go across the periods. The zig-zag line across the p-block in Figure 3 marks the change. The elements to the left of the

Synoptic link

You will find out more about intermolecular forces in Chapter 4, The ozone story.

zig-zag line display properties associated with metals, such as electrical conductivity. Elements to the right of the zig-zag line non-metals, displaying properties associated with them, for example, non-conductors of electricity. Other physical properties also show periodicity.

Melting points and boiling points

When elements are melted or boiled, the bonds between the individual atoms or molecules – the intermolecular forces – must be overcome. The strength of these bonds influences whether an element has a high or low melting point or boiling point.

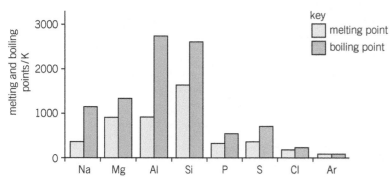

key
■ melting point
■ boiling point

▲ Figure 4 *Melting and boiling points of elements in Period 3*

In Period 3 (sodium to argon), both the melting and boiling points initially increase as you go across the period, and then fall dramatically from silicon, Si, to phosphorus, P (Figure 4). This means that the bonds between the particles in phosphorus must be much weaker and easier to overcome than those between the particles in silicon. Similar patterns in melting and boiling points are observed for the elements in other periods.

Electronic structure

The arrangement of elements by rows and columns in the periodic table is a direct result of the electronic structure of atoms. The number of outer shell electrons determines the group number. For example, lithium, sodium, and potassium all have one outer shell electron and are in Group 1. The noble gases all have filled outer shells – they have eight electrons in their outer shells, except helium which has two. They are in Group 0 (the zero refers to the shell beyond the filled outer one, which does not have any electrons in it yet).

The number of the shell that is being filled determines the period an element belongs to. From lithium to neon, the outer electrons are being placed in shell two, so these elements belong to Period 2.

Chemical properties

The chemical properties of an element are decided by the electrons in the incomplete outer shells – these are the electrons that are involved in chemical reaction.

Compare the electronic arrangement of the noble gases:

He	$1s^2$	Ar	$1s^2 2s^2 2p^6 3s^2 3p^6$
Ne	$1s^2 2s^2 2p^6$	Kr	$1s^2 2s^2 2p^6 3s^2 3p^6 3d^{10} 4s^2 4p^6$

All have sub-shells fully occupied by electrons. Such arrangements are called **closed shell** arrangements. These are particularly stable arrangements.

Now compare the electronic configurations of the elements in Group 1. They all have one electron in the outermost s-sub-shell and as a result show similar chemical properties. Group 2 elements all have two

Study tip

When writing the electronic configuration of an anion, (a negatively charged ion) electrons are added to the electron configuration of the neutral atom.

When writing the electronic configuration of a cation (a positively charged ion), electrons are taken away to the electron configuration of the neutral atom

electrons in the outermost s-sub-shell. Groups 1 and 2 elements are known as s-block elements.

In Groups 3, 4, 5, 6, 7, and 0 the outermost p-sub-shell is being filled. These elements are known as p-block elements. The elements where a d-sub-shell is being filled are called d-block elements, and those where an f-sub-shell is being filled are called f-block elements (Figure 3). Dividing it up in this way is very useful to chemists since it groups together elements with similar electronic configurations and similar chemical properties.

Electronic configuration of s- and p-block ions

- Group 1 elements form +1 ions.
- Group 2 elements form +2 ions.
- Group 7 elements usually form −1 ions.
- Group 6 elements form −2 ions.
- Aluminium forms +3 ions.

Knowing this, you can write the electronic configurations of ions formed by the s- and p-block elements in the first four periods.

 Worked example: Electronic configuration of an anion

What is the electronic configuration of S^{2-}?

Step 1: Write out the electronic configuration of a sulfur atom.

$$1s^2 2s^2 2p^6 3s^2 3p^4$$

Step 2: Use the charge on the ion to decide how many electrons to add to give the required charge.

S^{2-} has a charge of −2 so need to add two electrons.

Step 3: Add the required number of electrons to the electronic configuration.

$$1s^2 2s^2 2p^6 3s^2 3p^6$$

 Worked example: Electronic configuration of a cation

Draw out the electronic configuration of Na^+, showing the sub-shells and atomic orbitals.

Step 1: Write out the electronic configuration of a sodium atom.

$1s^2 \quad 2s^2 \quad 2p^6 \quad 3s^1$

| ↑↓ | ↑↓ | ↑↓ | ↑↓ | ↑↓ | ↑ |

Step 2: Use the charge on the ion to decide how many electrons to take away to give the required charge.

Na^+ has a charge of +1 so need to take away one electron.

Step 3: Add the required number of electrons to the electronic configuration.

$1s^2 \quad 2s^2 \quad 2p^6$

| ↑↓ | ↑↓ | ↑↓ | ↑↓ | ↑↓ |

1 The electron shell configuration for sodium can be written as 2.8.1. Use this notation to write down the electron shell configurations for:
   a lithium   (1 mark)
   b phosphorus   (1 mark)
   c calcium.   (1 mark)

2 The electron shell configurations of unknown elements **A** to **E** are given below. Which of these elements are in the same group?   (2 marks)
   A 2.8.2   B 2.6   C 2.8.8.2
   D 2.7   E 2.2

3 Copy and complete the following table that shows the electron shell configurations of some elements and the groups and periods to which they belong.   (3 marks)

Electron shell configuration	Group	Period
2.8.7	7	3
2.3		
2.8.6		
	4	2
	4	3
2.1		
2.8.1		
2.8.8.1		

4 Classify the following elements as s-, p-, d-, or f-block elements.
   a $[Kr]5s^1$ (1 mark)
   b $1s^2 2s^2 2p^6 3s^2 3p^4$ (1 mark)
   c $[Ar]3d^{10}4s^2 4p^6$ (1 mark)
   d $[Xe]6s^2$ (1 mark)

5 Classify the following elements as s-, p-, d-, or f-block elements.
   a chromium   (1 mark)
   b aluminium   (1 mark)
   c uranium   (1 mark)
   d strontium   (1 mark)

EL 5 The molecules of life

Specification reference: EL(i), EL(j), EL(k)

Learning outcomes

Demonstrate and apply knowledge and understanding of:

→ chemical bonding in terms of electrostatic forces – simple electron dot-and-cross diagrams to describe the electron arrangements in ions and covalent and dative covalent bonds

→ the bonding in simple molecular structure types; the typical physical properties (melting point, solubility in water, electrical conductivity) characteristic of these structure types

→ use of the electron pair repulsion principle, based on dot-and-cross diagrams, to predict, explain, and name the shapes of simple molecules (such as $BeCl_2$, BF_3, CH_4, NH_3, H_2O, and SF_6) and ions (such as NH_4^+) with up to six outer pairs of electrons (any combination of bonding pairs and lone pairs) and assigning bond angles to these structures.

Humans are not made up of single atoms but rather molecules and some ions. What are the molecules of life and how did they come into existence?

Cold chemistry and the molecules of life

Although hydrogen is the most common element in space, its atoms are spread out. There is about one atom per cubic centimetre (cm^3) in the space between the stars, so there is almost no chance that hydrogen atoms will come together to form hydrogen molecules.

However, dense gas clouds, or molecular gas clouds, do exist between stars and may contain between 100 and 1×10^6 particles per cm^3. The gas clouds are made up of a mixture of atoms and molecules, mainly of hydrogen, together with the dust of solid material from the breakup of old stars. They have been detected by radio and infrared telescopes on Earth and by spectroscopic instruments carried by rockets.

Table 1 shows some of the chemical species found in these gas clouds. Many of the substances contain carbon atoms bonded to elements other than just oxygen – they are **organic species**. These elements are major constituents of the human body.

Monatomic (1 atom)	Diatomic (2 atoms)	Triatomic (3 atoms)	Tetra-atomic (4 atoms)	Penta-atomic (5 atoms)
C^+	H_2	H_2O	NH_3	HCOOH
Ca^{2+}	OH	H_2S	H_2CO	NH_2CN
H^+	CO	HCN	HNCO	HC_3N
	CN	HNC	HNCS	C_4H
	CS	SO_2	C_3N	CH_2NH
	NS	OCS		CH_4
	SO	N_2H^+		
	SiO	HCS^+		
	SiS	HCO^+		
	C_2	NaOH		
	CH^+			
	NO			

Hexa-atomic (6 atoms)	Hepta-atomic (7 atoms)	Octa-atomic (8 atoms)	Nona-atomic (9 atoms)	Others
CH_3OH	CH_3CHO	$HCOOCH_3$	CH_3CH_2OH	HC_9N
NH_2CHO	CH_3NH_2		CH_3OCH_3	$HC_{11}N$
CH_3CN	H_2CCHCN		CH_3CH_2CN	
CH_3SH	CH_3C_2H		HC_7N	
CH_2CCH				

▶ Table 1 *Some chemical species in the dense gas clouds*

Where did the molecules of life come from?

In 1950 the scientist Stanley Miller put methane, ammonia, carbon dioxide, and water – simple molecules like those present in the dense gas clouds – into a flask. He heated and subjected the mixture to an electrical discharge to simulate the effect of lightning. Miller found that some of the reaction mixture had been converted into amino acids – the building blocks of proteins. Proteins are a group of compounds needed for correct cell functioning. This is one of the experiments that suggests life on Earth could have originated from the molecules in the dense gas clouds in outer space.

Analysis of dust, collected from a comet, showed the presence of polycyclic aromatic hydrocarbons as well as organic compounds rich in oxygen and nitrogen. These molecules are of interest to astrobiologists as these compounds could play important roles in terrestrial biochemistry. Temperatures in gas clouds are too low for chemical reactions to occur naturally, but ultraviolet light may penetrate the clouds and provide the energy to break covalent bonds. The simple molecules may then go on to form larger molecules. The evidence is building but the jury is still out.

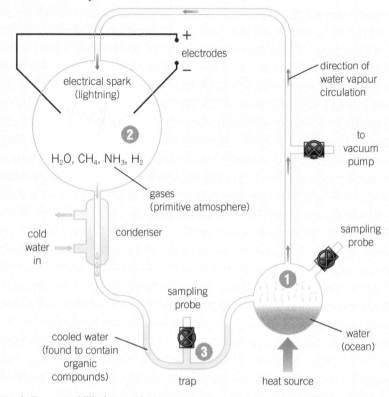

▲ Figure 1 *Miller's experiment*

> ### Study tip
> Hydrocarbons are molecules made of carbon and hydrogen.

Chemical ideas: Bonding, shapes, and sizes 3.1a

Chemical bonding

Covalent bonding

When scientists realised that the noble gases all have eight electrons in their outer shells (with the exception of helium) they linked this to the chemical stability of the noble gases. They suggested that other elements achieve eight outer shell electrons by losing or gaining electrons during reactions to form compounds. Furthermore, it seemed that some light elements achieve stability by reaching the helium configuration of two outer shell electrons. This was the basis for early ideas about why elements combine to form compounds. It is still useful today – though it is not the whole story.

> ### Synoptic link
> You will find out about ions and ionic bonding in Topic EL 7, Blood, sweat, and seas.

▲ Figure 2 *Electron sharing in the hydrogen molecule, H_2*

Dot-and-cross diagram

Diagrams used to represent the way that atoms bond together. In these diagrams, the outer shell electrons of one atom are represented by dots, with crosses for the other.

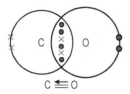

▲ Figure 4 *Dot-and-cross diagram for oxygen, O_2*

Study tip

The dot-and-cross model is an over-simplified model for bonding but it is nevertheless useful. It needs to be extended before it can do more than explain simple situations. Don't be surprised if dot-and-cross diagrams sometimes don't work, or if some molecules seem to break the rules.

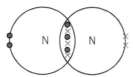

c ⇌ o

▲ Figure 7 *Dative covalent bonding in carbon monoxide, CO*

Covalent bonds

When non-metallic elements react it is not energetically favourable to form ions – electrons are *shared* between the atoms of the elements in these compounds. This sharing of electrons is called **covalent bonding**. The resulting compound is more stable than the individual elements. Shared electrons count as part of the outer shell of *both* atoms in the bond. Dot-and-cross diagrams can be used to show the bonding in covalent compounds (Figure 2).

The dot-and-cross diagrams show how sharing electrons gives the atoms more stable electron structures, like noble gases. Electron pairs that form bonds are called bonding pairs. Pairs of electrons not involved in bonding are called **lone pairs**. Ammonia, NH_3, has one lone pair of electrons and water, H_2O, has two lone pairs of electrons (Figure 3).

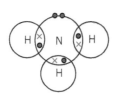

▲ Figure 3 *Dot-and-cross diagrams for ammonia, NH_3, and water, H_2O*

When two pairs of electrons form a covalent bond, it is called a *double bond*. The bonds in molecular oxygen are double covalent bonds (Figure 4).

When three pairs of electrons form a bond, it is called a *triple bond*. The bonds in nitrogen, N_2, and between carbon and nitrogen in hydrogen cyanide, HCN, are examples (Figure 5).

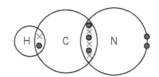

▲ Figure 5 *Dot–and-cross diagrams for nitrogen, N_2, and hydrogen cyanide, HCN*

Dot-and-cross diagrams are useful for representing individual electrons in a chemical bond. A simpler way of drawing molecules is by using lines to represent a pair of electrons shared between two atoms. A single line represents a single covalent bond, whilst double and triple lines represent double and triple covalent bonds (Figure 6).

H — H
hydrogen

O ═ C ═ O
carbon dioxide

H — C ≡ N
hydrogen cyanide

O ═ O
oxygen

N ≡ N
nitrogen

▲ Figure 6 *Covalently bonded molecules*

Dative covalent bonds

Carbon monoxide, CO, involves a triple bond. Two of the pairs of electrons are formed by the carbon and oxygen atoms each contributing one electron to the pair – these are ordinary covalent bonds. However both electrons in the third pair come from the oxygen atom – this is called a **dative covalent bond**. In a dative bond, *both* bonding electrons come from the *same* atom. A dative covalent bond is shown by an arrow, with the arrow pointing away from the atom that donates the pair of electrons (Figure 7).

Physical properties of covalently bonded simple molecules

The covalent intramolecular bonds *within* simple molecules such as water and ammonia are strong. However, the electrostatic attractions between different simple molecules are weak. This means relatively small amounts of energy are needed to separate one molecule from another. Therefore, elements and compounds with a simple molecular structure have relatively low melting and boiling points. Because there are no charged particles, covalent simple molecules do not conduct electricity. They mostly do not dissolve readily in water.

Chemical ideas: Bonding, shapes, and sizes 3.2

The shapes of molecules

Simple, covalently bonded molecules have a specific shape that can be determined using a theory called electron pair repulsion theory.

Electron pair repulsions

In methane, CH_4, there are *four groups* of electrons around the central carbon atom – the four covalent bonding pairs. Because similar charges repel, these groups of electrons arrange themselves so that they are as far apart as possible.

The farthest apart they can get is when the H—C—H bond angles are 109.5°. This corresponds to a tetrahedral shape, with the hydrogen atoms at the corners and the carbon atom in the centre. When drawing three-dimensional structures, bonds that lie in the plane of the paper are drawn as solid lines. Bonds that come towards you are represented by solid, wedge-shaped lines. Bonds that go away from you are shown as dashed wedges (Figure 8).

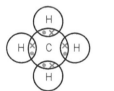

▲ Figure 8 *A dot-and-cross diagram showing the covalent bonding in methane (left), tetrahedral methane (middle), and a tetrahedron (right)*

All carbon atoms that are surrounded by four single bonds have a tetrahedral distribution of bonds, for example, in ethane both carbons have a tetrahedral arrangement (Figure 9). The same ideas apply to ions, for example, the ammonium ion, NH_4^+, is tetrahedral (Figure 10).

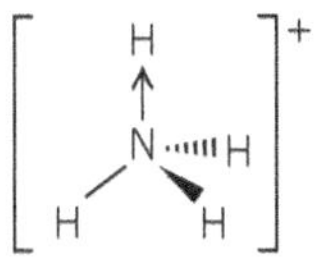

Lone pairs count too

Molecules such as ammonia and water have lone pairs in the outer shells of nitrogen and oxygen (Figure 3). These lone pairs repel

look along this bond

▲ Figure 9 *The tetrahedral bonding around the carbon atoms in ethane*

▲ Figure 10 *The tetrahedral arrangement of the ammonium ion*

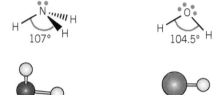

▲ **Figure 11** *The shapes of molecules of ammonia (left) and water (right)*

the bonding pairs of electrons, as in methane. The electron pairs in ammonia and water adopt a tetrahedral shape, but one or more of the corners are occupied by lone pairs of electrons.

The H—N—H and H—O—H bond angles are both close to 109° (Table 2). Ammonia is said to have a *pyramidal* shape and water is described as *bent* (Figure 11).

The shapes of covalent molecules are decided by a simple rule – groups of electrons in the outer shell repel one another and move as far apart as possible. It doesn't matter if they are groups of bonding electrons or lone pairs – they all repel one another. Lone pairs of electrons repel more strongly than electrons involved in bonding. This explains the different bond angles observed in methane, ammonia, and water even though all three molecules have four groups of electrons around a central atom (Table 2).

▼ **Table 2** *Summary of bond angles for compounds with different numbers of covalent bonds and lone pairs of electrons*

Number of groups of electrons	Types of groups	Example	Bond angle
4	4 × single covalent bonds	methane	109.5°
4	3 × single covalent bonds 1 × lone pair of electrons	ammonia	107.0°
4	2 × single covalent bonds 2 × lone pair of electrons	water	104.5°

Other shapes

There are several other types of shapes that molecule can have.

Linear molecules

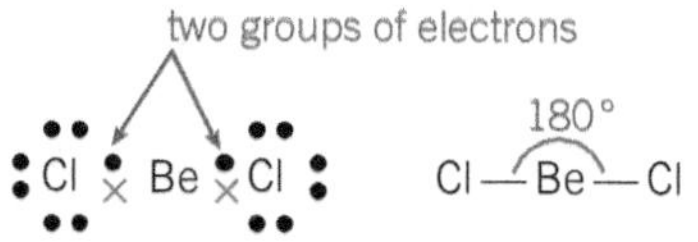

In $BeCl_2$, there are two groups of electrons around the central atom (Figure 12). Because there are fewer groups of electrons than in methane, they can get further apart. The furthest apart they can get is at an angle of 180°, so $BeCl_2$ is linear with a Cl—Be—Cl angle of 180°.

two groups of electrons

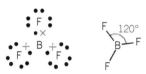

▲ **Figure 12** *The shape of the $BeCl_2$ molecule*

$O{=}C{=}O$ $H{-}C{\equiv}C{-}H$

▲ **Figure 13** *Carbon dioxide (left) and ethyne (right) both have linear molecules*

Carbon dioxide and ethyne are other examples of linear molecules – there are only two groups of electrons around the central atoms in these molecules (Figure 13).

Planar molecules

In BF_3 there are *three*, not four, groups of electrons around the central atom. The F—B—F bond angle is 120°. BF_3 is flat and shaped like a triangle with the boron atom at the centre and the fluorine atoms at the corners (Figure 14). It is described as planar triangular.

▲ **Figure 14** *The shape of the BF_3 molecule*

Methanal and ethane are other examples of planar structures (Figure 15). In methanal there are three groups of electrons around the carbon atom, and in ethene *both* carbon atoms have three groups of electrons. Remember it is the number of *groups* of electrons that determine the shape – there are four pairs of electrons around the carbon atoms in these molecules, but the bonds are *not* tetrahedrally directed because there are only three *groups* of electrons.

▲ Figure 15 *Methanal and ethene have planar molecules*

Bipyramidal molecules

Five groups of electrons around a central atom give rise to a trigonal bipyramidal shape, with electrons at the five corners of the shape. An example of such a molecule is phosphorus pentachloride, PCl_5. Bond angles approximate to 120° or 90°, depending on their position within the molecule (Figure 16).

▲ Figure 16 *The shape of the PCl_5 molecule. The lone pairs of electrons on chlorine have been omitted for clarity*

Octahedral molecules

Six groups of electrons around a central atom gives rise to an octahedral shape, where the electrons are directed to the six corners of an octahedron. Some molecules, such as SF_6, adopt this structure but it is more commonly found in the octahedral shapes of complexes of metal ions with six ligands (Figure 17).

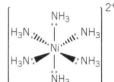

▲ Figure 17 *Six groups of electrons around a central atom (left) or an ion (right) give an octahedral shape (because there are eight faces). The lone pairs of electrons on fluorine have been omitted for clarity*

Summary questions

1 Draw electron dot-and-cross diagrams for the following covalent substances.
   a chlorine, Cl_2 *(1 mark)*
   b hydrogen chloride, HCl *(1 mark)*
   c methane, CH_4 *(1 mark)*
   d hydrogen sulfide, H_2S *(1 mark)*
   e aluminium bromide, $AlBr_3$ *(1 mark)*
   f silicon chloride, $SiCl_4$ *(1 mark)*

2 Draw electron dot-and-cross diagrams for the following covalent substances:
   a ethene *(1 mark)*
   b ethyne *(1 mark)*
   c methanol *(1 mark)*

3 Ammonia, NH_3, and boron trifluoride, BF_3, combine together to form the molecule NH_3BF_3. This molecule has a dative covalent bond between the nitrogen atom and the boron atom. Draw a dot-and-cross diagram for this molecule. *(1 mark)*

4 For each of the following molecules, draw an electron dot-and-cross diagram to show the bonding pairs and lone pairs of electrons in the molecule. Give the bond angles you would expect in each molecule.
   a SiH_4 *(1 mark)*   e SF_2 *(1 mark)*
   b H_2S *(1 mark)*   f BCl_3 *(1 mark)*
   c PH_3 *(1 mark)*   g C_2H_2 *(1 mark)*
   d CO_2 *(1 mark)*

5 Draw a diagram to show the bonds and lone pairs of electrons in each of the following molecules. Write in the bond angles you would expect to find in each structure.
   a CH_3CH_3 *(2 marks)*
   b CH_3OH *(2 marks)*
   c CH_3NH_2 *(2 marks)*
   d $CH_2{=}CH_2$ *(2 marks)*
   e $CH_3C{\equiv}N$ *(2 marks)*
   f NH_2OH *(2 marks)*
   g $COCl_2$ *(2 marks)*

6 What shapes are the following molecules?
   a NF_3 *(1 mark)*   c SF_6 *(1 mark)*
   b BF_3 *(1 mark)*

Learning outcomes

Demonstrate and apply knowledge and understanding of:

→ Avogadro constant N_A, relative formula mass and relative molecular mass M_r

→ the concept of amount of substance (moles) and its use to perform calculations involving – masses of substances, empirical and molecular formulae, percentage composition, percentage yields, water of crystallisation

→ balanced full chemical equations, including state symbols.

What are you made of? You could answer this biologically, and talk about organs and bones, or be more detailed and mention proteins, fats, and DNA. However, a chemist would talk about atoms and molecules or elements and compounds.

Elements and the body

A body is not really made up of a mixture of elements but rather a mixture of compounds, many of which appear quite complicated. The elements that make up these compounds are shown in Figure 1. You will be finding out more about some of these compounds later in this chapter. To begin with, however, you will look at the elements that are most likely to be in the compounds in your body.

Counting atoms of elements

Elements in the body are classified as major constituent elements, trace elements, and ultra-trace elements (Table 1). The ultra-trace elements are not given because their quantities are so small.

▼ Table 1 *The major constituent elements in the human body*

Element	Mass in a 60 kg person / g	Percentage of atoms	Element classification
oxygen	38 800	25.50	major constituent
carbon	10 900	9.50	major constituent
hydrogen	5990	63.00	major constituent
nitrogen	1860	1.40	major constituent
calcium	1200	0.31	trace
phosphorus	650	0.22	trace
potassium	220	0.06	trace
sulfur	150	0.05	trace
chlorine	100	0.03	trace
sodium	70	0.03	trace
magnesium	20	0.01	trace

The data gives conflicting interpretations of the importance of different elements in the body. For example, there are more atoms of hydrogen in your body than atoms of any other element, but hydrogen contributes far less than carbon or oxygen to the mass of your body. To decide which category an element belongs to, scientists determine how many atoms of an element there are in a body.

The mass of an element can be converted into the number of atoms by using a unit called the mole. One mole is the amount of an element

oxygen – 64.7%

potassium – 0.37%
hydrogen – 10%
phosphorus – 1.1%
chlorine – 0.17%
magnesium – 0.03%
calcium – 2.0%
nitrogen – 3.1%
iron – 0.006%
carbon – 18.2%

sulfur – 0.25%
sodium – 0.12%

▲ Figure 1 *The elements a human body is made of (% is by mass)*

that contains the same number of atoms as 12 g of carbon. Because atoms have exceedingly small masses, the number of atoms in one mole of an element is large: 6×10^{23} atoms per mole. Once you know the total number of atoms in the body and the number of atoms of each element in a body, you can work out the percentage of each element in the body. The mole is an important concept in chemistry and can be applied to both elements and compounds.

Chemical ideas: Measuring amounts of substance 1.1

Amount of substance

Imagine a bag containing 10 golf balls and 20 table tennis balls. The golf balls would make up most of the mass, but the table tennis balls would make up most of the volume. The composition of water has a similar situation. Oxygen makes up nearly 90% of the mass of water, but $\frac{2}{3}$ of the atoms are hydrogen atoms. However, you cannot pick up and count the number of oxygen and hydrogen atoms in water.

The formula of water is H_2O – there are *two* hydrogen atoms combined with *one* oxygen atom in each water molecule. The relative atomic mass A_r of an element links the mass of an element with the number of atoms.

Relative atomic mass

The relative atomic mass of an element is the mass of its atom relative to carbon-12, which is assigned a relative atomic mass of exactly 12. A_r values have no units. The relative atomic masses of some elements are listed in Table 2.

▼ Table 2 *Some relative atomic masses*

Element	Symbol	Approximate relative atomic mass A_r
hydrogen	H	1
helium	He	4
carbon	C	12
nitrogen	N	14
oxygen	O	16
magnesium	Mg	24
sulfur	S	32
calcium	Ca	40
iron	Fe	56
copper	Cu	64
iodine	I	127
mercury	Hg	200

Study tip

In this topic the A_r values are rounded to whole numbers. You will usually use A_r values to one decimal place.

▲ **Figure 2** *Two carbon-12 atoms have the same mass as one magnesium atom. So the relative atomic mass of magnesium is 2 × 12 = 24*

Chemical quantities

If you had two bottles containing equal masses of copper (A_r = 64) and sulfur (A_r = 32) you would know that you had twice as many sulfur atoms as copper atoms because sulfur atoms have only half the mass of copper atoms. If you had a bottle containing mercury (A_r = 200) that was five times heavier than a similar bottle containing calcium (A_r = 40), you would know that both bottles contained equal numbers of atoms because each mercury atom has five times the mass of each calcium atom.

12 g of carbon, 1 g of hydrogen, and 16 g of oxygen all contain equal numbers of atoms because these masses are in the same ratio as their relative atomic masses. This amount of each of these elements is called a **mole**, or mol for short. The mole is a unit for measuring the amount of substance. One mole of a substance contains as many particles (e.g., atoms, molecules, groups of ions) as there are atoms in 12 g of carbon-12.

Chemical amounts are defined so that the mass of one mole (the **molar mass**) is equal to the relative atomic mass in grams. So, the molar mass of carbon is 12 g mol^{-1}. If you had 6 g of carbon you would have 0.5 mol of carbon atoms and 3 g of hydrogen would contain 3 mol of hydrogen atoms.

$$\text{amount in moles } n \text{ (mol)} = \frac{\text{mass } m \text{ (g)}}{\text{molar mass } M \text{ (g mol}^{-1})}$$

 Worked example: Calculating the amount of substance

Calculate the amount of substance in 4 g of oxygen.

Step 1: Identify the molar mass of oxygen. 16 g mol^{-1}

Step 2: Substitute the values into the equation to calculate the amount of substance, in moles.

$$\frac{4 \text{ g}}{16 \text{ g mol}^{-1}} = 0.25 \text{ mol}$$

Relative formula mass

You can use the mole to deal with compounds as well as elements. A molecule of methane, CH_4, is formed when one carbon atom combines with four hydrogen atoms. Therefore one mole of methane is formed when one mole of carbon atoms combines with four moles of hydrogen atoms.

The **relative formula mass** M_r (or relative molecular mass) of a substance is the sum of the relative atomic masses of the elements making it up. The M_r of a substance can be worked out by writing the chemical formula of the substance then adding together the relative atomic masses of each of the atoms in the formula.

 Worked example: Relative formula mass

Calculate the relative formula mass of methane.

Step 1: Write out the chemical formula of methane. CH_4

Step 2: Use a periodic table to identify the relative atomic mass of each of the elements.

$$A_r(C): 12 \qquad A_r(H): 1$$

Step 3: Add together the relative atomic mass for each atom in the chemical formula.

$$1\ C \qquad 4\ H$$
$$(1 \times 12) + (4 \times 1) = 16$$

Like relative atomic mass, relative formula masses have no units and are on the same scale – relative to $A_r(^{12}C) = 12.000$. When a substance is made of discrete molecules, such as methane, the relative formula mass is often called the relative molecular mass. This also has the symbol M_r.

Formula units

Substances are made up of formula units. They are the basic units or building blocks, and match the formulae of the substances. Formula units can be single atoms, molecules, or groups of ions.

For example, the formula unit in metal copper is simply a copper atom, and the formula unit in non-metal carbon is a carbon atom. The formula unit of most elements is a single atom so their relative formula mass is identical to their relative atomic mass. There are some exceptions – in oxygen gas, the formula unit is the O_2 molecule so the relative formula mass is *twice* the relative atomic mass of oxygen.

In many covalent compounds, the formula unit is a molecule, for example, in methane the formula unit is the CH_4 molecule. However in ionic compounds the formula unit is a *group of ions*. In calcium nitrate the formula unit (or repeating unit) is $Ca(NO_3)_2$ and contains a group of three ions – one Ca^{2+} and two NO_3^-. These groups of ions are not labelled with any special name and so you just use the general name formula unit when referring to ionic compounds.

Moles of formula units

The relative formula mass in grams is equal to the molar mass, so the molar mass of methane is 16 g and calcium nitrate is 164 g. This means you can calculate the amount in moles of formula units.

amount in moles of formula units = mass ÷ molar mass

So, 8 g of methane contains 0.5 mol of CH_4 formula units and 41 g of calcium nitrate contains 0.25 mol of $Ca(NO_3)_2$ formula units.

As well as describing 16 g of methane as consisting of one mole of formula units (or molecules) of CH_4, you can also say that it contains one mole of carbon atoms and $(1 \times 4 =)$ four moles of hydrogen atoms. Similarly, 164 g of calcium nitrate contains one mole of formula units of $Ca(NO_3)_2$ – it also contains one mole of Ca^{2+} two moles 2 mol of NO_3^- ions.

For elements that exist as diatomic gases, such as oxygen, you must be especially careful. 32 g of oxygen gas contains one mol of O_2 molecules, but two moles of oxygen atoms. It is therefore essential to give the formula unit you are referring to when using moles, for example, moles of oxygen atoms or moles of O_2 molecules.

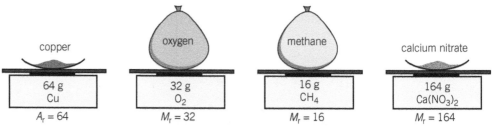

copper	oxygen	methane	calcium nitrate
64 g Cu	32 g O_2	16 g CH_4	164 g $Ca(NO_3)_2$
$A_r = 64$	$M_r = 32$	$M_r = 16$	$M_r = 164$

all contain the same number of formula units because their masses are in the same ratio as their relative formula masses

▲ Figure 3 *The important thing to remember is that equimolar amounts of substances contain equal numbers of formula units*

The Avogadro constant

The number of formula units in one mole of a substance is a constant. It is called the Avogadro constant N_A, after the Italian scientist Amedeo Avogadro. The value of the Avogadro constant is 6.02×10^{23} formula units per mole. One mole of atoms, molecules, and electrons will all contain $6.02 \times 10^{23}\,mol^{-1}$. The number is so huge that it is difficult to comprehend (Figure 4).

To work with specific numbers of formula units, you only have to use moles – easily done by using molar mass and mass of substance.

Working out chemical formulae

Moles can be used to work out chemical formulae. If a known mass of magnesium is reacted with oxygen to form magnesium oxide, you can find out the mass of oxygen that combined with the magnesium.

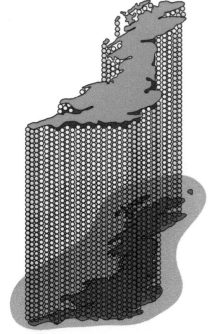

▲ Figure 4 *If the 6.02×10^{23} atoms in 12 g of carbon were turned into marbles, the marbles could cover Great Britain to a depth of 1500 km*

🔲 Worked example: Chemical formula using moles

In an experiment, 0.84 g of magnesium was burnt and combined with 0.56 g of oxygen. Using these results, work out the chemical formula of the magnesium oxide produced.

Step 1: Calculate the amount in moles of magnesium atoms in the reaction.

$$\frac{0.84\,g}{24\,g\,mol^{-1}} = 0.035\,mol^{-1}$$

Step 2: Calculate the amount in moles of oxygen atoms in the reaction.

$$\frac{0.56\,g}{16\,g\,mol^{-1}} = 0.035\,mol$$

Step 3: Calculate the ratio of moles of atoms of Mg:O in magnesium oxide. 1:1

Step 4: Write the formula of magnesium oxide. MgO

Activity EL 6.1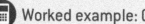

This activity allows you to practise writing formulae of ionic compounds.

Another way of analysing a compound is to find the *percentage mass* of each element it contains.

 Worked example: Chemical formula using percentage mass

The results from an experiment indicate that methane contains 75% mass of carbon and 25% mass of hydrogen. Work out the formula of methane.

Step 1: Calculate the mass of carbon in 100 g of methane.

100 g of methane contains 75 g of carbon.

Step 2: Calculate the amount of carbon atoms in 75 g.

$$\frac{75\,g}{12\,g\,mol^{-1}} = 6.25\,mol$$

Step 3: Calculate the mass of hydrogen in 100 g of methane.

100 g methane contains 25 g of hydrogen.

Step 4: Calculate the amount of hydrogen atoms in 25 g.

$$\frac{25\,g}{1\,g\,mol^{-1}} = 25\,mol$$

Step 5: Calculate the ratio of moles of atoms of C : H in methane. 1 : 4

Step 6: Write the formula of methane. CH_4

In a methane molecule a central carbon atom is surrounded by four hydrogen atoms – the simple ratio of atoms is the same as the formula of the molecule. This isn't the case for all substances. The **molecular formula** tells you the actual numbers of different types of atom. The molecular formula of ethane is C_2H_6, but the simplest ratio for the moles of atoms of C : H is 1 : 3 – a calculation from percentage masses would give a formula CH_3. This is the **empirical formula**. Table 3 gives examples of the two types of formula.

▼ Table 3 *Some molecular formulae and empirical formulae*

Substance	Molecular formula	Empirical formula
ethene	C_2H_4	CH_2
benzene	C_6H_6	CH
butane	C_4H_{10}	C_2H_5
phosphorus (V) oxide	P_4O_{10}	P_2O_5
oxygen	O_2	O
bromine	Br_2	Br

▲ **Figure 5** *When water is added to anhydrous copper(II) sulfate (white), hydrated copper(II) sulfate is formed (blue)*

Water of crystallisation

The crystals of some ionic lattices include molecules of water. These water molecules are fitted within the lattice in a regular manner, and are called water of crystallisation. The crystals are said to be hydrated, for example, $CuSO_4 \cdot 5H_2O$ is hydrated copper(II) sulfate and is blue coloured. When hydrated crystals are heated, the water of crystallisation is removed as steam leaving the anhydrous solid, for example, $CuSO_4$ is the anhydrous form of copper(II) sulfate and is white.

Calculations using compounds with a water of crystallisation

The formula of hydrated compounds can be determined using a simple experimental technique. The mass of a sample of the hydrated salt is measured. The salt is then heated, with the mass measured at regular intervals. When no more mass is lost, the formula can be determined.

 Worked example: Calculating the formula of a hydrated ionic compounds

2.53 g of hydrated magnesium chloride, $MgCl_2 \cdot xH_2O$, was heated to constant mass. 1.17 g of solid remained. What is the formula of the hydrated compound?

Step 1: Calculate the mass of water in 2.53 g $MgCl_2$

$$2.53 - 1.17 = 1.36\,g$$

Step 2: Calculate the number of moles of water.

$$\frac{1.36}{18} = 0.076 \text{ moles}$$

Step 3: Calculate the number of moles of $MgCl_2$ in the anhydrous solid. ($M_r(MgCl_2) = 95.9$)

$$\frac{1.17}{95.9} = 0.012 \text{ moles}$$

Step 4: Calculate the ratio of moles of magnesium chloride to moles of water by dividing both vales by the number of moles of magnesium chloride.

$$\frac{0.012}{0.012} = 1 \qquad \frac{0.076}{0.012} = 6.3 \qquad 1:6.3$$

Step 5: Use the ratio from Step 4 to determine the formula of the hydrated magnesium chloride.

$$MgCl_2 \cdot 6H_2O$$

Working out percentage yield

The mole can also be used to calculate the expected amount of product in a reaction carried out under ideal conditions. This is called the *theoretical yield*. However, certain factors can reduce the amount of products produced:

- loss of products from reaction vessels, particularly if there are several stages to the reaction
- side-reactions occurring, producing unwanted by-products
- impurities in the reactants
- changes in temperature and pressure
- if the reaction is an equilibrium system.

The reduced amount of products (called the *experimental* yield) can be expressed as a percentage of the theoretical (maximum) yield. This is the percentage yield.

$$\text{percentage yield} = \frac{\text{experimental yield}}{\text{theoretical yield}} \times 100$$

The theoretical yield can be calculated from the balanced equation for the particular reaction.

Synoptic link

You will find out more about equilibrium systems in Chapter 3, Elements from the sea.

 Worked example: Calculating percentage yield

0.84 g of magnesium was burnt in excess oxygen, producing 1.1 g of magnesium oxide. Using these results, work out the percentage yield of the reaction.

Step 1: Write the balanced equation for the reaction.

$$2Mg(s) + O_2(g) \rightarrow 2MgO(s)$$

Step 2: Calculate the amount of substance, in moles, of magnesium.

$$\frac{0.84\,g}{24\,g\,mol^{-1}} = 0.035\,mol$$

Step 3: The balanced equation tells us that the ratio of Mg to MgO produced is 2:2 or as the lowest whole number ratio 1:1. This means the maximum amount of magnesium oxide that can be produced from this amount of magnesium is also 0.035 mol.

Step 4: Calculate the formula mass of MgO.

$$A_r(Mg) = 24; A_r(O) = 16 \quad M_r(MgO) = 24 + 16 = 40$$

Step 5: Calculate the maximum mass of MgO you can expect to produce.

$$0.035\,mol\ of\ MgO \quad so,\ 40 \times 0.035 = 1.4\,g$$

Step 6: The experiment produced 1.1 g. Calculate the percentage yield.

$$\frac{1.1\,g}{1.4\,g} \times 100 = 79\%\ (2\ s.f.)$$

Study tip

You will need to use a table of relative atomic masses or the periodic table to answer the questions that follow. From now on, you will always see A_r and M_r figures quoted to one decimal place.

In the example above you were given the necessary balanced chemical equation. However, there will be many times where you will need to write your own equation.

Chemical ideas: Measuring amounts of substance 1.2

Balanced equations

A balanced chemical equation tells you the reactants and products in a reaction, and the relative amounts involved. The equation is balanced so that there are equal numbers of each type of atom on both sides of the arrow.

$$CH_4(g) + 2O_2(g) \rightarrow CO_2(g) + 2H_2O_{(l)}$$

This equation tells you that one molecule of methane reacts with two molecules of oxygen to form one molecule of carbon dioxide and two molecules of water. These are also the amounts in moles of the substances involved in the reaction, that is, one mole of methane reacts with two moles of oxygen.

The number written in front of each formula in a balanced equation tells you the number of formula units involved in the reaction. Remember that a formula unit may be a molecule or another species such as an atom or an ion. The small subscript numbers are part of the formulae and cannot be changed.

Writing balanced equations

The only way to be sure of the balanced equation for a reaction is to do experiments to find out what is formed in the reaction and what quantities are involved. However, equations are used a lot in chemistry and it isn't possible to do experiments every time. Fortunately, if you know the reactants and products, usually the formulae can be worked out and a balanced equation predicted. Equations can only be balanced by putting numbers in front of the formulae. You cannot balance them by altering the formulae because that would create different substances.

 Worked example: The steps for predicting balanced equations

What is the balanced equation for the reaction between calcium and water.

Step 1: Decide what the reactants and products are.

     calcium + water → calcium hydroxide + hydrogen

Step 2: Write the formula for the substances involved, including state symbols.

$$Ca(s) + H_2O(l) \rightarrow Ca(OH)_2(aq) + H_2(g)$$

Step 3: Balance the equation so that there are the same numbers of each type of atom on each side.

$$Ca(s) + 2H_2O \rightarrow Ca(OH)_2(aq) + H_2(g)$$

Study tip

State symbols are included in chemical equations to show the physical state of the reactants and products.

State	Symbol
gas	(g)
liquid	(l)
solid	(s)
aqueous solution	(aq)

Summary questions

1. One atom of element X is approximately 12 times heavier than one carbon atom.
   a What is the approximate relative atomic mass of this element? *(1 mark)*
   b Identify X. *(1 mark)*

2. Balance the following equations.
   a $Mg + O_2 \rightarrow MgO$ *(1 mark)*
   b $H_2 + O_2 \rightarrow H_2O$ *(1 mark)*
   c $CaCO_3 + HCl \rightarrow CaCl_2 + CO_2 + H_2O$ *(1 mark)*
   d $HCl + Ca(OH)_2 \rightarrow CaCl_2 + H_2O$ *(1 mark)*
   e $CH_3OH + O_2 \rightarrow CO_2 + H_2O$ *(1 mark)*

3. Write balanced equations, including state symbols for the following reactions.
   a zinc reacting with sulfuric acid, H_2SO_4, to form zinc sulfate, $ZnSO_4$, and hydrogen. *(2 marks)*
   b magnesium carbonate, $MgCO_3$, decomposing on heating to form magnesium oxide, MgO, and carbon dioxide. *(2 marks)*
   c barium oxide, BaO, reacting with hydrochloric acid to form barium chloride, $BaCl_2$, and water. *(2 marks)*

4. The empirical formula of a compound can be calculated when you know the masses of the elements in a sample of it.

   The steps to calculate the empirical formula of a compound if a 16.7 g sample of it contains 12.7 g of iodine and 4.0 g of oxygen are shown in Table 1.
   a Why do we need to know the mass of the sample as well as the masses of the elements in it? *(1 mark)*
   b In step 2, what do the numbers 0.1 and 0.25 represent? *(1 mark)*
   c Why, in step 3, do we divide both 0.1 and 0.25 by 0.1? *(1 mark)*
   d Why have we doubled the numbers in moving from step 3 to step 4? *(1 mark)*
   e Write down three possibilities for the molecular formula of this compound based on its empirical formula. *(3 marks)*
   f What additional information do you need to work out the actual molecular formula of the compound? *(1 mark)*

5. How many moles of atoms are contained in:
   a 32.1 g of sulfur *(1 mark)*
   b 31.8 g of copper *(1 mark)*
   c Explain why they are different, even though each question involves approximately 32 g of an element. *(2 marks)*

6. Calculate how many moles of each substance are contained in
   a 88 g of carbon dioxide *(1 mark)*
   b 2.92 g of sulfur hexafluoride, SF_6 *(1 mark)*
   c 0.37 kg of calcium hydroxide, $Ca(OH)_2$ *(2 marks)*
   d 18 tonnes of water (1 tonne $= 1 \times 10^6$ g) *(2 marks)*

▼ Table 1 *Steps involved in working out an empirical formula. Units have been omitted*

	Iodine	Oxygen
Step 1	12.7	4.0
Step 2	$\frac{12.7}{126.9} = 0.1$	$\frac{4.0}{16.0} = 0.25$
Step 3	$\frac{0.1}{0.1} = 1$	$\frac{0.25}{0.1} = 2.5$
Step 4	2	5
Step 5	The empirical formula of the compound is I_2O_5	

EL 7 Blood, sweat, and seas

Specification reference: EL(d), EL(l), EL(o), EL(s), EL(t)

Learning outcomes

Demonstrate and apply knowledge and understanding of:

→ balanced ionic chemical equations, including state symbols

→ the bonding in giant lattice (metallic, ionic, covalent network and simple molecular) structure types; the typical physical properties (melting point, solubility in water, electrical conductivity) characteristic of these structure types

→ structures of compounds that have a sodium chloride type lattice

→ the relationship between the position of an element in the s- or p-block of the Periodic Table and the charge on its ion; the names and formulae of NO_3^-, SO_4^{2-}, CO_3^{2-}, OH^-, NH_4^+, HCO_3^-, Cu^{2+}, Zn^{2+}, Pb^{2+}, Fe^{2+}, Fe^{3+}; formulae and names for compounds formed between these ions and other given anions and cations

→ the solubility of compounds formed between the following cations and anions: Li^+, Na^+, K^+, Ca^{2+}, Ba^{2+}, Cu^{2+}, Fe^{2+}, Fe^{3+}, Ag^+, Pb^{2+}, Zn^{2+}, Al^{3+}, NH_4^+, CO_3^{2-}, SO_4^{2-}, Cl^-, Br^-, I^-, OH^-, NO_3^-; colours of any precipitates formed; use of these ions as tests, for example, Ba^{2+} as a test for SO_4^{2-}; a sequence of tests leading to the identification of a salt containing the ions above

→ techniques and procedures for making soluble salts by reacting acids and bases and insoluble salts by precipitation reactions.

Other than water, what connects blood, sweat, and seas? An important link is ions.

Salts

Sea water contains dissolved salts. Salts are ionic compounds and many of the ions found present in sea water are important for a healthy body. For example, sodium and potassium ions are present in both blood and sea water, but in *very* different amounts (Table 1). Na^+ ions are lost in sweat and these must be replaced. Natural sea salt sometimes contains iodide ions, I^-, which are also needed for a healthy life. However, evidence shows that too much salt (in everyday language, salt usually means sodium chloride) can be harmful.

Salts get into our seas and rivers by dissolving out of rocks. Occasionally shallow seas form in hot climates, particularly over geological time, and evaporation of the water leaves large deposits of solid salts, called evaporites (Figure 1).

▼ Table 1 *Concentration of sodium and potassium ions in blood and sea water*

Substance	Concentration / mmol dm^{-3}	
	Na$^+$	K$^+$
sea water	470	10
blood cells	10	150

Salts of other metallic elements are also important in the body. Compounds of calcium are vital for healthy bones and teeth and these salts are insoluble. Calcium ions are precipitated from sea water as calcium carbonate, and make the shells and skeletons of many marine animals (Figure 2).

▲ Figure 1 *Lithium salt mining in Atacama desert, Chile*

▲ Figure 2 *Calcareous fossils called foraminifera*

Formation of salts

A consequence of the increase in atmospheric carbon dioxide is that more of the gas dissolves in sea water. This lowers the pH of the sea and could result in the calcium carbonate shells of marine animals dissolving, with disastrous consequences to the marine ecosystem.

The reaction of carbonate ions with acids to form a salt is an example of a neutralisation reaction. Table 2 gives some examples of other salts and their formation and uses.

▼ Table 2 *Some salts and their uses*

Salt	Use	Possible method (with reactants) to make salt in the lab		Formula of salt
		Reactant 1	Reactant 2	
magnesium sulfate	bath salts 	sulfuric acid **acid**	magnesium metal **metal**	$MgSO_4$
lithium chloride	lithium batteries 	hydrochloric acid **acid**	lithium oxide **metal oxide**	$LiCl$
barium sulfate	'barium meal' shows up soft tissue on X-ray 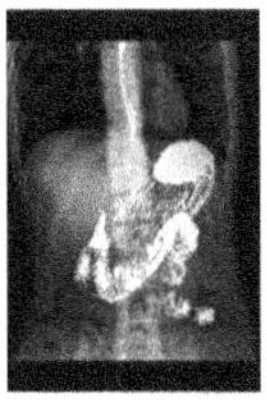	barium chloride $Ba^{2+}(aq)$	magnesium sulfate $SO_4{}^{2-}(aq)$	$BaSO_4$ this salt is insoluble and made by precipitation
sodium ethanoate	hand warmers 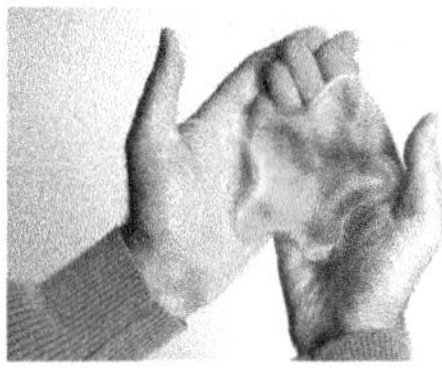	ethanoic acid **acid**	sodium carbonate **carbonate**	CH_3COONa
ammonium nitrate	fertiliser 	nitric acid **acid**	ammonium hydroxide (ammonia solution) **hydroxide**	NH_4NO_3

Acids

In almost all the methods in Table 2, one of the reactants is always an acid. Acids produce hydrogen ions, H^+, in solution. Some common acids are:

- sulfuric acid, H_2SO_4
- hydrochloric acid, HCl
- nitric acid, HNO_3
- ethanoic (or acetic) acid, CH_3COOH.

Chemical bonding

Formation of ions and electrostatic bonds

The elements in Groups 1 and 2 of the periodic table have only one or two outer shell electrons. They *lose* these electrons to form positively charged ions, called **cations**.

Most non-metal atoms have more than three outer shell electrons. They are able to *gain* electrons to form negatively charged ions, called **anions**.

Gaining one electron gives an anion with a single negative charge. A second electron is repelled by this anion because their charges are the same, so making a doubly charged anion is much harder. Getting a anion with a −3 charge is very difficult and does not often happen. It is also hard to remove three or more electrons from atoms – cations with a +4 charge are almost unknown.

- Group 1 elements form +1 ions
- Group 2 elements form +2 ions
- Group 6 elements form −2 ions
- Group 7 elements form −1 ions

Ionic bonding

When *metals* react with *non-metals* in a chemical reaction, ions are only formed if the *overall* energy change for the reaction is favourable. Electrons are *transferred* from the metal atoms to the non-metal atoms, often giving both the metal and the non-metal a stable electronic structure like that of a noble gas. The cations and anions formed are held together by their opposite charges in an **electrostatic bond**.

Figure 3 shows dot-and-cross diagrams for the formation of sodium chloride and magnesium fluoride. Each sodium atom loses one electron and each chlorine atom gains one electron, so the compound formed has the formula NaCl. Each magnesium atom loses two electrons but each fluorine atom gains only one electron, so the formula for magnesium fluoride is MgF_2.

Activity EL 7.1

In this activity you can check your understanding of why atoms form ions.

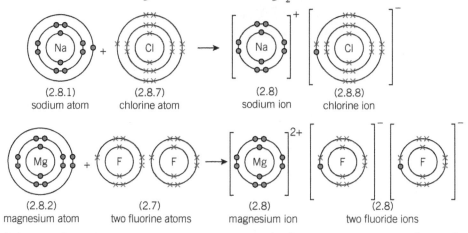

| (2.8.1) | (2.8.7) | (2.8) | (2.8.8) |
| sodium atom | chlorine atom | sodium ion | chlorine ion |

| (2.8.2) | (2.7) | (2.8) | (2.8) |
| magnesium atom | two fluorine atoms | magnesium ion | two fluoride ions |

▲ Figure 3 *Dot-and-cross diagrams for sodium chloride (top) and magnesium fluoride (bottom). The numbers show the arrangement of electrons in shells and only the outer electrons are shown*

The oppositely charged ions attract each other strongly in an ionic bond. In the solid compound, each ion attracts many others of opposite charge and the ions build up into a giant lattice (Figure 3).

Writing chemical formulae of ionic compounds

Table 3 lists some common cations and anions.

Study tip

When an anion name ends in –ate, it means there is one or more non-metals bonded to oxygen.

▼ Table 3 *The names and formulae of some common ions. For elements where there is more than one common ion, the oxidation state is given*

Cations (positive ions)			Anions (negative ions)	
+1	+2	+3	−1	−2
H^+ hydrogen	Mg^{2+} magnesium	Al^{3+} aluminum	F^- fluoride	O^{2-} oxide
Li^+ lithium	Ca^{2+} calcium	Fe^{3+} iron(III)	Cl^- chloride	CO_3^{2-} carbonate
Na^+ sodium	Ba^{2+} barium		Br^- bromide	SO_4^{2-} sulfate
K^+ potassium	Fe^{2+} iron(II)		I^- iodide	
NH_4^+ ammonium	Cu^{2+} copper(II)		OH^- hydroxide	
H^+ hydrogen	Zn^{2+} zinc		NO_3^- nitrate(V) or nitrate	
	Pb^{2+} lead(II)		HCO_3^- hydrogen carbonate	

Some ions contain more than one type of atom, such as hydroxide, OH^-, and sulfate, SO_4^{2-}. These are complex ions and consist of a group of atoms held together by covalent bonds, but the whole group carries an electric charge. The dot-and-cross diagram for sodium hydroxide is shown in Figure 4.

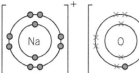 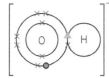

 Figure 4 *Dot-and-cross diagram for sodium hydroxide*

Once you know the formulae of the ions in an ionic compound, it is easy to write the formula of the compound. You need to make sure that the total positive charge is equal to the total negative charge – when you add all the charges on the ions, they add up to zero.

🖩 Worked example: Writing formulae for ionic compounds

Write the formula for sodium chloride.

Step 1: Identify the atoms and charges a compound is made of – sodium chloride is made up of sodium ions, Na^+, and chloride ions, Cl^-.

Step 2: Work out the formula so the charges balance.

(+1) + (−1) = 0, so the formula is NaCl.

Study tip

When brackets are used in a formula, everything inside the bracket is multiplied by the number outside, for example, $Fe(OH)_2$ means one Fe^{2+} and 2 × OH^-.

→

Write the formula for chromium(III) hydroxide.

Step 1: Identify the atoms and charges a compound is made of – chromium(III) hydroxide is made up of chromium(III) ions (Cr^{3+}) and hydroxide ions (OH^-).

Step 2: Work out the formula so the charges balance.

To balance the +3 charge you need three −1 charges. The formula is $Cr(OH)_3$.

Making ionic salts

The salts in Table 2 are of ionic compounds. There are several ways of making salts. You should already be familiar with these:

- acid + alkali → salt + water
- acid + base → salt + water
- acid + carbonate → salt + water + carbon dioxide
- acid + metal → salt + hydrogen

Chemical ideas: Structure and properties 5.1

Ionic substances in solution

Many ionic substances dissolve readily in water. There are however some ionic compounds that are not soluble or only sparingly soluble and these include:

- barium, calcium, lead, and silver sulfates
- silver and lead halides (chlorides, bromides, and iodides)
- all metal carbonates
- metal hydroxides (except Group 1 hydroxides and ammonium hydroxide).

When ionic substances dissolve, the ions become surrounded by water molecules and spread out through the solution. Once they are separated they behave independently of each other. This presence of hydrated ions in solution explains why aqueous solutions of salts can conduct electricity.

Ionic equations

When ions are in solution, the reactions of an ionic substance, such as sodium chloride, quite often involve only one of the two types of ion – the other ion does not get involved in the reaction. For example, if you add a solution of sodium chloride to silver nitrate solution, you get a white precipitate of silver chloride. Silver ions, Ag^+, and chloride ions, Cl^-, have come together to form insoluble silver chloride, which precipitates out. You can write an equation for this reaction showing only the ions that take part in the reaction.

$$Ag^+(aq) + Cl^-(aq) \rightarrow AgCl(s)$$

Synoptic link

You will find out more about neutralisation in EL 9, How salty?

Synoptic link 🧪

You can find out more about making soluble and insoluble salts in Techniques and procedures.

▲ **Figure 5** *Lead(II) chromate – also known as the pigment chrome yellow – precipitating from mixing lead nitrate and potassium chromate solutions*

Study tip

It is worth remembering that all Group 1 compounds, ammonium compounds, and all nitrates are soluble in water.

Activity EL 7.2 ↗

This activity allows you to apply your understanding of precipitation reactions.

The $Na^+(aq)$ and $NO_3^-(aq)$ ions are not involved in the reaction so can be left out of the equation – they are **spectator ions**. This is an **ionic equation** – it only shows the ions that take part in the reaction and excludes the spectator ions.

State symbols are very important in ionic equations. They help to identify that the above reaction involves ionic **precipitation** – a suspension of solid particles is produced by a chemical reaction in solution.

Sometimes precipitation reactions can be used to identify certain metal cations in solution (Table 5).

▼ Table 4 *State symbols*

State	State symbol
gas	(g)
liquid	(l)
solid	(s)
aqueous solution	(aq)

▼ Table 5 *Precipitation tests for some cations and anions*

Cation or anion being tested for	Solution added	Precipitate formed	Colour of precipitate	Overall ionic equation
Cu^{2+}	sodium hydroxide	copper hydroxide	blue	$Cu^{2+}(aq) + 2OH^-(aq) \rightarrow Cu(OH)_2(s)$
Fe^{2+}	sodium hydroxide	iron(II) hydroxide	'dirty' green	$Fe^{2+}(aq) + 2OH^-(aq) \rightarrow Fe(OH)_2(s)$
Fe^{3+}	sodium hydroxide	iron(III) hydroxide	orange/brown	$Fe^{3+}(aq) + 3OH^-(aq) \rightarrow Fe(OH)_3(s)$
Pb^{2+}	potassium iodide	lead iodide	bright yellow	$Pb^{2+}(aq) + 2I^-(aq) \rightarrow PbI_2(s)$
Cl^-	silver nitrate	silver chloride	white	$Ag^+(aq) + Cl^-(aq) \rightarrow AgCl(s)$
Br^-	silver nitrate	silver bromide	cream	$Ag^+(aq) + Br^-(aq) \rightarrow AgBr(s)$
I^-	silver nitrate	silver iodide	pale yellow	$Ag^+(aq) + I^-(aq) \rightarrow AgI(s)$
SO_4^{2-}	barium chloride	barium sulfate	white	$Ba^{2+}(aq) + SO_4^{2-}(aq) \rightarrow BaSO_4(s)$

Precipitation reactions have lots of practical applications, including water treatment, production of coloured pigments for paints and dyes, and in the identification of certain metal ions in solution.

Chemical ideas: Structure and properties 5.2

Bonding, structure, and properties
Ionic bonding

Ionic compounds are typically solids at room temperature and pressure and have lattice structures consisting of repeating positive and negative ions in all three dimensions. Because of this regular arrangement of ions, ionic compounds often form regularly shaped crystals (Figure 6).

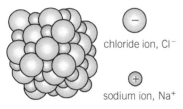

chloride ion, Cl^-

sodium ion, Na^+

▲ Figure 6 *The sodium chloride lattice, built up from oppositely charged sodium ions and chloride ions*

The electrostatic attraction of oppositely charged ions overcomes any repulsion between ions of the same charge, holding together the ionic lattice structures (including those of hydrated salts). This is called ionic bonding. The electrostatic attractions are strong. This means a large amount of energy is needed to pull the ions apart. This is the reason why ionic compounds have high melting points. Once melted the ions are free to move and this is why molten ionic compounds conduct electricity.

Metallic bonding

Metals also have a lattice structure. One simple model that accounts for some of the most important characteristics of metals is the electron-sea model. In this model the metal is pictured as a giant lattice structure of metal cations in a 'sea' of delocalised valence electrons. These electrons are free to move, so account for the flow of electricity in metals. The whole structure is held together by the attraction of the metal cations to the delocalised electrons (Figure 7). These electrostatic attractions are strong, so account for the relatively melting and boiling points of metals.

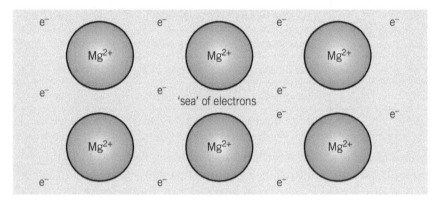

▲ Figure 7 *Metallic bonding – a sea of electrons*

Covalent networks

Some elements and compounds, such as carbon in the form of diamond (Figure 8) and silicon dioxide found in quartz, form large covalently bonded networks. These have strong covalent bonds between the atoms in the network. Because it takes a lot of energy to break these intramolecular electrostatic attractions, these networks have very high melting and boiling points. These networks are also insoluble in water and do not conduct electricity (graphite is an exception, due to its unique structure).

▲ Figure 8 *The structure of diamond.*
Each circle represents a carbon atom

Summary of bonding, structure, and properties

How does the type of the structure and bonding in a substance affect its properties? Table 6 summarises the similarities and differences, including covalent bonding in simple molecules.

▼ Table 6 *A summary of the various types of structures and bonding*

	Giant lattice			Covalent molecular	
	Ionic	Covalent network	Metallic	Simple molecular	Macromolecular
What substances have this type of structure?	compounds of metals with non-metals	some elements in Group 4 and some of their compounds	metals	some non-metal elements and some non-metal/ non-metal compounds	polymers
Examples	sodium chloride, NaCl, and calcium oxide, CaO	diamond and graphite (both C) and silica, SiO_2	sodium, Na, copper, Cu, and iron, Fe	carbon dioxide, CO_2, chlorine, Cl_2, and water, H_2O	poly(ethene), nylon, proteins, DNA
What type of particles does it contain?	ions	atoms	positive ions surrounded by delocalised electrons	small molecules	long-chain molecules
How are the particles bonded together?	strong ionic bonds; attraction between oppositely charged ions	strong covalent bonds – attraction of atoms' nuclei for shared electrons	strong metallic bonds – attraction of atoms' nuclei for delocalised electrons Figure 7	weak intermolecular bonds between molecules – strong covalent bonds between the atoms within molecules	weak intermolecular bonds between molecules – strong covalent bonds between the atoms within molecules
What are the typical properties? Melting point and boiling point	high	very high	generally high (except mercury)	low	moderate – often decompose on heating
Hardness	hard but brittle	very hard (if three-dimensional)	hard but malleable (except mercury)	soft	variable; many are soft but often flexible
Electrical conductivity	electrolytes conduct when molten or dissolved in water	do not normally conduct (except graphite)	conduct when solid or liquid	do not conduct	do not normally conduct
Solubility in water	often soluble	insoluble	insoluble (but some react)	usually insoluble, unless molecules contain groups which can hydrogen bond with water	usually insoluble
Solubility in non-polar solvents (e.g., hexane)	insoluble	insoluble	insoluble	usually soluble	sometimes soluble

Summary questions

Activity EL 7.3

In this activity you can check your understanding of the links between properties and structure.

1   Draw electron dot-and-cross diagrams for the following ionic compounds.
   a   lithium hydride, LiH   (*2 marks*)
   b   potassium fluoride, KF   (*2 marks*)
   c   magnesium oxide, MgO   (*2 marks*)
   d   calcium sulfide, CaS   (*2 marks*)

2   Which type of structure would you expect each of these substances to have?
   a   A white solid which starts to soften at 200 °C and can be drawn into fibres.   (*1 mark*)
   b   A white solid which melts at −190 °C.   (*1 mark*)
   c   A white solid which melts at 770 °C and conducts electricity when molten, but not in the solid state.   (*1 mark*)

3   Draw electron dot-and-cross diagrams for the following ionic compounds.
   a   calcium chloride, $CaCl_2$ (*2 marks*)
   b   sodium sulfide, Na_2S (*2 marks*)

4   An ammonia molecule, NH_3, forms a dative bond with a hydrogen ion, H^+, to produce an ammonium ion, NH_4^+. The other three hydrogen atoms are held to the nitrogen atom by conventional covalent bonds.
   a   What is the essential difference between a dative bond and a covalent bond?   (*2 marks*)
   b   Draw dot-and-cross diagrams for the ammonia molecule and the ammonium ion.   (*2 marks*)

5   Which type of structure would you expect each of the following substances to have?
   a   A hard grey solid which conducts electricity and melts at 3410 °C.   (*1 mark*)
   b   A liquid which conducts electricity and solidifies at −39 °C.   (*1 mark*)

6   Draw electron dot-and-cross diagrams for the following ionic compounds.
   a   sodium nitride, Na_3N (*2 marks*)
   b   aluminium fluoride, AlF_3 (*2 marks*)

EL 8 Spectacular metals

Specification reference: EL(p), EL(q), EL(r), EL(u)

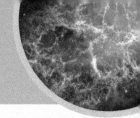

The metals of Groups 1 and 2 are the most reactive metals in the periodic table. Some of their reactions can be impressive (Figure 1).

Uses of Group 1 and 2 metals

Group 1 and 2 metals have some interesting uses. The **alloy** of the Group 1 metals sodium and potassium remains liquid at room temperature and is used in the nuclear industry as a heat transfer fluid.

The Group 2 metal beryllium is used in mirrors in satellites and magnesium uses range from pencil sharpeners to parts of car engines. The *compounds* of Group 2 find even more extensive use in our everyday lives (Table 1).

▼ Table 1 *Uses of Group 2 compounds*

Group 2 compound	Property	Use
beryllium oxide, BeO	high melting point	nose-cones of rockets
magnesium hydroxide, $Mg(OH)_2$	basic oxide	some indigestion tablets
calcium carbonate, $CaCO_3$	insoluble and reacts with acids	agriculture to neutralise acid soils
calcium sulfate, hydrated $CaSO_4$	insoluble	plaster to keep broken limbs in place
strontium carbonate, $SrCO_3$	red colour when heated	fireworks
barium sulfate, $BaSO_4$	opaque to X-rays and insoluble	barium meal to allow X-raying of the digestive tract
radium chloride, $RaCl_2$	radioactive	treat certain cancers

Learning outcomes

Demonstrate and apply knowledge and understanding of:

→ a description and comparison of the following properties of the elements and compounds of Mg, Ca, Sr, and Ba in Group 2: reactions of the elements with water and oxygen, thermal stability of the carbonates, solubilities of hydroxides and carbonates

→ the term ionisation enthalpy; equations for the first ionisation of elements; explanation of trends in first ionisation enthalpies for Periods 2 and 3 and groups and the resulting differences in reactivities of s- and p-block metals in terms of their ability to lose electrons

→ charge density of an ion and its relation to the thermal stability of the Group 2 carbonates

→ the basic nature of the oxides and hydroxides of Group 2 (Mg–Ba).

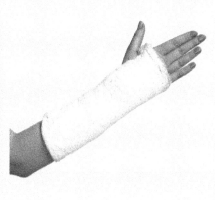

▲ Figure 1 *Uses of Group 2 compounds*

▲ Figure 2 *Caesium reacting with water– an indicator has been added to the water to show how the solution becomes alkaline*

3	4
Li	Be
lithium	beryllium
6.9	9.0
11	12
Na	Mg
sodium	magnesium
23.0	24.3
19	20
K	Ca
potassium	calcium
39.1	40.1
37	38
Rb	Sr
rubidium	strontium
85.5	87.6
55	56
Cs	Ba
caesium	barium
132.9	137.3
87	88
Fr	Ra
francium	radium
[223]	[226]

▲ Figure 3 *The elements of Group 1*   ▲ Figure 4 *The elements of Group 2*

First ionisation enthalpy

The first ionisation enthalpy of an element is the energy needed to remove one electron from every atom in one mole of isolated gaseous atoms of the element – a mole of gaseous ions with one positive charge are formed.

Chemical ideas: The periodic table 11.2

The s-block: Groups 1 and 2

The s-block contains two groups of reactive metals. Group 1 metals (Figure 3) are also called the alkali metals. Group 2 metals (Figure 4) are also called the alkaline earth metals. Groups 1 and 2 display two trends visible in other groups in the periodic table.

● Elements become *more* metallic *down* a group, for example, they more readily form cations in ionic compounds. For this reason, the most reactive metals in Groups 1 and 2 are found at the bottom of each group.

● Elements become *less* metallic *across* a period from left to right. For this reason, the Group 1 metals are more reactive than the Group 2 metals in the same period.

The Group 1 and 2 metals are not as widely used (in elemental form) as the familiar metals of the d-block, such as iron, copper, and chromium. The s-block metals tend to be soft, weak metals with low melting points, and the metals themselves are too reactive with water and oxygen to have many uses. However, the *compounds* of the s-block elements are very important (for examples, see Table 1).

Chemical reactivity

As Group 1 and 2 metals are all very reactive, they are never found naturally in their elemental form. However, compounds of s-block metals are very common throughout nature.

Like all groups in the periodic table, Groups 1 and 2 show patterns of reactivity down the group. There are *similarities* between the reactions of the elements within a group, and *differences* that show up as patterns, or trends. The *similarities* happen because the elements in a particular group all have similar arrangements of electrons in their atoms. The *differences* happen because the size of the atom increases down the group.

First ionisation enthalpy

If sufficient energy is given to an atom, an electron is lost and the atom becomes a positive ion – ionisation has taken place. An input of energy is *always* needed to remove electrons because they are attracted to the nucleus.

When one electron is pulled out of an atom, the energy required is called the **first ionisation enthalpy** (or first ionisation energy).

The general equation for the first ionisation process is

$$X(g) \rightarrow X^+(g) + e^-$$

where X represents the symbol for the element. For oxygen, the first ionisation enthalpy is $+1320\,kJ\,mol^{-1}$. This means that $1320\,kJ$ of energy is required to remove one electron from one mole of gaseous oxygen atoms.

$$O(g) \rightarrow O^+(g) + e^-$$

The first ionisation removes the most loosely held electron. This will be one of the outer shell electrons since they are furthest from the nucleus.

Figure 5 shows the first ionisation enthalpy for elements 1–56. The elements at the peaks are all in Group 0 (the noble gases). These elements have high first ionisation enthalpies – they are difficult to ionise and are very unreactive. The elements at the troughs are all in Group 1 (the alkali metals). These elements, with only one outer shell electron, have low ionisation enthalpies – they are easy to ionise and are very reactive.

This pattern provides chemists with data to support the idea of electron shells, for example, the outermost filled electron shell for neon is n = 2. But the outermost electron in sodium is in the n = 3 shell. In sodium, there is more shielding of the outermost electrons from the nucleus and therefore less energy needed to remove an outer electron from sodium than from neon. Trends in Period 2 and Period 3 are similar for similar reasons.

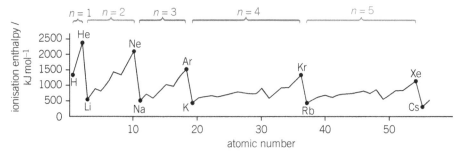

▲ Figure 5 *Variation of first ionisation enthalpy for elements 1–56*

Figure 6 shows the first ionisation enthalpies for elements 1–20 in more detail. There is a general trend that as you go across a period in the periodic table it becomes more difficult to remove an electron. Across a period, electrons are being added to the outer shell but, at the same time, protons are being added to the nucleus. As the nuclear charge becomes more positive, the electrons will be held more tightly and so it gets harder to pull one from the outer shell. This is an example of periodicity.

Ionisation energies and electron shells

The pattern of first ionisation enthalpies can be used as evidence to support the existence of sub-shells as well as shells. Look at Figure 6 for Period 2 (n = 2). Although the general trend across the period is an increase, there is some variation.

Between beryllium and boron there is a decrease. The electronic configuration of beryllium is $1s^2\ 2s^2$ and boron is $1s^2\ 2s^2\ 2p^1$. The s-sub-shell is lower in energy than the p-sub-shell, therefore less energy is needed to remove the outer electron of boron, in spite of increased nuclear charge.

Between nitrogen and oxygen there is a decrease in first ionisation enthalpies.

N is 2p [↑][↑][↑]   and   O is 2p [↑↓][↑][↑]

2s [↑↓]   2s [↑↓]

1s [↑↓]   1s [↑↓]

The extra repulsion from the paired electron sub-shelll in oxygen means less energy is needed to remove one of the paired electrons, despite the increased number of protons on going from nitrogen to oxygen.

The observed data (ionisation enthalpies) supports the theory of electron sub-shells.

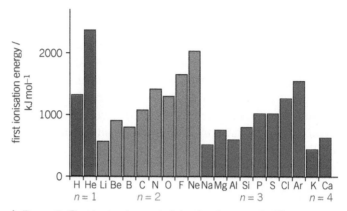

▲ Figure 6 *First ionisation enthalpies for elements 1–20*

Going down a group in the periodic table the first ionisation enthalpies *decrease.* This is because the attraction between the nucleus and the outermost electron decreases. There are more filled shells of electrons between the nucleus and the outermost electron and these shield the positively charged nucleus from the outermost electron, reducing the attraction the electron experiences. It is therefore easier for the outermost electron to be removed. The metals of Groups 1 and 2 react by losing their outer electrons, so the first ionisation energy trends also correspond to an increase in reactivity down the group.

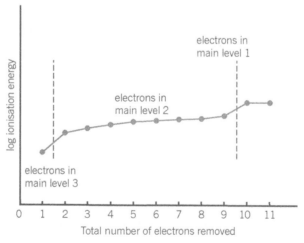

▲ Figure 7 *Successive ionisation enthalpies for sodium – a logarithmic scale is used for the ionisation enthalpy*

Successive ionisation enthalpies

More than one electron can be removed from an atom (except from a hydrogen atom, of course). So there are second, third, and fourth ionisation enthalpies, and so on. This is the energy required to

Activity EL 8.1

In this activity you will look at the reactions of Group 2 metals and their compounds.

remove further electrons. The general equations for these ionisation processes are:

first ionisation: $\quad Ca(g) \quad \rightarrow \quad Ca^+(g) + e^-$

second ionisation: $\quad Ca^+(g) \quad \rightarrow \quad Ca^{2+}(g) + e^-$

Each of the second and subsequent ionisation processes involves the removal of an electron from a *positive ion*, for example, the second ionisation enthalpy is the energy needed to remove one electron from an $X^+(g)$ ion. It is *not* the energy required to remove two electrons from an $X(g)$ atom.

Chemical properties of Group 2 elements and compounds

The elements of Group 2 are all reactive. They form compounds containing ions with a +2 charge, such as Mg^{2+} and Ca^{2+}.

Reactions with oxygen

All the Group 2 elements react with oxygen to produce the metal oxide.

$$2M(s) + O_2(g) \rightarrow 2MO(s)$$

Reactions with water

All the elements react with water to form hydroxides and hydrogen, with an increase in reactivity down the group. Magnesium (third period) reacts slowly, even when the water is heated. Barium (sixth period) reacts rapidly, giving a steady stream of hydrogen. The general equation, using M to represent a typical Group 2 metal reacting with water, is:

$$M(s) + 2H_2O(l) \rightarrow M(OH)_2(aq) + H_2(g)$$

The Group 2 metals do not react with water as vigorously as the Group 1 metals do.

Effect of heating carbonates

The general formula of Group 2 carbonates is MCO_3. When carbonates are heated, they decompose forming the oxide and releasing carbon dioxide.

$$MCO_3(s) \rightarrow MO(s) + CO_2(g)$$

The carbonates become more difficult to decompose down the group. For example, magnesium carbonate is easily decomposed by heating in a test tube over a Bunsen burner flame. Calcium carbonate needs much stronger heating directly in the flame before it will decompose. The **thermal stability** of calcium carbonate is greater than that of magnesium carbonate. The decomposition of calcium carbonate (limestone) is an important process, used to manufacture calcium oxide (quicklime).

The change in thermal stability down the group can be explained in terms of the **charge density** of the cations. Charge density is a measure of the concentration of charge on the cation. The smaller the +2 ion the higher the charge density. Cations with a higher charge density (those at the top of Group 2) can distort or **polarise** the negative charge cloud

▲ Figure 8 *Magnesium metal burning in air*

Study tip

Remember Group 1 metals form oxides with the general formula M_2O whilst Group 2 metals form oxides with the general formula MO.

Activity EL 8.2

Here you will be able to investigation the decomposition of Group 2 carbonates.

▲ Figure 9 *Effect of charge density on distortion of a carbonate ion*

around the carbonate ion making it less stable and easier to break up on heating (Figure 9).

Oxides and hydroxides

The general formula of the Group 2 oxides is MO, and that of the hydroxides is $M(OH)_2$. In water, the oxides and hydroxides form alkaline solutions, although they are not very soluble. Forming alkaline solutions is typical of metal oxides and hydroxides, in contrast to non-metals whose oxides are usually acidic. The most strongly alkaline oxides and hydroxides are those at the bottom of the group.

The oxides and hydroxides react with acids to form salts.

$$MO(s) + 2HCl(aq) \rightarrow MCl_2(aq) + H_2O(l)$$

$$M(OH)_2(s) + H_2SO_4(aq) \rightarrow MSO_4(aq) + 2H_2O(l)$$

This neutralising effect is used by farmers when they put lime (calcium hydroxide) on their fields to neutralise soil acidity.

Summary

The trends in properties of some compounds of Group 2 elements are summarised in Table 2.

▼ Table 2 *Summary of the reactivity trends of Group 2 compounds*

Element	Trend in reactivity with water	Trend in thermal stability of carbonate	Trend in pH of hydroxide in water	Trend in solubility of hydroxide	Trend in solubility of carbonate
Mg ↓ Ba ▼	increasing reactivity ↓	decomposes at increasingly higher temperature ↓	increasing pH ↓	increasing solubility ↓	decreasing solubility ↓

Summary questions

1 State the two factors that affect the charge density of an ion. (*2 marks*)

2 Write a word equation and a balanced chemical equation (with state symbols) for each of the following reactions.
   a the action of heated magnesium on steam (*2 marks*)
   b the neutralisation of hydrochloric acid with calcium oxide (*2 marks*)
   c the thermal decomposition of beryllium carbonate (*2 marks*)
   d the action of sulfuric acid on barium hydroxide. (*2 marks*)

3 The first, second, and third ionisation enthalpies of calcium are $+596\,kJ\,mol^{-1}$, $+1160\,kJ\,mol^{-1}$, and $+4930\,kJ\,mol^{-1}$ respectively.
   a Write equations corresponding to each of these three ionisation enthalpies. (*3 marks*)
   b Explain why the second ionisation enthalpy of calcium is larger than its first ionisation enthalpy. (*2 marks*)
   c Explain why there is a very sharp rise between the second and third ionisation enthalpies of calcium. (*2 marks*)

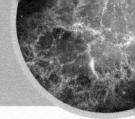

You have already looked at the different types of salts, where they are found, and how they are made. But how can you find out how much salt is in the sea, in our blood, or our sweat?

Calculating quantities of salts

The mole and certain practical techniques can be used to calculate the amounts of products and reactants formed and used up in reactions.

Titrations are used to calculate quantities or concentrations of substances in solution. In titrations a solution of known concentration (a standard solution) is reacted with another solution of unknown concentration. Along with the balanced equation for the reaction, it is possible to calculate the concentration of the unknown solution.

There are many applications of titrations, from calculating the amount of acid in acid rain and determining the amount of active ingredient in a medicine, to determining the amount of a pollutant ion in waste from mining. In the industries that do lots of titrations, automatic titration systems are used. These enable one operator to set up and do many titrations in a short time.

Chemical ideas: Measuring amount of substances 1.3

Using equations to work out reacting masses

A balanced chemical equation tells you the reactants and products in a reaction, and the relative amounts involved. The equation is balanced so that there are equal numbers of each type of atom on both sides. For example, in the combustion of methanol the theoretical equation is:

$$2CH_3OH(g) + 3O_2(g) \rightarrow 2CO_2(g) + 4H_2O(g)$$

The number written in front of each formula in a balanced equation tells you the number of formula units involved in the reaction. Remember that a formula unit may be a molecule or another species such as an atom or an ion. The small subscript numbers are part of the formulae and cannot be changed. Atoms are not created or destroyed in chemical reactions, they are simply rearranged, so equations must balance and the total mass on each side of the equation must always be the same. Chemists can use equations to work out the masses of reactants and products involved in a reaction, without having to do an experiment.

Learning outcomes

Demonstrate and apply knowledge and understanding of:

→ the techniques and procedures used in experiments to measure masses of solids

→ the terms acid, base, alkali, neutralisation

→ the use of the concept of amount of substance (moles) to perform calculations involving concentration (including titration calculations and calculations for making and diluting standard solutions)

→ the techniques and procedures used in experiments to measure volumes of solutions; the techniques and procedures used in experiments to prepare a standard solution from a solid or more concentrated solution and in acid–base titrations.

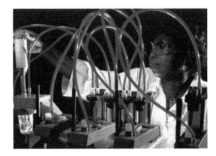

▲ Figure 1 An automatic titration device being used to perform quality controls on a pharmaceutical product

Study tip

The steps for working out reacting masses

Step 1: Write a balanced equation.

Step 2: In words, state what the equation tells you about the amount in moles of the substances you are interested in.

Step 3: Change amounts in moles to masses in grams.

Step 4: Scale the masses to the ones in the question.

 Worked example: Calculating masses from balanced equations

Calculate the mass of calcium oxide and carbon dioxide that would be produced if 100.0 g of calcium carbonate is heated.

Step 1: Write a balanced equation for the reaction.

$$CaCO_3(s) \rightarrow CaO(s) + CO_2(g)$$

Step 2: From the equation, work out the moles of reactants and moles of products.

1 mole of $CaCO_3$ produces 1 mole of CaO and 1 mole of CO_2.

Step 3: Calculate the molecular mass of the reactants and products from the atomic mass of the atoms that make up the substances.

$$A_r(Ca) = 40.0 \qquad A_r(C) = 12.0 \qquad A_r(O) = 16.0$$

$$M_r(CaCO_3) = 40.0\,g + 12.0\,g + (3 \times 16.0\,g) = 100.0\,g$$

$$M_r(CaO) = 40.0\,g + 16.0\,g = 56.0\,g$$

$$M_r(CO_2) = 12.0\,g + (2 \times 16.0\,g) = 44.0\,g$$

Step 4: Work out the masses of the products from the mass that was reacted

100.0 g of calcium carbonate gives 56.0 g of calcium oxide and 44.0 g carbon dioxide.

Chemical ideas: Acids and bases 8.1

Acids, bases, alkalis, and neutralisation

A key property of the Group 2 oxides and hydroxides is that they are bases – they react with acids to form salts.

What do we mean by acid and base?

Hydrochloric acid is an **acid** because of its properties. It turns litmus red, reacts with carbonates to give carbon dioxde, and is neutralised by bases – all properties that are expected of acids.

Chemists explain properties in terms of what goes on at the level of atoms, molecules, and ions. The properties of acids are due to the ability to transfer H^+ ions to something else. The substance which accepts the H^+ ion is called a **base**.

Take the reaction of hydrogen chloride with ammonia. The reaction forms a white salt, ammonium chloride

$$HCl(g) + NH_3(g) \rightarrow NH_4Cl(s)$$

Hydrochloric acid donates a proton, H^+, to the ammonia – it acts as an acid. The ammonia accepts a proton – it acts as a base. The general

Acid

A compound that dissociates in water to produce hydrogen ions.

Base

A compound that reacts with an acid – is a proton acceptor – to produce water (and a salt).

definition of an acid is a substance that donates H^+ in a chemical reaction. The substance that accepts the H^+ is a base. The reaction is an acid–base reaction. Since a hydrogen atom consists of only a proton and an electron, a H^+ ion corresponds to just one proton. Acids are sometimes referred to as *proton donors* and bases as *proton acceptors*. This theory of H^+ transfer is known as the Brønsted-Lowry theory of acids and bases.

Solutions of acids and bases

Hydrogen chloride is a gas containing HCl molecules. Water is almost totally made up of H_2O molecules. But hydrochloric acid, a solution of hydrogen chloride in water, readily conducts electricity so it must contain ions. A reaction between the hydrogen chloride molecules and the water molecules produces these ions.

$$HCl(aq) + H_2O(l) \rightarrow H_3O^+(aq) + Cl^-(aq)$$
$$\text{acid} \qquad \text{base} \qquad \text{oxonium ion}$$

In this reaction, water behaves as a base. The H_3O^+ ion is called the oxonium ion. It is a very common ion and is present in every solution of an acid in water – it occurs in every *acidic* solution. The acid donates H^+ to H_2O to form H_3O^+ (Figure 2).

▲ Figure 2 *The bonding in the H_3O^+ ion. A lone pair on the oxygen atom forms a dative covalent bond to H^+. (Once the bond has formed it is indistinguishable from the two other O—H bonds)*

The H_3O^+ ion can itself act as an acid, donating H^+ and becoming an H_2O molecule. The familiar properties of acidic solutions are all properties of the H_3O^+ ion. You will often see the formula $H_3O^+(aq)$ shortened to $H^+(aq)$ and the dissociation of HCl(aq) into ions represented by

$$HCl(aq) \rightarrow H^+(aq) + Cl^-(aq)$$

When an acid dissolves in water, $H^+(aq)$ ions form in solution.

An **alkali** is a base that dissolves in water to produce hydroxide ions, $OH^-(aq)$ (Figure 3).

Some alkalis, such as sodium hydroxide and potassium hydroxide, already contain hydroxide ions, whereas others, such as sodium carbonate and ammonia, form $OH^-(aq)$ ions when they react with water.

$$CO_3^{2-}(aq) + H_2O(l) \rightleftharpoons HCO_3^-(aq) + OH^-(aq)$$

$$NH_3(aq) + H_2O(l) \rightleftharpoons NH_4^+(aq) + OH^-(aq)$$

When an alkali reacts with an acid a salt is formed. This is a **neutralisation** reaction. The reaction involves the hydrogen ions in the acidic solution and the hydroxide ions in the alkali, so the ionic equation for a neutralization reaction is:

$$H^+(aq) + OH^- \rightarrow H_2O(l)$$

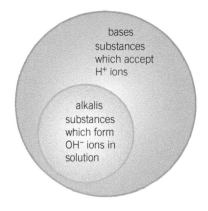

▲ Figure 3 *The relationship between alkalis and bases*

Alkali

A base that dissolves in water to produce hydroxide, OH^-, ions

Synoptic link

You will develop your understanding of acids in Chapter 8, Oceans.

Synoptic link 🧪

You can find out how to weigh solids and measure volumes of liquids in Techniques and procedures.

Activity EL 9.1

This activity gives you the opportunity to make a standard solution.

Synoptic link 🧪

You can find out more about these three practical skills in Techniques and procedures.

Activity EL 9.2

In this activity you will be able to find the concentration of an acid solution.

Activity EL 9.3

In this activity you will practise your titration technique and carry out accurate dilution of solutions.

Activity EL 9.4

In this activity you will make a compound that can be used in a fertiliser and calculate your percentage yield.

Study tip

You may see $mol\,dm^{-3}$ abbreviated to M. This abbreviation was once widely used, but now $mol\,dm^{-3}$ is used.

Ionic equations can also be written for insoluble salts formed by precipitation reactions.

$$Ba^{2+}(aq) + SO_4{}^{2-}(aq) \rightarrow BaSO_4(s)$$

Analysing acids and bases

It is often useful to know exactly when an acid and a base in solution have neutralised each other. From this, the amount of acid or base in the original solution can be calculated. By using a known concentration of one of the reactants – either acid or base – the exact reacting volume at neutralisation can be found out.

In order to successfully do this you will need to be able to do three things:

- dilute a solution
- make up a standard solution
- Carry out an acid-base titration.

The volume and concentration of the standard sodium hydroxide solution required for neutralisation can be used to calculate the concentration of the hydrochloric acid.

Chemical ideas: Measuring amounts of substance 1.4

Concentrations of solutions

Reactions are often carried out in solution. When using a solution of a substance, it is important to know how much of the substance is dissolved in a particular volume of solution. Concentrations are sometimes measured in $g\,dm^{-3}$. A solution containing 80 g of sodium hydroxide made up to $1\,dm^3$ of solution has a concentration of $80\,g\,dm^{-3}$.

However, chemists usually prefer to measure quantities in moles rather than in grams, because working in moles tells gives the number of particles present. The preferred unit for measuring concentration is $mol\,dm^{-3}$. To convert $g\,cm^{-3}$ to $mol\,dm^{-3}$, you need the molar mass of the substance involved. For example, the molar mass of sodium hydroxide, NaOH, is $40.0\,g\,mol^{-1}$, so a solution containing $80\,g\,dm^{-3}$ has a concentration of:

$$\frac{80.0\,g\,dm^{-3}}{40.0\,g\,mol^{-1}} = 2.0\,mol\,dm^{-3}$$

In general:

$$\text{concentration }(mol\,dm^{-3}) = \frac{\text{concentration }(g\,dm^{-3})}{\text{molar mass }(g\,mol^{-1})}$$

When a solution is made, its concentration will depend on:

- the amount of solute
- the final volume of the solution (Figure 6).

With the concentration of a solution, you can work out the amount of solute in a particular volume. In general:

amount (mol) = concentration of solution ($mol\,dm^{-3}$) × volume of solution

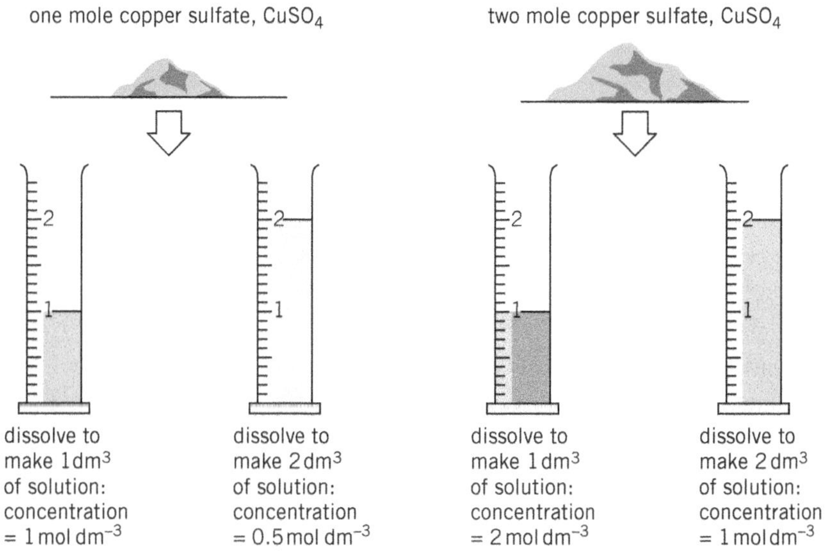

one mole copper sulfate, CuSO$_4$

two mole copper sulfate, CuSO$_4$

dissolve to make 1 dm^3 of solution: concentration = 1 mol dm^{-3}

dissolve to make 2 dm^3 of solution: concentration = 0.5 mol dm^{-3}

dissolve to make 1 dm^3 of solution: concentration = 2 mol dm^{-3}

dissolve to make 2 dm^3 of solution: concentration = 1 mol dm^{-3}

▲ Figure 4 *The concentration of a solution depends on the amount of solute and the final volume of the solution*

Worked example: The amount of a solute in a solution

Calculate how many moles of sodium hydroxide are in a 250 cm^3 solution of sodium hydroxide with a concentration of 2 mol dm^{-3}.

Step 1: Convert 250 cm^3 to dm^3.

$$\frac{250}{1\,000} = 0.25 \, \text{dm}^3$$

Step 2: Calculate the moles in 0.25 dm^3 of solution.

$$2 \, \text{mol dm}^{-3} \times 0.25 \, \text{dm}^3 = 0.5 \, \text{mol of sodium hydroxide}$$

Using concentrations in calculations

When carrying out a chemical reaction in solution, and you know the equation for the reaction, you can use the concentrations of the reacting solutions to predict the volumes you will need.

$$\text{concentration } c \text{ (mol dm}^{-3}) = \frac{\text{amount } n \text{ (moles)}}{\text{volume } V \text{ (dm}^3)} \qquad \textbf{Equation 1}$$

This equation can be rearranged easily using Figure 7. The horizontal line represent dividing n by c or V (Equation 1 and 2). The vertical lines represents multiplying c and V (Equation 3).

$$\text{volume } V \text{ (dm}^3) = \frac{\text{amount } n \text{ (moles)}}{\text{concentration } c \text{ (mol dm}^{-3})} \qquad \textbf{Equation 2}$$

$$\text{amount } n \text{ (moles)} = \text{concentration } c \text{ (mol dm}^{-3}) \times \text{volume } V \text{ (dm}^3) \qquad \textbf{Equation 3}$$

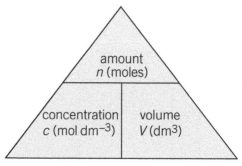

▲ **Figure 5** *A memory aid showing the relationship between concentration, the amount, and the volume of solution*

In general, the steps for working out the reacting volumes of solutions are as follows:

1   Write a balanced equation.

2   Write down what the equation tells you about the amount in moles of the substances you are interested in.

3   Use the known concentrations of the solutions to change amounts in moles to volumes of solutions.

4   Scale the volumes of solutions to the ones in the question.

 Worked example: Predicting volumes from chemical equations

Consider the reaction of sodium hydroxide with hydrochloric acid. The concentrations of both solutions are $2.0\,mol\,dm^{-3}$. If you have $0.25\,dm^3$ of sodium hydroxide solution, what volume of hydrochloric acid would be needed to neutralise it?

Step 1: Write the equation for the reaction.

$$NaOH(aq) + HCl(aq) \rightarrow NaCl(aq) + H_2O(l)$$

Step 2: Calculate the moles of sodium hydroxide.

$$2.0\,mol\,dm^{-3} \times 0.25\,dm^3 = 0.50\,mol$$

Step 3: Calculate the amount of hydrochloric acid needed.

From the equation, one mole of NaOH reacts with one mole of HCl, therefore $0.50\,mol$ of HCl needed.

Step 4: Calculate the volume of hydrochloric acid needed.

$$\text{volume of HCl needed} = \frac{0.50\,mol}{2.0\,mol\,dm^{-3}} = 0.25\,dm^3$$

Summary questions

1. Complete the following table that shows the concentrations of ions in water from a sample of Dead Sea water. *(7 marks)*

Ion	Concentration / $g\,dm^{-3}$	Concentration / $mol\,dm^{-3}$
Cl^-	183.0	
Mg^{2+}	36.2	
Na^+		1.37
Ca^{2+}		0.355
K^+	6.8	
Br^-	5.2	
SO_4^{2-}		0.00625

2. Calculate the mass of solute needed to make up the following solutions.
   a $1\,dm^3$ of a $2\,mol\,dm^{-3}$ solution of NaCl *(2 marks)*
   b $250\,cm^3$ of a $0.1\,mol\,dm^{-3}$ solution of $KMnO_4$ *(2 marks)*
   c $50\,cm^3$ of a $0.5\,mol\,dm^{-3}$ solution of KOH *(2 marks)*
   d $15\,dm^3$ of a $2\,mol\,dm^{-3}$ solution of $Pb(NO_3)_2$ *(2 marks)*
   e $10\,cm^3$ of a $0.01\,mol\,dm^{-3}$ solution of LiOH *(2 marks)*

3. In a titration, $25.00\,cm^3$ of a sodium hydroxide solution were pipetted into a conical flask. A $0.10\,mol\,dm^{-3}$ solution of sulfuric acid was run from a burette into the flask. An indicator in the flask changed colour at an average of $22.00\,cm^3$ of the acid added.

   $$H_2SO_4(aq) + 2NaOH(aq) \rightarrow Na_2SO_4(aq) + 2H_2O(l)$$

   a What is the concentration of the sodium hydroxide solution in $mol\,dm^{-3}$? *(2 marks)*
   b The student washes the conical flask with water between titrations and does not dry it. Explain the effect, if any, on the titre of the next titration. *(1 mark)*

4. A standard solution of sodium hydroxide cannot be made by direct weighing of the solid.
   a Explain the meaning of the term *standard solution*. *(1 mark)*
   b Suggest why sodium hydroxide cannot be made into a standard solution by direct weighing of the solid. *(2 marks)*

Practice questions

1   How many hydrogen atoms are there in a mole of methanol, CH_3OH?

    A   4   C 1.8×10^{24}

    B 6×10^{23} D 2.4×10^{24} (1 mark)

2   What mass of aluminium is contained in $5.10g$ of aluminium oxide?

    A $1.35g$ C $2.70g$

    B $4.26g$ D $3.66g$ (1 mark)

3   Which is the correct method for making a pure dry sample of barium sulfate?

    A   Mix barium carbonate with sulfuric acid, filter, and evaporate the filtrate.

    B   Mix barium chloride solution with sodium sulfate solution. Filter the precipitate and allow it to dry.

    C   Mix barium nitrate solution with sulfuric acid. Filter, wash and dry the precipitate.

    D   Add barium to sulfuric acid. Evaporate the solution.   (1 mark)

4   The change in structure and bonding in the elements across a period of the periodic table follows the following pattern:

    A   metallic to giant covalent to small molecules

    B   metallic to ionic to giant covalent

    C   ionic to giant covalent to small molecules

    D   giant covalent to small molecules to ionic.   (1 mark)

5   In the Geiger and Marsden experiment, the scientists fired alpha particles at a thin piece of gold foil. Which of the following happened to the alpha particles?

    A   Most of the particles were deflected.

    B   Some particles came almost straight back.

    C   The particles were mostly absorbed by the foil.

    D   The foil emitted neutrons.   (1 mark)

6   Which row of the table below contains two correct formulae for ionic substances?

A	$CaOH_2$	$CaSO_4$
B	$Fe_2(SO_4)_3$	$Fe(NO_3)_2$
C	AlN	AlS
D	KCO_3	KNO_3

(1 mark)

7   Which is a correct equation for a reaction from the chemistry of barium?

    A $BaCl_2(aq) + 2H_2O(l) \rightarrow$
        $Ba(OH)_2(aq) + 2HCl(aq)$

    B $Ba(s) + H_2O(l) \rightarrow BaO(s) + H_2(g)$

    C $Ba(OH)_2(s) + 2HCl(aq) \rightarrow$
        $BaCl_2(aq) + 2H_2O(l)$

    D $BaCO_3(s) + H_2SO_4(aq) \rightarrow$
        $BaSO_4(aq) + CO_2(g) + H_2O(l)$
        (1 mark)

8   Which of the following is/are true about atomic emission spectra?

    1   They consist of bright lines on a dark background.

    2   They are caused by electrons being excited to higher energy levels.

    3   The lines get closer at higher wavelength.

    A   1, 2, and 3 correct

    B   1 and 2 are correct

    C   2 and 3 are correct

    D   Only 1 is correct   (1 mark)

9   Which of the following is/are true about a sample of solid sodium chloride?

    1   There are equal numbers of sodium and chloride ions.

    2   The ions are arranged in a lattice.

    3   The ions cannot move through the lattice.

    A   1, 2, and 3 correct

    B   1 and 2 are correct

    C   2 and 3 are correct

    D   Only 1 is correct   (1 mark)

10 Which of the following is/are reasons why the bond angle in NH_3 is smaller than the bond angle in CH_4?

    1   Lone pair repulsion is greater than bond pair repulsion.

    2   The hydrogen atoms repel each other less in ammonia.

    3   Methane has more lone pairs than ammonia.

    A   1, 2, and 3 correct

    B   1 and 2 are correct

    C   2 and 3 are correct

    D   Only 1 is correct   (1 mark)

11 Helium is made in the Sun by reactions such as that shown below:

$$^2_1H + ^3_1H \rightarrow ^4_2He + ^1_0n$$

a Name this type of reaction that involves light nuclei joining to form heavier ones. (*1 mark*)

b Explain why ^2_1H and ^3_1H are described as *isotopes*. Give their similarities and differences in terms of nuclear particles. (*2 marks*)

c Characteristic lines in a star's absorption spectrum show the presence of hydrogen gas in the gas surrounding the star.

(i) Describe the appearance of an absorption spectrum. (*1 mark*)

(ii) Why do hydrogen atoms give an absorption spectrum whereas hydrogen nuclei do not? (*1 mark*)

(iii) Draw a diagram of the energy levels in a hydrogen atom. Draw arrows on this diagram to show the origin of two lines in the hydrogen absorption spectrum. (*2 marks*)

d Using a dot-and-cross diagram for a hydrogen molecule, explain how the atoms are held together. (*2 marks*)

e (i) Write the equation to represent the first ionisation enthalpy of hydrogen. (*1 mark*)

(ii) Explain how the value of this first ionisation enthalpy would compare with that of lithium. (*2 marks*)

12 Calcium carbonate rocks give off carbon dioxide when strongly heated. This can dissolve in water from underground springs making it fizzy. The fizzy water is sometimes sold as 'naturally carbonated spring water'.

a Write the equation for the decomposition of calcium carbonate, showing state symbols. (*1 mark*)

b Calcium hydroxide can be made from one of the products of the reaction. Describe tests that could be done to identify each of the ions in aqueous calcium hydroxide. (*2 marks*)

c (i) Calculate the volume of carbon dioxide that is given off when 0.35 g of calcium carbonate is fully decomposed. (*2 marks*)

(ii) Calculate the mass of barium carbonate that would be needed to produce the same volume of gas. (*1 mark*)

(ii) Draw a diagram of an apparatus you could use to check your answers to (i) and (ii). Explain how you could use your apparatus to show that barium carbonate was more thermally stable than calcium carbonate. (*6 marks*)

(iii) Explain in terms of the ions involved, why barium carbonate is more thermally stable than calcium carbonate. (*2 marks*)

d Draw a dot-and-cross diagram for carbon dioxide and use it to explain the shape and bond angle of the molecule. (*3 marks*)

13 Some students set out to make pure copper sulfate crystals from copper carbonate and sulfuric acid.

$$CuCO_3(s) + H_2SO_4(aq) \rightarrow$$
$$CuSO_4(aq) + CO_2(g) + H_2O(l)$$

a They plan to use the following methods:

Student A: React excess copper carbonate with sulfuric acid. Evaporate and allow to crystallise.

Student B: React excess sulfuric acid with copper carbonate. Filter, wash, and dry.

Student C: React excess copper carbonate with sulfuric acid. Filter. Evaporate the filtrate to dryness.

(i) None of these students would end up with pure crystals of copper sulfate? In each case, say what *would* be left at the end. (*3 marks*)

(ii) Outline the method that should be used. (*2 marks*)

(iii) Calculate the mass of copper carbonate that would react exactly with 20 cm³ of 2.0 mol dm⁻³ sulfuric acid. (*2 marks*)

(iv) How could the students test for the presence of sulfate in their crystals once they have made them? (*2 marks*)

b Draw a dot-and-cross diagram for the carbonate ion, CO_3^{2-}, using the minus symbol for the extra two electrons. Give the bond angle in the ion. (*2 marks*)

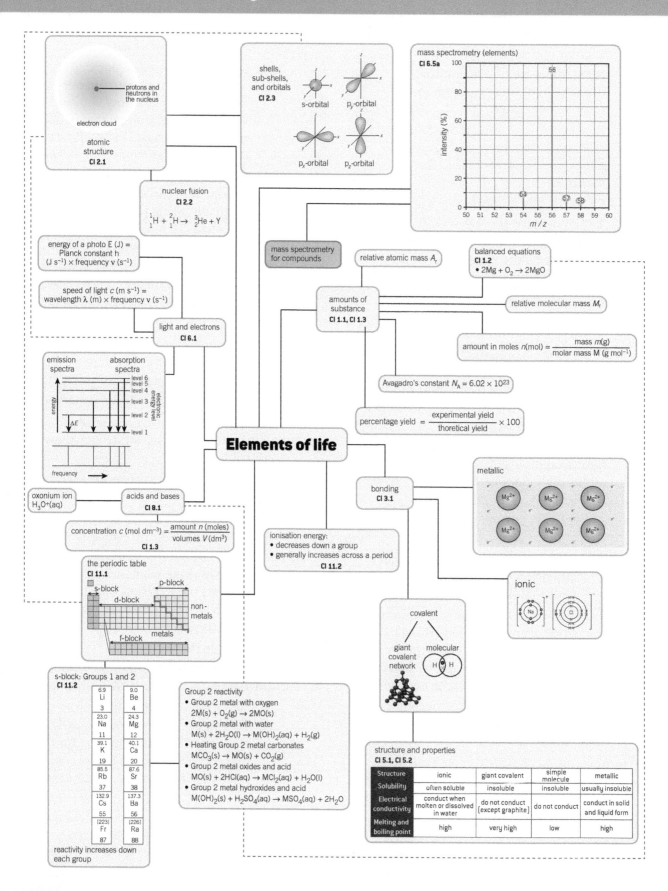

atomic structure
Cl 2.1

protons and neutrons in the nucleus

electron cloud

shells, sub-shells, and orbitals
Cl 2.3

s-orbital

p_y-orbital

p_x-orbital

p_z-orbital

mass spectrometry (elements)
Cl 6.5a

nuclear fusion
Cl 2.2

$${}^{1}_{1}H + {}^{2}_{1}H \rightarrow {}^{3}_{2}He + Y$$

energy of a photo E (J) = Planck constant h (J s^{-1}) × frequency ν (s^{-1})

speed of light c (m s^{-1}) = wavelength λ (m) × frequency ν (s^{-1})

light and electrons
Cl 6.1

mass spectrometry for compounds

relative atomic mass A_r

balanced equations
Cl 1.2
• $2Mg + O_2 \rightarrow 2MgO$

amounts of substance
Cl 1.1, Cl 1.3

relative molecular mass M_r

amount in moles n(mol) = $\dfrac{\text{mass } m(g)}{\text{molar mass M (g mol}^{-1})}$

Avagadro's constant $N_A = 6.02 \times 10^{23}$

percentage yield = $\dfrac{\text{experimental yield}}{\text{thoretical yield}} \times 100$

emission spectra
absorption spectra

level 6
level 5
level 4
level 3
level 2
level 1

energy

ΔE

electronic energy level

frequency

Elements of life

oxonium ion H_3O^+(aq)

acids and bases
Cl 8.1

concentration c (mol dm^{-3}) = $\dfrac{\text{amount } n \text{ (moles)}}{\text{volumes } V \text{ (dm}^3)}$
Cl 1.3

ionisation energy:
• decreases down a group
• generally increases across a period
Cl 11.2

bonding
Cl 3.1

metallic

Mg^{2+} Mg^{2+} Mg^{2+}
Mg^{2+} Mg^{2+} Mg^{2+}

the periodic table
Cl 11.1

s-block

p-block

d-block

non-metals

metals

f-block

ionic

Na   Cl

covalent

giant covalent network

molecular

H H

s-block: Groups 1 and 2
Cl 11.2

6.9 Li 3	9.0 Be 4
23.0 Na 11	24.3 Mg 12
39.1 K 19	40.1 Ca 20
85.5 Rb 37	87.6 Sr 38
132.9 Cs 55	137.3 Ba 56
[223] Fr 87	[226] Ra 88

reactivity increases down each group

Group 2 reactivity
• Group 2 metal with oxygen
  $2M(s) + O_2(g) \rightarrow 2MO(s)$
• Group 2 metal with water
  $M(s) + 2H_2O(l) \rightarrow M(OH)_2(aq) + H_2(g)$
• Heating Group 2 metal carbonates
  $MCO_3(s) \rightarrow MO(s) + CO_2(g)$
• Group 2 metal oxides and acid
  $MO(s) + 2HCl(aq) \rightarrow MCl_2(aq) + H_2O(l)$
• Group 2 metal hydroxides and acid
  $M(OH)_2(s) + H_2SO_4(aq) \rightarrow MSO_4(aq) + 2H_2O$

structure and properties
Cl 5.1, Cl 5.2

Structure	ionic	giant covalent	simple molecule	metallic
Solubility	often soluble	insoluble	insoluble	usually insoluble
Electrical conductivity	conduct when molten or dissolved in water	do not conduct (except graphite)	do not conduct	conduct in solid and liquid form
Melting and boiling point	high	very high	low	high

The transport of oxygen

One of the important elements of life is iron as it is an important component of haemoglobin – the molecule that carries oxygen in the blood. Haemoglobin is a complex protein, but part of its structure contains 'heme' groups (Figure 1).

▲ Figure 2 *A horseshoe crab (top) and a sea cucumber (bottom)*

▲ Figure 1 heme B, $C_{34}H_{32}O_4N_4Fe$, $M_r = 616\,g\,mol^{-1}$. One type of heme

Horseshoe crabs' blood contains a similar molecule called haemocyanin, which contains copper instead of iron. Sea cucumbers' blood contains vanabins, which contain vanadium, although the oxygen-carrying function of vanabins is uncertain as sea cucumbers also have haemoglobin. Haemocyanin and vanabins cause the blood to be blue and green respectively.

Other molecules with related structures include vitamin B-12 (cobalt-based) and chlorophyll (magnesium-based).

1   Write down the electron configurations, using s, p, d, and f notation, of magnesium, vanadium, iron, and cobalt.
2   Sketch a graph showing successive ionisation energies of magnesium and explain why it is evidence for the electron configuration of magnesium.
3   Explain why the bond angle in the CH_3 groups in heme B is 109°.
4   Show that the percentage by mass of carbon in heme B is 66.2%. Calculate the percentage by mass of iron in heme B.
5   Describe and write ionic equations for the precipitation tests for Fe^{2+} and Cu^{2+} ions in solution.

 Extension

1   Research the electron configuration of copper and chromium. Explain why they have slightly different configurations to other first row transition metals.
2   Prepare a summary of how to deduce shapes of molecules. For the molecules you choose, included details of the covalent bonds, repulsion of the groups of electrons and the shapes of the molecules. Give a wide range of examples.
3   On 12 November 2014 the *Philae* robotic lander detached from the European Space Agency *Rosetta* spacecraft and landed on the surface of comet 67P, orbiting near Jupiter, with the intention of studying the chemical composition of the comet. Research the findings of the mission and the instrumentation on board the lander.

CHAPTER 2

Developing fuels

Topics in this chapter

- → DF 1 Getting energy from fuels
  thermochemistry
- → DF 2 How much energy?
  enthalpy cycles and Hess' law
- → DF 3 What's in your tank?
  alkanes and shapes of molecules
- → DF 4 Where does the energy come from?
  bond enthalpies
- → DF 5 Getting the right sized molecules
  catalysis
- → DF 6 Alkenes – versatile compounds
  electrophilic addition, σ-bonds, and π-bonds
- → DF 7 Polymers and plastics
  addition polymerisation
- → DF 8 Burning fuels
  combustion of fuels, gas calculations, and the ideal gas equation
- → DF 9 What do the molecules look like?
  shapes of molecules, structures, and *E/Z* isomerism
- → DF 10 The trouble with emissions
  polluting gases
- → DF 11 Other fuels
  alternative fuels

Why a chapter on Developing fuels?

This chapter tells the story of fuels including petrol and diesel – what they are, how they are made, and the use of food as fuels for our bodies. It describes the work of chemists on improving fuels for motor vehicles, and in developing alternative fuels and sustainable energy sources for the future. Important ideas about vehicle pollutants and their control are also covered.

Some fundamental chemistry is introduced to achieve this. There are two main areas. First, it is important to understand where the energy comes from when a fuel burns. This leads to a study of enthalpy changes in chemical reactions, the use of energy cycles and the relationship between energy changes and the making and breaking of chemical bonds. Second, the module provides an introduction to organic chemistry. Alkanes and alkenes are studied in detail and other homologous series, such as alcohols and haloalkanes are introduced.

Isomerism is looked, and simple ideas about catalysis arise out of the use of catalytic converters to control exhaust emissions. All these topics will be developed and used in later modules.

Knowledge and understanding checklist

From your Key Stage 4 study you will have studied the following. Work through each point, using your Key Stage 4 notes and the support available on Kerboodle.

- [ ] Simple organic chemistry and homologous series.
- [ ] Useful products from crude oil.
- [ ] Combustion of alkanes.
- [ ] Exothermic and endothermic reactions.
- [ ] Addition polymerization.
- [ ] Sources of atmospheric pollutants.
- [ ] Catalysis.

You will learn more about some ideas introduced in earlier chapters:

- [ ] moles (**Elements of life**)
- [ ] empirical and molecular formulae (**Elements of life**)
- [ ] covalent bonding (**Elements of life**)
- [ ] polar bonds (**Elements of life**)
- [ ] molecular shape (**Elements of life**).

Maths skills checklist

In this chapter, you will need to use the following maths skills. You can find support for these skills on Kerboodle and through MyMaths.

- [ ] Recognise and make use of appropriate units in calculation.
- [ ] Use appropriate numbers of significant figures.
- [ ] Change to subject of an equation.
- [ ] Substitute numerical values into algebraic equations using appropriate units for physical quantities.
- [ ] Solve algebraic equations.
- [ ] Visualise and represent 2D and 3D forms including 2D representations of 3D objects.
- [ ] Understand the symmetry of 2D and 3D shapes.
- [ ] Plot data, lines of best fit, and extrapolate.

MyMaths.co.uk
Bringing Maths Alive

Fuel economy, or fuel efficiency, is the relationship between the distance travelled and the amount of fuel consumed by a vehicle. In the UK it is often measured in miles per gallon. Figure 1 shows a 1939 Mercedes car and a modern Mercedes car. The modern car has a fuel economy of nearly twice that of the 1939 vehicle. This shows the improvement in technology between 1939 and the present day. There are also hybrid and electric vehicles with even better consumption, but petrol-only or diesel-only engines will most likely be with us for many years yet.

The two cars in Figure 1 are petrol driven. Petrol is a highly concentrated energy source, meaning that the amount of energy released per gram of petrol is high when compared to many other fuels. To answer questions like 'How much energy can you get from a fuel?' and 'Which fuels store the most energy?' you have to know about thermochemistry.

Thermochemistry

Thermochemistry is the study of the energy and heat associated with chemical reactions. Different fuels give out different amounts of energy when one mole is burnt. Compare six important fuels (Figure 2).

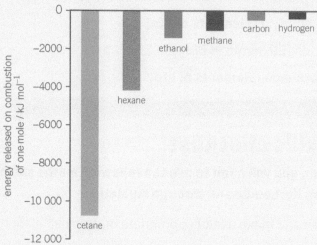

▲ **Figure 2** *The energy released on combustion of one mole of some important fuels*

The values vary widely. Why and what decides how much energy you get when you burn a mole of a particular fuel?

Fuels can be thought of as energy sources, but they can't release any energy until they have combined with oxygen. Therefore, the fuel–oxygen systems should be thought of as the energy sources. This is discussed in Topic DF 4. The following section looks at the energy changes during chemical reactions, such as combustion.

▲ **Figure 1** *A 1939 Mercedes (top) and a modern Mercedes (bottom)*

Chemical ideas: Energy changes and chemical reactions 4.1

Energy out, energy in

Energy changes are a characteristic feature of chemical reactions. Many chemical reactions give out energy and some take energy in. A reaction that gives out energy and heats the surroundings is described as **exothermic**. A reaction that takes in energy and cools the surroundings is **endothermic**.

During an exothermic reaction the chemical reactants are losing energy to their surroundings. This energy is used to heat the surroundings, for example, the air, the test tube, the laboratory, the car engine. The products end up with less energy than the reactants had but the surroundings end up with more, and get hotter. The energy transferred to and from the surroundings is measured as **enthalpy change**, ΔH. Enthalpy changes can be shown on an **enthalpy level diagram**, also called an energy level diagram (Figure 3).

> **Exothermic reaction**
>
> A reaction that gives out energy and heats the surroundings.

> **Endothermic reaction**
>
> A reaction that takes in energy and cools the surroundings.

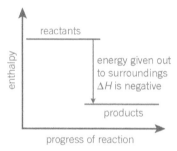

▲ Figure 3 *Enthalpy level diagram for an exothermic reaction such as burning methane:*
$CH_4(g) + 2O_2(g) \rightarrow CO_2(g) + 2H_2O(l)$

In an endothermic reaction, the reactants take in energy from the surroundings leaving the products at a higher energy level than the reactants (Figure 4).

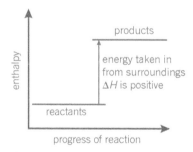

▲ Figure 4 *Enthalpy level diagram for an endothermic reaction such as decomposing calcium carbonate:*
$CaCO_3(s) \rightarrow CaO(s) + CO_2(g)$

Enthalpy change

You cannot measure the enthalpy H of a substance. What you can measure is the change in enthalpy when a reaction occurs. This is represented ΔH (pronounced 'delta H')

$$\Delta H = H_{\text{products}} - H_{\text{reactants}}$$

The enthalpy change in a chemical reaction gives the quantity of energy transferred to or from the surroundings, when the reaction is carried out in an open container.

For an exothermic reaction, ΔH is negative. This is because, from the point of view of the chemical reactants, energy has been lost to the surroundings. Conversely, for an endothermic reaction ΔH is positive – energy has been gained from the surroundings. Enthalpy changes are measured in kilojoules per mole ($kJ\,mol^{-1}$).

 ## Worked example: Making sense of ΔH

Example 1

The equation for the reaction of methane with oxygen (Figure 3) is:

$$CH_4(g) + 2O_2(g) \rightarrow CO_2(g) + 2H_2O(l) \qquad \Delta H = -890\,kJ\,mol^{-1}$$

How much energy would be released to the surroundings when two moles of methane undergo compete combustion?

Step 1: Calculate the energy transferred to heat the surroundings for one mole of methane.

From the equation $\Delta H = -890\,kJ$

Step 2: Multiply energy transfer for one mole by the number of moles in the question (in this case, two moles).

ΔH for two moles = $2 \times -890 = -1780\,kJ$

This assumes that all the methane is converted into products and that none is left unburnt.

Example 2

When calcium carbonate is heated, it decomposes. Energy is taken in – it is an endothermic reaction. How much energy would be taken in if 0.1 moles of calcium carbonate decomposed?

$$CaCO_3(s) \rightarrow CaO(s) + CO_2(g) \qquad \Delta H = +179\,kJ\,mol^{-1}$$

Step 1: Calculate the energy transferred to heat the surroundings for one mole of calcium carbonate.

From the equation = $+179\,kJ$

Step 2: Multiply energy transfer for one mole by the number of moles in the question (in this case, 0.1 moles).

ΔH for 0.1 moles = $0.1 \times +179 = +17.9\,kJ$

System or surroundings?

When chemists talk about enthalpy changes, they often refer to the system. This means the reactants and the products of the reaction they are interested in. The system may lose or gain enthalpy as a result of the reaction. The surroundings means the rest of the world – the test tube, the air, and so on.

Standard conditions

Like many physical and chemical quantities, ΔH varies according to the conditions. In particular, ΔH is affected by temperature, pressure, and the concentration of solutions. As such, certain **standard conditions** are chosen to compare enthalpy changes. Under standard conditions elements exist in their standard states.

If ΔH refers to these standard conditions, it is written as $\Delta H^{\ominus}_{298}$. You can use data sheets to look up $\Delta H^{\ominus}$ values.

A word about temperature

There are two scales of temperature used in science and you should be familiar with both. The kelvin K is the unit of absolute temperature and $0\,K$ is called absolute zero. The kelvin is the **SI (International System)** unit of temperature and should always be used in calculations involving temperature. However, you will usually measure temperature using a thermometer marked in degrees Celsius °C. You can convert temperatures from the Celsius scale to the kelvin scale by adding 273 to the Celsius reading. Similarly, you can convert from the Kelvin scale to the Celsius scale by subtracting 273 from the Kelvin reading.

 Worked example: Converting between Celsius and Kelvin

A liquid boils at 100 °C. What is this in K?

Step 1: Identify the conversion.

Converting from °C to K, so add 273.

Step 2: Add 273 to the Celsius value.

$$100 + 273 = 373\,K$$

A solid melts at 478 K. What is this in °C?

Step 1: Identify the converstion.

Converting from K to °C, so take away 273.

Step 2: Take away 273 from the kelvin value.

$$478 - 273 = 205\,°C$$

Different kinds of enthalpy change

There are several different kinds of enthalpy change that you need to be familiar with.

Standard enthalpy change for a reaction

The **standard enthalpy change for a** *reaction* is the enthalpy change when molar quantities of reactants, as stated in the equation, react together under standard conditions. This means at 100 kPa (pressure) and 298 K (temperature), with all the substances in their standard states. The symbol for the standard enthalpy change for a reaction is $\Delta_r H^{\ominus}_{298}$.

Standard conditions

Set conditions to allow us to compare enthalpy changes:

- a specified temperature normally chosen as 298 K (25 °C).
- a standard pressure of 100 kPa (equivalent to $1.01 \times 10^5\,N\,m^{-2}$)
- a standard concentration of 1 mol dm^{-3} for solutions.

Standard states

The physical state of a substance under standard conditions. This may be a pure solid, liquid, or gas.

Study tip

The kelvin *does not* have a degree symbol, °.

A change in temperature has the same numerical value on both scales.

Standard enthalpy change for a reaction $\Delta_r H^{\ominus}_{298}$

The enthalpy change when molar quantities of reactants as stated in the equation react together under standard conditions.

Worked example: Determining the standard enthalpy change of reaction

The equation for the reaction of hydrogen and oxygen is

$$2H_2(g) + O_2(g) \rightarrow 2H_2O(l) \qquad \textbf{Equation 1}$$

Determine the standard enthalpy of reaction given that

$$H_2(g) + \frac{1}{2}O_2(g) \rightarrow H_2O(l) \quad \Delta_r H^\ominus_{298} = -286\,\text{kJ mol}^{-1} \quad \textbf{Equation 2}$$

Step 1: Identify the difference in moles of the reaction the standard enthalpy value is given in (Equation 2) and the reaction you are calculating a value for (Equation 1).

$2H_2(g)$	$+$	$O_2(g)$	$\rightarrow$	$2H_2O(l)$	**Equation 1**
$H_2(g)$	$+$	$\frac{1}{2}O_2(g)$	$\rightarrow$	$H_2O(l)$	**Equation 2**
$2-1=1$		$1-\frac{1}{2}=\frac{1}{2}$		$2-1=1$	

Equation 1 is double the number of moles compared to Equation 2.

Step 2: Calculate how much energy is transferred under the reaction conditions, Equation 1.

Since 286 kJ are transferred to the surroundings when one mole of hydrogen reacts with oxygen and the equation you are interested in involves the reaction of two moles of hydrogen, the enthalpy change must also be doubled:

$$286 \times 2 = -572\,\text{kJ}$$

So, $2H_2(g) + O_2(g) \rightarrow 2H_2O(l)$ $\qquad \Delta_r H^\ominus_{298} = -572\,\text{kJ mol}^{-1}$

The following kinds of enthalpy change are particularly important and are given special names.

Measuring enthalpy changes

Many enthalpy changes can be measured in the laboratory by arranging for the energy involved in a reaction to be transferred to or from water surrounding the reaction vessel. If it is an exothermic reaction, the water gets hotter. If it is endothermic, the water gets cooler. By measuring the temperature change of the water, knowing the mass, and knowing the **specific heat capacity** of water (4.18 J g^{-1} K^{-1}), you can calculate the amount of energy that was transferred to or from the water during the chemical reaction. To do this, you need to use the relationship:

$$\begin{array}{ccccc} \text{energy} & & \text{specific heat} & & \\ \text{transferred} & = & \text{capacity} & \times & \text{mass} & \times & \text{temperature} \\ q\ (\text{kJ}) & & c\ (\text{J g}^{-1}\,\text{K}^{-1}) & & m\ (\text{g}) & & \text{change } \Delta T\ (\text{K}) \end{array}$$

Using a bomb calorimeter to accurately measure energy changes

In a bomb calorimeter, the fuel is ignited electrically and burns in oxygen inside the pressurised vessel. Energy is transferred to the surrounding water, where the temperature rise is measured. The experiment is done at constant volume in a closed container. Enthalpy changes are for reactions carried out at constant pressure, so the result needs to be modified accordingly.

1 Suggest why the sample is burnt in oxygen under pressure.
2 Suggest why the temperature measured is more accurate than when using a simple calorimeter.
3 Suggest why the heat transferred at constant volume might be different from that transferred at constant pressure.

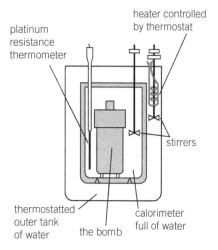

▲ Figure 3 *A bomb calorimeter*

Standard enthalpy change of combustion

The **standard enthalpy change** of *combustion* is the enthalpy change that occurs when one mole of a substance is burnt completely in oxygen. In theory, the substance needs to be burnt under standard conditions – 100 kPa and 298 K. In practice this is impossible, so the substance is burnt and then adjustments are made to allow for the non-standard conditions. The symbol for the standard enthalpy of combustion is $\Delta_c H^{\ominus}_{298}$.

For example, the enthalpy change of combustion of heptane, one of the alkanes found in petrol, is $-4187\,kJ\,mol^{-1}$. This is much bigger than for methane ($-890\,kJ\,mol^{-1}$) because burning heptane involves breaking and making more bonds than burning methane (see DF 4). Note that if no temperature is given with $\Delta_c H^{\ominus}$ then assume that the value refers to 298 K.

The equations for the combustion of methane and heptane are shown below.

$$CH_4(g) + 2O_2(g) \rightarrow CO_2(g) + 2H_2O(l) \qquad \Delta_c H^{\ominus}_{298} = -890\,kJ\,mol^{-1}$$

$$C_7H_{16}(l) + 11O_2(g) \rightarrow 7CO_2(g) + 8H_2O(l) \qquad \Delta_c H^{\ominus}_{298} = -4187\,kJ\,mol^{-1}$$

When writing an equation to represent an enthalpy change of combustion, the equation must always balance and show one mole of the substance reacting, even if this means having half a mole of oxygen molecules in the equation. You should always include state symbols.

Standard enthalpy change of combustion $\Delta_c H^{\ominus}_{298}$

The enthalpy change that occurs when one mole of a substance is burnt completely in oxygen under standard conditions in standard states.

Study tip

All combustion reactions are exothermic, so $\Delta_c H^{\ominus}_{298}$ is *always* negative.

Activity DF 1.1

In this activity you compare the energy given out by burning hexane and methanol, both compounds found in some types of petrol/

Standard enthalpy change of formation

The **standard enthalpy change of** *formation* is the enthalpy change when one mole of a compound is formed from its elements – again with both the compound and its elements being in their standard states under standard conditions. The symbol for standard enthalpy of formation is $\Delta_f H^\ominus_{298}$.

For example, the enthalpy change of formation of water, $H_2O(l)$, is $-286\,kJ\,mol^{-1}$. When you make one mole of water from hydrogen and oxygen, $286\,kJ$ are transferred to the surroundings. This is summed up as:

$$H_2(g) + \frac{1}{2} O_2(g) \rightarrow H_2O(l) \quad \Delta_f H^\ominus_{298} = -286\,kJ\,mol^{-1}$$

The equation refers to one mole of H_2O, so only $\frac{1}{2}$ mole of O_2 is needed in the equation.

It is often impossible to measure enthalpy changes of formation directly. For example, the standard enthalpy change of formation of methane is $-75\,kJ\,mol^{-1}$. This refers to the reaction

$$C(s) + 2H_2(g) \rightarrow CH_4(g) \quad \Delta_f H^\ominus_{298} = -75\,kJ\,mol^{-1}$$

However, this reaction doesn't actually occur under standard conditions. So how did anyone manage to measure the value of $\Delta_f H^\ominus_{298}$? It has to be done indirectly, making use of quantities that can be measured and incorporating these into an **enthalpy cycle**. You can find out about enthalpy cycles in Topic DF 2.

Standard enthalpy change of neutralisation

The **standard enthalpy change of** *neutralisation* can be measured from the energy given out when acids react with alkalis in aqueous solution. The symbol for this change is $\Delta_{neut} H^\ominus_{298}$.

The enthalpy changes can be calculated from these measurements:

$$NaOH(aq) + HCl(aq) \rightarrow NaCl(aq) + H_2O \qquad \Delta_{neut} H^\ominus_{298} = -58\,kJ\,mol^{-1} \qquad \textbf{Equation 3}$$

$$NaOH(aq) + HNO_3(aq) \rightarrow NaNO_3(aq) + H_2O \qquad \Delta_{neut} H^\ominus_{298} = -58\,kJ\,mol^{-1} \qquad \textbf{Equation 4}$$

$$NaOH(aq) + H_2SO_4(aq) \rightarrow NaHSO_4(aq) + H_2O \qquad \Delta_{neut} H^\ominus_{298} = -58\,kJ\,mol^{-1} \qquad \textbf{Equation 5}$$

$$2NaOH(aq) + H_2SO_4(aq) \rightarrow Na_2SO_4(aq) + 2H_2O \qquad \Delta_r H^\ominus_{298} = -115\,kJ\,mol^{-1} \qquad \textbf{Equation 6}$$

Notice that many of these values are the same. This is because the reaction that is occurring in Equations 3–5 can be represented as:

$$H^+(aq) + OH^- \rightarrow H_2O \qquad \Delta H^\ominus = -58\,kJ\,mol^{-1} \qquad \textbf{Equation 7}$$

So the enthalpy change of neutralisation is defined per mole of H_2O formed.

If one mole of sodium hydroxide, NaOH, reacts with one mole of sulfuric acid, H_2SO_4, the reaction in Equation 7 occurs and the enthalpy change is $-58\,kJ\,mol^{-1}$. If two moles of sodium hydroxide are available for each mole of sulfuric acid, then $\Delta_{neut} H^\ominus_{298}$ is virtually doubled. However, the enthalpy change of neutralisation (measured per mole of water produced) is virtually the same.

Summary questions

You will need to look up values for standard enthalpy changes when doing these problems.

1 Define the following enthalpy changes.
   a standard enthalpy change of combustion *(1 mark)*
   b standard enthalpy change of formation *(1 mark)*

2 Explain why standard enthalpy changes of formation may have a positive sign, but standard enthalpy changes of combustion are always negative. *(2 marks)*

3 The standard enthalpy change of formation of hydrogen chloride is $-92.3\,kJ\,mol^{-1}$ and that of hydrogen iodide is $+26.5\,kJ\,mol^{-1}$. Draw labelled enthalpy level diagrams to represent the reactions which occur when each of these compounds is formed from its elements. *(4 marks)*

4 The standard enthalpy change of combustion of carbon is equal to the standard enthalpy change of formation of carbon dioxide. Explain why this is so by referring to the equations for the two reactions. *(3 marks)*

5 a Write the equation to represent the formation of one mole of water from its elements in their standard states. *(2 marks)*
   b Look up and write down the standard enthalpy change of formation of water. *(1 mark)*
   c Calculate the enthalpy change when 1.0 g of hydrogen burns in oxygen. What assumptions have you made? *(3 marks)*
   d What is the standard enthalpy change for the following reaction?

   $$H_2O(l) \rightarrow H_2(g) + \frac{1}{2}O_2(g)$$ *(2 marks)*

6 Use Equations 3–7 to help you answer these questions.
   a $20\,cm^3$ of $0.10\,mol\,dm^{-3}$ NaOH reacts with $20\,cm^3$ $0.10\,mol\,dm^{-3}$ HCl. Calculate the temperature rise. *(2 marks)*
   b Calculate the temperature rise if both the concentrations were doubled in (a). *(2 marks)*
   c Calculate the temperature rise if $10\,cm^3$ of $0.10\,mol\,dm^{-3}$ NaOH is reacted with $20\,cm^3$ $0.10\,mol\,dm^{-3}$ HCl. *(2 marks)*
   d Estimate the temperature rise if $20\,cm^3$ of $0.10\,mol\,dm^{-3}$ NaOH is reacted with $20\,cm^3$ $0.10\,mol\,dm^{-3}$ HBr (another 'strong' acid, like HCl, that is present entirely as its ions in solution). Explain your reasoning. *(3 marks)*
   e Calculate the temperature rise when $20\,cm^3$ of $0.10\,mol\,dm^{-3}$ NaOH is reacted with $20\,cm^3$ $0.10\,mol\,dm^{-3}$ H_2SO_4. *(2 marks)*
   f Calculate the temperature rise when $40\,cm^3$ of $0.10\,mol\,dm^{-3}$ NaOH is reacted with $20\,cm^3$ $0.10\,mol\,dm^{-3}$ H_2SO_4. *(2 marks)*

Learning outcomes

Demonstrate and apply knowledge and understanding of:

→ the determination of enthalpy changes of reaction from enthalpy cycles and enthalpy level diagrams based on Hess' law

→ calculations involving enthalpy changes.

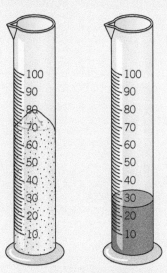

▲ Figure 1 *Fats and oils are more energy-rich than carbohydrates – the oil on the right will provide the same quantity of energy as the solid glucose on the left*

| 1 single measure of spirits | 1 glass of wine | $\frac{1}{2}$ pint of beer or lager |

▲ Figure 2 *Alcohol can be fattening: each of these drinks provides about 300 kJ of energy (about 70 Calories), equivalent to $1\frac{1}{2}$ slices of bread*

Important news for slimmers

Just as cars need fuels, so you need fuel for your body. When you eat too much of an energy-rich food, the excess energy gets stored in your body as fat. The more energy-rich the food, the more fattening it is.

Compare a carbohydrate such as glucose, $C_6H_{12}O_6$, with a fat such as glycerol trioleate, $C_{57}H_{104}O_6$, the main component of olive oil.

For each carbon atom, glucose has more oxygen atoms than olive oil, so glucose is much less energy-rich. From burning 1 g of a carbohydrate such as glucose you can get about 17 kJ. From burning 1 g of a fat such as olive oil you can get about 39 kJ. Gram for gram, fats contain twice as much energy as carbohydrates (Figure 1).

Alcohol is neither a fat nor a carbohydrate. In fact, there is a whole series of related compounds called alcohols, and the particular alcohol present in drinks is ethanol, C_2H_5OH. The same substance is used as an alternative to petrol for cars in some countries. It burns in the car engine releasing energy – and it also releases energy when metabolised in the body (Figure 2).

How is this energy measured? For some reactions, the enthalpy changes can be measured practically, but some cannot be measured directly and need to be calculated. The introductory activity in Topic DF 1 involved using a simple calorimeter to measure two enthalpy changes of combustion.

Chemical ideas: Energy changes and chemical reactions 4.2

Enthalpy cycles

For some reactions, measuring ΔH is very straightforward, for example, the burning of methane. For other reactions it is less simple. Decomposing calcium carbonate, $CaCO_3$, needs a temperature of over 800 °C. In cases like this, enthalpy changes can be measured indirectly, using enthalpy cycles.

Figure 4 shows an **enthalpy cycle**, also known as an energy cycle. There is both a direct and an indirect way to turn graphite, C, and hydrogen, H_2, into methane, CH_4. The enthalpy change for the direct route cannot be measured. The indirect route goes via carbon dioxide and water and involves two enthalpy changes both of which both can be measured. Since most organic compounds burn easily, cycles such as this can often be used, based on enthalpy changes of combustion, to work out indirectly the enthalpy change of an organic reaction.

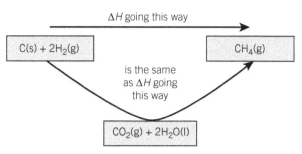

▲ Figure 4 *An enthalpy cycle for finding the enthalpy change of formation of methane, CH₄*

The same reaction can be represented using an enthalpy level diagram (Figure 5) instead of an enthalpy cycle.

In Hess' law, the enthalpy change for any chemical reaction is independent of the intermediate stages, so long as the initial and final conditions are the same for each route.

This is an alternative way of representing the same processes as in Figure 4, and the numerical values can be processed in exactly the same way as the worked examples that follow.

Using enthalpy changes of combustion in enthalpy cycles

The key idea is that the total enthalpy change for the indirect route is the same as the enthalpy change via the direct route. Energy cannot be created or destroyed – this is the law of conservation of energy. So as long as your starting and finishing points are the same, the enthalpy change will always be the same, irrespective of how you get from start to finish. This is one way of stating **Hess' law**, and an enthalpy cycle, like the one in Figure 5, is called a Hess' cycle or a thermochemical cycle.

If you know the enthalpy changes involved in two parts of the cycle, you can work out the enthalpy change in the third. So, referring to Figure 6, if you can measure ΔH_2 and ΔH_3, ΔH_1 can be calculated, which is the enthalpy change that cannot be measured directly.

$$\Delta H_1 = \Delta H_2 - \Delta H_3$$

It has to be minus ΔH_3 because the reaction to which ΔH_3 applies actually goes in the opposite direction to the way you want it to go in order to produce methane. To find the enthalpy change for the reverse reaction, you must reverse the sign.

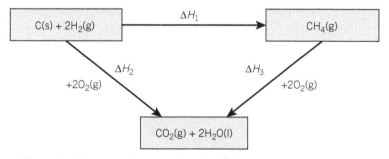

▲ Figure 6 *Using an enthalpy cycle to find ΔH_1*
$$\Delta H_1 = \Delta H_2 - \Delta H_3$$

Specific heat capacity *c*

The specific heat capacity of a substance is the amount of energy needed to raise the temperature of 1 g of a substance by 1 K.

Activity DF 2.1

In this activity you will examine the reaction of zinc with copper(II) sulfate. Because the reaction does not happen instantly, you will plot a cooling curve in order to estimate the maximum temperature more accurately. This activity also involves the concept of experimental errors.

Activity DF 2.2

In this activity you can work in small groups to discuss how you can make use of Hess' law to calculate an enthalpy change of reaction.

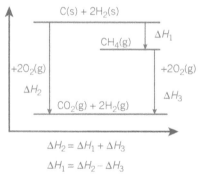

$$\Delta H_2 = \Delta H_1 + \Delta H_3$$
$$\Delta H_1 = \Delta H_2 - \Delta H_3$$

▲ Figure 5 *An enthalpy level diagram for the enthalpy of formation of methane*

Activity DF 2.3

In this activity you will determine the enthalpy change of a reaction and use your results along with Hess' law.

Worked example: Using Hess' cycles

Use the following standard enthalpy values to calculate the enthalpy of change of formation of methane (Figure 5).

$$\Delta_c H^{\ominus}(C) = -393 \text{ kJ mol}^{-1}$$

$$\Delta_c H^{\ominus}(H_2) = -286 \text{ kJ mol}^{-1}$$

$$\Delta_c H^{\ominus}(CH_4) = -890 \text{ kJ mol}^{-1}$$

Step 1: Identify the reactions involved from the Hess' cycle (Figure 6).

ΔH_2 is the sum of the enthalpy changes of combustion of one mole of carbon and two moles of hydrogen.

$$\Delta H_2 = \Delta_c H^{\ominus}(C) + \Delta_c H^{\ominus}(H_2)$$

ΔH_3 is the enthalpy change of combustion of methane.

$$\Delta H_3 = \Delta_c H^{\ominus}(CH_4)$$

ΔH_1 is the enthalpy change of formation of methane – the quantity you are trying to find.

Step 2: Write the equation to calculate the ΔH_1 value.

$$\Delta H_1 = \Delta H_2 - \Delta H_3$$

This is the same as:

$$\Delta H_1 = \Delta_c H^{\ominus}(C) + 2\Delta_c H^{\ominus}(H_2) - \Delta_c H^{\ominus}(CH_4)$$

Step 3: Substitute in the standard enthalpy values.

$$= -393 + 2(-286) - (-890)$$

$$= -75 \text{ kJ mol}^{-1}$$

Using enthalpy changes of formation in enthalpy cycles

Enthalpy changes for the formation of many compounds from their elements in their standard states are available in data books. Some of these have been measured directly and others indirectly, as shown above for methane. These data are very useful as they can be used to calculate the enthalpy change for a reaction, rather than doing experiments yourself.

 Worked example: Using enthalpy changes of formation to calculate an enthalpy change of reaction

Calculate the enthalpy change of reaction for the following process.

$$NH_3(g) + HCl(g) \rightarrow NH_4Cl(s)$$

Use the following data for the standard enthalpy of formation.

$$\Delta_f H^{\ominus}(NH_3) = -46.1 \text{ kJ mol}^{-1}$$

$$\Delta_f H^{\ominus}(HCl) = -92.3 \text{ kJ mol}^{-1}$$

$$\Delta_f H^{\ominus}(NH_4Cl) = -315 \text{ kJ mol}^{-1}$$

Step 1: Draw a Hess' cycle for the reaction.

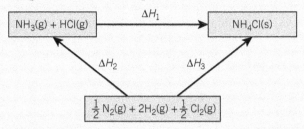

Step 2: Identify the reactions involved from the Hess' cycle.

ΔH_2 is the sum of the enthalpy changes of formation of one mole of ammonia and hydrogen chloride.

$$\Delta H_2 = \Delta_f H^\ominus (NH_3) + \Delta_f H^\ominus (HCl)$$

ΔH_3 is the enthalpy change of formation of ammonium chloride.

$$\Delta H_3 = \Delta_f H^\ominus (NH_4Cl)$$

ΔH_1 is the enthalpy change of reaction of ammonia and hydrogen chloride – the quantity you are trying to find.

Step 3: Write the equation to calculate the ΔH_1 value.

$$\Delta H_1 = -\Delta H_2 + \Delta H_3$$

This is the same as:

$$\Delta H_1 = -(\Delta_f H^\ominus (NH_3) - \Delta_f H^\ominus (HCl)) + \Delta_f H^\ominus (NH_4Cl)$$

Step 4: Substitute in the standard enthalpy of formation values.

$$= -(-46.1) - (-92.3) + (-315)$$
$$= -176.6 \, kJ \, mol^{-1}$$

You may find the following shortcut useful:

$$\Delta_r H^\ominus = \Sigma \Delta_r H^\ominus_{(products)} - \Delta_r H^\ominus_{(reactants)}$$

enthalpy change of reaction = sum of enthalpy changes of formation of products − sum of enthalpy changes of formation of reactants

Summary questions

1 You have been asked to measure the enthalpy change of combustion of methane. You have been given a gas cooker and a saucepan.
  a What other equipment would you need? *(1 mark)*
  b What measurements would you make? *(1 mark)*
  c What would be the main sources of error? *(1 mark)*

2 a Write an equation for the formation of one mole of butane, $C_4H_{10}(g)$, from its elements in their standard states. *(2 marks)*
  b Draw an enthalpy cycle to show the relationship between the formation of butane from carbon and hydrogen, and the combustion of these elements to give carbon dioxide and water (see Figure 6 if you need help). *(3 marks)*

c Use your enthalpy cycle to calculate a value for the standard enthalpy change of formation of butane. *(3 marks)*

You will need the following data:

$$\Delta_c H^{\ominus}(C) = -393 \text{ kJ mol}^{-1}$$
$$\Delta_c H^{\ominus}(H_2) = -286 \text{ kJ mol}^{-1}$$
$$\Delta_c H^{\ominus}(C_4H_{10}) = -2877 \text{ kJ mol}^{-1}$$

3 A student measures the enthalpy change of the reaction:

$$CuSO_4(s) \quad + \quad 5H_2O \quad \rightarrow \quad CuSO_4 \cdot 5H_2O$$
anhydrous   hydrated
copper sulfate   copper sulfate

The student uses two experimental steps and applies Hess' law to work out the answer.

Experiment 1

Dissolve 3.99 g (0.0250 mol) of anhydrous copper sulfate in 50 cm^3 water in a plastic cup. The temperature rose from 19.52 °C to 27.40 °C.

Experiment 2

Add 6.24 g (0.0250 mol) of hydrated copper sulfate to 48 cm^3 water in a plastic cup (to allow for the '5H$_2$O' making up the volume to 50 cm^3). The temperature changed from 19.56 °C to 18.28 °C.

a Why can't the enthalpy change of the reaction be measured directly? *(1 mark)*

b Show, by calculation, that 48 cm^3 was the correct volume to use for the second experiment. *(2 marks)*

c Calculate the energy change in each experiment in kJ mol^{-1}. (The specific heat capacity of water is 4.18 J g^{-1} K^{-1}) *(2 marks)*

d Explain any assumptions you made in these calculations. *(1 mark)*

e Draw a Hess' cycle which connects the enthalpy changes in the two experiments with the enthalpy change in the equation. *(3 marks)*

f Calculate a value for the enthalpy change of the reaction in the equation. *(2 marks)*

g Making suitable assumptions about the apparatus used, calculate the percentage uncertainty in each value in part **c**. Work these out as actual ($\pm$) uncertainty values. Add these uncertainties to give the overall uncertainty in your answer to part **f**. *(3 marks)*

h Which piece of apparatus generates the greatest percentage uncertainty? *(1 mark)*

i The accepted value for the enthalpy change is -77.4 kJ mol^{-1}. Comment on this compared with your value and uncertainty limits. *(2 marks)*

DF 3 What's in your tank?

Specification reference: DF(l), DF(m), DF(r)

What are petrol and diesel?

Petrol and diesel are both complex mixtures of many different compounds, carefully blended to give the right properties. The compounds are obtained from crude oil in several ways.

Crude oil is a mixture of many hundreds of hydrocarbons. It is a thick black liquid but dissolved in it are gases and solids. Oil from the North Sea is pumped along pipes on the seabed to UK refineries, and special tankers bring crude oil from distant oilfields, such as those in the Middle East and Alaska. These refineries are either close to the shore (such as at Fawley, near Southampton) or the oil is off-loaded into a pipeline leading to a refinery (such as from Finnart, on the west coast of Scotland, to the refinery at Grangemouth, near Edinburgh).

At the refinery the crude oil is heated to vaporise it and the vapour passes into a distillation column. This process is known as **fractional distillation**. The oil is separated into fractions, each having a specific boiling point range. The fractions do not have an exact boiling point because they are mixtures of many different hydrocarbons. For example, the gasoline fraction is a mixture of liquids, mostly alkanes with between five and seven carbon atoms, boiling in the range 25–75 °C.

The gasoline and gas oil fractions are the sources of petrol components. Another important fraction, naphtha, is also converted into high-grade petrol as well as being used in the manufacture of many organic chemicals. Table 1 contains information on some other fractions produced in the distillation of crude oil.

Learning outcomes

Demonstrate and apply knowledge and understanding of:

→ the nomenclature, general formulae, and structural formulae for alkanes and cycloalkanes (names up to 10 carbon atoms)

→ the terms aliphatic, aromatic, saturated, functional group, homologous series

→ structural formulae (full, shortened, and skeletal).

▲ Figure 1 *Grangemouth refinery*

▼ Table 1 *Fractions obtained from the fractional distillation of crude oil*

Name of fraction	Boiling point range / °C	Composition	% of crude oil	Use(s)
refinery gas	<25	C_1–C_4	1–2	liquid petroleum gas (propane, butane), blending in petrol, feedstock for organic chemicals
gasoline	25–75	C_5–C_7		car petrol
naphtha	75–190	C_6–C_{10}	20–40	production of organic chemicals, converted to petrol
kerosene	190–250	C_{10}–C_{16}	10–15	jet fuel, heating fuel (paraffin)
gas oil	250–350	C_{14}–C_{20}	15–20	diesel fuel, central heating fuel, converted to petrol
residue	>350	>C_{20}	40–50	fuel oil (e.g., power stations, ships), lubricating oils and waxes, bitumen or asphalt for roads and roofing

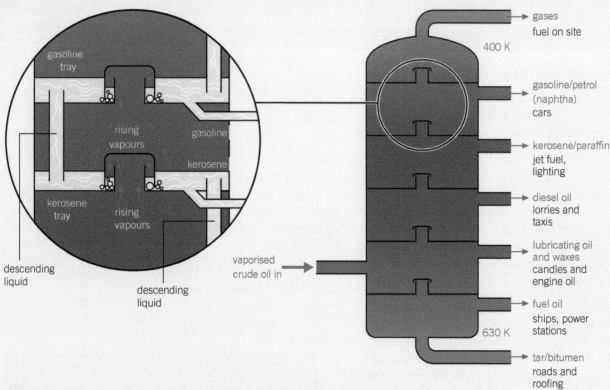

▲ Figure 2 *The primary fractional distillation of crude oil is a continuous process. Vapour rises up through the column and liquids condense and are run off at different levels, depending on their volatility*

Organic chemistry – alkanes

Structure and naming of alkanes

Many carbon compounds are found in living organisms, which is why their study got the name **organic chemistry**. Today, organic chemistry includes all carbon compounds whatever their origin – except carbon monoxide, CO, carbon dioxide, CO_2, and the carbonates, which are traditionally included in inorganic chemistry studies.

Only carbon can form the diverse range of compounds necessary to produce the individuality of living things.

Why carbon?

Carbon's electron structure is shown in Figure 3. This electron structure makes it the first member of Group 4 in the centre of the periodic table, and is responsible for its special properties.

A carbon atom has four electrons in its outer shell. It could achieve stability by losing or gaining four electrons but this is too many electrons to lose or gain. The resulting carbon ions would have charges of +4 or −4 respectively, and would be too highly charged. So when carbon forms compounds the bonds are covalent rather than ionic.

carbon (2,4)

▲ Figure 3 *The electron structure of carbon*

In methane, CH_4, the carbon atom achieves stability by sharing its outer electrons with four hydrogen atoms (Figure 4) forming four carbon–hydrogen covalent bonds.

Carbon forms strong covalent bonds with itself to give chains and rings of its atoms, joined by carbon–carbon covalent bonds. This property is called **catenation** and leads to the limitless variety of organic compounds possible.

Each carbon atom can form four covalent bonds, so the chains may be straight or branched, and can have other atoms or groups substituted on them.

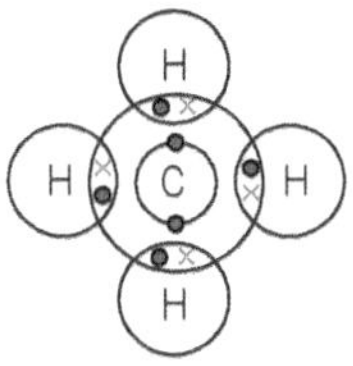

methane, CH_4

▲ Figure 4 *A dot-and-cross diagram showing covalent bonds in methane*

Hydrocarbons

Chemists cope with the vast number of organic compounds by dividing them into groups of related compounds. **Hydrocarbons** are compounds containing *only* carbon atoms and hydrogen atoms. They are represented by the general molecular formula C_xH_y. There are different types of hydrocarbons (Table 2).

▼ Table 2 *Some common hydrocarbons*

Name	Formula	Shape	Type of compound
methane	CH_4	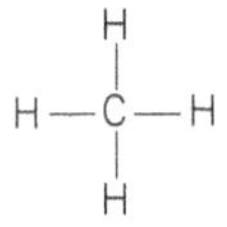	alkane
ethene	C_2H_4	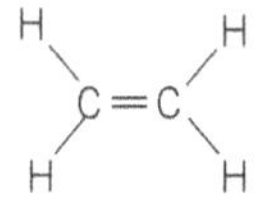	alkene
benzene	C_6H_6		arene
cyclohexane	C_6H_{12}		cycloalkane

The ring of six carbon atoms in benzene has special properties. Compounds that contain a benzene ring are called **aromatic compounds**. These compounds are called arenes. Compounds that do not contain a benzene ring are called **aliphatic compounds**.

Hydrocarbons are relatively unreactive – this is particularly true of alkanes and arenes. They form the unreactive framework of organic compounds. But when you attach other groups to the hydrocarbon framework its properties are modified. So you can think of organic compounds as having hydrocarbon frameworks, with modifiers attached.

A modifier such as the hydroxyl group, –OH, is also called a **functional group**. Compounds with the –OH group are called **alcohols**. Petrol usually contains alcohols in the blend and you will find out more about them later.

Aromatic compounds

Compounds that contain one or more benzene rings.

Aliphatic compounds

Compounds that do not contain any benzene rings.

Functional group

Modifiers that are responsible for the characteristic chemical reactions of molecules.

The double bond in an **alkene**, C=C, is much more reactive than a carbon–carbon single bond. It is often regarded as a functional group, even though it is part of the hydrocarbon framework.

Alkanes

Alkanes are **saturated** hydrocarbons. Saturated means that they contain the maximum number of hydrogen atoms possible, with no double or triple bonds between carbon atoms.

The general molecular formula of the alkanes is C_nH_{2n+2} where n is the number of carbon atoms and can be any whole number. The names of all the alkanes end in –ane. Table 3 shows the names and formulae of the first 10 alkanes.

Look at the names in Table 3. The first part of the name of an alkane indicates the number of carbon atoms in each molecule. The second part, –ane, indicates that the compounds are part of the class of compounds called alkanes. The names are irregular up to butane – they do not use the normal prefix associated with 1, 2, 3, and 4 – but after butane they are more predictable. You need to learn the prefix for 1–10 carbons (Table 3).

A series of compounds related to each other in this way is called an **homologous series**. All the members of the series have the same general molecular formula, and each member differs from the next by a $-CH_2-$ unit. All the compounds in a series have similar chemical properties, so chemists can study the properties of the series rather than those of individual compounds. However, physical properties such as melting point, boiling point, and density do change gradually in the series as the number of carbon atoms in the molecules increases.

Structure of alkanes

Figure 4 shows a dot-and-cross formula for methane. It shows all the outer electrons in each atom and how electrons are shared to form the covalent bonds. For larger molecules, dot-and-cross formulae can be complicated and the shared electron pairs can be replaced by lines representing the covalent bonds. Figure 5 is called a full structural formula – it shows all the atoms and all the bonds in the molecule.

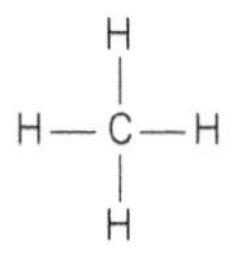

You can also write an abbreviated version known as a shortened structural formula. Take heptane, for example. Its full structural formula is shown in Figure 6.

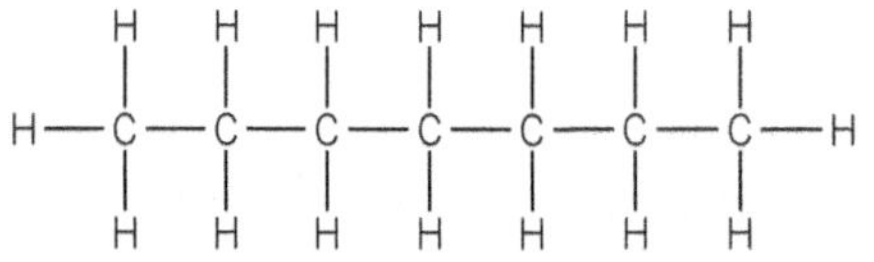

▲ Figure 6 *The full structural formula for heptane, C_7H_{16}*

Its shortened structural formula is

$$CH_3-CH_2-CH_2-CH_2-CH_2-CH_2-CH_3$$

This can be further shortened to

$$CH_3CH_2CH_2CH_2CH_2CH_2CH_3$$

Saturated

Hydrocarbons containing the maximum number of hydrogen atoms possible, no carbon–carbon double or triple bonds.

▼ Table 3 *The first 10 alkanes*

n	Molecular formula	Prefix	Name
1	CH_4	meth-	methane
2	C_2H_6	eth-	ethane
3	C_3H_8	prop-	propane
4	C_4H_{10}	but-	butane
5	C_5H_{12}	pent-	pentane
6	C_6H_{14}	hex-	hexane
7	C_7H_{16}	hept-	heptane
8	C_8H_{18}	oct-	octane
9	C_9H_{20}	non-	nonane
10	$C_{10}H_{22}$	dec-	decane

Homologous series

A series of compounds in which all members have the same general molecular formula

▲ Figure 5 *A full structural formula uses lines to represent covalent bonds*

Table 4 gives the full structural formulae and shortened formulae for some alkanes. You will see that each carbon atom is bonded to four other atoms. Each hydrogen atom is bonded to only one other atom.

▼ Table 4 *Structural formulae of alkanes*

Name	Molecular formula	Full structural formula	Shortened structural formula	Further shortened to
methane	CH_4	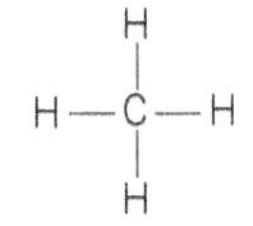	CH_4	
ethane	C_2H_6	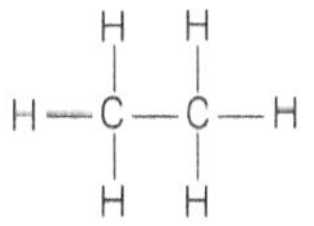	$CH_3{-}CH_3$	CH_3CH_3
propane	C_3H_8	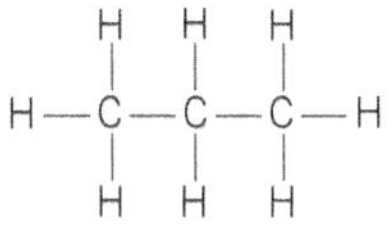	$CH_3{-}CH_2{-}CH_3$	$CH_3CH_2CH_3$

Branched alkanes

Alkanes may have straight or branched chains. It is often possible to draw more than one structural formula for a given molecular formula. There is often a straight-chain compound and one or more branched-chain compounds with the same molecular formula.

For a compound with the molecular formula C_4H_{10} there are two possible structural formulae:

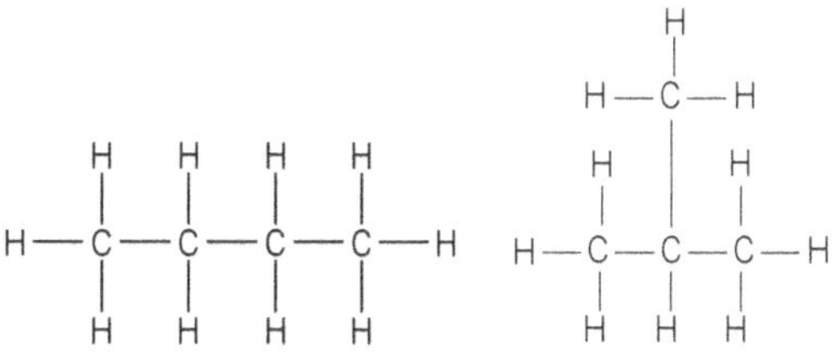

◀ Figure 7 *The full structural formulae of butane (left) and methylpropane (right). Both have the molecular formula C_4H_{10}*

These two compounds are **structural isomers** because they have the same molecular formulae but different structural formulae. There is more about this type of isomerism later in this section.

The branched-chain isomer is regarded as being formed from a straight-chain alkane (propane) with a $-CH_3$ group attached to the second carbon atom. The $-CH_3$ group is just methane with a hydrogen atom removed so that it can join to another atom. It is called a methyl group. The isomer is therefore called methylpropane.

Side groups of this kind are called alkyl groups. They have the general formula C_nH_{2n+1} (Table 5) and are often represented by the symbol R.

Butane and methylpropane are **systematic names**. Every organic compound can be given a systematic name derived from an internationally agreed set of rules. Many compounds also have common names. The systematic name is important because it allows for the full structural formula to be determined. The systematic name can also be determined from the full structural formula.

▼ Table 5 *Some common alkyl groups*

Alkyl group	Formula
methyl	$CH_3{-}$
ethyl	$CH_3CH_2{-}$
propyl	$CH_3CH_2CH_2{-}$
butyl	$CH_3CH_2CH_2CH_2{-}$
pentyl	$CH_3CH_2CH_2CH_2CH_2{-}$

Activity DF 3.1

In this activity you will practice naming alkanes and cycloalkanes.

Cycloalkanes

As well as open-chain alkanes, it is also possible for alkane molecules with cyclic structures to exist. These molecules are called cycloalkanes and have the general formula C_nH_{2n}. They have two fewer hydrogen atoms than the corresponding alkane, because there are no $-CH_3$ groups at the ends of the chain.

Table 6 shows some different ways of representing cycloalkanes. The skeletal formula shows only the shape of the carbon framework. Each line represents a carbon–carbon bond. The carbon atoms are at the corners. The carbon–hydrogen bonds are not shown but it is easy to work out how many there are – in saturated compounds, carbon always forms four covalent bonds

▼ Table 6 Cycloalkanes – the general formula is C_nH_{2n}

Cycloalkane	Shortened structural formula	Skeletal formula
cyclopropane, C_3H_6	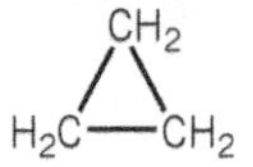 CH_2 / H_2C-CH_2	 △
cyclobutane, C_4H_8	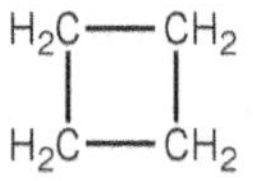 H_2C-CH_2 / H_2C-CH_2	▢
cyclohexane, C_6H_{12}	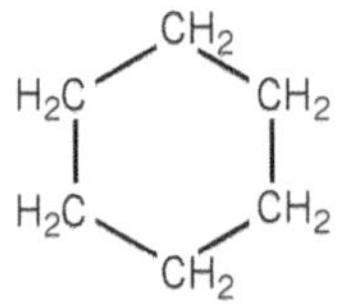 CH_2 / H_2C CH_2 / H_2C CH_2 / CH_2	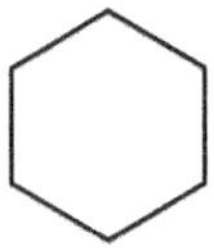 ⬡

Naming alkanes

Alkanes can be either straight chain or branched.

They have the general formula C_2H_{2n+2}. The rules for naming the first 10 straight chain alkanes are given in Table 3.

Naming branched alkanes

When alkanes are branched, the following rules apply.

1. Find the longest chain of carbons in the molecule and name the chain, for example, a chain five carbon atoms long would be pentane.

2. Identify any side chains off the main chain.

3. Name these side chains as substituents on the main chain by adding –yl to the appropriate prefix.

4. State the location of any side chains by prefixing the name of the side chain with the number of the carbon atom to which the side chain is attached.

5. Keep the number as low as possible, for example:

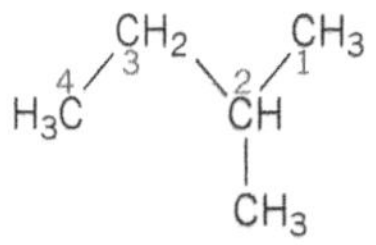

2-methylbutane   not   3-methylbutane

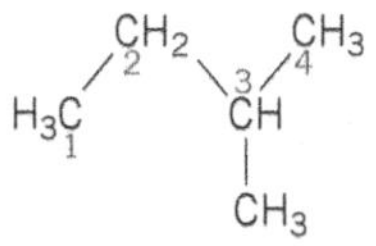

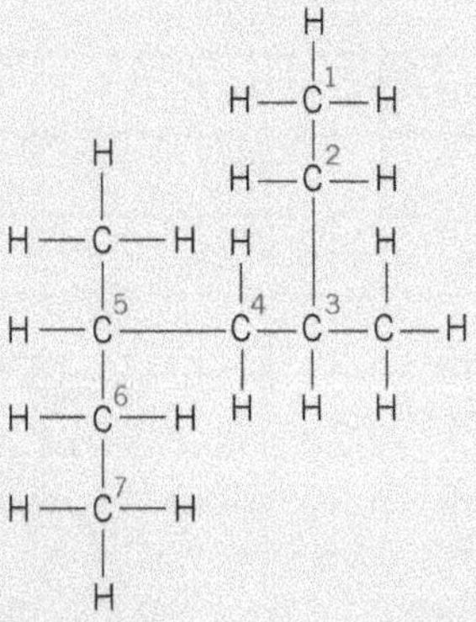

6 If there is more than one side chain, the rules still apply. Keep the numbering as low as possible, but when naming state the side chains in alphabetical order. If there are more than one of the same type of side group, use the prefixes di- (if there are two), tri- (if there are three), tetra- (if there are four).

7 Name the compound by starting the side chains and their location first, then the 'parent' (longest chain). Use hyphens between numbers and letters and commas between numbers.

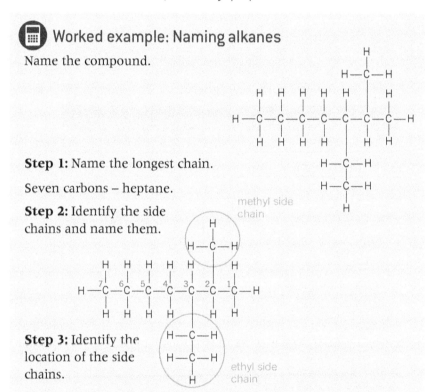

2,2-dimethyl propane

Worked example: Naming alkanes

Name the compound.

Step 1: Name the longest chain.

Seven carbons – heptane.

Step 2: Identify the side chains and name them.

Step 3: Identify the location of the side chains.

Numbering the heptane chain from right to left gives the lowest numbering with the methyl group on carbon 2 and the ethyl group on carbon 3.

Step 4: List the two side chains in alphabetical order. Therefore, 3-ethyl-2-methylheptane.

Naming cycloalkanes

Cycloalkanes have the general formula C_nH_{2n}. They can be names using the following rules:

1 Count the number of carbons in the cyclic structure and identify the appropriate prefix (Table 3) to name the alkane.

▲ Figure 8 Propane, C_3H_8, gas cylinders

▲ Figure 9 The hydrocarbon $C_{31}H_{64}$ is a typical component of paraffin wax

2 Add the alkane name to the end of cyclo.

There are five carbons so this is cyclopentane.

What are alkanes like?

Whether an alkane is solid, liquid, or gas at room temperature depends on the size of its molecules. The first four members of the series ($n = 1-4$) are colourless gases (Figure 8). Higher members ($n = 5-16$) are colourless liquids and the larger alkanes ($n = 17+$) are white waxy solids (Figure 9).

Alkanes mix well with each other but do not mix with water. The alkanes and water form two separate layers. This is because alkanes contain non-polar molecules but liquids such as water contain polar molecules that attract each other and prevent the alkane molecules mixing with them. There is more about the polarity of molecules in Topic DF 6.

Summary questions

1 Copy and complete the following table.

Empirical formula	Molecular formula	M_r
	C_3H_8	44.0
CH_2		168.0
	C_6H_6	
$C_{10}H_{21}$		282.0
	C_5H_{10}	
CH		26.0
	$C_{10}H_8$	

2 What is the molecular formula of each of the following alkanes?

  a heptane   *(1 mark)*

  b hexadecane – sixteen carbon atoms   *(1 mark)*

  c eicosane – twenty carbon atoms   *(1 mark)*

3 Draw dot-and-cross diagrams for:

  a ethane   *(1 mark)*

  b ethene   *(1 mark)*

  c propane   *(1 mark)*

4 A hydrocarbon contains 85.7% of carbon and 14.3% of hydrogen by mass. Its relative molecular mass is 28.0.

  a Find its empirical formula.   *(1 mark)*

  b Suggest a molecular formula for this compound.   *(1 mark)*

5 A hydrocarbon contains 82.8% by mass of carbon.

  a Work out its empirical formula.   *(1 mark)*

  b Suggest its molecular formula, explaining your reasons.   *(1 mark)*

DF 4 Where does the energy come from?

Specification reference: DF(e)

Carrying fuels around

The enthalpy change of combustion may not be the most important thing to consider for a practical fuel. What really matters is the **energy density** – how much energy you get per kilogram of fuel. This can be worked out from the enthalpy change of combustion using the relative molecular mass (Table 1).

▲ Figure 1 *The more energy this tanker can carry, the more cost effective the journey is*

Why do different fuels release different amounts of energy? All chemical reactions involve breaking and making chemical bonds.

▼ Table 1 *Energy densities of some important fuels*

Fuel	Formula	Standard enthalpy change of combustion $\Delta_c H^\ominus$ / kJ mol^{-1}	Relative molecular mass	Energy density – energy transferred on burning 1 kg of fuel / kJ kg^{-1}
hexadecane (cetane)	$C_{16}H_{34}(l)$	−10 700	226	−47 300
hexane	$C_6H_{14}(l)$	−4163	86	−48 400
methane	$CH_4(g)$	−890	16	−55 600
ethanol	$C_2H_5OH(l)$	−1367	46	−29 700
carbon	$C(s)$	−393	12	−32 800
hydrogen	$H_2(g)$	−286	2	−143 000

Bonds break in the reactants and new bonds form in the products. The energy changes in chemical reactions come from the energy changes that happen when bonds are broken and made.

Chemical ideas: Energy changes and chemical reactions 4.3

Bond enthalpies

A chemical bond is basically electrical attraction between atoms or ions. Breaking a bond involves overcoming these attractive forces. To break the bond completely, the atoms or ions need (theoretically) to be an infinite distance apart. Figure 2 illustrates this for the hydrogen–hydrogen bond in a molecule of hydrogen, H_2.

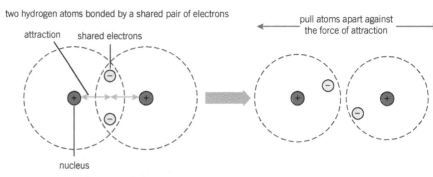

two hydrogen atoms bonded by a shared pair of electrons

attraction shared electrons

pull atoms apart against
the force of attraction

nucleus

Both nuclei are attracted to the same shared
pair of electrons. This holds the nuclei together.

▲ Figure 2 *Breaking a bond involves using energy to overcome the forces of attraction*

The quantity of energy needed to break a particular bond in a molecule is called the bond dissociation enthalpy, or **bond enthalpy**.

For the bond shown in Figure 2, the process involved is

$$H_2(g) \rightarrow 2H(g) \qquad\qquad \Delta H = +436\,\text{kJ mol}^{-1}$$

The bond enthalpy of the H—H bond is $+436\,\text{kJ mol}^{-1}$. ΔH has a positive value because breaking a bond is an endothermic process – it needs energy. Bond enthalpies are very useful because they indicate how strong bonds are. The stronger a bond, the more energy is needed to break it and the higher its bond enthalpy.

Bond enthalpy and bond length

When a bond like the one in Figure 2 forms, the atoms move together because of the attractive forces between nuclei and electrons. There are also repulsive forces between the nuclei of the two atoms. These get bigger as the atoms approach until the atoms stop moving together. The distance between them is now the equilibrium bond length (Figure 3). The shorter the bond length, the stronger the attraction between the atoms.

> **Study tip**
>
> Bond enthalpy is the energy needed to break one mole of a bond to give separate atoms all in the gaseous state. The units are kJ mol^{-1}.

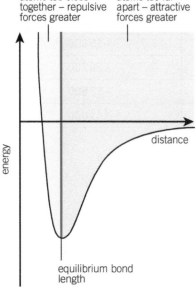

atoms too close together – repulsive forces greater

atoms too far apart – attractive forces greater

energy

distance

equilibrium bond length

▲ Figure 3 *In a chemical bond there is a balance between attractive and repulsive forces*

▼ Table 2 *Average bond enthalpies and bond lengths from a range of compounds (*except for C═O where the data is for carbon dioxide)*

Bond	Average bond enthalpy / kJ mol^{-1}	Bond length / nm
C—C	+347	0.154
C═C	+612	0.134
C≡C	+838	0.120
C—H	+413	0.108
O—H	+464	0.096
C—O	+358	0.143
C═O*	+805	0.116
O═O	+498	0.121
N≡N	+945	0.110

Table 2 gives some values for bond enthalpies and bond lengths. These are all **average bond enthalpies** because the exact value of a bond enthalpy actually depends on the particular compound in which the bond is found. From the table you can see that:

Average bond enthalpy
The average quantity of energy needed to break a particular bond.

● Double bonds have much higher bond enthalpies than single bonds. Triple bond enthalpies are even higher.

● In general, the higher the bond enthalpy, the shorter the bond – you can see this if you compare the lengths of the single, double, and triple bonds between carbon atoms. This is because there are more electrons between the atoms being attracted to the positive nuclei. More attraction makes shorter bonds.

Measuring bond enthalpies

It isn't easy to measure bond enthalpies because there is often more than one type of bond in a compound. It is also very difficult to make measurements when everything is in the gaseous state. For this reason, bond enthalpies are measured indirectly using enthalpy cycles.

Breaking and making bonds in a chemical reaction

Let's look again at the reaction that occurs when methane burns:

$$CH_4(g) + 2O_2(g) \rightarrow CO_2(g) + 2H_2O(l) \qquad \Delta H = -890 \, kJ \, mol^{-1}$$

The reaction involves both breaking bonds and making new bonds. By looking at the structures of both reactants and products it is possible to work out which bonds have been broken and which have been formed. In this case, four bonds between carbon and hydrogen in a methane molecule and a bond between oxygen atoms in two oxygen molecules have been broken. This bond-breaking requires energy. Once the bonds have been broken, the atoms can join together again to form new bonds – two carbon–oxygen double bonds, C═O, in a carbon dioxide molecule and four oxygen–hydrogen bonds in two water molecules.

Bond enthalpies always refer to breaking a bond in the gaseous compound. This means you can make fair comparisons between different bonds. You do not have to break all the old bonds before you can make new ones. New bonds start forming as soon as the first of the old bonds have broken.

In the combustion of methane, the energy taken in during the bond-breaking steps is less than the energy given out during the bond-making steps (Figure 4) so the overall reaction is exothermic. (If the reverse is true, the reaction is endothermic.)

As you need to break bonds before product molecules can begin to form, many reactions need heating to get them started. All reactions initially need energy to stretch and break bonds. Some reactions need only a little energy and there is enough energy available in the surroundings at room temperature. The reaction of acids with alkalis is an example.

Study tip
Bond-breaking is an endothermic process, so bond enthalpies are always positive.

Bond-making is exothermic.

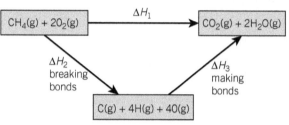

gaseous atoms

breaking bonds
(takes in energy)

making new bonds
(gives out energy)

energy

reactants

products

▲ Figure 4 *Breaking and making bonds in the reaction between methane and oxygen*

Other reactions need heating to get them started for example, applying a lit match to ignite methane on a gas cooker.

It isn't necessary for all the bonds to break before a reaction gets going. If it was, you would have to heat things to very high temperatures to make them react. Once a few bonds have broken, new bonds can start to form and this usually gives out enough energy to keep the reaction going. This is what happens when fuels burn. Some other reactions need continuous heating, for example, reactions that are only slightly exothermic.

Bonds and enthalpy cycles

You can represent bond-breaking and bond-making in an enthalpy cycle, such as the one given in Figure 5.

$CH_4(g) + 2O_2(g)$ — ΔH_1 → $CO_2(g) + 2H_2O(g)$

ΔH_2
breaking
bonds

ΔH_3
making
bonds

$C(g) + 4H(g) + 4O(g)$

▲ Figure 5 *An enthalpy cycle to show bond-breaking and bond-making in the combustion of methane*

🖩 Worked example: Enthalpy changes from bond enthalpies

Use the enthalpy cycle shown above to work out a value for the enthalpy change of combustion of methane.

Step 1: Calculate the enthalpy change when bonds are broken, ΔH_2.

$$= 4 \times (C—H) + 2 \times (O=O)$$
$$= +2648 \, kJ \, mol^{-1}$$

Step 2: Calculate the enthalpy change when bonds are made, ΔH_3.

$$= -[2 \times (C=O) + 4 \times (O—H)]$$
$$= -3466 \, kJ \, mol^{-1}$$

The minus sign is there because energy is *released* when bonds are made.

Step 3: Write the expression for the enthalpy change of combustion, ΔH_1.

$$\Delta H_1 = \Delta H_2 + \Delta H_3$$

Step 4: Substitute in the values for ΔH_2 and ΔH_3.

$$= +2648 \, kJ \, mol^{-1} + (-3466 \, kJ \, mol^{-1})$$
$$= -818 \, kJ \, mol^{-1}$$

Study tip

Some people use the following shortcut here:

$$\frac{\text{enthalpy}}{\text{change}} = \sum \frac{\text{bonds}}{\text{broken}} - \sum \frac{\text{bonds}}{\text{made}}$$

The value calculated in the Worked Example is different from the $-890\,kJ\,mol^{-1}$ given in data sheets for the standard enthalpy change of combustion of methane. There are two main reasons for this:

1 The value of ΔH_1 calculated here is not actually the standard value. In the equation the water product is gaseous, $H_2O(g)$, not liquid, $H_2O(l)$, as it would be under standard conditions. $H_2O(g)$ is used because when using bond enthalpies you have to work in the gaseous state.

2 The bond enthalpies given are often averages from several compounds. This makes them very useful as a 'toolkit' of values but it does mean that the results of such calculations are not always precise. However bond enthalpies are useful because they enable enthalpy changes to be measured when there is little specific data for a compound.

> **Study tip**
>
> The following are common errors made in these type of calculations:
>
> - forgetting that bonds in O_2 are broken
>
> - not noticing that there are two $C═O$ bonds per CO_2 and two $O—H$ bonds per H_2O.

Summary questions

You may need to look up values of bond enthalpies for these problems.

1 a Write an equation, with state symbols, for the combustion of one mole of propane, C_3H_8.
   (2 marks)
  b Write this equation again using full structural formulae, showing all the bonds within each molecule. *(2 marks)*
  c Make a list of the types and numbers of bonds that are broken in the reaction. *(1 mark)*
  d Make a list of the types and numbers of bonds that are made in the reaction. *(1 mark)*
  e Use bond enthalpies to calculate a value for the total enthalpy change involved in breaking all the bonds in the reaction of propane and oxygen. Decide whether bond breaking is an exothermic or endothermic process and put a negative or positive sign in front of the value calculated. *(2 marks)*
  f Use bond enthalpies to calculate a value for the total enthalpy change involved in making all the new bonds in the reaction of propane and oxygen. Decide whether bond making is an exothermic or endothermic process and put a negative or positive sign in front of the value calculated. *(2 marks)*
  g Add together the enthalpy changes involved to calculate the overall enthalpy change of the reaction. *(1 mark)*

2 Ethene reacts with bromine to form 1,2-dibromoethane.

$$H_2C═CH_2(g) + Br_2(g) \rightarrow BrH_2C—CH_2Br(g)$$

Use bond enthalpies to calculate a value for the enthalpy change of the reaction. Bond enthalpies: *(3 marks)*

$Br—Br = +193\,kJ\,mol^{-1}$; $C—Br = +290\,kJ\,mol^{-1}$.

3 Some apparently simple organic molecules do not exist because they are unstable and form another compound or compounds very easily. One such compound is ethenol, CH_2CHOH, which converts to ethanal, CH_3CHO.

Use bond enthalpies to calculate the enthalpy change of the reaction. *(3 marks)*

4 The difference between the enthalpy change of combustion of successive straight-chain alkanes is around $-650\,kJ\,mol^{-1}$. This is because each alkane differs from the previous one by a methyl fragment, $—CH_2$.
  a Write an equation for the combustion of the one carbon and two hydrogen atoms in this fragment. *(1 mark)*
  b Use bond enthalpies to calculate a value for the enthalpy change of the reaction in part a (include the breaking of the C—C bond). *(2 marks)*
  c Suggest why this value is not $-650\,kJ\,mol^{-1}$. *(2 marks)*

DF 5 Getting the right sized molecules

Learning outcomes

Demonstrate and apply knowledge and understanding of:

→ a simple model to explain the function of a heterogeneous catalyst

→ the term cracking; the use of catalysts in cracking processes; techniques and procedures for cracking a hydrocarbon vapour over a heated catalyst

→ the term unsaturated

→ the term catalyst, catalysis, catalyst poison, heterogeneous.

In Topic DF 3 you saw that gasoline is produced by fractional distillation of crude oil. This leaves two problems. The first is that the 'straight-run' gasoline from the primary distillation makes poor petrol. Some is used directly in petrol but most is treated further. The second is a problem of supply and demand. Crude oil contains a surplus of the high boiling fractions, such as the gas oil and the residue, and not enough of the lower boiling fractions, such as gasoline. Although demand for gas oil itself is comparatively low, it can be cracked and used in car petrol, therefore increasing demand. Figure 1 compares the supply and demand for different fractions of crude oil. The demand is greater than supply for both petrol and diesel.

The job of the refinery is to convert crude oil into useful components. In order to do this, the structure of the alkane molecules present must be altered to produce different alkanes. The alkanes are also converted into other types of hydrocarbon that are used in petrol. These include cycloalkanes, arenes (aromatic hydrocarbons), and alkenes. The products are blended to produce high-grade petrol.

Using the whole barrel

Cracking is one of the most important reactions in the petroleum industry. It starts with alkanes that have large molecules that are too big to use in petrol, for example, alkanes from the gas oil fraction. These large molecules are broken down to give alkanes with shorter chains that can be used in petrol. What's more, these shorter-chain alkanes tend to be highly branched so petrol made by cracking has a higher octane number. Another benefit of cracking is that it also helps to solve the supply and demand problem.

Cracking

The term **cracking** is used to describe any reaction in which a larger molecule is made into smaller molecules.

$$C_{11}H_{24} \rightarrow C_8H_{18} + C_3H_6$$

In this example a long-chain alkane from the kerosene fraction is made into octane (an alkane suitable for car petrol) and another compound, propene. Propene is **unsaturated** – it does not have as many hydrogen atoms as it could for the three carbon atoms. It has a carbon–carbon double bond (Figure 2) and is an **alkene**. Alkenes are described in more detail in DF 7.

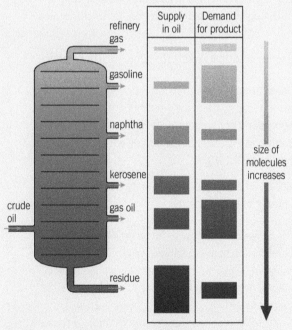

▲ Figure 1 *Supply and demand for different fractions of crude oil*

▲ Figure 2 *Propene*

Cracking: how is it done?

Much of the cracking carried out to produce petrol is done by heating heavy oils, such as gas oil, in the presence of a **catalyst**. You will be able to find out about catalysts later in this topic. This is catalytic cracking. The molecules in the feedstock can have 25–100 carbon atoms, although most will usually have 30–40 carbon atoms.

Cracking reactions are quite varied. Some of the types of reactions are:

- alkanes → branched alkanes + branched alkenes
- alkanes → smaller alkanes + cycloalkanes
- cycloalkanes → alkenes + branched alkenes
- alkenes → smaller alkenes.

The alkenes that are produced are important starting materials for other parts of the petrochemicals industry. Cracking always produces many different products, which need to be separated in a fractionating column.

In a modern catalytic cracker, the cracking takes place in a 60-metre high vertical tube about two metres in diameter. It is called a riser reactor because the hot vaporised hydrocarbons and zeolite catalyst are fed into the bottom of the tube and forced upwards by steam. The mixture is a moving fluidised bed where the solid particles flow like a liquid.

It takes the mixture about two seconds to flow from the bottom to the top of the tube – so the hydrocarbons are in contact with the catalyst for a very short period of time.

One of the problems with catalytic cracking is that, in addition to all the reactions you have already met, coke (carbon from the decomposition of hydrocarbon molecules) forms on the catalyst surface so that the catalyst eventually becomes inactive. The powdery catalyst needs to be regenerated to overcome this problem.

After the riser reactor, the mixture passes into a separator where steam carries away the cracked products leaving behind the solid catalyst. The catalyst goes into the regenerator, where it takes about 10 minutes for the coke to burn off in the hot air that is blown through the regenerator. The catalyst is then reintroduced into the base of the reactor ready to repeat the cycle.

The energy released from the burning coke heats up the catalyst. The catalyst transfers the energy to the feedstock so that cracking can occur without additional heating.

Catalytic crackers have been in operation since the late 1940s and have become very flexible and adaptable. They can handle a wide range of different feedstocks. The conditions and catalyst can be varied to give the maximum amount of the desired product – in this case branched alkanes for blending in petrol.

Cracking

Any reaction in which a larger molecule is made into smaller molecules.

Unsaturated

Any organic compound that has a double (or triple) bond between carbon atoms.

Synoptic link

You can find a general account of how to carry out cracking in Techniques and procedures.

Activities DF 5.1 and DF 5.2

You can find out more about cracking, and try cracking alkanes for yourself, in these activities.

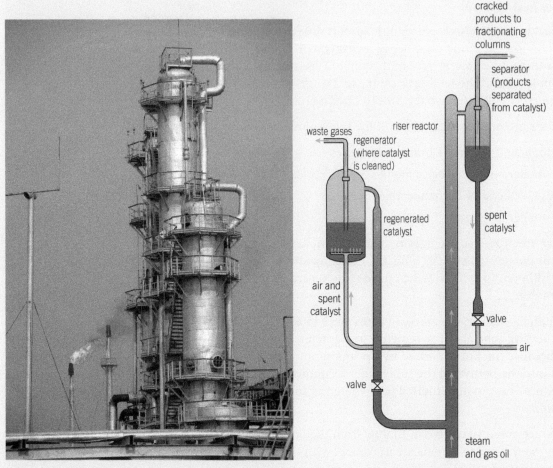

▲ **Figure 3** *A catalytic cracker (left) and how a catalytic cracker works (right). The feedstock is gas oil and the cracking reaction takes place in the riser reactor*

Chemical ideas: Rates of reactions 10.3

Catalysts

A **catalyst** is a substance that speeds up a reaction but can be recovered chemically unchanged at the end. The process of speeding up a chemical reaction using a catalyst is called **catalysis**.

Catalysts do not undergo any permanent *chemical* change, though sometimes they may be changed physically, for example, the surface of a solid catalyst may crumble or become roughened. This suggests that the catalyst is taking some part in the reaction, but is being regenerated.

Usually only small amounts of a catalyst are needed. The catalyst does not affect the amount of product formed, only the rate at which it is formed. A catalyst does not appear as a reactant in the overall equation for a reaction.

Homogenous catalysis

If the reactants and catalyst are in the same physical state (e.g., both are in aqueous solution) then the reaction is said to involve **homogeneous**

Catalyst

A substance which speeds up a reaction but can be recovered chemically unchanged at the end.

Catalysis

The process of speeding up a chemical reaction using a catalyst.

Synoptic link

Catalysts in living systems are called enzymes. You can find out more about enzymes in Chapter 7, Polymers and life.

catalysis. Enzyme-catalysed reactions in cells take place in aqueous solution and are examples of this type of catalysis.

Heterogeneous catalysis

Many important industrial processes involve **heterogeneous catalysis**, where the reactants and the catalyst are in different physical states. This usually involves a mixture of gases or liquids reacting in the presence of a solid catalyst.

When a solid catalyst is used to increase the rate of a reaction between gases or liquids, the reaction occurs on the surface of the solid (Figure 4). The reactants form bonds with atoms on the surface of the catalyst – they are **adsorbed** onto the surface. As a result, bonds in the reactant molecules are weakened and break. New bonds form between the reactants, held close together on the surface, to form the products. This in turn weakens the bonds to the catalyst surface and the product molecules are released.

It is important that the catalyst has a large surface area for contact with reactants. For this reason, solid catalysis are used in a finely divided form or as a fine wire mesh. Sometimes the catalyst is supported on a porous material to increase its surface area and prevent it from crumbling. This happens in the catalytic converters fitted to car exhaust systems.

Many of the heterogeneous catalysts used in industrial processes are transition metals (the metals in the central block of the periodic table) or transition metal compounds. You will meet a number of examples at different stages in the course.

Some examples of heterogeneous catalysis you will study more of are the use of platinum and rhodium in catalytic converters in cars (Topic DF 10) and the use of nickel powder in the hydrogenation of unsaturated oils to give saturated fats (Topic DF 6).

Catalyst poisoning

Catalysts can be poisoned so that they no longer function properly. Many substances which are poisonous to humans operate as a **catalyst poison**, blocking an enzyme-catalysed reaction.

In heterogeneous catalysis, the poison molecules are adsorbed more strongly to the catalyst surface than the reactant molecules. The catalyst cannot catalyse a reaction of the poison and so becomes inactive, with poison molecules blocking the active sites on its surface. This is the reason why leaded petrol cannot be used in cars fitted with a catalytic converter – lead is strongly adsorbed to the surface of the catalyst.

Catalyst poisoning is also the reason why it is not possible to replace the very costly metals (platinum and rhodium) in catalytic converters by cheaper metals (such as copper and nickel). These metals are vulnerable to poisoning by the trace amounts of sulfur dioxide present in car exhaust gases. Once the catalyst in a converter becomes inactive it cannot be regenerated. A new converter has to be fitted and this can be very costly.

Heterogeneous catalysis

Where the catalyst and the reactants are in different physical states.

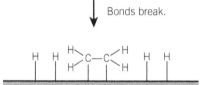

catalyst surface

Reactants get adsorbed onto catalyst surface. Bonds are weakened.

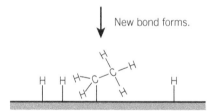

Bonds break.

New bond forms.

Second bond forms, and product diffuses away from catalyst surface, leaving it free to absorb fresh reactants.

▲ Figure 4 *An example of heterogeneous catalysis. The diagrams show a possible mechanism for nickel catalysing the reaction between ethene and hydrogen to form ethane*

Catalyst poison

A substance that stops a catalyst functioning properly.

Catalyst poisoning can be a problem in industrial processes. In the UK, nearly all the hydrogen for the Haber process is prepared by steam reforming of methane. Methane reacts with steam in the presence of a nickel catalyst.

$$CH_4(g) + H_2O(g) \xrightarrow{Ni(s)} CO(g) + 3H_2(g)$$

If the feedstock for the process contains sulfur compounds, these must be removed first to prevent severe catalyst poisoning.

Sometimes it is possible to clean or regenerate the surface of a catalyst. For example, in the catalytic cracking of long-chain hydrocarbons, carbon is produced and the surface of the catalyst becomes coated in a layer of soot. This blocks the adsorption of reactant molecules and the activity of the catalyst is reduced. The catalyst is constantly recycled through a separate container where hot air is blown through the powder. The oxygen in the air converts the carbon to carbon dioxide and cleans the catalyst surface.

Synoptic link

You will learn more about catalysts in Chapter 4, The ozone story.

Synoptic link

Catalysts are used in chemical industry and you will find out more about this in Chapter 6, The chemical industry.

Summary questions

1. Name a catalyst involved in each of the following industrial processes. In each case, state whether the process involves homogeneous or heterogeneous catalysis.
   a. Catalytic cracking of long-chain hydrocarbons. *(1 mark)*
   b. Oxidation of carbon monoxide and unburnt petrol in a car exhaust. *(1 mark)*

2. When carbon monoxide and nitrogen monoxide in car exhaust gases pass through a catalytic converter, carbon dioxide and nitrogen are formed.
   a. Write a balanced chemical equation for this reaction. *(2 marks)*
   b. Explain why it is important to reduce the quantities of carbon monoxide and nitrogen monoxide released into the atmosphere. *(1 mark)*
   c. Explain the meaning of the term adsorbed. *(1 mark)*
   d. Suggest why catalytic converters do not work effectively until a car engine has warmed up. *(2 marks)*

3. Figure 4 in this section shows a possible mechanism for the nickel-catalysed reaction between ethene and hydrogen. The reactants are adsorbed onto the surface of the nickel catalyst, where the reaction takes place. Using Ⓝ–Ⓞ to represent nitrogen monoxide and Ⓒ–Ⓞ to represent carbon monoxide, draw out a possible mechanism for the formation of carbon dioxide, Ⓞ=Ⓒ=Ⓞ as in the equation you have written in 2a. *(3 marks)*

DF 6 Alkenes – versatile compounds

Specification reference: DF(b), DF(m), DF(o), DF(q)

A source of energy is essential for human beings to function. The source of our energy, our fuel, is food. If you look at the labels on many foods, you will see references to fats, and these described as saturated or unsaturated. Unsaturated fats contain carbon-carbon double bonds and are related to alkenes.

▲ Figure 1 *Assorted fats and oils including olive oil, sunflower oil, butter, goose fat, duck fat, lard, and margarine*

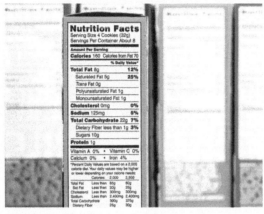

▲ Figure 2 *The nutrition label from a cookie box emphasizing that the product is high in sugar and fat content. The label states that the food contains 0 g of trans fat*

Animal fats, for example, butter and lard, contain a much higher proportion of saturated fat than oils and fats derived from plants. (Oils have the same chemical make-up as fats, just lower melting points.) These saturated fats were for a long time associated with heart disease, as they were thought to cause a build-up of cholesterol in the arteries. However, the modern villains are trans fats. *cis* and *trans* refer to the arrangement of the groups across a carbon–carbon double bond. Naturally occurring unsaturated fats are *cis*. However, these often have melting points that are slightly too low for use in the food industry, so partial hydrogenation was carried out to react the fats with hydrogen in order to saturate some of the unsaturated bonds. In the process, some *cis* compounds were turned to *trans* and these compounds are found to be very likely to lead to a build-up of cholesterol. *Trans* fats are therefore being phased out in most food areas despite the fact they did add to the crispiness of things like biscuits.

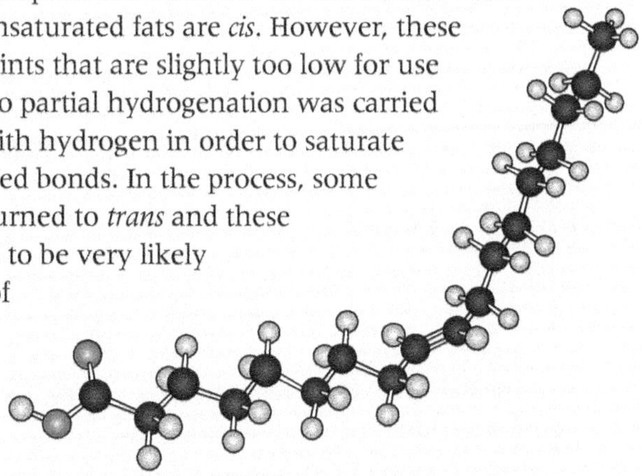

▲ Figure 3 *Molecular model of elaidic acid. Elaidic acid is derived from a major trans fat found in hydrogenated vegetable oils. The trans group can be seen in the middle of the molecule*

Learning outcomes

Demonstrate and apply knowledge and understanding of:

→ the nomenclature, general formula, and structural formulae for alkenes

→ the bonding in organic compounds in terms of σ- and π-bonds

→ the terms addition, electrophile, carbocation; the mechanism of electrophilic addition to alkenes using curly arrows; how the products obtained when other anions are present can be used to confirm the model of the mechanism

→ the addition reactions of alkenes with the following, showing the greater reactivity of the C=C bond compared with C—C:

- bromine to give a dibromo compound, including techniques and procedures for testing compounds for unsaturation using bromine water

- hydrogen bromide to give a bromo compound

- hydrogen in the presence of a catalyst to give an alkane (Ni with heat and pressure or Pt at room temperature and pressure)

- water in the presence of a catalyst to give an alcohol (concentrated H_2SO_4, then add water; or steam/ H_3PO_4/ heat and pressure).

▲ Figure 4 *Ethene*

▼ Table 1 *Other members of the alkene family*

Name of alkene	Formula
propene	CH_3—CH=CH_2
but-1-ene	CH_3—CH_2—CH=CH_2
but-2-ene	CH_3—CH=CH—CH_3
pent-1-ene	CH_3—CH_2—CH_2—CH=CH_2

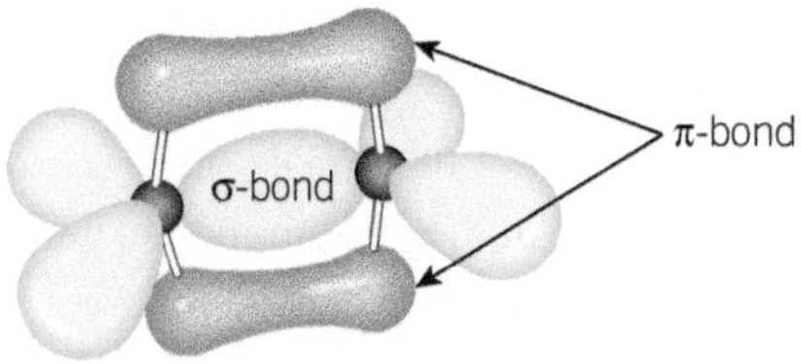

▲ Figure 5 *σ-bonds and π-bonds*

Electrophile

A positive ion, or a molecule with a partial positive charge, that will be attracted to a negatively charged region and react by accepting a lone pair of electrons to form a covalent bond.

Chemical ideas: Organic chemistry: frameworks 12.2a

Alkenes

Naming alkenes and electrophilic addition

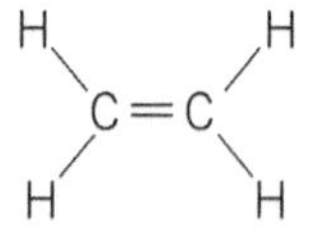

Ethene (Figure 4) is the simplest example of a class of hydrocarbons called **alkenes**.

Alkenes are distinguished from other hydrocarbons by the presence of the carbon–carbon double bond, C=C. The double bond implies that they are unsaturated hydrocarbons.

Examples of other members of the alkene family are listed in Table 1.

As with the alkanes, the boiling points of alkenes increase as the number of carbon atoms increases. Ethene, propene, and butene are gases. After that they are liquids and, eventually, solids.

All non-cyclic alkenes have the general formula, C_nH_{2n}, where n is the number of carbon atoms and can be any whole number. This is the same as the general formula of the cycloalkanes (e.g., cyclohexane) but the double bond makes alkenes react very differently from cycloalkanes.

Naming alkenes

Alkenes are named using similar rules to those used for alkanes.

1   Identify the longest chain and use the appropriate prefix to the number of carbons (Topic DF 3).

2   Follow the prefix with 'ene' to indicate the compound in an alkene.

3   For alkenes where the longest carbon chain is four carbon atoms or longer, insert a number before 'ene' to indicate the location of the carbon–carbon double bond.

4   For branched alkenes, you follow the same rules for branched alkanes (Topic DF 3).

For example, in but-1-ene the C=C is between C_1 and C_2 so the lowest number is used.

There are cycloalkenes, such as cyclohexene (an important intermediate in the production of some types of nylon) and there are dienes, such as penta-1,3-diene. When there is a diene, an 'a' is added to the prefix.

Bonding in alkenes

Previously, you have used dot-and-cross diagrams to illustrate the sharing of electrons in covalent bonds. In a single bond, for example, the C—C bond in alkanes, the two electrons are arranged between the atoms in an area of increased electron density, called a σ-bond. This symbol is the Greek letter *sigma*, the equivalent to s. It is so-called because the overlap of s-orbitals gives rise to a σ-bond, though there are other ways in which they can be formed. A double bond consists of one σ-bond and another type of bond, called a π-bond. *Pi* is the Greek letter for p. *One* π-bond consists of *two* areas of negative charge, one above and the other below the line of the atoms (Figure 5).

Chemical reactions of ethene

The four electrons in the double bond of ethene give the region between the two carbon atoms a higher than normal density of negative charge. Positive ions, or molecules with a partial positive charge on one of the atoms, will be attracted to this negatively charged region. They can react by accepting a pair of electrons from the C═C double bond. Substances that do this are called **electrophiles**.

Electrophilic addition reactions

Reaction with bromine

When ethene gas is bubbled through bromine, the red-brown bromine becomes decolorised – this is a good general test for unsaturation in an organic compound.

In order to interpret how a reaction occurs, chemists use reaction mechanisms. This involves logically deciding the movement of electrons using ideas such as bond polarity and charges. Look at the mechanism for the reaction of ethene with bromine. Chemists believe that the bromine molecule becomes **polarised** as it approaches the alkene. This means that the electrons in the bromine are repelled by the alkene electrons and are pushed back along the molecule. The bromine atom nearest the alkene becomes slightly positively charged and the bromine atom furthest from the alkene becomes slightly negatively charged. The positively charged bromine atom now behaves as an **electrophile** and reacts with the alkene double bond.

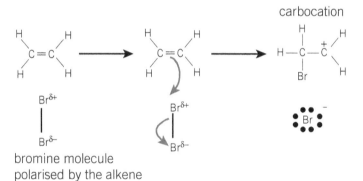

bromine molecule
polarised by the alkene

Remember, curly arrows like these represent the movement of a pair of electrons in chemical reactions. One of the carbon atoms now only has six outer electrons – it has become positively charged. It is a **carbocation**.

Carbocations react very rapidly with anything that has electrons to share – such as the bromide ion. A pair of electrons moves from the bromide ion to the positively charged carbon to form a new carbon–bromine covalent bond.

Study tip

An organic reaction mechanism represents the sequence of events in a reaction using the movement of electrons, represented by curly arrows.

Activity DF 6.1

This activity enables you to test for unsaturation used bromine.

Study tip

Make sure you clearly show precisely where electrons move to and where they have moved from using your curly arrows.

Study tip

The Br⁻ could attack from either side of the positively charged carbon atom – here it is shown attacking from below.

Carbocation

An ion with a positively-charged carbon atom.

Synoptic link ⚗

Adding bromine water to an alkene can be used as a test for unsaturation. You can find out more about this in Techniques and procedures.

Addition reaction

A reaction where two or more molecules react to form a single larger molecule.

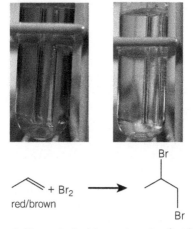

▲ **Figure 6** *Red-brown bromine (left) and decoloured bromine after propene has been passed through (right). The colourless 1,2-dibromoprapane has been formed*

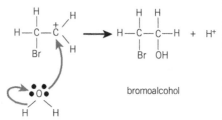

▲ **Figure 7** *Formation of a bromoalcohol*

Activity DF 6.2 ↗

This activity helps you to check your understanding of electrophilic addition mechanisms.

The overall reaction for the two steps in the mechanism is

This overall scheme is an example of an organic **reaction mechanism**.

An **addition reaction** is one where two or more molecules react to form a single, larger molecule. This equation represents an addition reaction and since the initial attack is by an electrophile the process is called an **electrophilic addition**.

How do you know the proposed mechanisms are correct? The mechanism for electrophilic addition via a carbocation is supported by experimental evidence. For example, if hex-1-ene reacts with bromine in the presence of chloride ions, two products are formed – 1,2-dibromohexane and 1-bromo-2-chlorohexane. However, 1,2-dichlorohexane is not formed as the electrophile must attack first, followed by an anion.

Often the test for an alkene involves shaking the alkene with bromine water rather than pure bromine. In this case, there is an alternative to the second stage in the reaction. Water molecules have lone pairs of electrons and can act as nucleophiles in competition with the bromide ions.

If the bromine water is dilute, there will be many more water molecules than bromide ions present and the bromoalcohol will be the main product of the reaction. This does not affect what you see – the bromine water is still decolorised (Figure 7).

Reaction with hydrogen bromide

The conditions under which a reaction is carried out can be very important in determining the mechanism. For example, ethene reacts readily at room temperature with a solution of hydrogen bromide, HBr, in a polar solvent. It is another example of electrophilic addition.

Alkenes also react with gaseous hydrogen bromide but ions are not involved and the mechanism involves a radical addition.

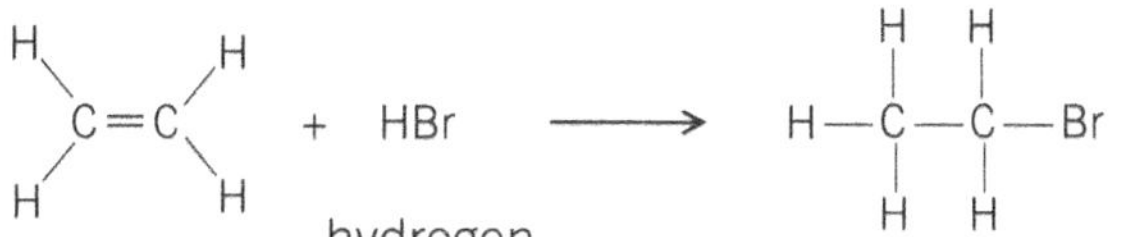

ethene | hydrogen bromide | bromoethane

Reaction with water

At high temperature, high pressure, and in the presence of a catalyst (phosphoric acid adsorbed onto solid silica), ethene and water (as steam) undergo an addition reaction. The process is used for the industrial manufacture of ethanol.

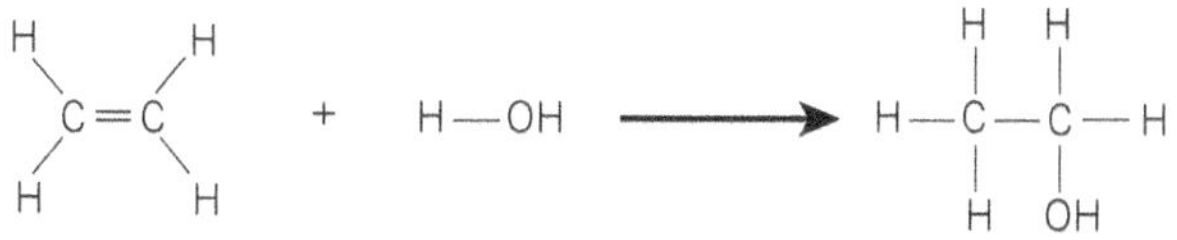

ethene | water | ethanol

Study tip

To help write electrophilic addition mechanisms, it helps to initially draw you electrophile in an X—Y format. For example, H—Br, H—OH, or H—OSO$_3$H.

In the laboratory, ethene can be converted to ethanol by first adding concentrated sulfuric acid, and then diluting with water.

Step 1

Step 2

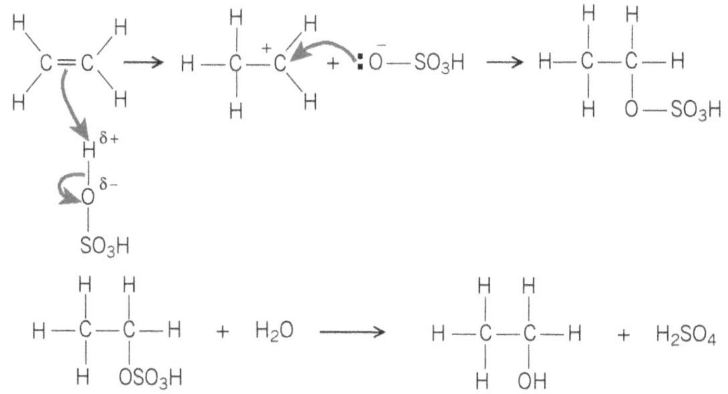

ethyl hydrogensulfate | ethanol

The overall reaction is addition of water across the double bond. The addition of water to an alkene is an example of a **hydration reaction**.

Reaction with hydrogen

The reaction of ethene with hydrogen is another example of an addition reaction, but here the mechanism involves hydrogen atoms, and takes place on the surface of a catalyst. A catalyst is needed to help break the strong hydrogen–hydrogen bond and form hydrogen atoms that can react with the alkene. If a platinum catalyst is used the process takes place under standard laboratory conditions. Nickel is a cheaper but less efficient catalyst. It needs to be very finely powdered and the gases need heating to approximately 150 °C under a pressure of 5 atm for hydrogenation to occur.

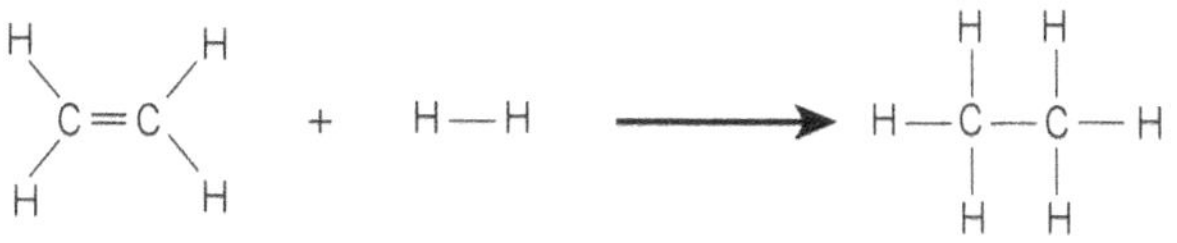

ethene | hydrogen | ethane

This hydrogenation reaction is the reaction used to make the unsaturated fats and oils more saturated. Ethene can also undergo addition reactions to form polymers, as you will see in Topic DF 7.

Summary questions

1 Complete the table of names, structural formulae, and
   skeletal formulae for the following alkenes. (12 marks)

Name	Structural formula	Skeletal formula
pent-2-ene		
		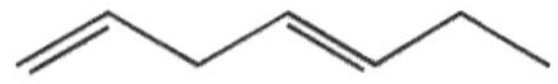
	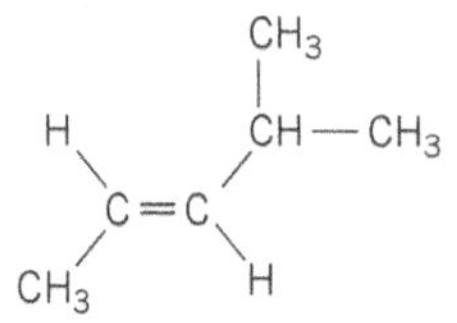	
3-ethylhept-1-ene		
		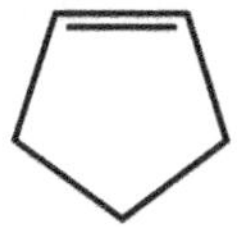
cyclopenta-1,3-diene		

2 Use full structural formulae to write an overall equation for:
   a the reaction of propene with bromine (2 marks)
   b the reaction of propene with hydrogen. (2 marks)

3 The reaction of propene with hydrogen bromide can give two different
   products. Draw the full structural formulae of these products. (2 marks)

4 Give the reagents and conditions needed to carry out each of the
   following reactions.
   a bromination of alkenes (2 marks)
   b industrial production of ethanol from ethene (2 marks)
   c laboratory hydrogenation of alkenes to alkanes (2 marks)
   d industrial production of margarine from sunflower oil (2 marks)

5 Describe a test you could carry out in the laboratory to
   show the presence of unsaturation in an organic compound.
   State what you would do and what you would see. (2 marks)

6 a Draw the mechanism for the reaction of propene with hydrogen
      bromide, HBr, to form $CH_3CHBrCH_3$. (4 marks)
   b Draw the structure of another compound that might be
      formed when propene reacts with HBr. (1 mark)

7 Ethene reacts with dilute aqueous bromine containing some
   dissolved sodium chloride.
   a Draw the mechanism for the formation of CH_2BrCH_2Cl using
      these reagents. (4 marks)
   b Give the formula of another compound that would be formed using
      these reagents and draw the mechanism for its formation. (4 marks)
   c Suggest why $CHBr_2CH_3$ is not formed. (2 marks)

DF 7 Polymers and plastics
Specification reference: DF(p)

You have seen how fuels can be produced from oil. There are many other useful products, including plastics, that can be made from oil based chemicals. These plastics are types of **polymers**. A polymer is a long molecule made up from lots of small molecules called monomers. Polymers are produced in abundance by nature – in plants, animals, and in our bodies. Synthetic polymers are so much part of our lives, both in terms of materials and culture, that it is difficult to believe that their development began as recently as the 1940s. Indeed, polymers have only been in widespread use since the 1950s.

▲ Figure 1 *Plastics are synthetic polymers that we use in many different ways such as plastic packaging, pipes, skis, and non-stick frying pans*

In the late nineteenth century, **plastics** were produced by modifying natural polymers. Celluloid, for example, was produced by reacting cellulose (from plants) with nitric acid. The first plastic to be made in significant quantities from manufactured chemicals was Bakelite, made from phenol and methanal (Figure 2). Bakelite is still used to make electrical fittings such as sockets and plugs. Although it was first made in 1872 by accident, it was not until 1910 that the process was patented and Bakelite was manufactured.

The polythene story

Imperial Chemical Industries (ICI) was formed in 1926 by the joining together of a number of smaller chemical companies. The prime aim of the merger was to form a strong competitor to the huge German chemical company IG Farben.

In 1930 Eric Fawcett, who was working for ICI, got the go-ahead to carry out research at high pressures and temperatures aimed at producing new dyestuffs. His results were disappointing and his project was eventually abandoned.

His team then moved into the field of high-pressure gas reactions and was joined by Reginald Gibson. On Friday 24th March 1933, Gibson and Fawcett carried out a reaction between ethene and benzaldehyde using a pressure of about 2000 atm. They were hoping to make the two chemicals add together to produce a ketone (Figure 3)

Their apparatus leaked and at one point they had to add extra ethene. They left the mixture to react over the weekend.

▲ Figure 2 *A celluloid film reel (top) and a Bakelite radio (bottom). Both are examples of polymers*

benzaldehyde + *ethene*

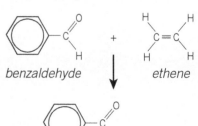

▲ Figure 3 *The reaction Gibson and Fawcett were attempting*

They opened the vessel on the following Monday and found a white waxy solid. When they analysed it they found that it had the empirical formula CH_2. They were not always able to obtain the same results from their experiment – sometimes they got the white solid, on other occasions they had less success, and sometimes their mixture exploded leaving them with just soot!

The work was halted in July 1933 because of the varied results and dangerous nature of the reaction.

Learning to control the process

In December 1935 the work was restarted. Fawcett and Gibson found that they could control the heat given out during the reaction if they added cold ethene at the correct rate. This kept the mixture cooler and prevented an explosion. They also found that they could control the reaction rate and relative molecular mass of the solid formed by varying the pressure.

A month later they had made enough of the material to show that it could be melted, moulded, and used as an insulator.

▲ Figure 4 *As well as more obvious uses, this tunnel uses poly(ethene) sheeting to protect the plants from the cold*

Most crucial of all was the identification of the role of oxygen in the process – this was done by Michael Perrin, who took charge of the programme in 1935. When oxygen was not present, the polymerisation did not occur. Too much oxygen caused the reaction to run out of control.

The trick was to add just enough oxygen. The leak in Fawcett and Gibson's original apparatus had accidentally let in a small amount of oxygen. If this had not happened then the discovery of poly(ethene) may not have been made. It was also Perrin who showed that even if benzaldehyde is left out of the reaction mixture, the polymer still forms.

Poly(ethene) is tough, durable, and has excellent electrical insulating properties. Unlike rubber, which had previously been used for insulating cables, poly(ethene) is not adversely affected by weather or water. It also has almost no tendency to absorb electrical signals. Its first important use was for insulating a telephone cable laid between the UK mainland and the Isle of Wight in 1939. Its unique electrical properties were again essential during the Second World War in the development of radar.

▲ Figure 5 *Poly(ethene) was essential in the development of radar equipment in the Second World War*

The first poly(ethene) washing-up bowls appeared in the shops in 1948 and were soon followed by carrier bags, squeezy bottles, and sandwich bags. Sadly, poly(ethene) and some other early polymer materials were over-exploited. They were used for all manner of novelty items and as cheap but poor substitutes for many natural materials. This gave plastic a bad name – the word plastic is often used to describe something which looks cheap and does not last. The reputation still sticks despite the durability and wide range of uses of modern polymers.

Elastomers, plastics, and fibres

Polymer properties vary widely. Polymers that are soft and springy, which can be deformed and then go back to their original shape, are called **elastomers**. Rubber is an elastomer.

Poly(ethene) is not so springy and when it is deformed it tends to stay out of shape, undergoing permanent or plastic deformation. Polymers like this can be incorporated into plastics.

Stronger polymers, which do not deform easily, are just what is needed for making clothing materials. Some can be made into strong, thin threads which can then be woven together. These polymers, such as nylon, can be used as fibres.

Poly(propene) is on the edge of the plastic/fibre boundary. It can be used as a plastic, like poly(ethene), but it can also be made into fibres for use in carpets.

▲ Figure 6 *This copy of the famous sculpture Venus Di Milo and was made from nylon using a 3D printer*

Chemical ideas: Organic chemistry: frameworks 12.2b

Addition polymerisation

What is a polymer?

A polymer molecule is a long molecule made up from lots of small molecules called monomers. If all the monomer molecules, A, are the same, an A–A polymer forms:

$$- - A + A + A + A - - \rightarrow - -A–A–A–A - -$$

Poly(ethene) and poly(chloroethene), PVC, are examples of A–A polymers. If two different monomers are used, an A–B polymer may be formed, in which A and B monomers alternate along the chain:

$$- - A + B + A + B - - \rightarrow - - A–B–A–B - -$$

Polyamides (nylons) and polyesters are examples of this type of A–B polymer.

Many polymers are formed in a reaction known as **addition polymerisation**. The monomers usually contain carbon–carbon double bonds, for example, in alkenes. The addition polymerisation of propene is typical example (Figure 7).

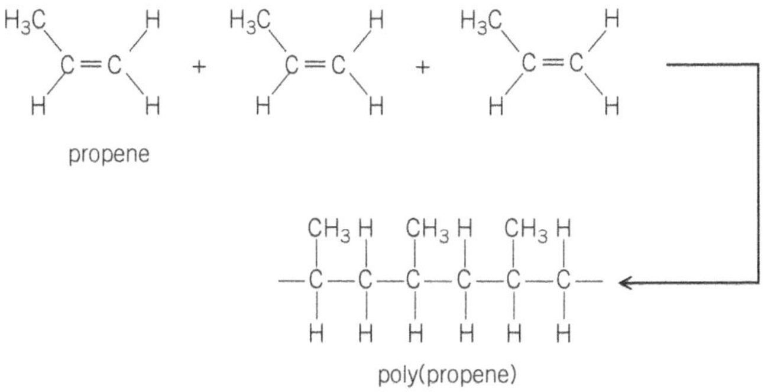

propene

poly(propene)

▲ Figure 7 *The polymerisation of propene*

> **Synoptic link**
>
> You will find out about another type of polymerisation (condensation) in Chapter 7, Polymers of life.

> **Polymerisation**
>
> small molecules called monomers join together to produce long chain polymers.

> **Activity DF 7.1**
>
> This activity allows you to learn about the properties and uses of some addition polymers.

> **Study tip**
>
> In addition polymerisation, when the monomers join together there is no other product except the polymer.

In the chain, the same basic unit is repeated over and over again, so the polymer structure can be shown simply as:

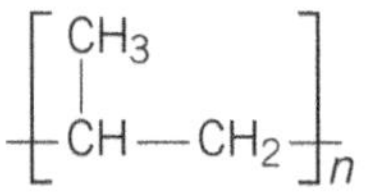

$$\left[\begin{array}{c} CH_3 \\ | \\ CH-CH_2 \end{array} \right]_n$$

The polymer is named by placing the name of the monomer in brackets and adding poly to the beginning, for example, poly(propene).

Sometimes more than one type of monomer is used in addition polymerisation. For example, if some ethene is added to the propene during the polymerisation process, both monomers become incorporated into the final polymer. This is called **copolymerisation**. A section of the copolymer chain could look like this:

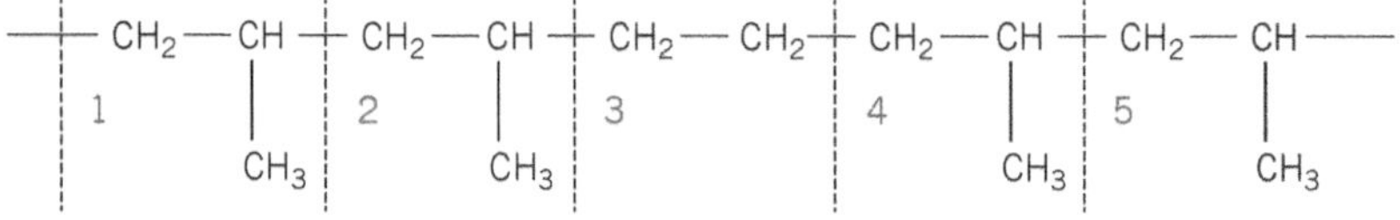

$$- CH_2-CH + CH_2-CH + CH_2-CH_2 + CH_2-CH + CH_2-CH -$$

Units 1, 2, 4, and 5 derive from propene, whilst unit 3 derives from ethene.

Summary questions

1. Write down the structure of a length of polymer formed from:
   a. three monomer units of poly(ethene)   (1 mark)
   b. three monomer units of poly(propene)   (1 mark)

2. The polymer PVC is made from a monomer with the structure $H_2C{=}CHCl$. Write down the structure of a length of polymer formed from 7 monomer units.   (1 mark)

3. Name the two alkenes used to make this copolymer.   (2 marks)

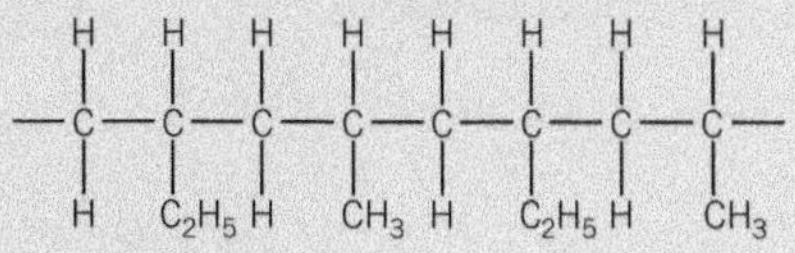

$$-\underset{H}{\overset{H}{C}}-\underset{C_2H_5}{\overset{H}{C}}-\underset{H}{\overset{H}{C}}-\underset{CH_3}{\overset{H}{C}}-\underset{H}{\overset{H}{C}}-\underset{C_2H_5}{\overset{H}{C}}-\underset{H}{\overset{H}{C}}-\underset{CH_3}{\overset{H}{C}}-$$

4. Three addition polymers have the general structure shown below.

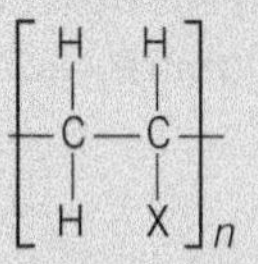

$$\left[\begin{array}{c} H \;\; H \\ | \;\;\; | \\ C-C \\ | \;\;\; | \\ H \;\; X \end{array} \right]_n \qquad X = H, CH_3, \text{ and } Cl.$$

   Name the three polymers.   (3 marks)

5. Draw a section of the polymer chain formed when chloroethene and ethenyl ethanoate copolymerise in an addition polymerisation reaction to form an A–B polymer.   (2 marks)

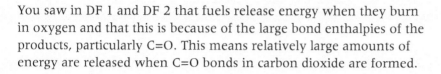

You saw in DF 1 and DF 2 that fuels release energy when they burn in oxygen and that this is because of the large bond enthalpies of the products, particularly C=O. This means relatively large amounts of energy are released when C=O bonds in carbon dioxide are formed.

Using the fuel

About 30–40% of each barrel of crude oil goes to make petrol, but it's not as simple as just distilling off the right bit at the refinery and sending it to the petrol stations. Petrol has to be blended to get the right properties. One important property is its volatility.

A mixture of petrol vapour and air is ignited in a cylinder in a car engine. The vapour-air mixture is provided via the fuel injection system. When the weather is very cold the petrol is difficult to vaporise, so the car is difficult to start.

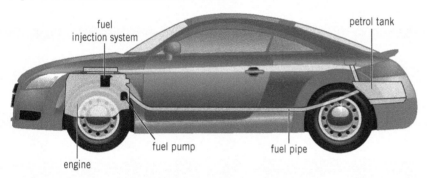

▲ Figure 1 *The fuel supply system. When you start a car engine, fuel is pumped to the electronic fuel injection system where a fine spray of the fuel is mixed with air in the correct proportions prior to ignition*

To get over this problem, petrol companies make different blends for different times of the year. During winter they put more volatile components in the petrol so it vaporises more readily. This means putting in more of the hydrocarbons with small molecules, such as butane and pentane.

In hot weather too many of these more volatile components will mean the petrol will vaporise too easily. Petrol would vaporise from the tank by evaporation – a process which is costly and polluting. Also, if the fuel vaporises too readily then pockets of vapour form in the fuel supply system. The fuel pump then delivers a mixture of liquid and vapour to the carburettor instead of mainly liquid. This means that not enough fuel gets through to keep the engine running – it's called vapour lock.

All petrols are a blend of hydrocarbons of high, medium, and low volatility. As well as altering the petrol blend for the different seasons in a particular country, the blend will be different in different countries depending on the climate. The colder the climate, then more volatile components are added to the blend. Petrol companies change their blend throughout the year – and you don't even notice.

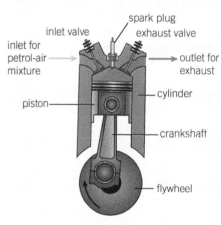

▲ Figure 2 *How a four-stroke petrol engine works. The petrol air mixture is drawn into the cylinder. The piston compresses the mixture, then a spark makes the mixture explode, pushing the piston down and turning the crankshaft. In a diesel engine, air alone enters via the inlet valve. There is no spark plug and fuel is injected via a separate inlet during the compression stroke. The high pressure causes the fuel–air mixture to ignite*

In both petrol and diesel engines, the fuel–air mixture has to ignite at the right time, usually just before the piston reaches the top of the cylinder.

Look at Figure 2. As the fuel–air mixture is compressed it heats up, and the more it is compressed the hotter it gets. Modern cars achieve greater efficiency than in the past by using higher compression ratios, often compressing the gases in the cylinder by about a factor of 10. These combustion reactions involve reacting gases.

Synoptic link

You will look at measuring volumes of gas in Chapter 4, The ozone story.

Activity DF 8.1

In this activity you will calculate the volume of one mole of hydrogen gas.

Synoptic link

You first encountered the concept of Avogadro's constant in Chapter 1, Elements of Life.

▲ Figure 3 *A mole of any gas at room temperature and pressure occupies 24.0 dm^3 (about the volume of a large biscuit tin)*

Chemical ideas: Measuring amounts of a substance 1.5a

Calculations involving gases

Volumes of gases

When carbon-based fuels burn, *complete* combustion forms carbon dioxide and water, though some carbon monoxide is also formed as a result of *incomplete* combustion.

For example:

$C_7H_{16} + 11O_2 \rightarrow 7CO_2 + 8H_2O$	complete combustion of heptane
$C_7H_{16} + 10O_2 \rightarrow 7CO_2 + 2CO + 8H_2O$	incomplete combustion of heptane
$C_2H_5OH + 3O_2 \rightarrow 2CO_2 + 3H_2O$	complete combustion of ethanol
$C_6H_6 + 7.5O_2 \rightarrow 6CO_2 + 3H_2O$	complete combustion of benzene

Chemists use chemical equations to work out the masses of reactants and products involved in a reaction. If one or more of these is a gas, it is sometimes more useful to know the volume of gas rather than its mass.

Measuring volumes

The units used to measure volume depend on how big the volume is. In chemistry, large volumes are usually measured in cubic decimetres (dm^3), with smaller volumes measured in cubic centimetres (cm^3). A decimetre is a tenth of a metre, 10 cm. A cubic decimetre is therefore 10 cm × 10 cm × 10 cm, or 1000 cm^3.

$$1 \, dm^3 = 1000 \, cm^3$$

Molar volume

The number of molecules in one mole of any gas is always equal to the Avogadro constant N_A, 6.02×10^{23}. The molecules in a gas are very far apart compared to the actual size of each molecule so the molecule has a negligible effect on the total volume the gas occupies. Therefore, one mole of any gas always occupies the same volume, no matter which gas it is. Avogadro realised this in 1811, when he put forward his famous law (sometimes called Avogadro's hypothesis).

> Equal volumes of all gases at the same temperature and pressure contain an equal number of molecules.

The volume occupied by one mole of any gas at a particular temperature and pressure is called the **molar volume**. At standard temperature and pressure (s.t.p.) the molar volume of a gas is 22.4 dm³. Standard temperature and pressure means a temperature of 0 °C (273 K) and a pressure of 1 atmosphere (101.3 kPa).

At room temperature, around 25 °C (298 K), and 1 atmosphere pressure, referred to as room temperature and pressure (RTP), the volume of a mole of any gas is about 24.0 dm³ (Figure 3).

The idea of molar volume allows you to calculate the amount in moles from the volume of a gas, and vice versa, provided you know the temperature and pressure of the gas. Here are two examples.

 ## Worked example: Calculating the volume of a gas from a given mass

Calculate the volume occupied by 4.4 g of carbon dioxide at room temperature and pressure. Assume that the molar volume of a gas at room temperature and pressure (RTP) is 24.0 dm³.

Step 1: Calculate the amount in moles from the mass. Relative formula mass of CO_2 = 44.0

$$\text{amount of } CO_2 = \frac{\text{mass}}{\text{molar mass}}$$

$$= \frac{4.4 \text{ g}}{44.0 \text{ g mol}^{-1}} = 0.1 \text{ mol}$$

Step 2: Calculate the volume.

1 mol CO_2 at r.t.p. has a volume of 24.0 dm³

$$24.0 \times 0.1 = 2.4$$

0.1 mol carbon dioxide at r.t.p. has a volume of 2.4 dm³

 ## Worked example: Calculating the mass of a gas from a given volume

Calculate the mass of 1.2 dm³ of methane gas, CH_4, at room temperature and pressure. Assume that the molar volume of a gas at room temperature and pressure (RTP) is 24.0 dm³.

Step 1: Calculate the amount in moles.

$$\text{amount of } CH_4 = \frac{\text{volume}}{\text{molar volume}}$$

$$= \frac{1.2 \text{ dm}^3}{24.0 \text{ dm}^3 \text{ mol}^{-1}} = 0.05 \text{ mol}$$

Step 2: Calculate the mass.

relative formula mass of CH_4 = 16.0

mass of CH_4 = amount in moles × molar mass
$$= 0.05 \text{ mol} \times 16.0 \text{ g mol}^{-1} = 0.8 \text{ g}$$

Reacting volumes of gases

Molar volumes can be used to work out the volumes of gases involved in a reaction. Consider the manufacture of ammonia from nitrogen and hydrogen:

$$N_2(g) + 3H_2(g) \rightarrow 2NH_3(g)$$

From the equation one mole of nitrogen reacts with three moles of hydrogen to produce two moles of ammonia. Using the idea that one mole of each gas occupies 24.0 dm³ at RTP:

$$24.0 \, dm^3 \, N_2 + (3 \times 24.0 \, dm^3) \, H_2 \rightarrow (2 \times 24.0 \, dm^3) \, NH_3$$

or $1 \, dm^3 \, N_2 + 3 \, dm^3 \, H_2 \rightarrow 2 \, dm^3 \, NH_3$

or   1 volume N_2 + 3 volumes H_2 → 2 volumes NH_3

So, if you had 10 cm³ of nitrogen it would react with 30 cm³ of hydrogen to form 20 cm³ of ammonia, provided all the measurements were taken at the same temperature and pressure.

For a reaction involving only gases, you can convert a statement about the numbers of moles of each substance involved to the same statement about volumes.

$N_2(g)$	+	$3H_2(g)$	→	$2NH_3(g)$
1 mole	+	3 moles	→	2 moles
1 volume	+	3 volumes	→	2 volumes

 Worked example: Calculations involving reacting volumes of gases

What volume of air is needed to completely burn 15 cm³ of hexane vapour? (Assume all volumes are measured at the same temperature and pressure and that air contains 21% oxygen)

Step 1: Write the equation.

$$C_6H_{14} + 9\frac{1}{2}O_2 \rightarrow 6CO_2 + 7H_2O$$

Step 2: Work out the volume of oxygen.

15 cm³ of hexane combines with 142.5 cm³ (15 × 9.5) oxygen from the ratio in the equation.

Step 3: Work out the volume of air required.

142.5 cm³ of oxygen is contained in $142.5 \times \dfrac{100}{21}$ cm³ air

= 680 cm³ of air (two significant figures, to match data)

Synoptic link

You can find a general description of how to measure the volumes of gases in Techniques and procedures.

Working with masses and volumes

You have seen how to calculate reacting masses using chemical equations and working with moles. Very similar calculations can be used to calculate the volumes of gases produced in reactions. This is because one mole of molecules of all gases occupies 24.0 dm³ at room temperature and pressure.

Study tip

The molar gas volume at room temperature and pressure (RTP) is 24.0 dm³ mol⁻¹.

Worked example: Calculation involving reacting masses and volumes

What volume of carbon dioxide is produced when 15 g of calcium carbonate completely decompose? Assume that one mole of gas occupies 24.0 dm^3 at r.t.p.

Step 1: Write the equation.

$$CaCO_3 \rightarrow CaO + CO_2$$

Step 2: Work out the moles of the calcium carbonate and carbon dioxide from the equation.

one mole of $CaCO_3$ decomposes to produce one mole of CO_2

Step 3: Write out the mass and volume of one mole for calcium carbonate and carbon dioxide.

mass of 1 mole of $CaCO_3$ = 100.1 g   volume of one mole of CO_2 = 24.0 dm^3

Step 4: Calculate the volume of carbon dioxide.

$$1\,g = \frac{24.0}{100.1}$$

$$15\,g = 15 \times \frac{24.0}{100.1}$$

$$= 3.6\,dm^3$$

The ideal Gas Equation

When doing calculations involving gas volumes, one mole of a gas occupies 24.0 dm^3 at room temperature and pressure. However, when temperature or pressure are different from room temperature and pressure the ideal gas equation is used.

pressure P (Pa) × volume V (m^3) =
amount of gas n (moles) × gas constant R (J K^{-1} mol^{-1}) × temperature T (K)

$PV = nRT$

The gas constant R has the value 8.31 J K^{-1} mol^{-1}.

 Worked Example: Calculating volume from the ideal gas equation

What is the volume of 1 mole of gas at 100 kPa pressure and 16 °C?

Step 1: Convert kPa to Pa.

1 kPa = 1000 Pa

So to convert from kPa to Pa, multiply the kPa value by 1000.

$100 \times 1000 = 1 \times 10^5\,Pa$

Step 2: Convert °C to K by adding 273 to the °C value.

$16 + 273 = 289\,K$

Step 3: Rearrange the ideal gas equation to make volume the subject and substitute in the values.

$$V = \frac{nRT}{P} = \frac{1 \times 8.31 \times 289}{1 \times 10^5} = 0.0240\,m^3 = 24.0\,dm^3$$

16 °C and 1×10^5 Pa are the values for room temperature and pressure.

 Worked example: Calculating pressure from the ideal gas equation

What is the pressure if 5.0 g of nitrogen is present in a volume of 50 cm³ at 300 K?

Step 1: Calculate the moles of nitrogen. $M_r(N_2) = 28$

$$n = \frac{mass}{relative\ molecular\ weight} = \frac{5.0}{28} = 0.179\,mol$$

Step 2: Convert cm³ to m³.

$$1\,cm^3 = 1 \times 10^{-6}\,m^3$$

So to convert from cm³ to m³, multiply the cm³ value by 1×10^{-6}.

$$50 \times 1 \times 10^{-6} = 5 \times 10^{-5}\,m^3$$

Step 3: Rearrange the ideal gas equation to make pressure the subject and substitute in the values.

$$P = \frac{nRT}{V} = \frac{0.179 \times 8.31 \times 300}{5.0 \times 10^{-5}} = 8.9 \times 10^6\,Pa$$

This is about nine times atmospheric pressure.

Summary questions

In the following problems, unless stated otherwise, you should assume that the volumes of all gases are measured at the same temperature and pressure.

1 The volumes of one mole of all gases are the same when measured at the same temperature and pressure. The volumes of one mole of liquids or solids are almost always different. Why do gases differ from liquids and solids in this way? *(2 marks)*

2 10 cm³ of hydrogen are burned in oxygen to form water.
   a Write a balanced equation, including state symbols for this reaction. *(2 marks)*
   b What volume of oxygen is needed to burn the hydrogen completely? *(3 marks)*

3 $10 \, cm^3$ of a gaseous hydrocarbon reacts completely with $40 \, cm^3$ of oxygen to produce $30 \, cm^3$ of carbon dioxide.

   a How many moles of carbon dioxide must have been formed from one mole of the hydrocarbon? *(2 marks)*

   b How many carbon atoms must there be in the formula of the hydrocarbon? *(2 marks)*

   c How many moles of oxygen were used in burning one mole of the hydrocarbon? *(2 marks)*

   d How many moles of water must have been formed in burning one mole of the hydrocarbon? *(2 marks)*

   e What is the formula of the hydrocarbon? *(1 mark)*

4 1.2 g of magnesium react with excess sulfuric acid. Calculate the volume of hydrogen produced at room temperature and pressure. *(3 marks)*

5 18 g of pentane, C_5H_{12}, are completely burnt in a car engine to form carbon dioxide and water.

   a How many moles of pentane are burned? *(2 marks)*

   b How many moles of oxygen are needed to burn all the pentane? *(1 mark)*

   c What volume of oxygen is needed, assuming that one mole of gas occupies $24.0 \, dm^3$? *(2 marks)*

   d What volume of air is needed (assume that air contains 21% oxygen by volume)? *(2 marks)*

   e What volume of carbon dioxide is formed? *(1 mark)*

6 7.5 g of a gas occupies $4.11 \, dm^3$ at 100 kPa and 290 K. Calculate the M_r of the gas. *(2 marks)*

7 a Calculate the amount (in moles) of air in a $1.7 \, dm^3$ car tyre at 1.5 bar and 0 °C. (1 bar = 100 kPa) *(2 marks)*

   b Assuming the volume does not change, calculate the pressure (in bar) in the tyre at 20 °C. *(2 marks)*

8 $10 \, cm^3$ of nitrogen is reacted with $20 \, cm^3$ hydrogen. Calculate the total volume of gases formed. *(4 marks)*

9 Two bulbs of $50 \, cm^3$ capacity are connected by a tube of negligible volume. Initially both bulbs are in melting ice and the pressure is $1.2 \times 10^5 \, Pa$.

   One bulb is then placed in boiling water whilst the other is left in melting ice. Calculate the new pressure in the bulbs. *(4 marks)*

DF 9 What do the molecules look like?

Specification reference: DF(c), DF(s), DF(t)

When considering fuels you have talked about the molecules. However, chemists cannot see molecules. Instead, chemists use models of molecular structures to explain how molecules behave in order to make predictions about further behaviour. When, in the past, these models led to correct predictions, chemists found they were *useful*. You cannot say they are *true*, however, and most models have their failings as well as successes.

For example, dot-and-cross diagrams are useful to describe bonding. However, try to draw such a diagram for nitrogen monoxide, NO, or diborane, B_2H_6, and you will see that they have failings.

You are going to look at some different ways of representing molecules, the concept of isomers, and naming conventions.

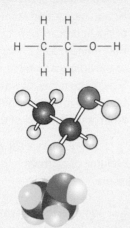

▲ Figure 1 *Three ways of representing an ethanol molecule – full structural formula, ball and stick, and space filling*

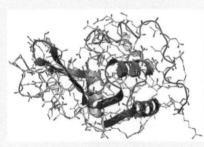

▲ Figure 2 *A representation of a more complicated molecule – the enzyme lysozyme*

Chemical ideas: Organic chemistry: frameworks 12.1b

Alkanes

Shapes of alkanes

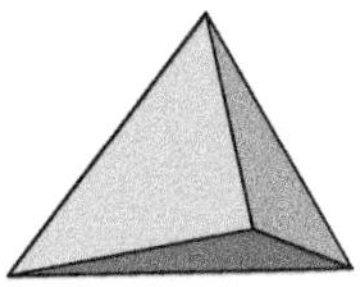

Representing structures in a two-dimensional way on paper can give a misleading picture of what the molecule looks like. The pairs of electrons in the covalent bonds repel one another, and so arrange themselves round the carbon atom as far apart as possible. Therefore, the carbon–hydrogen bonds in methane are directed so they point towards the corners of a regular tetrahedron (Figure 3). The carbon atom is at the centre of the tetrahedron, and the H—C—H bond angles are 109° (109° 28′ to be precise).

The best way to show this is to use a molecular model. Figure 4 shows how you represent the three-dimensional shape of the methane molecule on paper.

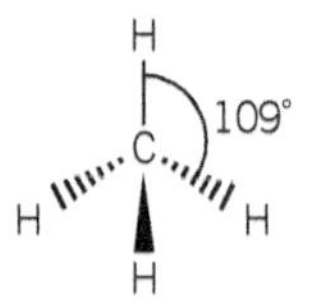

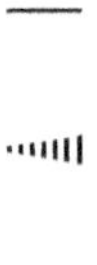

▲ Figure 3 *A regular tetrahedron*

— represents a bond in the plane of the paper

⠈⠈⠈⠈⠈ represents a bond in a direction behind the plane of the paper

◢ represents a bond in a direction in front of the plane of the paper

▲ Figure 4 *Three-dimensional shape of methane*

The structure of ethane in three dimensions is shown in Figure 5. Each carbon atom is at the centre of a tetrahedral arrangement.

Figure 6 shows the three-dimensional structure of butane. You can see that hydrocarbon chains are not really straight, but a zig-zag of carbon atoms. All the bond angles are 109°. The shape of a hydrocarbon chain is often represented by a **skeletal formula**, also shown in Figure 6.

Skeletal formulae show only the carbon skeleton and associated functional groups of the molecule (Figure 7), so do not assist in visualising shapes of molecules).

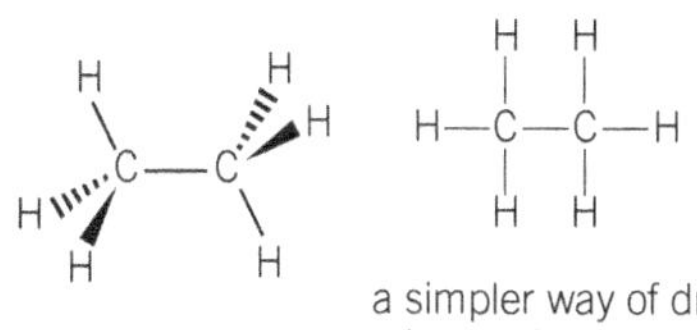

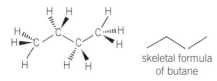

a simpler way of drawing ethane which shows the shape less accurately

▲ Figure 5 *The three-dimensional shape of ethane*

skeletal formula of butane

▲ Figure 6 *The three-dimensional shape of butane and its skeletal formula*

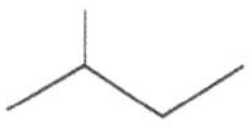

Chemical ideas: Bonding, shapes, and sizes 3.3

Structural and *E/Z* isomerism

Isomerism

Two molecules that have the same molecular formula but differ in the way their atoms are arranged are called **isomers**. Isomers are distinct compounds with different physical properties, and often different chemical properties too. The occurrence of isomers (isomerism) is very common in carbon compounds because of the great variety of ways in which carbon atoms can form chains and rings. However, you will also meet examples of isomerism in inorganic chemistry.

▲ Figure 7 *The skeletal formula of 2-methylbutane*

There are two ways in which atoms can be arranged differently in isomers.

1   The atoms are bonded together in a different order in each isomer – these are called structural isomers.

2   The order of bonding in the isomers is the same, but the arrangement of the atoms in space is different in each isomer – these are called stereoisomers.

Structural isomerism

Structural isomers have the same molecular formula but the atoms are bonded together in a different order. They have different *structural formulae*. There are various ways in which structural isomerism can arise.

Chain isomerism

There is only one alkane corresponding to each of the molecular formulae CH_4, C_2H_6, and C_3H_8. With four or more carbon atoms in a chain, different arrangements are possible (Figure 8). These are called **chain isomers**.

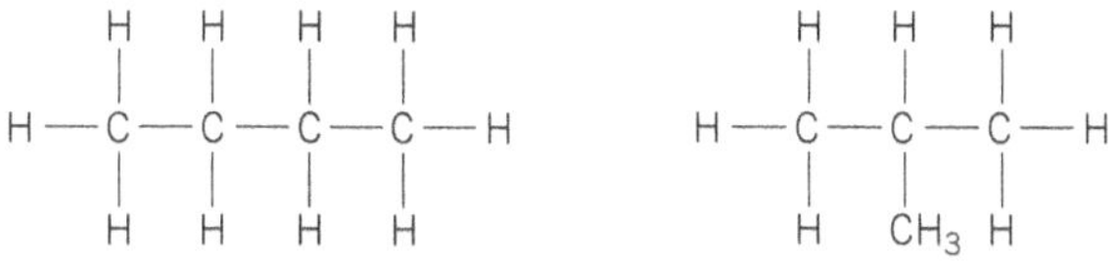

$CH_3CH_2CH_2CH_3$ $CH_3CH(CH_3)CH_3$

▲ Figure 8 *Two structures with the molecular formula of C_4H_{10}*

Both butane and methylpropane have the same molecular formula C_4H_{10}. Their different structures lead to different properties, for example, the boiling point of methylpropane is 12 K lower than that of butane.

Activity DF 9.1

This activity checks your understanding of structural isomers and structural isomerism.

As the number of carbon atoms in an alkane increases, the number of possible isomers increases. There are over four thousand million isomers with the molecular formula $C_{30}H_{62}$.

Position isomerism

Position isomerism can occur where there is an atom, or group of atoms, substituted in a carbon chain or ring. These are called functional groups. You first encountered functional groups in Topic DF 3. Position isomerism occurs when the functional group is situated in different positions in the molecules.

The isomers of C_3H_7OH are shown in Figure 9. The –OH functional group is situated at two different places on the hydrocarbon chains.

Study tip

In alcohols the functional group is –OH and the general formula is $C_nH_{2n+1}OH$.

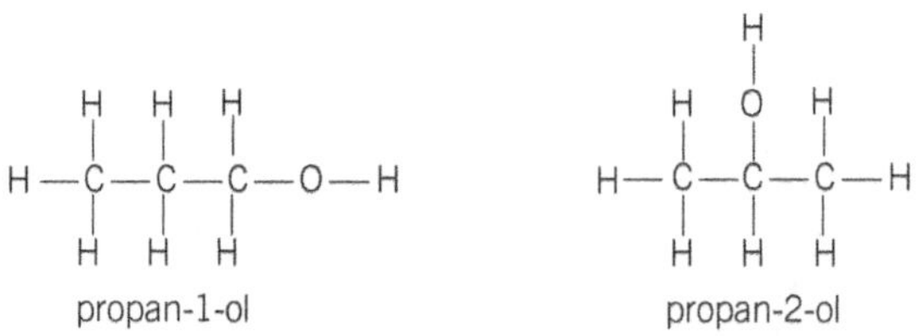

▲ Figure 9 *Two position isomers for the molecular formula C_3H_7OH*

Functional group isomerism

It is sometimes possible for compounds with the same molecular formula to have different functional groups, and because they have different functional groups they will belong to a different homologous series. These are called **functional group isomers**. The isomers of molecular formula C_3H_8O are shown in Figure 10.

Synoptic link

You will learn more about reactions of alcohols in Chapter 5, What's in a medicine?

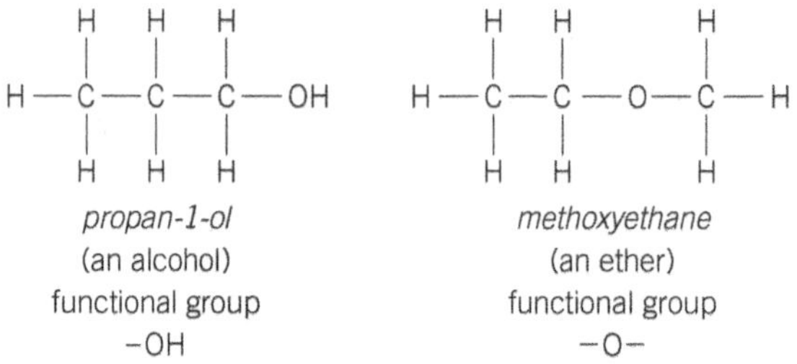

▲ Figure 10 *Two functional group isomers for the molecular formula C_3H_8O*

Study tip

As part of the course, you do not need to be able to name ethers.

The isomer on the left is called propan-1-ol. Its functional group is –OH and it is a member of the homologous series known as the alcohols. The isomer on the right is methoxyethane. Its functional group, –O–, is quite different from that of propan-1-ol and it belongs to another family of organic compounds, known as the ethers.
As well as having different physical properties, propan-1-ol and methoxyethane have very different chemical properties due to the different functional groups.

Synoptic link

You will study optical isomerism, along with other types of isomerism, in Chapter 8, Polymers and life.

Stereoisomerism

Stereoisomerism is when molecules have the same structural formula but differ in how their atoms are arranged in space. There are two types of stereoisomerism – *E*/*Z* isomerism and optical isomerism.

E/Z isomerism

E/Z **isomerism**, or *cis/trans* isomerism is one type of stereoisomerism. Stereoisomers have identical molecular formulae and the atoms are bonded together in the same order, but the arrangement of atoms in space is different in each isomer.

But-2-ene has two *E/Z* isomers (Figure 11).

To turn the second form into the first form you would have to spin one end of the molecule round in relation to the other end. This can only be done by first breaking the π-bond in the double bond. You can easily prove this by using molecular models.

The average bond enthalpy for the bond that has to be broken here is about $+270 \, kJ \, mol^{-1}$. This much energy is not available at room temperature. A covalent bond has to be broken and another reformed in order to interconvert the forms of but-2-ene. In other words, the process is an example of a chemical reaction and the two forms are different chemicals.

Naming stereoisomers of alkenes

The two different but-2-enes need different names. The older way of naming them uses *cis/trans* nomenclature. The form of but-2-ene with the same groups (in this case both are methyl groups) on the same side of the double bond is called *cis*-but-2-ene (Figure 12). When the substituents are on opposite sides of the double bond the molecule is said to be the *trans* form (Figure 13).

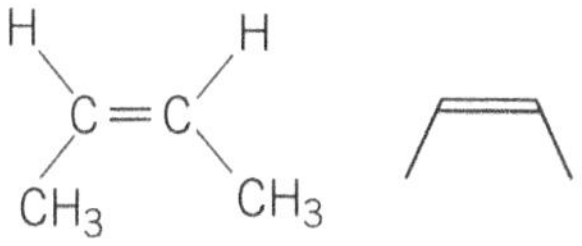

▲ Figure 12 *cis-but-2-ene*

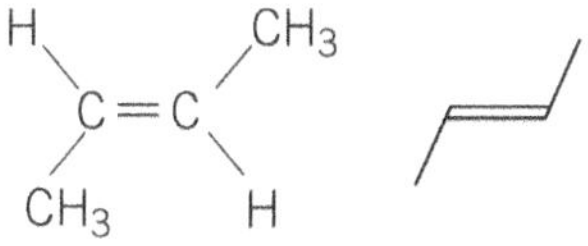

▲ Figure 13 *trans-but-2-ene*

There is a second nomenclature system for naming alkenes and this is the *E/Z* system. In the *E/Z* notation. *E* means opposite and corresponds generally to the term *trans* whilst *Z* means together and corresponds to the term *cis*. This means that *cis*-but-2-ene is also called (*Z*)-but-2-ene and *trans*-but-2-ene is also called (*E*)-but-2-ene.

This rule works when either the two groups of either end of the double bond are the same or there is a hydrogen on both carbons.

The *cis/trans* system has its limitations, for example, the two isomers in Figure 14 cannot be named using the *cis/trans* system. The *E/Z* system has its own set of rules to help inform the naming.

You will only be expected to name isomers with the same groups on either end of a double bond and name them as *E* or *Z* isomers.

Z and *E* isomers are different compounds, so they have different properties. Table 1 gives some information on some isomers which are clearly different substances.

▲ Figure 11 *E/Z isomers of but-2-ene*

Activity DF 9.2

This activity helps you check your understanding of *E/Z* isomerism.

Synoptic link

You will be looking at *cis/trans* isomers later in Chapter 10, Colours by design.

Study tip

If you have difficulty remembering which isomer is Z and which is E, the phrase 'Z-zame-zide' may help!

▲ Figure 14 *E/Z isomers of a haloalkane*

▼ Table 1 *Physical properties of some E/Z isomers*

E or *trans* isomer	Melting point / K	Density / g cm^{-3}	*Z* or *cis* isomer	Melting point / K	Density / g cm^{-3}
H, CH₃ / C=C / CH₃, H	168	0.604	H, H / C=C / CH₃, CH₃	134	0.621
Br, H / C=C / H, Br	267	2.23	Br, Br / C=C / H, H	220	2.25
H, COOH / C=C / HOOC, H	573	1.64	H, H / C=C / HOOC, COOH	412	1.59

Chemical ideas: Organic compounds: modifiers 13.1

Naming organic compounds

You have already learnt how to name alkanes and alkenes. Alcohols and haloalkanes follow similar rules for naming.

Naming alcohols

1 Name the alcohol using the root prefix of the longest chain.

2 Determine the position of the alcohol group, –OH, by counting the carbons to give the lowest number.

3 Add 'ol' onto the end of the name.

H H		
H—C—C—OH (with H atoms)	CH$_3$CH$_2$OH	ethanol
H—C—C—C—C—C—H (with OH on 2nd carbon)	CH$_3$CH$_2$CH$_2$(CHOH)CH$_3$	pental-2-ol

When an alcohol is branched, just follow the rules for a branched alkane, but replace the 'e' at the end of the name with 'ol' and include the number the –OH group is attached to.

2-methylpropan-2-ol

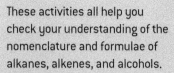

Activity DF 9.3, DF 9.4, and DF 9.5

These activities all help you check your understanding of the nomenclature and formulae of alkanes, alkenes, and alcohols.

General rules for naming alkanes, haloalkanes, and alcohols

A systematic naming system makes it possible to name compounds unambiguously, using a few simple rules. The following rules summarise what you have learnt so far about naming organic compounds.

1   Count the carbon atoms in the longest chain. This gives the root name, for example, an alkane with *five* carbons is called *pent*ane. Beware – occasionally the longest carbon chain goes 'round corners'.

2   Use alkyl group prefixes to indicate chain branches, for example, a –CH_3 group attached to a carbon chain is called a methyl group – methylpropane, $CH_3CH(CH_3)CH_3$.

3   Use chloro-, bromo-, iodo-, and so on to indicate a halogen atom substituted for a hydrogen atom, for example, bromoethane, CH_3CH_2Br. (These are called haloalkanes and you will meet them later in the course.)

4   Add suffixes to indicate other groups that have been substituted for a hydrogen atom, for example 'ol' for alcohols – ethanol, CH_3CH_2OH.

5   If necessary indicate the number(s) of the carbon atom(s) on which the substitution has occurred, keeping the numbers as low as possible. In none of the above examples is a number necessary, as they are unambiguous, but longer carbon chains need numbers to be unambiguous.

For example, 2-methylpentane $CH_3CH(CH_3)CH_2CH_2CH_3$ and not 4-methylpentane.

If there are two groups attached to the chain, the name would be 2,2-dimethylpentane $CH_3C(CH_3)_2CH_2CH_2CH_3$ not 4,4-dimethylpentane.

6   Prefixes are listed in alphabetical order, for example, 2-bromo-1-chloropropane, $CH_3CHBrCH_2Cl$.

There are other rules that occur in more complicated situations, and these will be introduced as you encounter them.

Study tip

The numbers are separated from each other by commas and from words by dashes. There are no gaps between prefixes or suffixes and the root name.
Also, if there are two of the same group, the number must be repeated (e.g., '2,2') and 'di' must be placed before the group.

Activity DF 9.1 and DF 9.2

You can practise naming organic compounds in these activities.

Summary questions

1   Write the molecular formula for each of the compounds in the following pairs. Use the molecular formulae to decide whether the compounds in the pair are isomers.

a $CH_3—CH_2—CH_2—CH_2—CH_2—CH_3$ and

$$\begin{array}{ccc} & CH_2 & \\ H_2C & & CH_2 \\ H_2C & & CH_2 \\ & CH_2 & \end{array}$$

(2 marks)

b $CH_3—CH_2—CH_2—CH_2—Cl$ and

$$H_3C—\underset{\underset{CH_3}{|}}{\overset{\overset{CH_3}{|}}{C}}—Cl$$

(2 marks)

c $CH_3—CH_2—CH_2—OH$ and

$$\underset{H_3C\qquad CH_3}{\overset{O}{\overset{||}{C}}}$$

(2 marks)

d

and

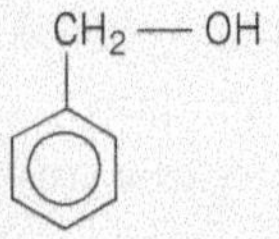

(2 marks)

e

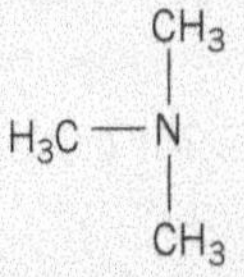

and

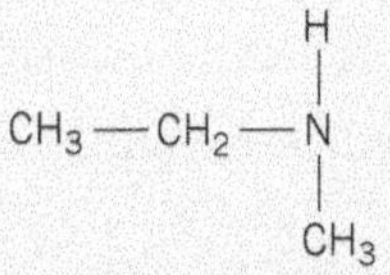

(2 marks)

2 There are two different compounds with the molecular formula C_4H_{10}. For each of them:

  a Draw full structural formulae. *(2 marks)*

  b Draw three-dimensional shapes using wedges and dotted lines. *(2 marks)*

  c Draw skeletal formulae. *(2 marks)*

  d Name each compound. *(2 marks)*

3 Draw the full structural formula of ethanol. Mark the values of a H—C—H bond angle and the C—O—H bond angle. Explain the values you have given. *(3 marks)*

4 There are several structural isomers with the molecular formula C_4H_9Br. Draw their skeletal formulae and name them. *(4 marks)*

5 There are four structural isomers with the molecular formula C_8H_{10} in which each isomer contains a benzene ring and at least one side chain. Draw their structures. *(4 marks)*

6 Draw the skeletal formulae of all the structural isomers which have the molecular formula $C_4H_{10}O$. Name all those that are alcohols. *(4 marks)*

7 **a** Draw and label the *E/Z* isomers of 1,2-dichloroethene. *(2 marks)*

  b Suggest which of the isomers has the higher boiling point. Explain your answer. *(2 marks)*

8 Nerol, which occurs in bergamot oil, geraniol (in roses), linalool (in lavender), and cilronellol (in geraniums) are four compounds from the lerpene family. Their skeletal structures are shown below.

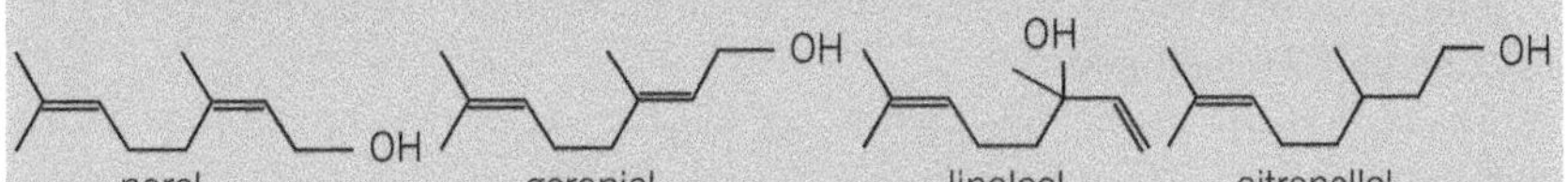

nerol   geraniol   linalool   citronellol

  a How are the structures of nerol and geraniol related? *(1 mark)*

  b How many moles of hydrogen (H_2) would be required to saturate one mole of geraniol? *(1 mark)*

  c How are nerol and geraniol related to citronellol? *(1 mark)*

  d How are the structures of nerol and geraniol related to linalool? *(1 mark)*

Figure 1 shows what goes into – and what comes out of – a car engine. Some of these exhausts are causing issues worldwide. There are slight differences between the emissions of petrol and diesel engines but many of the pollutants are similar. The oxides of nitrogen – nitrogen oxide and nitrogen dioxide – are grouped together as NO_x. Similarly, SO_x represents the oxides of sulfur – sulfur dioxide and sulfur trioxide – and C_xH_y represents the various hydrocarbons present in the exhaust fumes.

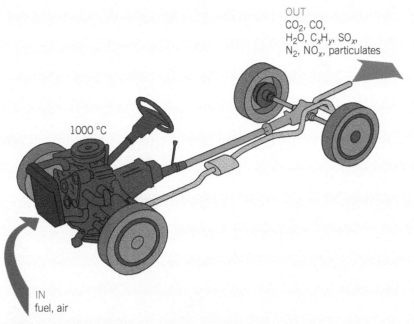

OUT
CO_2, CO,
H_2O, C_xH_y, SO_x,
N_2, NO_x, particulates

1000 °C

IN
fuel, air

▲ Figure 1 *What goes into – and comes out of – a car engine*

Learning outcomes

Demonstrate and apply knowledge and understanding of:

→ the origin of atmospheric pollutants from a variety of sources: particulates, unburnt hydrocarbons, CO, CO_2, NO_x, SO_x; the environmental implications and methods of reducing these pollutants

→ environmental implications of these pollutants

→ balanced equations for the complete and incomplete combustion (oxidation) of alkanes, cycloalkanes, alkenes, and alchols.

Furthermore, because petrol is so volatile, on a warm day a parked car gives off hydrocarbon fumes from the petrol tank and carburettor. This is **evaporative emission** and accounts for about 10% of emissions of volatile organic compounds from petrol vehicles.

The oxides of sulfur in vehicle exhausts come from sulfur compounds in the fuel. These combine with the oxygen from the air in the heat of the engine. Oxides of nitrogen are formed mainly from the components of the air itself. At the high temperatures in vehicle engines, nitrogen and oxygen in the air react to form nitrogen oxide. Some of this reacts with more oxygen to form nitrogen dioxide. Oxides of sulfur and nitrogen are acidic and give rise to acid rain in the atmosphere. This can cause health problems (particularly for asthmatics), corrode limestone buildings, and damage forests and lakes.

Carbon monoxide is formed by the incomplete combustion of hydrocarbon fuels. It is very toxic to humans and is oxidised to carbon dioxide in the atmosphere.

Particulates are very small carbon particles, not visible to the naked eye, that can get into our lungs and cause irritation and disease. Particulates are also produced by incomplete combustion of the hydrocarbon fuels in diesel.

Photochemical smog

The substances in Figure 1 are not the only pollutants caused by vehicles. Ozone is a *secondary pollutant* – it is not released directly into the atmosphere. Ozone is formed from chemical reactions that occur when sunlight shines on a mixture of *primary pollutants* (nitrogen oxides and hydrocarbons), oxygen, and water vapour. Other irritating compounds are formed by the breakdown and further reaction of the hydrocarbons. These reactions all occur in **photochemical smogs**, which are a great cause for concern. (A photochemical reaction occurs when a molecule absorbs light energy and then undergoes a chemical reaction.)

▲ Figure 2 *Photochemical smog in Beijing*

Ozone has a vital role in the troposphere, however it can also be an irritating, toxic gas and high concentrations near ground level are damaging to human health. It weakens the body's immune system and attacks lung tissue. Furthermore, ozone in the troposphere acts as a greenhouse gas, contributing to global warming.

Photochemical smogs contain a mixture of primary and secondary pollutants. The exact composition varies and depends on the nature of the primary pollutants, the local geography, weather conditions, the time of day, and the length of the smog episode. Photochemical smogs normally occur in the summer during high pressure (anticyclonic) conditions.

Synoptic link

You will find out more about the role of ozone in Chapter 4, The ozone story.

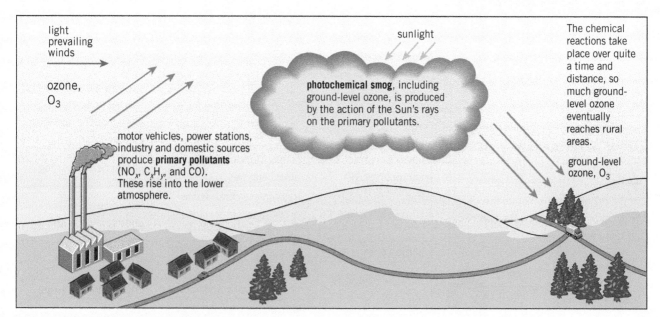

▲ Figure 3 *Formation of a photochemical smog*

What are the effects of photochemical smog?

Photochemical smogs cause haziness and reduced visibility in the air close to the ground. For many people they can cause eye and nose irritation and some difficulty in breathing. In vulnerable groups, such as asthmatics who already have respiratory problems, very young children, and many older people, this effect can be enhanced.

High ozone concentrations affect animals and plants too. Ozone is a highly reactive substance that attacks most organic matter. Compounds with carbon–carbon double bonds are particularly vulnerable so many materials such as plastics, rubbers, textiles, and paints can be damaged.

Chemical ideas: Human impacts 15.1

Atmospheric pollutants

The formation of pollutants

Table 1 summarises the sources and polluting effects of the primary pollutants formed by vehicle engines.

Complete and incomplete combustion

When a hydrocarbon (alkane, cycloalkane, or alkene) undergoes combustion the products of combustion will vary, depending on the amount of oxygen available. In a plentiful supply of air:

$$\text{hydrocarbon} + \text{oxygen} \rightarrow \text{carbon dioxide} + \text{water}$$

▼ Table 1 *Sources and polluting effects of primary pollutants from vehicle engines and other sources*

Pollutant	Major sources	Major polluting effects
particulates – particularly small carbon particles below 2.5×10^{-12} m	volcanoes, burning fuels, burning coal	penetrate deep into the human body causing heart attacks and lung cancer
volatile organic compounds, VOC	plants, unburnt fuel from petrol engines	photochemical smog
carbon monoxide, CO	Incomplete combustion of hydrocarbons in fossil fuels, burning biomass	toxic gas, photochemical smog
carbon dioxide, CO_2	combustion of fossil fuels	greenhouse effect
nitrogen oxides, NO_x	combustion of fuels in power stations and vehicles	acid rain, photochemical smog
sulfur oxides, SO_x	volcanoes, burning of fuels containing sulfur	toxic gas, acid rain

Study tip

Complete and incomplete combustion of hydrocarbons and alcohols always lead to the production of water. It is the carbon compounds that differ.

Activity DF 10.1

In this activity you can develop your research skills to learn about aspects of pollutants.

If the air (oxygen) supply is limited, combustion of hydrocarbons will lead to the production of water and either carbon monoxide or carbon (soot).

$$C_7H_{16} + 11O_2 \rightarrow 7CO_2 + 8H_2O \quad \text{complete combustion of heptane}$$

$$C_7H_{14} + 10\tfrac{1}{2}O_2 \rightarrow 7CO_2 + 7H_2O \quad \text{complete combustion of hept-1-ene}$$

$$C_8H_{18} + 7\tfrac{1}{2}O_2 \rightarrow 6CO + 2C + 9H_2O \quad \text{incomplete combustion of octane}$$

$$C_6H_{12} + 6O_2 \rightarrow 6CO + 6H_2O \quad \text{incomplete combustion of cyclohexene}$$

Alcohols can also be used as fuels. The same rules apply. For example:

$$C_2H_5OH + 3O_2 \rightarrow 2CO_2 + 3H_2O \quad \text{complete combustion of ethanol}$$

$$C_2H_5OH + 2O_2 \rightarrow 2CO + 3H_2O \quad \text{incomplete combustion of ethanol}$$

The burning of sulfur compounds in fuels produces sulfur dioxide, SO_2.

$$S + O_2 \rightarrow SO_2$$

Some nitrogen oxides, NO_x, are produced by burning nitrogen compounds in fuels, but these are present in very low proportions, especially in vehicle fuels. However, nitrogen and oxygen in the air react in the high temperatures of a vehicle engine.

$$N_2 + O_2 \rightarrow 2NO$$

Production of acid rain

Sulfur dioxide reacts with water in the atmosphere to form sulfuric(IV) acid, a weak acid.

$$SO_2 + H_2O \rightarrow H_2SO_3$$

However sulfur dioxide is oxidised to sulfur trioxide, SO_3, in the stratosphere, which then reacts with water in the atmosphere to form sulfuric(VI) acid, a strong acid.

$$SO_3 + H_2O \rightarrow H_2SO_4$$

▲ Figure 4 *A stone sculpture eroded by acid rain*

Nitrogen oxide, NO, and nitrogen dioxide, NO_2, react with water and oxygen in the atmosphere to form nitric(V) acid, a strong acid.

$$2NO + H_2O + 1\frac{1}{2}O_2 \rightarrow 2HNO_3$$

$$2NO_2 + H_2O + \frac{1}{2}O_2 \rightarrow 2HNO_3$$

Acid rain causes breathing difficulties, corrodes limestone buildings, and kills forests and life in lakes.

Photochemical smog

This is formed when primary pollutants are acted upon by sunlight to produce secondary pollutants, for example, ozone, nitrogen dioxide, and nitric acid. Photochemical smogs cause haziness and reduced visibility in the air (Figure 2) as well as causing respiratory problems in humans.

Tackling the emissions problem

Many countries are now bringing in legislation to limit exhaust emissions.

▼ Table 2 *European emission limits in mg km^{-1}. Diesel engines produce less carbon dioxide per km than petrol engines*

Pollutant	Petrol		Diesel	
	Emissions limit in year 2000	Emissions limit in year 2015	Emissions limit in year 2000	Emissions limit in year 2015
carbon monoxide	2300	1000	600	500
NO_x	500	60	200	80
particulates	–	5	50	5
hydrocarbons	200	100	–	–

There are important indirect methods for tackling the emissions problem, such as limiting the traffic entering towns and encouraging car-sharing schemes for people travelling to and from work. However, there are two ways of tackling the emissions problem directly. One involves changing the design of cars, and the other involves changing the fuel used by the car (Topic DF 11).

Using catalysts

Catalysts speed up reactions that involve pollutants in car exhausts. In the following reactions, pollutants are being converted to carbon dioxide, water, and nitrogen, all naturally present in the air.

1   using oxygen to turn carbon monoxide to carbon dioxide
2   using oxygen to turn hydrocarbons to carbon dioxide and water
3   reacting nitrogen oxide with carbon monoxide to form carbon dioxide and nitrogen

These reactions occur naturally, but under the conditions inside an exhaust system they go too slowly to get rid of the pollutants.

Catalytic converters in petrol cars contain catalysts of platinum or rhodium on a honeycomb structure. They are called three-way catalysts because they speed up the three reactions above.

This kind of catalyst system works *only* if the air–petrol mixture is carefully controlled so that it is exactly the stoichiometric mixture for the fuel. This means it has the exact calculated ratio of hydrocarbon to oxygen for complete combustion. If the mixture is too rich (too much fuel) then there is not enough oxygen in the exhaust fumes to remove carbon monoxide and the hydrocarbons. Cars fitted with three-way catalyst systems need to have oxygen sensors in the exhaust gases, linked back to electronically controlled fuel injection systems.

Catalytic converters work only when they are hot. A platinum catalyst starts working around 240 °C, but you can get the catalyst to start working at about 150 °C by alloying the platinum with rhodium. These catalysts are poisoned by lead, so the converters can only be used with lead-free fuel.

▲ Figure 5 *A catalytic converter*

The catalyst is used in the form of a fine powder spread over a ceramic support with a surface that has a network of tiny holes (Figure 4). The surface area of the catalyst exposed to the exhaust gases is about the same as two or three football fields.

Diesel engines have a higher concentration of oxygen, so any attempt to reduce NO_x to nitrogen would fail as the reducing agent would simply be oxidised by the oxygen. Diesel engines do have oxidation catalysts that turn carbon monoxide to carbon dioxide, and hydrocarbons to carbon dioxide and water. There have been attempts to reduce the amount of particulates in diesel exhaust gases by burning them at high temperatures. Unfortunately this reduces the fuel efficiency.

Catalytic converters

Petrol engines

The main pollutants in the exhaust are carbon monoxide, hydrocarbons, and nitrogen monoxide. These are removed by the following reactions:

- carbon monoxide $2CO + O_2 \rightarrow 2CO_2$
- hydrocarbons $C_7H_{16} + 11O_2 \rightarrow 7CO_2 + 8H_2O$
- nitrogen monoxide $2NO + CO \rightarrow N_2 + 2CO_2$

All three of these reactions take place in a three-way catalytic converter consisting of platinum or rhodium on a porous support.

Sulfur oxide pollutants are best prevented, by removing sulfur impurities from the fuels, before they are made available to the motorist.

Diesel engines

The main pollutants present in the exhaust are carbon monoxide, hydrocarbons, particulates, and nitrogen oxide compounds. Carbon monoxide and hydrocarbons are removed in the same way as in petrol engines. Particulates are removed by diesel particulate filters that contain a variety of materials, the most common being a ceramic. Regeneration (burning off the carbon particles) is accomplished by increasing the temperature at times decided by the vehicle's computer. This increases fuel consumption. Nitrogen oxides can by reduced by recycling some of the exhaust gases through the cylinder, lowering the temperature and thus the amount of NO_x formed. Alternatively, a reagent such as ammonia is used in the presence of a catalyst.

$$4NO + 4NH_3 + O_2 \rightarrow 4N_2 + 6H_2O$$

Again sulfur dioxide is best avoided in the first place by using ultra low-sulfur fuels.

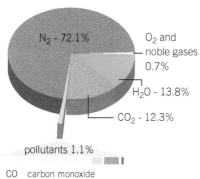

pollutants 1.1%

CO   carbon monoxide
HC   hydrocarbons
NO_x nitrogen oxide
PM   particles

▲ Figure 6 *Whilst the amount of harmful pollutants in a diesel engine exhaust is small, it is important they are removed*

Summary questions

1   a   Explain the difference between primary and secondary pollutants. *(2 marks)*
   b   Choose one secondary pollutant and say how it is made from a primary pollutant. *(2 marks)*
   c   What are the main disadvantages of tropospheric ozone? *(1 mark)*

2   a   Why is it important that catalytic converters start working at as low a temperature as possible? *(1 mark)*
   b   What is meant by a catalyst poison? *(1 mark)*
   c   Why are these catalysts used in the form of a fine powder? *(1 mark)*
   d   Suggest a reason why a catalytic converter has to be replaced eventually. *(1 mark)*
   e   Catalytic converters convert the pollutants carbon monoxide, CO, hydrocarbons, C_xH_y, and nitrogen oxides, NO_x, into less harmful gases. This is still only a partial solution to the emissions problem. Why? *(2 marks)*

3   a   Explain how oxides of nitrogen are removed from the exhaust of a petrol engine. *(2 marks)*
   b   Explain why this method does not work for a diesel engine. *(2 marks)*
   c   Describe a method that is used to reduce the oxides of nitrogen in a diesel exhaust. *(2 marks)*

4   Explain, with equations, how each of the following primary pollutants is made in a car engine. *(4 marks)*
   a   nitrogen oxide, NO
   b   carbon monoxide, CO
   c   hydrocarbons
   d   sulfur dioxide, SO_2

5   a   Use the data in Table 2 to decide how *you* would choose between a petrol and a diesel car. *(2 marks)*
   b   Write equations for the reactions involving a diesel oxidation catalyst which remove the following pollutants.
     i   carbon monoxide *(1 mark)*
     ii   cetane, $C_{16}H_{34}$ *(1 mark)*

DF 11 Other fuels

Specification reference: DF(k), DF(u)

Learning outcomes

Demonstrate and apply knowledge and understanding of:

→ the methods of reducing atmospheric pollutants

→ the benefits and risks associated with using fossil fuels and alternative fuels (biofuels and hydrogen); making decisions about ensuring a sustainable energy supply.

In addition to changing the design of the car (Topic DF 10), the other way to tackle the emissions problem is to change the fuel used by cars.

Aromatic hydrocarbons make up as much as 40% of petrol. Aromatic hydrocarbons may cause higher carbon monoxide, CO, hydrocarbon, C_xH_y, and nitrogen oxide, NO_x, emissions, and some may cause cancer. Benzene for example is strictly controlled, but others may be controlled in the future.

Butane content is also likely to be lowered in the future. Butane is volatile and responsible for evaporative emissions leading to ozone formation and photochemical smogs. However, both butane and aromatic compounds help petrol to perform well in modern engines. If they are removed then they must be replaced by something else. This is why the petrol companies looked to oxygenates (compounds such as ethanol that contain oxygen) as a possible solution.

There are a range of options of different fuel sources for the future, each of which has advantages and disadvantages.

Other hydrocarbon fuels

Liquified petroleum gas (LPG) comes from the distillation of crude oil and is often called autogas when it is used in cars. It is a mixture of propane and butane in varying quantities – often around 60% propane. It has to be kept under pressure to store the hydrocarbons as liquids. Petrol vehicles can be converted fairly easily to run on both fuels – one of the main changes needed is a larger fuel tank.

Autogas works in high performance engines and produces about 20% less carbon dioxide per mile than petrol. Because of the higher ratio of carbon to hydrogen it releases less carbon monoxide. It also produces fewer unburnt hydrocarbons and nitrogen oxides than petrol. For vehicle owners, road taxes and fuel taxes are lower than for petrol. The main disadvantage is that, although numbers are increasing, LPG filling stations are still relatively rare.

Liquid natural gas (LNG) is mainly methane and comes from oil and natural gas fields. Methane cannot be liquefied by pressure alone and it must be cooled to below −160 °C. LNG is therefore most suitable for large vehicles, especially as it works in modified diesel engines. Once again, there is a high carbon to hydrogen ratio so less carbon monoxide is produced, and much fewer NO_x than diesel.

▲ Figure 1 A bus powered by natural gas, pictured in Bristol

Replacements for fossil fuels

You have seen how fossil fuels are useful to humans. However, even though energy is extracted from fossil fuels more economically and with modern developments such as fracking (which produces shale gas), fossil fuels will not last for ever.

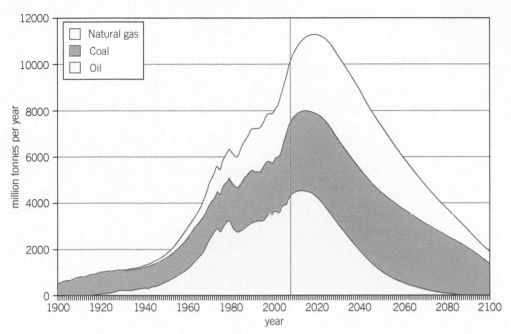

▲ Figure 2 *Supplies of conventional fossil fuels, according to a 2008 study. Their use will decline in the future as supplies run out*

Because of this potential supply problem, coupled with concerns about emissions (Topic DF 10), alternative fuels are being considered, such as biofuels.

Chemical ideas: Human impacts 15.2

Alternatives to fossil fuels

Biofuels

Biofuels are alternative fuels derived from renewable plant and animal materials. Examples of biofuels include ethanol and biodiesel.

Ethanol

Ethanol can be made by fermentation of carbohydrate crops such as sugar cane. Cars cannot easily run on ethanol alone as it is too volatile, but mixtures of petrol containing up to 15% ethanol are used.

Ethanol is sometimes said to be carbon neutral as the carbon dioxide produced in fermentation and burning the ethanol matches the carbon dioxide absorbed in the growing plant. However, energy is used to produce and distribute the ethanol, producing carbon dioxide if the energy comes from fossil fuels. Another issue, especially in developing countries, is that the land used to grow crops for fermentation could be used to produce food instead.

Biodiesel

Biodiesel is typically made by chemically reacting fats and oils, such as vegetable oil or animal fat, with an alcohol producing fatty acid esters. This process is called trans-esterification.

▲ Figure 3 *A biofuel pump in Belgium. The fuel, gasohol, is a mixture of petrol (85%) and ethanol*

Synoptic link

You will learn about esters and carboxylic acids in Chapter 5, What's in a medicine?

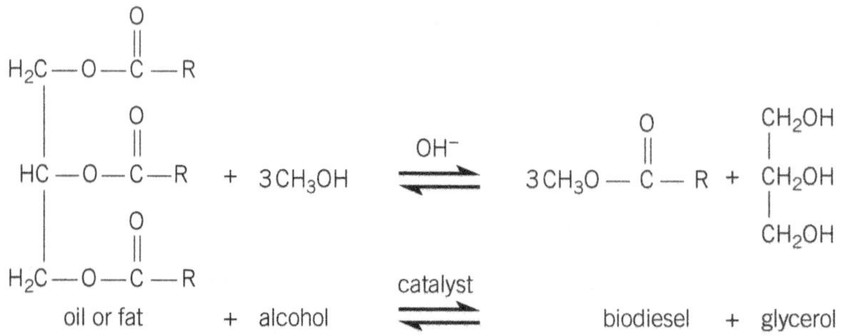

▲ Figure 4 *The trans-esterification of an oil or fat*

The advantages of biodiesel over diesel are:

- it can be made from waste oil rather than using fossil-fuel based oil
- it is carbon-neutral (except for the energy required to produce and distribute it)
- some diesel vehicles can run on pure biodiesel, but most on mixtures with regular diesel
- it is biodegradable if spilled
- it contains virtually no sulfur, so reduces oxides of sulfur in emissions
- it produces less particulates, carbon monoxide, and hydrocarbons than petrol and diesel (though these reductions are limited by the ways in which nitrogen oxides are removed).

The disadvantage is that it produces more nitrogen oxides than conventional fossil fuels.

Green diesel (often derived from algae), and biogas (methane derived from animal manure and other digested organic material) are other examples of biofuels.

Hydrogen

Hydrogen is another potential fuel for the future. On combustion, hydrogen produces just water.

$$2H_2(g) + O_2(g) \rightarrow 2H_2O(g)$$

Its advantages are:

- it is renewable and can be made by electrolysis of water
- it can be stored and sent down pipelines, in much the same way that methane currently is
- it can be used in internal combustion engines or in a fuel cell to generate electricity
- it produces no carbon dioxide, carbon monoxide, or hydrocarbons when burnt.

Its disadvantages are:

- its production from water often depends on the use of electricity from fossil fuel power stations (though alternative energy sources for electricity are being developed)

- it is less energy dense than petrol – it does not release as much energy per litre as petrol

- oxides of nitrogen are still produced at the high temperatures a hydrogen internal combustion engine runs at.

The hydrogen economy would use hydrogen as a way of storing and distributing energy. If systems are costed over whole lifetime use in terms of money and energy, then distributing hydrogen by pipeline may be cheaper than transmitting electricity.

Fuel cells are being used to generate electricity on a small scale in cars. These fuel cells convert the chemical energy from a fuel into electricity through a chemical reaction with oxygen or another oxidizing agent in an electrochemical cell. In the case of fuel cells for cars the fuel is usually oxygen. This means the main product is water.

The main problem in the design of cars run on hydrogen is that a large volume of gaseous hydrogen is required to get the mileage equivalent to a fuel tank of petrol. Some way of storing hydrogen more compactly is needed. One solution is to store it as a liquid in a high-pressure fuel tank.

Schemes for developing alternative fuel for use on a large scale depend on long-term and large-scale investments in new infrastructures, so success will depend on political as well as economic factors. The reward could be cleaner renewable fuels.

▲ Figure 5 *A hydrogen fuel station in Iceland*

▲ Figure 6 *Refilling a car with hydrogen*

Synoptic link

You will find out about redox reactions and electrolysis in Chapter 3, Elements from the sea.

Summary questions

1 a Write equations for the complete combustion of one mole of octane, cetane ($C_{16}H_{34}$), and ethanol. *(3 marks)*

 b Look up the enthalpy changes of combustion of these compounds. Work out the volume of carbon dioxide produced per kJ of energy transferred by the burning of each of the three fuels. (Assume one mole of CO_2 occupies 24.0 dm^3.) *(6 marks)*

 c Comment on the values you obtain. *(1 mark)*

2 The petrol tank of a typical car holds about 45 litres of petrol (approximately 10 gallons). Calculate the amount of energy released by burning 45 litres of petrol. Use the following information to help you.
 - assume that petrol is octane, C_8H_{18}
 - the standard enthalpy change of combustion of C_8H_{18} is −5500 kJ mol^{-1}
 - the density of octane is 0.70 g cm^{-3}
 - 1 litre is 1000 cm^3. *(2 marks)*

3 a Calculate the mass and volume (at 20 °C and 1 atmosphere pressure) of hydrogen needed to provide the same amount of energy as 45 litres of octane in **2**. *(2 marks)*
 - The standard enthalpy change of combustion of H_2 is −286 kJ mol^{-1}.
 - One mole of a gas at 20 °C and atmospheric pressure has a volume of about 24 litres.

 Hydrogen engines are efficient. In motorway driving conditions, a hydrogen engine can be over 20% more efficient than a petrol engine. In city 'stop-go' driving conditions, the hydrogen engine is about 50% more efficient than a petrol engine.

 b Taking efficiencies into account, what mass and volume of hydrogen are needed to give the same mileage as 45 litres of petrol in
  i motorway driving conditions *(2 marks)*
  ii city driving conditions. *(2 marks)*

Practice questions

1 The amount of energy needed to raise 1 g of a substance 1 °C, is called its:

A heat capacity

B thermal capacity

C molar heat capacity

D specific heat capacity *(1 mark)*

2 Which row in the table is correct?

A	bond-making	endothermic	$+\Delta H$
B	bond-breaking	exothermic	$-\Delta H$
C	bond-making	exothermic	$+\Delta H$
D	bond-breaking	endothermic	$+\Delta H$

(1 mark)

3 $10 \, cm^3$ of $2.0 \, mol \, dm^{-3}$ NaOH is mixed with $10 \, cm^3$ $2.0 \, mol \, dm^{-3}$ H_2SO_4. The temperature rises by y °C.

The enthalpy change of neutralisation of the reaction of NaOH with H_2SO_4 (in kJ) is given by:

A 4.18 y

C 20 × 4.18 y

B $4.18 \frac{y}{2}$

D $15 \times 4.18 \frac{y}{2}$ *(1 mark)*

4 Chloroethene has the formula of $CH_2{=}CHCl$. Poly(chloroethene) can be represented as:

A $-(CH_3CH_2Cl)-$ C $-(CHCHCl)-$

B $-(CHClCHCl)-$ D $-(CH_2CHCl)-$

(1 mark)

5 Which of the following *cannot* be formed from ethene in a one-step reaction?

A CH_2BrCH_2Br C CH_3CH_2OH

B CH_3CHBr_2 D CH_3CH_2Br *(1 mark)*

6 Which of the following might be found in the exhaust when hydrogen burns in an internal combustion engine?

A NO_x C hydrocarbons

B CO D SO_2 *(1 mark)*

7 A fuel is burnt in a small lamp and heats up a copper calorimeter containing water. Look at these three statements.

1 The specific heat capacity of the water is not accurately known.

2 Fuel may evaporate from the wick.

3 The fuel may not burn completely.

Which statements are limitations to accuracy in measuring the energy transferred?

A 1, 2, and 3 correct C 2 and 3 correct

B 1 and 2 correct D 1 correct *(1 mark)*

8 Which of the following are isomers of propan-1-ol?

1 $CH_3CH(OH)CH_3$

2 $CH_3CH_2OCH_3$

3 CH_3CH_2CHO

A 1, 2, and 3 correct C 2 and 3 correct

B 1 and 2 correct D 1 correct *(1 mark)*

9 Disadvantages of fossil fuels compared with biofuels include:

1 they give off CO_2 when burned

2 they have not (recently) absorbed CO_2 from the atmosphere

3 supplies are running out.

A 1, 2, and 3 correct C 2 and 3 correct

B 1 and 2 correct D 1 correct *(1 mark)*

10 Poly(propene) is a polymer which is now being used to make 'polymer banknotes'. One useful property is that it is unreactive to reagents such as acids.

a Draw **full** structural formulae for propene and poly(propene) *(2 marks)*

b Why is poly(propene) less reactive than propene? *(1 mark)*

c Propene reacts with bromine.

(i) Describe a test for propene based on this reaction. *(1 mark)*

(ii) Give the mechanism for the reaction of propene with bromine, using curly arrows and showing lone pairs and partial charges. Give the formula of the product. *(4 marks)*

(iii) Say what you understand by the term *electrophile* and identify an electrophile from your mechanism. *(2 marks)*

(iv) Some chloride ions are added to the reaction in (ii).

Give the formula of a product containing chlorine that would be formed. *(1 mark)*

d Butene is the next alkene in the homologous series after propene.

(i) What do you understand by the term *homologous series?* (*1 mark*)

(ii) What term is used to describe compounds like alkenes that have C=C bonds? (*1 mark*)

(iii) One structural isomer of butene has two stereoisomers.

Draw the structures for these two stereoisomers and name them.

(*2 marks*)

11 Sherbet sweets get their fizz from the reaction between sodium bicarbonate and citric acid.

$$C_6H_8O_7(aq) + 3NaHCO_3(s) \rightarrow$$
$$Na_3C_6H_5O_7 + 3CO_2(g) + 3H_2O$$
Equation 12.1

a (i) Some students decide to investigate this **endothermic** reaction and measure the enthalpy change of reaction.

They have available 10.0 g portions of sodium bicarbonate and 25.0 cm³ portions of a solution of citric acid. The citric acid portions represent an excess over the sodium bicarbonate portions in the reaction in **Equation 12.1.**

Describe how they would carry out their experiment and how they would work out ΔH per mole of sodium bicarbonate from their results.

(*6 marks*)

(ii) A student says that the temperature change they measure would be inaccurate because of heat losses. Comment on this statement.

(*2 marks*)

b Calculate the volume of carbon dioxide (measured at room temperature and pressure) that the students would collect if they reacted 10.0 g of sodium bicarbonate with excess citric acid. Assume none of the carbon dioxide dissolves.

Give your answer to an appropriate number significant figures. (*3 marks*)

12 Petrol cars produce less NO_x and particulates than diesel cars but more CO and hydrocarbons.

a (i) Suggest why diesel cars produce fewer hydrocarbons. (*1 mark*)

(ii) Write an equation to show how CO is formed from the combustion of hexane in a petrol engine. (*1 mark*)

(iii) Give a reason why particulates are a pollutant. (*1 mark*)

b NO and CO can be removed from the exhaust of a petrol engine by reacting them together over a catalytic converter. This uses a heterogeneous catalyst.

(i) Give the equation for this reaction.

(*1 mark*)

(ii) Explain the term *heterogeneous* in the context of catalysis and describe the first stage in the mechanism of this type of catalysis. (*2 marks*)

c Heterogeneous catalysts are also used for cracking hydrocarbons. Write the equation for a reaction in which nonane is cracked to produce ethene and one other product.

(*1 mark*)

d Ethanol is one example of a biofuel.

(i) Write the equation for the complete combustion of ethanol.

Show state symbols under standard conditions. (*1 mark*)

(ii) Use the bond enthalpy values in the table to calculate a value for the enthalpy change of combustion of ethanol. (*3 marks*)

Bond	Average bond enthalpy / kJ mol^{-1}
C—H	413
C—C	347
C—O	358
O—H	464
C=O	805
O=O	498

(iii) The enthalpy change of combustion of ethanol in a Data Book is different from your answer to (ii). Suggest **two** reasons for this. (*2 marks*)

(iv) Biofuels are said to be sustainable. Explain the word *sustainable* in this context. (*1 mark*)

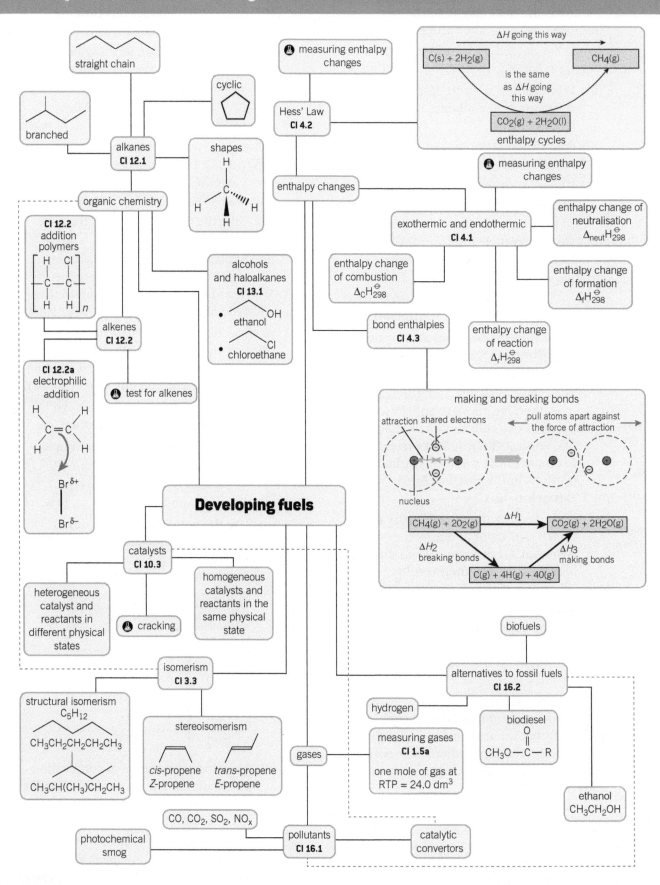

Carbon dioxide, enthalpy cycles, and alternative fuels

1 Below are two diagrams representing the bonding in carbon dioxide. The first is a simple dot-and-cross diagram whilst the second shows σ- and π-orbitals. Evaluate the two models of bonding, identifying phenomena that can be explained by each.

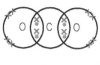

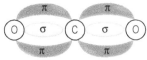

2 In this module you have learnt about different enthalpy cycles. Produce a summary of Hess's Law to include how enthalpy cycles can be used to calculate:

a enthalpy changes of combustion from enthalpy changes of formation;

b enthalpy changes of formation from enthalpy changes of combustion;

c enthalpy changes of reaction from enthalpy changes of formation or combustion;

d enthalpy changes of reaction from bond enthalpies.

Carry out research to include a wide range of worked examples in your summary.

3 Society is currently highly dependent on fossil fuels, yet the Intergovernmental Panel on Climate Change (IPCC) has stated that use of fossil fuels should be phased out by 2100 if the world is to avoid dangerous climate change. Discuss the challenges and opportunities of moving to a low-carbon economy and the role that chemists can play.

 ### Extension

Silanes and silenes are the silicon analogues of alkanes and alkenes. They contain silicon and hydrogen atoms only.

Silanes contain single bonds only. At standard temperature and pressure, the two simplest silanes are gases, whereas the next two (Si_3H_8 and Si_4H_{10}) are liquids. Silanes are highly reactive when mixed with air, spontaneously catching fire.

Silenes contain silicon–silicon double bonds. The simplest contain just two silicon atoms and are known as disilenes. They are extremely reactive and polymerise readily. The first stable disilene, tetramesityldisilene, was synthesised in 1972. It contains very bulky side groups to stabilise the molecule.

Bond	Average bond enthalpy / kJ mol⁻¹
Si—Si	222
Si=Si	113

1 What is the general formula for silanes and silenes?

2 Draw dot-and-cross diagrams for SiH_4 and Si_2H_4 and deduce the bond angles around each silicon atom.

3 Write a balanced chemical reaction for the oxidation of SiH_4 when it mixes with air.

4 Look up the structure of tetramesityldisilene. Suggest why silenes readily polymerise, and why bulky side groups in tetramesityldisilene help to stabilise the molecule.

5 What volume of hydrogen would react with 0.9 g of Si_2H_4? (one mole of H_2 occupies 24.0 dm³ at r.t.p.)

6 Would SiH_4 make a good fuel?

CHAPTER 3
Elements from the sea

Topics in this chapter

→ ES 1 The lowest point on Earth
chemistry of the halogens

→ ES 2 Bromine from sea water
oxidation states and redox

→ ES 3 Manufacturing chlorine
electrolysis

→ ES 4 From extracting bromine to making bleach
dynamic equilibrium and the equilibrium constant K_c

→ ES 5 The risks and benefits of using chlorine
redox, titration, and the risks and benefits of chlorine

→ ES 6 Hydrogen chloride in industry and the laboratory
atom economy and hydrogen halides

→ ES 7 The Deacon process solves the problem again
Le Chatelier's principle

Why a chapter on Elements from the sea?

The first chapter in the course – Elements of life – told the story of how the elements were formed. The theme is taken further in this chapter, which tells how extract some elements, such as bromine, from the natural resources that contain them and turn them into useful substances. In order to understand the extraction of bromine from sea water, redox chemistry is used. This leads on to looking at the manufacture of another useful halogen – chlorine and the chemistry of the halogens, needing an understanding of electrolysis. The risks and benefits of using chlorine are discussed. Chlorine is used to make bleach which leads to a study of dynamic equilibria, including a quantitative treatment of this subject. The industrial processes you look at in this chapter introduce the concept atom economy.

Knowledge and understanding checklist

From your Key Stage 4 study you will have studied the following. Work through each point, using your Key Stage 4 notes and the support available on Kerboodle.

- ☐ The halogens.
- ☐ Oxidation and reduction.
- ☐ Electrolysis.
- ☐ Dynamic equilibria.
- ☐ Atom economy.

You will learn more about some ideas introduced in earlier chapters:

- ☐ amounts in moles (**Elements of life**)
- ☐ electronic structure of atoms (**Elements of life**)
- ☐ covalent bonding (**Elements of life** and **Developing fuels**)
- ☐ the periodic table (**Elements of life**).

Maths skills checklist

In this chapter, you will need to use the following maths skills. You can find support for these skills on Kerboodle and through MyMaths.

- ☐ Recognise and make use of appropriate units in calculation.
- ☐ Recognise and use expressions in decimal and ordinary form.
- ☐ Use ratios, fractions, and percentages.
- ☐ Understand and use the symbols =, <, <<, >>, >, $\propto$, ~, $\rightleftharpoons$.
- ☐ Interpret data in tables and graphs and make estimations.

MyMaths.co.uk
Bringing Maths Alive

ES 1 The lowest point on Earth

Specification reference: ES(h), ES(i), ES(j), ES(k)

In Figure 1, the person is floating on water in the Dead Sea. They float because of the high **density** of the water there. The high density is due to the water containing about $350\,g\,dm^{-3}$ of salts compared with $40\,g\,dm^{-3}$ in water from the oceans. A coffee mug could hold about $350\,g$ of salts and $1\,dm^3$ is about four coffee mugs – the solution is very concentrated.

The Dead Sea is the lowest point on Earth, almost $400\,m$ below sea level in the Rift Valley that runs from East Africa to Syria (Figure 2). It is like a vast evaporating basin – water flows in at the north end from the River Jordan, but there is no outflow. The countryside around it is desert and in the scorching heat so much water evaporates that the air is thick with haze, making it hard to see across to the mountains a few kilometres away on the other side. This steady evaporation of water for thousands of years has resulted in huge accumulations of **salts**.

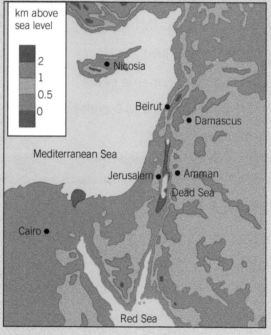

◀ **Figure 2** *The region around the Dead Sea in relief*

▲ **Figure 1** *The Dead Sea – humans float easily due to the high density of the water*

Salts found in the Dead Sea

Surveys of the salt concentration of the Dead Sea were conducted as early as the seventeenth century. Estimates suggest that there are about 43 billion tonnes of salts in the Dead Sea, and a particular feature is the relatively high proportion of bromides.

The sea is the major source of minerals in the region. A chemical industry has grown up around the Dead Sea in Israel and it has become one of the largest exporters of bromine compounds in the world. The annual production of bromine compounds in Israel exceeds 230 000 tonnes. Bromide, chloride, and iodide ions are colourless in solution and the crystalline salts from the Dead Sea look white. All of these ions are from Group 7. The elements in Group 7 look very different to their salts.

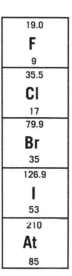

19.0
F
9
35.5
Cl
17
79.9
Br
35
126.9
I
53
210
At
85

Chemical ideas: The periodic table 11.3a

The p-block – Group 7

The **halogens** are the elements in Group 7 of the periodic table (Figure 3).

All halogen atoms have seven electrons in the outer shell. The halogens are the most reactive group of non-metals and none of them are found naturally in the elemental form. They are all found naturally in compounds such as calcium fluoride and sodium chloride. These two compounds contain the halide ions F^- and Cl^-, respectively.

Fluorine and chlorine are the most abundant halogens, bromine occurs in smaller quantities, iodine is quite scarce, and astatine is artificially produced, short-lived, and radioactive.

All the halogen elements occur as **diatomic molecules**, for example, fluorine, F_2, and bromine, Br_2. The two atoms are linked by a single covalent bond (Figure 4).

▲ **Figure 3** *The elements of Group 7*

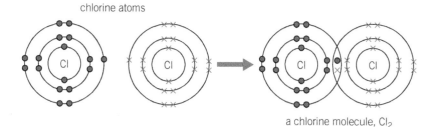

chlorine atoms

a chlorine molecule, Cl_2

▲ **Figure 4** *Covalent bonding in the chlorine, Cl_2, molecule*

In compounds, a halogen atom achieves stability by:

- gaining an electron from a metal atom, forming a halide ion in an ionic compound (Figure 5)

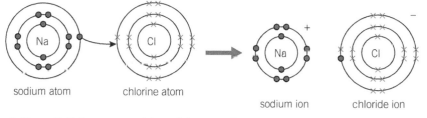

sodium atom   chlorine atom

sodium ion   chloride ion

▲ **Figure 5** *Halogens can achieve eight outer electrons by forming an ion, for example, Cl^-, via ionic bonding with a metal*

- sharing an electron from another non-metal atom in a covalently bonded compound (Figure 6).

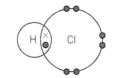

H Cl

▲ **Figure 6** *Halogens can achieve eight outer electrons by covalent bonding with a non-metal*

Physical properties of the halogens

Table 1 shows some properties of the halogens.

▼ **Table 1** *Some physical properties of the halogens*

Element	fluorine	chlorine	bromine	iodine
Formula of the element	F_2	Cl_2	Br_2	I_2
Appearance at room temperature	pale yellow gas	green gas	dark red volatile liquid	shiny grey/black solid – sublimes to give purple vapour on warming
Melting point / K	53	172.0	266.0	387.00
Boiling point / K	85	239.0	332.0	457.00
Solubility at 298 K / g per 100 g of water	reacts with water	0.6	3.5	0.03

Synoptic link

Chapter 4, The ozone story, explains how intermolecular bonds cause the trends in boiling points for the halogens.

▲ **Figure 7** *Halogens dissolved in water (the lower layer) – chlorine is pale green (left), bromine is orange/yellow (middle), and iodine is brown (right)*

▲ **Figure 8** *Halogens dissolved in cyclohexane – chlorine is pale green (left), bromine is orange/brown/red (middle), and iodine is violet (right)*

This demonstrates some trends of properties going down the group:

- become darker in colour
- melting point and boiling point increase
- change from gases to liquids to solids at room temperature
- become less volatile.

There is no trend for the solubility of the halogens in water as you go down the group, but the halogens are more soluble in organic solvents (such as hexane and cyclohexane) than they are in water. The solutions of halogens in organic solvents have a much more distinct colour and this makes it easier to say which halogen is present without ambiguity (Figure 7 and Figure 8).

Chemical properties of the halogens

Which is most reactive and why?

The halogens are a group of reactive elements. They tend to remove electrons from other elements so that they can complete their outer shell. For example:

$$2Na(s) + Cl_2(g) \rightarrow 2NaCl(s)$$

- Sodium, Na, has lost an electron to become Na^+. It has been **oxidised**.
- Each chlorine atom, Cl, has gained an electron to become Cl^-. It is an **oxidising agent**.
- The elements at the top of the group are the most reactive and are the strongest oxidising agents.

The core of an atom is made up of the nucleus and inner shell electrons (Figure 9). Both fluorine and chlorine have atoms that consist of a core with a charge of +7, surrounded by an outer electron shell containing seven electrons. Both react by gaining one electron to complete their outer shell. Fluorine atoms are very small, so the attraction between the core and the electron that completes its outer shell is very strong. In chlorine, the outer shell is further from the core and the attraction for the extra electron is weaker. This means that fluorine gains an extra electron more readily to become a negative ion than chlorine does. This trend continues down the halogen group, as each successive member of the group has one more complete electron shell than the previous one.

Activity ES 1.1

In this activity you will find out more about the physical properties of the halogens.

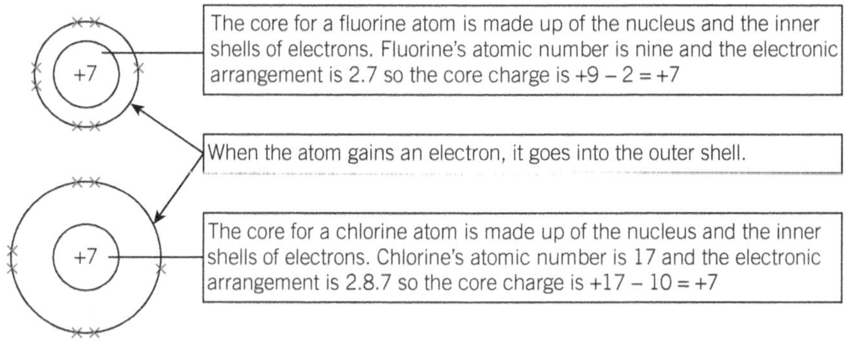

The core for a fluorine atom is made up of the nucleus and the inner shells of electrons. Fluorine's atomic number is nine and the electronic arrangement is 2.7 so the core charge is $+9 - 2 = +7$

+7

When the atom gains an electron, it goes into the outer shell.

+7

The core for a chlorine atom is made up of the nucleus and the inner shells of electrons. Chlorine's atomic number is 17 and the electronic arrangement is 2.8.7 so the core charge is $+17 - 10 = +7$

Activity ES 1.2

This activity helps you learn about the reactions of halogens and halides.

▲ **Figure 9** *Core charges and outer shells of electrons in fluorine (top) and chlorine (bottom)*

Overall, this means that fluorine is the most reactive member of the halogen group – it is the strongest oxidising agent. The reactivity decreases as you go down Group 7, for example, in the reaction with metals such as iron and sodium, chlorine reacts more vigorously than bromine.

Reactions of the halogens with halide ions

If you add a solution of chlorine (pale green) to a solution of potassium iodide ions (colourless) there is a chemical reaction. The solution turns brown and iodine is produced.

$$Cl_2 \text{ (aq)} + 2KI \text{ (aq)} \rightarrow 2KCl \text{ (aq)} + I_2 \text{(aq)}$$
pale green   brown

This equation can be simplified by writing an ionic equation – the K^+ ions can be left out because they are unchanged:

$$Cl_2\text{(aq)} + 2I^-\text{(aq)} \rightarrow 2Cl^-\text{(aq)} + I_2\text{(aq)}$$

Ions that are left out of ionic equations because they are unchanged are called **spectator ions**.

Here each iodide ion loses an electron and is oxidised. Each chlorine atom gains an electron and you call the chlorine an oxidising agent. In a reaction, the oxidising agent is said to have been **reduced**. You will learn more about reduction in Topic ES 2.

The **half-equations** make this clearer:

$$Cl_2\text{(aq)} + 2e^-\text{(aq)} \rightarrow 2Cl^-\text{(aq)}$$
$$2I^-\text{(aq)} \rightarrow I_2\text{(aq)} + 2e^-\text{(aq)}$$

Half-equations show more clearly what is happening in a redox reaction. In one of the half-equations a species gains electron(s) and is reduced (the Cl_2). In the other half-equation a species loses electron(s) and is oxidised (the I^-).

A similar thing happens if bromine solution is added to iodide ions:

$$Br_2(aq) \ + \ 2I^-(aq) \ \rightarrow \ 2Br^-(aq) \ + \ I_2(aq)$$
$$\text{orange/yellow} \qquad\qquad\qquad\qquad \text{brown}$$

These are examples of **displacement reactions** because a halogen displaces or pushes out a less reactive halogen from a compound. They are also called **redox reactions** because both oxidation and reduction have occurred in the same reaction. In the example above, the bromine gains electrons and is reduced whilst the iodine loses electrons and is oxidised.

Table 2 summarises the reactions of halogens and halide ions. You won't be surprised by the entries in the table if you remember that:

- reactivity decreases down the group
- the strongest oxidising agent is fluorine
- a halogen displaces a less reactive halide.

▼ **Table 2** *Does a halogen displace a halide ion?*

		Halide			
		F^-	Cl^-	Br^-	I^-
Halogen	F_2	no reaction	yes	yes	yes
	Cl_2	no reaction	no reaction	yes	yes
	Br_2	no reaction	no reaction	no reaction	yes
	I_2	no reaction	no reaction	no reaction	no reaction

When you carry out these reactions, the colour changes in the solutions can be difficult to detect. Adding an organic solvent can make the colour changes more distinct. The upper layer will be the organic layer.

Reactions of halide ions

At GCSE you may have met the reactions of silver ions with halide ions when studying **qualitative analysis**.

Reactions of halide ions with silver ions

When two solutions mix and a solid is formed, the solid is called a **precipitate**. Silver halides are precipitated when a solution of silver ions is added to a solution containing chloride, bromide, or iodide ions. The general equation is:

$$Ag^+(aq) + X^-(aq) \rightarrow AgX(s)$$

In the equation, X represents a halide. These are examples of precipitation reactions. Figure 10 shows the colours of these precipitates.

It can be difficult to distinguish between the colours of these precipitates so ammonia solution is sometimes added (Figure 11). The solubility of silver chloride is greater than that of silver bromide and silver iodide is insoluble in ammonia solution.

▲ **Figure 10** *The colours of the silver halides (from left to right) – white silver chloride, AgCl, cream silver bromide, AgBr, and pale yellow silver iodide, AgI*

▲ **Figure 11** *Distinguishing between the silver halides using ammonia solution – silver chloride (left), silver bromide (middle), and silver iodide (right) is insoluble*

Study tip

When you write an ionic equation for precipitation, first put the solid on the right.

$$\rightarrow AgCl(s)$$

Then add the aqueous ions on the left.

$$Ag^+(aq) + Cl^-(aq) \rightarrow AgCl(s)$$

Always include state symbols. In this example, no balancing is needed.

Summary questions

1 Write a stoichiometric equation with state symbols for the following reactions. Remember that Group 7 ions have a charge of -1, Group 1 ions have a charge of $+1$, and the molecules of Group 7 elements are diatomic.
   a Chlorine water is mixed with aqueous sodium iodide. (*2 marks*)
   b Bromine water is mixed with aqueous potassium iodide. (*2 marks*)
   c Bromine water is mixed with aqueous sodium chloride. (*1 mark*)

2 Write the two half-equations with state symbols for the following ionic equation.

$$Br_2(aq) + 2I^-(aq) \longrightarrow 2Br^-(aq) + I_2(aq) \qquad \text{(4 marks)}$$

3 Write balanced ionic equations with state symbols for the following precipitation reactions when silver nitrate solution is added to the following.
   a potassium iodide solution (*2 marks*)
   b sodium bromide solution (*2 marks*)
   c copper(II) chloride solution (*2 marks*)

4 Given that water in the Dead Sea has a bromide ion concentration of $5.2\,g\,dm^{-3}$ and a chloride ion concentration of $208\,g\,dm^{-3}$, calculate the following:
   a Concentration in $mol\,dm^{-3}$ for bromide ions in Dead Sea water. (*1 mark*)
   b Concentration in $mol\,dm^{-3}$ for chloride ions in Dead sea water. (*1 mark*)
   c The simplest ratio of bromide ions to chloride ions in the Dead Sea. One of the numbers in $Br^-{:}Cl^-$ should be 1. (*1 mark*)

5 Predict what you would observe after these chemicals have been added together and shaken.
   a Silver chloride solution and excess dilute ammonia solution. (*1 mark*)
   b Chlorine water, potassium iodide solution, and cyclohexane. (*2 marks*)
   c Sodium chloride solution, iodine solution, and cyclohexane. (*2 marks*)

Synoptic link

You carried out calculations where you converted concentration in $g\,dm^{-3}$ to concentration in $mol\,dm^{-3}$ in Chapter 1, Elements of life. You may decide to revisit this before attempting the summary questions.

ES 2 Bromine from sea water

Specification reference: ES(b), ES(d), ES(e), ES(f), ES(g)

The high levels of bromide ions in the Dead Sea make it an ideal source of the element bromine. The Dead Sea Bromine Group Ltd at S'Dom in Israel opened on the shore of the Dead Sea in the 1930s and still produces bromine. Bromine compounds have a wide range of uses in the pharmaceutical industry. Drugs that use bromine in their manufacture are undergoing trials for treatment of Alzheimer's disease. Most of the bromine produced is used in the manufacture of flame-retardants, which have been instrumental in saving many thousands of lives.

Getting bromine from dissolved bromide ions involves simple chemistry – in a laboratory a chlorine solution is added to a solution containing bromide ions. Industrially it is more complicated, and involves the use of some ingenious engineering (Figure 1).

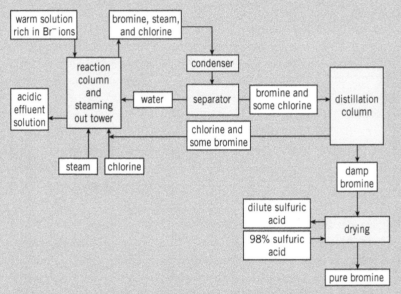

▲ **Figure 1** A scheme for the industrial manufacture of bromine

The industrial production of bromine

Chlorine is added to warm, partially evaporated, acidified sea water to displace bromine from the bromide ions. Bromine is volatile (boiling point 331 K) and bromine vapour and water is given off when steam is blown through the solution. The vapours are then condensed and two layers form because liquid bromine is not very soluble in water. The dense bromine layer is run off from the water layer that floats on top. The impure bromine is then distilled and dried.

The chlorine required for the process is produced by electrolysis on site. The reaction of chlorine with bromide ions is an example of a very important type of chemical reaction in which both oxidation and reduction take place – redox reactions.

Oxidation and reduction

Definitions of oxidation and reduction

At GCSE, you learnt that oxidation is gain of oxygen, and reduction is loss of oxygen. When magnesium and copper(II) oxide are heated together the following reaction happens:

$$Mg(s) + CuO(s) \rightarrow MgO(s) + Cu(s)$$

- Magnesium is oxidised because it gains oxygen – copper(II) oxide causes this to happen and so it is the oxidising agent.
- Copper(II) oxide is reduced because it loses oxygen – magnesium causes this to happen and so it is the **reducing agent**.

This is a redox reaction because it involves both oxidation and reduction.

Redox defined by electrons

Not all redox reactions involve oxygen, for example, the reaction between chlorine and bromide ions used to make bromine at the Dead Sea chemical plant:

$$Cl_2(aq) + 2Br^-(aq) \rightarrow 2Cl^-(aq) + Br_2(aq)$$

The half-equations are:

$$Cl_2(aq) + 2e^- \rightarrow 2Cl^-(aq)$$

$$2Br^-(aq) \rightarrow Br_2(aq) + 2e^-(aq)$$

- Chlorine is reduced because it gains electrons – bromide ions cause this to happen and so they are the reducing agent.
- Bromide ions are oxidised because they lose electrons – chlorine causes this to happen and so it is the oxidising agent.

This is also a redox reaction.

This is a much more useful definition. Half-equations can be written to show what happens to magnesium atoms and copper(II) ions.

$$Mg(s) \rightarrow Mg^{2+}(s) + 2e^-$$

$$Cu^{2+}(s) + 2e^- \rightarrow Cu(s)$$

- Magnesium is oxidised because it loses electrons.
- Copper(II) ions are reduced because they gain electrons.

Redox defined by oxidation state

A third definition is:

- when something is oxidised, its **oxidation state** (also called its oxidation number) increases
- when something is reduced, its oxidation state decreases.

Study tip

Remember, oxidising agents are reduced in chemical reactions and reducing agents are oxidised in chemical reactions.

Study tip

Some students find OIL RIG useful for remembering the definition involving electrons:

Oxidation **I**s **L**oss, **R**eduction **I**s **G**ain.

Synoptic link

You will develop your understanding of redox reactions in Chapter 9, Developing metals.

Study tip

Oxidation is:

- gain of oxygen
- loss of electrons
- an increase in oxidation number.

Reduction is:

- loss of oxygen
- gain of electrons
- a decrease in oxidation number.

Oxidation states

The oxidation state assigned to an element in a chemical tells you how many electrons have been lost or gained compared to the unreacted element. It is useful for:

- naming inorganic compounds
- deciding what has been oxidised and what has been reduced in a reaction
- identifying the oxidising agent and the reducing agent in a redox reaction
- balancing redox equations.

Working out the oxidation state for an element

The atoms in elements always have oxidation number 0. For example, oxygen, O_2, magnesium, Mg, and chlorine, Cl_2, all have the oxidation number 0.

Working out the oxidation state for a simple ion

In simple ions, the oxidation state is the same as the charge on the ion. For example, chlorine in Cl^- has the oxidation state –1, oxygen in O^{2-} has the oxidation state –2, and potassium in K^+ has the oxidation state +1.

Working out oxidation states in compounds

When compounds have no overall charge, the oxidation states of all the constituent elements must add up to zero. Some elements have oxidation states that never or rarely change. Table 1 shows the oxidation state for some elements in compounds.

▼ **Table 1** *Oxidation states of some elements in compounds*

Element	Oxidation state
fluorine	–1
oxygen	–2 (except when combined with F or in the peroxide ion O_2^{2-})
chlorine	–1 (except when combined with O or F)
bromine	–1 (except when combined with O, F, or Cl)
iodine	–1 (except when combined with O, F, Cl, or Br)
hydrogen	+1 (except when in a metal hydride e.g., NaH)
Group 1	+1
Group 2	+2
aluminium	+3

Worked example: Oxidation states in compounds

What is the oxidation state of bromine in BrF_3?

Step 1: Work out the oxidation state of fluorine in BrF_3.

Using the rules in Table 1, fluorine has an oxidation number of –1. There are three fluorine atoms in BrF_3.

$$(-1) \times 3 = -3$$

Step 2: Work out the oxidation state of bromine in BrF_3.

The overall charge of BrF_3 is 0. The oxidation state of bromine and fluorine must equal 0.

$$0 - (-3) = +3$$

The oxidation state of bromine is +3.

Working out oxidation state in more complicated ions

The same rules that are used above can also be used for the elements that make up complicated ions containing more than one element, such as PO_4^{3-}. This time the oxidation state of the constituent elements add up to the overall charge on the ion.

Worked example: Oxidation states in complex ions

What is the oxidation state of phosphorus in PO_4^{3-}?

Step 1: Work out the contribution of oxygen to the overall PO_4^{3-} ion in terms of oxidation state.

Using the rules in Table 1, oxygen has an oxidation state of −2. There are four oxygen atoms in PO_4^{3-}.

$$(-2) \times 4 = -8$$

Step 2: Work out the oxidation state of phosphorus in PO_4^{3-}.

The overall charge of PO_4^{3-} is −3. The oxidation states of phosphorus and oxygen when added together must equal −3.

$$\underset{\substack{\text{overall charge} \\ \text{on the ion}}}{(-3)} - \underset{\substack{\text{total oxidation numbers of} \\ \text{oxygen in the ion}}}{(-8)} = +5$$

The oxidation state of phosphorus is +5.

Systematic names

From a given formula you can work out the **systematic name**. A systematic name has a Roman numeral in brackets and the Roman numeral is the oxidation state of one of the elements in a compound.

At GCSE you may have met iron compounds in qualitative analysis. Iron can have a +2 or +3 oxidation state. The iron(II) ion is Fe^{2+} and the iron(III) ion is Fe^{3+}. As such, FeO is called iron(II) oxide and Fe_2O_3 is called iron(III) oxide. Copper also has two oxidation states, +1 and +2. Figure 2 shows the two oxides of copper – copper(II) oxide and copper(I) oxide.

The following rules are used for systematic names.

- Oxidation states are only used in systematic names for elements which have variable oxidation states, for example, transition metals, tin, lead, sulfur, nitrogen, and the halogens.
- The number shows the oxidation state of a preceding element. In iron(II) oxide the iron has oxidation state +2 but in potassium nitrate(V), KNO_3, it is the nitrogen that has the oxidation state +5.
- The number is placed close up to the element it refers to – there is no space between the name and the number.

 Worked example: Writing systematic names

What is the systematic name for $KMnO_4$?

Step 1: Identify the name of the compound.

$KMnO_4$ is potassium manganate.

Activity ES 2.1

This activity helps you learn the rules for assigning oxidation states.

▲ **Figure 2** *Two oxides of copper – copper(II) oxide (left) and copper(I) oxide (right)*

▼ **Table 2** *Roman numerals*

Number	Roman numeral
1	I
2	II
3	III
4	IV
5	V
6	VI
7	VII
8	VIII

Step 2: Identify the element that has a variable oxidation state.

Manganese, like iron, is a transition metal. Transition metals have variable oxidation states, so the Roman numeral sits next to the manganate.

Step 3: Work out the oxidation state of manganese.

Using the rules from Table 1, the oxidation state of oxygen is −2.

$$(-2) \times 4 = -8$$

The oxidation state of potassium, a Group 1 metal, is +1.

The overall charge of $KMnO_4$ is 0.

$$0 - (-8) - (+1) = +7$$

Step 4: Write out the systematic name for $KMnO_4$.

The Roman numeral for seven is VII (Table 2), so the systematic name must be potassium manganate(VII).

Naming oxyanions

An **oxyanion** is a negative ion with oxygen in. The name always ends in –ate to show that oxygen is present. The nitrate(V) ion has the formula NO_3^- and the nitrate(III) ion has the formula NO_2^-. Oxyanions need to have some indication of the oxidation state of the element other than oxygen.

Chlorine, bromine, and iodine usually have the oxidation state −1 but not when they are in oxyanions. Table 3 shows the oxidation state of chlorine in different chlorate ions.

▼ **Table 3** *Oxidation state of chlorine in different chlorate ions*

Ion	Oxidation state of clorine	Name of ion
ClO^-	+1	chlorate(I)
ClO_2^-	+3	chlorate(III)
ClO_3^-	+5	chlorate(V)
ClO_4^-	+7	chlorate(VII)

Study tip

Be careful when adding oxidation states. Although there are two I⁻ ions in the example, it is the oxidation state on an individual iodine that is required, so it is just −1.

Using oxidation states to decide what has been oxidised and what has been reduced

Oxidation states can be used to find what has been oxidised and what has been reduced in a redox reaction. Look at the below reaction and allocate oxidation states for each element.

$$Cl_2(aq) + 2I^-(aq) \rightarrow 2Cl^-(aq) + I_2(aq)$$

Cl	0	−1	decrease in oxidation state, Cl is reduced
I	−1	0	increase in oxidation state, I is oxidised

Here the Cl_2 is the oxidising agent and the I⁻ is the oxidising agent.

Using oxidation states to help balance equations

Sometimes **stoichiometric equations** (balanced equations) are easy to balance but sometimes they are very difficult. If you are trying to balance a redox reaction then using oxidation states can speed things up. In a redox reaction, the number of electrons lost must equal the number of electrons gained.

 Worked examples: Balancing equations using oxidation states

Example 1

What is the balanced equation for the following reaction?

$$Na + Mg^{2+} \rightarrow Na^+ + Mg$$

Step 1: Identify how the oxidation state for each element changes in the reaction.

$$Na + Mg^{2+} \rightarrow Na^+ + Mg$$

Na	0	+1	loses 1e⁻ so oxidised
Mg	+2	0	gains 2e⁻ so reduced

Step 2: Balance the equation so the number of electrons lost is equal to the electrons gained.

For every Mg^{2+} that gains two electrons, **two** Na must lose one electron.

$$2Na + Mg^{2+} \rightarrow 2Na^+ + Mg$$

Example 2

What is the balanced equation for the following reaction?

$$H^+ + I^- + H_2SO_4 \rightarrow H_2S + H_2O + I_2$$

Step 1: Identify how the oxidation state for each element changes in the reaction.

$$H^+ + I^- + H_2SO_4 \rightarrow H_2S + H_2O + I_2$$

H	+1	+1	+1	+1	does not change
I	−1				0 loses one electron so oxidised
S		+6	−2		gains eight electrons so reduced
O		−2		−2	does not change

Step 2: Balance the equation so the number of electrons lost is equal to the electrons gained.

One H_2SO_4 molecule gains eight electrons from **eight** I^- ions, forming **four** I_2 molecules and one H_2S

$$H^+ + 8I^- + H_2SO_4 \rightarrow H_2S + H_2O + 4I_2$$

Activity ES 2.2

In this activity you can carry out a number of test-tube redox reactions and check your understanding of them.

Step 3: Balance the number of hydrogens and oxygens by adjusting the number of H^+ ions and H_2O molecules.

The elements in these chemicals have been neither oxidised nor reduced. You cannot change the number of I^-, H_2SO_4, H_2S, or I_2 because they were involved in the loss or gain of electrons.

There are four oxygen atoms in H_2SO_4, so there must be four H_2O molecules.

$$H^+ + 8I^- + H_2SO_4 \rightarrow H_2S + 4H_2O + 4I_2$$

Therefore, there must be **eight** H^+ ions to balance the equation.

$$8H^+ + 8I^- + H_2SO_4 \rightarrow H_2S + 4H_2O + 4I_2$$

Summary questions

1 a Insert electrons, e^-, on the appropriate side of the following half-equations in order to balance and complete them, so that the electrical charges on both sides are equal.

   i $K \rightarrow K^+$

   ii $H_2 \rightarrow 2H^+$

   iii $O \rightarrow O^{2-}$

   iv $Cr^{3+} \rightarrow Cr^{2+}$ (4 marks)

   b Identify whether each process is oxidation or reduction.   (4 marks)

2 Write down the oxidation states of the elements in the following examples.

   a Ag^+ (1 mark)

   b Al_2O_3 (1 mark)

   c SO_4^{2-} (1 mark)

   d P_4 (1 mark)

   e SF_6 (1 mark)

   f PO_4^{3-} (1 mark)

3 Work out the oxidation state of the chlorine in each of these species.

   a ClO_2 (1 mark)

   b $HClO_4$ (1 mark)

   c $MgCl_2$ (1 mark)

   d Cl_2O_7 (1 mark)

   e HCl (1 mark)

   f Cl_2O (1 mark)

4 a In the process described for the manufacture of bromine from Dead Sea water, bromine is separated from other materials involved in the process. Which properties of bromine make it possible to separate it from:

   i water

   ii chlorine.   (4 marks) b Write an ionic equation with state symbols for the reaction of chlorine gas with aqueous bromide ions to produce aqueous chloride ions and bromine liquid. *(2 marks)*

c In the production of 1.0 tonne of bromine, what mass of chlorine is required in tonnes? Give your answer to 1 d.p. *(2 marks)*

d In the production of 5.0 g of bromine, what volume of chlorine is required at RTP? Give your answer to 2 s.f. and in dm^3. The volume of one mole of gas at room temperature and pressure is 24.0 dm^3. *(3 marks)*

5 a Some reactions of the halogens are shown below – they are all examples of redox reactions. In each case state which element is oxidised and which is reduced, and give the oxidation states of *each* atom or ion before and after the reaction.
  i $H_2 + Cl_2 \longrightarrow 2HCl$
  ii $2FeCl_2 + Cl_2 \longrightarrow 2FeCl_3$
  iii $2H_2O + 2F_2 \longrightarrow 4HF + O_2$ *(6 marks)*

 b For each of the redox reactions in part **a** identify by formula the:
  i oxidising agent
  ii reducing agent. *(6 marks)*

6 Use oxidation states to help you balance the following redox reactions.
 a $Br^- + H^+ + H_2SO_4 \longrightarrow Br_2 + SO_2 + H_2O$ *(2 marks)*
 b $I^- + H^+ + H_2SO_4 \longrightarrow I_2 + H_2S + H_2O$ *(2 marks)*

7 Use oxidation states to name the following ions and compounds.
 a SnO_2 *(1 mark)*
 b $FeCl_2$ *(1 mark)*
 c NO_3^- *(1 mark)*
 d $PbCl_4$ *(1 mark)*
 e $Mn(OH)_2$ *(1 mark)*
 f CrO_4^{2-} *(1 mark)*
 g VO_3^- *(1 mark)*
 h SO_3^{2-} *(1 mark)*

8 Write formulae for the following compounds. In each case, the negative ion has a charge of −1.
 a potassium chlorate(III) *(1 mark)*
 b sodium chlorate(V) *(1 mark)*
 c iron(III) hydroxide *(1 mark)*
 d copper(II) nitrate(V) *(1 mark)*

ES 3 Manufacturing chlorine

Specification reference: ES(c)

Learning outcomes

Demonstrate and apply knowledge and understanding of:

→ techniques and procedures in the electrolysis of aqueous solutions; half-equations for the processes occurring at electrodes in electrolysis of molten salts and aqueous solutions:

  - formation of oxygen or a halogen or metal ions at the anode

  - formation of hydrogen or a metal at the cathode.

Chlorine is needed for the extraction of bromine from Dead Sea water. The human demand for chlorine is high and it is manufactured worldwide on an enormous scale. Chlorine can be made by **electrolysis** (decomposing a compound using an electric current) of a concentrated solution of sodium chloride (brine).

The electrolysis of brine

Although the sea contains high concentrations of sodium chloride, rock salt is an even better source of sodium chloride. Rock salt can be recovered either by underground mining or by pumping water into the salt and collecting the salt solution at the surface. As well as chlorine, the electrolysis of brine generates hydrogen and sodium hydroxide (an alkali). As chlorine manufacture and sodium hydroxide manufacture are directly linked, the production of these chemicals is often referred to as the chlor–alkali industry.

Often in chemical reactions, additional products are made as well as the product you want. These are called **co-products**. The co-products, sodium hydroxide and hydrogen, can be sold and this helps to reduce the waste, as well as increasing profitability. The membrane cell is the most modern of the electrolysis methods for producing chlorine and has been used since the 1980s (Figure 1).

The half-equations involved in the electrolysis of sodium chloride solution are:

- at the positive electrode $2Cl^-(aq) \rightarrow Cl_2(aq) + 2e^-$
- at the negative electrode $2H_2O(l) + 2e^- \rightarrow 2OH^-(aq) + H_2(g)$

The equation representing the overall reaction occurring in the cell is:

$$2Cl^-(aq) + 2H_2O(l) \rightarrow Cl_2(aq) + 2OH^-(aq) + H_2(g)$$

Extracting iodine from seaweed

Just below the surface of the seas off the coast of Scotland lie hidden forests of a seaweed – kelp. This has historically been a valuable source of organic matter that local people used to fertilise their soil. In the early part of the nineteenth century, Napoleon Bonaparte was looking for a source of nitrate for making explosives. In 1811, the French entrepreneur Bernard Courtois responded to Napoleon's demands by attempting to make nitrate from rotting seaweed. He noticed some curious purple fumes and, by chance, discovered another of the halogens – iodine. This led to a whole industry developing in the coastal areas of Scotland.

The extraction of iodine from kelp can be carried out in the laboratory.

▲ **Figure 2** A mechanical seaweed harvester as used by Hebridean Seaweed Ltd off the north coast of Scotland

1. Heat the seaweed strongly on a tin lid using a blue Bunsen burner flame until only a small quantity of ash remains.
2. Boil the ash with distilled water. Filter whilst still hot and allow the filtrate to cool.
3. Add dilute sulfuric acid and then hydrogen peroxide solution to the filtrate. A deep brown coloured solution of iodine will be seen.
4. Transfer the mixture to a separating funnel and add cyclohexane. Shake for 30 seconds then clamp the separating funnel and allow the two layers to separate.
5. Discard the lower aqueous layer then run the upper cyclohexane layer into an evaporating dish. Leave this in a fume cupboard so that the cyclohexane evaporates leaving iodine crystals behind.

1. Why does the seaweed have to be heated strongly before adding the acid and hydrogen peroxide?
2. The filtrate contains chloride and bromide ions as well as iodide ions. Why are the iodide ions preferentially oxidised by the hydrogen peroxide?
3. Write a balanced equation with state symbols for the redox reaction which occurs between the iodide ions and the acidified hydrogen peroxide.
4. What colour will the upper cyclohexane layer be?
5. What does the final step tell you about the relative volatilities of iodine and cyclohexane?

Chemical ideas: Redox reactions 9.2

Electrolysis as redox reactions

When electricity is passed through a molten or aqueous ionic compound, the compound is broken down and the process is called electrolysis. The charged ions are free to move to the oppositely charged electrode and create a complete circuit.

Electrolysis of molten compounds

Solid ionic compounds do not conduct electricity since ions in a solid are not free to move. If the ionic compound is melted then the charged ions are free to move and carry a current. Electrons are lost or gained by ions at the **electrodes**.

Figure 3 shows the electrolysis of molten lead bromide to produce lead and bromine. The positive lead ions **migrate** to the negative electrode, called the **cathode**. The Pb^{2+} ions gain two electrons to form lead atoms. Since the lead bromide is hot, molten lead metal collects at the bottom of the container. Bromide ions have a negative charge and migrate to the positive electrode or **anode**. At the temperature needed to melt the lead bromide, bubbles of bromine gas will be seen. The molten lead bromide has been broken down into its elements.

Synoptic link
You found out about the structure of ionic lattices in Chapter 1, Elements of life.

Study tip
Cations migrate to the cathode.
Anions migrate to the anode.

cathode reaction: $Pb^{2+}(l) + 2e^- \rightarrow Pb(l)$ The lead ion gains electrons and is reduced.

anode reaction: $2Br^-(l) \rightarrow Br_2(g) + 2e^-$ The bromide ions lose electrons and are oxidised.

Study tip

Reduction occurs at the cathode.

Oxidation occurs at the anode.

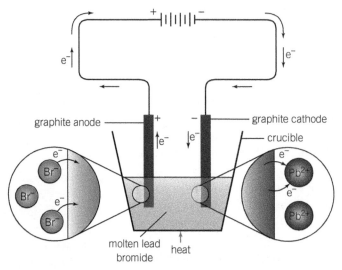

▲ **Figure 3** *Electrolysis of molten lead bromide*

It is easy to predict the products of the electrolysis of molten salt:

● product at cathode – metal

● product at anode – non-metal, apart from hydrogen.

Electrolysis of solutions

The electrolysis of solutions is much easier to carry out than electrolysis of molten compounds. When an ionic compound dissolves, the ions become free to move and carry the current. Figure 4 shows the apparatus that can be used to electrolyse a solution and collect any gases made so they can be identified.

Water can take part

When you electrolyse salt solutions, predicting the products at the electrodes is more difficult. In the electrolysis of molten salts there was no competition at the electrodes. In the electrolysis of solutions, water competes with the ions from the salt.

Water can be reduced at the cathode:

$$2H_2O(l) + 2e^- \rightarrow 2OH^-(aq) + H_2(g)$$ Gain of electrons so reduction.

Water can also be oxidised at the anode:

$$2H_2O(l) \rightarrow O_2(g) + 4H^+(aq) + 4e^-$$ Loss of electrons so oxidation.

Reduction at the cathode

At the cathode there will be both metal ions from the salt and the water the salt is dissolved in. More reactive metals such as sodium or potassium remain as ions and so hydrogen gas is produced by reduction of the water. Less reactive metals such as copper and zinc are plated (deposited) on the cathode.

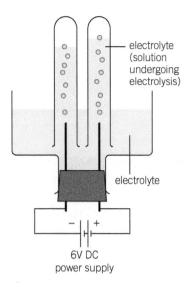

▲ **Figure 4** *Apparatus used for electrolysis of solutions*

Activity ES 3.1

In this activity you can carry out and check your understanding of electrolysis of aqueous solutions.

Oxidation at the anode

At the anode there will be negative ions from the salt and the water the salt is dissolved in. Halide ions have a greater tendency to be oxidised (or lose electrons) than water and so the halogen is produced at the anode. Other negative ions such as sulfates and nitrates have less tendency to be oxidised than water and so oxygen gas is produced at the anode.

There is another possibility at the anode. Usually in electrolysis the electrodes are themselves unreactive. Graphite and platinum are commonly used. A special case is when the anode and the metal ions in solution are the same. In the electrolysis of a copper compound with a copper anode, the metal anode loses mass because copper atoms change to copper ions and go into solution. An example would be the electrolysis of copper sulfate solution using copper electrodes.

▲ **Figure 5** *A copper plated oak leaf*

At the anode, oxidation occurs since electrons are lost by the copper electrode:

$$Cu(s) \rightarrow Cu^{2+}(aq) + 2e^-$$

At the cathode, copper is deposited and the concentration of the solution remains constant:

$$Cu^{2+}(aq) + 2e^- \rightarrow Cu(s)$$

Summary

You need to be able to predict the products of the electrolysis of solutions and be able to write half-equations. The information needed is summarised in Table 1.

> **Synoptic link** 🧪
>
> A summary of how to carry out electrolysis of aqueous solutions can be found in Techniques and procedures.

▼ **Table 1** *Products at electrodes from electrolysis of solutions*

Product at cathode (negative electrode)	Product at anode (positive electrode)	Lost at anode (positive electrode)
All electrodes	Unreactive electrode, e.g., graphite or platinum	Reactive anode, e.g., copper in $CuSO_4$(aq)
Reduction	Oxidation	Oxidation
Hydrogen if the metal in the salt comes from Group 1 or 2 or is aluminium $2H_2O(l) + 2e^- \rightarrow 2OH^-(aq) + H_2(g)$	Halogen if the salt is a halide $2Cl^-(aq) \rightarrow Cl_2(g) + 2e^-$	$Cu(s) \rightarrow Cu^{2+}(aq) + 2e^-$
Metal for all other salts $Cu^{2+}(aq) + 2e^- \rightarrow Cu(s)$	Oxygen if the salt is a sulfate or nitrate $2H_2O(l) \rightarrow O_2(g) + 4H^+(aq) + 4e^-$	
Hydrogen also made on electrolysis of acids, e.g., sulphuric acid $2H^+(aq) + 2e^- \rightarrow H_2(g)$	Oxygen also made on electrolysis of hydroxides, e.g., sodium hydroxide $4OH^-(aq) \rightarrow O_2(g) + 2H_2O(l) + 4e^-$	

Worked example: Half-equations for the electrolysis of zinc chloride

Write the half-equations for the reactions at the cathode and anode in the electrolysis of zinc chloride solution with graphite electrodes. Decide what is oxidised and what is reduced.

Step 1: Write the formula for zinc chloride – $ZnCl_2$

Step 2: Identify the cation – zinc, Zn^{2+}

Step 3: Identify the products of the reaction at the cathode – zinc is not in Group 1 or 2 (and is not aluminium) so hydrogen won't be produced.

Step 4: Write the half-equation for the reaction at the cathode.

Oxidation state of zinc in zinc chloride is +2. Zinc ions will *gain* two electrons at the cathode. Zinc is reduced.

$$Zn^{2+}(aq) + 2e^- \rightarrow Zn(s)$$

Step 5: Identify the anion – chloride, Cl^-

Step 6: Identify the products of the reaction at the anode – chloride is a halide so chlorine gas will be produced.

Step 7: Write the half-equation for the reaction at the anode.

Each chloride ion *loses* an electron to become an atom that pairs up with another atom to become a diatomic molecule. Chloride is oxidised.

$$2Cl^-(aq) \rightarrow Cl_2(g)$$

 Worked example: Half-equations for the electrolysis of zinc nitrate

Write the half-equations for the reactions at the cathode and anode in the electrolysis of zinc nitrate solution with zinc electrodes. Decide what is oxidised and what is reduced.

Step 1: Identify the products of the reaction at the cathode – Zinc is produced rather than hydrogen since the metal less reactive.

Step 2: Write the half-equation for the reaction at the cathode.

Each zinc ion gains two electrons to become an atom. The Zn^{2+} is reduced because it gains electrons.

$$Zn^{2+}(aq) + 2e^- \rightarrow Zn(s)$$

Step 3: Identify the products of the reaction at the anode – zinc electrode in a solution of a zinc compound so zinc atoms change into ions.

Step 4: Write the half-equation for the reaction at the anode.

Zinc atoms lose electrons to become zinc ions that go into solution. Zinc is oxidised because electrons are lost.

$$Zn(s) \rightarrow Zn^{2+}(aq) + 2e^-$$

Summary questions

1 Predict the products at the cathode and anode in the electrolysis of these molten compounds.
   a lead bromide   (1 mark)
   b sodium chloride   (1 mark)
   c zinc iodide.   (1 mark)

2 Look at the equation for the electrolysis of sodium chloride:

$$2Cl^-(aq) + 2H_2O(l) \rightarrow Cl_2(aq) + 2OH^-(aq) + H_2(g)$$

   a Calculate the amount (in moles) of sodium hydroxide, NaOH, in 1 tonne of solid sodium hydroxide.
     1 tonne is 1 000 000 g.   (3 marks)
   b What amount (in moles) of chlorine, Cl_2, is produced for each mole of NaOH?   (2 marks)
   c Calculate the mass of chlorine produced at the same time as 1 tonne of sodium hydroxide.   (2 marks)

3 Predict the products at the anode and cathode if the following solutions were electrolysed using the named electrodes.
   a sodium bromide with graphite electrodes   (2 marks)
   b aluminium nitrate with graphite electrodes   (2 marks)
   c zinc bromide with graphite electrodes.   (2 marks)

4 Write the half-equations for the cathode and anode in the electrolysis below. Say if they are reduction or oxidation.
   a zinc bromide solution with graphite electrodes   (2 marks)
   b sodium bromide solution with graphite electrodes   (2 marks)
   c sodium hydroxide solution with graphite electrodes   (2 marks)
   d nitric acid with platinum electrodes   (2 marks)
   e copper nitrate solution with copper electrodes.   (2 marks)

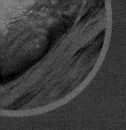

ES 4 From extracting bromine to making bleach

Specification reference: ES(o), ES(p), ES(q)

Demonstrate and apply knowledge and understanding of:

→ the characteristics of dynamic equilibrium

→ the equilibrium constant K_c for a given homogeneous reaction; calculations of the magnitude of K_c using equilibrium concentrations; relation of position of equilibrium to size of K_c, using symbols such as $>, <, >>, <<$

→ the use of K_c to explain the effect of changing concentrations on the position of a homogeneous equilibrium.

As well as its use in extracting bromine from sea water, chlorine is used to make bleaches. Chlorine gas is passed through a cold solution of sodium hydroxide. The sodium hydroxide solution reacts with the chlorine to form sodium chlorate(I).

$$Cl_2(aq) + 2NaOH(aq) \rightleftharpoons NaCl(aq) + NaOCl(aq) + H_2O(l)$$

In order to understand what the symbol $\rightleftharpoons$ means, you need to learn about equilibrium.

Chemical ideas: Equilibrium in chemistry 7.1

Chemical equilibrium

The general meaning of the term equilibrium is a state of balance where nothing changes. For example, a see-saw with two people of equal mass, sitting one on each side, is in a state of equilibrium. In chemistry, a state of equilibrium is also a state of balance, but it has a special feature – chemical equilibrium is **dynamic equilibrium**.

Why is equilibrium dynamic?

Figure 1 shows a sealed bottle of bromine. The bottle and its contents make up a **closed system**. Nothing can enter or leave the bottle. In this system, bromine is present in two states – as a liquid, $Br_2(l)$, and as a gas above the liquid, $Br_2(g)$.

When the bottle has been standing at a steady temperature for some time the depth of the orange colour above the liquid remains constant. The mass of bromine that is a gas and the mass of bromine that is a liquid in the closed system is constant.

▲ **Figure 1** *Bromine liquid and gas at equilibrium*

The system is at equilibrium and nothing appears to change on the macroscopic scale – the scale that you can see.

If you were able to see the individual molecules – the microscopic scale – the picture would be rather different. In the gas, the molecules are constantly moving rapidly in random directions, and inevitably collide with the molecules on the surface of the liquid (Figure 2). Some bounce back into the gas phase, but some enter the liquid phase.

At the same time, the molecules of bromine in the liquid are also constantly moving around, colliding with each other. Near the surface of the liquid, some of these molecules escape into the gaseous phase (Figure 2).

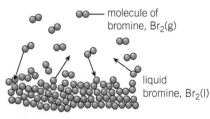

molecule of bromine, $Br_2(g)$

liquid bromine, $Br_2(l)$

▲ **Figure 2** *The bromine equilibrium on a microscopic scale*

There are molecules entering the liquid phase and molecules leaving the liquid phase – it is a **reversible change**. When the system is at equilibrium, the molecules enter and leave at the same rate. On the macroscopic scale it seems as though nothing is changing, but on the molecular scale molecules are constantly moving from one phase to the other. That is why the situation is described as dynamic equilibrium. It can be represented by the equation:

$$Br_2(g) \rightleftharpoons Br_2(l)$$

The $\rightleftharpoons$ sign represents dynamic equilibrium.

The road to equilibrium

Many chemical reactions are reversible reactions, for example, the reaction between hydrogen and iodine to produce hydrogen iodide.

$$H_2(g) \quad + \quad I_2(g) \quad \rightleftharpoons \quad 2HI(g)$$

colourless   purple   colourless

When reactions are represented using the $\rightleftharpoons$ symbol the reaction going from left to right is known as the **forward reaction** and the reaction going in the opposite direction is known as the **reverse reaction**. Even though the reaction is reversible, the substances on the left of the equilibrium sign are called the reactants and those on the right are called the products.

If the colourless hydrogen and the purple iodine are mixed in a closed container at 731 K, the purple colour becomes paler as the iodine reacts. The purple colour does not disappear and after a while the depth of the purple colour will stay constant (Figure 3).

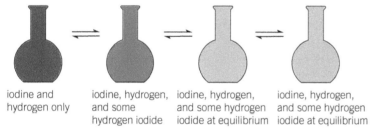

iodine and   iodine, hydrogen,   iodine, hydrogen,   iodine, hydrogen,
hydrogen only   and some   and some hydrogen   and some hydrogen
                 hydrogen iodide   iodide at equilibrium   iodide at equilibrium

▲ **Figure 3** *Changes in appearance of an iodine and hydrogen mixture until equilibrium is reached*

The rate of the forward reaction will decrease as hydrogen and iodine react to make hydrogen iodide and their concentrations decrease (Figure 4). The rate of the reverse reaction will be zero on first mixing the hydrogen and iodine because there will be no hydrogen iodide. As more hydrogen iodide is made, the rate of the reverse reaction increases because the hydrogen iodide concentration increases.

The system has reached equilibrium when the rates of the forward and reverse reactions are the same. Hydrogen, iodine, and hydrogen iodide are all present and their concentrations remain constant once equilibrium has been reached. Reactants are turning into products at the same rate as products are turning into reactants.

Activity ES 4.1

In this activity you can model a dynamic equilibrium.

Study tip

When talking about an equilibrium, it is important to state which reaction you are referring to. The easiest way to do this is to write an equation.

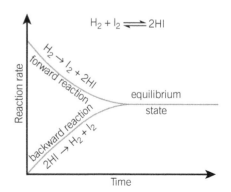

▲ **Figure 4** *Changes in the rate of reaction for the forward and reverse reaction*

So the full definition of dynamic equilibrium is:

● concentrations of reactants and products stay constant

● forward and reverse reactions are both happening (so dynamic)

● the rate of the forward and reverse reactions are equal to each other.

Although the concentrations of the reactants and products are constant, it is not true to say that the concentrations are the same at equilibrium. In Figure 5a, hydrogen and iodine were placed in a closed flask and allowed to reach equilibrium. The concentration of hydrogen iodide is greater than the concentration of the hydrogen and iodine at equilibrium. All the concentrations become constant at equilibrium as shown by the horizontal lines but the concentrations of reactants and products are not the same. In Figure 5b only hydrogen iodide was placed in the closed flask. At equilibrium the concentrations of reactants and products stay constant but are not the same.

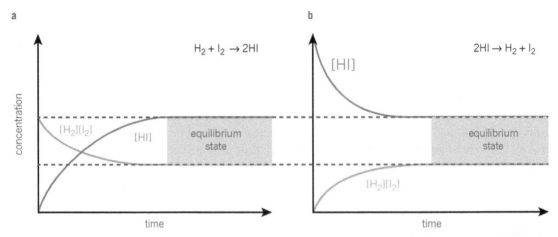

▲ **Figure 5** *Changes in concentration of a hydrogen and iodine mixture until equilibrium has been reached*

Figure 5 shows that under the same reaction conditions, the same equilibrium position is reached when the reaction starts with the products – hydrogen iodide – or the reactants – hydrogen and iodine. You cannot tell them apart.

Chemical ideas: Equilibrium in chemistry 7.2

The equilibrium constant K_c

Table 1 shows results obtained for the reaction between hydrogen and iodine to form hydrogen iodide. In the first three experiments, mixtures of hydrogen and iodine were put into sealed reaction vessels. In the final two experiments hydrogen iodide alone was sealed into the vessel. The mixtures were held at a constant temperature of 731 K until equilibrium was reached. The equilibrium concentrations of all three substances were then recorded.

▼ **Table 1** *Initial and equilibrium concentrations for the reaction $H_2(g) + I_2(g) \rightleftharpoons 2HI(g)$*

Experiment	Initial concentrations / mol dm⁻³			Equilibrium concentrations / mol dm⁻³			$K_c = \dfrac{[HI][HI]}{[H_2][I_2]}$
	$[H_2(g)]$	$[I_2(g)]$	$[HI(g)]$	$[H_2(g)]$	$[I_2(g)]$	$[HI(g)]$	
1	2.40×10^{-2}	1.38×10^{-2}	0.00	1.140×10^{-2}	0.120×10^{-2}	2.52×10^{-2}	46.4
2	2.44×10^{-2}	1.98×10^{-2}	0.00	0.770×10^{-2}	0.310×10^{-2}	3.34×10^{-2}	46.7
3	2.46×10^{-2}	1.76×10^{-2}	0.00	0.920×10^{-2}	0.220×10^{-2}	3.08×10^{-2}	46.9
4	0.00	0.00	3.04×10^{-2}	0.345×10^{-2}	0.345×10^{-2}	2.35×10^{-2}	46.9
5	0.00	0.00	7.58×10^{-2}	0.860×10^{-2}	0.860×10^{-2}	5.86×10^{-2}	46.4

For this reaction (looking at the last column) it was found that the following was constant:

$$\frac{[HI][HI]}{[H_2][I_2]}$$

This is the same as $\dfrac{[HI]^2}{[H_2][I_2]}$

This constant is called the **equilibrium constant K_c**. The square brackets [ ] mean concentration of.

The values of K_c are shown in Table 1. The equilibrium constant is the same whether the starting reaction mixture is $H_2 + I_2$ (Experiments 1 to 3) or HI (Experiments 4 and 5). The mean value of K_c for this reaction at 731 K is about 46.7.

K_c is greater than 1. This tells us that there must be more products than reactants – the top line must have a larger numerical value than the bottom line to calculate a value of 46.7 for K_c. The position of equilibrium lies to the products or to the right-hand side of the reaction. Since $K_c > 1$ (greater than 1), this means that, at equilibrium, most of the H_2 and I_2 has been converted to HI, but not all.

Writing the expression for K_c

The rules for writing K_c expressions have been discovered by using the results of many experiments. This makes it possible to write an expression for K_c for any reaction, without having to examine data. In the expression for K_c, the products of the forward reaction appear on the top line and the reactants on the bottom line. The power to which you raise the concentration of a substance is the same as the number which appears in front of it in the balanced equation.

In general, if an equilibrium mixture contains substances A, B, C, and D that react according to the equation:

$$aA + bB \rightleftharpoons cC + dD$$

then the expression for K_c is $K_c = \dfrac{[C]^c [D]^d}{[A]^a [B]^b}$

Once you have written the expression for K_c using the stoichiometric equation, use the concentrations of the substances to calculate a value for K_c.

Synoptic link

Acid–base equilibria are looked at in detail in Chapter 4, The ozone story.

Study tip

Always write a balanced stochiometric equation for a reaction before writing the equation for K_c. You will then be able to easily see the superscript values for K_c.

⊞ Worked example: Calculating K_c

Calculate the value for K_c (to 3 s.f.) for the reaction between nitrogen and hydrogen at 1000 K given the following equilibrium concentrations:

$$[N_2] = 0.142 \, mol \, dm^{-3}$$

$$[H_2] = 1.84 \, mol \, dm^{-3}$$

$$[NH_3] = 1.36 \, mol \, dm^{-3}$$

Step 1: Write the stoichiometric equation for the reaction.

$$N_2 + 3H_2 \rightleftharpoons 2NH_3$$

Step 2: Use the stoichiometric equation to write an expression for K_c.

$$K_c = \frac{[NH_3]^2}{[N_2][H_2]^3}$$

Step 3: Use the concentration values to calculate K_c.

$$K_c = \frac{(1.36)^2}{(0.142) \times (1.84)^3} = 2.09 \, mol^{-2} \, dm^6 \text{ at } 1000 \, k$$

Synoptic link

You will learn about units of K_c in Chapter 6, The chemical industry.

The size of K_c

Values of K_c vary enormously. Table 2 shows some values for two reactions at the same temperature. K_c is temperature-dependent so temperature is quoted alongside the equilibrium constant.

▼ **Table 2** *Some values of K_c*

Reaction	K_c
$H_2(g) + Br_2(g) \rightleftharpoons 2HBr(g)$	1010 at 550 K
$2H_2(g) + S_2(g) \rightleftharpoons 2H_2S(g)$	0.000 094 at 1020 K

For the reaction of hydrogen with bromine, the top line in the K_c relationship is greater than the bottom line. At equilibrium there are more products than reactants.

$$K_c = \frac{[HBr]^2}{[H_2][Br_2]}$$

For the reaction of hydrogen with sulfur, the top line in the K_c relationship is much smaller than the bottom line. At equilibrium there are more reactants than products.

$$K_c = \frac{[H_2S]^2}{[H_2]^2[S_2]}$$

All reactions are equilibrium reactions and even reactions that seem to go to completion actually have a little bit of reactant left in equilibrium with the product.

Study tip

$K_c > 1$ means there are more products than reactants at equilibrium.

$K_c \gg 1$ (greater than 10^{10}) means that the reaction appears to have gone to completion.

$K_c < 1$ means that there are more reactants than products.

$K_c \ll 1$ (less than 10^{-10}) means that the reaction appears not to have happened.

Chemical ideas: Equilibrium in chemistry 7.3

K_c and changed in concentration

Suppose a system is at equilibrium and you suddenly disturb it by adding more of a reagent. The composition of the system will change until equilibrium is reached again. The composition of the mixture will always adjust to keep the value of K_c constant, provided the *temperature stays constant.*

For example, in an experiment involving the formation of ethyl ethanoate the system was allowed to reach equilibrium. The equilibrium concentrations are shown in Table 3.

$$CH_3COOH(l) + C_2H_5OH(l) \rightleftharpoons CH_3COOC_2H_5(l) + H_2O(l)$$

▼ **Table 3** *Equilibrium is set up in Experiment 1, starting with equal concentrations of ethanoic acid and ethanol. In Experiment 2, the equilibrium is disturbed by adding extra ethanol. Both experiments are carried out at 298 K*

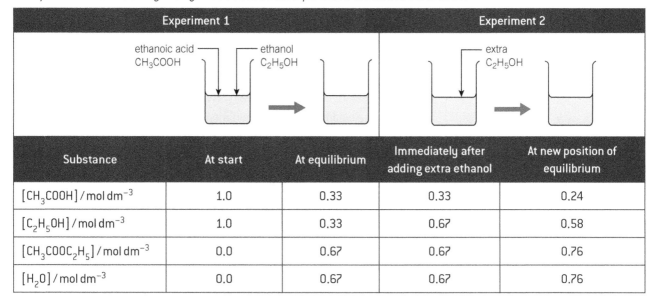

Substance	At start	At equilibrium	Immediately after adding extra ethanol	At new position of equilibrium
$[CH_3COOH]$ / mol dm^{-3}	1.0	0.33	0.33	0.24
$[C_2H_5OH]$ / mol dm^{-3}	1.0	0.33	0.67	0.58
$[CH_3COOC_2H_5]$ / mol dm^{-3}	0.0	0.67	0.67	0.76
$[H_2O]$ / mol dm^{-3}	0.0	0.67	0.67	0.76

Using the equilibrium concentrations from Experiment 1:

$$K_c = \frac{[CH_3COOC_2H_5][H_2O]}{[CH_3COOH][C_2H_5OH]}$$

$$= \frac{(0.67\,mol\,dm^{-3})(0.67\,mol\,dm^{-3})}{(0.33\,mol\,dm^{-3})(0.33\,mol\,cm^{-3})} = 4.1 \text{ at } 298\,K$$

In Experiment 2, one of the concentrations was deliberately changed by adding more C_2H_5OH to give a new concentration of $0.67\,mol\,dm^{-3}$. Immediately after adding the extra C_2H_5OH, before any changes occur, the new concentration ratio is:

$$\frac{(0.67\,mol\,dm^{-3})(0.67\,mol\,dm^{-3})}{(0.33\,mol\,dm^{-3})(0.67\,mol\,dm^{-3})} = 2.0$$

This value is smaller than K_c. In order to restore the value of K_c to 4.1 some C_2H_5OH and CH_3COOH must react (making the bottom line smaller) to produce $CH_3COOC_2H_5$ and H_2O (making the top line numerically larger). The system was left to reach equilibrium again, and the new equilibrium concentrations were measured.

$$K_c = \frac{(0.76\,mol\,dm^{-3})(0.76\,mol\,dm^{-3})}{(0.24\,mol\,dm^{-3})(0.58\,mol\,cm^{-3})} = 4.1 \text{ at } 298\,K$$

When the equilibrium was disturbed it moved in such a way that K_c remained constant. Some C_2H_5OH and CH_3COOH had to react to keep the K_c value at 4.1. More $CH_3COOC_2H_5$ and H_2O were produced. The equilibrium position had moved to the right, or moved to the product side to re-establish K_c.

Activity ES 4.2

In this activity you will be able to observe the effect of changing concentration on the position of equilibrium.

Study tip

Changing concentrations do not alter K_c once equilibrium as been reached, assuming the temperature remains constant.

Summary questions

1 Write expressions for K_c for the following reactions.
   a $2NO(g) + O_2(g) \rightleftharpoons 2NO_2(g)$ *(1 mark)*
   b $C_2H_6(g) \rightleftharpoons C_2H_4(g) + H_2(g)$ *(1 mark)*
   c $2HI(g) \rightleftharpoons H_2(g) + I_2(g)$ *(1 mark)*
   d $CO_2(aq) + H_2O(l) \rightleftharpoons HCO_3^-(aq) + H^+(aq)$ *(1 mark)*
   e $CH_3COOH(l) + C_3H_7OH(l) \rightleftharpoons CH_3COOC_3H_7(l) + H_2O(l)$ *(1 mark)*

2 The equilibrium constant K_c for a reaction is given by the expression:

$$K_c = \frac{[SO_3(g)]^2}{[SO_2(g)]^2\,[O_2(g)]}$$

   Write the balanced chemical equation for the reaction. *(1 mark)*

3 A mixture of nitrogen and hydrogen was sealed in a steel vessel and held at 1000 K until equilibrium was reached. The contents were then analysed. The results are given in the following table.

Substance	Equilibrium concentration / mol dm^{-3}
$N_2(g)$	0.142
$H_2(g)$	1.840
$NH_3(g)$	1.360

   a Write an expression for K_c for the reaction:
$$N_2(g) + 3H_2(g) \rightleftharpoons 2NH_3(g)$$ *(1 mark)*
   b Calculate a value for K_c. *(2 marks)*

4 When PCl_5 is heated in a sealed container and maintained at a constant temperature, an equilibrium is established. At 523 K, the following equilibrium concentrations were determined.

Substance	Equilibrium concentration / mol dm^{-3}
PCl_5	0.077
PCl_3	0.123
Cl_2	0.123

   a Write an expression for K_c for the reaction:
$$PCl_5(g) \rightleftharpoons PCl_3(g) + Cl_2(g)$$ *(1 mark)*
   b Calculate a value for K_c. *(2 marks)*

5 For the reaction of aqueous chloromethane with alkali the equilibrium constant has a value of 1×10^{16} at room temperature.

$$OH^-(aq) + CH_3Cl(aq) \rightleftharpoons CH_3OH(aq) + Cl^-(aq)$$

   What does this tell you about the concentration of chloromethane at equilibrium? *(2 marks)*

6 Use K_c to explain how the position of equilibrium would change if acid was added to the reaction between carbon dioxide and water.
$$CO_2(aq) + H_2O(l) \rightleftharpoons HCO_3^-(aq) + H^+(aq)$$ *(2 marks)*

ES 5 The risks and benefits of using chlorine

Specification reference: ES(f), ES(n)

There are a variety of risks associated with the production, storage, and transportation of chlorine, yet it makes a valuable contribution to improving our lives. The risks and benefits associated with toxic chlorine gas need to be weighed up before it is used.

Chlorine has a poor public image. It is associated with pollution – pollution of the land through pesticides that contain organochlorine compounds and pollution of the upper atmosphere through CFCs (chlorofluorocarbons). Chlorine has also been used as a poisonous gas. Both chlorine and phosgene, $COCl_2$, were used with deadly effect in the trenches in the First World War. They are also thought to have been used against civilian populations in recent years.

Learning outcomes

Demonstrate and apply knowledge and understanding of:

→ the risks associated with the storage and transport of chlorine; uses of chlorine which must be weighed against these risks, including: sterilising water by killing bacteria, bleaching

→ techniques and procedures in iodine–thiosulfate titrations.

Chemical ideas: Chemistry in industry 14.1

Risks and benefits of chlorine

Chlorine is a toxic gas detectable by smell at 1 part per million (ppm). Even in these small doses, chlorine can irritate the eyes, skin, and respiratory system. If inhaled at concentrations above 40 ppm, chlorine reacts in the lungs to form hydrochloric acid, HCl, which affects lung tissue and essentially causes drowning as liquid floods the lungs. Any leaks of chlorine during transport or storage cause danger for both workers and the general public.

Transporting chlorine

Although chlorine is usually prepared on site at the chemical plant requiring chlorine, this is not always the case and some chlorine is transported. Chlorine can be transported by road (Figure 1) or rail in specially designed pressurised tank containers. In some countries including the UK, a Hazchem warning plate (Figure 2) is attached to the tank during transport. In the event of an accident, this gives information for the fire brigade regarding what action is needed.

▲ **Figure 1** *A tank truck for transporting chlorine*

The chlorine is transported as a liquid as more chlorine can be stored in a fixed volume as a liquid under pressure than as a gas. If the temperature or pressure becomes too high, the tanks have pressure release devices designed to vent the tank and release some chlorine as gas. It is better to vent a small amount of chlorine to the atmosphere than to have a catastrophic explosion in case of a tank failure. Generally, the tanks are made and lined with steel. It is essential that the inside of the tank is dry as chlorine reacts with water to produce corrosive acids. Tanks have a cylindrical, protective housing at the top. This means that all loading and unloading is done through the protective housing at the top of the tank. Another safety feature

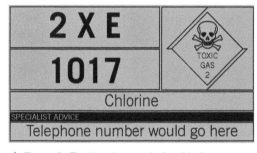

▲ **Figure 2** *The Hazchem code for chlorine*

Synoptic link

Aspects of the impact of the chemical industry on society are explored in Chapter 6, The chemical industry.

on large chlorine tanks is an excess flow valve, which is designed to close automatically if the angle valve which regulates the discharge of chlorine is broken or sheared off in the case of an accident in transport. It is activated if the discharge of liquid chlorine at the exit port exceeds some predetermined value.

Unloading chlorine

When chlorine is transferred on delivery from the rail tanker to a bulk trailer on site, a scrubber unit ensures that air being displaced from the bulk trailer has any chlorine removed from it (Figure 3). The scrubber has sodium hydroxide solution that reacts with the chlorine to produce sodium chlorate(I) – bleach. The bleach can be sold on.

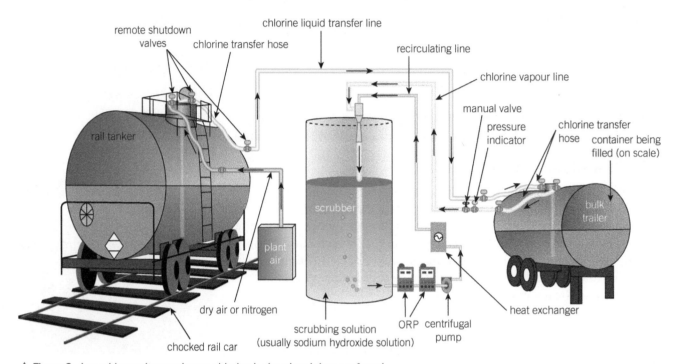

▲ **Figure 3** *A scrubber unit ensuring no chlorine leaks when it is transferred*

Storing chlorine

Chlorine may also be transported and then stored in cylinders. Workers at the chemical plant meet regulations regarding handling the cylinders by carefully moving the cylinders using a hoist to avoid damage to the outside (Figure 4). One method of routine checking of stored cylinders is to take a stick with cloth soaked in concentrated ammonia solution over the end. If a cylinder is leaking then a white cloud of ammonium chloride will be seen.

Uses and benefits of chlorine

In spite of the risks of its transport and storage, chlorine is used in many ways to make our lives safer and more comfortable. About 50 million tonnes of chlorine are produced worldwide annually. The best known use is in water treatment, where it is added to the water to

▲ **Figure 4** *Moving a chlorine cylinder carefully with a hoist*

kill bacteria and other pathogens. After its introduction for chlorination of water in the early twentieth century there was a rapid decline in the number of deaths from typhoid. Chlorine is also used in household bleach products, to kill bacteria on surfaces, or to remove stains from clothing. The bleach, which is an oxidising agent, removes stains by breaking bonds in coloured chemicals to form colourless products.

Determining the concentration of sodium chlorate in bleach

Chlorine reacts with sodium hydroxide to make sodium chlorate(I), NaClO. A solution of about 12% NaClO by mass is used in some water purifying plants to kill bacteria. A solution of about 5% is used in household bleach products. One way to determine the accurate concentration of sodium chlorate(I) is to carry out an iodine–thiosulfate titration.

Iodine–thiosulfate titrations

Not all titrations are acid–base reactions. Iodine–thiosulfate titrations involve redox reactions. They are used to find the concentration of a chemical that is a strong enough oxidising agent to oxidise iodide ions.

In the case of bleach, excess iodide ions are added to the chlorate(I) ions. The following redox reaction occurs:

$$ClO^- + 2I^- + 2H^+ \rightarrow I_2 + Cl^- + H_2O$$
$$\text{brown}$$

The iodine produced can be titrated using thiosulfate ions, $S_2O_3^{2-}$, in the following reaction:

$$2S_2O_3^{2-} + I_2 \rightarrow 2I^- + S_4O_6^{2-}$$
$$\text{brown} \quad \text{pale yellow}$$

The end point of the titration can be clearly identified by adding starch solution. The end point is determined when the final trace of blue/black colour is no longer visible.

> **Activity ES 5.1**
>
> In this activity you can work with other members of your class to investigate the cost-effectiveness of different comercially available bleaches.

> **Synoptic link**
>
> You met acid–base titrations in Chapter 1, Elements of life.

> **Synoptic link** 🧪
>
> Detail of how to carry out an iodine–thiosulfate titration can be found in Techniques and procedures.

 Worked example: The concentration of chlorate(I)

In a titration, 25.00 cm³ of sodium chlorate(I) solution was pipetted into a conical flask before excess potassium iodide and sulfuric acid were added. A 0.10 mol dm⁻³ solution of sodium thiosulfate was then run into the conical flask. The end point was reached when 22.0 cm³ of sodium thiosulfate had been added. Calculate the concentration of chlorate(I) ions in the original solution to three significant figures.

Step 1: Convert all cm³ readings to dm³

$$\frac{25.00\,\text{cm}^3}{1000} = 25.00 \times 10^{-3}\,\text{dm}^3$$

$$\frac{22.00\,\text{cm}^3}{1000} = 22.0 \times 10^{-3}\,\text{dm}^3$$

Step 2: Write the ionic equations for the titation.

$$ClO^- + 2I + 2H^+ \rightarrow I_2 + Cl^- + H_2O \quad \textbf{Equation 1}$$

$$I_2 + 2S_2O_3^{2-} \rightarrow 2I^- + S_4O_6^{2-} \qquad \textbf{Equation 2}$$

Step 3: Calculate the amount in moles of sodium thiosulfate.

$$\text{moles } S_2O_3^{2-} = \text{concentration } c \times \text{volume } v = 0.10 \times 22.00 \times 10^{-3} = 2.20 \times 10^{-3} \text{ moles}$$

Step 4: Calculate the amount in moles of iodine.

Looking at Equation 2, two moles of thiosulfate react with one mole of iodine. This is the same as one mole of thiosulfate reacting with 0.5 moles of iodine.

$$\text{moles } I_2 = 2.20 \times 10^{-3} \times 0.5 = 1.10 \times 10^{-3}$$

Step 5: Calculate the amount in moles of chlorate(I) ions.

Looking at Equation 1, one mole of chlorate ions produces one mole of iodine.

$$\text{moles } ClO^- = 1.10 \times 10^{-3}$$

Step 6: Calculate the concentration of chlorate(I) ions.

$$\text{concentration of } ClO^- = \frac{\text{moles } ClO^-}{\text{volume of } ClO^- \text{ used}} = \frac{1.10 \times 10^{-3}}{25.00 \times 10^{-3}} = 0.0440 \, \text{mol dm}^{-3}$$

Summary questions

1   Calculate the amount of sodium thiosulfate in the following solutions writing the answer in standard form and to an appropriate number of significant figures.
   a   20.0 cm³ of solution with a concentration of 1.00 mol dm⁻³. *(2 marks)*
   b   24.6 cm³ of solution with a concentration of 0.0100 mol dm³. *(2 marks)*

2   In a titration, 10.00 cm³ of sodium chlorate(I) solution was pipetted into a conical flask before excess potassium iodide and sulfuric acid were added. A 0.500 mol dm⁻³ solution of sodium thiosulfate was then run into the conical flask. The end point was reached when 11.2 cm³ of sodium thiosulfate had been added. Calculate the concentration of the sodium chlorate(I) solution to 3 s.f. *(3 marks)*

3   When chlorine gas reacts with water, it makes dilute hydrochloric acid and dilute chloric(I) acid, HCIO. This is a reversible reaction. Write a balanced equation with state symbols. *(2 marks)*

4   Household bleach is diluted by making 10.0 cm³ of bleach up to 100 cm³ in a volumetric flask. To a 10.0 cm³ aliquot of this, excess acid and potassium iodide were added. A mean titre of 9.80 cm³ of 0.0100 mol dm⁻³ sodium thiosulfate solution was required to change the starch indicator to colourless. Calculate the concentration of the undiluted bleach to 3 s.f. *(4 marks)*

ES 6 Hydrogen chloride in industry and the laboratory

Specification reference: ES(a), ES(l), ES(m)

Making hydrochloric acid

Hydrochloric acid can be made by a variety of methods. A simple method for its production is to start by making hydrogen chloride gas directly from the elements (Figure 1). This would be possible at a plant producing chlorine from brine by electrolysis, as described in Topic ES 3, as both chlorine and hydrogen are produced in the process.

$$H_2(g) + Cl_2(g) \rightarrow 2HCl(g)$$

This is a good example of an atom economy of 100%.

Chemical ideas: Chemistry in industry 14.2

Atom economy

These days, much more household waste is recycled, and less sent to landfill. The movement towards waste reduction is mirrored in the chemical industry, and is widely referred to as green chemistry. When deciding which reactions to use in a chemical plant, the percentage of reactant atoms ending up in the desired product is one factor that is taken into consideration. This percentage is called the **atom economy** and the greater the atom economy, the less the waste.

The following equation shows how to calculate the atom economy in a reaction.

$$\% \text{ atom economy} = \frac{\text{relative formula mass of the desired product}}{\text{relative formula mass of all reactants used}} \times 100$$

 Worked example: Calculating atom economy

What is the atom economy for the production of hydrogen chloride?

Step 1: Write the equation for the reaction.

$$H_2(g) + Cl_2(g) \rightarrow 2HCl(g)$$

Step 2: From the equation, work out the moles of reactants and products.

$$H_2(g) + Cl_2(g) \rightarrow 2HCl(g)$$
$$1\,mol + 1\,mol \rightarrow 2\,mol$$

Step 3: Calculate the mass of one mole for each substance using the relative molecular mass.

$M_r(H_2)$: 2; Mass of one mole = 2.0 g

$M_r(Cl_2)$: 71; Mass of one mole = 71.0 g

$M_r(HCl)$: 36.5; Mass of one mole = $2 \times 36.5 = 73.0$ g

Step 4: Calculate the percentage economy.

$$\% \text{ atom economy} = \frac{73.0}{2.0 + 71.0} \times 100 = 100\%$$

▲ **Figure 1** *A production unit for manufacturing hydrochloric acid*

Synoptic link

You met percentage yield in Chapter 1, Elements of life. Percentage yield is another way of considering the efficiency of a chemical reaction. However, a reaction might have a large percentage yield but a low atom economy.

Activity ES 6.1

This activity gives you the opportunity to calculate some atom economies.

Synoptic link

Atom economy is one aspect of the green chemistry principles. You will find out more about them in Chapter 5, What's in a medicine?

Making hydrochloric acid as a co-product

A large proportion of the hydrochloric acid that is made is a co-product from the chlorination of organic compounds. For example, the first stage in the manufacture of poly(chloroethene), also called poly (vinyl chloride) (PVC), is the reaction of ethene with chlorine. The 1,2-dichloroethane that is formed undergoes thermal cracking to give chloroethene and hydrogen chloride.

$$CH_2ClCH_2Cl \rightarrow CH_2CHCl + HCl$$

The hydrogen chloride can then be converted to hydrochloric acid by passing it through water. A solution of high concentration can be produced easily because hydrogen chloride has a very high solubility in water. Hydrogen chloride gas is made up of covalent molecules – when dissolved in water it forms the hydrated ions $H^+(aq)$ and $Cl^-(aq)$.

Chemical ideas: The periodic table 11.3b

Hydrogen halides
Preparing hydrogen halides

In Topic ES 1 you learnt that fluorine was the strongest oxidising agent in Group 7. Fluorine atoms have the greatest tendency to be reduced or gain electrons. In doing so, fluorine atoms become fluoride ions.

$$F_2 + 2e^- \rightarrow 2F^-$$

Fluoride ions have a low tendency to lose electrons and turn back into atoms. Fluoride ions are difficult to oxidise and so are poor reducing agents.

$$2F^- \rightarrow F_2 + 2e^-$$

stronger oxidising agents ⟵——————————

| F_2 | Cl_2 | Br_2 | I_2 |
| F^- | Cl^- | Br^- | I^- |

——————————⟶ stronger reducing agents

Sodium fluoride and sodium chloride

Sodium fluoride and sodium chloride both react with concentrated acid to make hydrogen fluoride or hydrogen chloride gas. In these experiments you see white fumes of hydrogen chloride as it meets the moist air. Tiny droplets of hydrochloric acid are being made.

$$NaCl(s) + H_2SO_4(aq) \rightarrow NaHSO_4(aq) + HCl(g)$$

This is not a redox reaction.

Sodium bromide

Sodium bromide first reacts with concentrated sulfuric acid to make hydrogen bromide.

$$NaBr(s) + H_2SO_4(aq) \rightarrow NaHSO_4(aq) + HBr(g)$$

However, the bromide ions produced are strong enough reducing agents to reduce the sulfuric acid to sulfur dioxide.

$$2H^+(aq) + 2Br^-(aq) + H_2SO_4(aq) \rightarrow SO_2(g) + Br_2(l) + 2H_2O(l)$$

Br	−1		0	increase in oxidation state – reducing agent
S		+6	+4	decrease in oxidation state – reduced

This means that adding concentrated sulfuric acid to sodium bromide would not be a good way to make hydrogen bromide gas because it won't be pure. The gas made will be a mixture of hydrogen bromide, sulfur dioxide, and bromine vapour (since the reaction is exothermic).

Sodium iodide

Sodium iodide first of all reacts with concentrated sulfuric acid to make hydrogen iodide.

$$NaI(s) + H_2SO_4(aq) \rightarrow NaHSO_4(aq) + HI(g)$$

However the iodide ions produced are even stronger reducing agents than the bromide ions above. The sulfuric acid is this time reduced further to make hydrogen sulfide gas.

$$8H^+(aq) + 8I^-(aq) + H_2SO_4(aq) \rightarrow H_2S(g) + 4I_2(s) + 4H_2O(l)$$

I	−1		0	increase in oxidation state – reducing agent
S		+6	−2	decrease in oxidation state – oxidising agent

With bromide ions, the oxidation state of sulfur decreased by two. With iodide ions, the oxidation state of sulfur is reduced by eight. (Iodide is a stronger reducing agent than bromide.) So adding concentrated sulfuric acid to sodium iodide would not be a good way to make hydrogen iodide gas because it won't be pure. The gas made will be a mixture of hydrogen iodide (white fumes) and hydrogen sulfide (smells of rotten eggs).

When preparing the hydrogen halides in the lab, the appropriate sodium halide is used. Concentrated sulfuric acid can be added to the sodium chloride when making hydrogen chloride. To make pure hydrogen bromide or hydrogen iodide concentrated phosphoric acid is used instead. Unlike sulfuric acid, the concentrated phosphoric acid will not be reduced and so a pure hydrogen halide can be collected.

> **Synoptic link**
>
> You first met bond enthalpies in Chapter 2, Developing fuels.

Similarities and differences in the properties of hydrogen halides

Thermal stability

The **thermal stability** of the hydrogen halides decreases as you go down Group 7 (Table 1) – hydrogen iodide, HI, is broken down into its elements at a lower temperature than hydrogen chloride, HCl. This is because the bond strength between hydrogen and the halogen decreases as you go down Group 7. Less energy is needed to break the bond for hydrogen iodide.

▼ **Table 1** Bond enthalpies for hydrogen–halogen bonds

Bond	Average bond enthalpy / kJ mol^{-1}	Bond length / nm
H–F	568.0	0.092
H–Cl	432.0	0.127
H–Br	366.3	0.141
H–I	298.3	0.161

When the hydrogen halides are heated in a laboratory:

- hydrogen fluoride isn't broken down into hydrogen and fluorine
- hydrogen chloride isn't broken down into hydrogen and chlorine
- some brown bromine gas is made when hydrogen bromide is strongly heated

$$2HBr(g) \rightarrow H_2(g) + Br_2(g)$$

- large amounts of purple gaseous iodine are made if a red hot needle is plunged into hydrogen iodide.

$$2HI(g) \rightarrow H_2(g) + I_2(g)$$

Acidity

In solution, the very soluble hydrogen halides are all acidic. Apart from HF they are strongly acidic. For HCl, HBr, and HI there is almost 100% **dissociation**. Remember that all acidic solutions have $H^+(aq)$ ions in. Another way of representing this is the **oxonium ion** $H_3O^+(aq)$:

$$HCl(aq) \rightarrow H^+(aq) + Cl^-(aq) \text{ or } H_2O(l) + HCl(aq) \rightarrow H_3O^+(aq) + Cl^-(aq)$$

Reaction with ammonia

All of the hydrogen halides react with ammonia to make salts. If a glass rod dipped in concentrated ammonia solution is placed in the hydrogen halide, a white cloud of ammonium halide is made. The following reaction is typical of the hydrogen halides:

$$NH_3(g) + HCl(g) \rightarrow NH_4Cl(s)$$

Reaction with sulfuric acid

The reactions of hydrogen halides with concentrated sulfuric acid are different. This is due to the increasing strength of the halide ions as reducing agents. Compare this with the reactions of solid halides with sulfuric acid.

- hydrogen fluoride, HF, and hydrogen chloride, HCl, do not react
- hydrogen bromide, HBr, makes sulfur dioxide, SO_2
- hydrogen iodide, HI, makes hydrogen sulfide, H_2S.

Synoptic link

You found out about acidic solutions in Chapter 1, Elements of life.

Activity ES 6.2

In this activity you will prepare and carry out reactions with hydrogen halides.

Summary questions

1 Propanol can be dehydrated to produce propene and water:

$$CH_3CH_2CH_2OH \rightarrow CH_3CH{=\!\!=}CH_2 + H_2O$$

a Calculate the relative formula mass of the starting material, propanol. (*1 mark*)

b Calculate the relative formula mass of the useful product, propene. (*1 mark*)

c Calculate the atom economy of this reaction. (*1 mark*)

2 1-bromobutane, C_4H_9Br, will react (rather slowly) with water to produce butan-1-ol, C_4H_9OH, and hydrogen bromide.

  a Write the equation for this reaction.   *(1 mark)*

  b Calculate the atom economy of this reaction.   *(2 marks)*

This reaction can be sped up by using sodium hydroxide, NaOH, instead of water. In this case, the waste product of the reaction is not hydrogen bromide, HBr, but sodium bromide, NaBr.

  c Write an equation for this reaction.   *(1 mark)*

  d What effect would changing the reactant in this way have on the atom economy?   *(1 mark)*

3 Write balanced equations with state symbols for the reaction of hydrogen iodide with the following:

  a ammonia   *(2 marks)*

  b concentrated sulfuric acid.   *(2 marks)*

4 1,2-dichloroethane undergoes thermal cracking to give chloroethene.

$$CH_2ClCH_2Cl \longrightarrow CH_2{=}CHCl + HCl$$

  a Calculate the percentage yield of this process if 10.0 tonnes of the 1,2-dichloroethane yield 2.0 tonnes of chloroethene.   *(2 marks)*

  b Use the percentage yield and atom economy of this reaction to calculate how much in tonnes of the 1,2-dichloroethane is actually converted into chloroethene.   *(2 marks)*

5 Explain why pure hydrogen chloride can be prepared by the addition of concentrated sulfuric acid to sodium chloride but the same method cannot be used to prepare hydrogen bromide from sodium bromide. Include any equations which help your explanation.   *(4 marks)*

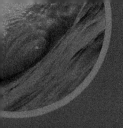

ES 7 The Deacon process solves the problem again

Specification reference: ES(q)

Hydrogen chloride gas has been produced by industrial chemical processes throughout history. Because it is highly toxic, it was important that early chemists could find a way of breaking it down or turning it into something useful. The solution has been built on and improved and is still being used today.

The Deacon process

With increased industrialisation in Britain in the 1800s, industrial pollution became a real problem particularly in areas such as Widnes in Cheshire. Demand for alkalis used to make soap and glass was high but unfortunately hydrogen chloride gas leaving the chimneys of the chemical plants producing the alkalis was devastating the land and killing farmers' crops. Parliament passed the first Alkali Act in 1863 and Victorian inspectors travelled the country to check the fumes from industrial chimneys. At first the hydrogen chloride was just dissolved in water and put into rivers where it killed all the fish. By 1874, Henry Deacon had developed what came to be known as the Deacon process. Hydrogen chloride was mixed with oxygen and passed over a catalyst. The products were chlorine and steam. Chlorine was in demand for bleaching paper and also fabrics.

▲ **Figure 1** *An industrial landscape from the nineteenth and early twentieth century*

$$4HCl(g) + O_2(g) \rightleftharpoons 2Cl_2(g) + 2H_2O(g) \ \Delta H = -114 \, kJ \, mol^{-1}$$

A high yield of chlorine requires that the equilibrium position is to the right. In theory this means that the best yield would be obtained using a high pressure, low temperature, and excess oxygen. In reality conditions in chemical plants are a compromise since factors such as rate of reaction, cost, and safety are taken into account.

The Deacon process today

The Deacon process is still important today. The Japanese Sumitoto Chemical Company have developed and improved the Deacon process so that it can produce almost pure chlorine whilst operating at lower temperature and at low cost.

In the production of the monomer for PVC (chloroethene) the following steps are involved.

Step 1 $CH_2{=}CH_2 \, (g) + Cl_2 \, (g) \rightarrow CH_2ClCH_2Cl \, (l)$

Step 2 $CH_2ClCH_2Cl \, (l) \rightarrow CH_2{=}CHCl(g) + HCl(g)$

Chlorine is used as a feedstock and hydrogen chloride is produced as a co-product. The hydrogen chloride can be used to make hydrochloric acid but this requires a great enough demand for it. The improved Deacon process developed by Sumitoto has been licensed to other chemical firms to solve the problem of hydrogen chloride as a product in polymer manufacture. For instance the hydrogen chloride gas made in Step 2 could be oxidised to chlorine using the Deacon process and recycled to use in Step 1. The costs for producing chlorine by this method are less than electrolysis of salt solution and since the chlorine recycling unit is on site the risks associated with the transport of liquid chlorine are eliminated.

Chemical ideas: Equilibrium in chemistry 7.4

Le Chatelier's principle

By studying data from many reactions, in 1888 Henri Le Chatelier (Figure 2) was able to propose a rule that enabled chemists to make qualitative predictions about the effect of a change on a system at equilibrium. He said that if a system is at equilibrium and a change is made in any of the conditions then the system will oppose the change. This is now known as **Le Chatelier's principle**.

Consider the reaction between ethanoic acid, CH_3COOH, and ethanol, C_2H_5OH. If the system was allowed to reach equilibrium, then the concentration of ethanol was increased, the equilibrium position will shift.

$$CH_3COOH(l) + C_2H_5OH(l) \rightleftharpoons CH_3COOC_2H_5(1) + H_2O(l)$$

The change in the equilibrium can be explained using le Chatelier's principle.

1 ethanol added – concentration of ethanol increased

2 The system opposes the change.

3 The system changes to decrease the concentration of ethanol

4 The forward reaction rate increases – ethanol reacts with ethanoic acid

5 More ethyl ethanoate, $CH_3COOC_2H_5$, and water are made

6 The equilibrium position moves to the right.

▲ **Figure 2** *Henri Le Chatelier*

This is exactly the same as the conclusion deduced using the equilibrium constant K_c in Topic ES 4.

Changing the concentration of any of the reactants or products will affect the equilibrium position.

- *Increasing* the concentration of *reactants* causes the equilibrium position to move to the *product* side.

- *Increasing* the concentration of *products* causes the equilibrium position to move to the *reactant* side.

- *Decreasing* the concentration of *reactants* causes the equilibrium position to move to the *reactant* side.

- *Decreasing* the concentration of *products* causes the equilibrium position to move to the *product* side.

Using Le Chatelier's principle for the effect of changes in pressure

Many important industrial processes involve reversible reactions that take place in the gas phase. For these processes it is essential that conditions are identified that ensure that the equilibrium is shifted as far to the right (the products) as possible. From the study of equilibria in gas-phase reactions, the following conclusions have been reached:

- increasing the pressure moves the equilibrium to the side of the equation with fewer gas molecules as this tends to reduce the pressure
- decreasing the pressure moves the equilibrium to the side of the equation with more gas molecules as this tends to increase the pressure.

In each case, the position of equilibrium shifts so as to oppose the change in pressure.

 Worked example: The effect of changes in pressure

In the first stage in the steam reforming of methane to make methanol, methane reacts with steam to form carbon monoxide, CO, and hydrogen, H_2. What would happen in this reaction if you *reduced* the pressure?

Step 1: Write the equation for the reaction.

$$CH_4(g) + H_2O(g) \rightleftharpoons CO(g) + 3H_2(g)$$

Step 2: Work out the number of molecules on each side of the equation.

$$CH_4(g) + H_2O(g) \rightleftharpoons CO(g) + 3H_2(g)$$
$$1 + 1 = 2 \qquad\qquad 1 + 3 = 4$$
$$\text{2 molecules} \qquad\qquad \text{4 molecules}$$

Step 3: Identify how the system will change.

Pressure is reduced. The system opposes the change, so the system increases the pressure.

Step 4: Identify which way the equilibrium position needs to shift to cause the system change identified in Step 3.

There are more molecules on the right-hand side of the equation than on the left-hand side of the equation. Therefore, making more carbon monoxide and hydrogen will increase the pressure. The equilibrium position moves to the right.

Synoptic link

You used the fact that if a forward reaction is exothermic then the reverse reaction is endothermic when you studied Hess' law in Chapter 2, Developing fuels.

Using Le Chatelier's principle for the effect of changes in temperature

Heating a reaction makes it go faster. However, how fast is not the same as how far. For a reversible reaction, if the forward reaction is exothermic then the reverse reaction will be endothermic to the same extent, and vice versa.

Worked example: The effect of changes in temperature

Nitrogen dioxide, NO_2, is a dark brown gas that exists in equilibrium with its colourless dimer, dinitrogen tetraoxide, N_2O_4.

$$2NO_2(g) \rightleftharpoons N_2O_4(g) \quad \Delta H \text{ is negative}$$
$$\text{brown} \qquad \text{colourless} \quad \text{(i.e., exothermic)}$$

The forward reaction forms $N_2O_4(g)$ and is exothermic, releasing thermal energy to the surroundings. The reverse reaction forms $NO_2(g)$ and is endothermic. Thermal energy is taken in from the surroundings.

If a sealed container of the brown equilibrium mixture is placed in iced water, it becomes paler. How can you explain this using the rule?

Step 1: Identify the temperature change – temperature is decreased.

Step 2: Identify how the system will change.

The system opposes the change. The system will act to increase the temperature.

Step 3: Identify which reaction will increase the temperature.

The forward reaction releases heat energy so will increase the temperature of the surroundings. This will increase the temperature of the sealed container. More N_2O_4 formed. The equilibrium position moves to the right. The colour becomes paler brown as N_2O_4 is colourless.

If the system in the worked example is put in a beaker of boiling water, the system would turn a darker brown. This is because the system would oppose the change of the increase in temperature, so the reverse reaction (the formation of the brown NO_2) would be favoured as this reaction takes in thermal energy. This reduces the temperature of the system.

- Heating a reversible reaction at equilibrium shifts the reaction in the direction of the endothermic reaction.

- Cooling a reversible reaction at equilibrium shifts the reaction in the direction of the exothermic reaction.

Summary questions

1 Which element is oxidised and which element is reduced in the Deacon process? *(2 marks)*

2 Ethanol is produced industrially at about 70 atmospheres pressure and 300 °C by the following reaction. The reaction needs a catalyst.

$$C_2H_4(g) + H_2O(g) \rightleftharpoons C_2H_5OH(g) \qquad \Delta H = -46 \, kJ \, mol^{-1}$$

Which of the following would move the position of equilibrium to the right?

A Increasing the temperature.

B Increasing the concentration of steam.

C Decreasing the pressure. *(1 mark)*

3 State the direction in which the position of equilibrium of each system would move (if at all) if the pressure was increased by compressing the reaction mixture. Give your answer as 'to the left' or 'to the right', or 'no change'.

a $2NO(g) + O_2(g) \rightleftharpoons 2NO_2(g)$ *(1 mark)*
b $C_2H_6(g) \rightleftharpoons C_2H_4(g) + H_2(g)$ *(1 mark)*
c $2HI(g) \rightleftharpoons H_2(g) + I_2(g)$ *(1 mark)*
d $2NO_2(g) \rightleftharpoons N_2O_4(g)$ *(1 mark)*
e $2CO(g) + O_2(g) \rightleftharpoons 2CO_2(g)$ *(1 mark)*

4 Consider the reaction between hydrogen and oxygen to produce steam.

a Write an equation for the reaction with state symbols. *(2 marks)*

b Write an expression for K_c. *(1 mark)*

c Describe and explain how the equilibrium position is affected by:

i an increase in temperature *(2 marks)*

ii an increase in the total pressure. *(2 marks)*

5 For the Deacon process explain why the following changes in conditions would increase the yield of chlorine.

$$HCl(g) + O_2(g) \rightleftharpoons 2Cl_2(g) + 2H_2O(g) \qquad \Delta H = -114 \, kJ \, mol^{-1}$$

a adding excess oxygen *(2 marks)*

b decreasing the pressure *(2 marks)*

c decreasing the temperature *(2 marks)*

6 This equilibrium exists in bleach:

$$Cl_2(aq) + 2NaOH(aq) \rightleftharpoons NaCl(aq) + NaOCl(aq) + H_2O(l)$$

Explain why you should never use another cleaning product that is acidic alongside the bleach. *(3 marks)*

Practice questions

1 Look at the two reactions of chlorine with ethene:

$C_2H_4 + Cl_2 \rightarrow C_2H_4Cl_2$ **Reaction 1**
$C_2H_4 + Cl_2 \rightarrow C_2H_3Cl + HCl$ **Reaction 2**

Which of the following rows is correct about the atom economies of these reactions in terms of the organic product?

	Reaction 1	Reaction 2
A	80%	50%
B	100%	37%
C	100%	63%
D	40%	100%

(1 mark)

2 Which row of the table is correct for the electrolysis of aqueous potassium iodide?

	Cathode	Anode
A	potassium	iodine
B	hydrogen	oxygen
C	iodine	potassium
D	hydrogen	iodine

(1 mark)

3 Which row in the table contains the correct half-equations for the electrolysis of molten sodium chloride?

	Cathode	Anode
A	$Na^+ + e^- \rightarrow Na$	$Cl^- \rightarrow Cl + e^-$
B	$2H^+ + 2e^- \rightarrow H_2$	$Cl^- \rightarrow \frac{1}{2}Cl_2 + e^-$
C	$Na^+ \rightarrow Na + e^-$	$Cl_2 \rightarrow 2Cl^- + 2e^-$
D	$Na^+ + e^- \rightarrow Na$	$2Cl^- \rightarrow Cl_2 + 2e^-$

(1 mark)

4 Which row in the table correctly describes the halogen elements at room temperature?

	Chlorine	Bromine	Iodine
A	green gas	brown gas	purple gas
B	colourless solution	brown solution	brown solution
C	green gas	red liquid	grey solid
D	yellow-green gas	brown liquid	purple solid

(1 mark)

5 Sodium bromide is reacted with silver nitrate solution.
The result is:

A a white precipitate that is soluble in dilute ammonia solution

B a cream precipitate that is soluble in concentrated ammonia solution

C a yellow precipitate that is insoluble in ammonia solution

D a white precipitate that is soluble in concentrated ammonia solution. (1 mark)

6 Which of the following will react with sodium iodide to produce the purest sample of hydrogen iodide?

A concentrated sulfuric acid
B dilute sulfuric acid
C dilute hydrochloric acid
D phosphoric acid (1 mark)

7 An aqueous solution of chlorine is added to an aqueous solution of sodium iodide. Some cyclohexane is added, forming the upper layer. Which of the following is the correct observation.

A There is no reaction.
B The upper layer goes purple and the aqueous layer goes brown.
C Both layers go brown.
D The lower layer is brown and the upper layer is yellow. (1 mark)

8 When sulfuric acid reacts with a bromide, which of the following are correct?

1 hydrogen bromide is produced
2 sulfur dioxide is produced
3 bromine is produced

A 1, 2, and 3 correct
B 1 and 2 are correct
C 2 and 3 are correct
D Only 1 is correct (1 mark)

9 Which of the following are true about hydrogen chloride, hydrogen bromide and hydrogen iodide?

   1 they all react with ammonia

   2 they are all acidic

   3 they all reduce sulfuric acid

   A 1, 2, and 3 correct

   B 1 and 2 are correct

   C 2 and 3 are correct

   D Only 1 is correct   *(1 mark)*

10 A solution of copper sulfate is electrolysed with copper electrodes. Which of the following is true?

   1 Copper is transferred from the anode to the cathode.

   2 Copper is plated on the anode.

   3 Sulfur dioxide is produced at the anode.

   A 1, 2, and 3 correct

   B 1 and 2 are correct

   C 2 and 3 are correct

   D Only 1 is correct   *(1 mark)*

11 In the manufacture of bromine from sea water:

   Step 1: Chlorine is bubbled through sea water containing a very dilute bromide solution to release bromine.

   Step 2: Air is blown through to produce bromine vapour.

   Step 3: The vapour is mixed with sulfur dioxide and passed into water:

$$Br_2 + SO_2 + ... H_2O \rightarrow 2HBr + ...$$

             Equation 11.1

   Step 4: Steam and chlorine are blown through to release bromine from the hydrogen bromide.

   Step 5: The bromine is dried using concentrated sulfuric acid.

   a (i) Write an ionic equation for the reaction that occurs in both Steps 1 and 4.   *(1 mark)*

      (ii) Which property of the halogens does this reaction illustrate? *(1 mark)*

   b Suggest why it is necessary to produce bromine in Step 1 and then again in Step 4.   *(1 mark)*

   c (i) On what property of bromine does Step 2 depend?   *(1 mark)*

      (ii) Suggest the appearance of the gas stream after Step 2.   *(1 mark)*

   d Use oxidation states to complete and balance Equation 11.1 and explain your reasoning.   *(3 marks)*

   e Chlorine is made by electrolysing an aqueous solution of sodium chloride. Give the half-equations for the reactions at the positive and negative electrodes during this electrolysis.   *(2 marks)*

   f Chlorine reacts with cold aqueous sodium hydroxide as follows:

$$Cl_2 + H_2O \rightleftharpoons HCl + HClO \quad \textbf{Equation 11.2}$$

      (i) Write the oxidation states under the chlorine atoms in Cl_2, HCl, and HClO in Equation 11.2.

          What is being reduced and what is being oxidised in this equation?   *(3 marks)*

      (ii) Give the systematic name of HClO.   *(1 mark)*

12 Hydrogen chloride is made industrially by the reaction:

$$H_2(g) + Cl_2(g) \rightleftharpoons 2HCl(g) \quad \textbf{Equation 12.1}$$

   a This reaction can reach dynamic equilibrium. Explain the meaning of the term *dynamic equilibrium*. *(2 marks)*

   b (i) Write the equation for the equilbrium constant K_c of the reaction in Equation 12.1. *(1 mark)*

      (ii) $K_c = 2 \times 10^{33}$ for this reaction at 298 K. What conclusion can be drawn about the composition of an equilibrium mixture at 298 K?   *(1 mark)*

      (iii) The reaction in Equation 12.1 is exothermic. Discuss the effect on an equilibrium mixture of, separately, changing the hydrogen concentration, varying the temperature and varying the pressure.   *(6 marks)*

13 The chemist Max Bodenstein investigated the equilibrium shown below:

$$H_2(g) + I_2(g) \rightleftharpoons 2HI(g) \qquad \text{Equation 13.1}$$

He allowed mixtures of known masses of hydrogen and iodine to react in sealed tubes at high temperatures until equilibrium had been established. Then he rapidly cooled the tubes and analysed the iodine present with sodium thiosulfate.

a Suggest why the flasks are rapidly cooled. *(1 mark)*

b Describe how a sealed tube could be investigated to measure the mass of iodine it contains. *(4 marks)*

c From such an experiment at 500 K, the masses of substances in a 100 cm³ tube were found to be:

Substance	Mass/g
H_2	0.20
I_2	25.38
HI	161.15

Calculate a value for K_c for the reaction in Equation 13.1 at 500 K. *(4 marks)*

d 20.0 cm³ of a solution of 0.002 mol dm⁻³ KIO_3 is reacted with excess iodide ions in the presence of acid.

$$KIO_3 + 5I^- + 6H^+ \rightarrow 3I_2 + 3H_2O$$

$$\text{Equation 13.2}$$

(i) Give the systematic name of KIO_3. *(1 mark)*

(ii) Calculate the volume of 0.50 mol dm⁻³ $Na_2S_2O_3$ that would react with the iodine formed. Give your answer to a *suitable* number of significant figures. *(4 marks)*

(iii) A teacher tells a student that this is not a very satisfactory titration result. Suggest why the teacher says this and suggest what the student could do to improve the titration result without changing the apparatus used. *(3 marks)*

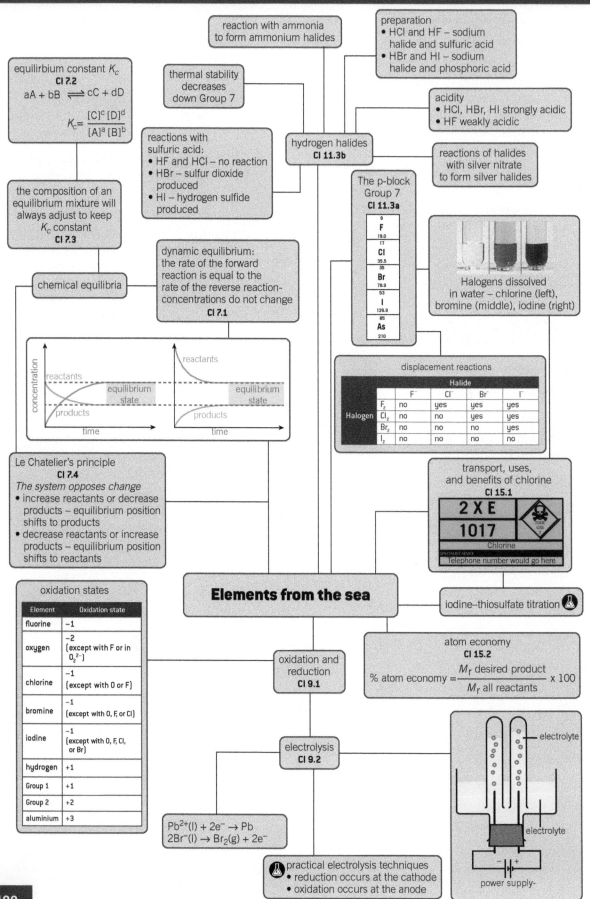

reaction with ammonia
to form ammonium halides

preparation
• HCl and HF – sodium
  halide and sulfuric acid
• HBr and HI – sodium
  halide and phosphoric acid

equilibrium constant K_c
CI 7.2
$$aA + bB \rightleftharpoons cC + dD$$
$$K_c = \frac{[C]^c [D]^d}{[A]^a [B]^b}$$

thermal stability
decreases
down Group 7

acidity
• HCl, HBr, HI strongly acidic
• HF weakly acidic

reactions with
sulfuric acid:
• HF and HCl – no reaction
• HBr – sulfur dioxide
  produced
• HI – hydrogen sulfide
  produced

hydrogen halides
CI 11.3b

reactions of halides
with silver nitrate
to form silver halides

the composition of an
equilibrium mixture will
always adjust to keep
K_c constant
CI 7.3

The p-block
Group 7
CI 11.3a

9
F
19.0
17
Cl
35.5
35
Br
79.9
53
I
126.9
85
As
210

Halogens dissolved
in water – chlorine (left),
bromine (middle), iodine (right)

dynamic equilibrium:
the rate of the forward
reaction is equal to the
rate of the reverse reaction-
concentrations do not change
CI 7.1

chemical equilibria

displacement reactions

		Halide			
		F⁻	Cl⁻	Br⁻	I⁻
Halogen	F_2	no	yes	yes	yes
	Cl_2	no	no	yes	yes
	Br_2	no	no	no	yes
	I_2	no	no	no	no

Le Chatelier's principle
CI 7.4
The system opposes change
• increase reactants or decrease
  products – equilibrium position
  shifts to products
• decrease reactants or increase
  products – equilibrium position
  shifts to reactants

transport, uses,
and benefits of chlorine
CI 15.1

2 X E
1017
Chlorine
Telephone number would go here

Elements from the sea

iodine–thiosulfate titration

oxidation states

Element	Oxidation state
fluorine	−1
oxygen	−2 (except with F or in O_2^{2-})
chlorine	−1 (except with O or F)
bromine	−1 (except with O, F, or Cl)
iodine	−1 (except with O, F, Cl, or Br)
hydrogen	+1
Group 1	+1
Group 2	+2
aluminium	+3

oxidation and
reduction
CI 9.1

atom economy
CI 15.2
$$\% \text{ atom economy} = \frac{M_r \text{ desired product}}{M_r \text{ all reactants}} \times 100$$

electrode
electrolyte

electrolysis
CI 9.2

$$Pb^{2+}(l) + 2e^- \rightarrow Pb$$
$$2Br^-(l) \rightarrow Br_2(g) + 2e^-$$

electrolyte

practical electrolysis techniques
• reduction occurs at the cathode
• oxidation occurs at the anode

power supply

Interhalogens

Interhalogens are compounds containing two or more different halogen atoms. There are many known interhalogens.

Interhalogen	Formula	Notes
chlorine monofluoride	ClF	colourless gas and the lightest interhalogen compound
iodine monofluoride	IF	gas which decomposes at 0 °C into iodine, I_2, and iodine pentafluoride
iodine monochloride	ICl	red crystalline substance that melts at 27°C
astatine monoiodide	AtI	heaviest diatomic interhalogen compound
bromine trifluoride	BrF_3	yellow liquid that is an electrical conductor
chlorine trifluoride	ClF_3	colourless gas
iodine trifluoride	IF_3	crystalline yellow solid
chlorine pentafluoride	ClF_5	colourless gas which reacts violently with water
iodine pentafluoride	IF_5	colourless highly reactive liquid

The three interhalogens containing chlorine and fluorine atoms are highly reactive oxidising and fluorinating agents. Examples of their reactions with metals are given below.

$$W + 6 \; ClF \longrightarrow WF_6 + 3 \; Cl_2 \qquad \text{Reaction A}$$

$$U + 3 \; ClF_3 \longrightarrow UF_6 + 3 \; ClF \qquad \text{Reaction B}$$

A dot-and-cross diagram of chlorine pentafluoride is shown below.

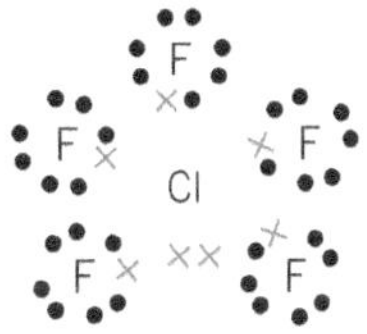

Some interhalogen molecules dissociate into halogens or other interhalogen molecules, as shown below.

$$ClF_3 \rightleftharpoons ClF + F_2 \qquad \text{Reaction C}$$

$$5 \; IF \rightleftharpoons IF_5 + 2 \; I_2 \qquad \text{Reaction D}$$

1. What is the oxidation state of chlorine in ClF, ClF_3, and ClF_5?
2. Use oxidation states to show that the interhalogens are acting as oxidising agents in Reactions A and B.
3. Comment on the number of electrons surrounding the chlorine atom in chlorine pentafluoride.
4. Write an expression for K_c for the dissociation of chlorine trifluoride (Reaction C).
5. Disproportionation occurs when different atoms of the same element are oxidised and reduced in the same reaction. Identify which of Reactions A–D is a disproportionation reaction.

Extension

1. Research further examples of equilibria in the chemical industry. Explain how careful control of the reaction conditions in each case can maximise the yield of product.
2. Prepare a summary of redox reactions. Include definitions of oxidation and reduction, and describe identification of oxidising and reducing agents, balancing equations, naming compounds, electrolysis, redox titrations and industrial applications of redox reactions.
3. There have been calls to ban the use of chlorine and chlorine compounds for environmental reasons. Research the risks and benefits of these chemicals and evaluate the opinions for and against a ban.

CHAPTER 4

The ozone story

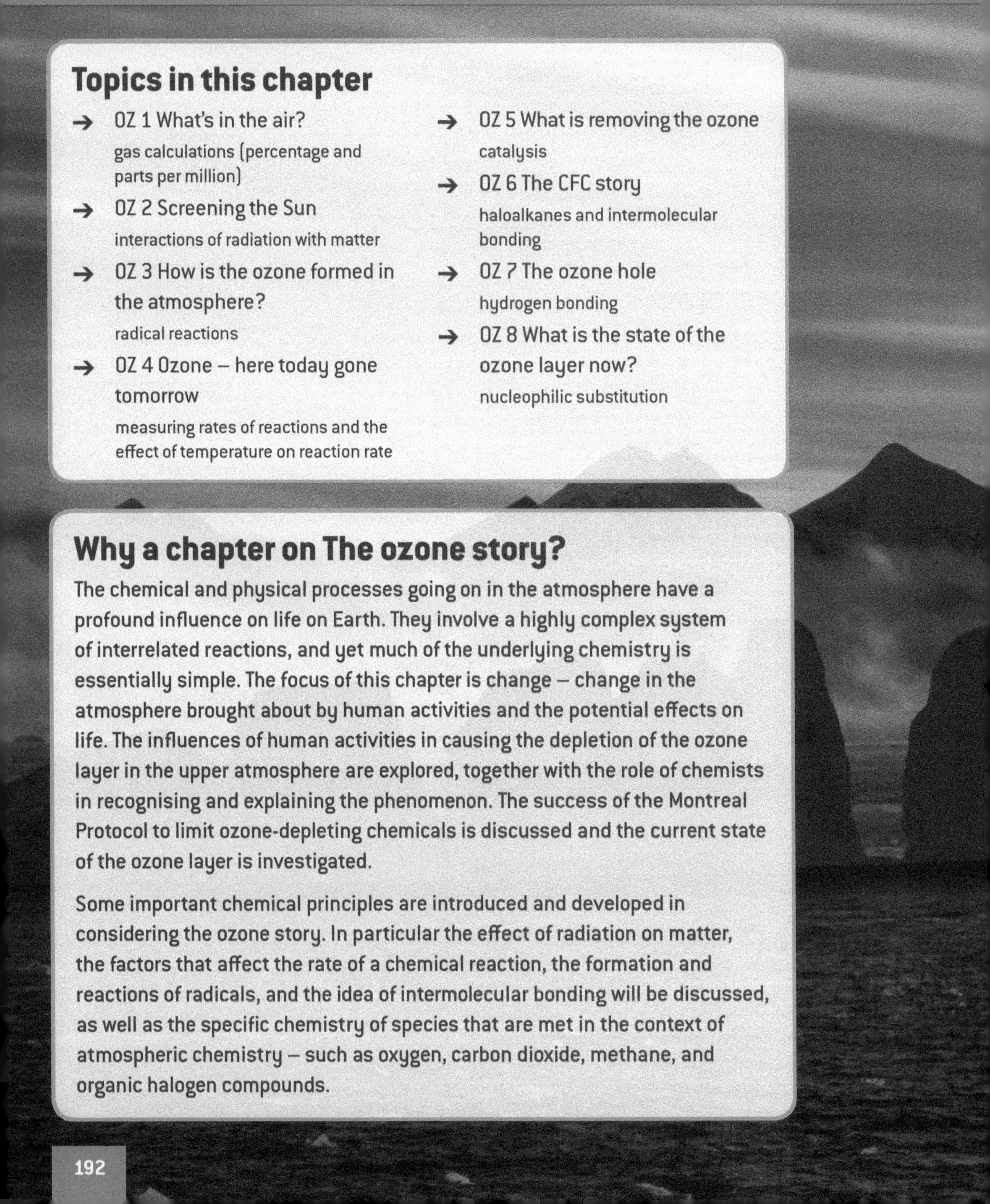

Topics in this chapter

- OZ 1 What's in the air?
  gas calculations (percentage and parts per million)
- OZ 2 Screening the Sun
  interactions of radiation with matter
- OZ 3 How is the ozone formed in the atmosphere?
  radical reactions
- OZ 4 Ozone – here today gone tomorrow
  measuring rates of reactions and the effect of temperature on reaction rate
- OZ 5 What is removing the ozone
  catalysis
- OZ 6 The CFC story
  haloalkanes and intermolecular bonding
- OZ 7 The ozone hole
  hydrogen bonding
- OZ 8 What is the state of the ozone layer now?
  nucleophilic substitution

Why a chapter on The ozone story?

The chemical and physical processes going on in the atmosphere have a profound influence on life on Earth. They involve a highly complex system of interrelated reactions, and yet much of the underlying chemistry is essentially simple. The focus of this chapter is change – change in the atmosphere brought about by human activities and the potential effects on life. The influences of human activities in causing the depletion of the ozone layer in the upper atmosphere are explored, together with the role of chemists in recognising and explaining the phenomenon. The success of the Montreal Protocol to limit ozone-depleting chemicals is discussed and the current state of the ozone layer is investigated.

Some important chemical principles are introduced and developed in considering the ozone story. In particular the effect of radiation on matter, the factors that affect the rate of a chemical reaction, the formation and reactions of radicals, and the idea of intermolecular bonding will be discussed, as well as the specific chemistry of species that are met in the context of atmospheric chemistry – such as oxygen, carbon dioxide, methane, and organic halogen compounds.

Knowledge and understanding checklist

From your Key Stage 4 study you will have studied the following. Work through each point, using your Key Stage 4 notes and the support available on Kerboodle.

☐ Factors affecting the rate of a chemical reaction.

☐ How catalysts work.

☐ Covalent bonding.

You will learn more about some ideas introduced in earlier chapters:

☐ the use of moles and quantitative chemistry (**Developing fuels**)

☐ the electromagnetic spectrum (**Elements of life**)

☐ electronic structure (**Elements of life**)

☐ the chemistry of simple organic molecules (**Developing fuels**)

☐ enthalpy changes and bond enthalpies (**Developing fuels**)

☐ oxidation states (**Elements from the sea**)

☐ catalysis (**Developing fuels**).

Maths skills checklist

In this chapter, you will need to use the following maths skills. You can find support for these skills on Kerboodle and through MyMaths.

☐ Recognise and make use of appropriate units in calculations.

☐ Recognise and use expressions in decimal and ordinary form.

☐ Use percentages.

☐ Understand and use the symbols $=$, $<$, $\ll$, $\gg$, $>$, $\propto$, $\sim$, $\rightleftharpoons$.

☐ Translate information between graphical and numerical forms.

☐ Plot variables for experimental data.

MyMaths.co.uk
Bringing Maths Alive

OZ 1 What's in the air?

▲ **Figure 1** *The Earth as seen from space – the blue haze is the atmosphere of the Earth*

The **atmosphere** is a relatively thin layer of gas that surrounds the Earth's surface. It extends about 100 km above the Earth's surface. If the world were a blown-up balloon, the rubber would be thick enough to contain nearly all the atmosphere. In Figure 1, the atmosphere is the thin blue haze you can see surrounding the Earth. Thin though it is, this layer of gas has an enormous influence on the Earth.

Structure and composition of the atmosphere

A simplified picture of the lower and middle parts of the atmosphere is shown in Figure 2. The two most chemically important regions are the **troposphere** and the **stratosphere**. In fact, 90% of all the molecules in the atmosphere are in the troposphere, and the atmosphere becomes less dense the higher you go. Figure 2 also shows the way that temperature changes with altitude. Mixing is easy in the troposphere because hot gases can rise and cold gases can fall. The reverse temperature gradient in the stratosphere means that mixing is much more difficult in the vertical direction. However, horizontal circulation is rapid in the stratosphere, particularly around circles of latitude.

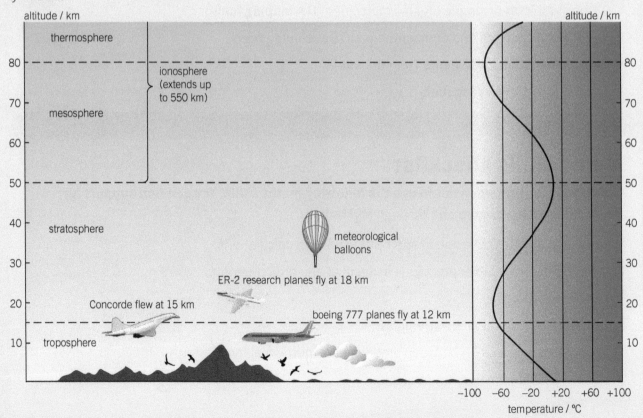

▲ **Figure 2** *The structure of the atmosphere and the change in temperature with altitude*

Table 1 shows the average composition by volume of dry air from an unpolluted environment. This is typical of the troposphere.

▼ Table 1 *Composition by volume of dry tropospheric air from an unpolluted environment. The concentrations of some of these gases are measured in parts per million (ppm) by volume. Gases marked with an asterisk are found naturally in the atmosphere, but their concentration is increased by human activity*

Gas	Concentration by volume
nitrogen, N_2	78%
oxygen, O_2	21%
argon, Ar	1%
carbon dioxide, CO_2	399 ppm *
neon, Ne	18.2 ppm
helium, He	5.2 ppm
methane, CH_4	1.8 ppm *
krypton, Kr	1.1 ppm
hydrogen, H_2	0.5 ppm
dinitrogen oxide, N_2O	0.3 ppm *
carbon monoxide, CO	0.1 ppm *
xenon, Xe	0.09 ppm
nitrogen monoxide, NO, and nitrogen dioxide, NO_2 (NO_x)	0.003 ppm *

Chemical ideas: Measuring amounts of substance 1.5b

Calculations involving gases

Gas concentrations

When dealing with a gas it is sometimes more useful to know its volume than its mass. When the concentration of a gas is small, it is more convenient to express the concentration as parts per million (ppm) by volume.

In Table 1 carbon dioxide has a concentration of 399 ppm. This means that of one million particles in a sample of air, 399 of them will be carbon dioxide molecules.

 Worked example: Calculating percentage composition

The concentration of carbon dioxide in a sample of air is 399 ppm. Calculate the percentage composition of carbon dioxide in this sample.

Step 1: Divide number of CO_2 molecules by 1 000 000.
$$\frac{399}{1\,000\,000} = 3.99 \times 10^{-4}$$
Step 2: Multiply answer by 100 to get the percentage.
$$3.99 \times 10^{-4} \times 100 = 0.04\%$$

Synoptic link

You have already looked at different types of gas calculations in Chapter 2, Developing fuels.

Study tip

This is the same as dividing by 10 000.

From the worked example, you can see that converting from per cent to parts per million involves a factor of 10 000. A concentration of 1% is equivalent to 10 000 ppm.

 Worked example: Calculating concentration in parts per million

A sample of air is 78% nitrogen. Calculate the concentration of nitrogen in ppm in this sample.

Step 1: Multiply the percentage by 10 000.

$$78 \times 10\,000 = 780\,000 \text{ ppm}$$

The early atmosphere

The atmosphere hasn't always had the same composition. The first atmosphere was lost during the upheavals in the early years of the Solar System. The next atmosphere consisted of compounds such as carbon dioxide, methane, and ammonia, which bubbled out of the Earth itself.

3000 million years ago there was very little oxygen in the atmosphere. When the first simple plants appeared they began to produce oxygen through photosynthesis. For more than 1000 million years very little of this oxygen reached the atmosphere. It was used up quickly as it oxidised sulfur and iron compounds, and other chemicals in the Earth's crust. It wasn't until this process was largely complete that oxygen began to collect in the atmosphere.

 The atmosphere past, present, and future

Researchers looking at the phenomenon of oxidative weathering believe that the Earth's atmosphere may have contained significant amounts of oxygen up to three billion years ago, earlier than previous estimates.

Their research involved studying the oxidation of two isotopes of chromium, chromium-53 and chromium-52. When oxidised, ^{53}Cr becomes slightly more soluble than ^{52}Cr. Rain water washes the heavier ^{53}Cr isotope from the soil into the sea more readily than ^{52}Cr. Consequently, soils should become depleted in ^{53}Cr and sea-bed sediments should become richer in ^{53}Cr.

This prediction was borne out when the researchers analysed ancient soils and sediments in Kwazulu-Natal Province, South Africa. The age of the sediments and the ratio of ^{52}Cr to ^{53}Cr indicates that 2.95 billion years ago the atmosphere could have had an oxygen concentration of 63 ppm.

Mauna Loa

From 1958 carbon dioxide levels in the atmosphere have been measured by continuous atmospheric monitoring devices at the Mauna Loa Observatory in Hawaii, USA. In 1958 the concentration of carbon dioxide was recorded as 315 ppm.

In early 1988 the level was recorded as 350 ppm, and on 10 May 2013 daily averages were recorded as 400 ppm. It is thought that this is the highest level for the last three million years of Earth's history. There is a natural annual variation of 3 to 9 ppm due to plants' growing seasons, but there has been a steady upward rise in average carbon dioxide levels.

March 2012   394.36 ppm

March 2013   397.27 ppm

March 2014   399.47 ppm

There is widespread concern at this increase in carbon dioxide levels, as they are linked to global warming.

1   State the number of protons, neutrons, and electrons in an atom of ^{52}Cr and an atom of ^{53}Cr.
2   Calculate the percentage composition of oxygen in the early atmosphere (63 ppm).
3   Suggest how plants' growing seasons affect the atmospheric concentration of carbon dioxide.

Synoptic link

You will learn more about increasing carbon dioxide levels and global warming in Chapter 8, Oceans.

When the oxygen concentration reached about 10%, there was enough for the first animals to evolve using oxygen for respiration. Eventually there was enough respiration and other processes going on to remove the oxygen as fast as it was formed. Since then, the oxygen concentration has remained at about 21%.

Look again at Table 1. All the gases listed are produced as a result of natural processes. Human activities add more gases to the atmosphere. Some of them, like carbon dioxide, are already present, but we increase their concentration. These gases are marked by an asterisk in Table 1. Their main sources as a result of human activities are shown in Table 2. Other gases in the atmosphere, like chlorofluorocarbons (CFCs) and hydrofluorocarbons (HFCs), are produced only as a result of human activity.

Synoptic link

You studied the origins of nitrogen oxides, carbon dioxide, and carbon monoxide in Chapter 2, Developing fuels.

▼ Table 2 *Sources of some of the gases in the atmosphere produced as a result of human activities*

Gas	Main source as a result of human activities
carbon dioxide	combustion of hydrocarbon fuels (e.g., in power stations, motor vehicles); deforestation
methane	cattle farming; landfill sites; rice paddy fields; natural gas leakage
nitrous oxide	fertilised soils; changes in land use (e.g., from the soil when land is ploughed up)
carbon monoxide	incomplete combustion of hydrocarbons (e.g., from car exhausts)
nitrogen oxides	internal combustion engines (from the reaction of N_2 and O_2 at high temperatures)

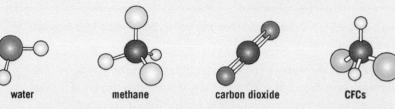

water   methane   carbon dioxide   CFCs

▲ Figure 3 *Molecular diagrams of some pollutant gases*

Given time, gases mix together and this natural diffusion process is greatly speeded up in the atmosphere by air currents and prevailing winds. So, in time, pollutant gases spread throughout the atmosphere. **Atmospheric pollution** is a global problem – it affects us all. In this chapter, we shall be concentrating on the depletion of the ozone layer in the stratosphere, looking at how chemists identified the problem, and at the international agreements to limit the damage.

Summary Questions

1 State the main gases in unpolluted air.   (*1 mark*)

2 State three human activities that add gases to the air.   (*1 mark*)

3 a Calculate how many parts per million by volume of argon are in a typical sample of tropospheric air.   (*1 mark*)
   b Calculate the percentage of neon in a typical sample of tropospheric air.   (*1 mark*)

4 a Calculate the volume of methane present in 1 dm³ of tropospheric air.   (*1 mark*)
   b Calculate the percentage of methane molecules in the sample.   (*1 mark*)

OZ 2 Screening the sun

Specification reference: OZ(r), OZ(s), OZ(t), OZ(u)

The sunburn problem

Until the 1920s a suntan was considered undesirable, as it showed that you had to work outdoors in the Sun. The clothes designer Coco Chanel made sunbathing fashionable after a cruise on the yacht belonging to the Duke of Westminster. As you discover more about the effects of the Sun's **electromagnetic radiation** on the chemical bonds in living material, sunbathing will seem like less of a good idea.

The Sun radiates a spectrum of electromagnetic radiation. Part of this spectrum corresponds to the energy required to break chemical bonds, including those in molecules such as DNA. This can cause damage to genes and lead to skin cancer. On a less serious level, sunlight can damage the proteins within skin, so that years of exposure can make people look wrinkly and leathery. Brief exposure to the Sun may irritate the blood vessels in the skin, making it look red and sunburnt.

▲ Figure 1 *The trend-setting clothes designer Coco Chanel*

Chemical sunscreens

Figure 3 shows the effects of different parts of the Sun's spectrum on the skin.

The most damaging part of this spectrum to the skin is the **ultraviolet** part. Fortunately, there are chemicals which absorb much of this radiation. You can sit by a window, or in a greenhouse, for hours on a sunny day without burning because the glass in the window lets through visible light but absorbs the high-energy ultraviolet radiation, so it never reaches your skin. On the other hand, water does let through some ultraviolet, so it is possible to burn under water.

Learning outcomes

Demonstrate and apply knowledge and understanding of:

→ the principal radiations of the Earth and the Sun in terms of the following regions of the electromagnetic spectrum: infrared, visible, ultraviolet

→ calculation of values for frequency, wavelength, and energy of electromagnetic radiation from given data

→ the formation and destruction of ozone in the stratosphere and troposphere; the effects of ozone in the atmosphere, including:

- ozone's action as a sunscreen in the stratosphere by absorbing high-energy UV (and the effects of such UV, including on human skin)

- the polluting effects of ozone in the troposphere, causing problems including photochemical smog

→ the effect of UV and visible radiation promoting electrons to higher energy levels, sometimes causing bond-breaking.

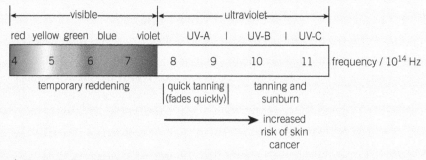

▲ Figure 3 *The effects of sunlight on the skin*

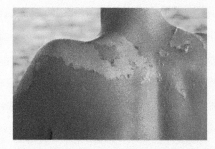

▲ Figure 2 *Over-exposure to sunlight causes severe sunburn and peeling skin, and increases the risk of developing skin cancers*

Activity OZ 2.1

This activity helps you develop your skills planning and investigating sunscreens.

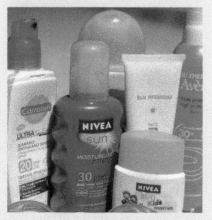

▲ **Figure 4** *The Sun protection factor (SPF) number on sunscreens indicates the time it will take for the Sun to produce a certain effect on your skin*

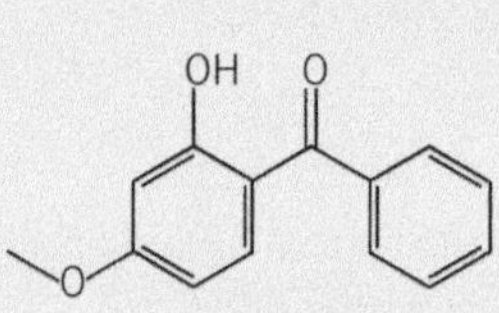

oxybenzone

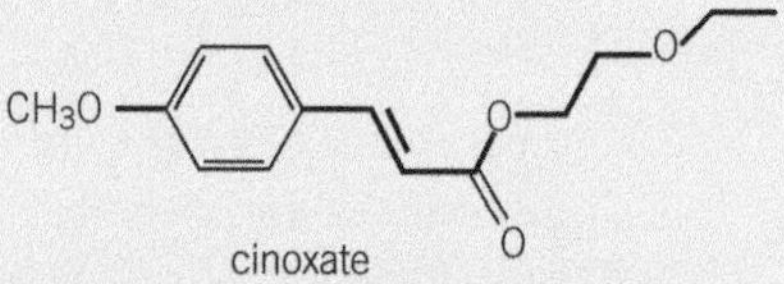

cinoxate

▲ **Figure 5** *Some examples of the molecules in sunscreen that are effective at absorbing ultraviolet*

Chemists have developed sunscreens which absorb high-energy ultraviolet radiation, and millions of pounds are spent on them in the UK every summer. The molecules in sunscreen often contain benzene rings or alternating double and single bonds (Figure 5). When ultraviolet light is absorbed, the electrons in the πorbitals in these bonds are promoted to higher energy levels.

However the best sunscreen of all is not made by chemists. It has always been with us – it is the atmosphere.

Why is the atmosphere such a good sunscreen?

Certain atmospheric gases absorb ultraviolet radiation strongly. They act as a global sunscreen, preventing much of the Sun's harmful radiation from reaching the Earth.

Most of this absorption goes on in the region of the upper atmosphere called the stratosphere (see Figure 2 in Topic OZ 1). Particularly important is the gas **ozone**, which is a form of oxygen with the formula O_3. It absorbs ultraviolet radiation in the region $10.1 \times 10^{14} - 14.0 \times 10^{14}$ Hz. This includes the UV-B and UV-C regions (Figure 3) which can damage DNA potentially leading to skin cancer, damage eyes leading to cataracts, and damage crops.

Although ozone in the stratosphere protects us from high energy ultraviolet radiation, ozone at ground level in the troposphere is a significant pollutant. Ozone is involved in reactions producing photochemical smog that causes haziness and reduced visibility, and irritation and respiratory problems for many people.

There is no life in the stratosphere because the high energy ultraviolet radiation would break down the delicate molecules of living things. Even simple molecular substances are broken down. Some of the covalent bonds break to give fragments of molecules called radicals.

Higher up in the atmosphere the radiation is powerful enough to knock electrons out of atoms, molecules, and radicals. Ions are produced, giving that part of the atmosphere its name – the ionosphere.

Synoptic link

In Chapter 10, Colour by design, you will learn more about how and why molecules absorb ultraviolet or visible radiation.

Study tip

You should use the term *high energy* ultraviolet radiation, rather than harmful UV radiation.

Chemical ideas: Radiation and matter 6.1b

Light and electrons

The wave theory and particle nature of light

The behaviour of light can be described using the wave model or the particle model. Like all electromagnetic radiation, light behaves like a wave with a characteristic wavelength λ and frequency ν. The speed of light c is the same for all kinds of electromagnetic radiation. It has a value of $3.00 \times 10^8\,m\,s^{-1}$ when the light is travelling in a vacuum.

Frequency and wavelength are related. If you multiply wavelength and frequency together you get a constant – the speed of light.

$$\text{speed of light } c\ (m\,s^{-1}) = \text{wavelength } \lambda\ (m) \times \text{frequency } \nu\ (s^{-1})$$

The behaviour of light can also be explained in some situations by thinking of it not as waves but as particles called photons.

Bringing the wave and particle models of light together

The two theories of light – the wave and photon models – are linked by a relationship:

the energy of a photon E (J) = Planck constant h (Js) × frequency v (s^{-1})

The energy of a photon is equal to the frequency of the light multiplied by the Planck constant. This is has a value of 6.63×10^{-34} Js.

 Worked example: Calculations using the particle theory

Calculate the frequency associated with a photon of red light with an energy of 3.00×10^{-19} J.

Step 1: Rearrange the equation the energy of a photon
E (J) = Planck constant h (Js) × frequency v (s^{-1})
to make frequency the subject.

$$v = \frac{E}{h}$$

Step 2: Calculate the frequency of the photon.

$$\frac{3.00 \times 10^{-19}}{6.63 \times 10^{-34}} = 4.52 \times 10^{14}\,\text{s}^{-1}$$

Study tip

You may see frequency with the units Hz. These are the same as s^{-1}.

Synoptic link

You first encountered the wave theory of light and defined wavelength, frequency, and speed in Topic EL 2, How do we know so much about outer space?

Activity OZ2.2

You can check your understanding of the electromagnetic spectrum in this activity.

Chemical ideas: Radiation and matter 6.2

What happens when radiation interacts with matter?

Energy interacts with matter

Electromagnetic radiation can interact with matter, transferring energy to the chemicals involved. This can cause changes in the chemicals, depending on the chemical and the amount of energy involved.

A molecule has energy associated with several different aspects of its behaviour, including:

- translation (the molecule moving around as a whole)
- rotation (of the molecule as a whole)
- vibration of the bonds
- electron energy.

These different kinds of energetic activities involve different amounts of energy as shown in Figure 6.

Electrons can occupy definite energy levels. The electronic energy of an atom or molecule changes when an electron moves from one level to another. Electronic energy is **quantised**; it has fixed levels. But

Study tip

If you are asked to calculate an energy value per mole of photon, remember to multiply the value per photon by the Avagadro constant N_A.

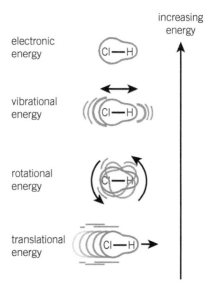

electronic energy

increasing energy

vibrational energy

rotational energy

translational energy

▲ Figure 6 *An HCl molecule has energy associated with different aspects of its behaviour*

here is a crucial point – *all* the other types of energy (translational, rotational, and vibrational) are quantised too.

Different energy changes for different parts of the spectrum

The spacing between vibrational energy levels corresponds to the infrared part of the spectrum. You sense **infrared radiation** as heat. The radiation makes bonds in the chemicals in your skin vibrate more energetically. The molecules have more kinetic energy and this is why you feel warmer.

Making molecules rotate requires less energy than making their bonds vibrate. Therefore, changes in rotational energy correspond to a lower energy, a lower frequency part of the electromagnetic spectrum, namely the microwave region. The spacing between translational energy levels is even smaller.

However, making electronic changes occur in a molecule requires *higher* energy than for vibrational changes. Exciting electrons to higher electronic energy levels requires energy corresponding to the **visible** and ultraviolet parts of the spectrum. See Table 1.

▼ Table 1 *Summary of molecular energy changes*

Change occurring	Size of energy change / J	Type of radiation absorbed
change of rotational energy level	1×10^{-22} to 1×10^{-20}	microwave
change of vibrational energy level	1×10^{-20} to 1×10^{-19}	infrared
change of electronic energy level	1×10^{-19} to 1×10^{-16}	visible and ultraviolet

Activity OZ2.3 and OZ2.4

You can check your understanding of how ozone and other substances act as a sunscreen in these activities.

The table gives *ranges* of energy. The particular value of the energy change depends on the substance involved. For example, the C—F bond is stronger than the C—Br bond, so it takes infrared of a higher energy to make a C—F bond vibrate than to make a C—Br bond vibrate. The electromagnetic radiation reaching the Earth's atmosphere is mainly in the visible and ultraviolet part of the electromagnetic spectrum. Radiation emitted by the Earth's surface is mainly in the infrared region.

What kind of electronic changes occur when molecules absorb ultraviolet radiation?

The electrons in a molecule, such as chlorine, Cl_2, occupy definite energy levels. The outer shell electrons are in the highest energy levels so can move most easily to higher levels.

When a chlorine molecule absorbs radiation, one of three things can happen depending on the amount of energy involved:

● Electrons may be excited to a higher energy level. Chlorine owes its colour to this process – Cl_2 absorbs visible light of such a frequency that the remaining, unabsorbed light looks green.

● If higher energy radiation is used the molecule may absorb so much energy that the bonding electrons can no longer bond the atoms together. This is **photodissociation** and **radicals** are formed. Radicals are molecules or atoms with at least one unpaired electron. They are usually very reactive. Their formation may lead to further chemical reactions. You will find out more about photodissociation and radicals later in this module.

● With very high energy photons, the molecules may acquire so much energy that an electron is able to leave it – the molecule is *ionised*.

Figure 8 illustrates these three possibilities:

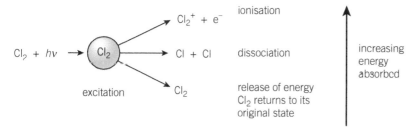

▲ **Figure 8** *When chlorine molecules absorb radiation (hν), they become excited – the excited molecules may then ionise, dissociate, or just release the energy*

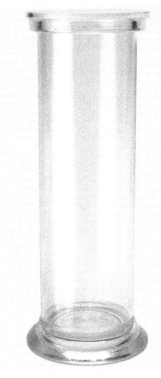

▲ Figure 7 *Chlorine gas*

Summary questions

1. A beam of infrared radiation has an energy of 3.65×10^{-20} J per photon. Calculate the frequency of the radiation. *(1 mark)*

2. A beam of X-rays has a frequency of $2.60 \times 10^{17}\,\text{s}^{-1}$. Calculate the wavelength of the X-rays. *(1 mark)*

3. Carbon dioxide is a greenhouse gas. It absorbs some of the infrared radiation given off from the Earth's surface.
   a. Explain why carbon dioxide molecules absorb only certain frequencies of infrared radiation. *(2 marks)*
   b. Explain why absorbing infrared radiation makes the atmosphere warmer. *(2 marks)*

4. To change 1 mole of molecular HCl from the lowest vibrational energy level (ground state) to the next vibrational level requires 32.7 kJ.
   a. Calculate the energy, in joules, gained by *one molecule* of HCl when energy is absorbed in this way. [Avogadro constant, $N_A = 6.02 \times 10^{23}\,\text{mol}^{-1}$] *(2 marks)*
   b. Calculate the corresponding frequency of radiation. State the type of radiation this corresponds to. *(2 marks)*
   c. Calculate the wavelength of this radiation in metres. *(1 mark)*

5. You want to heat up a cup of coffee in a microwave cooker. The cooker uses radiation of frequency 2.45×10^9 Hz. The cup contains 150 cm³ of coffee, which is mainly water.
   a. Calculate the energy needed to raise the temperature of the water by 30 °C. [Specific heat capacity of water = $4.18\,\text{J}\,\text{g}^{-1}\text{K}^{-1}$] *(2 marks)*
   b. Calculate the energy transferred to the water by each photon of microwave radiation. *(2 marks)*
   c. Calculate the energy transferred by one mole of photons. *(1 mark)*
   d. Calculate how many moles of photons are needed to supply the energy calculated in part **a**. *(1 mark)*

OZ 3 How is ozone formed in the atmosphere?

Specification reference: OZ(o), OZ(p)

If all the ozone in the atmosphere were collected and brought to the Earth's surface at atmospheric pressure, it would form a layer only 3 mm thick.

It isn't really surprising that there is so little ozone in the atmosphere – it reacts so quickly with other substances and gets destroyed. In fact, you might ask why the ozone in the atmosphere hasn't run out. Some reactions must be producing it too.

Many of the ideas in this part of the storyline are concerned with the formation and reactions of radicals.

Chemical ideas: Radiation and matter 6.3

Radiation and radicals

Ways of breaking bonds

In a covalent bond, a pair of electrons is shared between two atoms. For example, in the HCl molecule.

$$H \!\!-\!\!\overset{\bullet}{\underset{\bullet}{}}\!\!-\!\! Cl$$

During a reaction bonds break and get remade. Bond-breaking is sometimes called **bond fission** and it involves the redistribution of the electrons in the covalent bond. There are two ways this can happen – by **heterolytic fission** or by **homolytic fission**.

Heterolytic fission

In this type of fission, both of the shared electrons go to just one of the atoms when the bond breaks. This atom becomes negatively charged because it has one more electron than it has protons. The other atom becomes positively charged.

In the case of HCl:

$$H \!\!-\!\!\overset{\bullet}{\underset{\bullet}{}}\!\!-\!\! Cl \rightarrow H^+ + {\overset{\bullet}{\underset{\bullet}{}}}Cl^-$$

Heterolytic fission is common when a bond is already **polar**. You will find out more about polar bonds in Topic OZ 5.

Homolytic fission

In this type of bond fission, one of the two shared electrons goes to each atom, as in the case of Br_2.

$$Br \!\!-\!\!\overset{\bullet}{\underset{\bullet}{}}\!\!-\!\! Br \rightarrow Br^\bullet + Br^\bullet$$

The dot beside each atom shows the *unpaired electron* that the atom has inherited from the shared pair in the bond. The atoms have no overall charge because they have the electronic structure they had before they shared their electrons to form the bond. Sometimes the dot is omitted and the radical simply represented as, for example, Br.

The unpaired electron has a strong tendency to pair up again with another electron from another substance. Radicals are most commonly formed when the bond being broken is non-polar, but many polar bonds can break this way too – particularly when the reaction is taking place in the gas phase and in the presence of light.

The amount of energy it takes depends on the bond enthalpy of the bond being broken. For example, a pattern can be seen in the bond enthalpies of carbon–halogen bonds, with C—F strongest through to C—I being weakest. Photodissociation of a C—F bond requires higher energy (shorter wavelength) light than photodissociation of a C—I bond.

Radical reactions involving ozone

Ozone, O_3, is formed when an oxygen atom (an example of a radical) reacts with a dioxygen molecule.

$$O + O_2 \rightarrow O_3$$

One way to make oxygen atoms is by dissociating dioxygen molecules. This requires quite a lot of energy ($+498 \, kJ \, mol^{-1}$), which can be provided by ultraviolet radiation or by an electric discharge.

You can often smell the sharp odour of ozone near electric motors, photocopiers, or ultraviolet lamps used to kill bacteria in food shops. The electric discharges make some of the dioxygen molecules in the air dissociate into atoms, which then react to make ozone.

Some of the ozone in the troposphere is formed in reactions taking place in photochemical smogs. In this case, oxygen atoms are produced by the action of sunlight on the pollutant gas nitrogen dioxide.

In the stratosphere, oxygen atoms are formed by the photodissociation of dioxygen molecules when ultraviolet radiation of the right frequency (indicated by $h\nu$) is absorbed.

The reaction can be summarised as:

$$O_2 + h\nu \rightarrow O + O \qquad \textbf{Reaction 1}$$

The resulting oxygen atoms may collide and react with an O_2 molecule, another O atom or an O_3 molecule in reactions 2–4:

$$O + O_2 \rightarrow O_3 \qquad \textbf{Reaction 2}$$
$$O + O \rightarrow O_2 \qquad \textbf{Reaction 3}$$
$$O + O_3 \rightarrow O_2 + O_2 \qquad \textbf{Reaction 4}$$

Reaction 2 is the one that produces ozone.

When the ozone absorbs radiation in the 10.1×10^{14} to $14.0 \times 10^{14} \, Hz$ region, some molecules undergo photodissociation and split up again:

$$O_3 + h\nu \rightarrow O_2 + O \qquad \textbf{Reaction 5}$$

This reaction is responsible for the vital screening effect of ozone, since it absorbs the radiation responsible for sunburn.

▲ Figure 1 Helium-filled balloons are sent up into the stratosphere to measure ozone concentrations

Synoptic link

You first came across photochemical smogs in Chapter 2, Developing fuels.

Other radical reactions

Radical polymerisation uses initiators to produce radicals. The initiator radical attaches to the monomer, forming a new larger radical, which then attacks another monomer molecule. Consequently the polymer chain grows. Polymerisation stops when a termination step occurs, or when all of the monomer is used up.

Combustion is a series of radical chain reactions. Flammable materials require a lower concentration of radicals before initiation and propagation reactions cause combustion of the material. Termination reactions cause the flame to extinguish. Lead additives to petrol reduced propagation and promoted termination, thereby reducing auto-ignition.

▲ **Figure 2** *Combustion*

Radicals have been implicated in *age-related biological effects*. Radicals interact with biological molecules, removing an electron. This often causes the molecule to become a radical, resulting in changes to the biological molecules:

- Chains of DNA can become cross-linked, which may lead to cancer.
- Fats and proteins can become cross-linked, leading to wrinkles in the skin.
- Low-density lipoproteins can become oxidised, leading to arterial plaques which can cause heart disease or strokes.

Antioxidants such as vitamin C can reduce this damage because they give electrons to radicals but remain stable.

1 Suggest why radical polymerisation tends to produce highly-branched irregular polymer chains.
2 Suggest how initiation, propagation, and termination reactions may explain the process of combustion.

Radicals are reactive

Filled outer electron shells are more stable than unfilled ones. Radicals are reactive because they tend to try to fill their outer shells by grabbing an electron from another atom or molecule.

When Cl• collides with H_2, the chlorine grabs an electron from the pair of electrons in the bond between the hydrogen atoms. This makes a new bond between the chlorine and hydrogen atoms.

▲ **Figure 3** *The reaction mechanism of a chlorine radical with hydrogen*

The **curly arrow** indicates the movement of an electron. The 'tail' of the arrow shows where the electron starts and the 'head' shows where it finishes. Look carefully at the head of the arrow – it is drawn this way to show the movement of a *single* electron. Full-headed curly arrows indicate movement of a *pair* of electrons. Half-headed curly arrows indicate movement of a single electron (Figure 4).

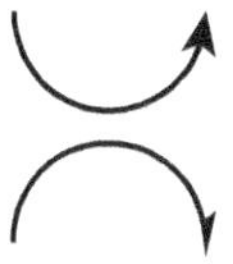

A hydrogen radical is formed in this reaction. This is also highly reactive and it will combine with another molecule, once again creating a new radical – and so it goes on. It is a **radical chain reaction** and it has three key stages:

Initiation Chlorine radicals are initially formed by the photodissociation of chlorine molecules:

$$Cl_2 + hv \rightarrow Cl• + Cl•$$

Cl• are so reactive that they soon react with something else – they *initiate* the reaction.

Propagation Chlorine radicals react with hydrogen molecules. The hydrogen radicals go on to react with chlorine molecules:

$$Cl• + H_2 \rightarrow HCl + H•$$
$$H• + Cl_2 \rightarrow HCl + Cl•$$

These reactions produce new radicals which keep the reaction going – they *propagate* the reaction.

Termination Occasionally two radicals collide with each other. When this happens the reaction is *terminated* because the radicals have been removed:

$$H• + H• \rightarrow H_2$$
$$Cl• + Cl• \rightarrow Cl_2$$
$$H• + Cl• \rightarrow HCl$$

The overall effect is to convert hydrogen and chlorine into hydrogen chloride:

$$H_2 + Cl_2 \rightarrow 2HCl$$

▲ **Figure 4** *Full-headed (top) and half-headed (bottom) curly arrows*

Synoptic link

You were first introduced to the use of curly arrows to represent the movement of electrons in Chapter 2, Developing fuels.

Activity OZ 3.1

This activity helps you check your understanding of radical mechanisms.

Methane and chlorine

Alkanes are generally considered to be unreactive, although they will react with halogens in the presence of light. This is another radical chain reaction, and has the same three key stages.

Initiation   Chlorine radicals are formed from chlorine molecules:

$$Cl_2 + h\nu \rightarrow Cl^\bullet + Cl^\bullet$$

Propagation Chlorine radicals react with methane molecules. The resulting methyl radicals can then react with chlorine molecules:

$$Cl^\bullet + CH_4 \rightarrow HCl + CH_3^\bullet$$
$$CH_3^\bullet + Cl_2 \rightarrow CH_3Cl + Cl^\bullet$$

These two steps *propagate* the reaction.

Termination The reaction ends when two free radicals combine:

$$Cl^\bullet + Cl^\bullet \rightarrow Cl_2$$
$$CH_3^\bullet + Cl^\bullet \rightarrow CH_3Cl$$
$$CH_3^\bullet + CH_3^\bullet \rightarrow C_2H_6$$

The effect is to produce hydrogen chloride, chloromethane, and small amounts of ethane. Further substitution may occur to form dichloromethane and trichloromethane.

> ### Study tip
>
> You should be able to identify whether radical reactions are initiation, propagation, or termination.

> ### Activity OZ 3.2 and OZ 3.3
>
> These activities involve worked with radical reactions in the laboratory.

Summary questions

1 State whether or not each of the following species is a radical. You may need to draw electron dot–cross diagrams to help you to decide.
   a F   (*1 mark*)   c H_2O (*1 mark*)   e NO_2 (*1 mark*)
   b Ar (*1 mark*)   d OH   (*1 mark*)   f CH_3 (*1 mark*)

2 The hydroxyl radical, $HO^\bullet$, is an important species in atmospheric chemistry. Reaction **A** shows one process in which $HO^\bullet$ is produced. The reaction is brought about by radiation with a wavelength below 190 nm.

   Reaction A $H_2O + h\nu \rightarrow H^\bullet + HO^\bullet$

   Hydroxyl radicals are very reactive and act as scavengers in the atmosphere. One set of reactions which involve stratospheric ozone is:

   Reaction B $HO^\bullet + O_3 \rightarrow HO_2^\bullet + O_2$
   Reaction C $HO_2^\bullet + O_3 \rightarrow HO^\bullet + 2O_2$

   a State whether reaction A is a homolytic or a heterolytic process.   (*1 mark*)
   b Explain whether reactions **A**, **B**, and **C** are initiation, propagation, or termination.   (*1 mark*)
   c i Write an equation which shows the overall result of reactions **B** and **C**.
      ii State the role of $HO^\bullet$ in this process.   (*2 marks*)

3 The creation of nitrogen monoxide from human activities is of concern because it is thought to lead to a loss of ozone from the stratosphere. Reactions **D** and **E** show how this loss can occur.

Reaction D $NO^{\bullet} + O_3 \longrightarrow NO_2^{\bullet} + O_2$ $\Delta H = -100\,kJ\,mol^{-1}$

Reaction E $NO_2^{\bullet} + O \longrightarrow NO^{\bullet} + O_2$ $\Delta H = -192\,kJ\,mol^{-1}$

  a State one human activity which leads to the production of a significant amount of NO.   (*1 mark*)
  b i Deduce the the overall effect of reactions **D** and **E**.
     ii State the role of NO in this process.
     iii Calculate the value of ΔH for the overall process.   (*3 marks*)

4 Reactions F to K show various processes involving ethane.

Reaction F $C_2H_6 \longrightarrow 2CH_3^{\bullet}$

Reaction G $CH_3^{\bullet} + C_2H_6 \longrightarrow CH_4 + C_2H_5^{\bullet}$

Reaction H $C_2H_5^{\bullet} \longrightarrow C_2H_4 + H^{\bullet}$

Reaction I $H^{\bullet} + C_2H_6 \longrightarrow H_2 + C_2H_5^{\bullet}$

Reaction J $2C_2H_5^{\bullet} \longrightarrow C_2H_4 + C_2H_6$

Reaction K $2C_2H_5^{\bullet} \longrightarrow C_4H_{10}$

         $(C_2H_4$ is ethene, $CH_2{=}CH_2)$

  a State whether these reactions are initiation, propagation, or termination.   (*3 marks*)
  b For most of these reactions, you would need information about bond enthalpies in order to decide whether the process is exothermic or endothermic. But two of the reactions can be classified by inspection.
     i Explain which reaction is endothermic and which reaction is exothermic.
     ii Explain your answer to i.   (*4 marks*)
  c State the names and formulae of the chemical species in the reaction sequence which are radicals.   (*3 marks*)

5 The radical chain reaction of methane (CH_4) with chlorine in the presence of sunlight is given below.

$$CH_4 + Cl_2 \longrightarrow CH_3Cl + HCl$$

  a Suggest a mechanism for the reaction showing clearly which reactions correspond to the initiation, propagation, and termination stages.   (*5 marks*)
  b Explain why the reaction product also contains some CH_2Cl_2, $CHCl_3$, and CCl_4.   (*3 marks*)

OZ 4 Ozone – here today and gone tomorrow

Learning outcomes

Demonstrate and apply knowledge and understanding of:

→ activation enthalpy and enthalpy profiles

→ the effect of concentration and pressure on the rate of a reaction, explained in terms of the collision theory

→ use of the concept of activation enthalpy and the Boltzmann distribution to explain the qualitative effect of temperature changes and catalysts on rate of reaction

→ techniques and procedures for experiments in reaction kinetics.

Reactions 1–5 in Topic OZ 3 show that ozone is being made and destroyed all the time. Left to themselves, these reactions would reach a point where ozone was being made as fast as it was being used up. This is called a steady state.

It's like the situation in Figure 1 when you are running water into a basin with the plug out of the waste pipe. When water is running out as fast as it's running in, the level of water stays constant. If you turned the tap on more, the water would rise, but it would run out faster because of the higher pressure. Before long you would reach a steady state again, but this time with more water in the basin.

The water coming out of the tap is like the reactions producing ozone, and the water going down the waste pipe is like the reactions that destroy it.

Chemists have studied the rates of the reactions that produce and destroy ozone and can write mathematical expressions for all the reactions involved. From these expressions, they can work out what the concentration of ozone should be in different circumstances. In the Ideas section below you will find out some of the ways that chemists calculate the rates of reactions.

▲ Figure 1 A steady state situation

Chemical ideas: Rates of reactions 10.1

Factors affecting reaction rates

Some chemical reactions, such as burning fuel in a car engine or precipitating silver chloride from solution, go very fast. Others, such as the souring of milk or the rusting of iron, are much slower. The study of the **rate of reaction**, or **reaction kinetics**, helps chemists to find ways of controlling and predicting chemical reactions, and to understand the mechanisms of chemical reactions. When you talk about the rate of something, you mean the rate at which some quantity changes. When you talk about the rate of reaction, you mean the rate at which reactants are converted into products.

In the decomposition of hydrogen peroxide, H_2O_2, the rate of the reaction means the rate at which $H_2O(l)$ and $O_2(g)$ are formed, which is the same as the rate at which $H_2O_2(aq)$ is used up.

$$2H_2O_2 \text{ (aq)} \rightarrow H_2O(l) + O_2 \text{ (g)}$$

You could measure the rate of this reaction in moles of water or oxygen formed per second, or moles of hydrogen peroxide used up per second.

Measuring rate of reaction

To determine the rate of a reaction you need to measure a property that changes during the reaction and that is proportional to the

▲ Figure 2 Decomposition of hydrogen peroxide

concentration of a particular reactant or product. The rate can then be calculated using:

$$\text{rate of reaction} = \frac{\text{change in property}}{\text{time taken}}$$

Measuring volumes of gases evolved

The reaction between calcium carbonate and hydrochloric acid produces carbon dioxide which can be collected. The volume produced can be used to follow the reaction rate. This method can be used for a range of experiments where a gas is evolved.

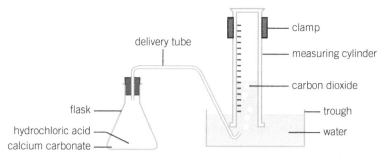

▲ Figure 3 *Measuring the rate of a gas produced*

Measuring mass changes

A different way of monitoring the same reaction is by recording the mass lost (in the form of carbon dioxide) from the reaction.

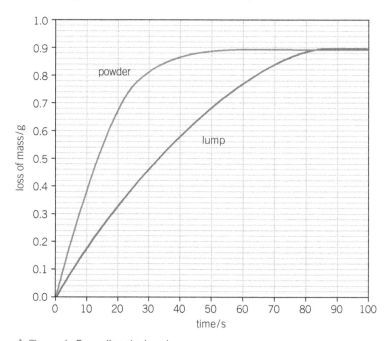

▲ Figure 4 *Recording the loss in mass*

Synoptic link

You first encountered pH in Chapter 1, Elements of life.

Synoptic link

You will learn about rate equations in Chapter 9, Developing metals.

pH measurement

As the above reaction proceeds, the hydrochloric acid concentration will fall, as it reacts with the calcium carbonate. Measuring the pH of the reaction mixture is another way of following the reaction. Again, this method is suitable for a range of reactions.

▲ **Figure 5** *When zinc reacts with copper sulfate solution the blue colour of CuSO₄ fades*

Synoptic link 🧪

How to use a colorimeter can be found in Techniques and procedures.

Colorimetry

A **colorimeter** measures the change in colour of a reaction. When zinc reacts with aqueous copper(II) sulfate (Figure 5) the blue coloration of the copper sulfate solution decreases and the reaction can be followed using a colorimeter.

$$Zn(s) + CuSO_4(aq) \rightarrow ZnSO_4(aq) + Cu(s)$$

Chemical analysis

All the techniques described so far have not interfered with the progress of the reaction. Chemical analysis, however, involves taking samples of the reaction mixture at regular intervals, and stopping the reaction in the sample (quenching it), before analysis.

For example, iodine and propanone react in the presence of an acid catalyst. A sample can be extracted from the reaction mixture and quenched by the addition of sodium hydrogen carbonate. This neutralises the acid catalyst, effectively stopping the reaction. The amount of unreacted iodine remaining can then be determined by titration.

What affects the rate of a reaction?

The rate of a chemical reaction may be affected by:

- The *concentration* of the reactants. In solutions, concentration is measured in $mol\,dm^{-3}$ – in gases, the concentration is proportional to the *pressure*.

- The *temperature*. Nearly all reactions go faster at higher temperatures.

- The *intensity of radiation*, if the reaction involves radiation. Dissociation of oxygen molecules happens faster when the intensity of the radiation increases.

- The *particle size* of a solid. A powder reacts faster than a lump of solid because there is a much larger *surface area* of solid exposed for reaction to take place on.

- The presence of a *catalyst*.

The collision theory of reactions

You can explain the effect of these factors using the **collision theory**. The idea is that reactions occur when particles of reactants collide with a certain minimum kinetic energy. An ozone molecule and a chlorine atom in the stratosphere must collide in order to react. This will happen more often if there are more particles in a given volume – so increasing concentration speeds up the reaction.

But not every collision causes a reaction. As the particles approach and collide. When a plot is drawn of reaction progress against enthalpy, this is known as an **enthalpy profile** (Figure 6).

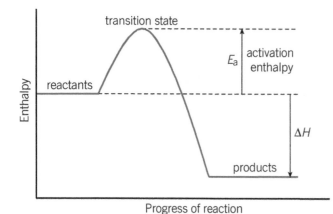

▲ **Figure 6** *Enthalpy profile for an exothermic reaction*

Existing bonds start to stretch and break and new bonds start to form. Only those pairs with enough combined energy on collision to overcome the energy barrier, or **activation enthalpy**, for the reaction will go on to produce products. At higher temperatures, a much larger proportion of colliding pairs have enough energy to react.

Enthalpy profiles

Plots like the one in Figure 6 are a useful way of picturing the energy changes that take place as a reaction proceeds. In going from reactants to products, the highest point on the pathway corresponds to the **transition state**, where old bonds stretch and new bonds start to form. This state exists for only a very short time. The curve in Figure 6 applies to a simple one-step reaction. Many reactions actually take place in a series of steps and there will be many curves – one for each step.

Catalysts provide an alternative reaction pathway of lower activation energy, thereby increasing the rate.

Chemical ideas: Rates of reactions 10.2

The effect of temperature on rate

Temperature has an important effect on the rate of chemical reactions. If you've measured the rates of reactions at different temperatures, you'll know that for many reactions the rate is roughly doubled by a temperature rise of just 10 °C.

The distribution of energies

At any temperature, the speeds of the molecules in a substance are distributed over a wide range. Just as the walking speeds of people in a street, some are moving slowly and some quickly, but the majority are moving at moderate speeds.

This distribution of kinetic energies in a gas at a given temperature is shown in Figure 7. The pattern is called the **Maxwell–Boltzmann distribution**.

Synoptic link

You first encountered enthalpy in Chapter 2, Developing fuels.

Activation enthalpy

Activation enthalpy is the minimum kinetic energy required by a pair of colliding particles before reaction will occur.

Synoptic link

You were introduced to catalysis in Chapter 2, Developing fuels.

Activity OZ 4.1

This activity you will have the opportunity to measure the rate of reactions at different temperatures.

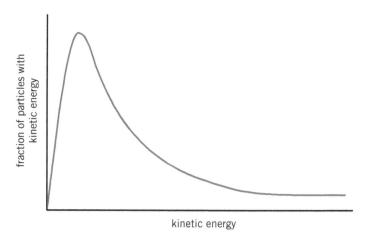

▲ **Figure 7** *Distribution curve for molecular kinetic energies in a gas*

As the temperature increases, more molecules move at higher speeds and have higher kinetic energies. Figure 8 shows how the distribution of energies changes when you increase the temperature from 300 to 310 K. There is still a spread of energies, but now a greater proportion of molecules have higher energies. The are under the curve is the same, but the distribution of energies is different.

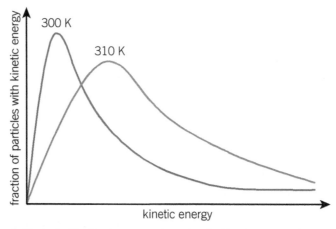

▲ **Figure 8** *Distribution curves for molecular kinetic energies in a gas at 300 K and 310 K*

Now let's look at the significance of this for reaction rates. Imagine a reaction whose activation enthalpy E_a is + 50 kJ mol^{-1}, a value typical of many reactions.

Figure 9 shows the number of collisions with energy greater than 50 kJ mol^{-1} for the reaction at 300 K – it's given by the shaded area underneath the curve. Only those collisions with energies in the shaded area can lead to a reaction.

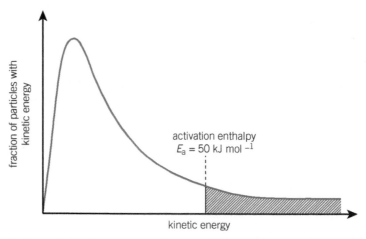

▲ **Figure 9** *Distribution curve showing collisions with energy 50 kJ mol^{-1} and above*

Now look at the graph in Figure 10, which shows the curves for both 300 K and 310 K. At the higher temperature a significantly higher proportion of molecules have energies above 50 kJ mol^{-1}.

Reactions go faster at higher temperatures because a larger proportion of the colliding molecules have the minimum activation enthalpy needed to react.

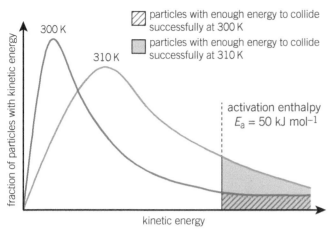

▲ Figure 10 *Distribution curves showing the effect of changing the temperature, from 300 K to 310 K, on the proportion of collisions with energy 50 kJ mol⁻¹ and above*

For many reactions, the rate is roughly doubled by a 10 °C temperature rise. It is only a rough rule, and in fact it only works exactly for reactions with an activation enthalpy of +50 kJ mol⁻¹ and for a temperature rise from 300 to 310 K, but it is a reasonable rough guide. The greater the activation enthalpy, the greater is the effect of increasing the temperature on the rate of the reaction.

Catalysts provide an alternative pathway for a reaction with a lower activation energy. This means that at any given temperature, there will be a larger proportion of particles that on collision produce a successful collision (Figure 11).

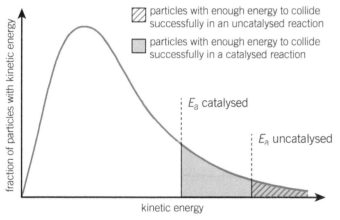

▲ Figure 11 *Distribution curve showing the effect of a catalyst on the proportion of particles with energy for successful collisions*

Synoptic link

You will find out more about the role of catalysts in Chapter 6, The chemical industry.

Synoptic link

You first came across catalysts in Chapter 2, Developing fuels

Summary questions

1 Use the collision theory to explain:
   a why coal burns faster when it is finely powdered than when it is in a lump   *(1 mark)*
   b why nitrogen and oxygen in the atmosphere do not normally react to form nitrogen oxides   *(1 mark)*
   c why reactions between two solids take place very slowly   *(1 mark)*
   d why flour dust in the air can ignite with explosive violence.   *(1 mark)*

2 The collision theory assumes that the rate of a reaction depends on:

A the rate at which reactant molecules collide with one another

B the proportion of reactant molecules that have enough energy to react once they have collided.

Which, out of **A** and **B**, explains each of the following observations?

a Reactions in solution go faster at higher concentration.   (*1 mark*)

b Solids react faster with liquids or gases when their surface area is greater.   (*1 mark*)

c Catalysts increase the rate of reactions.   (*1 mark*)

d Increasing the temperature increases the rate of a reaction.   (*1 mark*)

3 For each of the following reactions, state which of the following factors might affect the rate.

A temperature

B total pressure of gas

C concentration of solution

D surface area of solid.

a The reaction of magnesium with hydrochloric acid

$$Mg(s) + 2HCl(aq) \rightarrow MgCl_2(aq) + H_2(g)$$ (*1 mark*)

b The reaction of nitrogen with hydrogen in the presence of an iron catalyst

$$N_2(g) + 3H_2(g) \rightarrow 2NH_3(g)$$ (*1 mark*)

c The decomposition of aqueous hydrogen peroxide

$$2H_2O_2(aq) \rightarrow 2H_2O(l) + O_2(g)$$ (*1 mark*)

4 A mixture of hydrogen and oxygen doesn't react until it is ignited by a spark – then it explodes. The mixture also explodes if you add some powdered platinum.

a The energy of a spark is tiny, yet it is enough to ignite any quantity of a hydrogen/oxygen mixture, large or small. Suggest an explanation for this.   (*2 marks*)

b Explain why platinum makes the hydrogen/oxygen reaction occur at room temperature.   (*2 marks*)

5 The activation enthalpy for the decomposition of hydrogen peroxide to oxygen and water is $+36.4\,kJ\,mol^{-1}$ in the presence of an enzyme catalyst and $+49.0\,kJ\,mol^{-1}$ in the presence of a very fine colloidal suspension of platinum. The overall reaction is exothermic.

a Sketch, on the same enthalpy diagram, the enthalpy profiles for both catalysts.   (*2 marks*)

b How will the rate of the decomposition of hydrogen peroxide differ for the two catalysts at room temperature? Explain your answer.   (*2 marks*)

OZ 5 What is removing the ozone?

Specification reference: OZ(g), OZ(h), OZ(q)

When chemists compare their calculated concentrations of ozone from the expected reaction rates with the actual measured values, they find that the measured values are a lot lower than expected – ozone is being removed faster than expected. Going back to the analogy of the basin and the running tap from Topic OZ 4, it's as if the waste pipe had been made larger. But by what?

You have seen that ozone is very reactive and reacts with oxygen atoms. But oxygen atoms aren't the only radicals to be found in the stratosphere.

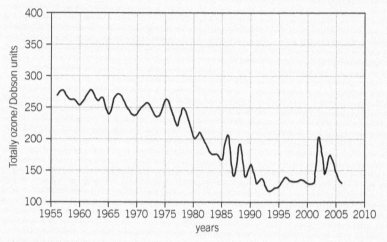

▲ **Figure 1** *Average levels of ozone at the Halley station in Antarctica in October between 1955 and 2006*

Chemical ideas: Rates of reaction 10.4

How do catalysts work?

In a chemical reaction, existing bonds in the reactants must first stretch and break. Then new bonds can form as the reactants are converted to products.

Bond breaking is an endothermic process. A pair of reacting particles must collide with an energy greater than the activation energy before any reaction can occur.

Catalysts speed up reactions by providing an alternative reaction pathway with a lower activation enthalpy for the breaking and remaking of bonds. Now that the energy barrier is lower, more pairs of molecules can react when they collide, so the reaction proceeds more quickly.

The enthalpy change ΔH is the same for the catalysed and uncatalysed reactions. Figure 2 shows the enthalpy profiles for an uncatalysed and a catalysed reaction. Figure 3 shows the Maxwell–Boltzmann distribution for an uncatalysed and a catalysed reaction.

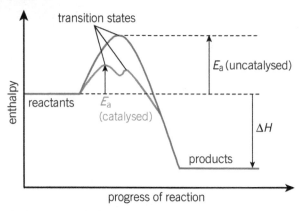

Figure 2 *The effect of a catalyst on the enthalpy profile for a reaction*

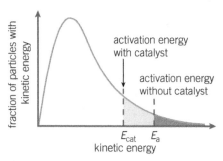

▲ **Figure 3** *A catalyst provides an alternative route with a lower activation energy, allowing a greater proportion of molecules to react*

Catalysts and equilibrium

Catalysts do not affect the position of equilibrium in a reversible reaction. They alter the *rate* at which the equilibrium is attained, but not the *composition* of the equilibrium mixture.

Homogeneous catalysts

Homogeneous catalysts are in the same physical state as the reactants in the reaction. They normally work by forming an intermediate compound with the reactants (sometimes called a transition state). That is why the enthalpy profile for the catalysed reaction in Figure 2 has two humps – one for each step. The intermediate compound then breaks down to give the product and reform the catalyst.

An example of homogeneous catalysis is the destruction of ozone in the stratosphere by chlorine and bromine atoms.

How do chlorine and bromine atoms delete ozone?

Small amounts of chloromethane, CH_3Cl, and bromomethane, CH_3Br, reach the stratosphere as a result of natural processes. Once in the stratosphere, their molecules are split up by solar radiation to give chlorine atoms and bromine atoms. CH_3Cl and CH_3Br are examples of compounds called haloalkanes. In the following reactions the unpaired electron of the radicals is not shown.

Chlorine atoms react with ozone like this:

$$Cl + O_3 \rightarrow ClO + O_2 \quad \textbf{Reaction 1}$$

and bromine atoms react in the same way.

The ClO formed is another reactive radical and can react with oxygen atoms:

$$ClO + O \rightarrow Cl + O_2 \quad \textbf{Reaction 2}$$

So now you have two radicals (Topic OZ 3) competing with each other to remove ozone from the stratosphere:

$$O + O_3 \rightarrow O_2 + O_2 \quad \textbf{Reaction 3}$$
$$Cl + O_3 \rightarrow ClO + O_2 \quad \textbf{Reaction 1}$$

The concentration of Cl atoms in the stratosphere is much less than the concentration of O atoms. So how significant is Reaction 1?

Homogeneous catalyst

Homogenous catalysts are catalysts that are in the same physical state as the reactants in the reaction. They normally work by forming an intermediate compound with the reactants.

The reaction of O_3 with Cl atoms would not matter much if it took place a lot more slowly than the reaction of O_3 with O atoms. Chemists have shown that, at temperatures and pressures similar to those in the stratosphere, the reaction of O_3 with Cl atoms is more than 1500 times faster than the reaction of O_3 with O atoms.

What's more, the Cl atoms used in Reaction 1 are regenerated in Reaction 2 – and can then go on to react with more O_3. Adding together the equations for Reactions 1 and 2 gives the equation for the overall reaction.

$$Cl + O_3 \rightarrow ClO + O_2 \qquad \textbf{Reaction 1}$$

and

$$ClO + O \rightarrow Cl + O_2 \qquad \textbf{Reaction 2}$$

overall reaction:

$$O + O_3 \rightarrow O_2 + O_2$$

The chlorine atoms act as a homogeneous catalyst for this reaction. By going through the catalytic cycle many times, a single chlorine atom can remove about one million ozone molecules. So you can see why even low concentrations of chlorine atoms can be devastating.

A similar catalytic cycle involves bromine atoms and, although the concentration of bromine atoms is much lower, bromine is about 100 times more effective in destroying ozone than chlorine.

Other ways ozone is removed

Chlorine and bromine atoms aren't the only radicals present in the stratosphere which can destroy ozone in a catalytic cycle in this way.

If you represent the radical by the generic symbol X, you can rewrite Reactions 1 and 2 as a general catalytic cycle:

$$X + O_3 \rightarrow XO + O_2$$

and

$$XO + O \rightarrow X + O_2$$

overall reaction:

$$O + O_3 \rightarrow O_2 + O_2$$

Two other important radicals (the hydroxyl radical and nitrogen monoxide) which can destroy ozone in this way are described below.

Hydroxyl radicals, HO

These are formed by the reaction of oxygen atoms with water in the stratosphere. They react with ozone like this:

$$HO + O_3 \rightarrow HO_2 + O_2$$

The HO_2 radicals then go on to react with oxygen atoms to reform the HO radicals:

$$HO_2 + O \rightarrow HO + O_2$$

So this is another example of a catalytic cycle, and the HO radicals released can go on to react with more O_3 molecules.

Nitrogen monoxide, NO

Nitrogen monoxide reacts with ozone to form nitrogen dioxide, NO_2, and dioxygen. Nitrogen dioxide can then react with oxygen atoms to release nitrogen monoxide and dioxygen to complete the catalytic cycle.

NO and NO_2 are both radicals. They are unusual radicals because they are relatively stable molecules and they can be prepared and collected like ordinary molecular substances. (It is important to remember that not all radicals are highly reactive.)

The radicals mentioned in this section (Cl, Br, HO, and NO) are important, but they are only part of the whole picture. Hundreds of reactions have been suggested which affect the gases in the stratosphere.

Many of these have been going on since long before there were humans on Earth. However human activities can have a serious effect on certain key reactions, and so lead to dramatic changes in the concentration of ozone in the stratosphere.

Summary questions

1 The enthalpy profiles **A–D** represent four different reactions – all the diagrams are drawn to the same scale.

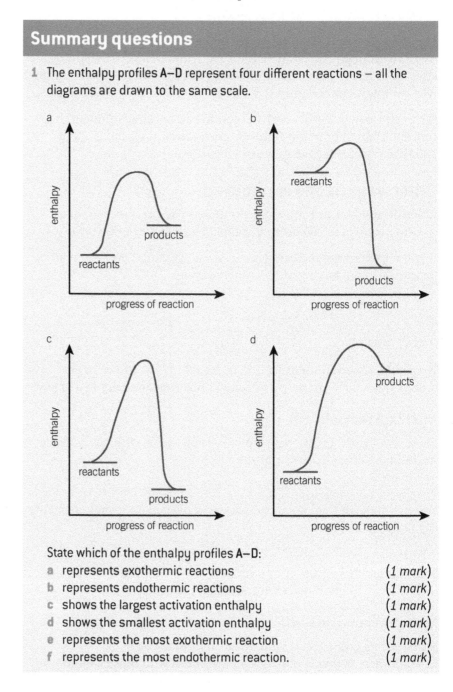

State which of the enthalpy profiles **A–D**:

a represents exothermic reactions (1 mark)
b represents endothermic reactions (1 mark)
c shows the largest activation enthalpy (1 mark)
d shows the smallest activation enthalpy (1 mark)
e represents the most exothermic reaction (1 mark)
f represents the most endothermic reaction. (1 mark)

2 The reaction of chlorine atoms with ozone in the upper atmosphere (stratosphere) is thought to be responsible for the destruction of the ozone layer:

$$Cl + O_3 \rightarrow ClO + O_2$$

The chlorine atoms are produced by the action of high energy solar radiation on CFCs.

a Use collision theory to explain how an increase in the concentration of chlorine atoms increases the rate of reaction with ozone. (2 marks)

b The activation enthalpy for the reaction of chlorine atoms with ozone is relatively small. Explain what effect you would expect a change of temperature to have on the rate of this reaction. (2 marks)

3 The diagram below shows the energy distribution for collisions between Cl atoms and O_3 molecules at temperature T_1. E_a is the activation enthalpy for the reaction of Cl atoms and O_3 molecules.

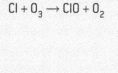

$$Cl + O_3 \rightarrow ClO + O_2$$

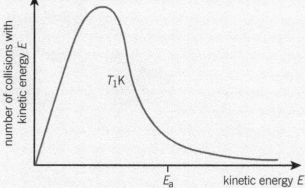

a Copy the diagram and shade the area of the graph to indicate the number of collisions with sufficient energy to lead to a reaction. (1 mark)

b On your graph, indicate the energy distribution for collisions between Cl atoms and O_3 molecules when the temperature of the reactants is increased by 10 K to T_2. Shade (using a different colour) the area under the graph to indicate the number of collisions with sufficient energy to lead to a reaction at T_2. (2 marks)

4 The breakdown of ozone in the stratosphere is catalysed by chlorine atoms. The overall reaction is:

$$O_3 + O \rightarrow 2O_2$$

a The Cl atoms act as a homogeneous catalyst in this process. State the meaning of the term homogeneous. (1 mark)

b The enthalpy profiles for the catalysed and uncatalysed reactions are shown in Figure 1. (2 marks)

  i Explain why the enthalpy profile for the catalysed reaction has two humps.

  ii Describe what is happening at the peaks and troughs on the two curves.

OZ 6 The CFC story

Specification reference: OZ(a), OZ(b), OZ(d), OZ(j), OZ(k)

Synoptic link

CFCs have similar structures to alkanes which you studied in Chapter 2, Developing fuels, with some hydrogen atoms substituted by halogen (Group 7) atoms.

CFCs: very handy compounds

In 1930 the American engineer Thomas Midgley, Jr demonstrated a new refrigerant to the American Chemical Society. He inhaled a lungful of dichlorodifluoromethane, CCl_2F_2, and used it to blow out a candle.

Midgley was flamboyantly demonstrating the lack of toxicity and lack of flammability of CCl_2F_2. Up to that time, ammonia had been the main refrigerant in use, but unfortunately it is toxic and very smelly.

Midgley had found what appeared to be the ideal replacement for ammonia. CCl_2F_2 belongs to a family of compounds called **chlorofluorocarbons** (CFCs) that contain chlorine, fluorine, and carbon. CFCs were used as:

- refrigerants and in air conditioning units

- aerosol propellants

- blowing agents for expanded plastics such as polystyrene

- dry cleaning solvents.

By the early 1970s, industry was producing about a million tonnes of CFCs a year.

▲ Figure 1 Thomas Midgley (1889–1944) developed CFCs and pioneered the use of lead compounds as 'anti-knock' agents in petrol

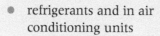

▲ Figure 2 The structures of two common CFCs

Haloalkanes

Properties and naming of haloalkanes

Organic halogen compounds have one or more halogen atoms (F, Cl, Br, or I) attached to a hydrocarbon backbone. Their occurrence in nature is limited, but they are very useful in chemical synthesis.

The halogen atom modifies the properties of the relatively unreactive hydrocarbon chain. The simplest examples are the **haloalkanes** (sometimes called halogenoalkanes), with the halogen atom attached to an alkane chain (Figure 3).

The haloalkanes are named after the parent alkanes, using the same basic rules as for naming alcohols – except that the halogen atom is added as a prefix to the name of the parent alkane.

Therefore, $CH_3CH_2CH_2Cl$ is called 1-chloropropane whilst $CH_3CHClCH_2Cl$ is called 1,2-dichloropropane. $CH_3CHBrCH_2CH_2Cl$ is called 3-bromo-1-chlorobutane.

▼ Table 1 *Naming haloalkanes*

Full structural formula	Skeletal formula	Name
H—C—C—C—Cl (with H atoms)	(skeletal with Cl)	1-chloropropane
H—C—C—C—Cl (with Cl, H atoms)	(skeletal with Cl, Cl)	1,2-dichloropropane
H—C—C—C—C—Cl (with Br, H atoms)	(skeletal with Br, Cl)	3-bromo-1-chlorobutane

The prefixes bromo- and chloro- are listed in alphabetical order. The numbers used to show the positions of the bromine and chlorine atoms are the lowest ones possible, for example, 3 and 1 rather than 2 and 4 in the case of 3-bromo-1-chlorobutane.

Physical properties of haloalkanes

The properties and reactions of haloalkanes are largely determined by the fact that carbon–halogen bonds are polar. Understanding the bond polarity helps explain the boiling points, intermolecular bonds, and reactions of haloalkanes.

Chemical ideas: Structure and properties 5.3

Bonds between molecules: temporary and permanent dipoles

Polar bonds

Figure 4 shows the way a hydrogen molecule is bonded.

The atoms are held together because both nuclei are attracted to the shared electrons between them. Both atoms are identical so the electrons are shared equally.

With larger atoms, it is the atomic cores (everything except the outer shell electrons) that are attracted to the shared electrons. When the two atoms bonded together are different sizes, the core of the smaller atom is closer to the shared electrons and exerts a stronger pull on them (Figure 5a). A similar situation arises when the atoms are from different groups in the periodic table and have different core charges (Figure 5b).

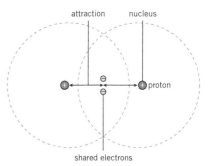

1-chloropropane

▲ Figure 3 *1-chloropropane is a typical haloalkane*

Synoptic link

You learnt about covalent bonding in Chapter 1, Elements of life, and Chapter 2, Developing fuels.

▲ Figure 4 *A hydrogen molecule – the electrons are held an equal distance from each of the two hydrogen nuclei because they are identical*

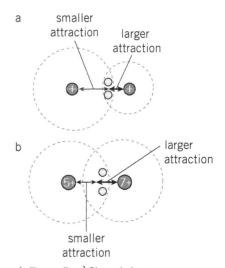

▲ Figure 5 *a) Shared electrons are attracted more strongly by the core of the smaller atom, which is closer b) Shared electrons are attracted more strongly by the atom with the greater core charge*

δ− 　　　　δ+
O ⁚——— H

More negative　　More positive
end of the bond　　end of the bond

because O has greater
share of electrons

▲ **Figure 6** *The O—H covalent bond is polar – the two electrons in the bond are drawn closer to the oxygen atom than the hydrogen atom*

Study tip

δ+ and δ− are referred to as *partial charges*.

Electronegativity

Electronegativity is a measure of the ability of an atom in a molecule to attract electrons in a chemical bond to itself.

▼ **Table 2** *Boiling points of some organic halogen compounds*

Compound	State at 298 K	Boiling point / K
CH_3F	gas	195
CH_3Cl	gas	249
CH_3Br	gas	277
CH_3I	liquid	316
CH_2Cl_2	liquid	313
$CHCl_3$	liquid	335
CCl_4	liquid	350
C_6H_5Cl	liquid	405

Activity OZ 6.1

In this activity you can investigate bond polarity.

When atoms attract bonding electrons unequally, one gets a slight negative charge because it has a greater share of the bonding electrons. The other atom consequently becomes slightly positively charged. Bonds like this are called **polar bonds** (Figure 6).

The small amounts of electrical charge are shown by δ− and δ+ where δ (lower case Greek letter delta) means a small amount of.

Some bonds are polar and some are not. The O—H bond is a particularly important example of a polar covalent bond. The polarity of the O—H bond has many consequences for the chemistry of water molecules, as you will learn in Topic OZ 7. For example, polar substances tend to be soluble in water whilst non-polar substances tend not to dissolve.

Electronegativity

To decide the polarity of a covalent bond, you need a measure of each atom's attraction for bonding electrons – its **electronegativity**. The electronegativity values in Figure 8 are derived from a method suggested by Linus Pauling.

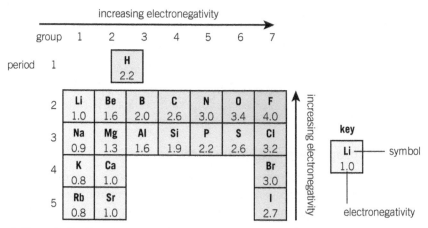

▲ **Figure 8** *Pauling electronegativity values for some main group elements in the Periodic Table, showing trends in groups and periods*

You can now predict how polar a particular covalent bond will be. In the C—F bond, fluorine has a higher electronegativity value (4.0) than carbon (2.6), so it attracts the shared electrons more strongly and the polarity of the bond is:

$$\begin{matrix} δ+ & δ- \\ C & - & F \end{matrix}$$

Electronegativity and haloalkanes

Carbon–halogen bonds are polar, but not enough to make a big difference to the physical properties of the compounds. For example, all haloalkanes are immiscible with water, like alkanes. Their boiling points depend on the halogen atoms present – the bigger the halogen atom and the more halogen atoms there are, Δthe higher the boiling point.

Intermolecular bonds

In a liquid or a solid there are **intermolecular bonds** causing molecules to be *attracted to one another*, otherwise they would move apart and become a gas. When a solid melts and boils (Figure 9), it is the intermolecular bonds that are broken, so boiling point gives a good indication of the strength of bonds *between* the molecules. Any covalent bonds *within* the molecules remain intact.

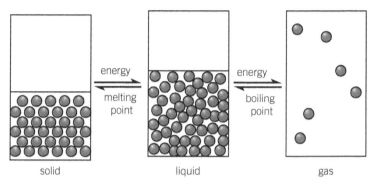

▲ **Figure 9** *Energy must be supplied to break intermolecular bonds. Covalent bonds within the molecules remain intact*

The low boiling points of noble gases (Table 3) show that the bonds *between the atoms* are very weak.

The boiling point trend of alkanes (Table 4) shows that the longer molecules have stronger bonds between them.

▼ **Table 3** *Boiling points for the noble gases*

Element	Boiling point / K
Helium	4
Neon	27
Argon	87
Krypton	121
Xenon	161

▼ **Table 4** *Boiling points of some alkanes*

Alkane	Structural formula	Boiling point / K	Alkane	Skeletal formula	Boiling point / K
methane	CH_4	111	hexane		342
ethane	CH_3CH_3	184	3-methylpentane		336
propane	$CH_3CH_2CH_3$	231	2-methylpentane		333
butane	$CH_3CH_2CH_2CH_3$	273	2,3-dimethylbutane		331
pentane	$CH_3CH_2CH_2CH_2CH_3$	309	2,2-dimethylbutane		323

For alkanes, the longer the chain, the stronger the intermolecular forces and the higher the boiling point. More energy is needed to break these stronger bonds and separate the molecules.

The table shows some branched isomers of hexane, C_6H_{14}. In straight-chain alkanes there are more contacts between molecules and more opportunities for intermolecular bonds to form, so straight-chain alkanes have higher boiling points than their branched isomers.

Dipoles

A **dipole** is a molecule (or part of a molecule) with a positive end and a negative end, because of its polar bonds. When a molecule has a dipole we say it is **polarised**. There are several ways a molecule can become polarised.

Permanent dipoles

Permanent dipoles occur when the two atoms in a bond have substantially different electronegativities. Hydrogen chloride has a permanent dipole because chlorine is much more electronegative than hydrogen, so attracts the shared electrons more.

When the atoms that form a molecule have only a small difference in electronegativity, any dipole will be very small. It is also possible for a molecule to have no overall dipole even though the bonds are polar. For example in CCl_4 each chlorine atom carries a small negative charge and the central carbon is positive.

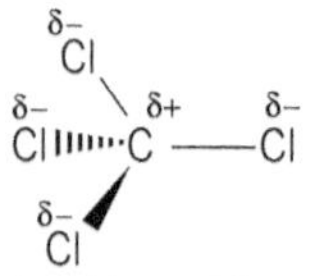

Because the chlorine atoms are distributed tetrahedrally around the carbon, the centre of negative charge is midway between all the chlorines. It is at the centre of the molecule and is superimposed on the positive charge on the carbon. The dipoles cancel due to the symmetry of the molecule and there is no overall dipole – the molecule is **non-polar** (Figure 10).

tetrachloromethane

▲ Figure 10 *The chlorine atoms are distributed symmetrically around the carbon atom, and CCl$_4$ has a tetrahedral shape, like that of methane. CCl$_4$ has polar bonds, but no overall dipole*

Instantaneous dipoles

Chlorine, Cl_2, does not have a permanent dipole, but a *temporary* dipole can arise. The electrons within the Cl—Cl bond are in constant motion, and at a particular instant they may not be evenly distributed. Then one end of the molecule has a greater negative charge than the other end – an **instantaneous dipole** (Figure 11).

Electron cloud evenly distributed–no dipole

At some instant, more of the electron cloud happens to be at one end of the molecule than the other–molecule has an instantaneous dipole

▲ Figure 11 *How a dipole forms in a chlorine molecule*

After an instant the electrons change position, changing the dipole.

Induced dipoles

An **induced dipole** occurs if an unpolarised molecule is next to a dipole. The dipole attracts or repels electrons in the unpolarised molecule, *inducing* a dipole in it (Figure 12).

In Figure 12, the dipole has been induced by a *permanent* dipole. A dipole can also be induced by an *instantaneous* dipole, so a whole series of dipoles can be set up in a substance that contains no permanent dipoles.

> **Study tip**
>
> Bond polarity depends on electronegativity differences.
>
> Molecular dipoles depend on electronegativity differences *and* the shape of the molecule. In symmetrical molecules such as CO_2 or CCl_4 the dipoles cancel.

> **Study tip**
>
> In the diagram for Figures 11, 12, and 13 red areas indicated electron deficiency (δ+) and blue areas indicated areas of higher electron density (δ–).

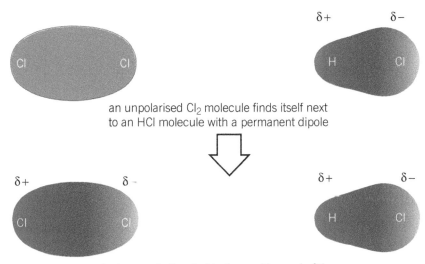

an unpolarised Cl_2 molecule finds itself next to an HCl molecule with a permanent dipole

electrons get attracted to the positive end of the HCl dipole, inducing a dipole in the Cl_2 molecule

▲ Figure 12 *How a dipole can be induced in a chlorine molecule*

Dipoles and intermolecular bonds

All intermolecular bonds arise from the attractive forces between dipoles. There are three kinds of bond:

- **instantaneous dipole–induced dipole bonds**, for example, in chlorine. These are the weakest type of intermolecular bonding. They act between *all* molecules because instantaneous dipoles can arise in molecules that already have a permanent dipole. However, they are most noticeable in non-polar substances. The more electrons the atoms have the greater the instantaneous dipole–induced dipole effect. This can be seen in the increased boiling points for the noble gases. The larger the atoms, the more electrons, the greater the instantaneous dipole-induced dipole bonds.

- **permanent dipole–permanent dipole bonds**, for example, in hydrogen chloride. These are relatively strong and such molecules are more likely to be liquids or solids.

- **permanent dipole–induced dipole bonds**, for example, between hydrogen chloride and chlorine molecules.

In the liquid state, the molecular dipoles are constantly moving. Sometimes opposite charges are next to each other causing attraction (Figure 13a) and sometimes like charges are next to each other causing repulsion (Figure 13b).

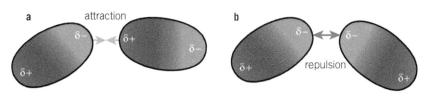

▲ Figure 13 *Molecular dipoles in the liquid state*

Study tip

Make sure that you can identify which substances have which type of intermolecular bond. You should be able to identify these in a range of different substances.

Melting and boiling points of the halogens

In the case of halogens (which form diatomic, non-polar molecules) the strongest type of intermolecular bond that can form is an instantaneous dipole–induced dipole. The smaller the atoms in the molecule the weaker the instantaneous dipole–induced dipole is.

Fluorine has the weakest instantaneous dipole–induced dipole bonds, having the fewest electrons, and therefore has the lowest melting and boiling points of the halogens.

▼ Table 5 *The melting and boiling points and physical appearance of the halogens*

Halogen	Melting point / K	Boiling point / K	Appearance at room temperature
fluorine	53	85	pale yellow gas
chlorine	172	239	green gas
bromine	266	332	dark red volatile liquid
iodine	387	457	black shiny solid

As the molecules get bigger and have more electrons, going down Group 7, the instantaneous dipole–induced dipole bonds increase and the melting and boiling points increase correspondingly.

Damage to the ozone layer

Chemists' concerns were raised in the early 1970s that nitrogen oxides from jet aircraft may be destroying the ozone layer. It turned out that this wasn't a significant problem because the number of aircraft was relatively small.

The issue arose again in 1974 when American professor Sherry Rowland published a paper predicting that CFCs would damage the ozone layer if they reached the stratosphere. For this work, Professor Rowland, together with Professors Mario Molina and Paul Crutzen, was awarded the Nobel Prize for Chemistry in 1995.

Chlorine reservoirs

Left to themselves, chlorine atoms would quickly destroy most of the ozone in the stratosphere, but fortunately there are other molecules there that react with chlorine atoms.

Methane, CH_4, is an important example. It removes chlorine atoms by reacting with them.

$$CH_4 + Cl \rightarrow CH_3 + HCl$$

This reaction is an example of a radical propagation reaction (Topic OZ 3).

HCl is a **chlorine reservoir molecule** because it stores chlorine in the stratosphere.

Another important reaction which produces a chlorine reservoir molecule is the reaction of nitrogen dioxide and chlorine monoxide.

$$NO_2 + ClO \rightarrow ClONO_2$$

These reservoir molecules are both soluble in water and can be removed in raindrops. Most, however, remain in the stratosphere with serious consequences.

Summary questions

1 Name the following haloalkanes.
   a $CHCl_3$ (*1 mark*)   d $CH_3—CHCl—CF_3$ (*1 mark*)
   b $CH_3CHClCH_3$ (*1 mark*)   e $CH_3—CHCl—CBr_2—CH_3$ (*1 mark*)
   c CF_3CCl_3 (*1 mark*)

2 Explain why noble gases have very low boiling points.   (*1 mark*)

3 Draw skeletal formulae showing how two molecules of pentane can approach close to one another. Now do the same for both of its structural isomers, 2-methylbutane, and 2,2-dimethylpropane.
   a Explain the boiling points of the isomers in terms of the strength of the intermolecular bonds present.   (*3 marks*)

Isomer	Boiling point / K
pentane	309
2-methylbutane	301
2,2-dimethylpropane	283

   b Explain the differences in strengths of the intermolecular bonds.   (*2 marks*)

4 Explain which of the following molecules possesses a permanent dipole, considering the shape of the molecule and the electronegativity of its atoms.
   a CO_2 (*1 mark*)   d CH_3OH (*1 mark*)
   b $CHCl_3$ (*1 mark*)   e $(CH_3)_2CO$ (*1 mark*)
   c C_6H_{12} (cyclohexane) (*1 mark*)   f benzene   (*1 mark*)

5 Silane, SiH_4, boils at 161 K whereas hydrogen sulfide, H_2S, boils at 213 K.
   a i State the number of electrons in each molecule.
   ii Explain the strengths of instantaneous dipole–induced dipole bonds in the two compounds.
   iii Explain whether either molecule possesses a permanent dipole.   (*6 marks*)

   b Use your answers in part **a** to explain the different boiling points of these two compounds.   (*3 marks*)

▲ **Figure 1** *Polar stratospheric clouds over the Antarctic. Years in which the temperature is low, with extensive formation of polar stratospheric clouds, correspond to particularly wide and deep ozone holes*

Why does the ozone hole develop over the poles?

Satellite measurements have shown that, whilst ozone concentrations have decreased in other parts of the globe too, the effect is particularly dramatic in the Antarctic spring.

The reasons for this are associated with the special weather conditions occurring in the Antarctic. Firstly, the very low temperatures (below $-80\,°C$) that occur in the polar winter lead to the formation of polar stratospheric clouds. These clouds are made up of tiny solid particles – some are mainly particles of ice, others are rich in nitric acid, HNO_3. These particles provide surfaces on which chemical reactions can occur.

Secondly, during the Antarctic winter a vortex of circulating air forms, which effectively isolates the air at the centre of the vortex. This cold core turns into a giant sealed reaction vessel (Figure 2).

Chlorine reservoir molecules – HCl and chlorine nitrate, $ClONO_2$ – are adsorbed onto the surface of the solid particles in the polar stratospheric clouds, where they react together:

$$ClONO_2 + HCl \rightarrow Cl_2 + HNO_3$$

The HNO_3 remains dissolved in the ice particle, but the chlorine molecules are released as a gas trapped in the isolated core of the vortex. In the Antarctic spring the vortex starts to break up and the chlorine molecules undergo homolytic fission to form chlorine atoms (Figure 3).

Ozone depletion also occurs over the Arctic. The effect is more variable here because temperatures are not usually as low as those above the Antarctic so polar stratospheric clouds are less abundant and a stable polar vortex does not form. However, ozone depletion here extended over more densely populated areas including Canada, the USA, and Europe, and is a cause for concern.

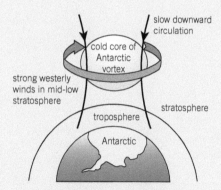

▲ **Figure 2** *The winter vortex over the Antarctic. The cold core is almost isolated from the rest of the atmosphere*

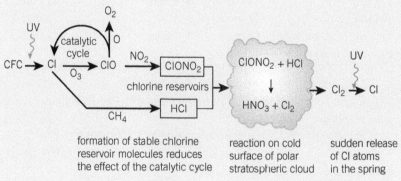

formation of stable chlorine reservoir molecules reduces the effect of the catalytic cycle

reaction on cold surface of polar stratospheric cloud

sudden release of Cl atoms in the spring

▲ **Figure 3** *How reactions in polar stratospheric clouds contribute to the dramatic loss of ozone in the Antarctic spring*

Ice particles are critical in the process above. Water has some interesting properties which can be explained by the bonds between its molecules.

Chemical ideas: Structure and properties 5.4

Bonds between molecules

Hydrogen bonding

Hydrogen bonding is the strongest type of intermolecular bond and can be thought of as a special case of permanent dipole–permanent dipole bonding.

What is special about hydrogen bonding?

For hydrogen bonding to occur, the molecules involved must have the following three features:

- a *large dipole* between a hydrogen atom and a highly electronegative atom such as oxygen, nitrogen, or fluorine
- a *small hydrogen atom* which can get very close to oxygen, nitrogen, or fluorine atoms in nearby molecules
- a *lone pair of electrons* on the oxygen, nitrogen, or fluorine atom that the positively charged hydrogen atom can line up with.

In hydrogen fluoride, HF, the hydrogen atoms have a strong positive charge because they are bonded to the highly electronegative fluorine atom. This positive charge lines up with a lone pair on another fluorine atom. The hydrogen and fluorine atoms can get very close, and therefore attract very strongly, because the hydrogen atom is so small.

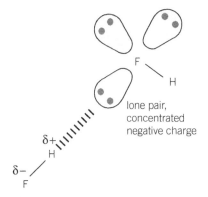

▲ Figure 5 *The positively charged hydrogen atom lines up with a lone pair on a fluorine atom*

> **Activity OZ 7.1**
>
> This activity visibly demonstrates the link between viscosity and hydrogen bonding.

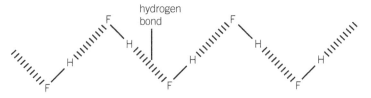

▲ Figure 4 *Hydrogen bonding in liquid hydrogen fluoride*

Dipole–dipole bonds are present in all the hydrogen halides (HF, HCl, HBr, and HI) but the greater strength of the hydrogen bonds in HF can be seen from a comparison of the boiling points. In Table 1, hydrogen fluoride has the highest boiling point despite having the lowest relative molecular mass. This is because the electronegativity of the halogens decreases down the group F>Cl>Br>I.

Water molecules can form twice as many hydrogen bonds as hydrogen fluoride. The oxygen atom possesses two lone pairs of electrons and there are twice as many hydrogen atoms as oxygen atoms (Figure 6).

Water is unique in this respect. In hydrogen fluoride, the fluorine has three lone pairs but there are only as many hydrogen atoms as fluorine atoms – only one-third of the available lone pairs can be used. In ammonia, NH_3, another hydrogen-bonded substance, there is only one lone pair on the nitrogen so, on average, only one of the three hydrogen atoms can form hydrogen bonds.

▼ Table 1 *Boiling points of the hydrogen halides*

Compound	Boiling point / K
HF	292
HCl	188
HBr	206
HI	238

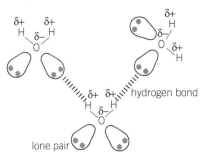

▲ Figure 6 *The positively charged hydrogen atoms line up with the lone pairs on the oxygen atoms*

▼ Table 2 *The effect of hydrogen bonding on boiling point*

Compound	Formula	Relative molecular mass	Hydrogen bonding?	Boiling point / K
Propane	$CH_3CH_2CH_3$	44	no	231
Ethanol	CH_3CH_2OH	46	yes	351
Heptane	$CH_3(CH_2)_5CH_3$	100	no	371
Glycerol	$CH_2(OH)CH(OH)CH_2(OH)$	92	yes	563

Synoptic link

Many covalent substances have N—H and O—H bonds in their molecules. Amino acids and proteins have both types of bonds and hydrogen bonding plays a major role in determining their properties. You will find out more about amino acids and proteins in Chapter 7, Polymers and life. You will also find out more about the properties of water in Chapter 8, The ocean.

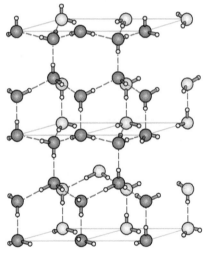

▲ Figure 7 *The arrangement of water molecules in ice*

In Table 2 you can see than the compounds with hydrogen bonding have higher boiling points that compounds with similar molecule masses that do not have hydrogen bonding. This is because more energy is needed to break hydrogen bonds than instantaneous dipole–induced dipole bonds.

Liquids that have hydrogen bonding have a high **viscosity** – viscosity is a measure of how easily a liquid flows. For a liquid to flow the molecules must be able to move past each other and this requires the constant breaking and forming of intermolecular bonds. The stronger the bonds between the molecules, the more difficult this becomes.

Substances with hydrogen bonding are also often soluble in water. Hydrogen bonds can form between water molecules and molecules of the substance, helping the dissolving process. When water freezes it forms ice crystals. The open structure, with four groups around each oxygen atom, maximises the hydrogen bonding between them (Figure 7)

Summary questions

1   a   Explain why the O—H bond is polar.
    b   State which of the covalent bonds in the following list will be significantly polar. In cases where there is a polar bond, show which atom is positive and which is negative using the δ+ and δ– convention.
        a   C—F   (*1 mark*)   e   H—N   (*1 mark*)
        b   C—H   (*1 mark*)   f   S—Br   (*1 mark*)
        c   C—S   (*1 mark*)   g   C—O   (*1 mark*)
        d   H—Cl   (*1 mark*)

2   Hydrogen bonding can occur between different molecules in mixtures. Draw diagrams to show where the hydrogen bonds form in the following mixtures.
    a NH_3 and H_2O (*2 marks*)
    b CH_3CH_2OH and H_2O (*2 marks*)

3   Explain why water exhibits a greater degree of hydrogen bonding than other substances.   (*3 marks*)

OZ 8 What is the state of the ozone layer now?

Specification reference: OZ(j), OZ(k), OZ(l), OZ(m), OZ(n), OZ(q)

Sherry Rowland's predictions come true

Science is all about making predictions, then testing them experimentally. The problem with Sherry Rowland's predictions (see OZ 6) was that they involved a long time scale. However in 1985, scientists examining the atmosphere above the Antarctic made a momentous discovery. Since the mid-1980s ongoing measurements have confirmed the presence of the 'hole' over the Antarctic.

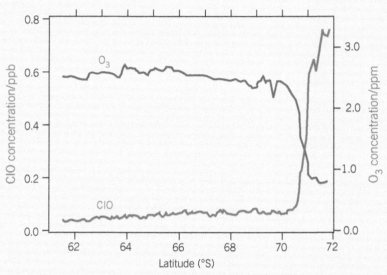

▲ **Figure 1** *These graphs show measurements of ClO radicals (in parts per billion) and ozone (in parts per million) at 18 km altitude. The measurements provided convincing evidence that chlorine radicals are involved in ozone depletion*

Could the trouble with CFCs have been foreseen?

When Thomas Midgley and other chemists developed CFCs, they found a family of compounds that had many useful applications. The trouble is that they are too unreactive.

In the 1930s, questions about environmental consequences of CFCs were simply not on the agenda. Atmospheric chemistry was in its infancy and sensitive instruments that could detect minute concentrations of compounds in the air did not exist.

Laboratory studies and computer modelling allowed scientists to run simulations and make predictions. In the 1980s and 1990s this contributed to an explosion of knowledge of the chemistry of the atmosphere.

It is now known that most organic compounds are broken down in the troposphere by species such as HO radicals and never reach the stratosphere. However the stability of CFCs, initially seen as a huge advantage, proved to be their downfall.

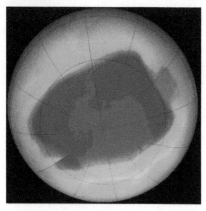

▲ **Figure 2** *The ozone hole over the Antarctic as recorded on 25th September 2006. This was the biggest recorded hole in the ozone layer. Blue represents areas of low ozone levels and green represents areas of high ozone levels*

Activity OZ 8.1

This activity provides an opportunity to develop your chemical literacy skills.

Substitution

A reaction in which one atom or group in a molecule is replaced by another atom or group.

Nucleophile

A nucleophile is a molecule or negatively charged ion with a lone pair of electrons that it can donate to a positively charged atom to form a covalent bond.

Haloalkanes

Chemical reactions of haloalkanes

Reactions of haloalkanes involve breaking the carbon–halogen bond. The bond can break homolytically or heterolytically (Topic OZ 3).

Homolytic fission

Homolytic fission of haloalkanes occurs when haloalkanes reach the stratosphere, where they are exposed to intense ultraviolet radiation. This is how ozone-depleting chlorine radicals are formed.

$$CH_3\text{—}Cl + h\nu \rightarrow CH_3{}^\bullet + Cl^\bullet$$

Haloalkanes can be *formed* by a radical halogenation mechanism (Topic OZ 3).

Heterolytic fission

Heterolytic fission is more common under normal laboratory conditions. The carbon–halogen bond is polar, and can break forming a negative halide ion and a positive **carbocation** (Figure 3). For example, with 2-chloro-2-methylpropane:

$$H_3C\text{—}\underset{\underset{CH_3}{|}}{\overset{\overset{CH_3}{|}}{C}}\text{—}Cl \longrightarrow H_3C\text{—}\underset{\underset{CH_3}{|}}{\overset{\overset{CH_3}{|}}{C}}{}^+ \ + \ Cl^-$$

2-chloro-2-methylpropane   carbocation   chloride ion

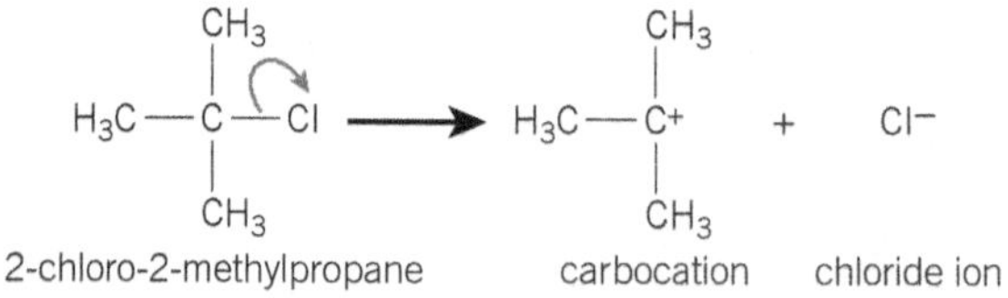

▲ **Figure 3** *2-chloro-2-methylpropane breaks down to a carbocation and a chloride ion*

Sometimes, heterolytic fission is caused by a negatively charged substance reacting with the positively polarised carbon atom, causing a **substitution** reaction.

Nucleophilic substitution reactions of haloalkanes

You have already seen radical reactions for haloalkanes. Another type of reaction haloalkanes take part in are substitution reactions. The halogen atom is substituted by another group. This group is called a **nucleophile**.

Consider the reaction between hydroxide ions, OH⁻, and 1-bromobutane. The haloalkane is heated under reflux with ethanolic sodium hydroxide

- The nucleophile, OH⁻, attacks the electron deficient carbon atom in the C—Br bond.

- The OH⁻ donates two electrons to form a new dative covalent bond.

- The C—Br bond breaks heterolytically and the bromine atom receives two electrons, producing a bromide ion. In this case the bromide ion is called the **leaving group**.

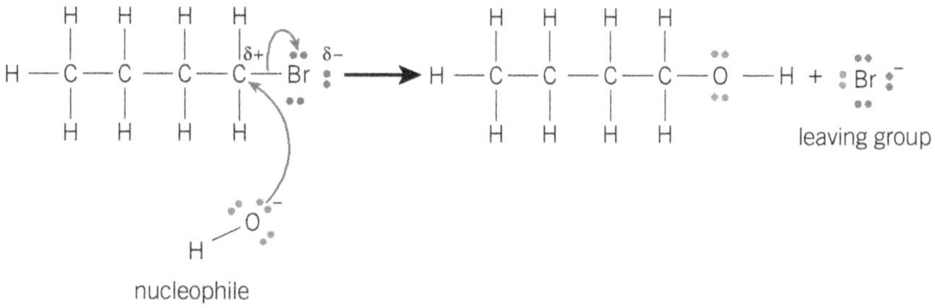

▲ **Figure 4** *The reaction mechanism of bromobutane and hydroxide ions*

Because the overall reaction involves a nucleophile replacing another group, this type of reaction is called a **nucleophilic substitution**.

This reaction involves heterolytic fission. A free carbocation is not formed because the OH^- ion attacks at the same time as the C–Br bond breaks.

Curly arrows show the movement of electrons. In this case a full-headed arrow shows the movement of *a pair* of electrons. Table 2 shows a number of nucleophiles. each nucleophile in Table 2 has a lone pair of electrons.

▼ **Table 2** *Some common nucleophiles*

Name and formula	Structure showing lone pairs
hydroxide ion, OH^-	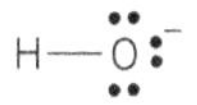 $H-\overset{\cdot\cdot}{\underset{\cdot\cdot}{O}}:^-$
cyanide ion, CN^-	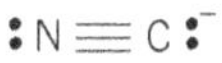 $:N \equiv C:^-$
ethanoate ion, CH_3COO^-	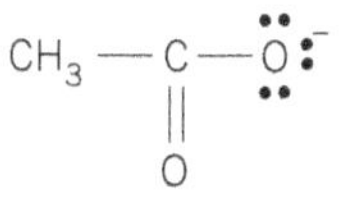 $CH_3-C-\overset{\cdot\cdot}{\underset{\cdot\cdot}{O}}:^-$ with $=O$ below
ethoxide ion, $C_2H_5O^-$	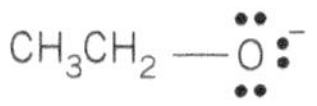 $CH_3CH_2-\overset{\cdot\cdot}{\underset{\cdot\cdot}{O}}:^-$
water, H_2O	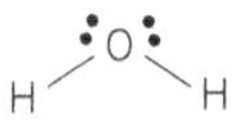 $H\diagdown\overset{\cdot\cdot}{O}\diagup H$
ammonia, NH_3	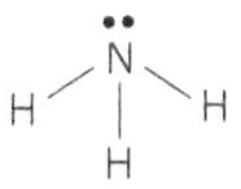 $H\diagdown\overset{\cdot\cdot}{N}\diagup H$ with H below

Using Nu^- as a general symbol for any nucleophile and X as a symbol for a halogen atom, the nucleophilic substitution process can be described by:

$$R{-}X + Nu^- \rightarrow R{-}Nu + X^-$$

The curly arrow moves from the lone pair of electrons to the electron deficient carbon.

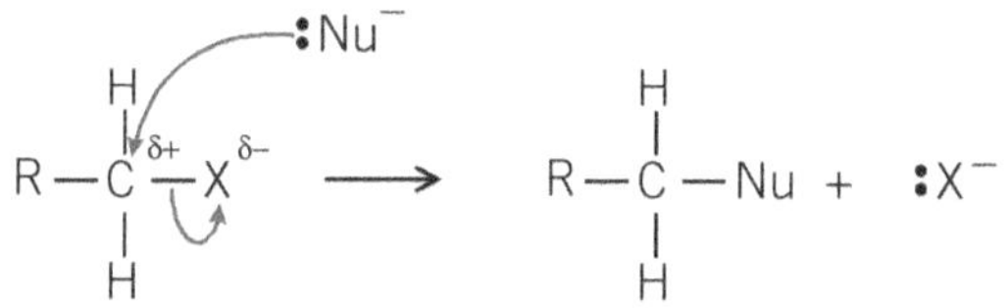

▲ **Figure 5** *General reaction mechanism and equation for nucleophilic substitution of haloalkanes.*

Water as a nucleophile

A water molecule has a lone pair on the oxygen atom, so water can act as a nucleophile – though this reaction is slower than the reaction with OH⁻ ions. When the two are heated together under reflux, water attacks the haloalkane.

Then the resulting ion loses H^+ to form an alcohol.

The general equation is known as a *hydrolysis* reaction:

$$R—X + H_2O \rightarrow R—OH + H^+ + X^-$$

Ammonia as a nucleophile

Ammonia, NH_3, can act in a similar way to water with the lone pair of electrons on the nitrogen atom attacking the haloalkane. The haloalkane is heated in a sealed tube with concentrated ammonia solution. The product is an **amine** with an NH_2 group. The overall equation is:

$$R—X + NH_3 \rightarrow R—NH_3^+ \, X^- \rightleftharpoons R—NH_2 + H^+ + X^-$$

Amines have the general formula $R—NH_2$. They are nitrogen analogues of alcohols ($R—OH$).

Some examples of amines are shown in Figure 6.

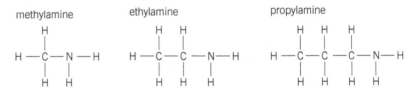

▲ **Figure 6** *Some examples of amines*

Using nucleophilic substitution to make haloalkanes

You can use the *reverse* of a hydrolysis reaction to produce a haloalkane from an alcohol. This time the nucleophile is a halide ion (X^-). The reaction is done in the presence of a strong acid, and the first step involves bonding between H^+ ions and the oxygen atom on the alcohol. For example:

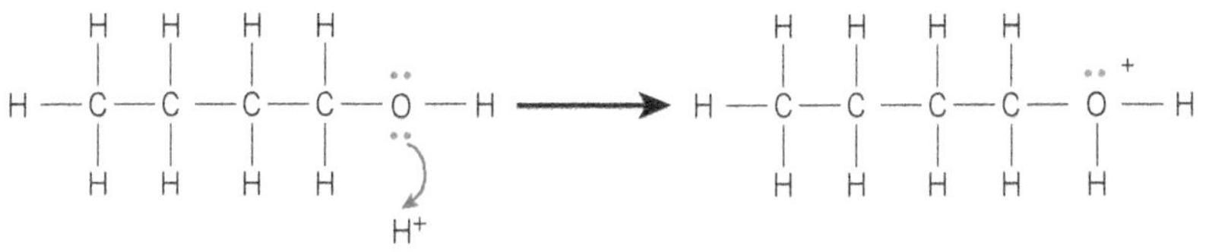

This gives the carbon atom to which the oxygen is attached a higher partial positive charge. It is now more readily attacked by halide ions:

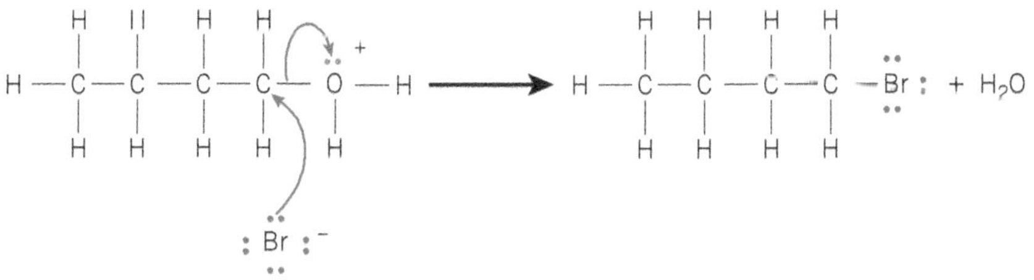

The overall equation for the reaction is:

$$CH_3CH_2CH_2CH_2OH + H^+ + Br^- \rightarrow CH_3CH_2CH_2CH_2Br + H_2O$$

Different halogens, different reactivity

Table 1 gives the bond enthalpies of the four different types of C—Hal bond.

▼ **Table 1** *Bond enthalpies of carbon–halogen bonds*

Bond	Bond enthalpy / $kJ\,mol^{-1}$
C—F	467
C—Cl	346
C—Br	290
C—I	228

You might imagine that a large bond polarity in a carbon–halogen bond would result in it breaking easily, for example, C—F bonds would break more easily than C—Cl bonds. However, it has been shown experimentally that bond enthalpy is the overriding factor in determining reactivity. Bond strength decreases in the order C—F > C—Cl > C—Br > C—I. On this basis, you would expect the C—I bond to be hydrolysed most easily because it is the weakest.

Bond polarity decreases in the order C—F > C—Cl > C—Br > C—I. On this basis, you would expect the C—F bond to be hydrolysed most easily because it is the most polar. Experimentally you find that C—I bonds are most easily hydrolysed, so bond strength is the most important factor.

The strength of the C—F bond makes it very difficult to break, so fluoro compounds are very unreactive. As you go down the halogen group the carbon–halogen bond gets weaker, so the compounds get more reactive. Bromo- and iodo-compounds are fairly reactive, so they are useful intermediates in organic synthesis.

Study tip

You need to be able to write balanced equations and reaction mechanisms for nucleophilic substitution reactions.

Activity OZ 8.3

You can investigate the relative reactivities of some haloalkanes in this activity.

What is the state of the ozone layer now?

It was clear in the 1980s that removing ozone from the stratosphere would have serious consequences, but it was not known what the full extent of these would be. One thing was certain – the numbers of cases of skin cancers and eye cataracts were increasing as the ozone was destroyed.

Another problem was whether increased ultraviolet radiation could affect species such as plankton in the oceans? That could affect other organisms involved in the food chain. Furthermore, could changes in the amount of radiation reaching the Earth affect its temperature and weather?

▲ Figure 7 *Bioluminescent plankton. There were concerns raised that plankton such as these could be directly affected by increases in UV radiation*

The Montreal Protocol

In 1987, Governments realised that it was not worth risking the global experiment needed to find the answers to these questions. In Montreal a procedure was agreed for restricting the release of CFCs into the atmosphere. This was known as the Montreal Protocol. Since then a series of amendments have strengthened the protocol, with the aim of eliminating emissions of ozone-depleting substances as a result of human activity.

By 1998 the developed nations had almost phased out their use of CFCs except for some specialised appliances such as asthma inhalers. A fund was set up to help developing countries to move away from CFCs by 2010.

CFC replacements

The chemical industry rapidly found replacements for CFCs that would have no significant damaging effect on the ozone layer. In the short term, replacement compounds were hydrochlorofluorocarbons (HCFCs), such as $CHClF_2$. The hydrogen–carbon bonds mean that HCFCs are

Synoptic link

You will learn more about greenhouse gases and global warming in Chapter 8, The oceans.

broken down in the troposphere. Nonetheless, some molecules will make it to the stratosphere where they will photodissociate to release chlorine atoms. It is hoped that HCFCs will be phased out in developed countries by 2020 and in the rest of the world by 2040.

For the longer term, hydrofluorocarbons, HFCs are a better option because they have no ozone-depleting effect, even if they make it to the stratosphere (Table 2).

Sadly, there is no perfect solution. Many of these compounds are greenhouse gases and there is concern about HFCs in the troposphere, since they can produce HF and trifluoroethanoic acid, CF_3COOH, but it is felt that the concentrations are too small to be a problem.

▼ Table 2 *Properties and uses of some CFCs, HCFCs, and HFCs. The ozone-depleting potential (ODP) is a measure of the effectiveness of the compound in destroying stratospheric ozone. CFC-11 is defined as having an ODP of 1.0*

Compound	Code	Ozone-depleting potential (ODP)	Lifetime / years	Uses
Main culprits				
CCl_3F	CFC-11	1.0	45	refrigeration, air conditioning, foams, cleaning solvents
CCl_2F_2	CFC-12	1.0	100	
$CBrF_3$	Halon-1301	10.0	65	firefighting
$CBrClF_2$	Halon-1211	3.0	16	
CCl_4	–	1.1	26	solvents
CH_3CCl_3	–	0.1	5	
Short-term solutions				
$CHClF_2$	HCFC-22	0.06	12	replaces CFC-11 and CFC-12
$CHCl_2CF_3$	HCFC-123	0.02	1	replaces CFC-11
Long-term solution				
CH_3CHF_2	HFC-152a	0	1.4	replaces CFC-11 and CFC-12

When will the ozone layer recover?

The concentrations of CFCs in the troposphere peaked around the beginning of this century and are now starting to decrease. The concentrations of HCFCs and HFCs are rising. Figure 7 shows the results from monitoring air samples around the world.

The long lifetimes of CFCs mean that recovery of the ozone layer will be slow – and will be subject to variations caused by weather conditions and solar activity. The lowest recorded ozone concentration was in 1992 when the level was 92 'Dobson units' but by 2002 this had risen to 131 units. It seemed that the ozone layer was recovering, but the ozone depletion in September 2006 was the most severe to date. The hole at that point covered 27 million km^2.

In 2013 the ozone concentration had risen to 133 Dobson units, and the size of the hole had receded to 21 million km². It seems that there is an established positive trend, but complete recovery is not expected until 2060–2070.

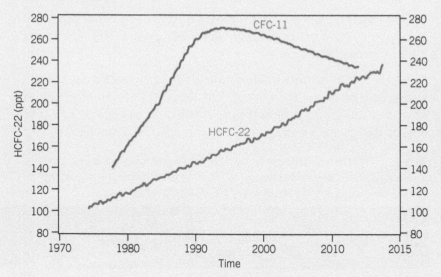

▶ Figure 7 *Results from air samples – the concentration of CFC-11 is decreasing but HCFC-22 is increasing*

Summary questions

1 When 1-chloropropane is heated under reflux with aqueous sodium hydroxide solution, a nucleophilic substitution reaction occurs, forming propan-1-ol.
   a Write a balanced equation for this reaction. (1 mark)
   b Explain why this is a *substitution* reaction. (1 mark)
   c Draw the mechanism of the reaction, showing the relevant lone pairs and partial charges. (3 marks)

2 Table 6 shows some common nucleophiles. Write balanced equations for each of the following nucleophilic substitution reactions, showing the structures of the reactants and products clearly.
   a iodoethane and OH^- ions (1 mark)
   b bromoethane and CN^- ions (1 mark)
   c chlorocyclopentane and OH^- ions (1 mark)
   d 2-chloro-2-methylpropane and H_2O (1 mark)
   e 1,2-dibromoethane and OH^- ions (1 mark)
   f bromomethane and $C_2H_5 O^-$ (ethoxide) ions (1 mark)
   g 2-chloropropane and CH_3COO^- (ethanoate) ions (1 mark)

3 Concentrated ammonia solution reacts with 1-bromoethane when heated in a sealed tube.
   a Write a balanced equation for this reaction. (1 mark)
   b Using the reaction of haloalkanes with water as a guide, draw out the mechanism for this reaction. (2 marks)
   c Write a few sentences to explain this mechanism to a fellow student. (2 marks)
   d Give definitions of all the terms in your mechanism. (2 marks)

Practice questions

1 The following substances all have similar M_r values.

   Which has the highest boiling point?

   A $CH_3CH_2CH_2CH_3$

   B CH_3CH_2Cl

   C $CH_3CH_2CH_2OH$

   D CH_3COCH_3 (*1 mark*)

2 Which of the following molecules would have the greatest overall dipole?

   A CCl_4

   B BF_3

   C NH_3

   D C_2H_6 (*1 mark*)

3 Nitrogen and hydrogen are mixed. Which of the following conditions would result in the equilibrium being set up fastest?

   $$N_2 + 3H_2 \; \rightarrow \; 2NH_3 \quad \text{exothermic}$$

	Pressure	Temperature
A	high	high
B	high	low
C	low	high
D	low	low

   (*1 mark*)

4 Which of the following are correct statements about the role of CFCs in ozone depletion?

	Troposphere	Stratosphere
A	no ozone to deplete	form Cl radicals that deplete ozone
B	do not form Cl radicals	form Cl radicals that deplete ozone
C	no ozone to deplete	react with ozone to deplete it
D	do not form Cl radicals	react with ozone to deplete it

   (*1 mark*)

5 Which of the following statements is correct about UV radiation in the stratosphere?

   A It is reflected from the Earth.

   B It comes from the Sun and is all absorbed.

   C It breaks down molecules.

   D It causes photochemical smog.   (*1 mark*)

6 The names of some bromoalkanes and their boiling points are shown below.

Name	1-bromobutane	2-bromobutane	2-bromo-2-methyl propane
Boiling point / K	375	364	346

   Which of the following is a correct statement?

   A The dipole of the C—Br bond is largest in 1-bromobutane.

   B The instantaneous dipole–induced dipole bonds get weaker as the molecules become more compact.

   C The molecules contain different numbers of electrons.

   D 1-bromobutane is the most volatile.

   (*1 mark*)

7 Hydrogen bonds can form between which of the following pairs of atoms in molecules?

A	Any hydrogen atom.	An oxygen atom in a molecule.
B	A hydrogen atom attached to a nitrogen atom.	A nitrogen atom in a molecule.
C	Any hydrogen atom.	A fluorine atom in a molecule.
D	A hydrogen atom attached to a chlorine atom.	A carbon atom attached to a chlorine atom.

   (*1 mark*)

8 Which of the following is a characteristic of a catalyst?

   1 It provides a route of lower activation enthalpy.

   2 It is unchanged at the end of the reaction.

   3 It does not take part in the reaction.

   A 1,2, and 3 correct

   B 1 and 2 are correct

   C 2 and 3 are correct

   D Only 1 is correct   (*1 mark*)

9 Which of the following are characteristics of a nucleophile?

1 It must have negative (or partial negative) charge.

2 It attacks positively charged atoms.

3 It has a lone pair of electrons.

A 1,2, and 3 correct

B 1 and 2 are correct

C 2 and 3 are correct

D Only 1 is correct   (*1 mark*)

10 Which of the following are termination steps in the radical chlorination of methane?

$$CH_4 + Cl_2 \rightarrow CH_3Cl + HCl$$

1 $2CH_3 \rightarrow C_2H_6$

2 $2Cl \rightarrow Cl_2$

3 $CH_3 + Cl \rightarrow CH_3Cl$

A 1,2, and 3 correct

B 1 and 2 are correct

C 2 and 3 are correct

D Only 1 is correct   (*1 mark*)

11 The hydroxyl radical OH has been described as 'the detergent of the atmosphere' as it is able to oxidise (and thus remove) most substances in the troposphere.

It is made by the reaction of oxygen atoms with water molecules:

$$O + H_2O \rightarrow 2OH \qquad \textbf{Equation 11.1}$$

a Draw a '*dot-and-cross*' diagram for a hydroxyl radical and explain why OH is called a *radical*.   (*2 marks*)

b Oxygen atoms are formed in the troposphere by the photolysis of ozone:

(i) Write an equation for the photolysis of ozone.   (*1 mark*)

(ii) 3% of oxygen atoms are removed from the atmosphere by the reaction in **Equation 11.1**. Suggest another equation (related to that from (i)) by which oxygen atoms are removed from the atmosphere.   (*1 mark*)

c In the stratosphere, oxygen molecules are broken down by photolysis.

(i) Given that the bond enthalpy of $O\!\!=\!\!O$ is 498 kJ mol^{-1}, calculate the minimum frequency of electromagnetic radiation needed to break this bond.   (*3 marks*)

(ii) Calculate the corresponding wavelength to this frequency.   (*2 marks*)

(iii) Suggest why this reaction only takes place in the stratosphere, whereas the photolysis of ozone takes place in the troposphere.   (*2 marks*)

d Carbon monoxide is one of the major substances removed from the troposphere by hydroxyl radicals.

$$CO + OH \rightarrow CO_2 + H \qquad \textbf{Equation 11.2}$$

Classify the reaction in **Equation 11.2** as initiation, propagation, or termination, giving a reason.   (*2 marks*)

12 Some students carry out experiments with 1-chlorobutane, 1-bromobutane, and 1-iodobutane.

a The students wish to discover the relative rates of hydrolysis of the three haloalkanes, using silver nitrate solution and ethanol.

Describe how they would carry out the experiment. Give equations for the reactions that occur and say what is seen.   (*6 marks*)

b Before the experiment, one student predicts that the haloalkane with the greatest permanent dipole would react fastest.

(i) Explain what is meant by *permanent dipole* and explain which of the haloalkanes used by the students has the greatest permanent dipole.   (*3 marks*)

(ii) The student uses the mechanism of the reaction to justify the prediction.

Draw out the mechanism of the reaction for the hydrolysis of 1-chlorobutane with water, showing curly arrows and partial charges.

(*3 marks*)

(iii) Indicate **one** *heterolytic* bond-breaking that occurs in the mechanism. (*1 mark*)

(iv) Explain why the haloalkane with the greatest dipole might have reacted fastest. (*1 mark*)

(v) State and explain the more important factor in determining the rate of hydrolysis. (*2 marks*)

c The students carry out a reaction in which they heat 1-bromobutane with concentrated ammonia in a sealed tube. Write an equation for the reaction that will occur and name the functional group in the product. (*2 marks*)

13 When aircraft started flying in or near to the stratosphere, there were concerns that nitrogen oxides in their exhausts would catalyse the breakdown of ozone by the reaction in **Equation 13.1**:

$$O + O_3 \rightarrow 2O_2 \qquad \textbf{Equation 13.1}$$

a The first reaction by which NO breaks down ozone is:

$$NO + O_3 \rightarrow NO_2 + O_2 \qquad \textbf{Equation 13.2}$$

(i) NO is a catalyst in the breakdown of ozone by the reaction in **Equation 13. 1.** Write the equation for the reaction that follows **Equation 13.2**. (*1 mark*)

(ii) What **type** of catalysis is involved here? (*1 mark*)

b In order to be a catalyst, the activation enthalpy of the catalysed reaction must be smaller than that for the uncatalysed reaction.

(i) Explain the meaning of the term *activation enthalpy*. (*1 mark*)

(ii) Label the Boltzmann distribution below to show how adding a catalyst makes a reaction go faster. (*2 marks*)

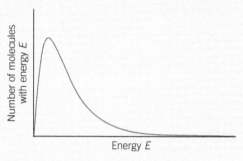

(iii) Complete the enthalpy profile below to show that a catalysed reaction has a lower activation enthalpy than the uncatalysed reaction. (*2 marks*)

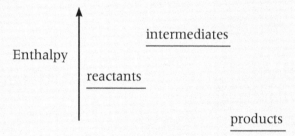

c (i) Why is there concern about the destruction of the ozone layer? (*2 marks*)

(ii) It was later discovered that the effect of NO from aircraft on the ozone layer was very small compared with the effect of another radical from human-derived sources.

Give the name of this radical and its source in the stratosphere. (*2 marks*)

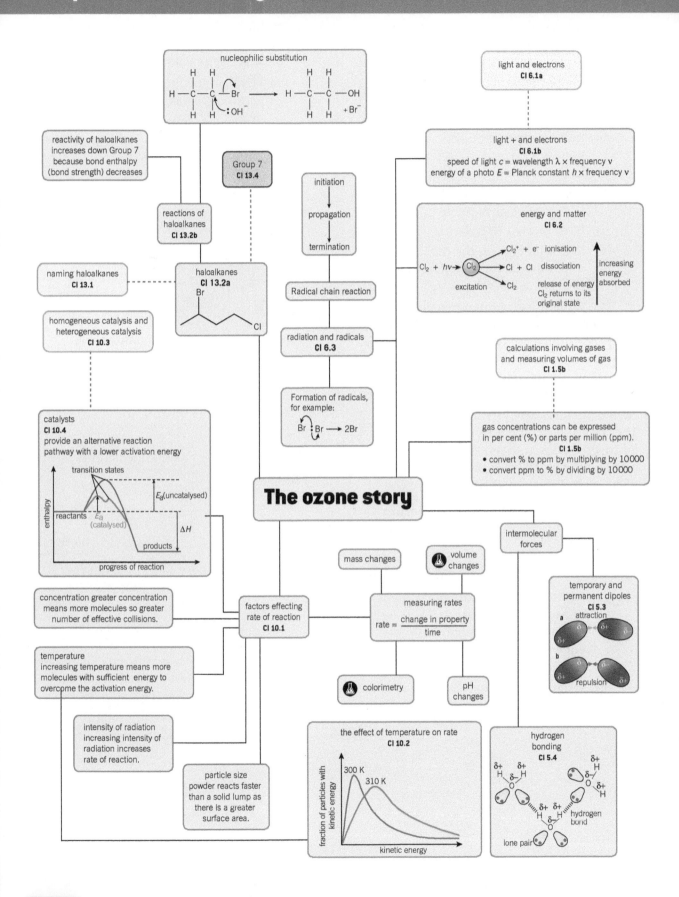

Anaesthetics

Many anaesthetics contain halogen atoms in their molecules. Anaesthetics are used to bring about a reversible loss of consciousness during operations, so that the patient does not feel any pain.

Isoflurane, enflurane, and halothane are all volatile liquids which readily evaporate. They have quite high M_r values (184–197 g mol^{-1}), but have low boiling points (around 50 °C) and very low solubility in water. Halothane is unstable in light and is packaged in dark bottles.

In contrast to some early anaesthetics such as diethyl ether, $C_2H_5OC_2H_5$, these three molecules are all completely non-flammable.

Isoflurane and enflurane are isomers of each other, and all three molecules contain a chiral carbon. This is a carbon with four different atoms or groups attached. The different positioning of the groups around the chiral carbon can have important biological effects, as you will learn in later topics.

1 What does the low boiling point of these molecules suggest about their intermolecular bonding? Identify the type(s) of intermolecular bonding they would contain.

2 Explain why these molecules are insoluble in water.

3 Write an equation for the complete combustion of diethyl ether. Suggest why the modern anaesthetics are non-flammable.

4 What might be the effect of light on halothane? Which bond is most likely to be affected?

5 Suggest the possible ozone depleting effect of these anaesthetics if they were to be released into the atmosphere.

6 Evaluate the advantages and disadvantages of these anaesthetics for doctors and patients.

Extension

1 Nucleophilic substitution reactions can occur by two main mechanisms, largely dependent on whether the starting molecule is a primary haloalkane or tertiary haloalkane. Research and contrast the two different mechanisms. (Note that only the mechanism described in the chapter will be tested in examinations.)

2 In the extension box, Other radical reactions, in Topic OZ 3, How is ozone formed in the atmosphere?, you learnt that radicals are involved in polymerisation, combustion, and age-related biological effects. Produce a summary of the reactions of radicals, including these three examples.

3 The challenge of ozone depletion appears to have been successfully addressed by politicians and chemists. Can the same approaches be applied to the challenge of global warming? What role can chemists play?

CHAPTER 5
What's in a medicine?

Topics in this chapter

→ WM 1 The development of modern ideas about medicine

reactions of alcohols

→ WM 2 Identifying the active chemical in willow bark

the –OH group in different environments and derivatives of carboxylic acids

→ WM 3 Infrared spectroscopy

Infrared spectroscopy

→ WM 4 Mass spectrometry

Mass spectrometry for compounds

→ WM 5 The synthesis of salicylic acid and aspirin

principles of green chemistry

Why a chapter on What's in a medicine?

This chapter introduces the pharmaceutical industry, which not only produces new and more effective medicines but is a net exporter of medicinal products and so contributes to the financial health of the UK. The molecule focussed on in this chapter is aspirin. Aspirin is a relatively simple drug which did not go through the drug testing regime carried out today but which is widely used as an over-the-counter painkiller, as well as being prescribed by doctors for heart disease and other conditions.

During the chapter you will study the application of instrumental methods for determining the structure of molecules and practise organic synthesis and use test-tube reactions to identify functional groups. You will also have the opportunity to see how green principles can be applied in the chemical industry.

The chemistry of alcohols, phenols, aldehydes, ketones, and carboxylic acids is studied in some detail. You will see how alcohols can react with carboxylic acids and acid anhydrides to produce esters, and that esters can also be made from phenols. You will also apply your understanding of acids and bases.

You will learn more about some ideas introduced in earlier chapters:

- ☐ hydrogen bonding (**The ozone story**)
- ☐ alcohols (**Developing fuels**)
- ☐ oxidation (**Elements from the sea**)
- ☐ alkenes (**Developing fuels**)
- ☐ equilibria (**Elements from the sea**)
- ☐ acids (**Elements of life**)
- ☐ the interaction of radiation with matter (**Elements of life and The ozone story**)
- ☐ bond polarity (**The ozone story**)
- ☐ mass spectrometry (**Elements of life**)
- ☐ atom economy (**Elements from the sea**).

Maths skills checklist

In this unit, you will need to use the following maths skills. You can find support for these skills on Kerboodle and through MyMaths.

- ☐ Translate information between graphical, numerical, and algebraic forms.
- ☐ Analyse and use data in a range of contexts.

MyMaths.co.uk
Bringing Maths Alive

WM 1 The development of modern ideas about medicines

Specification reference: WM(a), WM(b), WM(d), WM(f), WM(h)

Learning outcomes

Demonstrate and apply knowledge and understanding of:

→ the formulae of the following homologous series: carboxylic acids, acid anhydrides, esters, aldehydes, ketones, ethers

→ primary, secondary, and tertiary alcohols in terms of the differences in structures

→ the following reactions of alcohols and two-step syntheses involving these reactions and other organic reactions in the specification:

- with carboxylic acids, in the presence of concentrated sulfuric acid or concentrated hydrochloric acid (or with acid anhydrides) to form esters

- oxidation to carbonyl compounds (aldehydes and ketones) and carboxylic acids with acidified dichromate(VI) solution, including the importance of the condition (reflux or distillation) under which it is done

- dehydration to form alkenes using heated Al_2O_3 or refluxing with concentrated H_2SO_4

- substitution reactions to make haloalkanes

→ techniques and procedures for preparing and purifying a liquid organic product including the use of a separating funnel and of Quickfit or reduced scale apparatus for distillation and heating under reflux

→ the term elimination reaction.

When there is something wrong with your body, a **medicine**, such as aspirin or penicillin, prevents things getting worse and can help to bring about a cure. The active ingredients of medicines are **drugs** – substances that alter the way your body works. If your body is already working normally the drug will not be beneficial, and if the drug throws the body a long way off balance it may even be a poison. Not all drugs are medicines – alcohol and nicotine are not medicines, but they *are* drugs. Some drugs, such as opium, may or may not be medicines depending on your state of health.

The study of drugs and their action is called **pharmacology**. Making and dispensing medicines is called **pharmacy**.

People have been using medicines for thousands of years – most of that time with no idea how they worked. Their effectiveness was discovered by trial and error and sometimes there were disastrous mistakes. Today's medicines are increasingly designed to have specific effects – this is becoming easier as scientists learn more about the body's chemistry and begin to understand the intricate detail of the complex molecules from which people are made. Work at this level comes into the field of molecular pharmacology.

Medicines from nature

Modern pharmacy has its origins in folklore, and the history of medicine abounds with herbal and folk remedies. Many of these can be explained in present day terms and the modern pharmaceutical industry investigates 'old wives' tales' to see if they lead to important new medicines.

▲ **Figure 1** *Feverfew has been used since ancient times for the treatment of migraine – research in the 1970s confirmed that it was an effective medicine for this disorder*

▲ **Figure 2** *Foxglove contains the compound digitalin, which is active against heart disease*

In 400 BC, Hippocrates recommended a brew of willow leaves to ease the pain of childbirth, and in 1763 the Reverend Edward Stone, an English clergyman living in Chipping Norton, Oxfordshire, used a willow bark brew to reduce fevers.

Unknown at the time, it was the substance salicin that was extracted from the willow bark. Salicin has no pharmacological effect by itself. The body converts it by hydrolysis and oxidation into the active chemical, salicylic acid. Salicin can be oxidised to salicylic acid because it contains a primary alcohol group. You need to know about the structure of alcohols, and their reactions, to fully understand this oxidation.

▲ Figure 3 *Extracts from willow trees have been used in medicine for thousands of years*

Chemical ideas: Organic chemistry: modifiers 13.3

Alcohols

Physical properties of alcohols

Like water molecules, alcohol molecules are polar because of the polarised O—H bond. In both water and alcohols, there is a special sort of strong attractive force between the molecules due to hydrogen bonds.

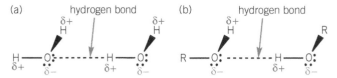

▲ Figure 4 *Hydrogen bonding in water (a) and alcohols (b)*

Hydrogen bonds are not as strong as covalent bonds, but are stronger than other attractive bonds between covalent molecules. When a liquid boils, these forces must be broken so the molecules escape from the liquid to form a gas. This explains why the boiling points of alcohols are higher than those of corresponding alkanes with similar relative molecular mass M_r. For example, ethanol ($M_r = 46.0$) is a liquid, whilst propane ($M_r = 44.0$) is a gas at room temperature.

Hydrogen bonding between alcohol and water molecules (Figure 5) also explains why the two liquids mix together.

Table 1 shows the solubility of some alcohols in water. As the hydrocarbon chain becomes longer and the molecule becomes larger, the influence of the –OH group on the properties of the molecule becomes less important. So the properties of the higher alcohols get more and more like those of the corresponding alkane.

▼ Table 1 *Solubility of primary alcohols in water*

Name	Formula	Solubility / g per 100 g water
methanol	CH_3OH	fully soluble
ethanol	CH_3CH_2OH	fully soluble
propan-1-ol	$CH_3CH_2CH_2OH$	fully soluble
butan-1-ol	$CH_3CH_2CH_2CH_2OH$	8.0
pentan-1-ol	$CH_3CH_2CH_2CH_2CH_2OH$	2.7
hexan-1-ol	$CH_3CH_2CH_2CH_2CH_2CH_2OH$	0.6

Synoptic link

You studied the naming of alcohols in Chapter 2, Developing fuels.

Synoptic link

You studied hydrogen bonding, as well as other intermolecular bonds, in Chapter 4, The ozone story.

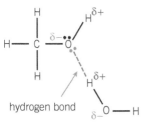

▲ Figure 5 *Hydrogen bonding between water molecules and alcohol molecules*

Reactions of alcohols

There are of three types of alcohols – **primary alcohols**, **secondary alcohols**, and **tertiary alcohols** – named according to the position of the –OH group (Table 2).

▼ Table 2 *Primary, secondary, and tertiary alcohols*

Type of alcohol	Position of –OH group	General formula	Example
primary	–OH bonded to a carbon bonded to *one* other carbon atom	RCH_2OH	butan-1-ol
secondary	–OH bonded to a carbon bonded to *two* other carbon atoms	$RCH(OH)R$	butan-2-ol
tertiary	–OH bonded to a carbon bonded to *three* other carbon atoms	$R_2C(OH)R$	2-methylpropan-2-ol

The reactions of alcohols depend on the type of alcohol involved.

Oxidation

The –OH group can be oxidised by strong oxidising agents, such as acidified potassium dichromate(VI), $K_2Cr_2O_7$, solution. The orange dichromate(VI) ion, $Cr_2O_7^{2-}$ (aq), is reduced to green chromate(III) ions, Cr^{3+}(aq), in the reaction. On being oxidised, the –OH group is converted into a carbonyl, C=O, group, and the reaction mixture turns from orange to green.

In this reaction two atoms of hydrogen are being removed – one from the oxygen atom, and one from the carbon atom. Oxidation of the –OH group will not take place unless there is a hydrogen atom on the carbon atom to which the –OH is attached.

The product is a **carbonyl** compound – an aldehyde or a ketone. The type of product you get depends on the type of alcohol you start with and the reaction conditions used.

Primary alcohols, such as ethanol, are initially oxidised to **aldehydes**. The aldehyde is then oxidised to a carboxylic acid, in the presence of excess oxidising agent.

Synoptic link

You encountered oxidation and reduction in Chapter 3, Elements from the sea.

Study tip

$Cr_2O_4^{2-}/H^+$ does not exist so when asked to give an example of such an oxidising agent it should be $K_2Cr_2O_7$(aq)/H^+(aq) *or* acidified potassium chromate(VII) solution.

Study tip

You do not need to know the reaction mechanism for this type of reaction.

If the aldehyde is required it can be distilled in situ out from the reaction mixture before it is oxidised further. In this case, having the alcohol in excess reduces the likelihood of the aldehyde being further oxidised.

Secondary alcohols, such as propan-2-ol, are oxidised to **ketones** by heating under reflux, in the presence of an oxidising agent.

Activity WM 1.1

In this activity you will have the opportunity to oxidise an alcohol.

$$H-\overset{\overset{\displaystyle H}{|}}{\underset{\underset{\displaystyle H}{|}}{C}}-\overset{\overset{\displaystyle H}{|}}{\underset{\underset{\displaystyle OH}{|}}{C}}-\overset{\overset{\displaystyle H}{|}}{\underset{\underset{\displaystyle H}{|}}{C}}-H \xrightarrow[\text{reflux}]{[O]} H-\overset{\overset{\displaystyle H}{|}}{\underset{\underset{\displaystyle H}{|}}{C}}-\overset{\overset{\displaystyle}{\underset{\underset{\displaystyle O}{||}}{C}}}{}-\overset{\overset{\displaystyle H}{|}}{\underset{\underset{\displaystyle H}{|}}{C}}-H \ + \ H_2O$$

propan-2-ol   propanone (a ketone)

Study tip

Sometimes an oxidising agent is represented as [O] in organic reactions.

The ketone is not oxidised further as this would involve breaking a strong, covalent C—C bond.

Tertiary alcohols, such as 2-methylpropan-2-ol, do not oxidise because they do not have a hydrogen atom on the carbon atom to which the –OH group is attached.

▼ Table 3 *Oxidation of alcohols*

Type of alcohol	Product(s) of oxidation with acidified potassium dichromate(VI) solution	Final colour of reaction mixture
primary	aldehyde (with no excess oxidising agent), carboxylic acid (with excess oxidising agent)	green
secondary	ketone	green
tertiary	does not oxidise	orange

Heating under reflux

Refluxing is a safe method for heating reactions involving volatile and flammable liquids. The liquid is boiled with a vertically mounted condenser so the vapour condenses and returns back into the reaction mixture (Figure 6).

Aldehydes and ketones

Both aldehydes and ketones contain a carbonyl group, C=O. In an aldehyde, the carbonyl group is at the end of an alkane chain – the functional group is:

$$\overset{\displaystyle R}{\underset{\displaystyle H}{\diagdown}}C=O$$

Aldehydes are named using the suffix –al (Figure 7).

In a ketone, the carbonyl group is within an alkane chain – the functional group is:

$$R-\overset{\overset{\displaystyle O}{||}}{C}-R$$

Ketones are named using the suffix –one (Figure 8).

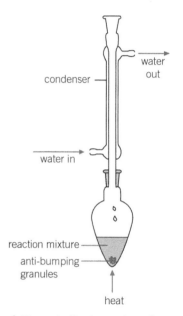

water out

condenser

water in

reaction mixture
anti-bumping granules

heat

▲ Figure 6 *Heating under reflux*

$$H-\overset{\overset{\displaystyle H}{|}}{\underset{\underset{\displaystyle H}{|}}{C}}-\overset{\overset{\displaystyle O}{\diagup\diagup}}{\underset{\underset{\displaystyle H}{\diagdown}}{C}}$$

▲ Figure 7 *ethanal*

$$H-\overset{\overset{\displaystyle H}{|}}{\underset{\underset{\displaystyle H}{|}}{C}}-\overset{\overset{\displaystyle O}{||}}{C}-\overset{\overset{\displaystyle H}{|}}{\underset{\underset{\displaystyle H}{|}}{C}}-H$$

▲ Figure 8 *propanone*

Synoptic link

The chemistry of aldehydes and ketones is developed in Chapter 10, Colour by design.

Synoptic link 🧪

You can find out more about reflux and distillation in Techniques and procedures.

Elimination reaction

A reaction where a small molecule is removed from a larger molecule leaving an unsaturated molecule. In the case of alcohols the small molecule is water.

Synoptic link

You studied alkenes and addition reactions in Chapter 2, Developing fuels.

Synoptic link

You studied haloalkanes and nucleophilic substitution reactions in Chapter 4, The ozone story.

Dehydration of alcohols

Many alcohols can lose a molecule of water to form an alkene. The conditions needed for this type of reaction are a heated catalyst of alumina, Al_2O_3, at 300 °C or reflux with concentrated sulfuric acid. Under either conditions the corresponding alkene and water are produced.

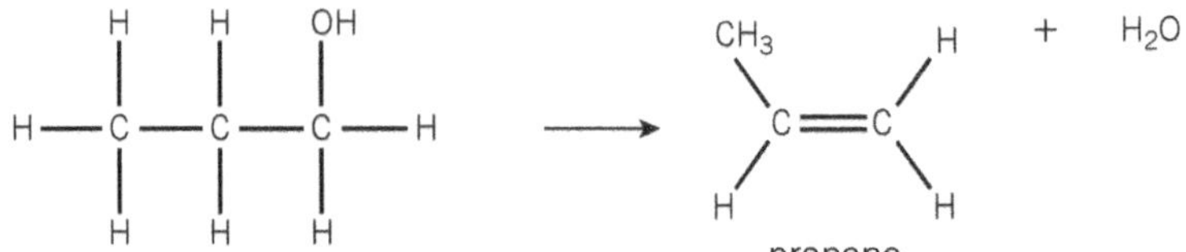

propan-1-ol → propene + H_2O

It can be easier to follow a reaction using the skeletal formulae:

The reaction is described as **dehydration** since it involves the removal of a water molecule from a molecule of the reactant. Dehydration is an example of an **elimination** reaction.

You can also think of an elimination reaction as being the reverse of an addition reaction.

Substitution reactions

Alcohols undergo nucleophilic substitution reactions with halide ions, in the presence of a strong acid, to produce haloalkanes. For example:

$$CH_3CH_2CH_2CH_2OH + H^+ + Br^- \rightarrow CH_3CH_2CH_2CH_2Br + H_2O$$

Formation of esters

There are two ways of converting alcohols into esters – esterification using an acid anhydride and esterification using a carboxylic acid.

Esterification

Esterification is the reaction of an alcohol with a carboxylic acid. For example, ethanol reacts with ethanoic acid to form the ester, ethyl ethanoate.

$$CH_3CH_2OH + CH_3COOH \rightleftharpoons CH_3COOCH_2CH_3 + H_2O$$

ethanol   ethanoic acid   ethyl ethanoate   water

You can write the same reaction using skeletal formulae:

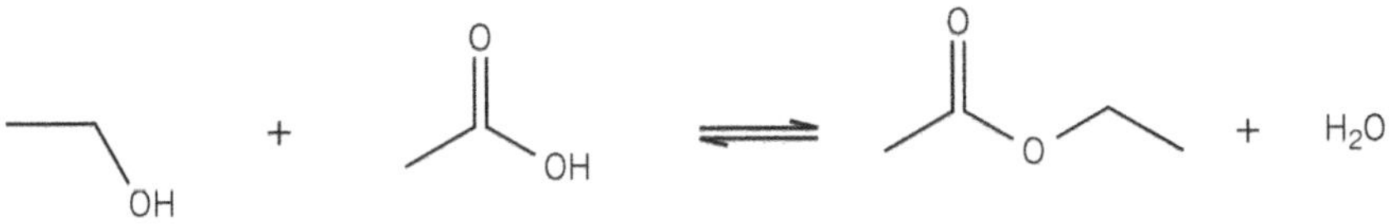

The reaction occurs extremely slowly unless a strong acid catalyst is present. A small amount of either concentrated sulfuric acid or concentrated hydrochloric acid is generally used, and the reaction

Activity WM 1.2

In this activity you can practise making a liquid haloalkane from an alcohol and purifying the product.

mixture is heated under reflux. The ⇌ symbol in the equation means the reaction is reversible and comes to an equilibrium, where both reactants and products are present. The ester would then have to be separated from the mixture using distillation and purified.

Look at the structure of ethyl ethanoate. Esters are named from the alcohol and acid that form them – 'ethyl' from ethanol and 'ethanoate' from ethanoic acid.

Using an acid anhydride

Acid anhydrides, derivatives of carboxylic acids, are more reactive than a carboxylic acid, and react completely with an alcohol on warming to give a much higher yield of ester.

$$(CH_3CO)_2O \quad + \quad CH_3CH_2OH \quad \rightarrow \quad CH_3COOCH_2CH_3 \quad + \quad CH_3COOH$$

ethanoic anhydride ethanol   ethyl ethanoate   ethanoic acid

Ethers

Be careful not to confuse alcohols with **ethers**. Ethers have the same molecular formula as alcohols but have a different structure – they are structural isomers. The general formula of ethers is R—O—R. They are derived from alkanes by substituting an alkoxy group (–OR) for a hydrogen atom. For example:

$$CH_3CH_2OCH_3CH_2$$

When naming an ether, the longer hydrocarbon chain is chosen as the parent alkane.

Purification of organic liquid products

Most common haloalkanes, esters, and ethers are organic liquids. During synthesis only a crude product is produced. This has to be purified before it can be used.

The main techniques are use of a separating funnel, use of drying agents, and simple distillation.

> **Synoptic link**
>
> You first encountered equilibrium in Chapter 3, Elements from the sea.

> **Synoptic link**
>
> You will learn more about naming esters in Chapter 7, Polymers and life.

> **Activity WM 1.3**
>
> This activity allows you to check your understanding of alcohol reactions.

> **Synoptic link** 🜂
>
> You can find out more about the purification of organic liquid products in Techniques and procedures.

Summary questions

1 a For each of the skeletal formula below give the type of function group present.

i

(1 mark)

iii

(1 mark)

ii

(1 mark)

iv

(1 mark)

v

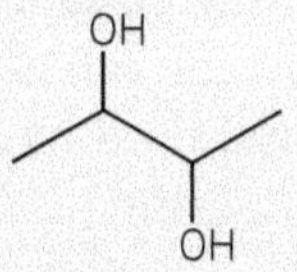

OH

OH

(1 mark)

vi

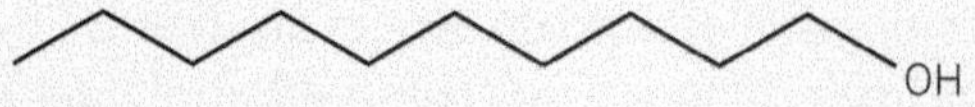

OH

(1 mark)

vii

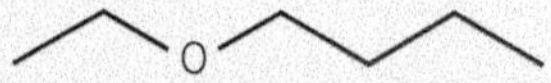

O

(1 mark)

b Give the systematic names of compounds **i** to **vi**.   *(7 marks)*

2 Why does ethanol mix with water but hexanol does not?   *(2 marks)*

3 Look at the following compounds:

A

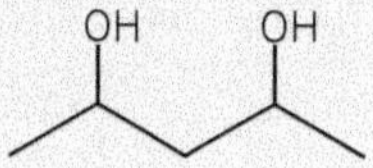

OH   OH

B

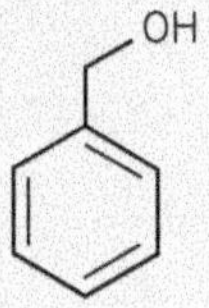

OH

C

O

D

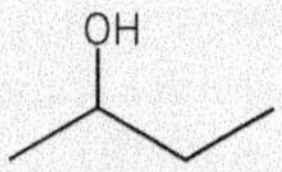

OH

E

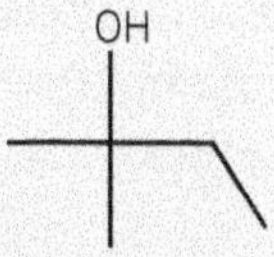

OH

a Which compound(s) is(are) alcohols?   *(1 mark)*
b Which compound(s) is(are) ethers?   *(1 mark)*
c Which compound(s) is(are) diols?   *(1 mark)*
d Which compounds are isomers?   *(1 mark)*
e Which compound do you think will be the most volatile?   *(1 mark)*
f Which compound would you expect to be the most soluble in water?   *(1 mark)*
g Which compound(s) would form carboxylic acid on refluxing with excess acidified potassium dichromate(VI)?   *(1 mark)*
h Which compound would not react on refluxing with excess acidified potassium dichromate(VI)?   *(1 mark)*

4 Here are the boiling points and relative molecular masses (M_r) of a number of substances:

Substance	Boiling point / °C	M_r
water, H_2O	100.0	18.0
ethane, CH_3CH_3	−88.5	30.0
ethanol, CH_3CH_2OH	78.0	46.0
butan-1-ol, $CH_3CH_2CH_2CH_2OH$	117.0	74.0
ethoxyethane, $CH_3CH_2OCH_2CH_3$	35.0	74.0

Use ideas about bonds between molecules to explain why
a ethanol has a higher boiling point than ethane   *(2 marks)*
b water has a higher boiling point than ethanol   *(2 marks)*
c butan-1-ol has a higher boiling point than ethanol   *(2 marks)*
d butan-1-ol has a higher boiling point than ethoxyethane.   *(2 marks)*

5 a Identify a compound that would give but-1-ene on dehydration.
(1 mark)

   b Write an equation for any dehydration identified.   (2 marks)

6 For each of the compounds i to v, what would you see on:
   a addition of sodium hydroxide solution   (1 mark)
   b reflux with excess potassium chromate(VII) solution.   (1 mark)

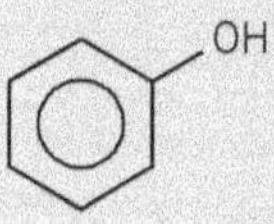

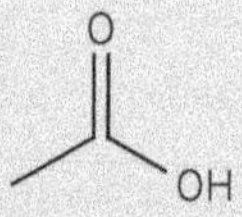

7 For each of the following reactions decide:
   a what type of reaction will occur.
   b what type of product will be formed.
      i   propan-1-ol warmed with ethanoic anhydride   (2 marks)
      ii   butan-1-ol refluxed with excess potassium
           dichromate(VI) solution   (2 marks)
      iii butan-2-ol refluxed with concentrated sulfuric acid   (2 marks)
      iv   2-methyl-propan-2-ol refluxed with excess potassium
           dichromate(VI) solution   (2 marks)
      v   butan-1-ol refluxed with propanoic acid with a few drops
           of concentrated sulfuric acid added   (2 marks)
      vi   ethanol (in excess) heated with sodium dichromate(VI)
           solution, with products distilled out as the reaction
           proceeds   (2 marks)
      vii butan-2-ol heated to 300 °C over aluminium oxide,
           Al_2O_3 (2 marks)
      viii 2-methyl-propan-2-ol shaken together with concentrated
           hydrochloric acid   (2 marks)
      ix   butan-2-ol refluxed with potassium dichromate(VI)
           solution   (2 marks)

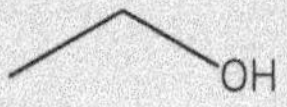

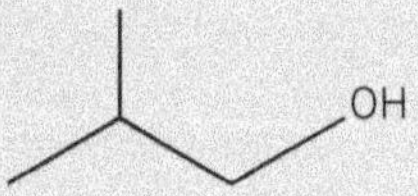

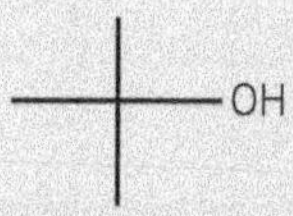

WM 2 Identifying the active chemical in willow bark

Specification reference: WM(a), WM(c)

Learning outcomes

Demonstrate and apply knowledge and understanding of:

→ the formulae of the following homologous series: phenols

→ the following properties of phenols:

- acidic nature, and their reaction with alkalis but not carbonates (whereas carboxylic acids react with alkalis and carbonates)

- test with neutral iron(III) chloride solution, to give a purple coloration

- reaction with acid anhydrides (but not carboxylic acids) to form esters.

How do chemists find out the chemical structure of compounds like salicylic acid? One way is to use chemical reactions to test for the presence of particular functional groups.

Relatively simple test-tube experiments can often be used effectively in the identification of unknown substances. For example, orange/brown bromine solutions go colourless on the addition of molecules with a double bond between carbon atoms.

Three chemical tests are particularly helpful in providing clues about the structure of salicylic acid:

1 An aqueous solution of the compound is weakly acidic.

2 Salicylic acid reacts with alcohols (such as ethanol) to produce compounds called esters. Esters have strong pleasant odours, often of fruit or flowers.

3 A neutral solution of iron(III) chloride turns an intense pink colour when salicylic acid is added.

Tests 1 and 2 are characteristic of carboxylic acids (compounds containing the –COOH functional group). Test 3 indicates the presence of a phenol group (an –OH group attached to a benzene ring).

Synoptic link

You will learn how to name carboxylic acids in Chapter 7, Polymers of life.

▲ Figure 1 *Carboxylic acid functional group*

Chemical ideas: Organic chemistry: modifiers 13.4

Carboxylic acids and phenols

Carboxylic acids and their derivatives

Carboxylic acids contain the carboxyl group (Figure 1). This formula is often abbreviated to –COOH, although the two oxygen atoms are not joined together. The structure of the rest of the molecule can vary widely and this gives rise to a large number of different carboxylic acids. When the remainder of the molecule is an alkyl group, the acids can be represented by the general formula R—COOH.

A carboxyl group can also be attached to a benzene ring (Figure 4).

The –OH group in the carboxyl group can be replaced by other groups to give a whole range of carboxylic acid derivatives (Table 1).

▼ **Table 1** *Some examples of carboxylic acid derivatives*

Acid derivative	Functional group	Example
ester RCOOR	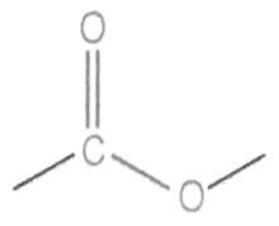 *(structure shown)*	ethyl ethanoate *(structure shown)* 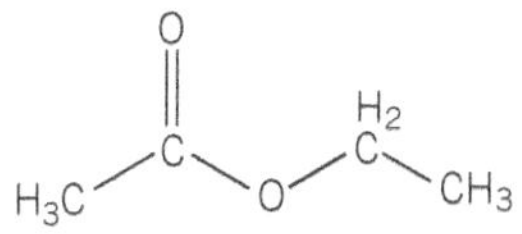
acid anhydride $(RCO)_2O$	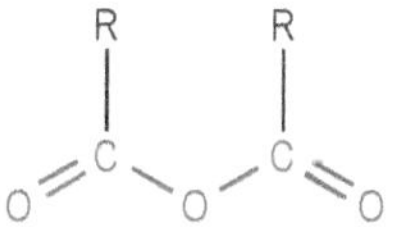 *(structure shown)*	ethanoic anhydride *(structure shown)* 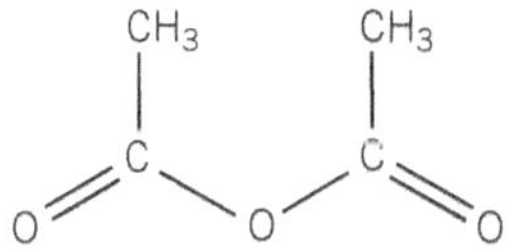

The –OH group in alcohols, phenols, and carboxylic acids

The hydroxyl group, –OH, can occur in three different environments in organic molecules.

- As part of a carboxyl group in carboxylic acids.

- Attached to an alkane chain in alcohols. There are three types of alcohols – primary, secondary, and tertiary – according to the position of the –OH group (Topic WM 1).

- Attached to a benzene ring in phenols (Figure 5).

Although phenols look similar to alcohols, they behave very differently. It is generally true that functional groups behave differently when attached to an aromatic ring from when they are attached to an alkyl group.

Acidic properties of the –OH group

The –OH group reacts with water.

$$R—OH + H_2O \rightleftharpoons R—O^- + H_3O^+$$

R stands for the group of atoms which makes up the rest of the molecule.

Water dissociates to a very small extent, so, at any one time a small number of water molecules donate H^+ ions to other water molecules – water behaves as a weak acid.

$$H—OH + H_2O \rightleftharpoons H—O^- + H_3O^+$$

A similar reaction occurs with ethanol, but to a lesser extent. The equilibrium lies further to the left, and ethanol is a weaker acid than water.

With phenol, the equilibrium lies further to the right than in water – phenol is slightly more acidic than water. Carboxylic acids are even more acidic, though still weak. The order of acid strength is

ethanol < water < phenol < carboxylic acids

Phenols and carboxylic acids are strong enough acids to react with strong bases such as sodium hydroxide, NaOH, to form salts (Figure 7).

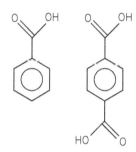

▲ **Figure 2** *Benzenecarboxylic acid or benzoic acid (left) and benzene-1,4-dicarboxylic acid (right)*

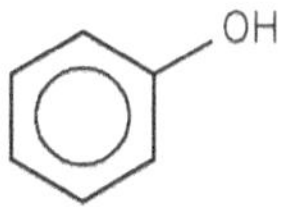

▲ **Figure 3** *Phenol*

Activity WM 2.1

This activity helps you recognise the formulae and structure of a range of different homologous series you have met so far.

Synoptic link

You first looked at the properties of acids in Chapter 1, Elements of life.

Synoptic link

You first encountered equilibrium in Chapter 3, Elements from the sea.

Synoptic link

Acid–base equilibria are explored more in Chapter 8, Oceans.

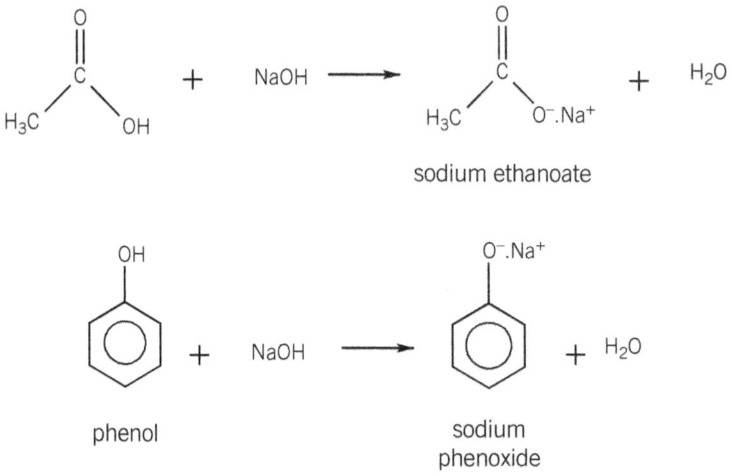

The reaction of ethanoic acid:

$$CH_3COOH + NaOH \longrightarrow CH_3COO^-.Na^+ + H_2O$$

sodium ethanoate

The reaction of phenol:

phenol $+$ NaOH $\longrightarrow$ sodium phenoxide $+ H_2O$

▲ **Figure 7** *The reaction of ethanoic acid and phenol with sodium hydroxide – a strong base*

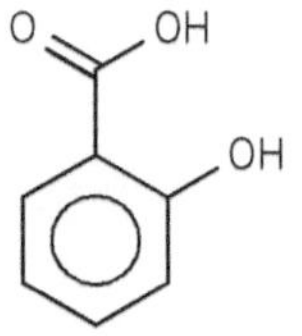

▲ **Figure 8** *2-hydroxybenzoic acid (salicylic acid)*

Aspirin structure

▲ **Figure 9** *Aspirin*

The salts produced are ionic and remain in solution after the reaction. Only carboxylic acids have high enough concentrations of H^+(aq) ions to produce carbon dioxide on reaction with carbonates.

$$CO_3^{2-}(aq) + 2H^+(aq) \rightarrow CO_2(g) + H_2O(l)$$

As such carboxylic acids make carbonates fizz, but alcohols and phenols do not. (The H_3O^+ ion is often written as H^+(aq).)

The iron(III) chloride test for phenol and its derivatives

Some groupings of atoms can become attached to metal ions and form **complexes**. The $-C{=}C-OH$ group (called the enol group) can form a purple complex with Fe^{3+} ions in neutral solution. Only phenol and its derivatives have such an arrangement of atoms and they are the only ones to give a colour with neutral iron(III) chloride solution. This is used as a test for phenol and its derivatives. Similar complexes are used to make the colours of some inks.

Ester formation

The formation of esters using alcohols was described in Topic WM 1. Esters can also be made using phenols and acid anhydrides in alkaline conditions.

$$\begin{array}{ccccc} C_6H_5OH & + & (CH_3CO)_2O & \rightarrow & CH_3COOC_6H_5 & + & CH_3COOH \\ \text{phenol} & & \text{ethanoic anhydride} & & \text{phenyl ethanoate} & & \text{ethanoic acid} \end{array}$$

Phenols *do not* react with carboxylic acids to produce esters.

Esters from salicylic acid

Figure 8 shows the structure of 2-hydroxybenzoic acid (salicylic acid). There are two ways of esterifying it. The phenol –OH group can react with an acid anhydride, or the –COOH group can react with an alcohol.

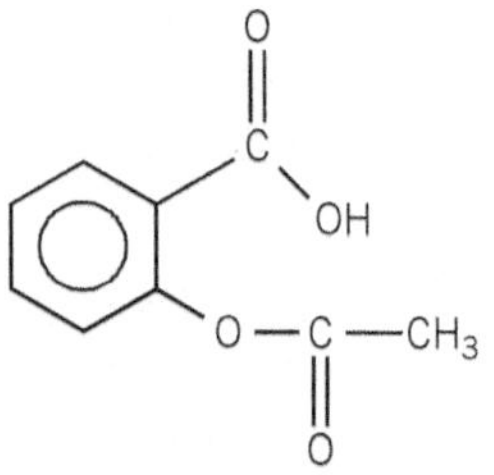

Aspirin is the product of esterifying the phenol –OH group to form 2-ethanoyloxybenzoic acid (Figure 9). It is quite soluble in water, so it can be absorbed into the bloodstream through the stomach wall.

The product of reacting the –COOH group with methanol is called methyl 2-hydroxybenzoate (Figure 10). This is better known as oil of wintergreen and is used as a linament. It is soluble in fats rather than water so it is absorbed through the skin – like aspirin, it reduces pain and swelling.

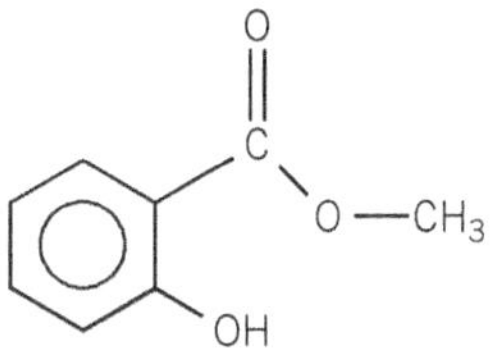

Summary questions

1 a Name the following compounds.

i OH

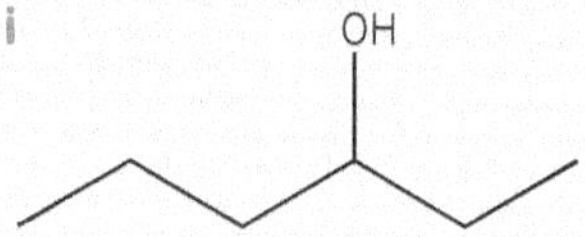

ii OH   iii OH Cl Cl

(2 marks)   (2 marks)   (2 marks)

iv
$$H-\overset{\overset{\displaystyle H}{|}}{\underset{\underset{\displaystyle H}{|}}{C}}-\overset{\overset{\displaystyle H}{|}}{\underset{\underset{\displaystyle H}{|}}{C}}-\overset{\overset{\displaystyle H}{|}}{\underset{\underset{\displaystyle H}{|}}{C}}-\overset{\overset{\displaystyle O}{\|}}{C}-H$$

v O

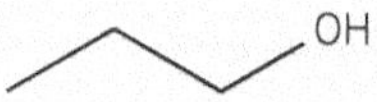

(2 marks)   (2 marks)

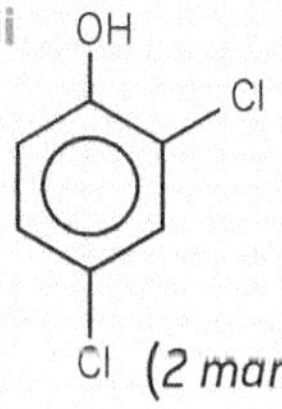

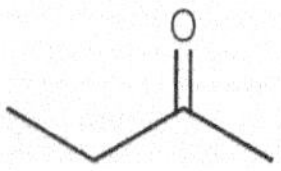

vi $CH_3C(CH_3)(OH)CH_2CH_3$ (2 marks)

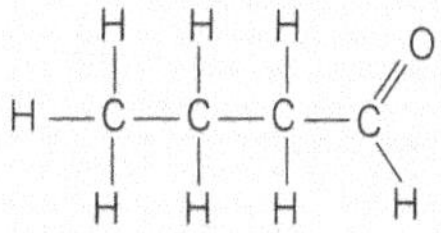

vii $HCOOH$ (2 marks)

b From compounds **i** to **vii** identify:

a a secondary alcohol (1 mark)   c a phenol (1 mark)

b an aldehyde (1 mark)   d a ketone (1 mark)

e an aliphatic alcohol that is not easily oxidised on heating with acidified potassium dichromate(VI) (1 mark)

f produces a purple colour with neutral aqueous iron(III) chloride (1 mark)

g gives carbon dioxide with sodium carbonate (1 mark)

h produces a carboxylic acid on refluxing with excess acidified potassium dichromate(VI) (1 mark)

i can be produced by the oxidation of methanol. (1 mark)

2 Draw the structure of the organic product when each of the following reacts with excess NaOH.

a OH OH

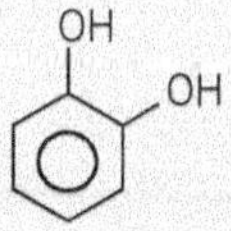

b O OH OH O

(2 marks)   (2 marks)

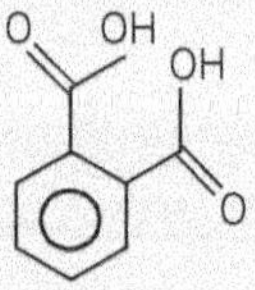

3 Phenols and carboxylic acids are weak acids and show typical acid properties. Write a balanced equation for each of the following reactions.

a phenol and sodium hydroxide (2 marks)

b propanoic acid and potassium hydroxide (2 marks)

c butanoic acid and sodium carbonate (2 marks)

4 Draw the structure of the ester produced in the following reactions.

a OH CH₃ + H₃C—C(=O)—O—C(=O)—CH₃

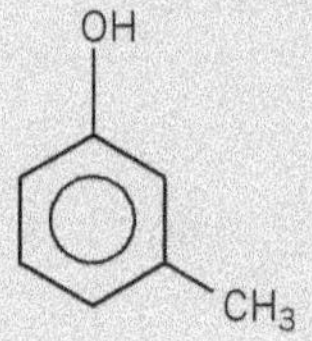

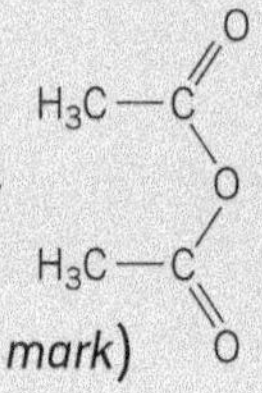

(1 mark)

b OH CH₃ + H₃C—C(=O)—OH

(1 mark)

▲ **Figure 10** *methyl 2-hydroxybenzoate*

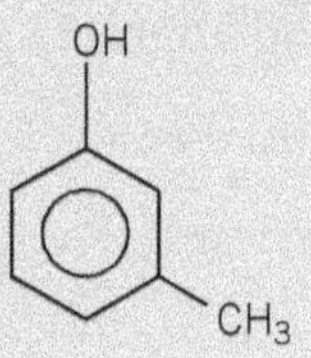

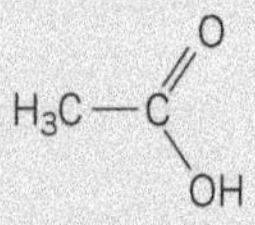

Study tip

You may need to look back at Topic WM 1 to answer question 2.

WM 3 Infrared spectroscopy

Specification reference: WM(j)

Learning outcomes

Demonstrate and apply knowledge and understanding of:

→ the effect of specific frequencies of infrared radiation making specific bonds in organic molecules vibrate (more); interpretation and prediction of infrared spectra for organic compounds, in terms of the functional group(s) present.

Although chemical tests provide evidence for the presence of carboxylic acid and phenol groups in salicylic acid, instrumental techniques are today's most efficient research tools. Three frequently used instrumental techniques are:

- infrared (IR) spectroscopy
- nuclear magnetic resonance (nmr) spectroscopy
- mass spectroscopy.

This section focuses on infrared spectroscopy.

Making use of infrared spectroscopy

One of the very first things done with any unidentified substance is to record an infrared (IR) spectrum. Figure 1 shows the IR spectrum of salicylic acid. Different function groups give different peaks on an IR spectrum.

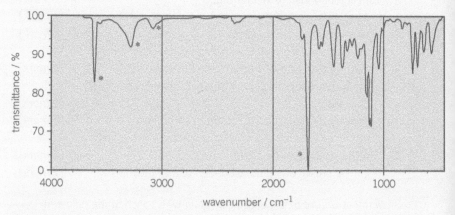

▲ Figure 1 Infrared spectrum of salicylic acid

Chemical ideas: Radiation and matter 6.4

Infrared spectroscopy

The energy possessed by molecules is **quantised** – molecules must take a small number of definite energy values rather than any energy value.

In **infrared spectroscopy** substances are exposed to radiation in the frequency range 10^{14}–10^{13} Hz, that is, wavelengths 2.5–15 μm. This makes vibrational energy changes occur in the molecules, which absorb infrared radiation of specific frequencies.

Frequency and wavelength are related.

speed of light c (m s^{-1}) = wavelength λ (m) × frequency v (s^{-1})

Synoptic link

You first encountered the equation $c = \lambda v$ in Chapter 1, Elements of life.

The speed of light c is a constant $-3.00 \times 10^8 \, \mathrm{m\,s^{-1}}$. From this equation, a direct measure of frequency is $\frac{1}{\lambda}$. This is the wavenumber of the radiation and usually measured in $\mathrm{cm^{-1}}$. This is the unit recorded on an **infrared spectrum**. Figure 2 shows the relationship between wavenumber, wavelength, and frequency. Note the directions of the arrows.

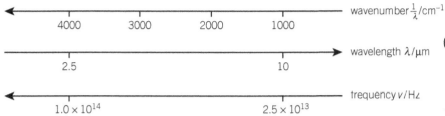

▲ Figure 2 *The relationship between wavenumber, wavelength, and frequency*

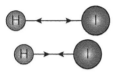

▲ Figure 3 *Stretching in the HI molecule*

Bond deformation

Simple diatomic molecules such as HCl, HBr, and HI can only vibrate in one way – stretching – where the atoms pull apart and then push together again (Figure 3). For these molecules there is only one vibrational infrared absorption. This corresponds to the molecules changing from their lowest vibrational energy state to the next higher level, in which the vibration is more vigorous.

The frequencies of the absorptions are different for each molecule. This is because the energy needed to excite a vibration depends on the strength of the bond holding the atoms together – weaker bonds require less energy. Table 1 shows how the bond strength is related to infrared absorption.

In more complex molecules, more bond deformations are possible. Most of these involve more than two atoms. For example, carbon dioxide can vibrate as shown in Figure 4.

More complex molecules can have many vibrational modes, descriptions such as rocking, scissoring, twisting, and wagging. Add these to the fact that the molecules also contain bonds with different bond enthalpies and you may have a very complicated spectrum.

▼ Table 1 *Bond enthalpies and infrared absorptions for the hydrogen halides*

Compound	Bond enthalpy / $\mathrm{kJ\,mol^{-1}}$	Infrared absorption / $\mathrm{cm^{-1}}$
HCl	+432	2886
HBr	+366	2559
HI	+298	2230

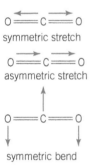

▲ Figure 4 *Vibrations in the CO_2 molecule*

The important point to remember about infrared spectroscopy is that you do not try to explain the whole spectrum – you look for one or two signals that are characteristic of particular bonds.

Figure 5 shows the infrared spectrum of ethanol and some of the vibrations that give rise to the signals.

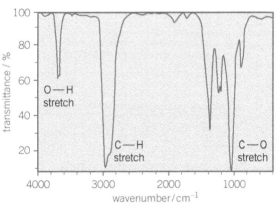

▲ Figure 5 *The infrared spectrum of ethanol in the gas phase (left) and the ethanol molecule, showing the vibrations which give rise to some characteristic absorptions*

Interpreting the spectra

An infrared spectrometer detects absorptions and produces an infrared spectrum. The infrared spectrum of but-1-ene (Figure 7) shows most of the characteristic absorptions of an unsaturated hydrocarbon containing a C=C double bond. These have been marked on the spectrum to show the bonds which are responsible.

Activity WM 3.1

In this activity you will be able to use your knowledge to interpret infrared spectra

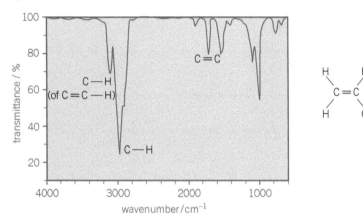

▲ Figure 7 Infrared spectrum of but-1-ene

In general, you can match a particular bond to a particular absorption region. Table 2 gives some examples. The precise position of an absorption depends on the environment of the bond in the molecule, so the wavenumber is only quoted for regions where the absorptions are expected to arise.

Figure 8 shows the infrared spectrum of propanone. The absorption at around $1740\,cm^{-1}$ is characteristic of the carbonyl, C=O, group. The absorption is very intense compared with the C=C absorption in a similar region for but-1-ene (Figure 7).

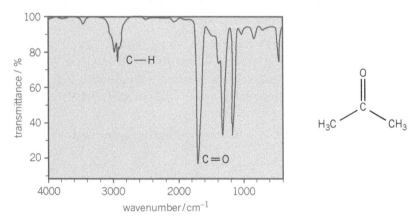

▲ Figure 8 Infrared spectrum of propanone

Why are some absorptions intense while others are weaker? The strongest infrared absorptions arise when there is a large change in bond polarity associated with the vibration. Therefore O—H, C—O, and C=O bonds, which are very polar, give more intense absorptions than the non-polar C—H, C—C, and C=C bonds.

Hydrogen bonding affects the absorption due to the O—H stretching vibration. Figure 5 showed the infrared spectrum of ethanol in the gas phase. There is little hydrogen bonding and the O—H absorption is a sharp peak at $3670\,cm^{-1}$.

Now look at Figure 9, which shows the infrared spectrum of a liquid film of ethanol. Hydrogen bonding between the hydroxyl groups changes the O—H vibration and the absorption becomes much broader. It is also shifted to a lower wavenumber and shows maximum absorption at $3340\,cm^{-1}$.

Synoptic link

You found out about bond polarity in Chapter 4, The ozone story.

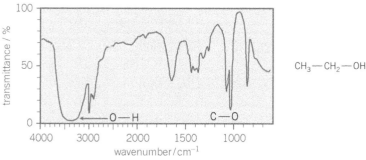

▲ **Figure 9** *Infrared spectrum of ethanol (liquid film)*

Study tip

The spectrum in Figure 9 has a non-linear scale for the horizontal axis. This is common for many instruments, so always check the scale carefully when reading off values of wavenumbers for absorptions.

Using the combination of wavenumber and intensity it is possible to interpret simple infrared spectra. Reference tables, like Table 2, are used when interpreting simple infrared spectra.

▼ **Table 2** *Characteristic infrared absorptions in organic molecules*

Bond	Location	Wavenumber/cm^{-1}	Intensity
C—H	alkanes	2850–2950	M—S
	alkenes, arenes,	3000–3100	M—S
	alkynes	*ca* 3300	S
C=C	alkenes	1620–1680	M
	arenes	several peaks in range 1450–1650	variable
C≡C	alkynes	2100–2260	M
C=O	aldehydes	1720–1740	S
	ketones	1705–1725	S
	carboxylic acids	1700–1725	S
	esters	1735–1750	S
	amides	1630–1700	M
C—O	alcohols, ethers, esters	1050–1300	S
C≡N	nitriles	2200–2260	M
C—F	fluoroalkanes	1000–1400	S
C—Cl	chloroalkanes	600–800	S
C—Br	bromoalkanes	500–600	S

Study tip

Remember the infrared spectrum helps you decide the type of bonds present. It cannot tell you where they are in the molecule or how many of each type there are.

▼ Table 2 (continued)

Bond	Location	Wavenumber/cm⁻¹	Intensity
O—H	alcohols, phenols	3600–3640	S
	*alcohols, phenols	3200–3600	S (broad)
	*carboxylic acids	2500–3200	M (broad)
N—H	primary amines	3300–3500	M—S
	amides	*ca* 3500	M

* hydrogen-bonded   M = medium   S = strong

It is helpful to divide the infrared spectrum into four regions (Table 3).

▼ Table 3 *Regions in the infrared spectrum where typical absorptions occur*

Absorption range / cm⁻¹	Bonds responsible	Examples
4000–2500	single bonds to hydrogen	O—H, C—H, N—H
2500–2000	triple bonds	C≡C, C≡N
2000–1500	double bonds	C=C, C=O
below 1500	various	C—O, C—X (halogen)

Below 1500 cm⁻¹ the spectrum can be quite complex and it is more difficult to assign absorptions to particular bonds. This region is characteristic of the particular molecule and is often called the **fingerprint region**. It is useful for identification purposes, for example, if you need to compare two spectra to find out if they are spectra of the same compound. It is only rarely used to identify functional groups. Figure 10 summarises the information in Table 3.

Aromatic compounds often exhibit complex absorption patterns in the fingerprint region (see Figures 13 and 14). Such compounds can often be identified by comparing their infrared spectra with reference spectra.

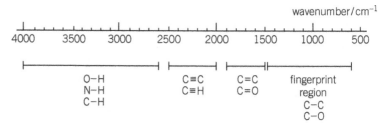

▲ Figure 10 *Typical regions of absorption in the infrared spectrum*

Worked example: Using infrared spectra – butane, $CH_3CH_2CH_2CH_3$

What are the key features shown in the infrared spectrum in Figure 11 that indicates the molecule is butane?

1 A strong absorption at 2970 cm^{-1} characteristic of C—H stretching in aliphatic compounds.

2 No indication of any functional groups.

3 An alkane fits with these features.

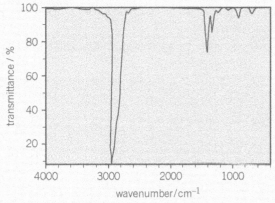

▲ **Figure 11** *Infrared spectrum of butane*

 Worked example: Using infrared spectra – methylbenzene,

What are the key features shown in the infrared spectrum in Figure 12 that indicates this molecule is methylbenzene?

1 C—H absorptions just above 3000 cm^{-1} for the C—H on the benzene ring.

2 C—H absorption just below 3000 cm^{-1} for C—H on the methyl group.

3 An absorption pattern around 700 cm^{-1} is typical of a benzene ring with one substituted group.

4 No indication of any functional groups.

5 An alkylbenzene (arene) matches these features.

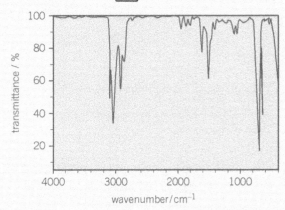

▲ **Figure 12** *Infrared spectrum of methylbenzene*

 Worked example: Using infrared spectra – benzoic acid,

What are the key features shown in the infrared spectrum in Figure 13 that indicate this is benzoic acid?

1 A sharp absorption at 3580 cm^{-1} characteristic of an O—H bond (not hydrogen-bonded).

2 A strong absorption at 1760 cm^{-1} shows the presence of a C=O group.

3 The position of the C—H absorption suggests it is an aromatic compound.

4 An aromatic carboxylic acid fits these features. The identity of the sample could be confirmed by comparing its spectrum with that of an authentic sample of benzoic acid, checking the fingerprint region carefully.

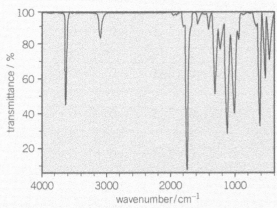

▲ **Figure 13** *Infrared spectrum of benzoic acid*

Summary questions

1 The infrared spectrum of carbon dioxide shows a strong absorption at $2360\,cm^{-1}$.

   a Calculate the wavelength of the radiation absorbed. Give your answer in µm. *(1 mark)*

   b Use $c = \lambda v$ to calculate the frequency of the radiation absorbed. *(1 mark)*

   $(c = 3.00 \times 10^8\,m\,s^{-1} \qquad 1\,\mu m = 1 \times 10^{-6}\,m)$

2 Figure 14 shows the infrared spectrum of phenol.

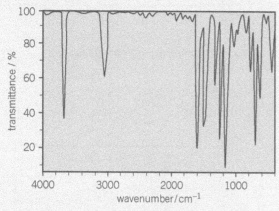

▲ **Figure 14** *Infrared spectrum of phenol*

Identify the key peaks in the spectrum, and the bond to which each corresponds. *(3 marks)*

3 The infrared spectra in Figure 15 (Spectrum A and Spectrum B) represent butan-2-ol and butanone.

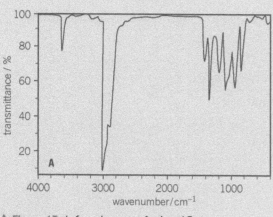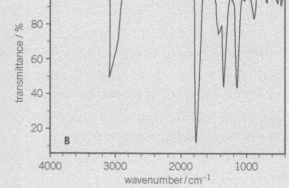

▲ **Figure 15** *Infrared spectra for A and B*

   a Draw structures for butan-2-ol and butan-2-one. *(2 marks)*

   b Identify the key peaks in each spectrum, and the bond to which each corresponds. Give your answer in the form of a table. *(4 marks)*

   c Decide which spectrum represents butan-2-ol and which represents butan-2-one. *(2 marks)*

4 The infrared spectra in Figure 16 represents three compounds – C, D, and E. The compounds are an ester, a carboxylic acid, and an alcohol, though not necessarily in that order.

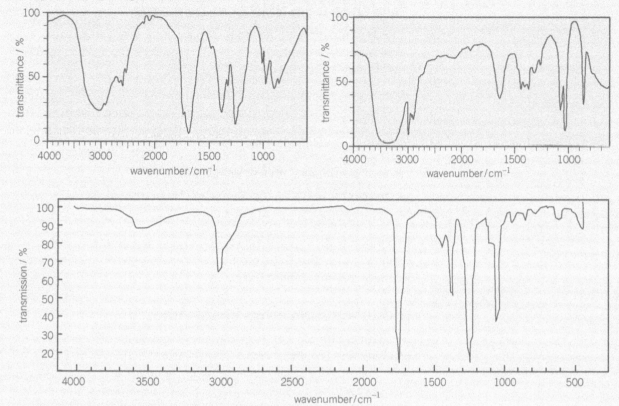

▲ Figure 16 *Infrared spectra for compounds C, D, and E*

a Identify the key peaks in each spectrum, and the bond to which each corresponds. Give your answer in the form of a table. (*6 marks*)

b Decide which spectrum represents which type of compound. (*3 marks*)

5 Oil of wintergreen has mild pain-killing properties. Its structure is shown below.

Draw up a table to show the key peaks you would expect to see in the infrared spectrum of oil of wintergreen, and the bond that each absorption corresponds to. (*8 marks*)

WM 4 Mass spectrometry

Specification reference: WM(i)

Learning outcomes

Demonstrate and apply knowledge and understanding of:

→ interpretation and prediction of mass spectra:

- the M+ peak and the molecular mass
- that other peaks are due to positive ions from fragments
- the M+1 peak being caused by the presence of ^{13}C.

Synoptic link

You will learn more about nmr spectroscopy in Chapter 7, Polymers and life.

Synoptic link

In Chapter 1, Elements of life, you saw how information about the relative abundance of isotopes of an element can be obtained from a mass spectrum.

Evidence from NMR spectrometry

A second instrumental technique that could be applied to an unidentified compound such as salicylic acid is **nuclear magnetic resonance spectroscopy** (NMR).

NMR spectroscopy measures the surrounding chemical environment of the nuclei of one particular element – most often hydrogen. The nucleus of a hydrogen atom consists of just one proton. The proton NMR spectrum for salicylic acid shows that salicylic acid contains:

- one proton in a –COOH environment
- one proton in a phenolic –OH environment
- four protons attached to a benzene ring.

Although NMR spectroscopy is of limited value in this case, it generally provides a powerful technique for determining the structure of many organic compounds.

Evidence from mass spectrometry

A combination of IR and NMR spectroscopy shows that salicylic acid has an –OH group and a –COOH group both attached to a benzene ring. In other words, a better name for salicylic acid is hydroxybenzoic acid.

However, there are three possible isomeric hydroxybenzoic acids – 2-hydroxybenzoic acid, 3-hydroxybenzoic acid, and 4-hydroxybenzoic acid. A decision about which isomer salicylic acid is can be made by analysis of the mass spectrum of the compound.

Chemical ideas: Radiation and matter 6.5b

Mass spectrometry with compounds

Mass spectrometry can be used to find out the atomic mass of elements and the relative abundances of isotopes in an element. In the case of more complex molecules, these can be fragmented, ionised, and detected in a mass spectrometer.

Using mass spectra to investigate the structure of molecules

A typical mass spectrum from an organic compound, such as the one for 2-ethoxybutane (Figure 1), can be quite complex. The spectrum is dislayed as percentage intensity against a value of *m/z*. You can assume this is equivalent to the molecular mass of the ion causing the peak.

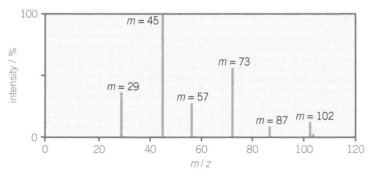

▲ Figure 1 *Simplified mass spectrum for 2-ethoxybutane, showing the six largest peaks*

The heaviest ion (m = 102) is the one corresponding to the ethoxybutane molecule with just one electron removed – this is called the **molecular ion** M^+. The m/z value for the M^+ ion helps chemists determine the molecular mass of the compound being analysed.

However, this is not the only ion reaching the detector. The M^+ ion fragments into smaller ions and the mass spectrometer detects and analyses these too. The way in which a parent ion breaks down is characteristic of that compound – the **fragmentation pattern**. This is why there are many peaks with a m/z value below the molecular mass of the unfragmented compound.

The most abundant ion gives the strongest detector signal. This is set to 100% in the spectrum and is referred to as the **base peak**. The intensities of all the other peaks are expressed as a percentage of its value. It is not uncommon for the molecular ion peak to be so weak as to be unnoticeable – the spectrum obtained then consists entirely of fragments.

Often a peak appears at a value of M+1. This is because a small proportion of carbon (approximately 1.1%) exist as ^{13}C. If a molecule has incorporated one of these carbon-13 atoms, the molecular mass of the compound will be M+1 (where M is the molecular mass of carbon-12). Because there are proportionally a small number of carbon-13 atoms compared to carbon-12 atoms, the M+1 peak will be small when compared to the peak caused by the M^+ ion.

Interpreting the mass spectrum of 2-ethoxybutane (Figure 1)

The peaks at 29, 45, 57, 73, and 87 are caused by fragmentation of the ion of the parent molecule. This produces a number of different fragment ions of different masses.

The molecular mass of the sample can be easily seen to be 102 and is a result of ionisation of the parent molecule (M^+), in this case $[CH_3CH(CH_3)OCH_2CH_3]^+$.

The small peak at 103 is caused by the M^+ ion with a carbon-13 atom present in each ion being detected.

Synoptic link

You will learn how to interpret fragmentation patterns in Chapter 7, Polymers and life.

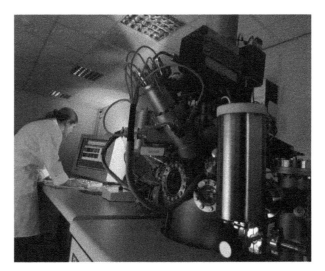

▲ Figure 2 *Mass spectrometer*

The mass spectrum of salicylic acid

Figure 3 shows the mass spectrum of salicylic acid. The signal at mass 138 is from the parent ion (molecular ion). Modern mass spectrometers give very accurate mass values for the signals in the spectrum. For salicylic acid, the high-resolution spectrum (Figure 3, top) gives a mass of 138.0317 for the molecular ion. This confirms that the substance has an empirical formula of $C_7H_6O_3$.

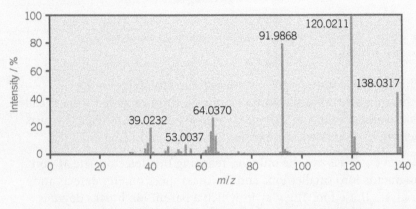

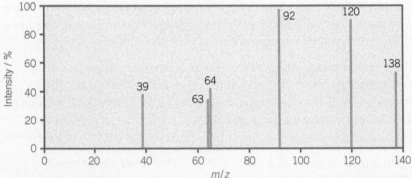

▲ **Figure 3** *Mass spectra of salicylic acid at high-resolution (top) and low-resolution (bottom). The relative heights are not the same because different experimental conditions have been used*

Considering the way that the 3-hydroxybenzoic acid and the 4-hydroxybenzoic acid isomers would break down suggests that that these isomers could not form some of the fragments observed in the mass spectrum of salicylic acid. For example, the signal at mass 120 could not be formed from either of these isomers. Therefore salicylic acid must be 2-hydroxybenzoic acid. Comparison of the fragmentation pattern with a database of known mass spectra identifies salicylic acid as 2-hydroxybenzoic acid.

Drawing the evidence about the structure of salicylic acid together

Chemical tests showed the presence of a phenolic, –OH, group and a carboxylic acid, –COOH, group. Infrared spectroscopy showed that the –OH and –C=O groups were present. Nuclear magnetic resonance spectroscopy confirmed that there were hydrogen atoms in three types of environment: –COOH, –OH attached directly to a benzene ring and

–H attached directly to a benzene ring. Mass spectroscopy showed that salicylic acid was the same compound as that stored in the database as 2-hydroxybenzoic acid (Figure 4).

The mass spectrum fragmentation pattern showed that the structure could not be 3-hydroxybenzoic acid or 4-hydroxybenzoic acid (Figure 5).

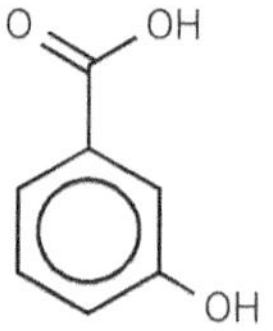

▲ Figure 4 The structure of 2-hydroxybenzoic acid

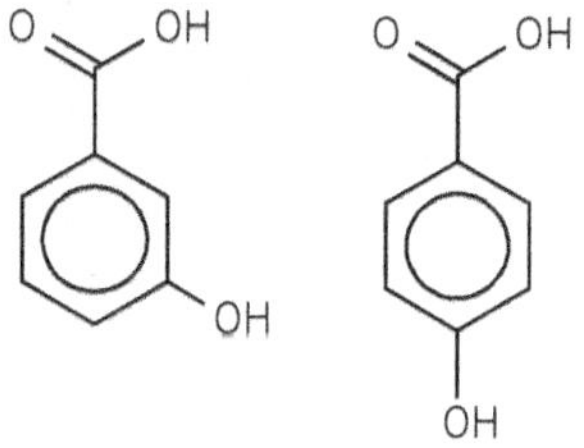

▲ Figure 5 The structure of 3-hydroxybenzoic acid (left) and 4-hydroxybenzoic acid (right)

Summary questions

1 Look at the mass spectrum of butan-2-one.

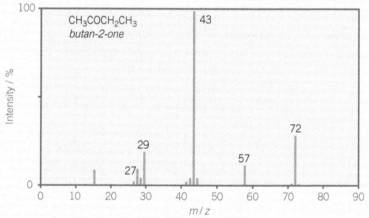

$CH_3COCH_2CH_3$
butan-2-one

a Identify the molecular ion peak and write the formula for it. (*1 mark*)
b What causes the peak at $m/z = 73$? (*1 mark*)
c Why are there peaks at m/z values of 27, 29, 43, 57, and 72? (*2 marks*)

2 Butan-2-ol has a number of different structural isomers. How would the mass spectra of these isomers:
a be the same as the mass spectrum for butan-2-ol? (*2 marks*)
b differ from the mass spectrum for butan-2-ol? (*2 marks*)

Activity WM 4.1

In this activity you can practise the use of both mass spectra and IR spectra to determine the structure of organic compounds.

WM 5 The synthesis of salicylic acid and aspirin

Specification reference: WM(e), WM(g)

Medicines that are natural products – those that come directly from plants – may be difficult to obtain when needed. Supply may be seasonal, depend on weather conditions, and be liable to contamination. Collecting plants from their natural habitat is also usually not environmentally sustainable.

Synthesising salicylic acid

Chemists do not want to rely on willow trees as their source of salicylic acid (2-hydroxybenzoic acid). Once the chemical structure of the active compound in a plant is known, chemists begin to search for ways of producing it artificially.

Simple inorganic substances, such as aluminium chloride, can be synthesised directly from their elements, but larger, more complex molecules cannot be made directly in this way. Instead, a known compound with a similar structure is identified, and the structure modified.

At the end of the nineteenth century, the compound phenol was already well known in the pharmaceutical industry – it has germicidal properties. It was also readily available as a product from heating coal in gas-works. Its molecular structure differs from that of 2-hydroxybenzoic acid by only one functional group. The problem in synthesis is to introduce this extra group in the right position without disrupting the rest of the molecule.

In this particular case, carbon dioxide can be combined directly with phenol to give 2-hydroxybenzoic acid by careful control of the conditions.

$$C_6H_5OH \quad + \quad CO_2 \quad \rightarrow \quad C_6H_4(OH)COOH$$

This general method is known as the Kolbe synthesis and an industrial version of this addition reaction was developed by the German chemist Felix Hoffmann. This is an early example of 'green chemistry' with an atom economy of 100% because there are no leaving atoms or molecules.

Chemical ideas: Greener industry 14.3

Key principles of green chemistry

Although green chemistry can mean different things to different people it is simply developing chemical products and especially chemical processes that are as sustainable and as environmentally friendly as possible. There are twelve key principles of green chemistry (Table 1).

▼ Table 1 *The 12 principles of green chemistry·*

Principles	Explanation
Better atom economy	Means more of the feedstock is incorporated into the product and less waste products are produced.
Prevention of waste products	This is better than treating and disposing of waste.
Less hazardous chemical synthesis	Using less hazardous chemicals in the chemical reaction.
Design safer chemical products	Less toxic and hazardous chemical products.
Use safer solvents	Minimise the use of organic solvents.
Lower energy usage	Lower temperature and pressure processes.
Use renewable feedstocks	Instead of depleting natural resources.
Reduce reagents used and the number of steps	As these can generate waste.
Use catalysts and more selective catalysts	These generally reduce energy usage and waste products.
Design chemical products for degradation	When released into the environment should break down into innocuous products.
Employ real time process monitoring	Better monitoring of chemical processes reduces waste products.
Use safer chemical processes	Choose processes that minimise the potential for releasing gases, fires and explosions.

Synoptic link

You first encountered atom economy in Chapter 3, Elements from the sea.

It follows that in any chemical process the key factors to consider are cost, impact on the environment, and health and safety. All individual chemical processes involve a balance of all of these. It cannot be assumed that application of the twelve principles will automatically results in a less expensive (therefore more profitable) process.

Examples of use if the green chemistry principles include

- The original manufacture of Taxol (a chemotherapy drug) was made by extraction from the tree bark of an extremely slow growing Pacific yew. This needed vast quantities of organic solvent and often killed the tree. A newer method involves growing tree cells in a fermentation vat followed by recrystallization, so reducing the amount of solvent and preventing the death of the yew trees. There are developments in using the bacterium *E. coli* to increase synthetic yields.

- The cholesterol reducing drug atorvastatin is now synthesised using an enzyme that catalyses chemical reactions in water, reducing the need for potentially polluting organic solvents.

- The painkiller Ibuprofen was originally synthesised in a 12 step synthesis with a 40% atom economy. It is now synthesised in a four stage process with an atom economy of of 77%.

▲ Figure 1 *The industrial production of medicines*

Synoptic link

You will look at more industrial applications of chemistry in Chapter 6, The chemical industry.

Each one of these changes will follow one or more of the green chemistry principles outlined above. However, there will always be other aspects to take into account. For example:

- By changing the reagents to improve atom economy, does this increase the cost of the new reactants too much?
- Has the replacement of one reagent with another raised new health as safety issues?
- Even though the number of steps in a synthesis have been reduced what is the overall yield?
- Maybe a lowerd temperature (thereby reducing energy costs) increases the time for the reaction to reach equilibrium (thereby increasing the overall cost).

Study tip

You do not need to learn the example below, but you should be able to analyse and use given information in a range of given examples.

Chemists are continually balancing out costs, health and safety, and the impact on the environment. It would be rare to meet a synthesis that holds to all 12 green principles.

An example of using some of these principles can be seen in the approaches taken in the manufacture of 4-aminodiphenylamine.

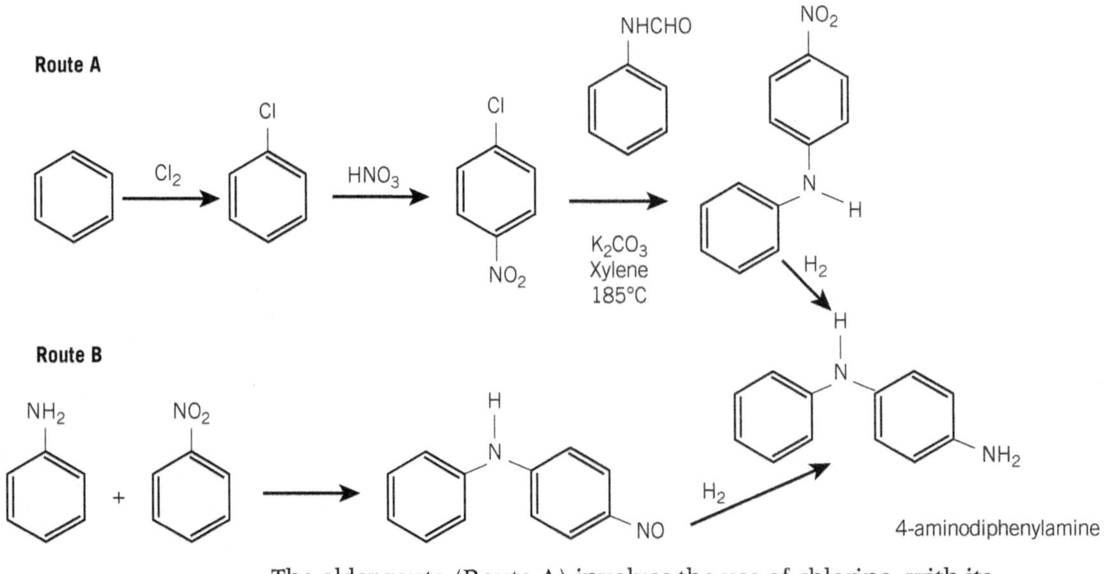

The older route (Route A) involves the use of chlorine, with its associated hazards and is a four step process. The newer route (Route B) has an improved atom economy, reduces the amount of waste and improved the risks associated with the process.

Making aspirin

Through Hoffmann's work, synthetic 2-hydroxybenzoic acid of reliable purity became available. Synthetic 2-hydroxybenzoic acid was widely used for curing fevers and suppressing pain, but reports began to accumulate of irritating effects on the mouth, gullet, and stomach. Clearly the new wonder medicine had unpleasant side-effects. Chemists had a new problem – could they modify the structure to reduce the irritating effects, whilst still retaining the beneficial ones?

Hoffmann prepared a range of compounds by making slight modifications to the structure of 2-hydroxybenzoic acid. His father was a sufferer from chronic rheumatism and Hoffmann tried out each of the new preparations on him to test its effects. This was a bit more primitive than the modern testing of medicines. It is not recorded what Hoffmann's father thought of all this, but he survived long enough for his son to prepare a derivative that was as effective as 2-hydroxybenzoic acid and had less unpleasant side effects.

The effective product was 2-ethanoxybenzoic acid (sometimes called 2-ethanoylhydroxybenzoic acid or acetylsalicylic acid). This is now known as aspirin and it is both an ester and a carboxylic acid (Figure 2).

However aspirin is not very soluble in water. It was first available as a powder in sachets – Bayer then decided to pellet the powder and aspirin became the first medicine to be sold as tablets.

Drug purity

When an organic solid product is synthesised it is in the crude form and it contains impurities, such as reactants, reagents, and solvents. The product needs to be purified and its purity checked, particularly if it is to be used as a drug. This is often more time consuming than the synthesis. There are several techniques for purifying products.

Techniques for purifying organic solids

The three main techniques for purifying and/or identifying organic solids are recrystallisation, thin layer chromatography, and melting point determination.

Recrystallisation

This technique is used to purify solid, crude organic products. Only the desired compound dissolves (to an appreciable extent) in the chosen hot solvent, leaving insoluble impurities to be filtered off. On cooling, the desired compound will **crystallise** out, leaving any soluble impurities in solution. The pure crystals can be filtered off, washed, and dried.

Thin layer chromatography

Thin layer chromatography is used to separate small quantities of organic compounds, purify or check the purity of organic substances, and follow the progress of a reaction over time.

A suitable solvent must be chosen. Different organic compounds have different affinities for a particular solvent, and so will be carried through the chromatography medium (plate) at different rates. When chromatography is carried out using a silica plate, it is known as **thin layer chromatography**.

Melting point determination

Measuring melting points is used as evidence of a solid organic compound's identity and purity. The value obtained is compared to the published value. A pure compound should melt within 0.5 °C of its true melting point.

▲ Figure 1 *Meadowsweet (Spiraea ulmaria), from which salicylic acid was first extracted in 1835 – aspirin got its name from 'a' for acetyl (an older word for ethanoyl) and 'spirin' for spirsaüre (the German word for salicylic acid)*

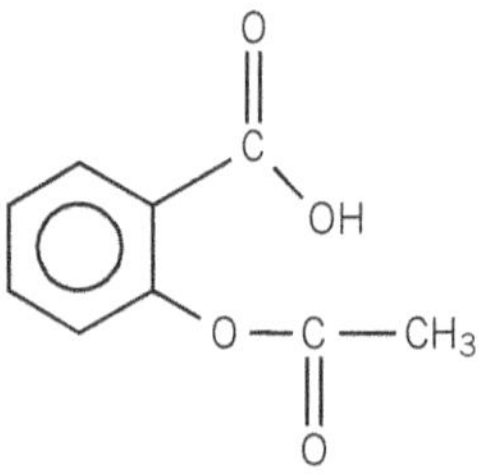

▲ Figure 2 *The structure of aspirin*

Synoptic link

You will find out more detail on recrystallisation, thin layer chromatography, and melting point determination in Techniques and procedures.

Activity WM 5.1

This activity gives you the opportunity to synthesis aspirin in a two stage process starting from oil of wintergarden.

Activity WM 5.2

In this activity you will recrystallise aspirin and check its purity.

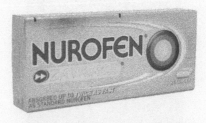

▲ **Figure 3** *Ibuprofen is available in a number of different formulations and trade names. It is relatively cheap now as the patent has expired*

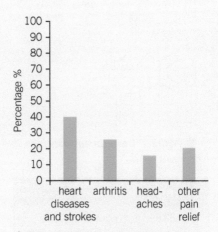

▲ **Figure 4** *Current uses of aspirin*

Developing a new medicine costs an enormous amount of money. The selling price charged by the pharmaceutical company must be sufficient not only to cover the costs of production and marketing, but also to recover the development costs. If other companies could simply copy the medicine, they would be able to sell it at a much lower price. This is where patents become important.

Protecting the discovery

When a pharmaceutical company discovers a new medicine, it takes out patents to protect the discovery. Patents apply in only one country, so several patents must be taken out to prevent companies in other countries manufacturing the medicine. Patents last only for a specific amount of time, but whilst the patent is in force no other company can manufacture the medicine in that country. By the time a patent runs out, the company that discovered the medicine will hopefully have sold enough of it to cover its development costs. Afterwards, any company can produce and sell the medicine. Ibuprofen – a drug used to treat pain, inflammation, and fever – is sold under a variety of trade names as tablets and in pain relief gels (Figure 3).

New uses of aspirin

The life cycle of modern medicines is often short because medical and pharmaceutical research offers better remedies all the time. Aspirin is now over 100 years old but new uses continue to emerge all the time. For example, the discovery of its potential in treating heart disease came from observations by doctors on their patients. By careful analysis, Dr Laurence Craven in the USA noticed that his male patients who took aspirin suffered fewer heart attacks. Similar work is being carried out to observe its effects on certain cancers, diabetes, deep vein thrombosis, and Alzheimer's disease, among others. This research is being carried out by doctors and universities using statistical analysis of medical records, and not by pharmaceutical companies, who have nothing to gain by it.

Summary questions

1 Write a step by step method for recrystallisation, using your practical work as a guide. *(3 marks)*

2 What principles could be used to reduce waste products in industrial processes for manufacturing chemicals? *(2 marks)*

3 List three different types of organic reaction and comment on their atom economy. *(3 marks)*

4 Research different techniques for determining melting points and give a brief description. What advantages has the technique chosen by your educational institution? *(3 marks)*

Practice questions

1 How many esters are there with formula $C_4H_8O_2$?

   A 1

   B 2

   C 3

   D 4   *(1 mark)*

2 Which pair of compounds would react to form CH_3COOCH_3?

A	CH_3CH_2OH	HCOOH
B	CH_3OH	$(CH_3CO)_2O$
C	CH_3COOH	CH_3CH_2OH
D	CH_3Cl	CH_3COOH

                                                      (1 mark)

3 What will be the main product when propan-1-ol is distilled with acidified dichromate?

   A CH_3COCH_3

   B CH_3CH_2COOH

   C CH_3CH_2CHO

   D $CH_3CH_2CH_2OH$ *(1 mark)*

4 A student is recrystallising a solid using water as solvent. The solid is dissolved in the minimum of hot water and allowed to crystallise.

   Which is the correct sequence that follows?

   A wash with water, filter, and dry the crystals

   B remove the wet crystals and let them dry on a hotplate

   C filter the crystals and dry them

   D filter the crystals, wash, and dry. *(1 mark)*

5 Which of the following sequences would work for purifying a liquid organic product?

   A wash, separate, dry, distil

   B distil, wash, separate, dry

   C separate, dry, wash, distil

   D dry, distil, wash, separate   *(1 mark)*

6 Which of the following is **not** true about a mass spectrum?

   A The M+1 peak is caused by impurities.

   B The M$^+$ peak indicates the M_r of the compound.

   C Peaks of smaller mass are caused by fragments of the molecule.

   D Only positive ions are detected.   *(1 mark)*

7 In which of the following rows does the name correctly describe the formula?

A	$(CH_3CO)_2O$	ester
B	CH_3CHO	ketone
C	CH_3OCH_3	ether
D	CH_3COCH_3	acid anhydride

                                                      (1 mark)

8 CH_3COOH can be made by the oxidation of which of the following:

   1 CH_3CHO

   2 CH_3CH_2OH

   3 CH_3COCH_3

   A 1,2, and 3 correct

   B 1 and 2 are correct

   C 2 and 3 are correct

   D Only 1 is correct   *(1 mark)*

9 A substance that fizzes when added to aqueous sodium carbonate could be:

   1 a carboxylic acid

   2 a phenol

   3 an alcohol.

   A 1,2, and 3 correct

   B 1 and 2 are correct

   C 2 and 3 are correct

   D Only 1 is correct   *(1 mark)*

10 Which of the following is/are correct about a tertiary alcohol?

   1 It cannot be oxidised using acid dichromate.

   2 It will form esters.

   3 It can be dehydrated.

   A 1,2, and 3 correct

   B 1 and 2 are correct

   C 2 and 3 are correct

   D Only 1 is correct   *(1 mark)*

11 Some students set out to make a sample of aspirin.

They react salicylic acid with ethanoic anhydride, using a phosphoric acid catalyst.

a Complete the equation for the reaction by drawing out the skeletal structures of the products. *(2 marks)*

$$C_7H_6O_3 \quad + \quad C_4H_6O_3 \longrightarrow$$

b The impure product is filtered off, washed, and dried, and its melting point taken.

   (i) Name the method of purification that the students would use. *(1 mark)*

   (ii) How would the melting point change (if at all) after this purification? *(1 mark)*

   (iii) The students start with 10.0 g of salicylic acid and finish with 5.0 g of purified product. Calculate the percentage yield. *(2 marks)*

   (iv) The students use thin layer chromatography to check they have made aspirin and to see if there is any unreacted salicylic acid in the sample of aspirin. Give details of their method. *(6 marks)*

c (i) Which of aspirin and salicylic acid will react with neutral iron(III) chloride? Give your reason and state the colour of a positive test. *(3 marks)*

   (ii) Write the equation for the reaction of salicylic acid with sodium carbonate, using structural formulae for organic substances. *(2 marks)*

d The students titrate 1.05 g of their purified aspirin with sodium hydroxide. 24.70 cm^3 of 0.100 mol dm^{-3} NaOH was required.

   (i) Calculate the purity of the aspirin, assuming that one mole of aspirin reacts with one mole of NaOH under these conditions. *(2 marks)*

   (ii) How might NaOH react with aspirin under other conditions? *(2 marks)*

   (iii) Which impurity, if present in the aspirin, would make the results of your calculation in (i) invalid? *(1 mark)*

12 Butan-2-ol is an intermediate in the industrial formation of 'MEK', a widely-used solvent.

$$CH_2CH_2CH(OH)CH_3 \xrightarrow{\text{oxidation}} MEK$$
$$\text{butan-2-ol}$$

a (i) Classify butan-2-ol as a primary, secondary, or tertiary alcohol, giving a reason. *(2 marks)*

b (i) Give the reagents and conditions that would be used in a laboratory to make MEK from butan-2-ol. *(2 marks)*

   (ii) Draw the structure of MEK and name its functional group. *(2 marks)*

c A chemist takes a sample from the plant making MEK from butan-2-ol and runs the infrared spectrum shown below.

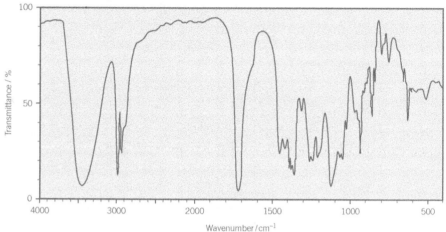

Explain the conclusions the chemist can draw about the progress of the reaction.

(*4 marks*)

d (i) A chemist runs a mass spectrum of butan-2-ol. Two of the peaks obtained are shown in the table below:

Give the type of ion responsible for each peak

Peak	Ion
74	
75 (very small)	

(*3 marks*)

(ii) Other peaks are found at m/z values below 74. How do these form?

(*1 mark*)

e A compound **A** gives just butan-2-ol when it reacts with water.

Give the full structural formula for compound **A**.　(*1 mark*)

f Butan-2-ol can be dehydrated in the laboratory to a mixture of compound **A** and one other compound.

(i) Give the reagents and conditions for this reaction.　(*2 marks*)

(ii) Give the skeletal formula of the other compound that is formed.　(*1 mark*)

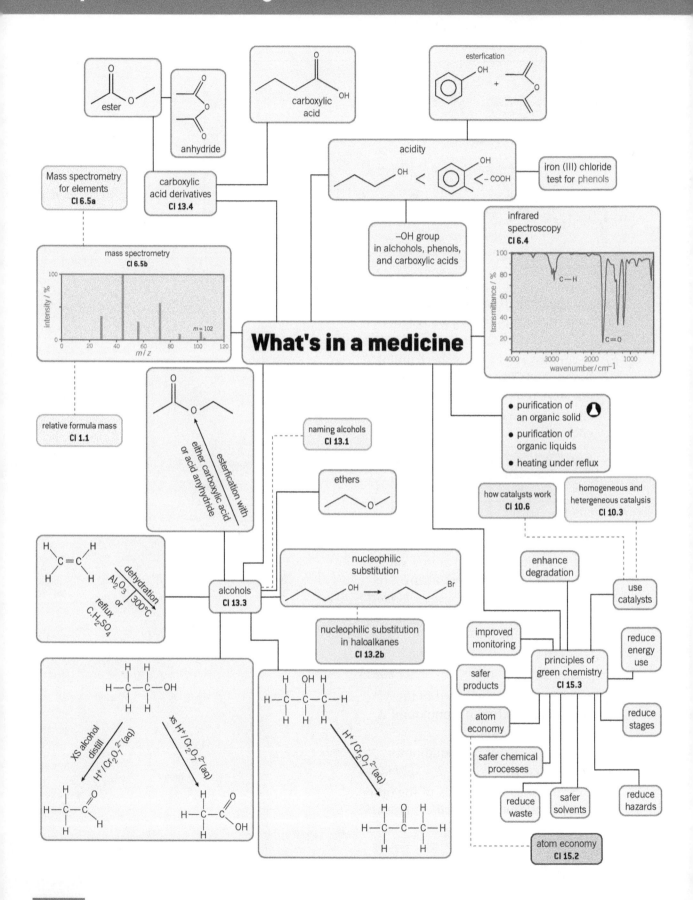

ester

anhydride

carboxylic acid

esterfication

OH +

Mass spectrometry for elements
CI 6.5a

carboxylic acid derivatives
CI 13.4

acidity

OH < OH COOH

iron (III) chloride test for phenols

mass spectrometry
CI 6.5b

intensity / %

m = 102

m / z

−OH group in alchohols, phenols, and carboxylic acids

infrared spectroscopy
CI 6.4

transmittance / %

C—H

C=O

wavenumber / cm^{-1}

What's in a medicine

relative formula mass
CI 1.1

esterfication with either carboxylic acid or acid anhydride

naming alcohols
CI 13.1

ethers

- purification of an organic solid
- purification of organic liquids
- heating under reflux

how catalysts work
CI 10.6

homogeneous and hetergeneous catalysis
CI 10.3

H$_2$C=CH$_2$

dehydration
Al$_2$O$_3$ | 300°C
or
reflux
C.H$_2$SO$_4$

alcohols
CI 13.3

nucleophilic substitution

OH → Br

enhance degradation

use catalysts

nucleophilic substitution in haloalkanes
CI 13.2b

improved monitoring

safer products

principles of green chemistry
CI 15.3

reduce energy use

H—C—C—OH

XS alcohol distill
H$^+$/Cr$_2$O$_7$$^{2-}$(aq)

xs H$^+$/Cr$_2$O$_7$$^{2-}$(aq)

H—C—C—C—H
OH

H$^+$/Cr$_2$O$_7$$^{2-}$(aq)

atom economy

reduce stages

safer chemical processes

H—C—C=O
H

H—C—C—C
OH

H—C—C—C—H
O

reduce waste

safer solvents

reduce hazards

atom economy
CI 15.2

Salbutamol and ibuprofen

salbutamol   ibuprofen

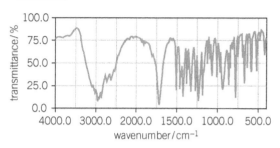

Salbutamol is used for the treatment of asthma and other bronchial diseases. It relaxes the muscles in the air passages of the lungs, making it easier to breathe by keep the airways open. Salbutamol is often prescribed as an inhaled preparation to be breathed in by the patient. This makes it fast acting.

Ibuprofen is an over-the-counter painkiller which reduces inflammation. It can be taken in tablet, capsule, or gel form and particularly works to improve movement by reducing muscle pain. Ibuprofen works by blocking production of molecules causing swelling and pain.

1 Name the oxygen-containing functional groups in salbutamol and ibuprofen. Choose from the following functional groups.

primary alcohol, secondary alcohol, tertiary alcohol, phenol, aldehyde, ketone, acid anhydride, carboxylic acid, ether, ester

2 Explain why salbutamol will react with sodium hydroxide but not sodium carbonate, whilst ibuprofen will react with both sodium hydroxide and sodium carbonate.

3 Describe and explain the effect of adding iron(III) chloride to separate samples of salbutamol and ibuprofen.

4 Identify whether the infrared spectrum is that of salbutamol or ibuprofen. Give the wavenumbers of key peaks and relate them to bonds in the molecule.

5 Salbutamol and ibuprofen will react with various reagents.

a State what you would observe when salbutamol reacts with acidified dichromate(VI) solution.

b Name the **type** of molecule made when salbutamol reacts with ethanoic acid in the presence of concentrated sulfuric acid.

c Draw the product when ibuprofen reacts with ethanol to form an ester.

d Draw the product when salbutamol undergoes a dehydration reaction using Al_2O_3. Classify the type of reaction mechanism occurring.

Extension

1 Research how salbutamol and ibuprofen can be made in the laboratory. Identify the starting materials, and describe as many of the steps as you can. Look for the conversion of one functional group into another in each step and suggest what reagents could be used to bring about each functional group change.

2 Draw a flowchart of organic reactions to summarise the reactions that functional groups covered in this topic can undergo.

3 Find out about the processes involved in testing a possible new drug before it is permitted for commercial use. Summarise the steps and explain why each is necessary.

CHAPTER 6

The chemical industry

Topics in this chapter

- → CI 1 Nitrogen chemistry
  Redox reactions in the soil and the laboratory
- → CI 2 Manufacturing nitric acid
  Equilibrium constants and the effect of changes in conditions
- → CI 3 Manufacturing sulfuric acid
  Measuring rates of reactions
- → CI 4 The importance of rate equations
  Orders of reactions and the use of the Arrhenius equation
- → CI 5 The treatment of diseases and the half-life of reactions
  Finding orders using a half-life method and the link between rate equations and mechanisms
- → CI 6 Manufacturing ethanoic acid
  A case study of an industrial process

Why a chapter on The chemical industry?

The origin of virtually every object that encountered in everyday life can be traced back in some way to the chemical industry. Food provides a good example of this. Food crops may grow naturally in soil, but without the fertilisers, pesticides, and preservatives that are used at various stages in food production we could not produce food in the quantity, quality, and reliability that modern society appears to demand.

In this chapter you will look at how the chemical industry manufactures some simple, yet important compounds – nitric acid, sulfuric acid, and ethanoic acid. Each of these plays a significant role in the story of food production. The factors that affect the position of chemical equilibrium and the rates of chemical reactions are studied and are used to explain how chemists in industry make processes more efficient and cost effective.

Many important mathematical skills are developed in this chapter. Equilibrium constant are used to predict concentrations of products that are formed at equilibrium, data about rates at different concentrations are used to deduce equations that describe how rate depends on concentration of different reactants. The relationship between rate and temperature involves the use of exponential functions and the techniques for handling these functions are explained in the chapter.

Analysis of experimental data using these mathematical techniques is a powerful tool for chemists, allowing them to deduce important facts about the mechanism of reactions and the activation enthalpies for the steps in these mechanisms.

Knowledge and understanding checklist

You will learn more about some ideas introduced in earlier chapters:

- ☐ Bond enthalpies (**Developing fuels**)
- ☐ Redox reactions and oxidation numbers (**Elements from the sea**)
- ☐ Equilibrium and equilibrium constants (**Elements from the sea**)
- ☐ Measuring rates of reactions and factors affecting reaction rates (**The ozone story**)
- ☐ Calculations involving amount of substance (**Elements of life**)
- ☐ Catalysts (**The ozone story**).

Maths skills checklists

In this chapter, you will need to use the following maths skills. You can find support for these on Kerboodle and through MyMaths.

- ☐ Recognise and make use of appropriate units in calculations.
- ☐ Recognise and use expressions in decimal and ordinary form.
- ☐ Use ratios, fractions, and percentages.
- ☐ Estimate results.
- ☐ Use calculators to find and use power and exponential functions.
- ☐ Use an appropriate number of significant figures.
- ☐ Understand and use the symbols $=$, $<$, $\ll$, $\gg$, $>$, $\propto$, $\sim$, $\rightleftharpoons$.
- ☐ Change the subject of an equation.
- ☐ Substitute numerical values into algebraic equations, using appropriate units for physical quantities.
- ☐ Solve algebraic equations.
- ☐ Translate information between graphical, numerical, and algebraic forms.
- ☐ Plot two variables from experimental or other data.
- ☐ Determine the slope and intercept of a graph.
- ☐ Calculate rate of change from a graph showing a linear relationship.
- ☐ Draw and use the slope of a tangent to a curve as a measure of rate of change.

MyMaths.co.uk
Bringing Maths Alive

CI 1 Nitrogen chemistry

Industry

Industry involves the production of goods or services, from the food you eat to medicine you take.

Nitric acid is one of many different acids that is essential to industry, and therefore to our lifestyles.

Nitric acid

Ammonia, a colourless gas with a strong smell, is an important nitrogen compound and is used to make fertilisers as well as nitric acid. Considering that ammonia is nitrogen-based, and nitrogen gas is abundant in the air (about 78%), you might assume that the manufacture of ammonia is an easy job for chemists. However, nitrogen gas is remarkably unreactive. Thousands of litres of it pass unreacted through our lungs every day and it even dissolves in our blood without reaction. Fritz Haber discovered in the 1890s that ammonia could only be made using high pressures, together with temperatures of around 450 °C and iron catalysts. For this he was awarded the 1918 Nobel Prize. The pressure dials on his prototype apparatus can be seen in Figure 2.

▲ Figure 1 *A fertiliser factory*

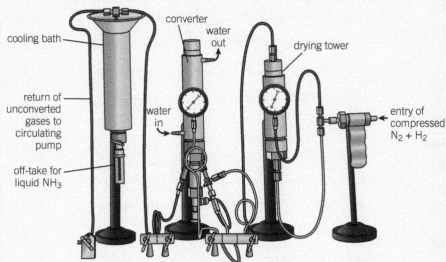

▲ Figure 2 *The Haber process*

▲ Figure 3 *Lightning generates enough energy to convert nitrogen and oxygen in the air to nitrogen oxide*

Nitrogen chemistry (Group 5)

Molecular structures of nitrogen gas, ammonia, and the ammonium ion

The low reactivity of nitrogen molecules arises from the strong triple bond holding the atoms together (Figure 4).

Bond enthalpy of N≡N bond is $+945\,kJ\,mol^{-1}$
(Bond enthalpy of N——N bond is $+158\,kJ\,mol^{-1}$)

▲ Figure 4 *The triple bond in nitrogen needs a lot of energy to break it*

Before nitrogen gas, N_2, can react, the triple bond between the atoms must be broken or partly broken. The **bond enthalpy** of N≡N is very large, at $+945\,kJ\,mol^{-1}$. Most reactions of molecular nitrogen have high activation enthalpies and therefore require high temperatures and catalysts to make them occur. Consider the following reaction:

$$N_2(g) + 3H_2(g) \rightleftharpoons 2NH_3(g) \qquad \Delta H = -92\,kJ\,mol^{-1}$$

As can be seen from the equation, ammonia is nitrogen hydride, NH_3. The bonding of the ammonia molecule is shown in Figure 5. The lone pair of electrons on the nitrogen atom is not involved in the bonding, so it is available to form dative covalent bonds.

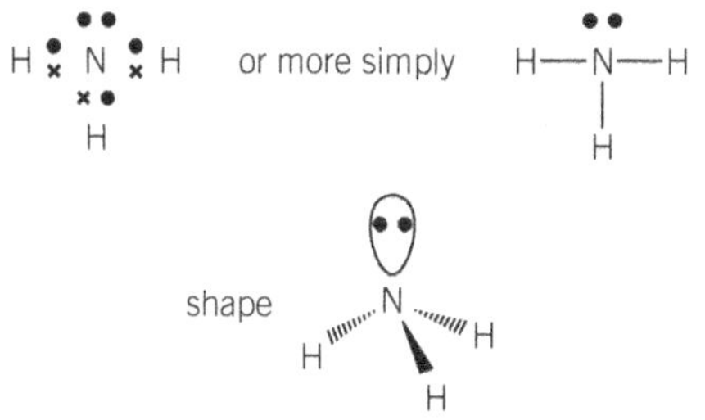

▲ Figure 5 *Bonding in the ammonia molecule*

This lone pair of electrons explains why ammonia acts as a base to form dative covalent bonds to hydrogen ions. The reaction gives the ammonium ion, NH_4^+, as a product:

$$NH_3(g) + H^+(aq) \rightleftharpoons NH_4^+(aq)$$

Figure 6 shows the bonding in the ammonium ion.

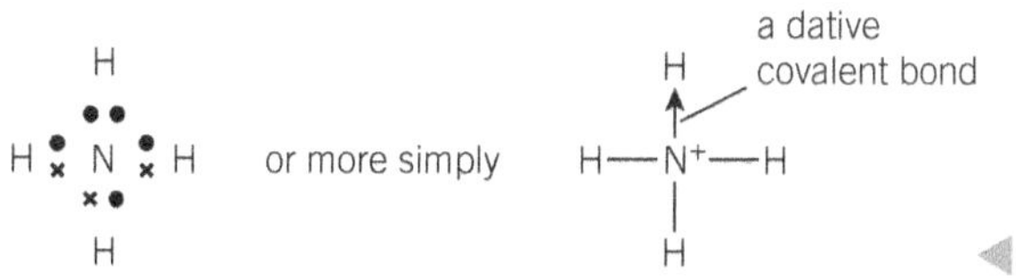

◀ Figure 6 *Bonding in the ammonium ion*

However, once formed, all four bonds are equivalent, so the ammonium ion is usually represented as shown in Figure 7.

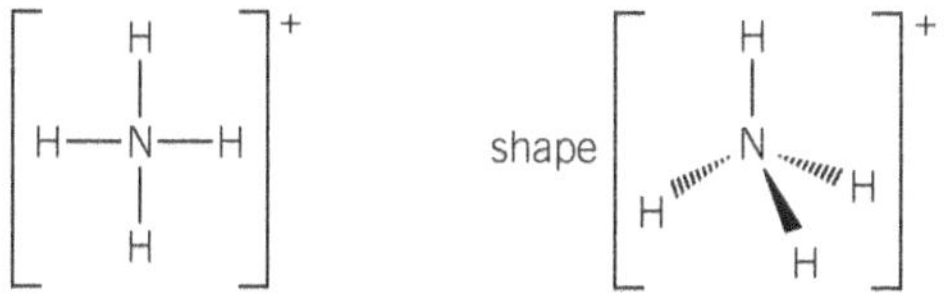

◀ Figure 7 *Representing an ammonium ion, NH_4^+*

Synoptic link

You first met the periodic table in Chapter 1, Elements of life.

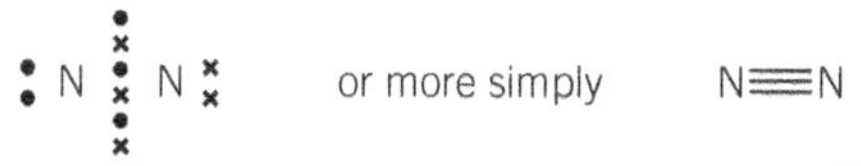

Synoptic link

For more on bond enthalpies, see Topic DF 4, Where does the energy come from?

The nitrogen cycle

Nitrogen is very unreactive which means it cannot be used directly by plants to make proteins. In addition, almost all the nitrogen in the soil is present in complex organic compounds and so is not readily available to plants. As such, various processes are used to convert gaseous nitrogen and organic nitrogen compounds into the soluble ammonium and nitrate(V) ions that plants can use. Figure 8 shows the main processes in the nitrogen cycle.

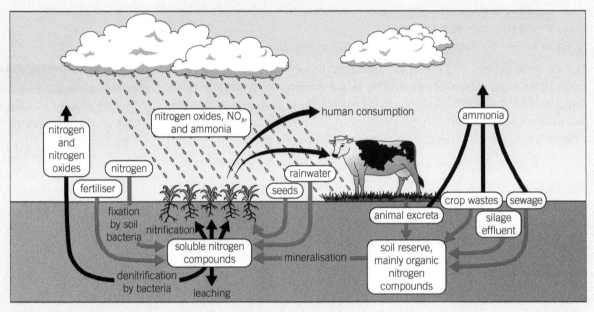

▲ Figure 8 The nitrogen cycle

Nitrogen oxides

Nitrogen forms several oxides, all of them gases. The most important ones are shown in Table 1.

▼ Table 1 Oxides of nitrogen

Name and formula	Appearance	Where it comes from
nitrogen oxide, NO	colourless gas, turns to brown NO_2 in air	combustion processes, especially vehicle engines, thunderstorms, formed in the soil by denitrifying bacteria
nitrogen dioxide, NO_2	brown gas	from oxidation of NO in atmosphere
dinitrogen oxide, N_2O	colourless gas	formed in the soil by denitrifying bacteria

Nitrogen(II) oxide, formed from combustion processes, turns to brown nitrogen(IV) dioxide in air:

$$N_2(g) + O_2(g) \rightarrow 2NO(g)$$

$$2NO(g) + O_2(g) \rightarrow 2NO_2(g)$$

Activity CI 1.1

This activity provides demonstrations to compare the properties of NO, NO_2, and N_2O

When oxygen content is low, anaerobic bacteria reduce nitrate(V) ions in the sequence:

$$NO_3^-(aq) \rightarrow NO(g) \rightarrow N_2O(g) \rightarrow N_2(g)$$

Two kinds of nitrate ion are involved in the nitrogen cycle – nitrate(III), NO_2^-, and nitrate(V), NO_3^-. They are both named nitrates, but they are distinguished from one another by showing the oxidation state of the nitrogen. Figure 9 shows the bonding in nitrate(III) and nitrate(V) ions.

Synoptic link

You met oxidation states and naming inorganic compounds containing elements with varying oxidation states in Topic ES 2, Bromine from sea water.

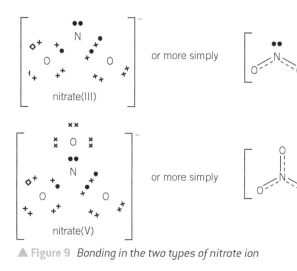

or more simply

The charge is delocalised over the two N—O bonds, which are equivalent.

or more simply

By using its lone pair to form a dative covalent bond to an oxygen atom, the nitrogen increases its oxidation state to +5.

▲ Figure 9 *Bonding in the two types of nitrate ion*

Both of these ions, NO_3^- and NO_2^-, are very soluble in water – they are made through oxidation of ammonium ions by certain aerobic bacteria in the soil. The bacteria carry out the reactions as a means of obtaining respiratory energy. The overall process is called nitrification as the end product is the nitrate(V) ion, formed via the nitrate(III) ion. Nitrification occurs in several stages, but the formation of nitrate(III) can be represented by:

Synoptic link

Variable oxidation states are revisited in Chapter 9, Developing metals.

$$NH_4^+(aq) + 1\frac{1}{2}O_2(g) \rightarrow NO_2^-(aq) + 2H^+(aq) + H_2O(l)$$

The nitrate(III) ion, NO_2^-, produced is rapidly oxidised further:

$$NO_2^-(aq) + \frac{1}{2}O_2(g) \rightarrow NO_3^-(aq)$$

Testing for nitrate(V) ions

To test for nitrate(V) ions, NO_3^-, sodium hydroxide solution, NaOH, and Devarda's alloy, Cu/Al/Zn, are added to a test solution and gently heated. Aluminium acts as the reducing agent. If the test solution has nitrate(V) ions, ammonia gas will be evolved. This can be shown in the following equation:

$$3NO_3^- + 8Al + 5OH^- + 18H_2O \rightarrow 3NH_3 + 8[Al(OH)_4]^-$$

Ammonia has a characteristic sharp, choking smell and makes damp red litmus paper turn blue. It also forms white fumes of ammonium chloride when it comes into contact with hydrogen chloride gas fumes from concentrated hydrochloric acid.

Activity CI 1.2

In this activity you will use tests for ammonium and nitrate(V) ions to identify unknown compounds.

Testing for ammonium ions

To test for ammonium ions, NH_4^+, sodium hydroxide solution is added to the test solution and gently heated. If ammonium ions are present, ammonia gas will be given off. This can be shown in the following equation:

$$NH_4^+(aq) + OH^-(aq) \rightarrow NH_3(g) + H_2O(l)$$

 Worked example: Balancing equations using oxidation states

What is the balanced equation for the following reaction?

$$NO_3^- + Al + OH^- + H_2O \rightarrow NH_3 + [Al(OH)_4]^-$$

Step 1: Identify how the oxidation state for each element changes in the reaction.

$$NO_3^- + Al + OH^- + H_2O \rightarrow NH_3 + [Al(OH)_4]^-$$

N	+5			−3		gains eight electrons so reduced
O	−2	−2	−2		−2	does not change
Al		0			+3	loses three electrons so oxidised
H			+1 +1	+1	+1	does not change

Step 2: Balance the equation so the number of electrons lost equals the electrons gained.

$$3NO_3^- + 8Al + OH^- + H_2O \rightarrow 3NH_3 + 8[Al(OH)_4]^-$$

Step 3: Balance the equation so the total charges of the compounds on two sides of the reaction are equal to zero.

$$3NO_3^- \quad nOH^- \quad 8[Al(OH)_4]^-$$

$$3 \times (-1) + n \times (-1) = 8 \times (-1)$$

$$n = 5$$

$$3NO_3^- + 8Al + 5OH^- + H_2O \rightarrow 3NH_3 + 8[Al(OH)_4]^-$$

Step 4: Balance the number of hydrogens and oxygens by adjusting the number of H^+ ions and H_2O molecules.

There are 32 oxygens in the products and 14 in the reactants, excluding water, so in order to balance the overall equation there must be 18 water molecules:

$$3NO_3^- + 8Al + 5OH^- + 18H_2O \rightarrow 3NH_3 + 8[Al(OH)_4]^-$$

Summary questions

1 Nitrogen, N_2, at the top of Group 15, is very unreactive. The next member of Group 15, phosphorus, is highly reactive – white phosphorus, P_4, catches fire spontaneously in air. The shape of the P_4 molecule is shown in Figure 10.

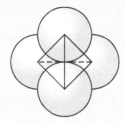

▲ Figure 10 *White phosphorus, P_4, is tetrahedral*

Suggest an explanation for the difference in reactivity between these two elements of Group 15. *(2 marks)*

2 One of the stages in the denitrification process brought about by bacteria in the soil involves the conversion of NO into N_2O.
   a What is the change in the oxidation state of nitrogen in this conversion? *(1 mark)*
   b Write a balanced half-equation for the conversion:

   $NO \rightarrow N_2O$

   You will need to add electrons, H_2O molecules, and H^+ ions to balance the equation. *(2 marks)*

3 Look at the following pairs of nitrogen species – the conversions from the first species to the second are part of the nitrogen cycle:

   A $NO_2^- \rightarrow NO_3^-$ nitrification
   B $NO_3^- \rightarrow N_2$ denitrification
   C $NH_4^+ \rightarrow NH_3$ putrefaction
   D $N_2 \rightarrow NH_3$ ammonia synthesis
   E $NO \rightarrow NO_2$ in the atmosphere

   For each conversion:
   a Give the oxidation state of nitrogen in each of the two species. *(5 marks)*
   b Say whether the conversion of the first compound to the second involves oxidation, reduction, or neither oxidation or reduction. *(5 marks)*

4 Write a balanced half-equation for each of the conversions A–E in Question 3. *(5 marks)*

5 A common nitrate(V) test, known as the brown ring test, can be performed by adding iron(II) sulfate to a solution of a nitrate then slowly adding concentrated sulfuric acid such that the acid forms a layer below the aqueous solution. Reactions are given below:

   $$NO_3^- + Fe^{2+} + H^+ \rightarrow Fe^{3+} + NO + H_2O$$
   $$[Fe(H_2O)_6]^{2+} + NO \rightarrow [Fe(H_2O)_5(NO)]^{2+} + H_2O$$

   a Balance the equations. *(2 marks)*
   b This test is not used if nitrate(III) ions may be present. Explain why. *(3 marks)*

CI 2 Manufacturing nitric acid

Specification references: CI(f), CI(h)

Learning outcomes

Demonstrate and apply knowledge and understanding of:

→ the effect of changes of pressure (if any) and temperature on the magnitude of the equilibrium constant; the fact that addition of catalysts has no effect on the position of equilibrium or the magnitude of the equilibrium constant

→ calculations, including units, involving K_c and initial and equilibrium concentrations for homogeneous equilibria; techniques and procedures for experiments to determine equilibrium constants.

The world population is growing. A rapid increase in human numbers began in about 1850 and, whilst the rate of increase has slowed down, numbers are set to continue to increase throughout the 21st century. The world population doubled from 3 billion in 1961 to 6 billion in 1999, and the United States Census Bureau forecasts that the world population will reach 9 billion by 2042 – an increase of 50% in 43 years.

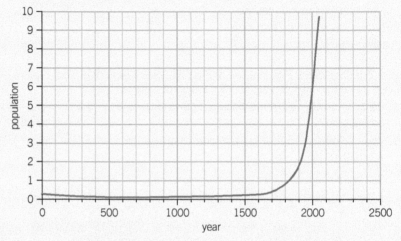

▲ Figure 1 *The global population and predicted population growth*

This rapid increase in human population poses a huge challenge to agriculture. How can food production be increased without encroaching on the world's remaining forests and wildernesses? It might be done by making the most efficient use of existing agricultural land – particularly by improving crop varieties and planting techniques, and making sensible use of added plant nutrients (in manure and fertilisers) and pesticides.

Plant nutrients

When crops are harvested, the natural nutrient cycles are disturbed by removing large quantities of plants before the natural processes of decay take place – these decay processes would return nutrients to the soil, as seen in the nitrogen cycle. The nutrients removed when plants are harvested need to be replenished before a crop with similar nutrient needs can be grown in the same soil. The organic store is replenished by organic manure (animal excretions) and by the death and decay of living organisms. However, in many cases the speed of harvest is too fast to be replenished by only organic manure. As such, fertilisers are used to replenish the soil.

For healthy growth, plants require many different nutrients, for example, nitrogen is needed for leaf growth and protein production. It is taken up by most plants as nitrate, NO_3^-, and ammonium, NH_4^+, ions from the soil solution. Deficiency of these ions in a particular soil can be treated by the addition of ammonium nitrate, NH_4NO_3, in the form of a nitrogen a fertiliser. The ammonium nitrate is produced from

▲ Figure 2 *Fertilisers are added to agricultural fields to help crops grow healthily*

nitric acid. Since there is huge demand on fertilisers to catch up with the population growth and food production, there is huge demand on production of nitric acid – a colourless, fuming, and highly corrosive liquid. Every year approximately 60 million tonnes of industrial nitric acid, HNO_3, are produced in the world.

Nitric acid production involves two stages:

1   oxidation of ammonia
2   absorption of the resulting nitrogen oxides.

Oxidation of ammonia

This part of the process involves the oxidation of ammonia to nitrogen monoxide (nitric oxide):

$$4NH_3(g) + 5O_2(g) \rightleftharpoons 4NO(g) + 6H_2O(g) \qquad \Delta H = -900 \, kJ \, mol^{-1}$$

Absorption of the nitrogen oxides

Air is added and the gases compressed again:

$$2NO(g) + O_2(g) \rightleftharpoons 2NO_2(g) \qquad \Delta H = -115 \, kJ \, mol^{-1}$$

$$2NO_2(g) \rightleftharpoons N_2O_4(g) \qquad \Delta H = -58 \, kJ \, mol^{-1}$$

The gases are then passed through one or more towers to meet a stream of water:

$$3N_2O_4(g) + 2H_2O(l) \rightleftharpoons 4HNO_3(aq) + 2NO(g) \qquad \Delta H = -103 \, kJ \, mol^{-1}$$

All the reactions above are equilibrium reactions.

Study tip

Remember, the symbol $\rightleftharpoons$ shows that the reaction is in equilibrium.

Chemical Ideas: Equilibrium in chemistry 7.5

Equilibrium constant K_c, temperature, and pressure

You can write equilibrium constants for all equilibrium processes. Table 1 shows data obtained for the hydrolysis of an ester, ethyl ethanoate:

$$CH_3COOC_2H_5(l) + H_2O(l) \rightleftharpoons CH_3COOH(l) + C_2H_5OH(l)$$

▼ Table 1 *Equilibrium concentrations for the hydrolysis of ethyl ethanoate at 293 K*

Experiment	Equilibrium concentrations / mol dm⁻³			
	$[CH_3COOC_2H_5(l)]_{eq}$	$[H_2O(l)]_{eq}$	$[CH_3COOH(l)]_{eq}$	$[C_2H_5OH(l)]_{eq}$
Experiment 1	0.090	0.531	0.114	0.114
Experiment 2	0.204	0.118	0.082	0.082
Experiment 3	0.151	0.261	0.105	0.105

For a reaction that occurs in solution, the quantities that matter are concentrations. Table 1 shows the equilibrium concentrations of different reaction mixtures, all at 293 K. Equilibrium concentration means the concentration when the reaction has reached equilibrium.

Synoptic link

For the equilibrium reactions, equilibrium constants, and shifting the position of equilibrium, look back to Topic ES 4, From extracting bromine to making bleach, and Topic ES 7, The Deacon process solves the problem again.

Synoptic link

You were first introduced to esters in Chapter 5, What's in a medicine?

You will study hydrolysis of esters in Topic PL 1, The polyester story.

Study tip

Equilibrium concentration can be indicated by the subscript eq, although this is often omitted.

Synoptic link

Equilibrium constants were covered in Topic ES 4, From extracting bromine to making bleach.

Remember that the expression for the equilibrium constant can be identified from the reaction equation.

$$aA + bB \rightleftharpoons cC + dD$$

$$K_c = \frac{[C]^c \, [D]^d}{[A]^a \, [B]^b}$$

Activity CI 2.1

In this activity values for K_c are calculated using your own experimental data.

Activity CI 2.2

This activity uses a microscale technique to calculate the value of K_c for a reaction between silver(I) and iron(II) ions.

If you look at the data in Table 1, you will find that the expression is constant for the three experiments carried out at the same temperature.

$$K_c = \frac{[CH_3COOH(l)][C_2H_5OH(l)]}{[CH_3COOC_2H_5(l)][H_2O(l)]}$$

The constant K_c is the **equilibrium constant** for this reaction.

The concentrations in the expression for K_c must be those at equilibrium. From the data given in Table 1, you can work out the value of K_c is constant at about 0.28 for the hydrolysis of ethyl ethanoate at 293 K. This value, which is less than 1, tells you that, at equilibrium, a substantial proportion of the reactants is left unreacted – the reaction is incomplete.

Units of K_c

The units of K_c vary – it depends on the expression for K_c for the particular reaction you are studying and needs to be calculated for each reaction.

Example 1

$$H_2(g) + Br_2(g) \rightleftharpoons 2HBr(g)$$

$$K_c = \frac{[HBr(g)]^2}{[H_2(g)][Br_2(g)]}$$

The unit of K_c are given by $\dfrac{(mol\,dm^{-3})^2}{(mol\,dm^{-3})\,(mol\,dm^{-3})}$

$$= \frac{\cancel{(mol\,dm^{-3})^2}}{\cancel{(mol\,dm^{-3})} \times \cancel{(mol\,dm^{-3})}}$$

So K_c for this reaction has no units, since they cancel out on the top and bottom of the expression.

For any value of K_c you must quote the units and the temperature for which it applies.

Example 2

Insulin can exist as dimers, In_2, or monomers, In, and are linked by the following equation:

$$In_2(aq) \rightleftharpoons 2In(aq)$$

$$K_c = \frac{[In(aq)]^2}{[In_2(aq)]}$$

The unit of K_c are given by $\dfrac{(mol\,dm^{-3})^2}{(mol\,dm^{-3})} = \dfrac{(mol\,dm^{-3})^{\cancel{2}}}{\cancel{(mol\,dm^{-3})}} = mol\,dm^{-3}$

You can see that the units of K_c vary from reaction to reaction, and need to be worked out from the equilibrium expression. When you quote a value for K_c you must always show a balanced equation for the reaction.

Worked example: Calculating K_c and its units

Calculate the value for K_c, and its unit, for the oxidation of ammonia at 890 K, given the following equilibrium concentrations:

$[NH_3] = 1.230\,mol\,dm^{-3}$ $\qquad$ $[O_2] = 0.432\,mol\,dm^{-3}$

$[NO] = 0.752\,mol\,dm^{-3}$ $\qquad$ $[H_2O] = 1.010\,mol\,dm^{-3}$

Step 1: Write the stoichiometric equation for the reaction.

$$4NH_3(g) + 5O_2(g) \rightleftharpoons 4NO(g) + 6H_2O(g)$$

Step 2: Use the stoichiometric equation to write an expression for K_c.

$$K_c = \frac{[NO]^4\,[H_2O]^6}{[NH_3]^4\,[O_2]^5}$$

Step 3: Use the concentration values to calculate K_c.

$$K_c = \frac{(0.752)^4 \times (1.010)^6}{(1.230)^4 \times (0.432)^5} = 9.857$$

Step 4: Use the units and the expression for K_c to calculate the units.

The units of K_c are given by $\dfrac{(mol\,dm^{-3})^4 \times (mol\,dm^{-3})^6}{(mol\,dm^{-3})^4 \times (mol\,dm^{-3})^5}$

$$= \frac{(\cancel{mol\,dm^{-3})^4} \times (mol\,dm^{-3})^{\cancel{6}}}{(\cancel{mol\,dm^{-3})^4} \times \cancel{(mol\,dm^{-3})^5}}$$

$$= 9.857\ mol\,dm^{-3}\ at\ 890\,K$$

Table 2 shows the results obtained for another reaction, the equilibrium mixture of nitrogen dioxide and dinitrogen tetroxide:

$$2NO_2(g) \rightleftharpoons N_2O_4(g)$$

▼ Table 2 *The equilibrium mixture of nitrogen dioxide and dinitrogen tetroxide at 293 K*

Experiment	Initial concentrations / mol dm^{-3}		Equilibrium concentrations / mol dm^{-3}		K_c
	$[NO_2]$	$[N_2O_4]$	$[NO_2]$	$[N_2O_4]$	$[N_2O_4]/[NO_2]^2$
1	0.0000	0.0400	0.0125	0.0337	215.68
2	0.0800	0.0000	0.0125	0.0337	215.68
3	0.0000	0.0600	0.0156	0.0524	215.32
4	0.0600	0.0000	0.0107	0.0247	215.74
5	0.0600	0.0200	0.0141	0.0429	215.78

In Experiment 1 and Experiment 2, the equilibrium mixtures have identical compositions, because the initial concentration of N_2O_4 in Experiment 1 is half the initial concentration of NO_2 in Experiment 2 – that is, the total number of nitrogen and oxygen atoms is the same in both experiments. In Experiments 3 to 5, different initial concentrations of N_2O_4 and NO_2 give different equilibrium concentrations. In all the experiments, however, the equilibrium concentrations are related. The last column of Table 2 shows that, at equilibrium, the expression $[N_2O_4]/[NO_2]^2$ has a value of around $215.5\,mol^{-1}\,dm^3$ at 293 K.

Pressure, temperature, and the magnitude of the equilibrium constant

Changing the pressure

Many important industrial processes involve reversible reactions that take place in the gas phase. For these processes, it is essential that conditions are identified that ensure that the equilibrium position is shifted as far to the right as possible in order to maximise the yield of the desired product.

From the study of equilibria in gas-phase reactions, a simple rule has emerged.

● Increasing the pressure moves the equilibrium to the side of the equation with fewer gas molecules, so to reducing the pressure.

The position of equilibrium shifts so as to counteract the change in pressure as much as possible, so the magnitude of K_c remains constant. This rule is part of **Le Chatelier's principle** and is a useful tool for predicting the effect that changing conditions will have on the equilibrium position.

An example is the steam reforming of methane. This process is used to make methanol, an important industrial chemical. In the first stage, methane reacts with steam to form carbon monoxide and hydrogen:

$$CH_4(g) + H_2O(g) \rightleftharpoons CO(g) + 3H_2(g)$$
$$\text{2 molecules} \qquad \text{4 molecules}$$

There are more gas molecules on the right hand side of the equation, so a *reduction* in the pressure results in the position of equilibrium shifting to the right. The lower the pressure, the more product will be obtained.

The second stage of methanol manufacture involves the reaction:

$$CO(g) + 2H_2(g) \rightleftharpoons CH_3OH(g)$$
$$\text{3 molecules} \qquad \text{1 molecule}$$

For this reaction, *increasing* the pressure will move the equilibrium to the right hand side – the *higher* the pressure, the more methanol will be obtained.

Le Chatelier's principle can be used to predict the *qualitative* effect of changing the total pressure – the position of equilibrium shifts so as to counteract the change in pressure. However, as with changes of concentration, the equilibrium will move in such a way that K_c remains constant.

You can demonstrate this by looking at the result of compressing the following reaction at equilibrium:

$$N_2(g) + 3H_2(g) \rightleftharpoons 2NH_3(g)$$

Start with a system at equilibrium that initially contains 2.5 mol dm^{-3} of nitrogen, 7.5 mol dm^{-3} of hydrogen, and 0.12 mol dm^{-3} of ammonia at 773 °C. When the system is compressed, and equilibrium re-established, you get the following results:

Before compression	After compression
$[N_2]$ = 2.5 mol dm^{-3}	$[N_2]$ = 21 mol dm^{-3}
$[H_2]$ = 7.5 mol dm^{-3}	$[H_2]$ = 62 mol dm^{-3}
$[NH_3]$ = 0.12 mol dm^{-3}	$[NH_3]$ = 8.4 mol dm^{-3}

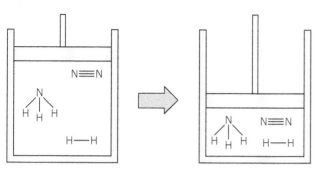

$$K_c \text{ before} = \frac{[NH_3]^2}{[N_2] \times [H_2]^3} = \frac{(0.12\,\text{mol dm}^{-3})^2}{(2.5\,\text{mol dm}^{-3}) \times (7.5\,\text{mol dm}^{-3})^3} = 1.4 \times 10^{-5}\,\text{mol}^{-2}\text{dm}^6$$

$$K_c \text{ after} = \frac{[NH_3]^2}{[N_2] \times [H_2]^3} = \frac{(8.4\,\text{mol dm}^{-3})^2}{(21\,\text{mol dm}^{-3}) \times (62\,\text{mol dm}^{-3})^3} = 1.4 \times 10^{-5}\,\text{mol}^{-2}\text{dm}^6$$

The results show that K_c has remained constant before and after the compression. However, the ratio of reactants to products has changed significantly. The compression subjected the equilibrium reaction to an increase in the total pressure on the system. In order to reestablish K_c the position of the equilibrium shifted in the direction that minimised the effect of this pressure. The reaction shifted towards the right hand side of the equation, $2NH_3(g)$, as this reduces the number of particles, thereby reducing the total pressure on the system.

Changing the temperature

Changes in concentration and total pressure may alter the composition of equilibrium mixtures, but they do not alter the value of the equilibrium constant itself, provided the temperature does not change. The proportions of reactants and products alter in such a way as to keep the ratio in the K_c expression unchanged. However, changes in temperature alter the value of the equilibrium constant itself.

Table 3 shows some experimental measurements of how temperature changes affect K_c for two reactions.

▼ Table 3 *The effect of temperature on the equilibrium constants of two reacting systems – the value of ΔH is for the forward reaction in each case*

$N_2(g) + 3H_2(g) \rightleftharpoons 2NH_3(g)$ $\Delta H = -92\,\text{kJ mol}^{-1}$		$N_2O_4 \rightleftharpoons 2NO_2(g)$ $\Delta H = +57\,\text{kJ mol}^{-1}$	
T/K	$K_c/\text{mol}^{-2}\text{dm}^6$	T/K	$K_c/\text{mol dm}^{-3}$
400	4.39×10^4	200	5.51×10^{-8}
600	4.03	400	1.46
800	3.00×10^{-2}	600	3.62×10^2
$K_c = \dfrac{[NH_3(g)]^2}{[N_2(g)]\,[H_2(g)]^3}$		$K_c = \dfrac{[NO_2(g)]^2}{[N_2O_4(g)]}$	

Table 3 shows that for the endothermic reaction, a rise in temperature favours the products and increases K_c. For the exothermic reaction, a rise in temperature favours the reactants and decreases K_c.

Using a catalyst

Catalysts do not affect the position of equilibrium or the equilibrium constant. They alter the rate at which equilibrium is attained, but not the composition of the equilibrium mixture.

Synoptic link 🧪

Techniques for determining equilibrium constants are covered in Techniques and procedures.

Techniques for experiments to determine equilibrium constants

There are various experimental methods to determine equilibrium constants. Some examples are pH measurements, colorimetric methods, and titration.

Summary

The effect of changing conditions on equilibrium mixtures is summarised in Table 4. The important thing to remember is that K_c is constant unless the temperature is changed.

Synoptic link

Acid–base equilibriums and solubility products K_{sp} are studied in Chapter 8, Oceans.

▼ Table 4 *The effect of changing conditions on equilibrium mixtures*

Reaction condition changed	Composition	K_c
concentration	may change	unchanged
total pressure	may change	unchanged
temperature	changed	changed
catalyst	unchanged	unchanged

Summary questions

1. Write expressions for K_c for the following reactions. In each case, give the units of K_c.
   a. $2NO(g) + O_2(g) \rightleftharpoons 2NO_2(g)$ (2 marks)
   b. $C_2H_6(g) \rightleftharpoons C_2H_4(g) + H_2(g)$ (2 marks)
   c. $2HI(g) \rightleftharpoons H_2(g) + I_2(g)$ (2 marks)
   d. $CO_2(aq) + H_2O(l) \rightleftharpoons HCO_3^-(aq) + H^+(aq)$ (2 marks)
   e. $In_6(aq) \rightleftharpoons 3In_2(aq)$ (2 marks)
   f. $CH_3COOH(l) + C_3H_7OH(l) \rightleftharpoons CH_3COOC_3H_7(l) + H_2O(l)$ (2 marks)

2. The equilibrium constant K_c for a reaction is given by the expression:
$$K_c = \frac{[SO_3(g)]^2}{[SO_2(g)]^2[O_2(g)]}$$
   Write the balanced chemical equation for the reaction. (2 marks)

▼ Table 5 *Results for the reaction between nitrogen and hydrogen*

Substance	Equilibrium concentration / $mol\,dm^{-3}$
$N_2(g)$	0.142
$H_2(g)$	1.84
$NH_3(g)$	1.36

3. A mixture of nitrogen and hydrogen was sealed in a steel vessel and held at 1000 K until equilibrium was reached. The contents were then analysed. The results are given in Table 5.
   a. Write an expression for K_c for the reaction:
   $$N_2(g) + 3H_2(g) \rightleftharpoons 2NH_3(g)$$ (3 marks)
   b. Calculate a value for K_c, and remember to give the units. (2 marks)

4. Consider the reaction between hydrogen and oxygen to produce steam.
   a. Write an equation for the reaction with state symbols. (2 marks)
   b. Write an expression for K_c. (1 mark)
   c. State how the equilibrium position is affected by:
      i. an increase in temperature (1 mark)
      ii. an increase in the total pressure. (1 mark)

d State how the value of K_c is affected by:
  i an increase in temperature (*1 mark*)
  ii an increase in the total pressure. (*1 mark*)

5 When PCl_5 is heated in a sealed container and maintained at a constant temperature, an equilibrium is established. At 523 K, the concentrations given in Table 6 were determined.
  a Write an expression for K_c for the reaction $PCl_5(g) \rightleftharpoons PCl_3(g) + Cl_2(g)$ (*1 mark*)
  b Calculate a value for K_c with units. (*3 marks*)

▼ Table 6 *Concentrations for PCl_5 at 523 K*

Substance	Initial concentration / mol dm^{-3}	Equilibrium concentration / mol dm^{-3}
PCl_5	0.2	
PCl_3		0.123
Cl_2		0.123

6 For the reaction $2H_2(g) + S_2(g) \rightleftharpoons 2H_2S(g)$, K_c was found to be $9.4 \times 10^5\,mol^{-1}\,dm^3$ at 1020 K.
Equilibrium concentrations were measured as:

$[H_2(g)] = 0.234\,mol\,dm^{-3}$ $[H_2S(g)] = 0.442\,mol\,dm^{-3}$
  a Write an expression for K_c for the reaction. (*1 mark*)
  b What is the equilibrium concentration of $S_2(g)$? (*2 marks*)

7 The equilibrium constant K_c for the reaction $2NO_2(g) \rightleftharpoons 2NO(g) + O_2(g)$ is $9.0\,mol\,dm^{-3}$ at 683 K:
  a Write an expression for K_c for the reaction. (*1 mark*)
  b Equilibrium concentrations were measured as:

$[NO(g)] = 0.50\,mol\,dm^{-3}$ $[O_2(g)] = 0.25\,mol\,dm^{-3}$

What is the equilibrium concentration of $NO_2(g)$ in the above mixture? (*2 marks*)

8 Ethanol and ethanal react together according to the equation:

$2C_2H_5OH(l) + CH_3CHO(l) \rightleftharpoons CH_3CH(OC_2H_5)_2(l) + H_2O(l)$
ethanol   ethanal

An excess of ethanol was mixed with ethanal, and the system was allowed to reach equilibrium at 298 K. The equilibrium concentrations were measured and are shown in Table 7.
  a Write an expression for K_c for this equilibrium. (*1 mark*)
  b Examine the equilibrium concentrations. What do you think predominates at equilibrium – reactants or products? (*2 marks*)
  c What does this lead you to expect for the magnitude of K_c? (*1 mark*)
  d Calculate a value for K_c. (*2 marks*)

▼ Table 7 *Equilibrium concentrations for the reaction between ethanol and ethanal*

Substance	Equilibrium concentration / mol dm^{-3}
$C_2H_5OH(l)$	13.24
$CH_3CHO(l)$	0.133
$CH_3CH(OC_2H_5)_2(l)$	1.311
$H_2O(l)$	1.311

9 Consider the reaction of aqueous chloromethane with alkali:

$OH^-(aq) + CH_3Cl(aq) \rightleftharpoons CH_3OH(aq) + Cl^-(aq)$

The equilibrium constant has a value of 1×10^{16} at room temperature. What does this tell you about the concentration of chloromethane at equilibrium? (*2 marks*)

10 A sample of NO_2 is allowed to reach equilibrium in a flask of volume $2.0\,dm^3$ at 293 K.

$2NO_2 \rightleftharpoons N_2O_4$ $K_c = 215.5\,dm^3\,mol^{-3}$ at 293 K

The equilibrium amount of N_2O_4 is found to be 0.1048 mol.
  a If the initial amount of NO_2 is x mol, write the amount of NO_2 at equilibrium in terms of x.
  b Write the expression for K_c and use this to calculate the value of x.

CI 3 Manufacturing sulfuric acid

Specification references: CI(a), CI(c)

▲ Figure 1 *Leaves from barley showing a phosphorus deficiency*

Diseases in plants can have a negative economic impact on individual farmers and on the entire agricultural industry. One reason for plant diseases is nutrient deficiency. Crop yields are reduced if there is a shortage of even one nutrient.

Deficiency of phosphorus leads to serious diseases in plants (Figure 1). Today, phosphorus deficiency is commonly treated with phosphate fertilisers such as ammonium phosphate. The chemical industry manufactures phosphate fertilisers from phosphoric acid, which is produced from sulfuric acid.

Sulfuric acid is one of the most important compounds made by the chemical industry. It is used to make hundreds of compounds needed by almost every industry (Figure 2). Every year, approximately 200 million tonnes of sulfuric acid is manufactured globally.

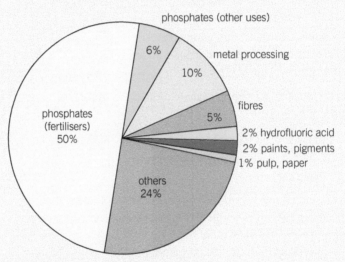

▲ Figure 2 *Pie chart of the uses of sulfuric acid*

The process for producing sulfuric acid has four stages:

1 extraction of sulfur

2 conversion of sulfur to sulfur dioxide

3 conversion of sulfur dioxide to sulfur trioxide

4 conversion of sulfur trioxide to sulfuric acid.

Making any of these stages faster has massive importance for the chemical industry throughout the world. Chemists investigate the speed of reactions – the rate of reaction.

Rates of reactions

Rate is how quickly a quantity of something changes, for example, speed is rate of change of distance. Whenever rate is measured, you need to be clear about the units used. Speed is often measured in metres per second ($m\,s^{-1}$).

The rate of a reaction is the rate at which reactants are converted into products. Scientists often use the concentration of reactants in order to calculate the rate of a reaction. Often this means the concentration of a solution, but you can also apply the idea to gases.

It is useful to know rates of reactions for all sorts of reasons. You might want to use changes of concentration to alter the rate of a reaction in an industrial process. Knowing how rate of reaction changes can also tell a great deal about the way that reactions occur – their **reaction mechanisms**.

Consider the decomposition of hydrogen peroxide, H_2O_2, to form water, H_2O, and oxygen, O_2:

$$2H_2O_2(aq) \rightarrow 2H_2O(l) + O_2(g)$$

The rate of the reaction is the rate at which $H_2O(l)$ and $O_2(g)$ are formed, which is the same as the rate at which $H_2O_2(aq)$ is used up. The rate of this reaction can be measured in moles of product (water or oxygen) formed per second, or moles of hydrogen peroxide used up per second.

Suppose $0.0001\,mol$ of oxygen are formed per second. The rate of the reaction is:

$0.0001\,mol(O_2)\,s^{-1}$ or
$0.0002\,mol(H_2O)\,s^{-1}$ or
$-0.0002\,mol(H_2O_2)\,s^{-1}$

The rate in terms of moles of water is twice the rate in terms of moles of oxygen. This is because two moles of water are formed for every one mole of oxygen.

The rate in terms of hydrogen peroxide has a minus sign. This is because hydrogen peroxide is getting used up instead of being produced.

In this case, the units for the rate of reaction are $mol\,s^{-1}$ (moles per second).

Synoptic link

You first came across factors affecting rates of reaction in Chapter 2, Developing fuels, and Chapter 4, The ozone story.

▲ Figure 3 *A lack of oxygen (caused by overwatering) can cause root rot in plants. Using a small concentration of hydrogen peroxide with water helps prevent root rot as the hydrogen peroxide decomposes to produce oxygen*

Study tip

The rate of a reaction can also be measured in $mol\,min^{-1}$ (moles per minute) or even $mol\,h^{-1}$ (moles per hour).

Measuring rate of a reaction

To determine the rate of a reaction, you need to be able to measure the rate at which one of the reactants is used up or the rate at which one of the products is formed. You can then find the rate of the reaction in terms of $\dfrac{\text{change in property}}{\text{time taken}}$.

Practically, a property is measured that changes during the reaction that is proportional to the concentration of a particular reactant or product. There are many experimental methods that can be used to follow the rate of a reaction. For example:

- measuring volumes of gases evolved
- measuring mass changes
- pH measurement
- colorimetry
- chemical analysis including titration.

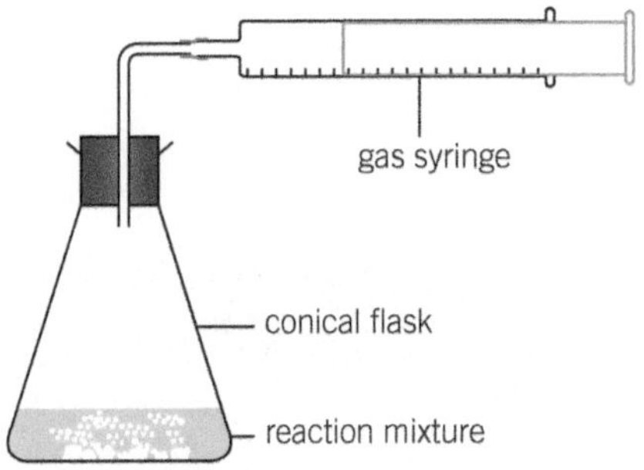

Measuring volumes of gases evolved

The reaction between calcium carbonate and hydrochloric acid produces carbon dioxide as one of the products.

$$CaCO_3(s) + 2HCl(aq) \rightarrow CaCl_2(aq) + H_2O(l) + CO_2(g)$$

The carbon dioxide can be collected in a gas syringe. The volume produced can be used to follow the reaction rate. The apparatus that could be used for this is shown in Figure 4. The unit for the rate of reaction will be $cm^3\,s^{-1}$.

Measuring mass changes

Figure 5 shows a different way of monitoring the same reaction. Instead of measuring the volume of gas produced, the rate is being monitored by recording the mass lost (in the form of carbon dioxide gas) from the reaction. The units of the rate of reaction will be $g\,s^{-1}$.

pH measurement

The reaction can also be measured using pH. In the reaction between hydrochloric acid and calcium carbonate, as the reaction proceeds, the hydrochloric acid concentration will fall and this can be measured by monitoring the pH of the reaction mixture.

Colorimetry

A colorimeter can be used to measure the change in colour of a reaction, such as in the reaction between zinc and aqueous copper(II) sulfate:

$$Zn(s) + CuSO_4(aq) \rightarrow ZnSO_4(aq) + Cu(s)$$

The blue colouration of the copper sulfate solution decreases as the reaction proceeds (Figure 6), and the reaction can be followed using a colorimeter.

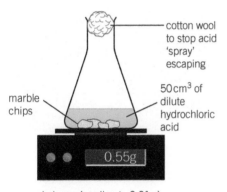

▲ Figure 4 *Apparatus for measuring the rate of a gas produced using a gas syringe*

gas syringe

conical flask

reaction mixture

cotton wool to stop acid 'spray' escaping

$50\,cm^3$ of dilute hydrochloric acid

marble chips

0.55g

balance (reading to 0.01 g)

▲ Figure 5 *Apparatus for measuring the rate of reaction by recording the loss in mass of a gaseous product*

▲ Figure 6 *The oxidation of zinc by copper(II) sulfate*

Chemical analysis and titration

All the techniques described so far have not interfered with the progress of the reaction. However, chemical analysis involves taking samples of the reaction mixture at regular intervals, and stopping the reaction in the sample before analysis, by a process known as quenching.

Example 1

In the acid-catalysed reaction between iodine and propanone, a sample can be extracted from the reaction mixture and quenched by the addition of sodium hydrogencarbonate. This neutralises the acid catalyst, effectively stopping the reaction. Any unreacted iodine can then be analysed by titration with a thiosulfate solution.

Example 2

The hydrolysis of methyl methanoate using sodium hydroxide is:

$$HCOOCH_3 + NaOH \rightarrow HCOO^-Na^+ + CH_3OH$$

The rate of the reaction can be followed by monitoring the concentration of sodium hydroxide as it is used up during the course of the reaction. Samples taken from the reaction mixture as it proceeds are quenched by dilution with a known volume of ice-cold water. The concentration of the sodium hydroxide remaining in each reaction sample is determined by titrating it with an acid such as dilute hydrochloric acid. The unit of the rate of reaction will be $mol\,dm^{-3}\,s^{-1}$.

Synoptic link

You covered iodine–thiosulfate titrations in Topic ES 5, The risks and benefits of using chlorine. Iodine–thiosulfate titrations are also covered in Techniques and procedures.

Summary questions

1   The reaction between calcium carbonate and dilute hydrochloric acid is given below.

$$2HCl(aq) + CaCO_3(s) \rightarrow CaCl_2(aq) + CO_2(g) + H_2O(l)$$

Describe how you would follow the rate of this reaction.   (1 mark)

2   The equation for the oxidation of bromide ions by bromate in an acidic solution is:

$$6H^+ + BrO_3^- + 5Br^- \rightarrow 3Br_2 + 3H_2O$$

Suggest an experimental way of following the rate of the reaction and explain your answer.   (2 marks)

3   The equation for the reaction for the iodination of propanone in acidic solution is:

$$I_2(aq) + CH_3COCH_3(aq) \rightarrow CH_2ICOCH_3(aq) + H^+(aq) + I^-(aq)$$

If you wished to remove a sample during the reaction for analysis, how could you quench the reaction?   (2 marks)

CI 4 The importance of rate equations

Learning outcomes

Demonstrate and apply knowledge and understanding of:

→ the terms:

- order of reaction (both overall and with respect to a given reagent); use of $\propto$

- rate equations of the form: rate = $k[\mathbf{A}]^m[\mathbf{B}]^n$, where m and n are integers;

calculations based on the rate equation; the rate constant k increasing with increasing temperature

→ the Arrhenius equation and the determination of E_a and A for a reaction, given data on the rate constants at different temperatures.

As you have previously seen, the process for the manufacture of sulfuric acid has four stages and making any of these stages faster has great importance for the chemical industry throughout the world. Leaving reactions to take place under standard conditions is often not very useful for chemists. Some chemical reactions do not take place under standard conditions at all. Some chemical reactions are too slow or too fast. For example, consider the oxidation, or rusting, of an iron gate. Rusting is a very slow reaction and it might take years for it to be completed under standard conditions. In contrast, explosions are often extremely fast reactions.

One of the jobs of chemists is to control chemical reactions and the rates at which they occur. This allows the yield to be increased and provides the large quantities of products (such as sulfuric acid) needed in much shorter times. There are various factors that affect the rates of reactions including temperature, catalysts (activation energies), surface area, and concentrations (or pressure for gases). For instance, vanadium(V) oxide is a catalyst commonly used for the manufacture of sulfuric acid in order to increase the reaction rate of the conversion of sulfur dioxide, SO_2, to sulfur trioxide, SO_3.

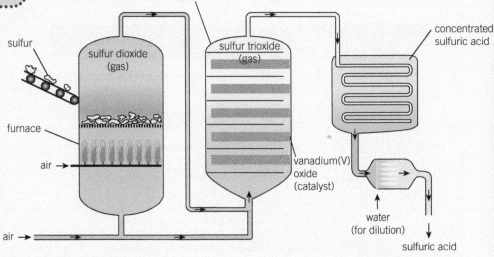

▲ Figure 1 *Rusting is a very slow reaction*

▲ Figure 2 *Vanadium(V) oxide is used as a catalyst for the manufacture of sulfuric acid. Sulfur is burnt in oxygen (from air) to form sulfur dioxide. Sulfur dioxide is oxidised in the presence of vanadium(V) oxide to form sulfur trioxide, which is then dissolved in sulfuric acid to form very concentrated sulfuric acid. This is then diluted to form sulfuric acid that is used in industry*

The impact of these factors can be measured through various experimental methods. Chemists use the data generated from experiments to create equations which represent the impact of certain factors on the rates of reactions. Rate equations allow estimations to be made about the impact of certain changes on the rate of a chemical reaction, without having to carry out long and expensive chemical experiments.

Chemical ideas: Rates of reactions 10.6

Rate equation of a reaction

The decomposition of hydrogen peroxide in solution proceeds slowly under normal conditions, but it is greatly sped up by catalysts. A particularly effective catalyst is the enzyme catalase.

Figure 3 shows the apparatus used to carry out an experiment on the enzyme-catalysed decomposition of hydrogen peroxide solution. The volume of oxygen produced is measured in the inverted measuring cylinder or burette.

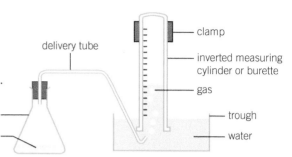

▲ Figure 3 *Apparatus for measuring oxygen produced from decomposition of hydrogen peroxide*

By measuring the total volume of oxygen given off at different times from the start of the experiment, a graph can be plotted of volume of oxygen released against time. From this graph, you can work out the rate of the reaction, in $cm^3 s^{-1}$, at any time during the reaction.

Figure 4 shows a graph of the results that were obtained by starting with hydrogen peroxide of concentration $0.4\,mol\,dm^{-3}$.

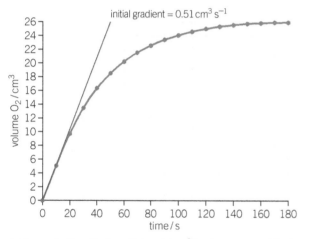

▲ Figure 4 *Results from $0.4\,mol\,dm^{-3}$ hydrogen peroxide decomposition*

Synoptic link 🧪

Measuring volumes of gas in experimental techniques is covered in Techniques and procedures.

Study tip

You can convert $cm^3 s^{-1}$ to $mol\,s^{-1}$, as 1 mol of oxygen occupies about $24\,000\,cm^3$ at room temperature, but you only need to compare rates. For this purpose, you can use cm^3 of oxygen without bothering to convert to moles.

From the graph, you can see two things:

1 The graph is steep at first.
The gradient of the graph gives you the rate of the reaction – the steeper the gradient, the faster the reaction. The reaction is at its fastest at the start, when the concentration of hydrogen peroxide in solution is high, before any has been used up.

2 The graph gradually flattens out.
The hydrogen peroxide is used up and its concentration falls. The lower the concentration, the slower the reaction. Eventually, the graph is horizontal – the gradient is zero – and the reaction has come to a stop.

The rate of the reaction at the start is called the **initial rate**. The initial rate can be found by drawing a tangent to the curve at the point t = 0 and measuring the gradient of this tangent.

The gradient of the tangent in Figure 4 is 0.51, so the initial rate is $0.51\,cm^3 s^{-1}$.

Figure 5 shows some results that were obtained when the same experiment was done using hydrogen peroxide solutions of different concentrations.

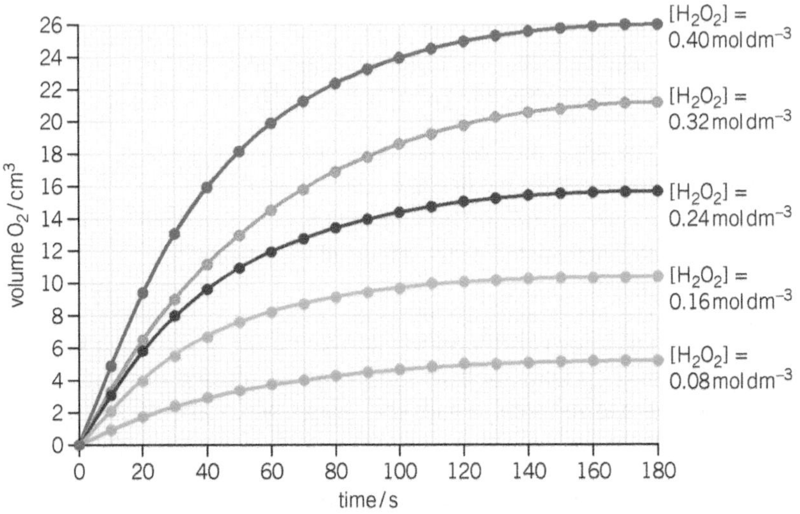

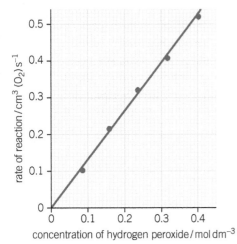

▲ **Figure 5** *Graph of decomposition of different concentrations of hydrogen peroxide*

In each case, the concentration of the catalase enzyme was kept constant, as were all other conditions such as temperature. The graphs start off with differing gradients, depending on the initial concentration of hydrogen peroxide. Table 1 shows the initial rates of the experiments in Figure 5.

▼ **Table 1** *Rates of decomposition of different concentrations of hydrogen peroxide*

Concentration of hydrogen peroxide at start / mol dm^{-3}	Initial rate / cm^3 s^{-1}
0.08	0.10
0.16	0.21
0.24	0.32
0.32	0.41
0.40	0.51

The initial rates plotted against concentration of hydrogen peroxide give a straight line graph (Figure 6).

A straight line graph means that the rate is directly proportional to the concentration of hydrogen peroxide, in other words:

$$\text{rate} \propto [H_2O_2(aq)] \qquad \text{or} \qquad \text{rate} = k[H_2O_2(aq)]$$

The rate of the reaction of the decomposition of hydrogen peroxide is also affected by the concentration of the catalase enzyme. To find the effect of changing the concentration of catalase on the rate of the reaction you could do another set of experiments – this time keeping the concentration of hydrogen peroxide constant.

▲ **Figure 6** *Initial rate of the decomposition of hydrogen peroxide plotted against the concentration of hydrogen peroxide*

Experimental results show that the rate of the reaction is also proportional to the concentration of catalase. In other words:

$$\text{rate} = k[\text{catalase}]$$

Combining this with the equation involving H_2O_2 gives:

$$\text{rate} = k[H_2O_2(aq)][\text{catalase}]$$

This is called the **rate equation** for the reaction – the constant k is called the **rate constant**. The value of k varies with temperature, so you must always say at what temperature the measurements were made when you give the rate, or the rate constant, of a reaction.

Study tip

k is the rate constant.

Do not confuse the rate constant, a lower case k, with the equilibrium constant, an upper case K.

Unit of the rate constant

The unit of the rate constant is dependent on the terms in the rate equation. To find the unit of the rate constant you have to rearrange the rate equation. Look at the rate equation for the decomposition of hydrogen peroxide.

$$\text{rate} = k\,[\text{H}_2\text{O}_2][\text{catalase}]$$

Rearrange this equation to make k the subject.

$$k = \frac{\text{rate}}{[\text{H}_2\text{O}_2][\text{catalase}]}$$

You can then rewrite this equation using only the units of the different terms. The unit of rate in this example is $\text{mol}\,\text{dm}^{-3}\,\text{s}^{-1}$, and the unit of concentration is $\text{mol}\,\text{dm}^{-3}$.

$$k = \frac{\text{mol}\,\text{dm}^{-3}\,\text{s}^{-1}}{(\text{mol}\,\text{dm}^{-3})(\text{mol}\,\text{dm}^{-3})}$$

You can now cancel out any common units to give the units of the rate constant k.

$$k = \frac{\cancel{\text{mol}\,\text{dm}^{-3}}\,\text{s}^{-1}}{\cancel{(\text{mol}\,\text{dm}^{-3})}(\text{mol}\,\text{dm}^{-3})}$$

The unit of the rate constant for the decomposition of hydrogen peroxide is $\text{dm}^3\,\text{mol}^{-1}\,\text{s}^{-1}$.

 ## Worked example: The unit of the rate constant

The rate equation for the reaction between nitrogen dioxide and ozone is:

$$\text{rate} = k[\text{NO}_2][\text{O}_3]$$

What is the unit of the rate constant?

Step 1: Rearrange the rate equation to make the rate constant k the subject.

$$k = \frac{\text{rate}}{[\text{NO}_2][\text{O}_3]}$$

Step 2: Insert units into the rearranged equation.

$$k = \frac{\text{mol}\,\text{dm}^{-3}\,\text{s}^{-1}}{(\text{mol}\,\text{dm}^{-3})(\text{mol}\,\text{dm}^{-3})}$$

Step 3: Cancel out common units.

$$k = \frac{\cancel{\text{mol}\,\text{dm}^{-3}}\,\text{s}^{-1}}{\cancel{(\text{mol}\,\text{dm}^{-3})}(\text{mol}\,\text{dm}^{-3})}$$

Step 4: Write out the unit for the rate constant.

$$\text{dm}^3\,\text{mol}^{-1}\,\text{s}^{-1}$$

The unit of the rate constant for the decomposition of hydrogen peroxide and of the reaction between nitrogen dioxide and ozone is the same. This is because the order of both these reactions is second order overall.

Study tip

You have to reverse the sign of the units in the denominator. So $\text{mol}\,\text{dm}^{-3}$ becomes $\text{mol}^{-1}\,\text{dm}^3$.

By convention, the positive indices are written first, so giving $\text{dm}^3\,\text{mol}^{-1}\,\text{s}^{-1}$.

Synoptic link

The role of catalysts in living systems (enzymes) is explored further in Chapter 7, Polymers and life.

Order of reaction

A rate equation can be written for any chemical reaction, provided an experiment can be carried out to find out how the rate depends on the concentration of the reactants. For a general reaction in which **A** and **B** are the reactants:

$$\mathbf{A} + \mathbf{B} \rightarrow \text{products}$$

The general rate equation is:

$$\text{rate} = k[\mathbf{A}]^m[\mathbf{B}]^n$$

m and n are the powers to which the concentrations need to be raised – they usually have values of 0, 1, or 2. The terms m and n are called the **order of the reaction** with respect to **A** and **B**. For example, in the decomposition of hydrogen peroxide:

$$\text{rate} = k[H_2O_2][\text{catalase}]$$

In this case, m and n are both equal to 1. The reaction is first order with respect to H_2O_2 and first order with respect to catalase. The overall order of the reaction is given by $m + n$, so in this case the reaction is overall second order.

Activity CI 4.1

This activity uses the iodine clock method to find the order of a reaction using experimental results.

Activity CI 4.2

This activity will help you interpret experimental data in order to find orders of reaction and derive a rate equation.

 ### Worked example: Order of a reaction 1

Iodide ions, I^-, react with peroxodisulfate(VI) ions, $S_2O_8^{2-}$. What is the order of the reaction?

The equation for the reaction is:

$$S_2O_8^{2-}(aq) + 2I^-(aq) \rightarrow 2SO_4^{2-}(aq) + I_2(aq)$$

Experimental data show that the rate equation for the reaction is:

$$\text{rate} = k[S_2O_8^{2-}(aq)][I^-(aq)]$$

So this reaction is first order with respect to $S_2O_8^{2-}$, first order with respect to I^-, and second order overall.

The order of the reaction with respect to I^- is one, even though there are two I^- ions in the equation for the reaction. This demonstrates an important point:

You cannot predict the rate equation for a reaction from its balanced equation.

The only way to find the rate equation for a reaction is by doing experiments to find the effect of varying the concentrations of reactants.

Worked example: Order of a reaction 2

Propanone, CH_3COCH_3, reacts with iodine, I_2, in a reaction that is catalysed by acid, H^+. What is the order of the reaction?

$$CH_3COCH_3(aq) + I_2(aq) \rightarrow CH_3COCH_2I(aq) + H^+(aq) + I^-(aq)$$

The rate equation found by experiments is:

$$\text{rate} = k[CH_3COCH_3][H^+]$$

The reaction is first order with respect to both CH_3COCH_3 and H^+.

Iodine does not appear in the rate equation, even though iodine is one of the reactants. The reaction is zero order with respect to I_2.

The reaction is second order overall.

Study tip

The acid catalyst, H^+, appears in the rate equation, even though it is not used up.

Units of the rate constant and the order of a reaction

All reactions with the same overall order of a reaction have the same unit of the rate constant.

- zero order   unit of $k = mol\,dm^{-3}\,s^{-1}$
- first order   unit of $k = s^{-1}$
- second order   unit of $k = dm^3\,mol^{-1}\,s^{-1}$

Study tip

Each of these units can be worked out using the rate equation, as shown earlier in this topic.

The Arrhenius equation

As you have seen, the rate equation for a reaction between two substances **A** and **B** looks like this:

$$\text{rate} = k[\mathbf{A}]^m[\mathbf{B}]^n$$

The equation shows the effect of changing the concentrations of the reactants on the rate of the reaction, but other variables such as temperature and the presence of catalysts can also change rates of reaction. Where do these fit into the rate equation?

These are all included in the rate constant k. The rate constant increases with increasing temperature. These changes are shown in the Arrhenius equation:

$$k = Ae^{-\frac{E_a}{RT}}$$

T is temperature in kelvin.

R is the gas constant $8.314\,J\,K^{-1}\,mol^{-1}$.

E_a is the activation energy with the unit of joules per mole.

e is a mathematical constant.

A is the frequency factor, which is a term that includes factors such as the frequency of collisions and their orientation. A is also taken as constant across small temperature ranges.

Synoptic link

You met the gas constant before when you studied the ideal gas equation $(PV = nRT)$ in Topic DF 8, Burning fuels.

From the Arrhenius equation, A, e, and R are constants, but E_a and T are variables which can change. The Arrhenius equation can be used to calculate the impact of changes in T and E_a on the rate constant k, and therefore the impact on the rates of reactions. T is the temperature, so you can calculate the impact of temperature on the rates of reactions. E_a is the activation energy which is affected by catalysts, so you can calculate the impact of catalysts on the rates of reactions.

Study tip

You do not need to learn the Arrhenius equation but you do need to know how to use it.

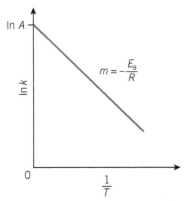

▲ **Figure 7** *Graph of ln k against $\frac{1}{T}$*

Activity CI 4.3

In this activity you can design and carry out work to find the activation enthalpy E_a for a reaction.

Study tip

Remember that the unit of E_a is $J\,mol^{-1}$, rather than $kJ\,mol^{-1}$.

Temperature is in K.

Determining E_a and A

The Arrhenius equation can be used to determine the activation energy and A for a reaction.

If you take the natural logarithm of both sides of the equation you get the following equation.

$$\ln k = \ln A - \frac{E_a}{RT}$$

This equation can be rearranged so that the different terms fit the equation for a straight line.

$$\ln k = -\frac{E_a}{R} \times \frac{1}{T} + \ln A$$

$$y = \quad m \times x + \quad c$$

You can now plot a graph of $\ln k$ versus $\frac{1}{T}$, which should give a straight line with a gradient of $-\frac{E_a}{R}$ (Figure 7) and intercept $\ln A$.

Worked example: Determination of E_a and A

Use the following data to determine the activation energy for a reaction.

▼ **Table 2** *Rate constant at different temperatures*

Temperature / K	Rate constant / s^{-1}
573	2.91×10^{-6}
673	8.38×10^{-4}
773	7.65×10^{-2}

Step 1: Calculate the $\ln k$ and $\frac{1}{T}$ values.

▼ **Table 3** *Values for $\frac{1}{T}$ and ln k*

$\frac{1}{T}$ / K^{-1}	$\ln k$
0.00175	−12.75
0.00149	−7.08
0.00129	−2.57

Step 2: Draw the graph of $\ln k$ against $\frac{1}{T}$ (Figure 8). Calculate the gradient from the graph.

$$m = \frac{y_2 - y_1}{x_2 - x_1}$$

$$= \frac{-13.9 - (-0.9)}{0.0018 - 0.0012} = \frac{-12.9}{0.0006}$$

$$= -21500$$

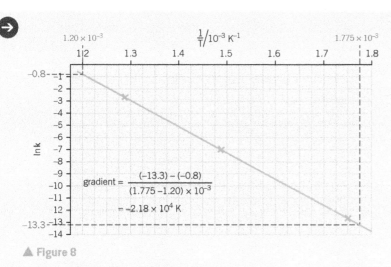

▲ Figure 8

Step 3: Calculate E_a using the gradient.

$$\text{gradient} = -\frac{E_a}{R}$$

$$-2.18 \times 10^4 = -\frac{E_a}{8.314}$$

$$E_a = -2.18 \times 10^4 \times 8.314$$

$$= 1.81 \times 10^5 \, \text{J mol}^{-1}$$

$$= 181 \, \text{kJ mol}^{-1}$$

Step 4: Calculate the value of A.

$$k = A\text{e}^{-\frac{E_a}{RT}}$$

$$\ln k = \ln A - \frac{E_a}{RT}$$

$$\ln A = \frac{E_a}{RT} + \ln k$$

Using the values in Table 3 for 573 K and the value of E_a calculated in Step 3:

$$\ln A = \frac{181\,000}{8.314 \times 573} + (-12.75)$$

$$\ln A = 37.99 - 12.75 = 25.24$$

$$A - 9.15 \times 10^{10} \, \text{s}^{-1}$$

Alternatively, if you are given the intercept of the graph of $\ln k$ against $\frac{1}{T}$, the intercept gives you the value of $\ln A$ (Figure 7). From this you can calculate A.

$$A = \text{e}^{\ln A}$$

The effect of a change of temperature

You can use the Arrhenius equation to show the effect of a change of temperature on the rate constant, and therefore on the rate of the reaction.

Many reaction rates are roughly doubled for a 10 °C temperature rise, even though everything else stays the same. Since the concentrations are not changed, the rate equation shows that it must be the value of k that has roughly doubled:

$$\text{rate} = k[\mathbf{A}]^m[\mathbf{B}]^n$$

Changing the temperature almost always changes the value of k, although the 10 °C rule is only a rough one.

 Worked example: Calculating the effect of a change of temperature

Calculate the change in the rate of a reaction with an activation energy of 50 kJ mol⁻¹, if the temperature is increased by 10 °C from 20 °C to 30 °C? ($R = 8.31$ J K⁻¹ mol⁻¹).

Step 1: Write the equations for the rates of reactions at the two different temperatures.

temperature = 20 °C   temperature = 30 °C

$\text{rate}_1 = k_1[\mathbf{A}]^m[\mathbf{B}]^n$ $\text{rate}_2 = k_2[\mathbf{A}]^m[\mathbf{B}]^n$

Since the rest of the equations are the same, the rates will only be affected by the differences between k_1 and k_2.

Step 2: Write expressions for k_1 and k_2 using the Arrhenius equation.

$$k_1 = Ae^{-\frac{E_a}{RT}}$$

$$k_2 = Ae^{-\frac{E_a}{RT}}$$

As the frequency factor A is the same, the rate constants will only be affected by the differences between the rest of the equations.

Therefore k_1 and k_2 are proportional to $e^{-\frac{E_a}{RT}}$.

Step 3: Calculate the values for the rest of the equations.

The unit of temperature (°C) needs to be converted to match the unit of temperature in R (K).

20 °C = 293 K   30 °C = 303 K

$$k_1 \propto e^{-\frac{50\,\text{kJ mol}^{-1}}{8.31\,\text{J K}^{-1}\text{mol}^{-1} \times 293}}$$

$$k_2 \propto e^{-\frac{50\,\text{kJ mol}^{-1}}{8.31\,\text{J K}^{-1}\text{mol}^{-1} \times 303}}$$

Step 4: Check the units and make the required conversions.

The unit of E_a (kJ) needs to be converted to match the unit of R (J):

50 kJ mol⁻¹ = 50 000 J mol⁻¹

$$k_1 \propto e^{-\frac{50\,000\,\text{J mol}^{-1}}{8.31\,\text{J K}^{-1}\text{mol}^{-1} \times 293\,\text{K}}} = 1.21 \times 10^{-9}$$

$$k_2 \propto e^{-\frac{50\,000\,\text{J mol}^{-1}}{8.31\,\text{J K}^{-1}\text{mol}^{-1} \times 303\,\text{K}}} = 2.38 \times 10^{-9}$$

The number of collisions with energy equal to or greater than the activation energy E_a has almost doubled by increasing the temperature by 10 °C. That causes the rate of reaction to approximately double.

Study tip

You can remove A from the equation so that you can calculate the ratio of k_1 to k_2. This can tell you the effect temperature will have on the number of collisions with an energy equal to or greater than the activation energy E_a.

However, by removing A from the equation, the equation is no longer an expression for k. You should use the proportional symbol $\propto$.

The effect of a catalyst

A catalyst provides a route for the reaction with a lower activation energy.

As an example, in the presence of a catalyst the activation energy falls from $50\,kJ\,mol^{-1}$ to $40\,kJ\,mol^{-1}$ in a reaction that takes place at 50 °C.

activation energy = $50\,kJ\,mol^{-1}$ activation energy = $40\,kJ\,mol^{-1}$

$$k_1 = Ae^{-\dfrac{E_{a1}}{RT}} \qquad k_2 = Ae^{-\dfrac{E_{a2}}{RT}}$$

$$k_1 \propto e^{-\dfrac{50\,kJ\,mol^{-1}}{8.31\,J\,K^{-1}mol^{-1} \times 323\,K}} \quad k_2 \propto e^{-\dfrac{40\,kJ\,mol^{-1}}{8.31\,J\,K^{-1}mol^{-1} \times 323\,K}}$$

$$\propto e^{-\dfrac{50\,000\,J\,mol^{-1}}{8.31\,J\,K^{-1}mol^{-1} \times 323\,K}} \quad \propto e^{-\dfrac{40\,000\,J\,mol^{-1}}{8.31\,J\,K^{-1}mol^{-1} \times 323\,K}}$$

$$\propto 8.13 \times 10^{-9} \qquad\qquad \propto 3.37 \times 10^{-7}$$

A decrease of $10\,kJ\,mol^{-1}$ in the activation energy significantly increases the number of collisions with energy equal to or greater than the activation energy, and therefore increases the rate of reaction.

Synoptic link
You covered catalysts in Topic OZ 4, Ozone – here today and gone tomorrow.

Study tip
Remember to convert temperatures to K by adding 273.

Synoptic link
The catalytic properties of transition metals are covered in Topic DM 2, Transition metals as catalysts.

Summary questions

1 Why does increasing the concentration of a reactant not necessarily mean that the rate of the reaction will increase? *(1 mark)*

2 Use the rate equations for the following reactions to write down the order of the reaction with respect to each of the reactants and catalyst where present.

 a The elimination of hydrogen bromide from bromoethane:

 $$CH_2CH_2Br + OH^- \rightarrow CH_2{=}CH_2 + Br^- + H_2O$$
 rate = $k[CH_3CH_2Br]$ *(2 marks)*

 b The acid-catalysed hydrolysis of methyl methanoate:

 $$HCOOCH_3 + H_2O \rightarrow HCOOH + CH_3OH$$
 rate = $k[HCOOCH_3][H^+]$ *(2 marks)*

 c The hydrolysis of urea, NH_2CONH_2, in the presence of the enzyme urease:

 $$NH_2CONH_2(aq) + H_2O(l) \xrightarrow{\text{urease}} 2NH_3(aq) + CO_2(g)$$
 rate = $k[NH_2CONH_2][urease]$ *(2 marks)*

 d One of the propagation steps in the radical substitution of an alkane by chlorine:

 $$CH_3{\bullet}(g) + Cl_2(g) \rightarrow CH_3Cl(g) + Cl{\bullet}(g)$$
 rate = $k[CH_3{\bullet}][Cl_2]$ *(2 marks)*

 e The formation of the poisonous gas phosgene, used in the First World War, from carbon monoxide and chlorine:

 $$CO(g) + Cl_2 \rightarrow COCl_2(g)$$

 rate = $k[CO]^{\frac{1}{2}}[Cl_2]$ *(2 marks)*

 f The decomposition of nitrogen dioxide to oxygen and nitrogen monoxide:

 $$2NO_2(g) \rightarrow 2NO(g) + O_2(g)$$
 rate = $k[NO_2]^2$ *(2 marks)*

3 Write down the rate equations for the following reactions:

 a Experiments show that the reaction of 1-chlorobutane with aqueous sodium hydroxide is first order with respect to 1-chlorobutane and first order with respect to hydroxide ions.

 $$CH_3CH_2CH_2CH_2Cl + OH^- \rightarrow CH_3CH_2CH_2CH_2OH + Cl^-$$
     (2 marks)

 b The hydrolysis of sucrose, $C_{12}H_{22}O_{11}$, is first order with respect to sucrose and first order with respect to an acid catalyst, $H^+(aq)$:

 $$C_{12}H_{22}O_{11} + H_2O \rightarrow 2C_6H_{12}O_6$$ *(2 marks)*

4 Assuming A is constant over this range, work out how much faster a reaction with an activation energy of $40\,kJ\,mol^{-1}$ would be at 80 °C rather than at 40 °C, assuming the concentrations of everything are the same. $R = 8.31\,J\,K^{-1}\,mol^{-1}$ *(3 marks)*

5 Suppose you added a catalyst to a reaction at 27 °C with an uncatalysed activation energy of $50\,kJ\,mol^{-1}$. The catalyst provides a route with an activation energy of $35\,kJ\,mol^{-1}$. How many times faster will the reaction occur on the lower activation energy route? $R = 8.31\,J\,K^{-1}\,mol^{-1}$ *(3 marks)*

Learning outcomes

Demonstrate and apply knowledge and understanding of:

→ the use of given data to calculate half-lives for a reaction

→ techniques and procedures for experiments in reaction kinetics; use of experimental data [graphical methods (including rates from tangents of curves), half-lives or initial rates when varying concentrations are used] to find the rate of reaction, order of a reaction (zero, first or second order)

→ the term rate-determining step; relation between rate-determining step and the orders and possible mechanism for a reaction.

▲ Figure 2 *One use of Streptomycin is in the control fire blight in apple, pear, and plum trees*

Nutrition deficiency is not the only cause of diseases in plants. Another cause of diseases in plants that would affect the yield of crops (and therefore impact the agricultural industry negatively) is pathogens. Pathogenic diseases can be prevented and treated with the application of certain drugs. For example, Streptomycin (Figure 1) is an antibiotic drug which was used in tuberculosis treatment in the past. However, it is also used as a pesticide to combat the growth of bacteria, fungi, and algae in plants. Streptomycin controls bacterial and fungal diseases of certain fruit, vegetables, seeds, and crops.

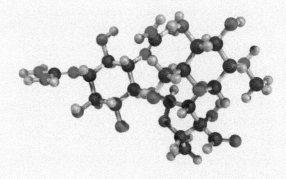

▲ Figure 1 *Streptomycin – a bactericide and fungicide*

As in the antibacterial treatment of human diseases, the antibacterial drug should be given to plants at the correct time to be effective. In order to do so, you need to know on average how long it will take for the drug to be used up in the plant so that you know when would be the appropriate time to treat the plant again.

The minimum concentration of Streptomycin for the effective treatment of bacterial blight disease in a plant has to be at least $20\,mg\,l^{-1}$. Once the Streptomycin is applied to the plant and it is absorbed, its concentration will decrease over time as it is metabolised by the plant. To ensure its effectiveness, the concentration should be kept above $20\,mg\,l^{-1}$. If you know how long it takes for the concentration of the drug to halve, you can apply the treatment of $40\,mg\,l^{-1}$ Streptomycin every time its concentration decreases by half. For instance, if it takes eight hours for $40\,mg\,l^{-1}$ of Streptomycin to fall to half of its concentration, you have to treat the plant with $40\,mg\,l^{-1}$ of Streptomycin every eight hours to ensure its effectiveness. By doing so, you can avoid wasting extra medicinal costs as well as avoiding possible side-effects. Chemists calculate half-lives of reactions.

Finding the order of reaction with experiments

To find the order of a reaction you have to do experiments. Most reactions involve more than one reactant, and you have to do several experiments to find the order with respect to each reactant separately. You have to control the variables, so that the initial concentration of only one substance is changed at a time, and you must make all your measurements at the same temperature.

Look again at the decomposition of hydrogen peroxide in the presence of the enzyme catalase:

$$2H_2O_2(aq) \xrightarrow{\text{catalase}} 2H_2O(l) + O_2(g)$$

If you are looking at the effect of changing the concentration of hydrogen peroxide, there is no need to worry about the catalase – it is an enzyme so it does not get used up and its concentration does not change. However, if you want to look at the effect of varying the concentration of catalase, you must control the concentration of hydrogen peroxide to keep it constant, otherwise two variables will change at the same time. One way is to do several experiments to measure the initial rate of the reaction, keeping the concentration of hydrogen peroxide the same each time but varying the concentration of catalase.

Another way of controlling the concentration of a reactant is to have a large excess of it so that over the course of the experiment the concentration does not change significantly.

Once a set of data has been collected for the effect of changing the concentration of a particular reactant, there are several methods you can use to find the order with respect to that reactant.

The progress curve method

A **progress curve** shows how the concentration of a reactant (or product) changes as the reaction proceeds. Figure 3 shows a progress curve.

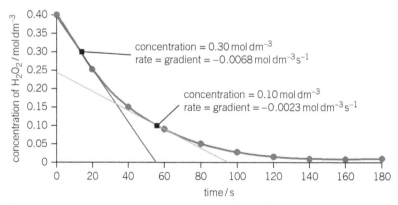

▲ Figure 3 *A progress curve for the decomposition of hydrogen peroxide with tangents to calculate the rates at various concentrations*

The gradient of each tangent gives the rate of reaction for a particular concentration of hydrogen peroxide. You can then find the order with respect to hydrogen peroxide, using the initial rate method.

The initial rate method

The initial rate method involves drawing tangents at the origin of different progress curves. This is the method used in the hydrogen peroxide investigation. Several experiments are run at different concentrations. For each run, the initial rate is found graphically (Figure 4).

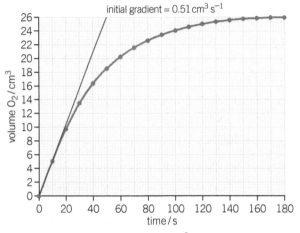

▲ **Figure 4** *Results from 0.4 mol dm^{-3} hydrogen peroxide decomposition*

Once you know the initial rates for different concentrations, you can find the order (Figure 5). If a graph of initial rate against concentration is a straight line then the reaction is first order with respect to that reactant. If a graph of initial rate against (concentration)^2 is a straight line then the reaction is second order with respect to that reactant. If the rate does not change with changing concentration at all then the reaction is zero order with respect to that reactant.

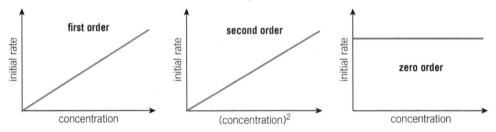

▲ **Figure 5** *The shapes of a graph of initial rate against concentration of a reactant tell you if a reaction if first, second, or zero order*

The process can be repeated, changing the concentration of other reactants. The resultant data can be used to determine the rate equation.

Using the reciprocal of the reaction time as a measure of the rate

One way of measuring the initial rate of a reaction is to measure how long the reaction takes to produce a small, fixed amount of one of the products – the time taken is called the reaction time. If the rate is high (the reaction proceeds quickly) then the reaction time will be small. If the rate is low (the reaction proceeds slowly) then the reaction time will be large.

A good example of this method is the reaction between sodium thiosulfate solution and hydrochloric acid:

$$Na_2S_2O_3(aq) + 2HCl(aq) \rightarrow 2NaCl(aq) + SO_2(aq) + S(s) + H_2O(l)$$

sodium   sulfur
thiosulfate

As the reaction proceeds, solid sulfur forms as a suspension of fine particles and the mixture becomes cloudy. The reaction flask is placed on a cross drawn on a piece of white paper and viewed from above. The cross becomes less visible and is obscured when a certain amount of sulfur has formed in the reaction mixture (Figure 6). The reaction time to reach this point can be measured using different starting concentrations of sodium thiosulfate solution. The volume of each solution and the concentration of the hydrochloric acid must be kept constant in each experiment.

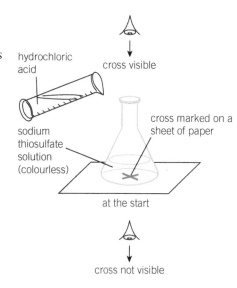

For each starting concentration:

$$\text{average rate for this stage of the reaction} = \frac{\text{amount of sulfur needed to obscure the cross}}{\text{reaction time for sulfur to form}}$$

The amount of sulfur needed to obscure the cross will be the same for each experiment, so the rate of reaction is proportional to $\frac{1}{t}$:

$$\text{average rate} \propto \frac{1}{t}$$

The shorter the time taken, the faster the reaction. The longer the time taken, the slower the reaction.

If a graph of $\frac{1}{t}$ is plotted against the concentrations of the thiosulfate solutions used, a straight line plot is obtained showing that the rate of the reaction is proportional to the concentration of sodium thiosulfate – it is first order with respect to $[Na_2S_2O_3]$.

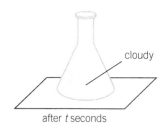

▲ Figure 6 *Using a cross to measure reaction time*

Half-life of reactions

Alternatively, you can use the progress curve to find half-lives for the reaction. If the half-lives are constant then the reaction is first order.

Consider again the decomposition of hydrogen peroxide.

$$2H_2O_2(aq) \xrightarrow{\text{catalase}} 2H_2O(l) + O_2(g)$$

Figure 4 shows the volume of oxygen produced in the decomposition of hydrogen peroxide. For every one mole of oxygen produced, two moles of hydrogen peroxide are used up. Figure 7 shows the same experiment but measuring the concentration of hydrogen peroxide remaining, instead of measuring the volume of oxygen released.

> **Study tip**
>
> Remember:
> - T is temperature
> - t is time

> **Activity CI 5.1**
>
> This activity provides the opportunity to experimentally find the kinetic half-life for a reaction.

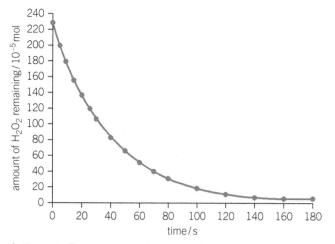

▲ Figure 7 *Progress of the decomposition of hydrogen peroxide measured by the amount of remaining hydrogen peroxide*

You can use Figure 7 to find the **half-life** $t_{\frac{1}{2}}$ of hydrogen peroxide in this experiment. The half-life means the time taken for half of the hydrogen peroxide to get used up.

In Figure 8 this has been done in three cases, using exactly the same graph as in Figure 7.

Activity CI 5.2

This activity checks your skills in interpreting kinetics graphs.

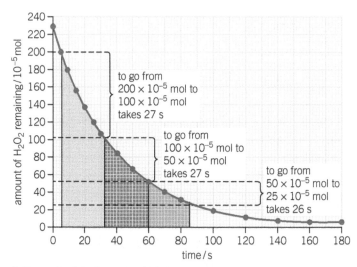

to go from
200×10^{-5} mol to
100×10^{-5} mol
takes 27 s

to go from
100×10^{-5} mol to
50×10^{-5} mol
takes 27 s

to go from
50×10^{-5} mol to
25×10^{-5} mol
takes 26 s

▲ Figure 8 *Calculating half-life from a graph*

Looking at Figure 8, you can see that to go from 200×10^{-5} mol of hydrogen peroxide to 100×10^{-5} mol takes 27 seconds. In other words, starting with 200×10^{-5} mol of hydrogen peroxide, the half-life $t_{\frac{1}{2}}$ is 27 s. Starting with 100×10^{-5} mol, $t_{\frac{1}{2}}$ is again 27 s. Starting with 50×10^{-5} mol, $t_{\frac{1}{2}}$ is 26 s. In fact, allowing for experimental error, $t_{\frac{1}{2}}$ is around 27 s, whatever the starting amount for this reaction at the stated temperature.

The decomposition of hydrogen peroxide is a first order reaction with respect to hydrogen peroxide. For first order reactions, the half-life is constant whatever the starting amount. This characteristic gives a useful way of deciding whether a reaction is first order or not – zero order and second order reactions do not have constant half-lives.

Rate equations and reaction mechanisms

Once you know the rate equation, you can then link it to the reaction mechanism. Most reaction mechanisms involve several individual steps. The rate equation gives information about the slowest step in the mechanism – the **rate-determining step**.

The rate equation for a reaction *cannot* be predicted from the balanced chemical equation.

One way of using the rate equation to find information about the rate-determining step and the reaction mechanism is:

1   Identify the substances from the chemical equation that are *not* involved in the rate equation.
    These substances are *not* involved in the rate-determining step. They will be involved in a step that is faster than the rate-determining step.

2   Identify the substances from the chemical equation that *are* in the rate equation. These substances are the ones involved in the rate-determining step.

3   Identify the order of reaction for each of the substances.
    This will tell you the relative number of moles of each substance involved in the rate-determining step. For example, if a reaction is first order with respect to a substance, then there will be one molecule, atom, or ion of this substance involved in the rate-determining step.

You now know what substances and the ratios of the substances that are involved in the rate-determining step. It is therefore possible to make predictions about the mechanism of the rate-determining step. Once you know the rate-determining step, it is often possible to deduce the full mechanism for a reaction. For example, consider the reaction of 2-bromo-2-methylpropane with hydroxide ions. The chemical equation is:

$$(CH_3)_3CBr + OH^- \rightarrow (CH_3)_3COH + Br^-$$

The rate equation is:

$$\text{rate} = k[(CH_3)_3CBr]$$

There is no [OH^-] term in the rate equation, so this reaction is found to be zero order with respect to OH^-. Therefore, the reaction cannot take place by direct reaction of 2-bromo-2-methylpropane with OH^- ions.

The rate equation contains [(CH_3)_3CBr] and this is the only substance in the rate equation, so you can deduce that $(CH_3)_3CBr$ is the only substance in the rate-determining step. Also, the reaction is first order with respect to $(CH_3)_3CBr$, therefore there is only one molecule involved in the rate-determining step. From this information, and your own knowledge of mechanisms, you may now be able to make a prediction of the mechanism for this step – and possibly for the rest of the reaction too.

Chemists have studied this reaction in detail and have found that it takes place in two steps. First, the C—Br bond breaks heterolytically:

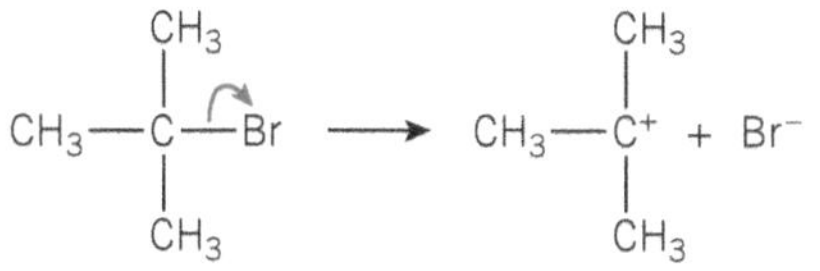

Step 1

Because this step only involves $(CH_3)_3CBr$, its rate depends only on $[(CH_3)_3CBr]$, not on [OH^-].

Activity CI 5.3

This activity uses two different methods to follow the progress of a reaction and uses the results to comment on the rate-determining step.

Study tip

Remember, rate equations cannot be determined from the chemical equation. They can only be determined experimentally.

Synoptic link

You have studied some reaction mechanisms in Topic DF 6, Alkenes – versatile compounds, and Topic OZ 8, What is the state of the ozone layer now?

df6

oz8

The second step involves reaction of the carbocation, $(CH_3)_3C^+$, with OH^-:

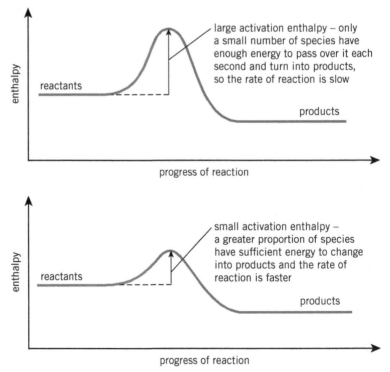

Step 2

Like most ionic reactions, the process in Step 2 is very fast – faster than Step 1. So the rate of Step 1 controls the rate of the whole reaction. That is why the overall rate of the reaction depends only on the concentration of $(CH_3)_3CBr$. The reaction is first order with respect to $(CH_3)_3CBr$ but zero order with respect to OH^-. Step 1 is said to be the rate-determining step and its rate equation becomes the rate equation for the whole reaction.

The mechanism in this example involves two steps. Some simple reactions occur in a single step, whilst more complex ones may involve more than two steps. However, in every case, once you have broken the reaction down into steps, you can write the rate equation for each step from its chemical equation – the overall rate equation for the reaction can only be found experimentally.

Synoptic link

Activate energy and collision theory was covered in Topic OZ 4, Ozone – here today and gone tomorrow.

Rate-determining steps

Why are some steps in a reaction slow and others fast? One reason is that the steps have different energy barriers (activation enthalpies). If the energy barrier is large, the reacting species (either a single species or pairs of colliding species) have enough energy to pass over it and the rate of conversion of reactants into products is slow. Figure 9 compares the enthalpy profiles for reactions with large and small activation enthalpies.

▲ **Figure 9** *The rate of reaction depends on the size of the activation enthalpy*

The enthalpy profile will be different for a reaction involving two steps. There will be two activation enthalpies – one for each step.

The rate-determining step may be the only step in the complete reaction mechanism. However, in a multi-step mechanism the rate-determining step can come at any point in the reaction. It may be the first step, in which case every step after the first step will happen more quickly and the mechanism will not be held up in forming the end product.

Alternatively, the rate-determining step may be the final step, causing a bottleneck as the previous steps happen quickly and are waiting for this step to form the product.

Finally, it may come somewhere in the middle.

In the case of the reaction of 2-bromo-2-methylpropane with hydroxide ions, Step 1 (the rate-determining step) has the larger activation enthalpy (Figure 10). Usually, the chemicals in the middle of a reaction mechanism – the intermediates – are at a higher energy than the reactants or products, because they have unusual structures or bonding, like the carbocation $(CH_3)_3C^+$.

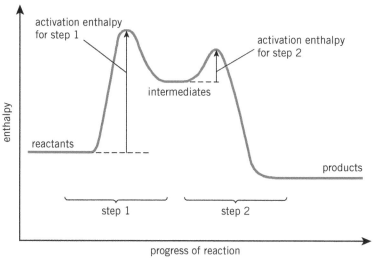

▲ Figure 10 *Progress of a reaction with two steps and an intermediate substance*

> **Activity CI 5.4**
>
> This activity aids chemical literacy and helps you asses your understanding of rates of reaction mechanisms as equilibrium constants.

In any reaction with several steps, the rate-determining step will be the one with the largest activation enthalpy.

Summary questions

1 Define what the half-life of a reaction is and explain how it could be used to decide about the order of reactions. *(2 marks)*

2 The decomposition of nitrogen(V) oxide, carried out at 318 K, was investigated:

$$N_2O_5 \rightarrow 2NO_2 + \frac{1}{2}O_2$$

The results obtained are shown in Table 1.

▼ Table 1 Results for the decomposition of nitrogen(V) oxide

Time / s	$[N_2O_5]$ / mol dm^{-3}
0	2.33
184	2.08
319	1.91
526	1.67
867	1.36
1198	1.11
1877	0.72
2315	0.55
3144	0.34
3500	0.21

a Plot a graph of $[N_2O_5]$ against time.   (3 marks)
b Find the first three half-lives in these results. What do these values suggest about the order of the reaction with respect to N_2O_5?
                                                               (2 marks)

c Using your graph, find the reaction rate for five of the values of $[N_2O_5]$.   (5 marks)
d Plot a graph of reaction rate against $[N_2O_5]$. What order with respect to $[N_2O_5]$ is suggested by this graph? Does this agree with the order deduced in part b?   (3 marks)
e Write the rate equation for the reaction.   (2 marks)
f Calculate the value of the rate constant, including the units, by finding the gradient of the graph plotted in part d.   (3 marks)

3 When cyclopropane gas is heated it isomerises to propene gas. The following data were obtained by heating cyclopropane in a sealed container. The temperature was 700 K and the initial pressure was 0.5 atm.

Time / 10^3 s	0	13	36	56	83	108
% cyclopropane	100	84	63	49	35	25

a Plot a graph of percentage cyclopropane against time. (2 marks)
b Use your graph to measure some half-life values, and so determine the order of the reaction with respect to cyclopropane.   (2 marks)

4 The initial rate method was used to investigate the reaction:

$$2H_2(g) + 2NO(g) \rightarrow 2H_2O(g) + N_2(g)$$

The results are shown below (temperature = 973 K).

$[H_2] / 10^{-2} \, mol \, dm^{-3}$	$[NO] / 10^{-2} \, mol \, dm^{-3}$	$Rate / 10^{-6} \, mol \, dm^{-3} \, s^{-1}$
2.0	2.50	4.8
2.0	1.25	1.2
2.0	5.00	19.2
1.0	1.25	0.6
4.0	2.50	9.6

a What is the order of reaction with respect to:
  i $H_2(g)$
  ii NO(g)   (2 marks)
b Write down the rate equation for this reaction.   (2 marks)
c Calculate a value for the rate constant for this reaction at 973 K.
                                                   (3 marks)

5 The equation for the reaction for the iodination of propanone in acidic solution is:

$$I_2(aq) + CH_3COCH_3(aq) \rightarrow CH_2ICOCH_3(aq) + H^+(aq) + I^-(aq)$$

The rate equation for the reaction has been determined experimentally – it is first order with respect to propanone, zero order with respect to iodine, and first order with respect to hydrogen ions. The following mechanism for the reaction has been proposed:

  Step 1 $CH_3COCH_3 + H^+ \rightarrow CH_3COCH_3^+$
  Step 2 $CH_3COCH_3^+ + I_2 \rightarrow CH_2ICOCH_3 + I^- + H^+$

a Write a rate equation for the reaction.   (2 marks)
b Is this proposed mechanism consistent with the rate equation? Justify your answer.   (2 marks)
c Which step is the rate-determining step?   (1 mark)
d If you wished to remove a sample during the reaction for analysis, how would you quench the reaction?   (2 marks)

6 Consider the following reaction between substances **A** and **B**:

$$A + 2B \rightarrow AB_2$$

The rate equation for the reaction is rate $= k[B]^2$.
Consider the two proposed mechanisms:
Mechanism 1
  Step 1 $B + B \rightarrow B_2$
  Step 2 $A + B_2 \rightarrow AB_2$
Mechanism 2
  Step 1 $A + B \rightarrow AB$
  Step 2 $AB + B \rightarrow AB_2$
Which mechanism do you think is correct? Explain your answer.
                                                   (2 marks)

CI 6 Manufacturing ethanoic acid

Specification references: CI(g), CI(k)

Learning outcomes

Demonstrate and apply knowledge and understanding of:

→ the determination of the most economical operating conditions for an industrial process using principles of equilibrium and rates of reaction

→ given examples of industrial processes:

- costs of raw materials, energy costs, costs associated with plant, co-products and by-products

- the benefits and risks associated with the process in terms of benefits to society of the product(s) and hazards involved.

Pesticides

Keeping crops healthy is not the only concern for the agricultural industry. Crops also have to be protected – both before harvest and in storage – from other organisms that compete for the crops grown. In 2006, the Institute for Plant Diseases in Bonn, Germany, reported that the total global potential loss of crops due to pests varied from about 50% in wheat to more than 80% in cotton production.

One natural method for reducing the loss of crops due to pests is using pest controller animals. For example, vinegaroon spiders are carnivorous and feed on organisms which act as pests to crops.

When a vinegaroon is threatened, it sprays a mist from an opening under its tail. The mist is about 85% ethanoic acid, which is the same acid that gives vinegar its taste and smell (this is where the name vinegaroon comes from). Vinegar is roughly 3–5% ethanoic acid by volume, making ethanoic acid the main component of vinegar apart from water. When sprayed by a vinegaroon spider, it affects the eyes and the skin of a predator, and gives the vinegaroon spider a chance to scamper away.

▲ Figure 1 A vinegaroon spider

Manufacturing ethanoic acid

Ethanoic acid is mainly manufactured in industry from methanol. Methanol and carbon monoxide react together in the liquid phase, with some water to keep the catalyst in solution, at moderate temperatures of about 450 K and a pressure of 30 atm:

$$CH_3OH(l) + CO(g) \rightarrow CH_3COOH(l) \qquad \Delta H = -137\,kJ\,mol^{-1}$$

This reaction is relatively straightforward compared with many others in industry. It works with around 99% efficiency.

▲ Figure 2 Water-based emulsion paints

Finding the right conditions for an industrial process

The production of chemicals on an industrial scale is an expensive business. In order to keep costs down and therefore maximize profits operating costs need to be kept to a minimum. This can be done, amongst other ways, by considering economy of the operating system (getting the best balance between rate and yield) and considering outgoing costs.

Ethanoic acid is also manufactured in bulk for a wide number of uses. Most of the ethanoic acid produced is used to make ethenyl ethanoate – the monomer for poly(ethenyl ethanoate) – and ethanoic anhydride. These chemicals are mainly used to produce water-based emulsion paints. This is a good example of a recent improvement in manufacturing a process. Until recently, much of the ethanoic acid produced was manufactured by the non-catalytic oxidation of naphtha, which gave large quantities of co-products. It is now usually manufactured from methanol, with yields of over 99%.

Chemical ideas: Chemistry in industry 14.4

The operation of a chemical manufacturing process

In order to determine the most economical operating conditions for industrial reactions you need to use the principles of equilibrium and rates of reaction.

Consider the production of methanol from synthesis gas (carbon monoxide and hydrogen). Traditionally, synthesis gas is made by passing steam over hot coke. Coke is made by heating coal at high temperatures and is mainly carbon.

$$C(s) + H_2O(g) \rightleftharpoons CO(g) + H_2(g) \qquad \Delta H = +131\,kJ\,mol^{-1}$$

$$CO(g) + 2H_2(g) \rightleftharpoons CH_3OH(g) \qquad \Delta H = -90\,kJ\,mol^{-1}$$

$$CH_3OH(l) + CO(g) \rightarrow CH_3COOH(l) \qquad \Delta H = -137\,kJ\,mol^{-1}$$

For the first reaction, high temperature will favour the endothermic production of hydrogen and carbon monoxide. Increasing the temperature also increases the rate of reaction considerably. As such, this reaction is carried out at 800–1200 °C. The reaction is carried out at atmospheric pressure because increasing the pressure would not increase rate of formation of carbon monoxide and hydrogen. There is only one mole of gas in the reactants but two moles of gas in the products. As such, increasing the pressure would favour the reverse reaction.

For the second reaction, because it is an exothermic reaction low temperature will favour the products. However, since low temperatures decrease the rates of reactions, a balance is found. This reaction is carried out at 250 °C in industry. Furthermore, the reaction is favoured by high pressure, since there are three moles of gases in the reactants for one mole of gas in the products. The reaction in industry is carried out at a pressure of around 10×10^6 Pa (around 100 atm).

Synoptic link

The production of synthesis gas and the production of methanol are reversible reactions. Look back at Topic ES 7, The Deacon process solves the problem again, for how changing temperature and pressure affect the position of equilibrium.

Study tip

You do not need to learn any of these specific examples but you will be expected to apply the ideas in a range of contexts.

You can apply similar ways of thinking to the manufacture of ammonia, nitric acid, and sulfuric acid met earlier in this chapter.

The costs of a chemical process

Raw materials

Raw materials are the starting point for any industrial chemical process. They are the materials from which **feedstocks** are made – the reactants that go into a chemical process. The raw materials usually have to be prepared or treated to ensure that they are sufficiently pure and present in the correct proportions to use as feedstock. For example, the largest part of an ammonia plant is concerned with feedstock preparation – making the nitrogen and hydrogen mixture for direct conversion to ammonia. Methane is reacted with steam to produce hydrogen, and nitrogen is obtained by fractional distillation of liquefied air. The raw materials for the manufacture of ammonia are therefore air, water, and methane.

An important part of feedstock preparation is getting the feedstock into a form in which it is easy to handle. Transferring gases and liquids is relatively easy, because they can be transported by pipes within the chemical plant, or even across country. Even so, the cost of pumping may be high and so every effort is made to keep the number of pumps and length of piping down to a minimum.

Solids are expensive to handle. Sometimes they are melted and maintained as hot liquids to reduce transportation costs. For example, the sulfur used in the manufacture of sulfuric acid is often transported and used as a molten liquid. Another way of handling solids is to mix them with a liquid to form a slurry – they can then be transported along pipelines.

Often there is little choice of feedstock and raw materials for a chemical process. You can see some examples in Table 1.

▼ Table 1 *Feedstocks and raw materials for some inorganic chemicals*

Product	Feedstock	Raw materials
ammonia	methane, air, water	natural gas, air, water
nitric acid	ammonia, air, water	natural gas/oil, air
sulfuric acid	sulfur, air, water	natural gas/oil (desulfurisation yields the sulfur), air, water
ethanoic acid	methanol, carbon monoxide, water	coal/natural gas, air, water

A vast proportion of the organic chemicals produced today are derived from oil and natural gas. Natural gas is mainly methane but ethane, propane, and butane are often present as well. These are steam-cracked to produce ethene and propene (Figure 4).

▲ Figure 3 *Steam cracking at an oil refinery to produce feedstocks for the petrochemical industry*

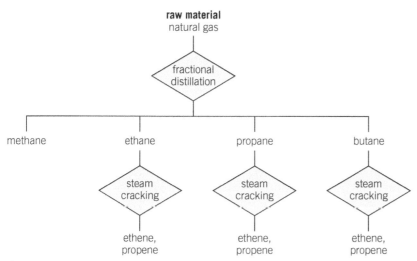

▲ Figure 4 *Feedstocks from natural gas*

Distillation of crude oil produces a variety of fractions, in particular LPG (liquefied petroleum gas), naphtha, and gas (or diesel) oil. These fractions are converted into a variety of building blocks from which many chemicals are made. These building blocks are often alkenes. However, branched-chain alkanes, cycloalkanes, and aromatic hydrocarbons are also produced for use in unleaded petrol (Figure 5).

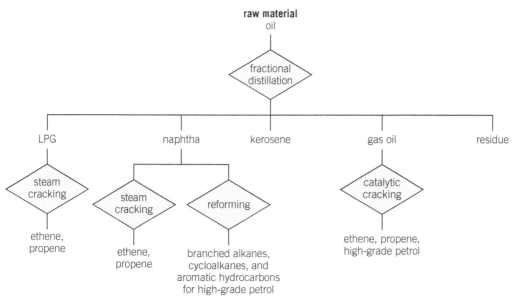

▲ Figure 5 *Feedstocks from oil*

Co-products and by-products

When feedstock is passed through a reactor, a number of things might happen to it.

1   The reaction may form only one product. For example, ammonia:

$$N_2(g) + 3H_2(g) \rightleftharpoons 2NH_3(g)$$

2 The reaction may form two products. For example, when phenol is manufactured from 1-methylethylbenzene (often called cumene), propanone is also produced. The overall reaction can be represented as:

$$C_6H_5CH(CH_3)_2 + O_2 \rightarrow C_6H_5OH + (CH_3)_2CO$$

cumene   phenol   propanone

The propanone formed in this reaction is an example of a **co-product**. The ratio of product to co-product is always fixed – the more desired product produced, the more co-product is also produced. In this example, six tonnes of propanone are produced at the same time as 10 tonnes of phenol. Proceeds from the sale of propanone make a significant contribution to profits. The route would become uncompetitive if demand for propanone were to fall.

3 A reaction other than the one that was intended may occur. For example, epoxyethane, $(CH_2)_2O$, can be made in a one-step process in which ethene is mixed with oxygen and passed over a silver catalyst at 300 °C and a pressure of about 3 atmospheres:

$$2C_2H_4 + O_2 \rightarrow 2(CH_2)_2O \qquad \text{desired reaction}$$

Under these conditions, there is also the possibility of ethene being completely oxidised:

$$C_2H_4 + 3O_2 \rightarrow 2CO_2 + 2H_2O \qquad \text{unwanted side reaction}$$

The carbon dioxide and water formed in this unwanted reaction are called **by-products**. Chemists and chemical engineers work to try to reduce the amount of by-products produced, by reducing the amount of side reactions occurring.

4 Some feedstock may remain unreacted.

Costs and efficiency

Fixed and variable costs

Many factors contribute to the cost of a chemical process. Major costs such as research and development, plant design and construction, and initial production are incurred before any product is sold into the market. Sales of the product have to generate enough return to offset these initial costs and generate a profit for the company (Figure 6).

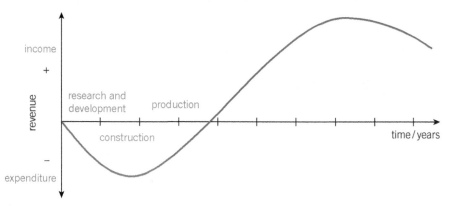

▲ Figure 6 *The profit of a product must eventually surpass the money invested in the development or the industrial process. Eventually demand for the product or increasing maintenance costs as a plant gets older may reduce the profit*

The profit generated is the difference between the selling price of the product and the costs of production. Production costs are made up of two elements – fixed costs and variable costs.

Fixed (or indirect) costs are those incurred by the company, whether they produce one tonne or tens of thousands of tonnes of product. For example, as soon as it has been built the production plant starts to lose value regardless of how much product is made. Other fixed costs include labour costs, land purchase or rental, sales expenses, and so on.

The fixed cost element in the production cost is calculated by spreading the total annual charges over the number of units of the product produced per year. This means that if only one tonne of product is produced per year, then the fixed cost element of production cost will be significantly higher than if 100 tonnes are produced with the same fixed costs.

Variable (or direct) costs relate specifically to the unit of production. Raw materials are the most obvious variable cost, along with costs of effluent treatment and disposal, and the cost of distributing the product. If no production occurs then these variable costs will not be incurred, whereas fixed costs will still have to be paid.

Efficiency

The efficiency of a chemical process will also affect total costs and depends on various physical factors such as temperature, pressure, and rate of mixing. When choosing the temperature and pressure for a particular reaction step, it might seem appropriate to go for the most extreme reactions conditions possible (e.g., high temperature for an endothermic reaction) to maximise the rate at which the product is formed. However, it is not always as simple as this and it is necessary to find the conditions that give the most economical conversion. For example, very high temperatures and pressures require a very specialised, expensive chemical plant that is costly to maintain, and add to the difficulty of controlling the chemical reactions.

Green chemistry aims to, amongst other things, reduce the use of feedstocks to a minimum, recycle unused reactants and solvents, and reduce energy consumption to a minimum. This usually results in minimum waste. Cost control often utilises the ideas of green chemistry.

Choice of reaction conditions can have a significant impact on **yield**. Industry aims to maximise the output of product in a given time in order to maximise profits. Achieving this involves a balancing act between rate and position of equilibrium.

The costs of a process can be reduced considerably if an effective catalyst can be found. Huge savings can also be made by *recycling* unreacted feedstock, and in saving energy that would otherwise be lost to the environment. Sometimes the ability to sell on a co-product can be the difference between making a profit or not.

Saving energy

Efficient use of energy is a significant aim in most chemical processes. During the past few years, often in response to high energy costs and environmental considerations, chemical companies have markedly reduced their energy consumption.

> **Synoptic link**
>
> You previously met green chemical principles in Topic WM 5, The synthesis of salicylic acid and aspirin.

Many chemical reactions involve the release of thermal energy. This thermal energy can be conserved by lagging pipes and by using heat exchangers (Figure 7). Energy from exothermic parts of the process can be used to supply energy to endothermic parts. The energy released in an exothermic reaction can be used to raise the temperature of the reactants, where this is safe to do so.

Water or steam is often used to transfer energy within a chemical plant. Steam, which is generated in a separate chemical plant away from the process, is a much safer alternative to oil and gas, or to using electricity where flammable substances are involved.

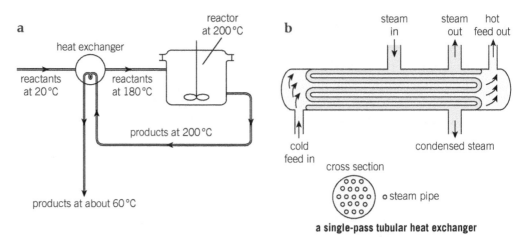

▲ Figure 7 *Conserving thermal energy with heat exchangers*

In an integrated plant, energy (transferred in steam, hot liquid, or gas streams) from one process can be used in a completely different part of the plant. This reduces the quantity of oil, gas, and electricity that has to be purchased. By contrast, a small site would have to raise steam in a special boiler using purchased fuel oil, gas, or electricity. The price of energy in the small plant would therefore be considerably greater in relative terms than that in an integrated plant.

Health and safety

As with every industrial process, there are certain risks involved in industrial manufacture of chemicals. A balance is needed between the risk and the likelihood of that risk occurring against the benefits the chemical industry can bring to quality of life.

Safety legislation

Safety is a major consideration in all operations in the chemical industry. All aspects of safety are affected by national and European Union legislation. Legislation and planning are essential but the key factor in ensuring safety is for everyone within the chemical plant to recognise that it is in their interest to work safely. Some of the main pieces of legislation affecting the chemical industry are:

1   *The Health and Safety at Work Act, 1974*

UK legislation places responsibility for health and safety with the employer. Personal safety is rated very highly, and on a typical visit to a production site you may see eye-baths (Figure 8),

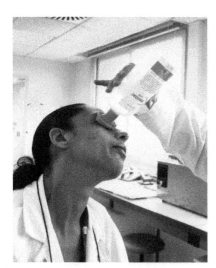

▲ Figure 8 *Laboratories and other production sites have facilities for washing your eyes if any chemicals are spilt onto them*

showers, toxic gas refuges, breathing apparatus, emergency control rooms, and (on larger sites) the company's fire brigade, ambulance service, and a well-equipped medical centre with its own qualified doctors and nurses.

2   *The Control of Substances Hazardous to Health (COSHH) Regulations, 2002*

These regulations control the amount of exposure that employees have to chemicals that can cause a **hazard**. There are various means of minimising exposure, depending on the chemical process concerned. Where possible, the production of these hazardous materials will already have been minimised by altering the reaction conditions, for example, altering the temperature might reduce the amount of a toxic by-product being produced. Steps taken might also include the use of extractors on site and the implementation

> **Hazard**
>
> A **hazard** is the potential that a hazardous substance has to cause harm.

> **Risk**
>
> **Risk** is the likelihood the hazardous substance will cause harm under the conditions of its use.

▼ Table 1 *Examples of how the occurrence of risks can be minimised*

Hazard	Risk	Methods taken to reduce the risk occurring
flammable gases	explosion/fire	stored in flameproof, pressurised cylinders incinerate under controlled conditions extractor fans
acidic gases	burns	neutralise by passing through a scrubber containing alkaline material regular checking of plant for leaks
toxic emissions	various, depending on the material	check protocol for process to minimise emissions monitor levels of emissions ensure that all personnel are familiar with evacuation procedure

> **Activity CI 6.1**
>
> This activity helps you interpret and apply information about an industrial process.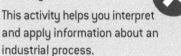

of safe handling and storage of hazardous chemicals. The hazards could be raw materials, reactants, products, or by-products. Table 1 shows some simplified examples of the risks caused by certain materials that are commonly found on chemical sites, and the methods that can be taken to reduce these **risks** occurring.

3   *The Control of Major Accident Hazards (COMAH) Regulations, 1999*

Some reagents used by chemical companies could become a hazard to people living near the plant in the event of an accident. This applies particularly to poisonous gases or volatile liquids. If a plant requires the use of a reagent such as chlorine then the company works with the local authority and emergency services. Emergency procedures are carefully planned and rehearsed at regular intervals. Warning siren is tested at regular intervals and this test ensures that everyone recognises the warning. The most common emergency procedure requires everyone to go indoors and close all windows and doors until the all-clear is given.

▲ Figure 9 *Bioaccumulation is a big environmental issue. The pesticide DDT was widely used before it was discovered to cause thinning in eggshells of birds of prey, such as the Bald Eagle, through bioaccumulation (build up in food chains). Since DDT was banned in the USA, the Bald Eagle's population has recovered significantly*

4   *Registration, Evaluation, Authorisation, and Restriction of Chemicals, 2007*

EU regulation addresses the production and use of chemicals, and any potential effects on human health and the environment. The REACH regulation puts responsibility on the company and requires that any manufactured or imported chemical substances must be registered with the European Chemicals Agency (ECHA). The ECHA must also be informed if the use of chemicals classified as *substances of very high concern* (SVHC) are used in any significant quantities. Substances are considered SVHC if they are carcinogenic (cause cancers), mutagens (cause genetic mutations), interfere in normal reproduction, or bioaccumulative (build up in food chains).

Risks versus benefits

Although manufacture of chemicals on a large scale involves a level of risk, this needs to be kept in perspective especially when considering the benefits gained from the products.

The chemical industry in the UK generated 60 billion US dollars worth of chemical sales, provided the world with biofuels, colourants, foodstuffs, paints, fertilisers, and so on.

From hydropolymers for contact lenses to waste water treatments chemicals helping provide sanitary living conditions, from nanoparticle technology used in suncream to the ever expanding range of pharmaceutical products, the chemical industry improves the day to day quality of life for millions of humans and animals every day.

Synoptic link

Risks and benefits were introduced with respect to chlorine in Topic ES 5, The risks and benefits of using chlorine.

Summary questions

1   Ethane is manufactured from the reaction of methane with chlorine gas. Chloroethane is produced as an unwanted side product of the reaction.
   a   Write the equation for the production of ethane and chloroethane from methane and chlorine gas. *(1 mark)*
   b   Classify products of the reaction as a *by-product* or *co-product* of ethane manufacture, explaining the difference between these terms. *(2 marks)*

2   Sulfur trioxide is an intermediate in the manufacture of sulfuric acid and is made with the reaction below:
$$2SO_2(g) + O_2(g) \rightleftharpoons 2SO_3(g) \qquad \Delta H = -192\,kJ\,mol^{-1}$$
   Discuss the conditions of temperature, pressure, and catalyst that might be used to make sulfur trioxide in a good yield as quickly and economically as possible. Use your knowledge of equilibrium reactions and rates of reactions. *(6 marks)*

Practice questions

1   The Birkeland–Eyde process was once used to make NO by combining the nitrogen and oxygen of the air, using an electrical arc.

$$N_2 + O_2 \rightleftharpoons 2NO \quad \Delta H = +180 \text{ kJ mol}^{-1}$$

Which of the following is correct about this process:

A The best equilibrium yield will be obtained at the highest pressure.

B The temperature should be as high as possible, limited only by cost.

C Use of a catalyst will improve the equilibrium yield.

D Increasing the pressure will have no effect on the rate at which the equilibrium is set up.   (*1 mark*)

2   It is possible to calculate the activation enthalpy of a reaction from the gradient of the line obtained by plotting:

A k against T

B $\ln k$ against $\dfrac{1}{T}$

C k against $\dfrac{1}{T}$

D $\ln k$ against T (*1 mark*)

3   Which row is correct?

	Rate constant k	Equilibrium constant K_c
A	always increases when temperature increases	never varies with temperature
B	sometimes decreases when the temperature increases	always increases as the temperature increases
C	always increases when temperature increases	can either increase or decrease with increasing temperature
D	sometimes decreases when the temperature increases	never varies with temperature

(*1 mark*)

4   A reaction is first order with respect to reagent **A** and second order with respect to reagent **B**.

Which of the following is correct?

A The rate-determining step has two molecules of **B** and one of **A** reacting.

B The reaction is second order overall.

C The units of k are $dm^6 mol^{-2} s^{-1}$.

D The rate goes up by a factor of 16 when the concentrations of both **A** and **B** are doubled.   (*1 mark*)

5   Which row is the correct description of the bonding in the species shown?

	N_2	NH_3	NH_4^+
A	triple bond and no lone pairs	three single bonds and a lone pair	four single bonds
B	triple bond and two lone pairs	three single bonds and a lone pair	four single bonds
C	triple bond and no lone pairs	four single bonds	three single bonds
D	triple bond and two lone pairs	four single bonds	three single bonds

(*1 mark*)

6   Which of the following rows is correct?

	N_2O	NO	NO_2
A	systematic name: nitrous oxide	systematic name: nitric oxide	systematic name: nitrogen dioxide
B	oxidation number of N: −1	oxidation number of N: −2	oxidation number of N: +4
C	brown gas	colourless gas	brown gas
D	possible name: nitrogen(I) oxide	possible name: nitrogen(II) oxide	possible name: nitrogen(IV) oxide

(*1 mark*)

7 The graph shows a first order decay process.

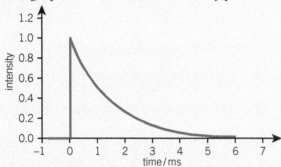

Which of the following statements are correct.

1 A graph of tangents to the curve plotted against time would give a straight line.

2 The half-life from an intensity of 0.8 is 1.0 ms.

3 The line reaches zero after 7 ms.

A 1, 2, and 3 correct

B 1 and 2 are correct

C 2 and 3 are correct

D Only 1 is correct   (*1 mark*)

8 Which of the following methods could be used to follow the reaction of calcium carbonate with hydrochloric acid?

1 Measuring the mass of the reacting flask.

2 Connecting the flask to a gas syringe.

3 Sampling from time to time and titrating with standard NaOH.

A 1, 2, and 3 correct

B 1 and 2 are correct

C 2 and 3 are correct

D Only 1 is correct   (*1 mark*)

9 It takes the concentration of a reagent in a first order reaction 2.0 minutes to fall from 1.0 mol dm^{-3} to 0.5 mol dm^{-3}.

Which of the following are correct?

1 The half-life is 2.0 minutes.

2 It takes 4.0 minutes for the concentration to fall from 1.0 mol dm^{-3} to 0.25 mol dm^{-3}.

3 It takes 4.0 minutes for the concentration to fall from 0.8 mol dm^{-3} to 0.2 mol dm^{-3}.

A 1, 2, and 3 correct

B 1 and 2 are correct

C 2 and 3 are correct

D Only 1 is correct   (*1 mark*)

10 A chemist generates some NO_2 gas by reacting copper with concentrated nitric acid. The chemist collects the gas in a syringe.
$$Cu + 4HNO_3 \rightarrow Cu(NO_3)_2 + 2NO_2 + 2H_2O$$

a Give the systematic name for:

(i) NO_2

(ii) $Cu(NO_3)_2$ (*2 marks*)

b The chemist isolates the $Cu(NO_3)_2$ and wishes to test for the anion.

Describe his method and the expected result.   (*2 marks*)

c The following equilibrium exists in the gas syringe:

$$N_2O_4 \rightleftharpoons 2NO_2 \quad \textbf{Equation 1}$$
colourless   brown

The chemist compresses the brown gas in the syringe. The colour becomes a darker brown and then fades to a paler colour than originally.

(i) Explain why the colour eventually becomes paler.

(ii) Suggest why the colour goes darker initially.   (*4 marks*)

d The chemist places the syringe in a beaker of hot water. The gas becomes darker in colour.

What information does this give about the reaction in **Equation 1**? Explain your answer.   (*2 marks*)

e (i) Write the equation for the K_c of the equilibrium in **Equation 1**.

(ii) In which of **c** and **d** does the value of K_c change? If it changes, does it get larger or smaller?   (*2 marks*)

f The chemist prepares 4.00×10^{-3} mol of N_2O_4 and places it in a 100 cm^3 syringe under the same conditions. Using colorimetry, the chemist calculates that the concentration of NO_2 in the syringe is $0.019 \text{ mol dm}^{-3}$. Calculate a value for K_c under these conditions.   (*3 marks*)

11 Some students investigated the kinetics of the hydrolysis of 2-bromo-2-methylpropane and 1-bromobutane by hydroxide ions.

$$RBr + OH^- \rightarrow ROH + Br^-$$

a Suggest why they chose to compare these two compounds, rather than, say 1-bromobutane and 1-bromopentane.

(*1 mark*)

b The students mixed 1-bromobutane and hydroxide ions in equal molar proportions and followed the reaction by withdrawing portions of the reaction mixture and titrating. They found that the half-life was not constant.

Suggest what they put in the burette for the titration. (*1 mark*)

c Sketch the graph they would have plotted and show how they would have measured one half-life. (*3 marks*)

d What could they deduce from the fact that the half-life was not constant?

(*1 mark*)

The students then used their method to measure the initial concentration of the reaction under various starting conditions. Their results were:

Experiment	Initial concentrations / mol dm^{-3}		Initial rate / mol dm^{-3} s^{-1}
	[OH$^-$]	[C$_4$H$_9$Br]	
A	0.10	0.25	3.2×10^{-6}
B	0.10	0.50	6.6×10^{-6}
C	0.50	0.50	3.3×10^{-5}

e Suggest how they could have measured one of the initial rates of reaction using a graph like that from c. (*1 mark*)

f Use these results to work out the orders of reaction, to calculate a value (with units) for the rate constant, and to write the rate equation for the reaction. (*4 marks*)

g When the students use 2-methyl-2-bromopropane, they discover that the reaction is first order with respect to the haloalkane but that varying the concentration of the hydroxide ion has no effect.

(i) What can be concluded about orders of reaction from the fact that varying the concentration of hydroxide ions has no effect?

(ii) Write the rate equation for this hydrolysis reaction.

(iii) One of the hydrolyses referred to above is thought to proceed by the following mechanism:

$$RBr \rightarrow R^+ + Br^- \qquad slow$$
$$R^+ + OH^- \rightarrow ROH \qquad fast$$

Which haloalkane hydrolyses by this mechanism? Explain how you decide. (*4 marks*)

12 Methanol is made as an intermediate in the manufacture of many other chemicals. One way of making it is to use the reaction:

$$CO(g) + 2H_2(g) \rightleftharpoons CH_3OH(g)$$
$$\Delta H = -129 \, kJ \, mol^{-1}$$

a Discuss the conditions of temperature, pressure, and catalyst that might be used to make methanol in a good yield as quickly and economically as possible.

(*6 marks*)

b A little methane is produced at the same time as the methanol.

(i) Write an equation for the production of methane from CO and H$_2$.

(ii) Classify methane as a *by-product* or *co-product* of methanol manufacture, explaining the difference between these terms.

(iii) Suggest a way of removing the methane from the methanol.

(*4 marks*)

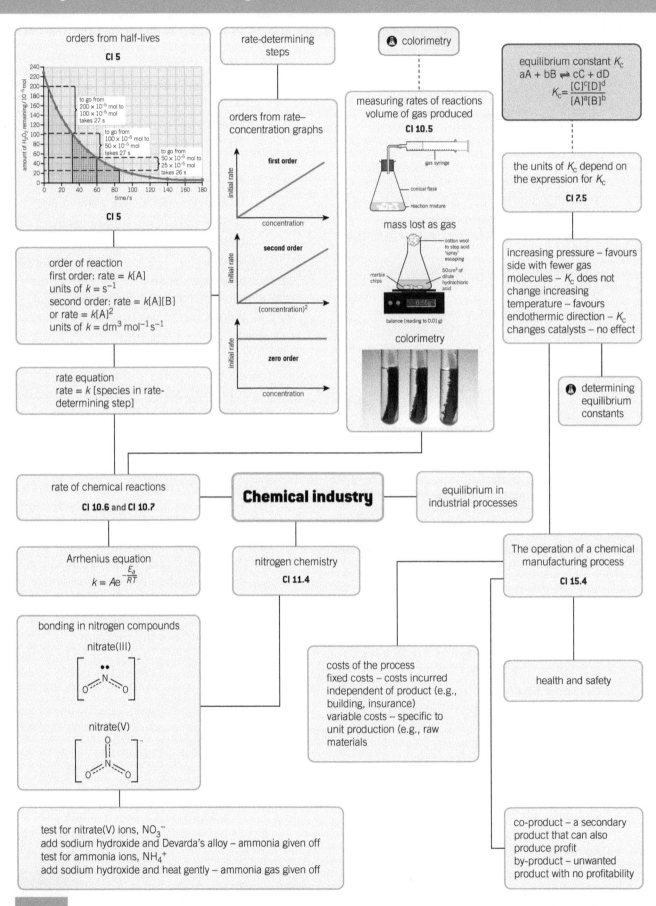

orders from half-lives
CI 5

to go from
200 × 10⁻⁵ mol to
100 × 10⁻⁵ mol
takes 27 s

to go from
100 × 10⁻⁵ mol to
50 × 10⁻⁵ mol
takes 27 s

to go from
50 × 10⁻⁵ mol to
25 × 10⁻⁵ mol
takes 26 s

CI 5

rate-determining steps

orders from rate–concentration graphs

first order

second order

zero order

colorimetry

measuring rates of reactions volume of gas produced
CI 10.5

gas syringe
conical flask
reaction mixture

mass lost as gas
cotton wool to stop acid 'spray' escaping
50cm³ of dilute hydrochloric acid
marble chips
0.55g
balance (reading to 0.01 g)

colorimetry

equilibrium constant K_c
$aA + bB \rightleftharpoons cC + dD$
$$K_c = \frac{[C]^c[D]^d}{[A]^a[B]^b}$$

the units of K_c depend on the expression for K_c
CI 7.5

increasing pressure – favours side with fewer gas molecules – K_c does not change increasing temperature – favours endothermic direction – K_c changes catalysts – no effect

determining equilibrium constants

order of reaction
first order: rate = k[A]
units of $k = s^{-1}$
second order: rate = k[A][B]
or rate = k[A]²
units of $k = dm^3\,mol^{-1}\,s^{-1}$

rate equation
rate = k [species in rate-determining step]

rate of chemical reactions
CI 10.6 and CI 10.7

Chemical industry

equilibrium in industrial processes

Arrhenius equation
$$k = Ae^{-\frac{E_a}{RT}}$$

nitrogen chemistry
CI 11.4

The operation of a chemical manufacturing process
CI 15.4

bonding in nitrogen compounds
nitrate(III)

nitrate(V)

costs of the process
fixed costs – costs incurred independent of product (e.g., building, insurance)
variable costs – specific to unit production (e.g., raw materials

health and safety

test for nitrate(V) ions, NO_3^-
add sodium hydroxide and Devarda's alloy – ammonia given off
test for ammonia ions, NH_4^+
add sodium hydroxide and heat gently – ammonia gas given off

co-product – a secondary product that can also produce profit
by-product – unwanted product with no profitability

The production of phenol

Phenol is a widely-used industrial chemical, used for making polymers, dyes, and antiseptics. World production is around 10 million tonnes per year.

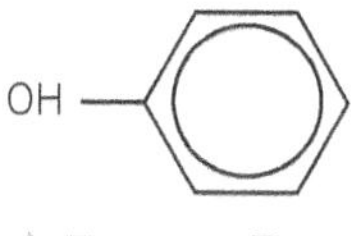

▲ **Figure 1** *The structure of phenol*

The main method of production of phenol involves several steps and produces 6 tonnes of propanone for every 10 tonnes of phenol. It takes place at around 450 K and the yield of phenol is 85%.

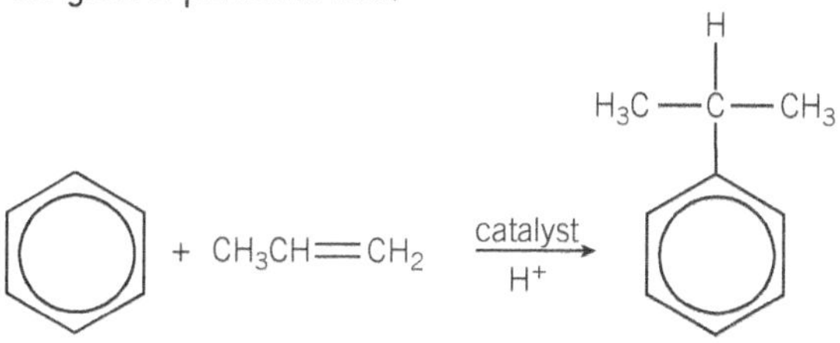

$$\bigcirc + CH_3CH{=}CH_2 \xrightarrow[H^+]{catalyst} \bigcirc$$

▲ **Figure 2** *Step 1*

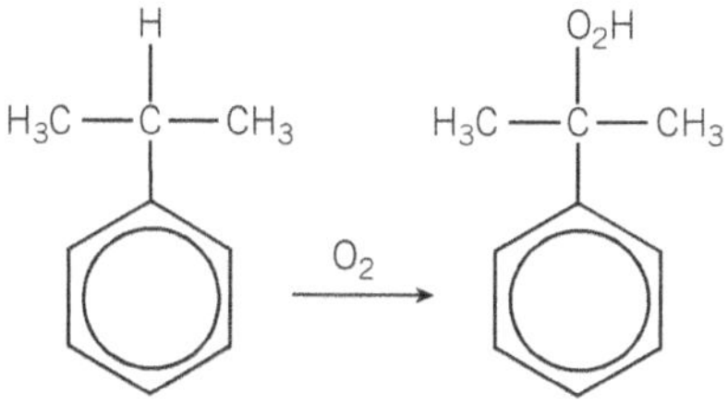

$$H_3C-\underset{\underset{\text{(benzene ring)}}{\mid}}{\overset{\overset{H}{\mid}}{C}}-CH_3 \xrightarrow{O_2} H_3C-\underset{\underset{\text{(benzene ring)}}{\mid}}{\overset{\overset{O_2H}{\mid}}{C}}-CH_3$$

▲ **Figure 3** *Step 2*

$$H_3C-\underset{\underset{\text{(benzene ring)}}{\mid}}{\overset{\overset{O_2H}{\mid}}{C}}-CH_3 \xrightarrow{H^+} \underset{\text{(phenol)}}{OH} + H_3C-\overset{\overset{O}{\parallel}}{C}-CH_3$$

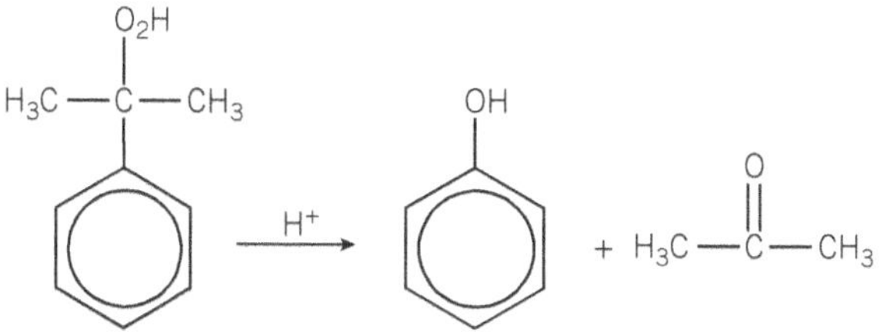

▲ **Figure 4** *Step 3*

A new method involves oxidation of benzene by N_2O at about 650 K:

$$C_6H_6(g) + N_2O(g) \rightleftharpoons C_6H_5OH(g) + N_2(g)$$
$$\Delta H^{\ominus} = -286 \text{ kJ mol}^{-1}$$

N_2O is a waste product in the formation of polyamides and is a greenhouse gas. The yield of phenol is 95%.

1 Calculate the atom economy of the final stage of the main method of making phenol.

2 Evaluate the two methods in terms of costs of raw materials, energy costs, risks and benefits to society, environmental concerns, and hazards involved.

3 For the new method, the oxidation of benzene by N_2O, explain the effect of increasing temperature, increasing pressure, and using a catalyst on (i) the rate of reaction; (ii) the magnitude of the equilibrium constant.

4 Write an expression for K_c for the oxidation of benzene by N_2O. Given the equilibrium concentrations $[N_2O] = 0.05 \text{ mol dm}^{-3}$, $[C_6H_6] = 0.05 \text{ mol dm}^{-3}$, $[C_6H_5OH] = 0.95 \text{ mol dm}^{-3}$ and $[N_2] = 0.95 \text{ mol dm}^{-3}$, calculate K_c and give the units.

5 The following initial rates data were obtained for the reaction of C_6H_6 with N_2O.

$[C_6H_6]$ / mol dm^{-3}	$[N_2O]$ / mol dm^{-3}	rate / mol dm^{-3} s^{-1}
0.012	0.008	0.001
0.024	0.008	0.002
0.024	0.024	0.006

Deduce the rate equation under these conditions and calculate the rate constant k, giving the units.

A student proposed the following mechanism for the oxidation of benzene by N_2O. Explain whether this mechanism is consistent with the rate equation deduced in question 5.

Step 1 $N_2O \rightarrow N_2 + O$ *slow*
Step 2 $C_6H_6 + O \rightarrow C_6H_5OH$ *fast*

✚ Extension

1 Find out about recent developments in the chemical industry. These could include novel procedures, new products or developments in 'green' chemistry.

2 This topic has considered the equilibria and rates of several examples of industrial processes. Produce a summary of equilibria and rates, explaining how the choice of conditions affects the economics of the process because of the effect on the position of equilibrium and rate.

3 What are the top ten products of the chemical industry and what are their uses? Research the contribution of chemical industry to the national and international economy.

CHAPTER 7
Polymers and life

Topics in this chapter

Why a chapter on Polymers and life?

Proteins and nucleic acids are two types of organic molecule that are fundamental to the existence of life. Proteins enable and regulate the chemical processes in living cells, whilst nucleic acids enable the synthesis of these proteins and store the information necessary for their manufacture. Despite the complexity of the roles that they play, both of these molecules are built up from relatively simple sub-units, bonded together in long chains – they are biological polymers

This chapter begins by looking at polymer structures that have been produced artificially by chemists – polyesters and polyamides. The condensation reactions used to form these polymers from monomers are described and some of the properties of the monomer molecules themselves are also explored.

The monomers that make up proteins – amino acids – are described, and a new type of stereoisomerism is introduced. The structure of the proteins formed from amino acids is then discussed, and the ability of some of these proteins to act as enzymes (biological catalysts) is explained. A knowledge of the three-dimensional structure of enzymes enables new drug molecules to be designed.

The key features of the structure of DNA are then described and related to the way in which DNA is able to copy itself when cells divide. DNA carries information that causes specific proteins to be synthesised, and the mechanism of this process is explained.

Finally, two important spectroscopic techniques are introduced. Information from these techniques help chemists to deduce the structure of organic molecules, and strategies for analysing this information are described.

Knowledge and understanding checklist

In this chapter you will learn more about some ideas introduced in earlier chapters:

- ☐ Addition polymers (**Developing fuels**)
- ☐ Carboxylic acid, aldehydes, ketones, and phenols (**What's in a medicine?**)
- ☐ The formation of esters (**What's in a medicine?**)
- ☐ Stereisomerism (**Developing fuels**)
- ☐ Chromatography (**What's in a medicine?**)
- ☐ Intermolecular bonds (**The ozone story**)
- ☐ Catalysts (**Developing fuels, The ozone story,** and **The chemical industry**).
- ☐ Mass spectrometry (**Elements of life** and **What's in a medicine?**).

Maths skills checklists

In this chapter, you will need to use the following maths skills. You can find support for these on Kerboodle and through MyMaths.

- ☐ Visualise and represent 2-D and 3-D forms, including 2-D representations of 3-D objects.
- ☐ Understand the symmetry of 2-D and 3-D shapes.

MyMaths.co.uk
Bringing Maths Alive

PL 1 The polyester story

Specification references: PL(h), PL(k), PL(o), PL(p)

Polyesters – condensation polymers

In the 1940s, UK chemist Rex Whinfield knew that another scientist, Wallace Carothers, had tried to produce polyesters. Carothers had produced the award-winning nylon in 1935, but Whinfield felt that Carothers had not used the acid most likely to give the desired properties for a polyester. Whinfield used 1,4-benzenedicarboxylic acid (terephthalic acid) and reacted it with ethane-1,2-diol. The structure of the polyester formed is:

$$\left[\begin{array}{c} \text{polyester structure} \end{array}\right]_n$$

The old name for this polyester is polyethylene terephthalate, often known as PET.

The polyester is produced as small granules. These are melted and squeezed through fine holes. The resulting filaments are spun to form a continuous fibre. The fibre, known as Terylene or Dacron, is widely used in making clothes – such as suits, shirts, and skirts – and for filling anoraks and duvets because it gives good heat insulation (Figure 1). The polyester can also be made into sheets for X-ray films.

A newer use of PET is for packaging. The granules of the polyester are heated to about 240 °C and further polymerisation takes place – a process known as curing. When the polymer is then stretched and moulded, the molecules are orientated in three dimensions. The plastic has great strength and is impermeable to gases – no wonder PET is widely used for bottling carbonated drinks (Figure 2).

The versatility of the polymer arises because the molecules can be aligned in one, two, or three dimensions, giving rise to different properties.

Polymers are important in our everyday lives. Not only are synthetic polymers such as plastics an integral material that is widely used but organisms (humankind included) are made of biological polymers such as proteins and DNA. Enzymes are proteins that control bodily functions and are also used in industrial applications, such as food production.

▲ Figure 1 Alligator raincoat advertisement from the 1960s

▲ Figure 2 Carbonated drinks' bottles are often made of PET

Carboxylic acids

Naming carboxylic acids

Carboxylic acids contain the carboxyl group:

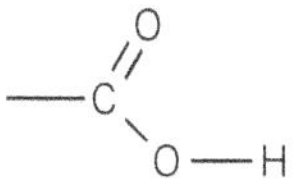

This is often abbreviated to –COOH, although the two oxygen atoms are not joined together.

The structure of the rest of the molecule can vary widely and this gives rise to a large number of different carboxylic acids. When the remainder of the molecule is an alkyl group, the acids can be represented by the general formula R–COOH.

Carboxylic acids are named from the parent alkane by omitting the final -e (since the suffix oic starts in a vowel) and adding the ending -oic acid (Table 1).

▼ Table 1 *Naming three simple carboxylic acids*

Number of carbon atoms	Alkane	Carboxylic acid
1	methane, CH_4	methanoic acid, HCOOH
2	ethane, CH_3CH_3	ethanoic acid, CH_3COOH
3	propane, $CH_3CH_2CH_3$	propanoic acid, CH_3CH_2COOH

Branched-chain carboxylic acids are named in a similar way to alkanes:

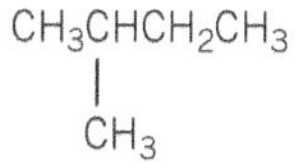

$CH_3CHCH_2CH_3$
        |
       CH_3
2-methylbutane

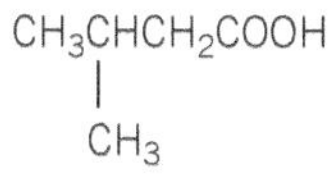

CH_3CHCH_2COOH
        |
       CH_3
3-methylbutanoic acid

The skeletal formula of 3-methylbutanoic acid is:

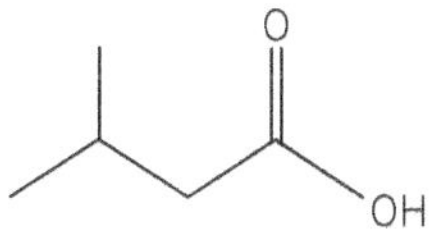

When two carboxyl groups are present, the ending -dioic acid is used. For example:

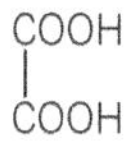

COOH
|
COOH

ethanedioic acid

COOH
|
CH_2
|
COOH

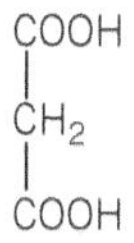

propanedioic acid

> **Study tip**
>
> The carbon atom in the –COOH group counts as one of the carbon atoms for the name.

> **Synoptic link**
>
> You will learn more about carboxylic acids in Chapter 9, Colour by design.

> **Study tip**
>
> When naming double carboxylic acids, the e remains on the prefix, for example, ethanedioic acid, unlike naming single carboxylic acids, for example, ethanoic acid.

A carboxyl group can be attached to a benzene ring. For example:

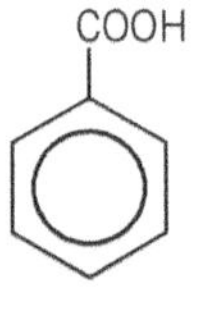

benzenecarboxylic acid

benzene-1,4-dicarboxylic acid

Benzenecarboxylic acid is also known as benzoic acid.

Naming phenols

When the –OH group is attached to an arene ring, it is called a **phenol** group. Phenol itself is:

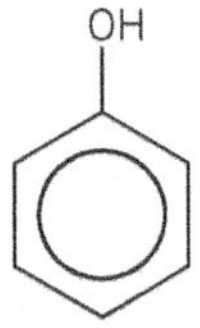

Other simple phenols are named as shown:

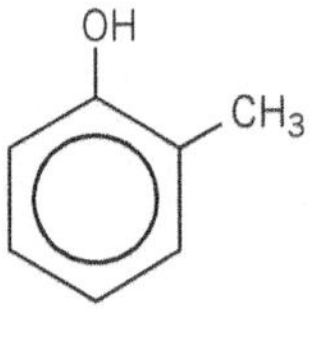

2-methylphenol

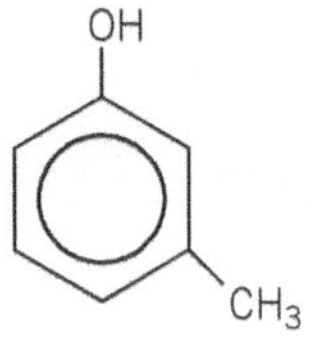

3-methylphenol

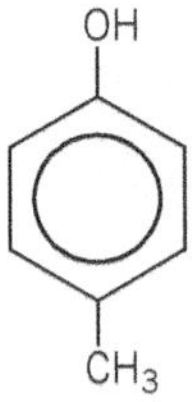

4-methylphenol

Naming aldehydes and ketones

Ketones have the ending -one and aldehydes have the ending -al after a stem which indicates the length of the carbon chain, for example:

- $CH_3CH_2COCH_3$ is butanone and $CH_3CH_2CH_2COCH_3$ is pentan-2-one, because another isomer (pentan-3-one) is possible. $(CH_3)_2CH_2COCH_3$ is 4-methylpentan-2-one, with the lower number being determined by the position of the carbonyl (C=O) group.

- $CH_3CH_2CH_2CHO$ is butanal. Since the CHO group is always on the end of the molecule, a number before the -al is never necessary.

Reactions of carboxylic acids

Carboxylic acids are strong enough acids to react with strong bases to form salts. Phenols also react in this way, but alcohols do not.

Carboxylic acids react with carbonates to form water and carbon dioxide.

$$CO_3^{2-}(aq) + 2H^+(aq) \rightarrow CO_2(g) + H_2O(l)$$

This is an acid–base reaction, where the carbonate ion, $CO_3^{2-}(aq)$, is the base.

The overall equation for the reaction of ethanoic acid with calcium carbonate is:

$$CaCO_3 + 2CH_3COOH \rightarrow Ca(CH_3COO)_2 + CO_2 + H_2O$$

Synoptic link

You learnt about the formulae of carboxylic acids, aldehydes, and ketones in Topic WM 2, Identifying the active chemical in willow bark.

Synoptic link

You learnt about the reactions of phenol and some of the reactions of carboxylic acids in Topic WM 1, The development of modern ideas about medicines.

Study tip

Soluble bases, such as sodium hydroxide, are called alkalis.

Study tip

The H_3O^+ ion is often written as $H^+(aq)$.

The salt formed is called calcium ethanoate.

Phenols and alcohols do not have a great enough concentration of $H^+(aq)$ ions to react in this way. So when carboxylic acids react with carbonates, the solution will fizz (because of the production of carbon dioxide) whereas solutions of alcohols and phenols with carbonates will not.

Carboxylic acids, phenols, and alcohols also form salts with metals and the reaction is *redox* rather than acid–base, for example:

$$Mg + 2CH_3COOH \rightarrow Mg(CH_3COO)_2 + H_2$$

Esters

What are esters and how are they made?

Esters are formed when an alcohol reacts with a carboxylic acid. The reaction occurs extremely slowly unless an acid catalyst is present – a small amount of concentrated sulfuric acid or concentrated hydrochloric acid is often used and the reagents are heated under reflux.

$$ROH + CH_3COOH \rightleftharpoons CH_3COOR + H_2O$$
(R is an alkyl group, e.g., CH_3 or C_2H_5.)

This process is called a **condensation reaction**. A condensation reaction can be thought of as two molecules reacting together to form a larger molecule with the elimination of a small molecule such as water.

The ester link is formed by the condensation reaction of the hydroxyl group in the alcohol and the carboxyl group in the acid – the process is known as **esterification**.

The reaction is reversible and eventually comes to equilibrium. To improve the yield of ester from a given amount of acid, an excess of alcohol is added or the water can be distilled off as it is formed. These help to drive the equilibrium position to the right. The sulfuric acid catalyst also absorbs some of the water formed, again helping to move the equilibrium position to the right.

The reaction is easily reversed – the reverse reaction is called **ester hydrolysis**.

Esters have strong, sweet smells which are often floral or fruity. Esters can be used in food flavourings and perfumes. Many naturally occurring esters are responsible for well-known fragrances. Many organic compounds dissolve readily in esters, so esters are widely used as solvents, for example, in some glues.

Naming esters

Esters are named after the alcohol and acid from which they are derived.

The ending -oate is used instead of the -oic from the parent acid to show clearly that the compound is an ester. For example, ethanol and ethanoic acid react together to form ethyl ethanoate. Phenol and benzoic acid react to form phenyl benzoate.

Synoptic link

Esterification was first introduced in Topic WM 1, The development of modern ideas about medicines.

Synoptic link

Equilibria was first covered in Chapter 3, Elements from the sea.

▲ Figure 3 *The ester ethyl ethanoate is responsible for the smell and flavour of pear drops*

The names of esters always consist of two words, the second ending in -oate. For example:

ethyl — this part comes from the alcohol
ethanoate — this part comes from the acid

You have to be careful to get the groups the right way round. For example, ethyl ethanoate is very different from its isomer, methyl propanoate:

$C_2H_5OCOCH_3$ ethyl ethanoate

$CH_3OCOC_2H_5$ methyl propanoate

However, when writing the formula of an ester, either the acid part or the alcohol part can be written first. For example, ethyl ethanoate may be represented by either $C_2H_5OCOCH_3$ or $CH_3COOC_2H_5$. You will need to recognise the structure written both ways. The important thing is to identify the group attached to the $C{=}O$, as this is from the acid, whilst the group attached to $-O-$ is from the alcohol.

Polyesters

Polymers can be formed by addition reactions and also by condensation reactions – the same type of reaction that is used to make esters. Polymers formed from condensation reactions are known as **condensation polymers**. The monomers must have at least two suitable functional groups per molecule for a condensation polymer to be produced. During condensation polymerisation, the monomers react together to give a longer chain polymer *and* a small stable molecule – usually water or hydrogen chloride.

Terylene, a **polyester**, is a typical condensation polymer. It is formed from two monomers, a diol (containing two −OH groups) with a dicarboxylic acid (containing two −COOH groups):

ethane-1,2-diol

benzene-1,4-dicarboxylic acid (terephthalic acid)

the repeating unit of terylene

Making esters from phenols

The −OH group in phenol is less reactive to esterification than the −OH of ethanol, so it needs a more vigorous reagent to esterify it, for example, an acid anhydride.

Summary questions

1 Name the following compounds:
   a C_2H_5COOH (1 mark)
   b $HOOCCH_2CH_2COOH$ (1 mark)
   c $(C_3H_7CO)_2O$ (1 mark)
   d

   (1 mark)

   e $HCHO$ (1 mark)
   f $CH_3CH(CH_3)COCH_3$ (1 mark)

2 Write structural formulae for:
   a phenol (1 mark)
   b methanoic acid (1 mark)
   c methanoic anhydride (1 mark)
   d propane-1,3-diol (1 mark)
   e hexanedioic acid (1 mark)
   f pentanal (1 mark)
   g 4-methylpentan-2-one. (1 mark)

3 Write the formula for the monomer used to form
   $$-(OCH(CH_3)CO)OCH(CH_3)CO-$$
   (1 mark)

4 Give the systematic names of the following esters:
   a CH_3COOCH_3 b $HCOOC_2H_5$ c $C_3H_7COOCH_3$ d $C_2H_5OCOCH_3$ (4 marks)

5 Write balanced equations for the reactions that occur when the following pairs of compounds are heated under reflux with a few drops of sulfuric acid catalyst:
   a propan-2-ol and propanoic acid (2 marks)
   b ethanoic acid and ethane-1,2-diol. (2 marks)

6 Draw *two* repeating units of the polymer formed between $HOCH_2CH_2OH$ and $HOOCCH_2COOH$. (2 marks)

7 Write equations for the reactions of:
   a phenol with sodium hydroxide (1 mark)
   b propanoic acid with calcium carbonate (2 marks)
   c methanoic acid with magnesium (1 mark)
   d ethanol with ethanoic anhydride. (2 marks)

8 There are four esters and two acids with the molecular formula $C_4H_8O_2$. Draw their structures and name them. (12 marks)

9 Lactic acid, $CH_3CH(OH)COOH$, forms a compound with molecular formula $C_6H_8O_4$ when warmed with concentrated sulfuric acid. Identify the compound and write an equation for its formation from lactic acid. (2 marks)

10 When malonic acid, *Z*-ethenedicarboxylic acid, is heated it loses water to form compound A:
   a Draw the structure of *Z*-ethenedicarboxylic acid. (1 mark)
   b Write an equation for the reaction of malonic acid to form compound A. (1 mark)
   c Name the functional group in compound A. (1 mark)
   d Suggest why *E*-ethenedicarboxylic acid has to be heated to a higher temperature before it forms compound A. (1 mark)

PL 2 The nylon story

Specification references: PL(j), PL(k), PL(l), PL(n)

The invention of nylon

Wallace Carothers joined the US chemical company DuPont in 1928. He led a team investigating the production of polymers that might be used as fibres. This was at a time when scientists were beginning to understand more about the structure of polymers, so Carothers had a scientific basis for his work.

It was already known that wool and silk have protein structures and are polymers involving the peptide linkage –CONH–. Chemists had also begun to discover that many natural fibres are composed of molecules that are very long and narrow – like the fibres themselves.

Carothers did not make his discoveries by accident – he set about systematically trying to create new polymers. In one series of experiments he decided to try to make synthetic polymers in which the polymer molecules were built up in a similar way to the protein chains in silk and wool. Instead of using amino acids (the starting materials for proteins), Carothers began with amines and carboxylic acids.

Amines are organic compounds that contain the amine functional group, $-NH_2$. When an $-NH_2$ group reacts with the $-COOH$ group in a carboxylic acid, an amide group $-CONH-$ is formed. A molecule of water is eliminated in the process – this is an example of a condensation reaction.

Carothers used diamines and dicarboxylic acids that contained reactive groups in two places in their molecules, so they could link together to form a chain. In this way he was able to make polymers in which monomer units were linked together by amide groups. The process is called condensation polymerisation because the individual steps are condensation reactions.

▲ Figure 1 *Formation of a polyamide*

Because the group linking the monomer groups together is an amide group, these polymers are called polyamides. More commonly they are known as nylons.

Carothers discovered nylon in the spring of 1935 and by 1938 the first product using nylon, Dr. West's Miracle-Tuft Toothbrush, appeared. Nylon stockings were seen for the first time in 1939. Most of the nylon produced at this time was used for parachute material instead of silk (Figure 2), so nylon stockings did not become generally available in Britain until the end of the Second World War.

The first aramids

After the invention of nylon, chemists began to make sense of the relationship between a polymer's structure and its properties. They were able to predict strengths for particular structures, and research was directed at inventing a 'super fibre'. In the early 1960s, DuPont were looking for a fibre with the 'heat resistance of asbestos and the stiffness of glass'.

The aromatic polyamides seemed promising candidates – the planar aromatic rings should result in rigid polymer chains and, because of the high ratio of carbon to hydrogen, they require relatively large concentrations of oxygen before they burn.

The first polymeric aromatic amide – an aramid – was made from 3-aminobenzoic acid. The polymer could be made into fibres and was fire-resistant but it was not particularly strong. The zigzag nature of the chains prevented the molecules from aligning themselves properly.

Stephanie Kwolek and her team at DuPont, investigated the properties of fibres made from other polymeric aromatic amides. One of these fibres was found to be extremely strong, fire-resistant, and flexible. The new aramid fibre was called Kevlar. It also has a low density because it is made from light atoms – carbon, hydrogen, oxygen, and nitrogen. Weight for weight, Kevlar is about five times stronger than steel. One of its early uses was to replace steel cords in car tyres – Kevlar tyres are lighter and last longer than steel-reinforced tyres.

Kevlar is strong because of the way the rigid, linear molecules are packed together (Figure 3). The chains line up parallel to one another, held together by hydrogen bonds. This leads to sheets of molecules. The sheets then stack together regularly around the fibre axis to give an almost perfectly ordered structure.

The important thing to realise about Kevlar is that it adopts this crystalline structure because of the way the polymer is processed to produce the fibre.

▲ Figure 2 *Nylon parachutes*

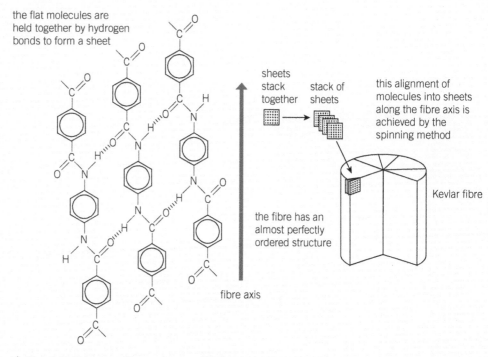

▲ Figure 3 *The crystalline structure of Kevlar*

Amines

What are amines and how are they named?

The structure of amines resembles ammonia molecules in which alkyl groups replace one, two, or all three hydrogen atoms.

Amines with one alkyl group are called primary amines. Lower primary amines are named by adding the alkane group as a prefix to amine, for example, methylamine, CH_3NH_2, and ethylamine, $C_2H_5NH_2$.

Higher homologues are often named using the prefix amino- and the name of the alkane from which they appear to be derived, for example, 2-aminopropane, $CH_3CH(NH_2)CH_3$.

Amines used to make condensation polymers have two amino groups attached and are called diamines. They are named after the parent alkane, for example, ethanediamine, $NH_2CH_2CH_2NH_2$, and 1,6-diaminohexane, $NH_2CH_2CH_2CH_2CH_2CH_2CH_2NH_2$.

Amines with low relative molecular masses are gases or volatile liquids. The volatile amines also resemble ammonia in having strong smells. The characteristic smell of decaying fish comes from amines such as ethylamine. Rotting animal flesh gives off the diamines $H_2N(CH_2)_4NH_2$ and $H_2N(CH_2)_5NH_2$, which are sometimes called by the names putrescine and cadaverine, respectively.

Properties of amines

The properties of amines are similar to those of ammonia but modified by the presence of alkyl groups. Most of the properties are due to the lone pair of electrons on the nitrogen atom.

The bonding around the nitrogen atom of an amine is similar to that in ammonia – three pairs of electrons form localised covalent bonds, whilst the other two electrons form a lone pair:

The lone pair of electrons is responsible for amines being soluble in water and acting as bases.

Solubility of amines

Like ammonia, amines can form hydrogen bonds with water (Figure 4).

Because of this strong attraction between amine molecules and water molecules, amines with small alkyl groups are soluble. Amines with larger alkyl groups are less soluble in water. The large alkyl groups are unable to break the hydrogen bonds between the water molecules. The enthalpy change to break these hydrogen bonds is greater than the enthalpy change in the formation of the new intermolecular forces between

Synoptic link

You first met hydrogen bonding in Topic OZ 7, The ozone hole.

Synoptic link

You met ammonia in Chapter 6, The chemical industry.

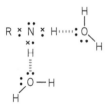
▲ Figure 4 *Hydrogen bonding between amines and water*

the alkyl group and water. As such, the formation of a solution is less energetically feasible for larger-chain amines than smaller-chain amines.

Amines as bases

The lone pair on the nitrogen atom can take part in **dative covalent bonding**. When the electron pair is donated to an H^+, an amine acts as an H^+ acceptor and is a base, for example:

$$CH_3NH_2(aq) + H_2O(l) \rightarrow \quad CH_3NH_3^+(aq) \quad + OH^-(aq)$$
$$\text{methylammonium ion}$$

The bonding in the cation formed has a dative bond, like that in the ammonium ion:

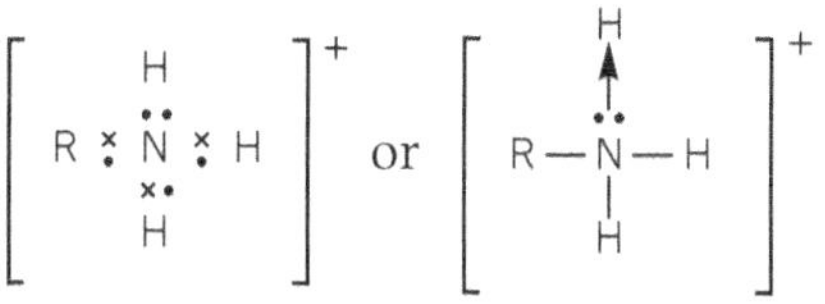

The presence of hydroxide ions makes the solution alkaline, so solutions of amines are alkaline.

Like ammonia, amines also react with acids. The H_3O^+ ions in acidic solutions are more powerful H^+ donors than water. Their reaction with amines goes to completion and the solution therefore loses its strong amine smell, for example:

$$CH_3CH_2NH_2(aq) + H_3O^+(aq) \rightarrow CH_3CH_2NH_3^+(aq) + H_2O(l)$$
$$\text{ethylamine} \qquad\qquad\qquad \text{ethylammonium ion}$$

Amides

Primary amides have the formula:

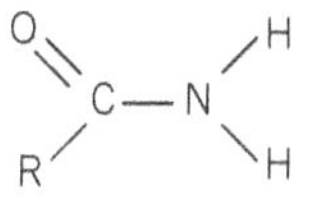

They are derivatives of carboxylic acids made by replacing the –OH group with $-NH_2$.

However, carboxylic acids do not react with ammonia to give amides, so acyl chlorides are used, for example:

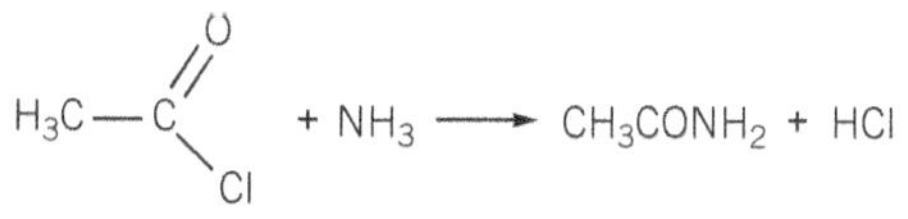

ethanoyl chloride   primary amide

Amines react with acyl chlorides in a similar way, for example:

$$CH_3COCl + RNH_2 \rightarrow CH_3CONHR + HCl$$

This is a condensation reaction – here the small molecule produced is hydrogen chloride, HCl, rather than water.

The product is a secondary amide:

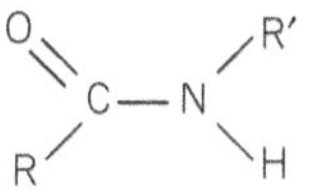

Synoptic link

Enthalpy was covered in Chapter 2, Developing fuels.

Synoptic link

Acids and bases were first introduced in Topic EL 9, How salty?

Synoptic link

Dative covalent bonding and the structure of the ammonium ion was covered in Topic CI 1, Nitrogen chemistry.

Activity PL 2.1

Using simple tests you can find out about the properties and reactions of amines.

Polyamides

By using diamines and dicarboxylic acids, which have reactive groups in two places in their molecules, a polymer chain can be made in which monomer units are linked together by amide groups.

The process is called condensation polymerisation because the individual steps are condensation reactions.

Examples of a diamine and a dicarboxylic acid that can be made to polymerise in this way are:

1,6-diaminohexane $H_2NCH_2CH_2CH_2CH_2CH_2CH_2NH_2$

hexanedioic acid $HOOCCH_2CH_2CH_2CH_2COOH$

The polymer is:

$$\left(\text{NH}-(\text{CH}_2)_6-\text{NH}-\overset{\displaystyle O}{\overset{\|}{\text{C}}}-(\text{CH}_2)_4-\overset{\displaystyle O}{\overset{\|}{\text{C}}}\right)_n$$

The repeating unit of nylon-6,6

Because the group linking the monomer groups together is a secondary amide group, these polymers are called **polyamides**. More usually, though, they are known as nylons.

Activity PL 2.2

In this card activity you can check your ability to name nylons.

The industrial preparation of nylon from a diamine and a dicarboxylic acid is quite slow. It is easier to demonstrate the process in the laboratory if an acyl chloride derivative of the acid is used, for example, 1,6-diaminohexane and decanedioyl dichloride react readily. The equation is:

$n\text{NH}_2(\text{CH}_2)_6\text{NH}_2$ + $n\text{ClCO}(\text{CH}_2)_8\text{COCl} \longrightarrow$ —$\{\text{NH}(\text{CH}_2)_6\text{NHCO}(\text{CH}_2)_8\text{CO}\}_n$— + $2n\text{HCl}$

1,6-diaminohexane decanedioyl chloride nylon-6,10

▲ Figure 5 *Making nylon in the laboratory*

Naming nylons

A nylon is named according to the number of carbon atoms in the monomers. If two monomers are used, then the first digit indicates the number of carbon atoms in the *diamine* and the second digit indicates the number of carbon atoms in the corresponding *dicarboxylic acid*. So nylon-6,6 is made from 1,6-diaminohexane (six carbon atoms) and hexanedioic acid (six carbon atoms). Nylon-6,10 is made from 1,6-diaminohexane (six carbon atoms) and decanedioic acid (10 carbon atoms):

It is also possible to make nylon from a single monomer containing an amine group at one end and an acid group at the other. For example, nylon-6 is —$\{\text{NH}-(\text{CH}_2)_5-\text{CO}\}_n$— and is made from molecules of $H_2N(\text{CH}_2)_5\text{COOH}$.

Acyl chlorides

Acyl chlorides are reactive forms of carboxylic acids. They are named by replacing the -ic acid from the acid name with -yl chloride, for example:

$$H_3C-\overset{\displaystyle O}{\overset{\diagup}{\underset{\diagdown}{\text{C}}}}-\text{Cl}$$

ethanoyl chloride

They react vigorously with amines to form amides:

$$RCOCl \;\; + \;\; R'NH_2 \rightarrow RCONHR' + HCl$$

acyl chloride   amine   amide

They also react vigorously with alcohols to form esters, for example:

$$C_2H_5COCl \;\; + CH_3OH \;\; \rightarrow \;\; C_2H_5COOCH_3 \;\; + HCl$$

propanoyl chloride methanol   methyl propanoate

Activity PL 2.3

This activity allows you to practice naming a range of organic compounds.

Summary questions

1　Write full structural formulae for *one* repeating unit of:
　a　nylon-6,6　　(*1 mark*)　b　nylon-6,10　(*1 mark*)　c　nylon-6　(*1 mark*)

2　Give the systematic names for:
　a　$NH_2CH_2NH_2$　　　　　(*1 mark*)　　b　$CH_3CH_2CH_2NH_2$　(*1 mark*)
　c　$CH_3CH(NH_2)CH_2CH_3$　(*1 mark*)　　d　C_3H_7COCl　　　(*1 mark*)

3　Name the nylons that contain the following repeating units:
　a　$-HN-(CH_2)_6-NH-CO-(CH_2)_6-CO-$　　　　　　　　　(*1 mark*)
　b　$-CO-(CH_2)_7-CO-HN-(CH_2)_9-NH-$　　　　　　　　　(*1 mark*)
　c　$-CO-(CH_2)_2-CO-HN-(CH_2)_4-NH-$　　　　　　　　　(*1 mark*)

4　Write out the repeating units and give the names for the polymers formed from the following molecules:
　a　$HOOC(CH_2)_5NH_2$　　　　　　　　　　　　　　　　(*1 mark*)
　b　$H_2N(CH_2)_5NH_2$ and $HOOC(CH_2)_5COOH$　　　　　(*1 mark*)

5　Nylon-5,10 can be formed by the reaction of a diamine with a diacyl dichloride:
　a　Write down the structures of the two monomers.　(*2 marks*)
　b　What small molecule is lost when the monomers react?　(*1 mark*)
　c　Draw the structure of the repeating unit in nylon-5,10.　(*1 mark*)

6　Draw structures for the products formed when 2-aminopropane reacts with:
　a　water　　(*1 mark*)　　　　b　hydrochloric acid　　(*1 mark*)
　c　ethanoyl chloride　　(*2 marks*)

7　a　Draw the repeating unit of the polymer that could be obtained from 3-aminobenzoic acid.　(*1 mark*)
　b　Suggest why this polymer is not a strong as Kevlar.　(*1 mark*)

8　Caprolactam has the formula:

Under certain conditions it undergoes ring polymerisation to form a nylon.
　a　Name the functional group in caprolactam.　(*1 mark*)
　b　Write an equation showing *n* units of caprolactam forming a nylon. (*1 mark*)
　c　Name the nylon formed.　(*1 mark*)
　d　Classify the polymerisation as addition or condensation, giving a reason.　(*1 mark*)
　e　A cyclic compound called caprolactone undergoes similar polymerisation to form a polyester.
　　Suggest formulae for caprolactone and the polyester it forms. (*2 marks*)

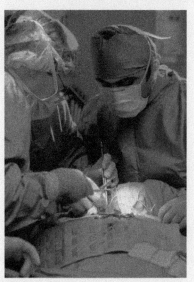

▲ **Figure 1** *Surgeons using stitches which hydrolyse in the body*

Surgical stitches

Polymers made from compounds in crude oil, such as nylons, require heating under reflux with acid or alkali to hydrolyse.

Since the 1980s, threads made from a special polyester have been used by surgeons to stitch together the sides of wounds. The thread is special because it 'disappears' as the wound heals.

The polyester is made from monomers that contain both a hydroxyl group, $-OH$, and a carboxylic acid group, $-COOH$, so only one type of monomer is needed.

The monomer used may be lactic acid (2-hydroxypropanoic acid), glycolic acid (2-hydroxyethanoic acid), or a mixture of the two.

The polymers form strong threads but water in the body slowly hydrolyses the ester bonds. The products of the hydrolysis (the monomers lactic acid or glycolic acid) are non-toxic. Indeed, lactic acid is a normal breakdown product of glucose in the body.

Another use for the degradable polyesters is the controlled delivery of a medicine. The medicine is dispersed throughout a tablet of the polymer, and this is then implanted in a suitable part of the body. The medicine is released at a rate determined by the rate of hydrolysis of the polyester.

The rate of hydrolysis of the polymer is very important. It is affected by the relative molecular mass of the polymer and its crystallinity. Tuning and balancing these features in polyesters made from lactic acid and glycolic acid can achieve a precise rate of hydrolysis.

Chemical ideas: Organic chemistry: modifiers 13.7

Hydrolysis of esters and amides

Hydrolysis of esters

The reverse of esterification corresponds to the breakdown of an ester by water – in other words, it is a hydrolysis.

$$C_2H_5COOCH_3 + H_2O \rightleftharpoons C_2H_5COOH + CH_3OH$$

On their own, water and an ester react very slowly but the process can be sped up by a catalyst.

A catalyst is effective for both directions of a reversible reaction, and sulfuric acid (or any other acid) is used. Excess water displaces the equilibrium position to the right and the yield of products is improved.

Another way of hydrolysing an ester is to add an alkali, such as sodium hydroxide solution. When alkali is used, the hydrolysis does not produce a carboxylic acid but a carboxylate salt. For ethyl ethanoate the reaction is:

$$C_2H_5COOCH_3 + OH^- \rightarrow C_2H_5COO^- + CH_3OH$$

Alkaline hydrolysis (unlike acid hydrolysis) goes to completion and so is usually preferred.

Hydrolysis of amides

The C—N bond in amides also can be broken by **hydrolysis** (reaction with water). The reaction is catalysed by either acid or alkali.

Acid hydrolysis

Acid hydrolysis occurs when an amide is heated with moderately concentrated sulfuric or hydrochloric acid, for example:

Overall equation:

$$CH_3CONH_2 + H_2O + H^+ \rightarrow CH_3COOH + NH_4^+$$

In this example, the hydrolysis of a primary amide is shown. As well as acting as a catalyst, the acid also forms a cation with the ammonia.

If a secondary amide is hydrolysed under similar conditions, an amine salt is formed, for example:

$$CH_3CONHCH_3 + H_2O + H^+ \rightarrow CH_3COOH + CH_3NH_3^+$$

Alkaline hydrolysis

The amide is heated with a moderately concentrated alkali, normally sodium hydroxide, for example:

Overall equation:

$$CH_3CONHR + OH^- \rightarrow CH_3COO^- + RNH_2$$

In this example, the hydrolysis of a secondary amide is shown. Here the alkali reacts with the acid to form a carboxylate ion and the amine is also formed.

> **Activity PL 3.1**
>
> This activity checks your understanding of how the products of hydrolysis of esters and amides depend on the conditions used.

> **Study tip**
>
> There will be a corresponding anion for the amine salt, for example, $CH_3NH_3^+Cl^-$ would be formed if hydrochloric acid was used as the catalyst.

> **Study tip**
>
> There will be a corresponding cation for the carboxylic acid salt, for example, $H_3COO^-Na^+$ would form is sodium hydroxide solution was used.

Summary questions

1 Write equations for:

a the acid hydrolysis of ethyl propanoate (*1 mark*)

b the alkaline hydrolysis of $C_2H_5CONHCH_3$ (*2 marks*)

c the acid hydrolysis of $C_2H_5CONH_2$. (*2 marks*)

2 Name the products of the alkaline hydrolysis of $CH_3CONHC_2H_5$ by sodium hydroxide. (*2 marks*)

3 Draw structures for one repeating unit of the polymers formed by the following:

a lactic acid (*1 mark*)

b glycolic acid (*1 mark*)

4 PHBV is a useful biodegradable polymer:

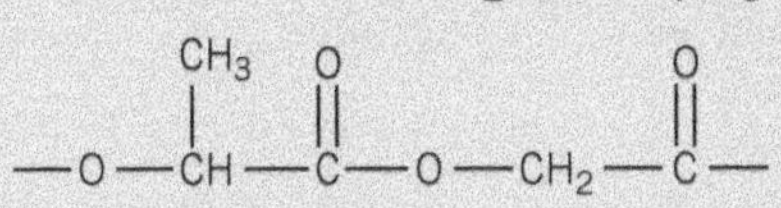

Give the formulae of the hydrolysis products of PHBV and give their systematic names. (*4 marks*)

5 Copy and complete the following reaction schemes by inserting the structures of the missing reactants or products, or by writing the reaction conditions over the arrow: (*6 marks*)

a $H_3C—CH_2—CO—NH_2 \longrightarrow H_3C—CH_2—COOH + NH_4Cl$

b 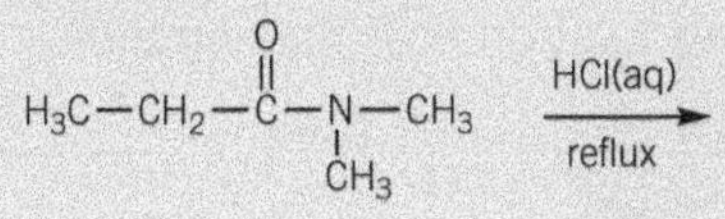

c 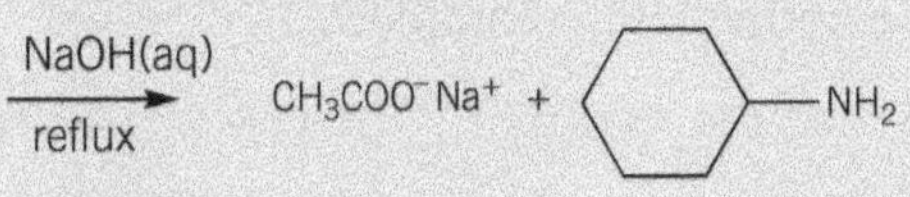

d 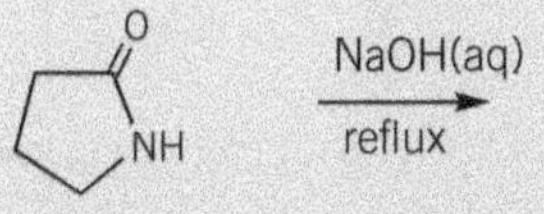

6 Give the formulae of the products of hydrolysis of the following polymers:

a acid hydrolysis of Terylene (*2 marks*)

b acid hydrolysis of nylon-6,10 (*2 marks*)

c alkaline hydrolysis of nylon-6 (*1 mark*)

d acid hydrolysis of Kevlar. (*2 marks*)

7 When ethyl ethanoate is hydrolysed with water enriched with oxygen-18, $H_2{}^{18}O$, the oxygen-18 appears in the ethanoic acid and not in the ethanol:

$$C_2H_5OCOCH_3 + H_2{}^{18}O \rightarrow C_2H_5OH + CH_3CO^{18}OH$$

a Use this information to identify which ester bond is broken during hydrolysis of the ester. (*1 mark*)

b Suggest why it is this bond that is broken in preference to any other bond. (*1 mark*)

c What role do you think the catalyst might have in this reaction? (*1 mark*)

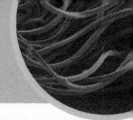

PL 4 What are proteins?

Specification references: PL[a], PL[b], PL[i], PL[q]

What are proteins?

The name protein was coined by Berzelius in 1838. He had little idea what proteins were but he recognised their importance because they were so widespread in living things.

Proteins are big molecules – they are natural polymers with relative molecular masses up to about 100 000. They play a key role in almost every structure and activity of a living organism (Figure 1). That is why they are regarded as some of the most important constituents of our bodies.

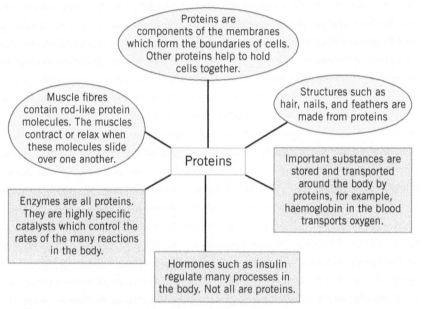

▲ Figure 1 *Proteins perform many functions in our bodies. Fibrous proteins (top, purple boxes) are the major structural materials. Globular proteins (bottom, green boxes) are involved in maintenance and regulation of processes and include hormones and enzymes*

▲ Figure 2 *Hair, fingernails, and feathers are all made up of proteins*

Learning outcomes

Demonstrate and apply knowledge and understanding of:

→ amino acid chemistry – the general structure of amino acids, proteins as condensation polymers formed from amino acid monomers, the formation and hydrolysis of the peptide link between amino acid residues in proteins

→ techniques and procedures for paper chromatography

→ the acid–base properties of amino acids and their existence as zwitterions

→ optical isomerism:

  ● diagrams to represent optical stereoisomers of molecules

  ● the use of the term chiral as applied to a molecule, and identifying carbon atoms that are chiral centres in molecules

  ● enantiomers as non-superimposable mirror image molecules

→ the primary structure of proteins.

▲ Figure 3 *Muscle is made up of protein*

Amino acids – the building blocks of proteins

Insulin is a hormone and is one of the simplest proteins (Figure 4).

The abbreviations in the circles of Figure 4 represent the α-amino acids that have combined to form insulin. There are two short chains of these amino acid residues – the parts of the original amino acid molecules that are joined together to form the protein. Just like some other polymers you have met, proteins are joined together by condensation reactions. All the proteins in the living world are made from just 20 α-amino acids.

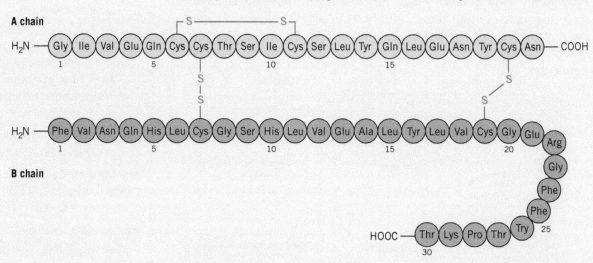

▲ Figure 4 *Human insulin – the two chains are joined together by –S—S– links*

In living organisms, proteins need to be replaced continuously – even in adults who have stopped growing. In some instances this is obvious, like when your hair and nails grow after they have been cut. It is also important that hormones and some enzymes (e.g., the digestive enzymes) are made when needed, and then destroyed once they have done their job so that they do not go on producing their effects after the need for them has passed.

Proteins are replaced from the food humans eat. However, the proteins consumed do not need to be identical to the proteins being replaced – you do not need to eat hair and fingernails for your own to grow. In the body, the proteins are broken down into the amino acids that the proteins are made of. About half of these amino acids are classified as essential amino acids – the human body cannot synthesis these and they have to be taken in from food. Non-essential amino acids can be made from carbohydrates and other amino acids in the body.

The amino acids are then reassembled into the proteins needed in your body. Most of your proteins will be identical to those found in other humans but some will be different. Human proteins are also different from the proteins of other animals. However, in some cases very similar proteins are found, not just throughout the animal kingdom but in plants and microorganisms as well, for example, the enzymes that oxidise glucose in cell metabolism.

All of these millions of proteins are built up from the same small number of amino acids. What makes each protein different is the order in which the amino acids are joined to one another. This is called the primary structure of the protein.

Each amino acid has a different side chain. Table 1 shows the structures of these R-groups and the names of the amino acids, together with their abbreviated symbols.

▼ Table 1 *The 20 amino acids that make up proteins*

Formula	Name and abbreviation	Formula	Name and abbreviation
H_2NCHCO_2H | H	glycine (Gly)	H_2NCHCO_2H | CHOH | CH_3	threonine (Thr)
H_2NCHCO_2H | CH_3	alanine (Ala)	H_2NCHCO_2H | CH_2SH	cysteine (Cys)
H_2NCHCO_2H | $CHCH_3$ | CH_3	valine (Val)	H_2NCHCO_2H | CH_2 | $CONH_2$	asparagine (Asn)
H_2NCHCO_2H | CH_2 | $CH_3(CH_3)_2$	leucine (Leu)	H_2NCHCO_2H | CH_2 | CH_2CONH_2	glutamine (Gln)
H_2NCHCO_2H | CHC_2H_5 | CH_3	isoleucine (Ile)	H_2NCHCO_2H | CH_2—⬡—OH	tyrosine (Tyr)
HN—CHCO_2H H_2C CH_2 CH_2	proline (Pro) (Proline is a secondary amine)	H_2NCHCO_2H | CH_2 | C=CH HN N CH	histidine (His)
H_2NCHCO_2H | CH_2 C CH NH (indole ring)	tryptophan (Try)	H_2NCHCO_2H | $(CH_2)_3$ | NH | $NH=C—NH_2$	arginine (Arg)
H_2NCHCO_2H | CH_2 | CH_2SCH_3	methionine (Met)	H_2NCHCO_2H | $(CH_2)_3$ | CH_2NH_2	lysine (Lys)
H_2NCHCO_2H | CH_2—⬡	phenylalanine (Phe)	H_2NCHCO_2H | CH_2CO_2H	aspartic acid (Asp)
H_2NCHCO_2H | CH_2OH	serine (Ser)	H_2NCHCO_2H | CH_2 | CH_2CO_2H	glutamic acid (Glu)

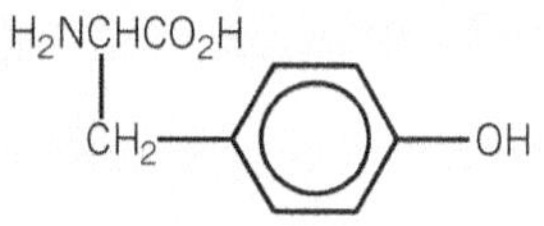

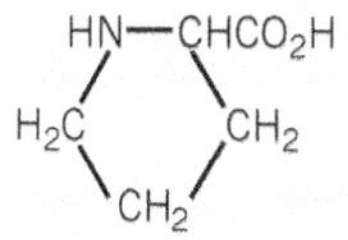

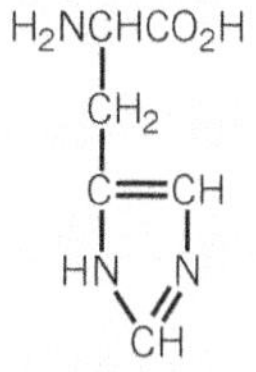

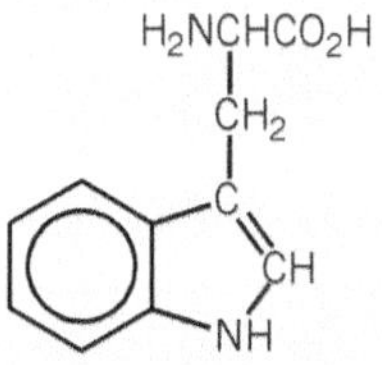

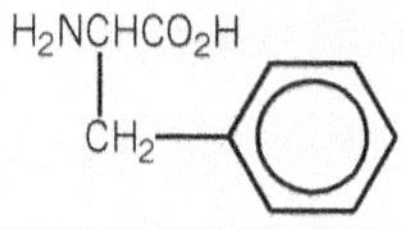

Amino acids, peptides, and proteins

Amino acids

Amino acids contain at least one amino group and one carboxylic acid group. The α-amino acids are particularly important in living systems. The diagram shows the general structure of an α-amino acid.

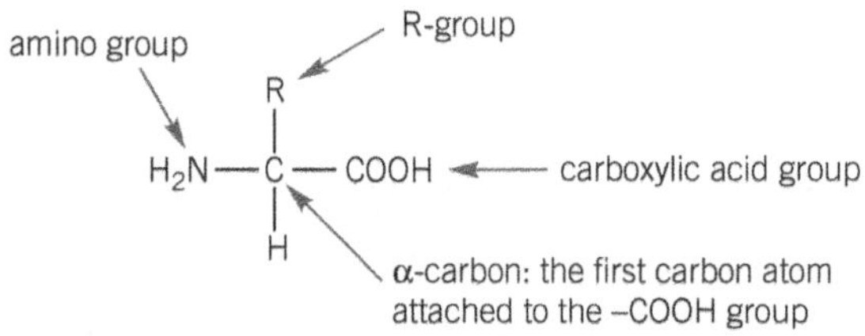

Amino acids are examples of bifunctional compounds – compounds with two functional groups. The properties of bifunctional compounds are sometimes simply the same as the properties of the two separate functional groups. Amino acids show some of the properties of amines (e.g., they form salts with acids) and some of the properties of carboxylic acids (e.g., they form salts with alkalis). However, the functional groups also interact. The proton-donating −COOH and proton-accepting −NH₂ groups can react with one another, forming **zwitterions**. Zwitterions contain both negatively charged groups and positively charged groups:

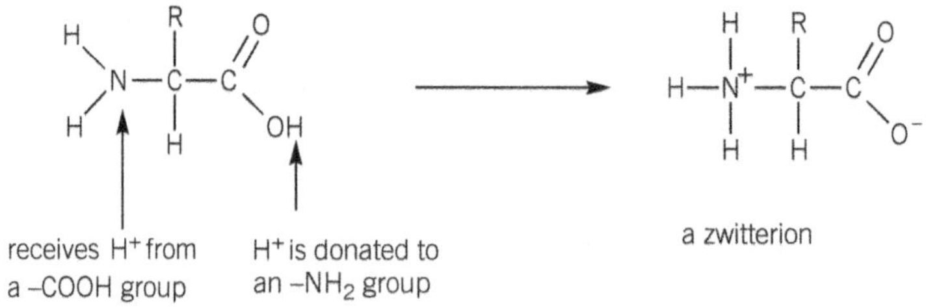

receives H⁺ from a −COOH group H⁺ is donated to an −NH₂ group

a zwitterion

An aqueous solution of an amino acid consists mainly of zwitterions, with very few molecules containing the un-ionised groups. Amino acids are very soluble in water, because they are effectively ionic.

Unless there is an extra −COOH or −NH₂ group in the R group of the molecule (as there is in some naturally occurring amino acids), they are neutral in aqueous solution.

Adding small quantities of acid or alkali to an amino acid solution causes little change to the pH because the zwitterions buffer the effect of the addition:

$$HO^- + H_3N^+-CHRCOO^- \longrightarrow H_2O + H_2N-CHRCOO^-$$
$$H_3O^+ + H_3N^+-CHRCOO^- \longrightarrow H_2O + H_3N^+-CHRCOOH$$

So, amino acids exist in three different ionic forms, depending on the pH of the solution they are in:

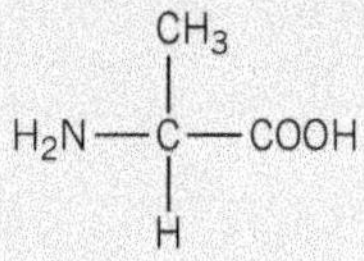

$H_3N^+-CHR-COOH$	$H_3N^+-CHR-COO^-$	$H_2N-CHR-COO^-$
in acid solution	in neutral solution	in alkaline solution

Solutions that can withstand the addition of small amounts of acid or alkali are called **buffer** solutions.

Optical isomerism

Since amino acids usually have four different groups around a carbon atom, they show **optical isomerism**.

Optical isomerism arises because of the different ways in which you can arrange four groups around a carbon atom. It is a type of **stereoisomerism**.

The four single bonds around a carbon atom are arranged tetrahedrally. When four different atoms or groups are attached to these four bonds, the molecule can exist in two isomeric forms. These two forms for the amino acid alanine (2-aminopropanoic acid) are shown in Figure 5.

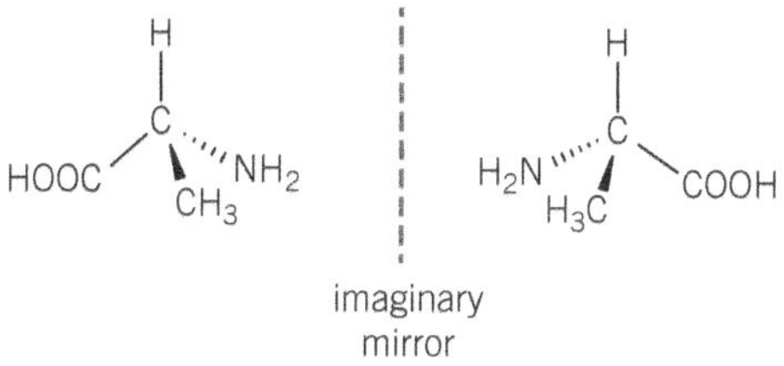

imaginary mirror

▲ Figure 5 *Alanine has two forms*

If you rotate the right-hand image so that the –COOH group is in the same position as the –COOH group in the left hand image, you will see that the –NH_2 group and –CH_3 group are not in the same places as in the left hand image (Figure 6). The right-hand structure is the mirror image of the left-hand structure.

All molecules have mirror images but they do not all exist as two isomers. For example, glycine (aminoethanoic acid) has only one form (Figure 7). If you rotate the right-hand image so that the –NH_2 group is in the same position as in the left-hand image, the two molecules are identical.

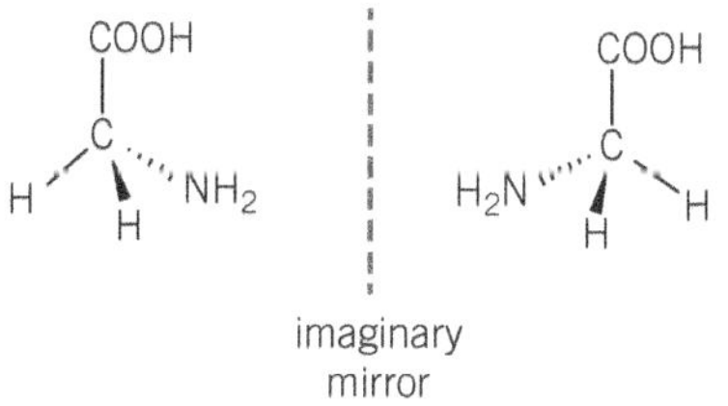

imaginary mirror

▲ Figure 7 *Glycine has only one form*

The two forms of alanine are **non-superimposable**. The only way you can make them superimpose is to break the C—NH_2 and C—COOH bonds and swap the groups round. Breaking and reforming bonds corresponds to a chemical reaction, and in such a reaction a compound is turned into a new compound. In this example, the new compound is a different isomer from the original one.

Synoptic link

You will learn more about buffer solutions in Chapter 8, Oceans.

Synoptic link

Isomerism and stereoisomerism was introduced in Topic DF 9, What do the molecules look like?

Activity PL 4.2

You can use modelling to reinforce your ideas about shapes of molecules and optical isomerism.

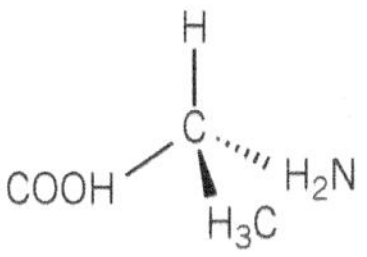

▲ Figure 6 *When you rotate the right hand image in Figure 5, you can see that it is not the same as the left hand image*

Study tip

You may need to build models of these two structures to convince yourself that they are different. If you have built models, then you can use a mirror to prove this to yourself.

Study tip

The mirror images of alanine are non-superimposable.

The mirror images of glycine are superimposable.

Synoptic link

Determining the shapes of molecules was covered in Topic EL 5, The molecules of life.

Enantiomers

Molecules such as the two forms of alanine are called optical isomers or **enantiomers**. You do not always have to build models to find them – whenever a molecule contains a carbon atom that is surrounded by four different atoms or groups, there will be optical isomerism.

Molecules that are non-superimposable on their mirror images are called **chiral molecules**. A carbon atom that is bonded to four different groups is called a chiral centre. There are lots of other chiral things in the world, for example, your shoes and your hands. If you look at the reflection of your right hand in a mirror, the mirror image is superimposable on your left hand but not on your right hand.

The proteins in our bodies are built up from only one enantiomer of each amino acid. These are called the L-enantiomers and they have the same arrangement of the four groups around the central carbon atom.

How do enantiomers differ?

Enantiomers behave identically in ordinary test-tube chemical reactions. Most of their physical properties, such as melting point, density, and solubility, are also the same. But enantiomers behave differently in the presence of other chiral molecules and they sometimes smell or taste different.

▲ Figure 8 *The different smell of oranges and lemons is caused by the two enantiomers of limonene*

Making peptides and proteins

When an $-NH_2$ group reacts with the $-COOH$ group in a carboxylic acid, a secondary amide group, $-CONH-$, is formed. In this process, a molecule of water is eliminated, so it is a condensation reaction.

When two amino acids join together in this way, the secondary amide group formed is called a **peptide link**. The dipeptide formed from glycine (Gly) and alanine (Ala) is:

$$H_2N-CH_2-C-N-CH-COOH$$

peptide link

This dipeptide is abbreviated to GlyAla – the amino acid with the free $-NH_2$ group is written first.

The amino acids that make up some proteins found in humans, and their abbreviations, are shown in Table 1. You do not need to learn these but you will need to consult the table.

From a dipeptide, scientists can make a tripeptide by adding another α-amino acid molecule. A polypeptide can contain up to 40 amino acid residues.

Proteins are naturally occurring condensation polymers made from amino acid monomers joined by peptide links. A protein contains more than about 40 amino acid residues, although the distinction between a polypeptide and a protein is an arbitrary one (the term residue is used for an α-amino acid which has lost the elements of water in forming a peptide or protein).

Insulin, for example, is made from 51 amino acid monomers, 14 of them being different. In fact, all proteins are constructed from just 20 amino acids. What makes each protein different is the order in which the amino acids are joined to one another – this is called the primary structure. Each protein has its own unique primary structure.

Hydrolysis of proteins

The peptide link in peptides and proteins can be hydrolysed to release the individual amino acids. Peptides are secondary amides, and hydrolysis can be carried out by heating with moderately concentrated acid or alkali.

The breakdown of proteins in the laboratory is routinely carried out by boiling with moderately concentrated hydrochloric acid to hydrolyse the amide C—N bonds. In living organisms, the hydrolysis of proteins is catalysed by enzymes rather than acid or alkali.

Paper chromatography can be used to identify the individual amino acids present in a peptide. The peptide is hydrolysed under reflux and the product compared to known samples of pure amino acids using chromatography.

Activity PL 4.4

In this practical activity you hydrolyse aspartame and use the resultant paper chromatogram to identify amino acids.

Synoptic link

Paper chromatography and reflux are covered more in Techniques and procedures.

Summary questions

1 The general structure of an α-amino acid is:

$$H_2N—\overset{\overset{\displaystyle R}{|}}{\underset{\underset{\displaystyle H}{|}}{C}}—COOH$$

Draw the structure of the zwitterion present in solutions of the amino acid. (*1 mark*)

2 A protein chain contains the following section:

$$—NH—\overset{\overset{\displaystyle CH_3}{|}}{CH}—CO—NH—\overset{\overset{\displaystyle H_3C—CH}{\overset{\displaystyle |}{}}}{\underset{}{CH}}—CO—$$

Draw the structures of the amino acids obtained from the hydrolysis of this section of the protein chain. (*2 marks*)

3 a What structural feature must a molecule have in order to form optical isomers (enantiomers)? (*1 mark*)

 b Which of the following compounds can have optical isomers? Give reasons.
  i CH_2Cl_2
  ii CH_2ClBr
  iii $CHClBrI$ (*2 marks*)

4 a Draw the structural formula of 3-methylhexane. (*1 mark*)

 b Identify the chiral carbon atom in your structural formula and mark it with an asterisk (*). (*1 mark*) c   Use three-dimensional diagrams to show the structures of the two optical isomers of this compound and how they are related.   (2 marks)

d   Draw the structural formula of 2-bromobutane.   (1 mark)

e   Use three-dimensional diagrams to show the structures of the two optical isomers of this compound.   (2 marks)

5   Limonene and carvone are examples of compounds for which both enantiomers occur naturally. One enantiomer of limonene smells of oranges, the other of lemons. Carvone can smell like caraway seeds or like spearmint.

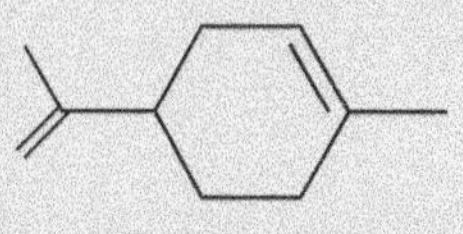

limonene 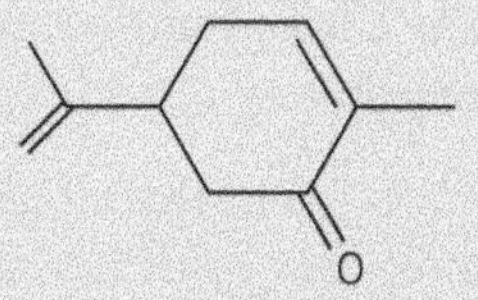   carvone

a   Draw the full structural formulae of limonene and carvone. Use an asterisk to show the chiral carbon atom in each molecule.   (2 marks)

b   Draw the structure of the compound which is formed when one mole of limonene undergoes addition with two moles of hydrogen molecules.   (1 mark)

c   Does the product of the reaction in part b exhibit optical isomerism? Explain your answer.   (2 marks)

6   Write equations for the following reactions. Look up the formulae of the amino acids in Table 1:

a   alanine with hydrochloric acid   (1 mark)

b   serine with sodium hydroxide solution   (1 mark)

c   lysine with excess hydrochloric acid   (2 marks)

d   aspartic acid with excess sodium hydroxide solution.   (2 marks)

7   Write equations for the following reactions, showing organic substances as structural formulae:

a   The hydrolysis of the dipeptide AlaGly using moderately concentrated hydrochloric acid.   (2 marks)

b   The hydrolysis of the dipeptide SerGly using dilute sodium hydroxide solution.   (2 marks)

c   Describe how you would use paper chromatography to investigate the amino acids formed when a dipeptide is hydrolysed.   (5 marks)

8   Tartaric acid (2,3-dihydroxybutanedioc acid) has more than one chiral centre.

a   Draw the structure of tartaric acid, marking the chiral centres with asterisks.   (2 marks)

b   There are three stereoisomers all with the structural formula of tartaric acid, two of which are enantiomers. Suggest reasons for this.   (2 marks)

c   A structural isomer of tartaric acid, with the same type and number of functional groups, has no chiral centres. Suggest a structure for this isomer.   (1 mark)

Folded chains

Scientists need to know about more than just a primary structure before they understand how a protein works. One thing they now know is that most proteins have a precise shape that arises from the folding together of the chains. The action of many proteins is critically dependent on this shape.

As long as different molecules fold to the same shape, they may have similar actions. Chain folding gives proteins their three-dimensional shape – it also places chemical groupings in positions where they can bond most effectively.

The chains in a protein are often folded or twisted in a regular manner as a result of hydrogen bonding. This results in the formation of a helix or a sheet, which is called the secondary structure of a protein. The figures show ribbon molecules with the parts colour-coded. The proteins are not really these colours. The ribbon model of insulin shows helical sections quite clearly (Figure 1). The more complicated molecule of ATPase (Figure 2) shows sheets (blue) and helices (red). Keratin, the structural protein that makes our hair, consists mainly of helices (Figure 3).

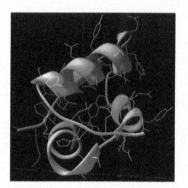

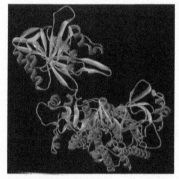

▲ Figure 1 *A ribbon model of insulin* ▲ Figure 2 *A ribbon model of the enzyme ATPase*

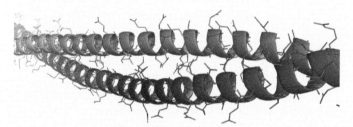

▲ Figure 3 *A ribbon model of keratin*

The chains may then fold up further. The overall shape of a protein is called its tertiary structure and this is stabilised by intermolecular and other bonds between the R-groups of the amino acids in the chains. The overall folding of secondary structures is clear from the ribbon molecules shown. A 'space-filling' model of insulin is shown for comparison (Figure 4).

▲ Figure 4 *A molecular model of an insulin molecule*

Protein structure

There are three levels of protein structure:

- primary structure – the order of the amino acid residues
- secondary structure – the coiling of parts of the chain into a helix or the formation of a region of sheet
- tertiary structure – the folding of the secondary structure.

Secondary structures

The helices and sheets are held together by hydrogen bonds between –NH groups on one peptide link and –C=O groups on another peptide link (Figure 5).

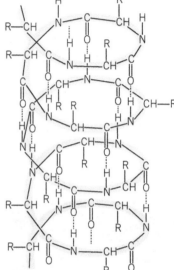

▲ Figure 5 *A sheet and a helix from a secondary protein structure*

Tertiary structures

Tertiary structures are held together by intermolecular bonds, ionic bonds, and covalent bonds.

Instantaneous dipole–induced dipole bonds form between non-polar side chains on amino acids such as phenylalanine and leucine. The centres of protein molecules tend to contain amino acids such as these, so that the non-polar groups do not interfere with the hydrogen bonding between the surrounding water molecules.

Hydrogen bonds form between polar side chains (e.g., $-CH_2OH$ in serine and $-CH_2CONH_2$ in asparagine). If amino acids with polar side chains are situated on the outside of proteins, then hydrogen bonds can also form to water molecules surrounding the protein, so allowing the proteins to dissolve.

Ionic bonds form between ionisable side chains, such as $-CH_2COO^-$ in aspartic acid and $-CH_2CH_2CH_2CH_2NH_3^+$ in lysine.

Covalent bonds form, for example, where the –SH groups on neighbouring cysteine residues are oxidised to form –S—S– links.

Synoptic link

You learnt about instantaneous dipole–induced dipole bonds and hydrogen bonds in Topic OZ 6, The CFC story, and Topic OZ 7, The ozone hole.

The shape of the tertiary structure varies with the function of the protein.

Structural proteins such as muscles and hair are fibrous (long and thin) and consist mainly of helices.

Proteins that control metabolism (e.g., enzymes and hormones) are globular, for example, insulin, and have both sheets and helices.

> **Activity PL 5.1**
>
> In this modelling activity you will check your understanding primary, secondary, and tertiary structures in proteins.

Summary questions

1. Use Table 1 in Topic PL 4 to answer this question.
   a. Illustrate the formation of a hydrogen bond between two different R-groups on different protein chains (other than serine and asparagine). *(2 marks)*
   b. Name two amino acids (other than aspartic acid and lysine) that would form ionisable side chains that would attract one another. Draw the ionised side chains. *(2 marks)*
   c. Name two amino acids with non-polar side chains (apart from phenylalanine and leucine). Describe the intermolecular bonds that might form between these side chains. *(2 marks)*

2. Describe how the amino acids are joined in:
   a. a helical structure *(2 marks)*
   b. a sheet *(1 mark)*
   c. the tertiary structure of a protein. *(2 marks)*

3. Washing and combing straightens hair. Perming involves chemically reducing the disulfide bridges between keratin helices and re-making them by oxidation.
   a. Suggest the bonds that are broken by washing and why these are different from those broken in perming. *(2 marks)*
   b. Describe the chemical process that occurs when disulfide bridges are reduced and what happens to the proteins in hair as a result. *(2 marks)*
   c. Describe the chemical process that occurs when the disulfide bridges are re-made and the effect this has on the hair. *(1 mark)*

4. A ribbon model of a protein is shown below:

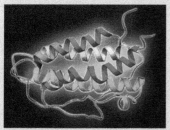

   a. Suggest the function of this protein, giving a reason. *(2 marks)*
   b. Which parts of this model can be described as the secondary structure? *(1 mark)*
   c. Which part can be described as the tertiary structure? *(1 mark)*
   d. Describe the chemical structure of the green, yellow, red, and blue twisted ribbons. *(1 mark)*
   e. Suggest what links these ribbons together. *(1 mark)*

PL 6 Enzymes

Specification references: PL(f), PL(g)

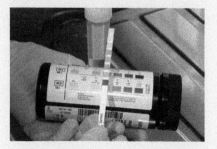

▲ Figure 1 Testing for glucose in urine

▲ Figure 2 Enzymes are important in the production of bread, cheese, and wine

What are enzymes?

People with diabetes are unable to control the level of glucose in their blood. When glucose builds up to a high level in the blood, it is excreted in the urine. They can buy reagent strips from a pharmacy to test for glucose in their urine. The strips contain an enzyme that produces a colour reaction when glucose is present. The fresh reagent strip has a small coloured square at one end that turns a different colour in the presence of glucose (Figure 1). The square is impregnated with four reagents:

- glucose oxidase – an enzyme that catalyses the reaction:
  glucose + oxygen → gluconic acid + hydrogen peroxide

- an indicator – this is present in its reduced form (XH_2) which is colourless, but it turns into a coloured form (X) when it is oxidised

- peroxidase – an enzyme that catalyses the oxidation of the indicator by hydrogen peroxide

- a buffer – a mixture of chemicals that keeps the reagents at a fixed pH during the test.

The manufacturer's instruction sheet recommends storing the test strips below 30 °C but not in a refrigerator.

Enzymes are proteins. The urine test strips illustrate that enzymes are catalysts that are specific to certain chemicals (in this case, glucose). They are also sensitive to pH and temperature.

Enzymes at work

The use of an enzyme in a diabetes testing strip to detect glucose is just one of many applications of enzymes. Many medical diagnostic kits are based on the use of enzymes.

However, medical applications use only small quantities of enzymes. Larger amounts are used in the food industry and in washing powders. Almost all of these are hydrolases – enzymes that hydrolyse fats, proteins, or carbohydrates.

In the food processing industry, two of the major uses of enzymes are:

- producing glucose syrup (used as a sweetener in food products) by breaking down starch with enzymes such as α-amylase

- making cheese using rennet enzymes. These break down the milk protein casein, and cause the separation of the curds (solid) from the whey (liquid).

Other uses include baking, brewing, and processing fruit in order to make fruit juices.

Many biological washing powders contain one or more enzymes to assist in the removal of stains. The enzyme is usually a protease to hydrolyse proteins in blood and food but a lipase may also be added to break down fats. More recently, cellulases have been added too – these break down the tiny surface fibres that give older clothes a fluffy, dull look. Protein engineering is being used to make enzymes that are more stable in hot washes, or to create a wider range of active enzymes that will do their job at lower temperatures.

Enzymes are finding applications in wider areas of waste treatment, for example, an enzyme is being used to destroy cyanide ions that are left over after gold extraction or after the production of some polymers. Enzymes can also be used to help break up oil spillages. They also find their way into fashion where they are used, for example, to give the stonewashed appearance in jeans.

Enzymes are often used because they are considered more environmentally friendly than other catalysts. They allow reactions to occur at lower temperatures, conserving energy, and they often reduce the number of steps needed to make a product. This is important because it increases the atom economy of the process.

> **Synoptic link**
>
> Atom economy was covered in Topic ES 6, Hydrogen chloride in industry and the laboratory.

Chemical ideas: Rates of reaction 10.8

Enzymes

Active sites

You can use the theory of the active site to explain why enzymes show their characteristics. Enzymes are:

* catalysts
* highly specific, for example, the test strips work only with glucose – they give no response with other sugars
* sensitive to pH – many work best at a particular pH and become inactive (denatured) if the pH becomes too acidic or too alkaline
* sensitive to temperature – many enzymes work best at temperatures close to body temperature and most are destroyed above 60–70 °C
* subject to competitive inhibition.

Enzymes have a precise tertiary structure. There is a cleft in the enzyme surface formed by the way the protein chain folds. Within the cleft are chemical groups, some of the R-groups on the amino acid residues, that bind molecules and possibly react with them. This cleft is known as the **active site** and it is where the catalysis takes place.

Figure 3 shows a space-filling model of the enzyme lysozyme, which catalyses the breakdown of the cell walls of bacteria to treat bacterial infections. The cleft for the active site (on the left) can be clearly seen.

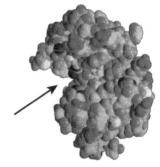

▲ Figure 3 *Lysozyme, showing the cleft for the active site*

> **Synoptic link**
>
> You learnt about catalysis in Chapter 2, Developing fuels, and Chapter 4, The ozone story.

Catalysis

The shape of the active site is tailored for the **substrate** molecules to fit into. Some of the R-groups on the amino acid residues bind the substrate. The bonds that bind the substrate to the active site have to be weak so that the binding can be readily reversed when the products need to leave the active site after the reaction. The bonds are usually hydrogen bonds or interactions between ionic groups. The binding may cause other bonds within the substrate to weaken, or it may alter the shape of the substrate so that atoms are brought into contact to help them to react.

After the reaction, the product leaves the enzyme, which is then free to start again with another molecule of substrate (Figure 4).

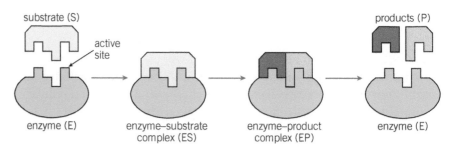

▲ Figure 4 *A model of enzyme catalysis*

Enzymes, like other catalysts, provide a route of lower activation enthalpy (Figure 5). When the activation enthalpy is lower, the reaction takes place more quickly.

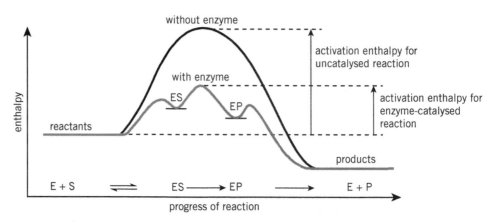

▲ Figure 5 *How an enzyme provides a route of lower activation enthalpy*

Synoptic link

You learnt about rates of reaction, orders of reactions, and the rate-determining step in Chapter 6, The chemical industry.

Enzyme kinetics

Enzymes are usually present only in small amounts and their relative molecular masses are very large, so their molar concentrations are very low. If the substrate concentration is high enough, all the enzyme active sites will have substrate molecules bound to them (the enzyme–substrate complex is written as ES). If the substrate concentration is now increased further, no more enzyme–substrate complexes can be formed, and the rate at which substrate molecules pass through the reaction pathway and change into products remains constant. In this situation the reaction rate does not depend on the substrate concentration – the reaction is zero order with respect to substrate. The rate-determining step is EP → E + P.

When the substrate concentration is low enough, not all the enzyme active sites will have a substrate molecule bound to them. The overall reaction rate will now depend on how frequently enzymes encounter substrates, which will depend on how much substrate there is – if there is twice as much substrate, there will be twice as many encounters. So the reaction is now first order with respect to the substrate. The rate-determining step is $E + S \rightarrow ES$. The graph in Figure 6 indicates the variation of the kinetics of an enzyme reaction with respect to substrate concentration.

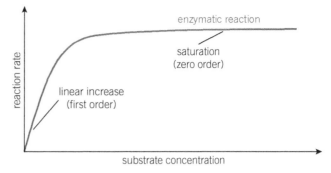

▲ Figure 6 *Graph showing how the initial rate of a typical enzyme reaction varies with substrate concentration*

Inhibitors

Molecules that fit on to the active site but cannot be catalysed are called **competitive inhibitors**. They compete with the substrate molecules for active sites but, once there, do not react further.

For example, the enzyme succinate dehydrogenase catalyses one of an important series of reactions in our bodies by which glucose is oxidised to carbon dioxide and water. The dehydrogenation reaction is:

$$HOOCCH_2CH_2COOH \rightarrow HOOCCH{=}CHCOOH + H_2$$

succinic acid
(butanedioic acid)

Propanedioic acid, $HOOCCH_2COOH$, is found to be a competitive inhibitor. It fits into the active site of the enzyme (because of the two carboxylic acid groups) but it cannot lose hydrogen in the same way as succinic acid because it does not have two $-CH_2-$ groups, therefore remains bound to the active site preventing succinic acid molecules entering the active sites.

Such inhibitor studies are used by biochemists to study the nature of the active site of enzymes. Here, the two carboxylic acid groups are clearly important for the substrate to bind to the active site.

Specificity

Inorganic catalysts will often catalyse a variety of similar reactions. Enzymes show much more **specificity** in which substrate they will catalyse. Enzymes are so specific because they have a precise tertiary structure that exactly matches the structure of the substrate. It is an example of molecular recognition. The active site is usually three-dimensional, so sometimes one stereoisomer of a molecule will fit when another will not. This again helps biochemists find out more about the nature of enzyme active sites.

Other factors that affect the performance of enzymes

Many industrial catalysts consist of inorganic substances that are relatively unaffected by changes in temperature and pH. However, the complex organic structure of enzymes means they are much more sensitive to such conditions.

Activity PL 6.1

In this data-handling exercise you can check your understanding of enzyme kinetics.

pH

An enzyme's active site usually contains ionisable groups and the enzyme's action will be affected by a change in pH. For example, if there is a –COOH group that acts by donating an H^+ to the substrate, raising the pH will turn it into $-COO^-$ and the enzyme will not be able to function correctly. Figure 7 shows a graph of enzyme activity plotted against pH. You can see that an enzyme has an optimum pH at which it works best. This is usually around pH 7 but do not forget that stomach bacteria, for example, work at much lower pH values. Even small differences from the optimum can reduce the activity, because the ionisation of –COOH and $-NH_2$ groups is changed by these small differences.

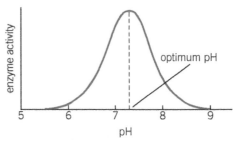

▲ Figure 7 *Graph showing how the rate of a typical enzyme reaction varies with pH*

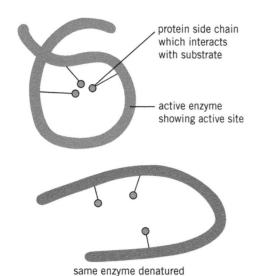

▲ Figure 8 *The shape of an enzyme is lost when it is denatured and its active site is destroyed*

An enzyme will also become inactive if its shape is destroyed. Figure 8 illustrates how the active site of an enzyme can be made up from the side chains of amino acids in different parts of the protein molecule. They are held close together by the enzyme's tertiary structure.

If the tertiary structure is broken, the enzyme loses its shape and the side chains are no longer close together. The active site is destroyed and the enzyme is said to be **denatured**. This happens at pH values that are further from the optimum, since ionic bonds holding the tertiary structure in shape will be broken.

Temperature

Figure 9 shows enzyme activity plotted against temperature. You can see that the enzyme activity rises with temperature at first, as for any other reaction. This is because more molecules have enough energy to collide with their combined energy being greater than the activation enthalpy.

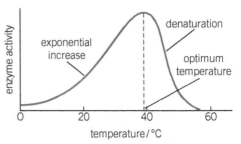

▲ Figure 9 *Graph showing how the rate of an enzyme reaction varies with temperature*

Some of the bonds holding the tertiary structure together are weak dipole–dipole bonds and hydrogen bonds. These can be broken easily by raising the temperature, which causes them to vibrate more vigorously and weaken or break.

At higher temperatures the enzyme is denatured and its activity falls. Enzyme systems have evolved to operate at close to their optimum. For example, most human enzymes have an optimum temperature between 35 and 40 °C – body temperature is 37 °C.

Summary questions

1 When vegetables are cooked, some vitamins are lost by enzyme action. At high temperatures, these enzymes have less time in which to act during cooking.
  a Suggest why the enzymes have less time to act during cooking at high temperatures. (*1 mark*)
  b Explain your answer to part a in terms of enzyme structure, and name the bonds involved. (*2 marks*)
  c Sometimes vegetables are blanched with boiling water before cooking at low temperatures. Suggest the effect of this on vitamin loss. (*2 marks*)

2 Pickling in vinegar is an ancient method of preserving food. Vinegar is acidic and affects the enzymes of food spoilage bacteria:
  a Explain the effect of acid on the enzymes. (*1 mark*)
  b Explain your answer to part a in terms of enzyme structure, and name the bonds involved. (*2 marks*)

3 This question refers to the dehydrogenation of succinic acid.
  a Name the product of dehydrogenation of butanedioic acid. (*1 mark*)
  b This product exists as two *E/Z* isomers. Draw the two structures and label them *E* and *Z*. (*2 marks*)
  c Write the structure of pentanedioic acid. Consider whether succinate dehydrogenase would catalyse the dehydrogenation of pentanedioic acid, and what the product might be. (*3 marks*)

4 When a slice of apple is exposed to air, it turns brown quickly. This is because the enzyme o-diphenol oxidase catalyses the oxidation of phenols in the apple to dark-coloured products. The first stage in the reaction is to convert the catechol to o-quinone:

catechol   o-quinone

Suggest why the conversion of catechol to o-quinone is significantly reduced when 3-hydroxybenzoic acid is added into a mixture of catechol and o-diphenol oxidase:

3-hydroxybenzoic acid   (*2 marks*)

5 *Canavalia ensiformis* (jack bean) is a common bean in America. It contains the enzyme urease. Urea, $(NH_2)_2CO$, is decomposed into ammonia and carbon dioxide in the presence of water and urease.
  a Write a balanced equation for the reaction. (*1 mark*)
  b Draw a line graph to show the change in rate with urea concentration. (*1 mark*)
  c Explain the shape of the graph in terms of the mechanism of the enzyme reaction. (*4 marks*)
  d Suggest an experimental method you could use to follow this reaction. (*2 marks*)

Some medicines work by inhibiting the action of enzymes. A good example is captopril which is widely used to treat high blood pressure. If high blood pressure is left untreated it can cause strokes or heart attacks.

An enzyme called ACE is known to convert an inactive 10-amino-acid peptide in the body called angiotensin I into an eight-amino-acid peptide called angiotensin II. This second peptide causes high blood pressure if too much is produced, so ACE inhibitors such as captopril make useful medicines to control high blood pressure.

Angiotensin I is thought to fit into the ACE active site (Figure 2). Three interactions, labelled 1, 2, and 3, between the three amino acids at the –COOH end of the peptide angiotensin I hold it in place. Hydrolysis occurs as shown, to release a dipeptide and form angiotensin II.

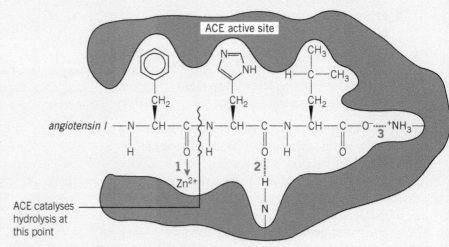

▲ **Figure 2** *Angiotensin I fitting into the active site of ACE*

Captopril is thought to fit into the ACE active site as shown in Figure 3:

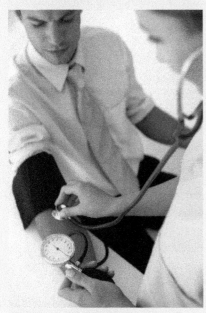

▲ **Figure 1** *Measuring a patient's blood pressure*

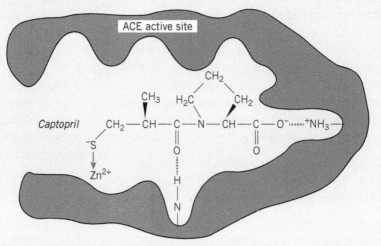

▲ **Figure 3** *Captopril in the ACE active site*

You will see that captopril makes the same three interactions with the active site but has no peptide link to be hydrolysed. So it is a competitive inhibitor – it blocks the enzyme from accepting substrate molecules.

A more recent medicine zofrenopril has the same pharmacophore.

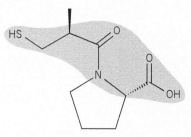

▲ Figure 5 *The active part of captopril, called the pharmacophore*

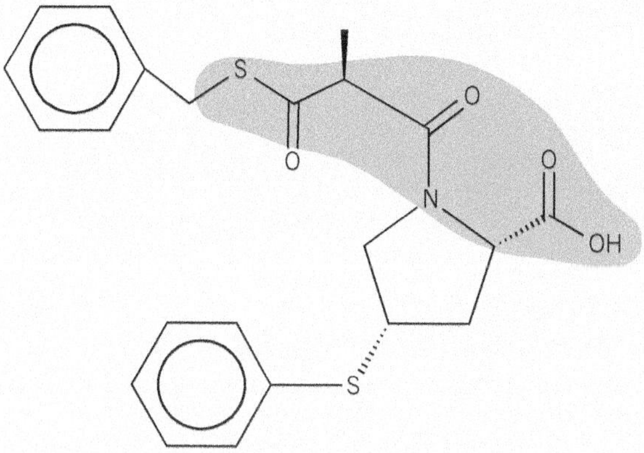

▲ Figure 4 *Zofrenopril has the same pharmacophore as captopril*

Its structure has been modified to remove the –SH group from captopril, which gave it unpleasant side-effects.

Chemical ideas: Structure and bonding 5.6

Molecular recognition

Molecular recognition refers to the ways in which molecules interact in terms of intermolecular and other non-covalent bonds.

Examples you have met have all been to do with substrates and inhibitors binding to enzyme active sites. Other molecules bind in the same way to receptor sites in the body and control how it works.

The pharmacophore of a medicine is the part of its molecule that gives that medicine, and others like it, its particular pharmacological effect. Pharmacophores:

- can be modified by changing functional groups attached to them to make the medicine more effective or to reduce side-effects
- interact with receptor sites by forming weak interactions, such as hydrogen bonding, metal coordination, ionic bonds, and dipole–dipole bonds
- fit into receptor sites with the correct size, shape, and orientation and with which they can form the correct weak bonds.

The example of captopril illustrates all of these points.

Study tip

You do not need to learn the action of captopril. However, you do need to have an understanding of molecular recognition and apply the concepts to a range of examples.

Synoptic link

Intermolecular bonds, including hydrogen bonding, were introduced in Topic OZ 6, The CFC story, and Topic OZ 7, The zone hole.

Summary questions

1 Use Table 1 in Topic PL 4 to identify the two amino acids at the right-hand end of the angiotensin I molecule (Figure 2).
  Write down the primary structure of the dipeptide that is produced when angiotensin I is hydrolysed by the ACE enzyme. *(2 marks)*

2 Captopril is a much more active ACE inhibitor than its isomer, with the methyl group in the other configuration.
  a What does 'the methyl group in the other configuration' mean? Draw a structure for the other isomer of captopril based on Figure 4. *(2 marks)*
  b Suggest why this isomer is less effective. *(1 mark)*
  c What other group in captopril is like the methyl group in being able to exist in another configuration in another isomer? *(1 mark)*

3 Enalapril is another ACE inhibitor:

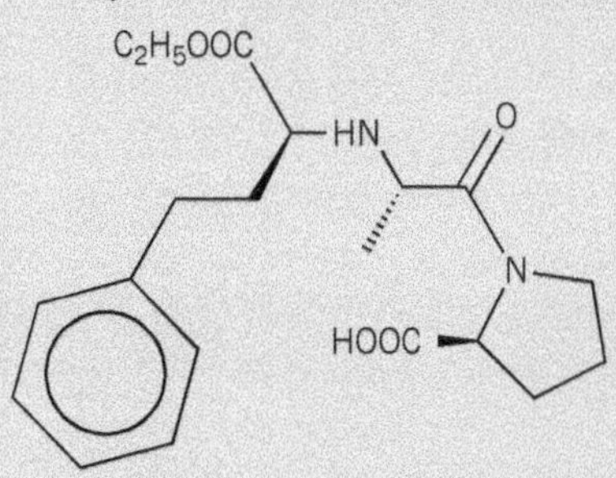

  a Draw the enalapril structure and circle the pharmacophore. *(1 mark)*
  b What do you understand by the term pharmacophore? *(1 mark)*

PL 8 The thread of life

DNA for life

Proteins are important molecules in the body but how do cells make proteins?

As a chemist you would need three things before you could synthesise a protein in the laboratory:

1 a set of instructions for the protein – in other words, something that told you its primary structure

2 supplies of the pure amino acids ready for you to use in the appropriate steps

3 a way of making the amino and carboxylic acid groups react with one another more easily.

If you look at how a cell makes its proteins, you can see a close parallel with what a chemist would do.

The instructions specifying the primary structure of a protein are carried by molecules of deoxyribonucleic acid (DNA). The building blocks for DNA are deoxyribose (a sugar), phosphate groups, and four different bases.

DNA consists of two strands twisted in a helical form with the sugar–phosphate backbone around the outside, held together by hydrogen bonding involving the internal base pairs. It is the bases that form the code for protein synthesis. DNA is a nucleic acid, not a protein. Although many proteins have helical secondary structures, these are not double helices.

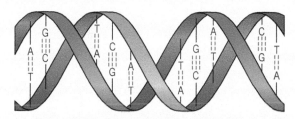

▲ Figure 1 *Illustration of a portion of a DNA molecule showing the double helix. You can think of the sugar–phosphate backbone as the sides of a twisted ladder with the rungs formed by pairs of bases*

DNA passes on the code to RNA, which codes for the amino acids in a protein.

Making use of DNA

DNA fingerprinting

No two people (except genetically identical twins) share the same DNA sequence. DNA fingerprinting is regularly used to help to investigate very serious crimes.

A solution of an enzyme is added to the sample (e.g., a trace of blood, semen, or skin) that 'cuts' the DNA at particular sites into a specific pattern of fragments. Fragments from genes are not used in this process because they do not vary much between individuals. However, areas of junk DNA are used. These nearly always differ (apart from within families, where there are similarities), especially if at least four DNA sites are sampled.

The resulting solution is applied to a gel, which is then subjected to an electric field. The DNA fragments are electrically charged owing to the negative charges carried by the phosphate groups, and so the different-sized DNA fragments move at different speeds through the gel towards the positive electrode. The process is known as gel electrophoresis.

To see the pattern of a DNA profile test, radioactive tracers are added that bind to DNA fragments. The plate on which the gel is spread is then exposed to a photographic film. A series of bands are seen that are compared with the DNA sample from the suspect (Figure 2).

DNA profiling is also used in paternity disputes (to prove who is the father), in medical analysis of genetic relationships between different people and populations, and in the identification of body remains (Figure 3).

▲ Figure 2 *DNA profiling*

▲ Figure 3 *The identity of the bones thought to be those of the last Tsar of Russia was confirmed by comparing DNA from the bones with DNA from some of the Tsar's living relatives*

The Human Genome Project

The Human Genome Project set out in 1993 to map the complete DNA of human beings. With the technological advances that occurred after its inception, the project was completed in 2003. The coordinators were the US Department of Energy and the National Institutes of Health. There were major contributions from the Wellcome Trust (UK), and additional contributions from Japan, France, Germany, China, and other nations.

The goals of the project were:

- to identify all the genes in human DNA, approximately 20 000–25 000
- to determine the sequences of the three billion chemical base pairs that make up human DNA
- to store this information in databases
- to improve tools for data analysis
- to transfer related technologies to the private sector
- to address the ethical, legal, and social issues arising from the project.

Chemical ideas: Molecules in living systems 16.1

DNA and RNA

The building blocks

The building blocks for DNA are shown in Figure 4. These building blocks are deoxyribose (a sugar), phosphate groups ($H_2PO_4^-$), and four different bases. These bases are adenine (A), thymine (T), cytosine (C), and guanine (G).

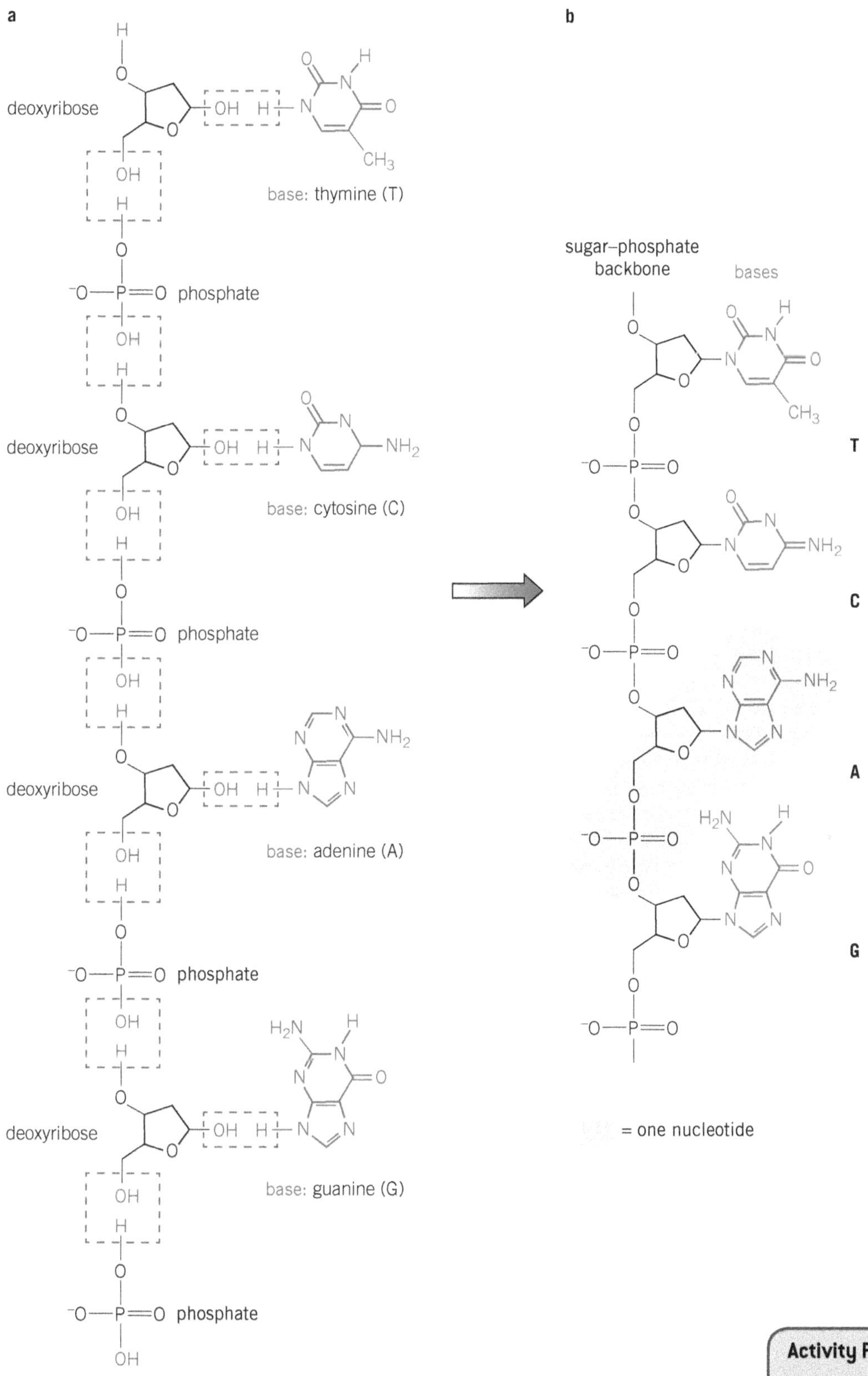

a

deoxyribose

OH
H

base: thymine (T)

OH
H

phosphate

OH
H

deoxyribose

OH H

base: cytosine (C)

OH
H

phosphate

OH
H

deoxyribose

OH H

base: adenine (A)

OH
H

phosphate

OH
H

deoxyribose

OH H

base: guanine (G)

OH
H

phosphate

OH

b

sugar–phosphate
backbone   bases

T

C

A

G

= one nucleotide

▲ Figure 4 *Representation of the structure of a single strand of DNA:*
a) the individual parts
b) the parts condensed together

Activity PL 8.1

This activity allows you to
check your understanding of DNA
structure and how two strands of
DNA form a double helix structure
(known as base pairing).

Figure 4b shows that a single strand of DNA consists of sugar (deoxyribose) molecules and phosphate groups condensed together to form a long chain of alternating sugar–phosphate groups. This is often referred to as the sugar–phosphate backbone. One of four bases (represented by A, T, C, and G) is condensed to a deoxyribose unit in the sugar–phosphate backbone. Each sugar/phosphate/base group is called a **nucleotide**. The order of bases shown in Figure 4 is for illustration only – the sequence is different in different DNA molecules.

Base pairing

Crucial to the double-helix model (Figure 1) was the understanding that pairs of bases in DNA can form hydrogen bonds together – adenine (A) with thymine (T), and cytosine (C) with guanine (G) (Figure 5, and Figure 6).

Synoptic link

Hydrogen was introduced in Topic OZ 7, The ozone hole.

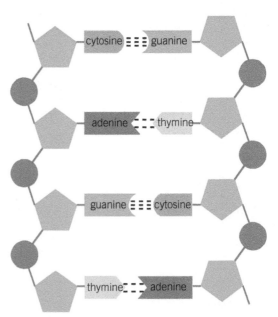

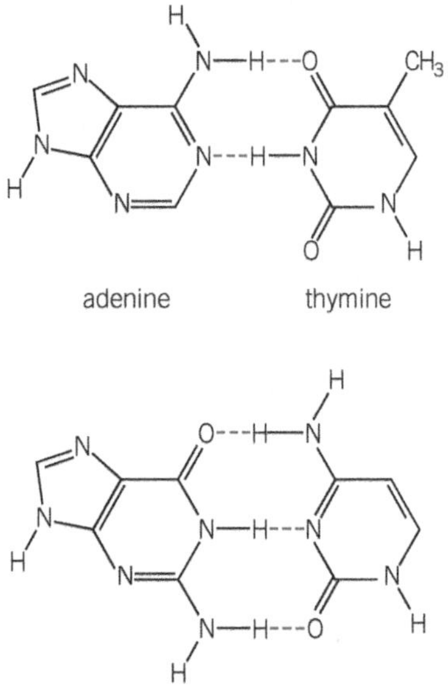

▲ Figure 5 *A simple model of base pairing in DNA*

adenine   thymine

guanine   cytosine

▲ Figure 6 *Base pairing in DNA – the molecular shapes of T and A allow them to fit together to form two hydrogen bonds between them, whilst the molecular shapes of G and C allow them to fit together to form three hydrogen bonds between them*

Protein synthesis

Chemists have found that the base pairing in DNA enables it to do two things.

- One strand, with the aid of enzymes, can synthesise a complementary copy of itself (i.e., identical to the other strand with which it is normally paired). For example, TCGAT on the original strand will appear as AGCTA on the complementary replicated strand. This replication of DNA happens prior to cell division, so the same information can be found in each new cell.

- One strand can make a complementary copy of another nucleic acid, called messenger RNA (mRNA). It does this by a process known as transcription. mRNA is used in the process of protein synthesis.

Transcription

Using enzymes, a section of the DNA relating to one protein unzips – this strand of DNA acts as a template for copying. RNA nucleotides present in the cell nucleus are joined together using enzymes. The order in which the RNA nucleotides join together is determined by the order of bases on the DNA template. When one section of DNA has been copied, the double helix then reforms (zips back up) and the mRNA produced is able to pass out of the cell nucleus (Figure 7).

mRNA differs from DNA in the following ways:

- It has ribose as the sugar, not deoxyribose, so it is called a ribonucleic acid rather than a deoxyribonucleic acid.
- It has the base uracil (U) rather than thymine (T) – these differ by a methyl group, $-CH_3$, at a place that does not affect base pairing, so U pairs with A, just as T pairs with A in DNA.
- It exists only as a single strand and does not pair up.

The similarities and differences between RNA and DNA components can be seen in Table 1.

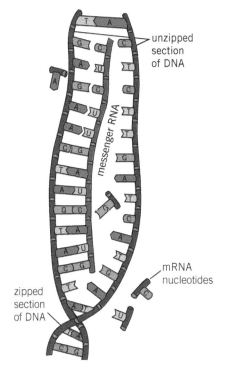

▲ Figure 7 *Diagram representing transcription. T – note how the mRNA nucleotide bases pair with the bases on the unzipped section of the DNA*

▼ Table 1 *Comparison of DNA and RNA components*

RNA	DNA
ribose	deoxyribose (note the absence of one OH group)
uracil	thymine

mRNA carries a code in its sequence of bases that corresponds to particular amino acids. The code is a triplet code, with three bases coding for each amino acid. The group of three bases is called a codon. Table 2 shows the codes, for example, GUG codes for valine and AAG codes for lysine. Because there are 64 arrangements of three bases and only 20 amino acids, the code is said to be degenerate – there is more than one code for each amino acid. The start code is AUG – that always puts the amino acid methionine at the start of the protein sequence. Sometimes the methionine is hydrolysed off later. There are three stop codons – UAA, UGA, and UAG – that end the synthesis of a chain of amino acids.

▼ Table 2 *Triplet base codes (codons) for each amino acid used in mRNA*

First base	Second base				Third base
	U	C	A	G	
U	UUU Phe	UCU Ser	UAU Tyr	UGU Cys	U
	UUC Phe	UCC Ser	UAC Tyr	UGC Cys	C
	UUA Leu	UCA Ser	UAA Stop	UGA Stop	A
	UUG Leu	UCG Ser	UAG Stop	UGG Trp	G
C	CUU Leu	CCU Pro	CAU His	CGU Arg	U
	CUC Leu	CCC Pro	CAC His	CGC Arg	C
	CUA Leu	CCA Pro	CAA Gln	CGA Arg	A
	CUG Leu	CCG Pro	CAG Gln	CGG Arg	G
A	AUU Ile	ACU Thr	AAU Asn	AGU Ser	U
	AUC Ile	ACC Thr	AAC Asn	AGC Ser	C
	AUA Ile	ACA Thr	AAA Lys	AGA Arg	A
	AUG Met	ACG Thr	AAG Lys	AGG Arg	G
G	GUU Val	GCU Ala	GAU Asp	GGU Gly	U
	GUC Val	GCC Ala	GAC Asp	GGC Gly	C
	GUA Val	GCA Ala	GAA Glu	GGA Gly	A
	GUG Val	GCG Ala	GAG Glu	GGG Gly	G

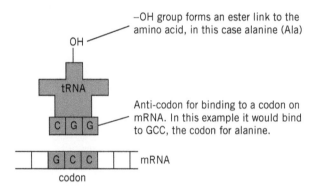

−OH group forms an ester link to the amino acid, in this case alanine (Ala)

Anti-codon for binding to a codon on mRNA. In this example it would bind to GCC, the codon for alanine.

▲ Figure 8 *Schematic representation of a tRNA molecule showing the three bases that form the anti-codon*

messenger RNA (mRNA) carries the code for protein synthesis

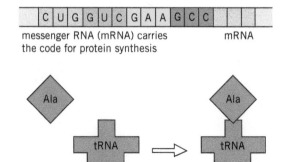

transfer RNA (tRNA) collects an amino acid and takes it to the mRNA strand

▲ Figure 9 *Messenger RNA and transfer RNA*

Translation

Protein synthesis takes place in structures called ribosomes within a cell. mRNA passes here from the nucleus after being synthesised by transcription. During protein synthesis, amino acids need to be joined together in the correct primary structure – mRNA provides a template through which this can happen. Amino acids are taken to the mRNA in the ribosomes and joined to small lengths of RNA called transfer RNA (tRNA) (Figure 8).

The code on the tRNA that fits to the mRNA is called an anti-codon, as it is the complementary sequence of bases to the codon. So GCC on the mRNA codes for alanine, and one of the tRNAs that attaches to alanine has the anti-codon CGG.

Figure 9 summarises the roles of mRNA and tRNA in protein synthesis. It shows how the codons on the mRNA are read as they are in contact with the ribosome and translated into a protein chain.

You can think of the ribosome as rolling along the mRNA chain. tRNA molecules, each carrying an amino acid, feed into the front of the ribosome and the protein chain grows from the back (Figure 10).

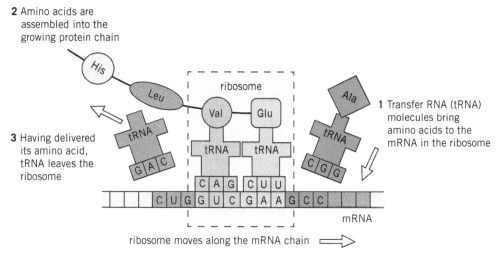

2 Amino acids are assembled into the growing protein chain

3 Having delivered its amino acid, tRNA leaves the ribosome

1 Transfer RNA (tRNA) molecules bring amino acids to the mRNA in the ribosome

ribosome moves along the mRNA chain

▲ Figure 10 *Protein synthesis and the reading of codons on mRNA*

Another important contrast between DNA and RNA is that each DNA molecule contains the information for the production of *many* different mRNA molecules but, in general, each mRNA molecule is a set of instructions for just *one* protein. A DNA segment responsible for a particular protein is called a gene.

Summary questions

1 **a** In the formation of a strand of DNA, what type of reaction is responsible for the linking of:
  i the deoxyribose and phosphate
  ii the deoxyribose and base? *(2 marks)*
  b The skeletal formula of deoxyribose is shown in Table 1. Draw a full structural formula for deoxyribose. *(1 mark)*

2 Table 2 shows that only the first two bases of the RNA codon are important for some amino acids.
  a Make a list of those amino acids where the identity of the third base does not matter. *(1 mark)*
  b There are two amino acids that have just one codon. Which amino acid are they? *(1 mark)*

3 **a** Use the codons from Table 2 to predict the peptides obtained if mRNA molecules with the following patterns of bases were used:
  i AAAAAA **ii** CGCGCGCGC
  iii UACCUAACU *(3 marks)*
  b Predict the anti-codons for these amino acids:
  i Trp **ii** Asp. *(2 marks)*

4 One strand of DNA in a cell nucleus carries the sequence of bases CAGT.
  a Write down the corresponding sequence of bases on the mRNA strand that copies from it. *(1 mark)*
  b Write down the corresponding sequence of bases on the other DNA strand in the double helix. *(1 mark)*

5 A nucleotide consists of a base–sugar–phosphate group. Draw the skeletal formulae of:
  a A nucleotide containing adenine from DNA hydrogen-bonded to a nucleotide containing thymine from DNA. *(2 marks)*
  b A nucleotide containing guanine joined to a nucleotide containing cytosene as part of the RNA structure. *(2 marks)*

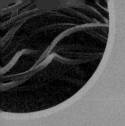

PL 9 Spectroscopic analysis of organic compounds

Specification references: PL(r), PL(s), PL(t)

Synoptic link

In Topic WM 4, Mass spectrometry, you discovered how the M_r of a substance could be found from the M^+ peak in its mass spectrum and that molecules broke up in the spectrometer to produce fragments.

Mass spectrometry

Mass spectrometry has a variety of uses, both in chemical analysis and more widely used techniques.

Use in chemical analysis

High-resolution values of the M^+ peak can be used to measure the M_r to greater accuracy. This allows the molecular formula to be found. Fragments from the molecule can also be used to determine its structure.

Wider uses

Protein analysis

In one method, a protein is broken by enzymes into smaller peptides. These are analysed sequentially by a mass spectrometer. The peptides are identified by matching against a database of known peptide spectra.

Respired gas monitor

Mass spectrometers can give instant readings of the gases in the breath of a patient undergoing anaesthesia to ensure all is well.

▲ Figure 1 *Using a respired gas monitor*

Nuclear magnetic resonance spectrometry

The theory

Nuclear magnetic resonance spectroscopy (NMR) is one of the most widely used analytical techniques available to chemists. NMR provides you with very detailed information about the nuclei of certain atoms. These atoms have to have an odd number of protons or neutrons. Of the elements present in most organic compounds, the common isotopes carbon-12 and oxygen-16 cannot be used. However, hydrogen can be. There is a small proportion of the carbon-13 isotope that is also used for NMR.

Such nuclei behave like tiny magnets and when they are placed in a strong magnetic field, they align themselves with or against the magnetic field. If they are aligned against the magnetic field, they have a higher energy than if they are aligned with it. If the correct frequency of radiation is applied, some of these nuclei move up to the higher level, absorbing some energy, ΔE, as they do so. The energy absorbed corresponds to radio frequencies.

For every type of molecular arrangement (molecular environment), there is a very slightly different magnetic field. The ^1H and ^{13}C nuclei in these different environments have different energy gaps ΔE between their high and low energy levels, and so absorb different frequencies of radiation. Therefore, they give different NMR absorption peaks, and it is possible to find out how many hydrogen or carbon atoms of different types there are in a molecule.

Uses of NMR

MRI scanners

Proton NMR (^1H NMR) is the principle behind magnetic resonance imaging (MRI) scanners used routinely in hospitals (Figure 2). The word nuclear was omitted, since some patients thought the technique would make them radioactive.

Chemical analysis

By looking at the number of carbon or hydrogen nuclei in different environments, NMR can give more information about the structure of a compound and often determine it uniquely.

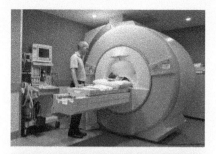
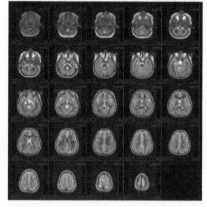

▲ Figure 2 *MRI scanner in use (top) and an MRI scan of a female child's brain (bottom)*

Chemical Ideas: Radiation and matter 6.5c

High-resolution mass spectrometry

Modern mass spectrometers exist that can measure the relative molecular masses of ions to four decimal places. At this resolution, the exact relative atomic masses become important. These are:

- ^1H 1.0078
- ^{14}N 14.0031
- ^{16}O 15.9949
- ^{12}C 12.0000

Molecules that have the same molecular mass to one decimal place have different molecular masses when measured more accurately. For example, propane, C_3H_8, and ethanal, CH_3CHO, both have a molecular mass of 44.0. However, to four decimal places the molecular masses become 44.0624 and 44.0261, so these two compounds could be distinguished by a high-resolution mass spectrometer.

If the M_r of a molecule is known to four decimal places, its molecular formula can be deduced directly. To determine the structural formula, other methods have to be used.

 Worked example: Finding molecular formulae from accurate M_r values

C_2H_4, CO, and N_2 all have M_r of 28. Substance **X** has an accurate M_r of 27.9949.

Identify substance **X**.

Step 1: Calculate the exact relative atomic mass of all three substances.

$C_2H_4 = (12.0000 \times 2) + (1.0078 \times 4) = 28.0312$

$CO = 12.0000 + 15.9949 = 27.9949$

$N_2 = 14.0031 \times 2 = 28.0062$

Step 2: Identify the compound

CO is substance **X**

Synoptic link

You have used mass spectrometry in Topic EL 1, Where do the chemical elements come from?, and Topic WM 4, Mass spectrometry.

Fragmentation

The mass spectrum of 2-ethoxybutane, $CH_3CH_2CH(CH_3)OCH_2CH_3$, is shown in Figure 3.

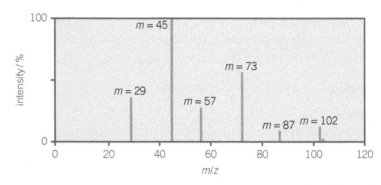

▲ Figure 3 *A simplified mass spectrum for 2-ethoxybutane, showing the six largest peaks*

Activity PL 9.1

You can use molecular models to check your understanding of how mass spectrometry is used to determine the structures of organic compounds.

The positive ions formed in a mass spectrometer can undergo some strange chemistry, unlike the reactions you are used to with laboratory chemicals, so mass spectra can be quite complicated to interpret. However, simple ideas can often help to make enough sense of what is going on to allow you to identify the sample.

If you imagine the molecules of 2-ethoxybutane so that can be pulled apart into their constituent building blocks, you can identify nearly all the peaks in its mass spectrum (Figure 4).

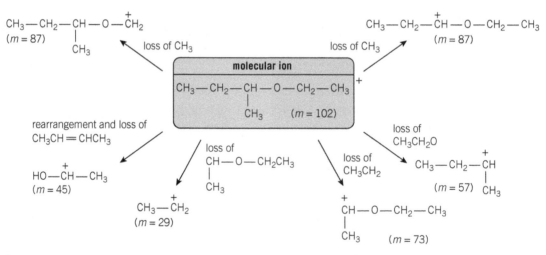

▲ Figure 4 *Fragmentation of 2-ethoxybutane – the fragments may be formed in several steps*

For each fragmentation, one of the products keeps the positive charge. This means that there are always two possibilities:

$$M^+ \rightarrow A^+ + B \qquad \text{or} \qquad M^+ \rightarrow A + B^+$$

Usually the fragment that gives the most stable positive ion is formed but you may see products from both routes in a spectrum.

Study tip

There is always a positive charge because a negatively charged electron has been removed during the process.

This approach enables you to make use of the mass differences between peaks (the masses of the bits which have fallen off) to help to make sense of the spectrum. For example, in Figure 3 peaks 102 and 87 differ by 15, corresponding to the loss of CH_3. This is a common process in mass spectrometry so be on the lookout for gaps of 15 in your spectra – they probably mean there is a methyl group in the substance you are investigating. Table 1 gives some other common differences and the groups that they suggest.

 Worked example: Fragmentation

A ketone has the molecular formula $C_5H_{10}O$. The mass spectrum is shown in Figure 5.

Suggest formulae for the ions that give the labelled peaks and suggest a structure for the substance.

The peak with the highest m/z value is 86, so it could be the M^+ peak, caused by knocking one electron off the whole molecule. This can be confirmed from the molecular formula: $C_5H_{10}O^+$ = $(12.0000 \times 5) + (1.0078 \times 10) + 15.9949 = 86.0729$. So 86 = $C_5H_{10}O^+$ (do not forget the plus).

86 − 57 = 29 that corresponds to C_2H_5 being lost from the molecule. There is also a peak at 29.

So 57 = $C_3H_5O^+$ and 29 = $C_2H_5^+$.

If the ketone has C_2H_5 in it and the molecular formula given, it must be $C_2H_5COC_2H_5$.

So the ion at 57 is $C_2H_5CO^+$ and the ketone is pentan-3-one.

The other fragments you would expect for pentan-2-one are:

- 71 for $C_3H_7CO^+$
- 15 for CH_3^+
- 43 for CH_3CO^+ and $C_3H_7^+$

Mass difference	Group suggested
15	CH_3
17	OH
28	C=O or C_2H_4
29	C_2H_5
43	$COCH_3$
45	$COOH$
77	C_6H_5

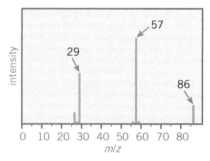

▲ Figure 5 *Mass spectrum of the ketone $C_5H_{10}O$*

Chemical Ideas: Radiation and matter 6.6

Nuclear magnetic resonance

^{13}C NMR

A ^{13}C NMR spectrum for ethanol is shown in Figure 6.

The zero of the spectrum is the TMS peak, given by the absorptions of the ^{13}C in tetramethylsilane, $Si(CH_3)_4$. There is only one peak, since all the carbon atoms are in the same environment.

TMS has been chosen as a standard reference, because it gives a sharp signal well away from most of the ones of interest to chemists. The extent to which a signal differs from TMS is called its chemical shift.

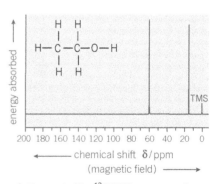

▲ Figure 6 *The ^{13}C NMR spectrum for ethanol*

Activity PL 9.2

This data-handling exercise enables you to check your understanding of ^{13}C NMR.

The spectrum of ethanol shows two peaks, corresponding to carbon atoms in different environments. Being attached to an oxygen atom causes the chemical shift of one carbon to be greater than the other.

The regions in which chemical shifts occur are shown in Figure 7:

^{13}C NMR chemical shifts relative to TMS

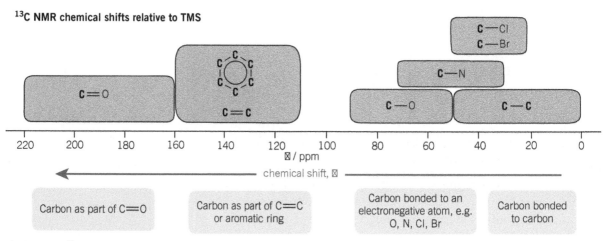

| Carbon as part of C=O | Carbon as part of C=C or aromatic ring | Carbon bonded to an electronegative atom, e.g. O, N, Cl, Br | Carbon bonded to carbon |

▲ Figure 7 ^{13}C chemical shifts

Study tip

The heights of the peaks in a ^{13}C NMR spectrum give no information about the number of carbon atoms in a molecule.

^{13}C NMR can be used to:

● work out the number of carbon atoms that have different environments in a molecule

● work out (sometimes only roughly) the groups to which these carbon atoms are attached.

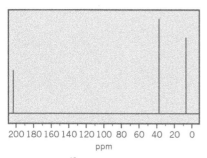

▲ Figure 8 ^{13}C NMR spectrum of a C_3H_6O compound

Activity PL 9.3

You can practise analysing ^1H NMR in order to identify organic compounds.

🖩 Worked example: The use of ^{13}C NMR

The ^{13}C NMR of a compound C_3H_6O is shown in Figure 8. Suggest the structure of the compound.

Possible structures are: CH_3COCH_3, $CH_2=CHCH_2OH$, CH_3CH_2CHO

Of these, the first (propanone) has carbon atoms in only two environments, so it can be eliminated as the spectrum shows three environments. The second would have a peak around 120-140 for C=C, which is not there, so it must be the third (propanal) with three carbon environments, one of which is for a C=O.

^1H NMR spectra

The ^1H NMR spectrum of ethanol is shown in Figure 9.

The protons in TMS provide the zero here and the protons in different environments have different chemical shifts relative to this zero.

The spectrum shows three peaks, showing that there are three proton environments in ethanol. The diagram in Figure 10 enables you to work out which is which.

This shows that the peak at δ1.2 is caused by the CH_3–C protons and the peak at δ3.7 is caused by the CH_2–O protons. The OH proton can be almost anywhere and, in the spectrum, it shows at δ2.7.

This time the number of protons can be deduced from the peak height (or more accurately the peak *area*). NMR spectrometers can work out the peak areas and the results are often shown on the spectrum, as in Figure 9, making it even easier to identify the three environments, and how many hydrogens are in each environment.

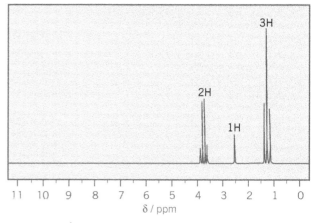

▲ Figure 9 ^1H NMR spectrum of ethanol

^1H NMR chemical shifts relative to TMS

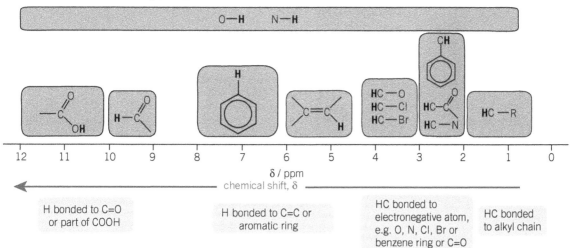

▲ Figure 10 Chemical shifts in ^1H NMR

Another important feature is the splitting patterns (Figure 11). The 3H peak has three sub-peaks and is known as a triplet, the 1H peak is a singlet, and the 2H peak is a quartet, as it consists of four peaks very close together.

The splitting pattern depends on the number of hydrogen atoms on the *adjacent* carbon atoms by the $n + 1$ rule. So, the CH_3–C hydrogens have two hydrogens on the adjacent carbon, CH_2. So $n = 2$ and the peak is split into three ($2 + 1 = 3$). The C–CH_2–O hydrogens have three hydrogens on the adjacent carbon, so $n = 3$ and the peak is split into a quartet. The hydrogen of the –OH does not have an adjacent carbon, so its peak is not split.

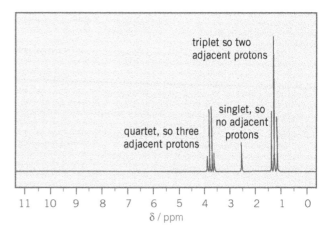

▲ Figure 11 ^1H NMR spectrum showing splitting patterns for ethanol

So, from a ^{1}H NMR spectrum, the following can be deduced:

● the number of different hydrogen environments in the molecule
● the number of hydrogen atoms in each environment (if this data is given)
● the nature of these environments (what atoms are connected to the carbon that carries these hydrogens)
● the number of hydrogen atoms attached to *neighbouring carbon atoms* (from the splitting of a peak).

✚ Why does the *n* + 1 rule work?

The ^{1}H nuclei behave like tiny magnets and they can be in one of two orientations depending on whether they are in the low energy level or the high energy level. In a methyl, $-CH_3$ group, there are four combinations in which the three tiny proton magnets can be arranged:

● all aligned with the external field
   S–N   S–N   S–N
● two with and one against the external field
   S–N   S–N   N–S
● one with and two against the external field
   S–N   N–S   N–S
● all aligned against the external field
   N–S   N–S   N–S

The magnetic effect of these arrangements in a methyl group is transmitted to the hydrogen atoms on the neighbouring carbon so that they sense one of four magnetic fields, or *n* + 1 fields, where *n* is the number of hydrogen atoms bonded to the neighbouring carbon atom. They can absorb four different frequencies and give four signals. These are centred on the expected chemical shift.

The peak intensities are in the ratio 1 : 3 : 3 : 1 because there are three possible combinations leading to each of the two central peaks:

S–N/S–N/N–S   S–N/N–S/S–N   N–S/S–N/S–N

and

N–S/N–S/S–N   N–S/S–N/N–S   S–N/N–S/N–S

> A methyl group is attached to a carbon with one hydrogen. Use the above to write out the explanation as to why the CH_3 hydrogen peak will exist as a doublet.

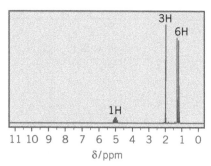

▲ **Figure 12** *^{1}H NMR spectrum of an ester*

🖩 Worked example: Interpreting a ^{1}H NMR spectrum

The ^{1}H NMR spectrum of an ester is shown in Figure 12.

Work out the structure of the ester.

Step 1: Identify the number of hydrogen environments in the molecule.

There are three hydrogen environments in the molecule.

Hydrogen atoms in these environments are in the ratio 1 : 3 : 6.

Step 2: Identify the hydrogen environments from the peaks.

The peak at δ1.2 has six hydrogens in the same environment, so it is $(CH_3)_2C-$. The carbon that the two methyl groups are attached to has one hydrogen on it (*n*=1, hence a doublet).

The peak at δ2.0 is for the hydrogens in $CH_3-C=O$ (right place on the spectrum and the adjacent carbon has no hydrogen atoms).

The peak at δ5.0 consists of one hydrogen. The carbon atom it is attached to is next to carbons with lots of hydrogen atoms, so it must be the hydrogen on the carbon with two methyl groups attached.

Step 3: Identify the structure.

So the ester is $CH_3COOCH(CH_3)_2$. The only peak that does not quite fit is that at 5.0 for the single hydrogen in $(CH_3)_2CH–O$. This has a slightly higher shift value than shown in the chart, but this sometimes happens. It could not be CH=C, as there would not be enough hydrogen atoms to make the 1:3:6 ratio.

> **Activity PL 9.4**
>
> In this activity you will use infrared, ^1H NMR, and mass spectra to identify the organic molecules involved in the synthesis of two medicines.

Putting it all together (including IR)

Table 2 is a summary of the evidence for the structure of a substance that can be obtained through mass spectrometry, IR, and NMR.

▼ Table 2 *Evidence for the structure of a substance that can be obtained from spectroscopy*

Low-resolution mass pectrometry	• M_r from the M⁺ ion • fragments which can be used to suggest structure
High-resolution mass spectrometry	molecular formula from the M⁺ ion
IR	certain functional groups
^{13}C NMR	the number and types of carbon environments
^1H NMR	• the number and types of hydrogen environments • the number of hydrogen atoms in each environment • the number of hydrogen atoms on the carbon atom adjacent to the carbon atom carrying the hydrogen.

 Worked example: Combining spectroscopic evidence to identify a compound

A compound **X** has the following spectra.
Identify the compound.

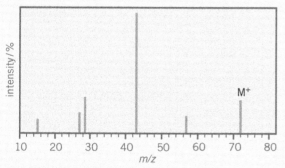

▲ Figure 13 *Mass spectrum of compound X*

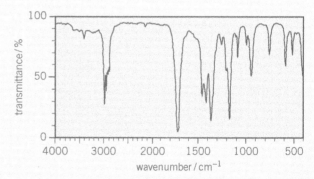

▲ Figure 14 *IR spectrum of compound X*

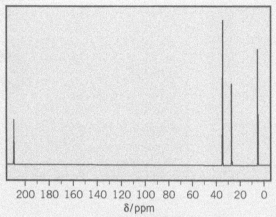

▲ Figure 15 ^{13}C NMR spectrum of compound **X**

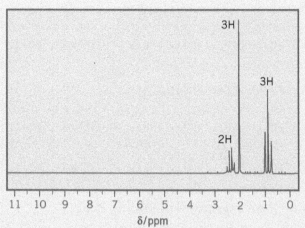

▲ Figure 16 ^{1}H NMR spectrum of compound **X**

Mass spectrum: M_r of 72 from M⁺ peak

IR: C=O (probably aldehyde, ketone or acid, since around 1720 cm⁻¹) but no OH (no *broad* peak 2500–3200 and no peaks around 3000), so probably aldehyde or ketone

^{13}C NMR: four carbon environments (one C=O)

^{1}H NMR: CH₃CH₂ to give triplet and quartet, CH₃ attached to a carbon with no hydrogens

All this points to CH₃CH₂COCH₃ (butanone). This is confirmed by the fragments from the mass spectrum.

Peak at *m/z* 57: CH₃CH₂CO⁺, peak at *m/z* 43: CH₃CO⁺, peak at *m/z* 29: C₂H₅⁺; peak at *m/z* 15: CH₃⁺

Summary questions

1 Figure 17 shows the mass spectra for ethanol, C₂H₅OH, and ethanoic acid, CH₃COOH, not necessarily in that order. The spectra are labelled A and B.

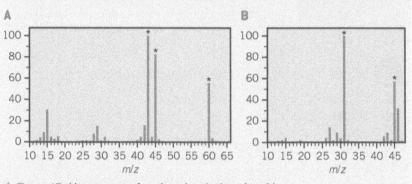

▲ Figure 17 *Mass spectra for ethanol and ethanoic acid*

a Assign the spectra to the correct compounds.   (*1 mark*)
b For each spectrum, suggest a possible ion responsible for the peaks marked with an asterisk (*).   (*3 marks*)

2 How many peaks would you expect in the ^{13}C NMR spectrum of:
   a ethane   (*1 mark*)   b propane   (*1 mark*)
   c propan-1-ol   (*1 mark*)   d propan-2-ol?   (*1 mark*)

3 Tartaric acid, HOOCCH(OH)CH(OH)COOH, and succinic acid,
   HOOCCH$_2$CH$_2$COOH, are commonly present in wine.
   a For each compound, state how many signals you would expect
      in its ^1H NMR spectrum. Give your reasons.   (*3 marks*)
   b Give the ratio of the peak areas in each compound.   (*2 marks*)
   c For each compound, state how many signals you would expect
      in its ^{13}C NMR spectrum. Give your reasons.   (*3 marks*)

4 The mass spectrum of a hydrocarbon C$_4$H$_{10}$ has peaks at 58, 43, 29, and 15.
   a Suggest ions that would give rise to each of these peaks.   (*4 marks*)
   b Draw two possible structures for C$_4$H$_{10}$.   (*1 mark*)
   c Decide which of these two structures would give rise to the
      given ions. Give your reasoning.   (*1 mark*)

5 Compounds **C** and **D** both have molecular formula C$_3$H$_6$O and
   a C=O group but they have different structures.
   a Draw out two possible structures for C$_3$H$_6$O.   (*1 mark*)
   b Compound **C** has a high intensity peak for a fragment ion at 43 in its mass
      spectrum. Which group of atoms is lost when this forms?   (*1 mark*)
   c Compound **D** has a fragment ion at 57 in its mass spectrum.
      What is lost in forming this ion?   (*1 mark*)
   d Compound **D** also has a fragment ion at 29 in its mass spectrum.
      Identify this ion.   (*1 mark*)
   e Identify **C** and **D**, giving reasons.   (*2 marks*)

6 An alcohol **E** has peaks in its ^1H NMR at δ1.3 (9H) and δ2.1 (1H).
   a Draw the structure of the alcohol.   (*1 mark*)
   b Explain, with reasons, whether either of the peaks would
      be split.   (*2 marks*)
   c Predict the number of peaks in the ^{13}C NMR spectrum and
      give their approximate ranges.   (*2 marks*)

7 Compound **F** has IR absorptions at 1740 cm^{-1}, 1100 cm^{-1}, no broad
   absorption around 3000 cm^{-1}, and no absorptions above 3000 cm^{-1}. Its
   ^1H NMR spectrum is below.

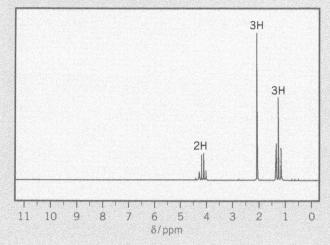

a Identify compound **F**, giving your reasons.   (*6 marks*)

b How many peaks would you expect in the ^{13}C NMR spectrum of compound **F**? Give your reasons.   (*1 mark*)

c Compound **F** has an isomer, compound **G**, with the same functional group and the same alkyl groups. Sketch a possible ^{1}H NMR spectrum for this compound.   (*4 marks*)

8 DIMP is an insect repellent developed to try to protect forestry workers from insects such as midges. Its structure is shown.

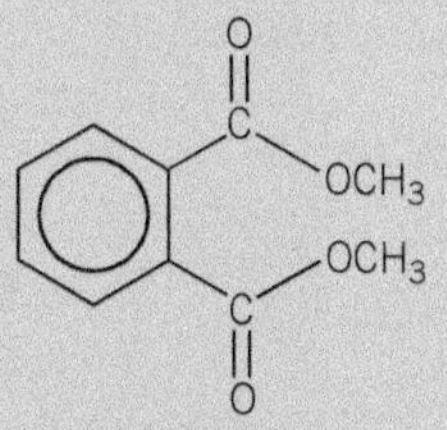

a Identify the different types of proton in the molecule, the expected chemical shifts, and the relative numbers of each type of proton.   (*2 marks*)

b Sketch the ^{1}H NMR spectrum you would expect for DIMP.   (*2 marks*)

c Sketch the ^{13}C NMR spectrum you would expect for DIMP.   (*4 marks*)

9 Compound **H** contains only carbon, hydrogen and oxygen. Its accurate molecular mass was found to be 72.0573. A database gave four compounds with masses in this region: $C_2H_4N_2O$, $C_3H_8N_2$, $C_3H_4O_2$, and C_4H_8O.

a Use these accurate atomic masses to work out the formula of compound **H**:

H = 1.0078, O = 15.9949, N = 14.0031, C = 12.0000   (*1 mark*)

b The low-resolution mass spectrum of compound **H** has peaks at 72, 57, 43 and 29. Use this data to work out the structure of compound **H**, giving your reasons.   (*2 marks*)

Practice questions

1   The dipeptide Gly-Lys is hydrolysed using dilute hydrochloric acid.

R-group	
Gly	$-H$
Lys	$-(CH_2)_4NH_2$

The organic ions present in the hydrolysed mixture will be:

A H_2NCH_2COOH and $H_2NCH((CH_2)_4NH_2)COOH$

B $^+H_3NCH_2COOH$ and $^+H_3NCH((CH_2)_4NH_2)COOH$

C $H_2NCH_2COO^-$ and $H_2NCH((CH_2)_4NH_2)COO^-$

D $^+H_3NCH_2COOH$ and $^+H_3NCH((CH_2)_4NH_3^+)COOH$ *(1 mark)*

2   Which of the following describes a nucleotide from RNA.

A   uracil joined to deoxyribose joined to a phosphate group

B   thymine joined to ribose joined to a phosphate group

C   guanine joined to ribose joined to a phosphate group

D   lysine joined to ribose joined to a phosphate group   *(1 mark)*

3   At high substrate concentrations, the order of an enzyme reaction is:

A   zero order with respect to both substrate and enzyme concentration

B   first order with respect to both substrate and enzyme concentration

C   zero order with respect to substrate concentration and first order with respect to enzyme concentration

D   second order overall   *(1 mark)*

4   Enzymes do not function well slightly away from their optimum pH because:

A   ionic bonds are less available to attach the substrate

B   the active site of the enzyme is destroyed

C   hydrogen bonds are less available to attach the substrate

D   the tertiary structure of the enzyme breaks down   *(1 mark)*

5   The diamine $H_2NCH_2CH_2NH_2$:

A   will react with ethanoic acid to form a polymer

B   will react with HCl in the ratio diamine: acid of 1:2

C   will be neutral in solution

D   will form an ionic compound when reacted with NaOH   *(1 mark)*

6   The compound HOCCHCHCOOH could form:

A   just addition polymers

B   just condensation polymers

C   both addition and condensation polymers

D   neither condensation nor addition polymers   *(1 mark)*

7   The mass spectrum of ethanol has peaks at (among others) 47, 45, and 31.

Which of the following is a correct statement about one of these peaks?

A   The peak at 45 is the M^+ peak

B   The peak at 47 is caused by isotopes of hydrogen

C   The peak at 31 is due to the loss of CH_3^+

D   The peak at 31 is caused by $CH_3CH_2^+$ *(1 mark)*

8   Ethanoic acid will form sodium ethanoate when reacted with:

1   sodium carbonate solution

2   sodium

3   sodium hydroxide solution

A   1, 2, and 3 correct

B   1 and 2 are correct

C   2 and 3 are correct

D   Only 1 is correct   *(1 mark)*

9   A competitive inhibitor of an enzyme reaction:

1   has a shape resembling the substrate

2   binds irreversibly with the active site

3   changes the shape of the active site

A   1, 2, and 3 correct

B   1 and 2 are correct

C   2 and 3 are correct

D   Only 1 is correct   *(1 mark)*

10 A peptide is hydrolysed. Amino acids can be identified in the hydrolysed mixture by paper chromatography, followed by:

1 using the colours of the spots with ninhydrin

2 measuring the R_f values of the spots and consulting a database

3 comparing the heights the spots have risen with spots from known amino acids.

A 1, 2, and 3 correct

B 1 and 2 are correct

C 2 and 3 are correct

D Only 1 is correct (*1 mark*)

11 A tripeptide that can be extracted from algae is Glu-Gln-Ala.

The R-groups of these three amino acids are:

Name	Abbreviation	R-group
glutamic acid	Glu	$-CH_2CH_2COOH$
glutamine	Gln	$-CH_2CH_2CONH_2$
alanine	Ala	$-CH_3$

a Name the functional groups in the R-groups of glutamic acid and glutamine. (*2 marks*)

b A student suggests that glutamine could be made from glutamic acid by the following route:

$$-CH_2CH_2COOH$$
$$\downarrow SOCl_2$$
$$-CH_2CH_2COCl$$
$$\downarrow NH_3$$
$$-CH_2CH_2CONH_2$$

(i) Name the functional group on the intermediate compound.

(ii) Comment on the student's method of converting Glu to Gln in this way. (*3 marks*)

c Give the structure of the tripeptide with the free NH_2 at the Glu end. (*2 marks*)

d Give the structure of the ion that would be obtained if the tripeptide were dissolved in dilute alkali. (*2 marks*)

e A codon for glutamic acid is GAG. Use this and information from the *Data Sheet* to write an mRNA sequence that would produce the tripeptide Glu-Gln-Ala and the DNA strand that would have produced this mRNA. (*2 marks*)

f Use the structures on the *Data Sheet* to draw structures to show how a cytosine molecule hydrogen bonds to a guanine molecule. (*2 marks*)

g Draw three-dimensional structures to show how the two enantiomers of alanine are related. (*2 marks*)

h Longer peptides and proteins have secondary and tertiary structures. Describe these and the bonds holding them together. (*6 marks*)

12 Noradrenaline is a 'flight or fight' hormone that is produced when scared or excited. It has several effects on the body including dilating the airways in the lungs and increasing heart rate. Asthmatics need to dilate their airways and their inhalers often contain salbutamol, which has the same effect as noradrenaline.

noradrenaline

salbutamol

a (i) Name a functional group that is present in noradrenaline but not salbutamol.

(ii) Name a functional group that is present in salbutamol but not adrenaline.

(iii) Would either compound fizz when added to a solution of a carbonate? Give a reason. (*3 marks*)

b **(i)** Suggest how salbutamol mimics the function of adrenaline, using the term pharmacophore in your answer.

(ii) Draw a line round the pharmacophore on salbutamol.

(3 marks)

13 Ketones are produced on a massive scale in industry as solvents, polymer precursors, and pharmaceuticals. There are three ketones with molecular formula $C_5H_{10}O$. Two of these are the 'straight-chained' pentan-2-one and pentan-3-one.

a These ketones will all have one characteristic peak in their IR spectra. Give details of this peak. *(1 mark)*

b One of the straight-chained pentanones has three peaks in its ^{13}C NMR spectrum, at 8, 35, and 212 ppm. Draw the structure of this pentanone and indicate the carbon atoms giving rise to each peak. *(2 marks)*

c Pentan-2-one has a singlet peak in its ^1H NMR spectrum.

(i) Give the region of the spectrum in which this peak would be found and explain why it is a singlet.

(ii) Give the total number of peaks in this spectrum. *(3 marks)*

d There is one branched-chain ketone with formula $C_5H_{10}O$.

(i) Give its skeletal formula and name

(ii) The ^1H NMR spectrum of this isomer has a peak at a chemical shift of 1.0 ppm. Give and explain the number of protons for this peak. *(4 marks)*

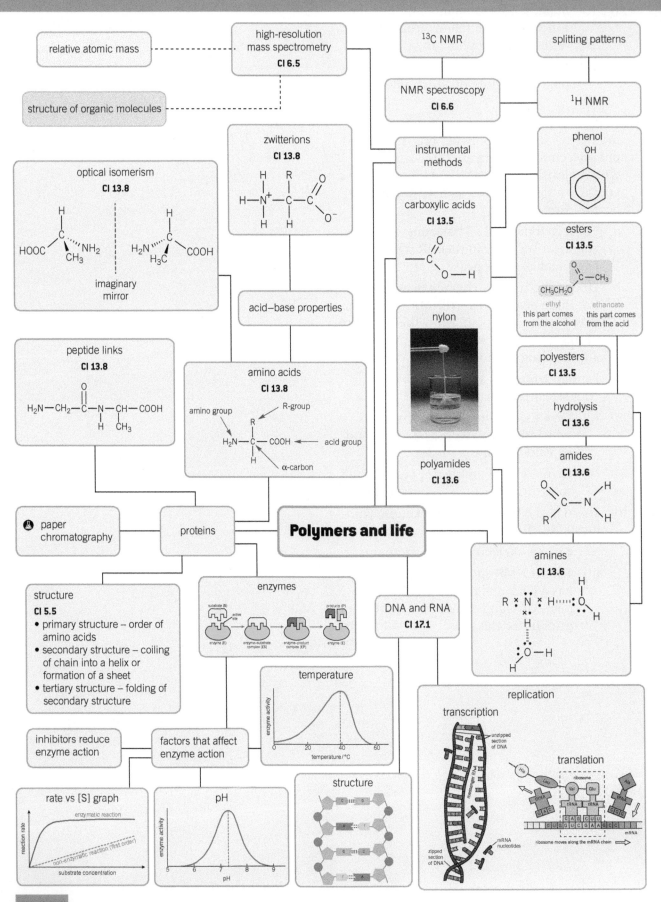

relative atomic mass

high-resolution mass spectrometry
CI 6.5

structure of organic molecules

^{13}C NMR

splitting patterns

NMR spectroscopy
CI 6.6

^{1}H NMR

instrumental methods

phenol
OH

zwitterions
CI 13.8

carboxylic acids
CI 13.5

esters
CI 13.5

CH$_3$CH$_2$O

ethyl
this part comes
from the alcohol

ethanoate
this part comes
from the acid

optical isomerism
CI 13.8

HOOC–C–NH$_2$
CH$_3$

H$_2$N–C–COOH
H$_3$C

imaginary
mirror

acid–base properties

nylon

polyesters
CI 13.5

hydrolysis
CI 13.6

peptide links
CI 13.8

H$_2$N–CH$_2$–C–N–CH–COOH
H   CH$_3$

amino acids
CI 13.8

amino group   R-group

H$_2$N–C–COOH ← acid group
H

α-carbon

amides
CI 13.6

polyamides
CI 13.6

🜨 paper
chromatography

proteins

Polymers and life

amines
CI 13.6

R × N × H ⋯⋯ O
H

O–H

structure
CI 5.5
• primary structure – order of
amino acids
• secondary structure – coiling
of chain into a helix or
formation of a sheet
• tertiary structure – folding of
secondary structure

enzymes

substrate (S)
active site
enzyme (E)   enzyme-substrate complex (ES)   enzyme-product complex (EP)   enzyme (E)
products (P)

DNA and RNA
CI 17.1

replication

transcription

translation

temperature

inhibitors reduce
enzyme action

factors that affect
enzyme action

rate vs [S] graph

enzymatic reaction
non-enzymatic reaction (first order)

reaction rate

substrate concentration

pH

enzyme activity

structure

Tripeptides

Tripeptides consist of three amino acids joined by peptide bonds. Their properties depend on the amino acids they contain and the order in which the amino acids are joined. For example, there are three distinct tripeptides formed from two glycine units and one alanine unit. These are GLyGlyAla, GlyAlaGly, AlaGlyGly.

Glutathione (Figure 1) is a tripeptide that is an important antioxidant in many plants and animals. It is formed from the amino acids glutamic acid, cysteine, and glycine.

Melanocyte-inhibiting factor (Figure 2) is a tripeptide that is produced by the hypothalamus. It is formed from the amino acids glycine-NH_2, leucine, and proline.

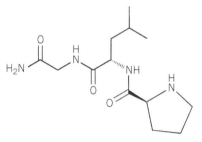

▲ Figure 1   *Glutathione*

1   Identify the peptide links and put an asterisk (*) next to the chiral carbon atoms in glutathione and melanocyte-inhibiting factor.
2   Explain why GlyGlyAla, GlyAlaGly, and AlaGlyGly are three different molecules. Include their structures in your answer.
3   Draw the products of hydrolysis of glutathione under acid conditions.
4   The amino acid proline is formed from the hydrolysis of melanocyte-inhibiting factor. It contains a five-membered ring. At pH < 6.30 it is protonated, at pH = 6.30 it is a zwitterion and at pH > 6.30 it is deprotonated. Draw the structure of proline under these different conditions.
5   A student hydrolysed glutathione. Explain how the student could use paper chromatography to identify the components and to prove that the hydrolysis had been successful.
6   Suggest the main peaks you would expect in the mass spectrum, infra-red spectrum, ^{1}H-NMR spectrum, and ^{13}C-NMR spectrum of glycine.

▲ Figure 2   *Melanocyte-inhibiting factor*

 Extension

1   The amino acids found in proteins are all α-amino acids. Research what an α-amino acid is and the occurrence and biological effects of β- and γ-amino acids such as β-alanine and γ-aminobutyric acid.
2   DNA and proteins both feature condensation polymerisation, hydrogen bonding, instantaneous dipole–induced dipole and permanent dipole–permanent dipole bonds. Compare and contrast the occurrence and influence of each of these types of bonding in DNA and proteins.
3   The human genome project, completed in 2003, aimed to work out the sequence of the three billion base pairs in the human genome and to identify all the genes. Research the methods used to sequence the genome, identifying the role of chemists in the project.

CHAPTER 8
Oceans

Topics in this chapter

Why a chapter on Oceans?

To many people, the term ocean probably conjures up an image of a seemingly endless expanse of water, of some biological interest but chemically inert – yet this is far from true. Oceans play an essential part in the cycles of many chemicals (e.g., sulfur and nitrogen compounds) throughout the Earth. The oceans absorb and store carbon dioxide and must be considered together with the atmosphere in any study of the greenhouse effect. The oceans play a further role in controlling our climate through the absorption of solar energy, and the consequent production of water vapour and flow of warm water that helps to drive currents in the air and the seas.

The oceans help to make the Earth hospitable to life, and have kept our planet that way for over 3.5 billion years. Despite their importance, our understanding of ocean processes is far from complete and they are one of the major sources of uncertainty in scientists' attempts to model future global conditions.

This chapter attempts to raise awareness of the importance of the oceans to life on Earth, and to bring out some of the fundamental chemistry which lies behind some ocean processes. Major chemical ideas are developed and linked (perhaps unexpectedly) to familiar objects such as seashells and to the behaviour of water itself.

Knowledge and understanding checklist

In this chapter you will learn more about some ideas introduced in earlier chapters:

☐ ionic bonding (**Elements of life**)

☐ acids and bases (**Elements of life**)

☐ enthalpy changes (**Developing fuels**)

☐ intermolecular bonds (**The ozone story**)

☐ chemical equilibrium (**Elements from the sea**)

☐ equilibrium constants (**The chemical industry** and **Elements from the sea**).

Maths skills checklists

In this chapter you will need to use the following maths skills. You can find support for these skills on Kerboodle and through MyMaths.

☐ Recognise and make use of appropriate units in calculations.

☐ Recognise and use expressions in decimal and ordinary form.

☐ Use calculators to find and use power and exponential functions.

☐ Use an appropriate number of significant figures.

☐ Understand and use the symbols $=$, $<$, $<<$, $>>$, $>$, $\propto$, $\sim$, $\rightleftharpoons$.

☐ Change the subject of an equation.

☐ Substitute numerical values into algebraic equations, using appropriate units for physical quantities.

☐ Solve algebraic equations.

☐ Use logarithms in relation to quantities that range over several orders of magnitude.

MyMaths.co.uk
Bringing Maths Alive

The deep oceans

Scientists know less about the deepest points on Earth than they do about the surface of Mars. On 26 March 2012, the *Deepsea Challenger* was successfully piloted to the oceans' deepest point. This historic expedition to the Mariana Trench's lowest point – the Challenger Deep, which lies 10.99 km below the ocean surface – was the first extensive scientific exploration in a manned submersible of the deepest spot on Earth.

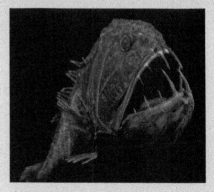

▲ **Figure 1** *The Fangtooth fish is one of the deepest-living fishes to be discovered, usually found up to 2000 metres below sea level (although it has been found up to 5000 metres below sea level)*

Surveying the seas

Although humans have been exploring the Earth throughout the ages, it is only relatively recently that detailed investigations of the deep oceans have taken place. The historic *HMS Challenger* voyage (1872–1876) was the first ever dedicated marine science expedition. The warship had undergone extensive alterations, including the construction of two laboratories on board – one for chemistry and one for biology – and the crew included a team of six scientists. It was assigned to investigate 'everything about the sea'.

After three and a half years, during a journey of almost 130 000 km, the expedition had collected information from most of the oceans, as well as samples of water, sediments, and marine life. More than 4000 new species of marine animals had been discovered, the extent of the Mid-Atlantic Ridge had been measured (although its significance was not recognised until the 1960s), valuable nodules rich in manganese, copper, nickel, and cobalt had been found on the sea floor, and the scientists had even managed to measure the depth of the Mariana Trench using only a weight tied to rope. The *HMS Challenger* scientists measured the depth of the trench as 8200 m, which is reasonably close to modern measurements that put it at about 11 000 m. However, some of the most important discoveries have arisen from the measurements the *HMS Challenger* scientists made of the temperature and salinity of deep-ocean water. Their data provided vital evidence for working out how water circulates around the oceans and how the oceans control the global climate.

The voyage of *HMS Challenger* provided the scientific basis of modern oceanography. Today an ocean survey ship, equipped with much more sensitive equipment, usually makes measurements at only one site per day, travelling 400 km between sites. Accurate information may be of little use if the system you are studying has changed significantly by

▲ **Figure 2** *HMS Challenger – her voyage from 1872 to 1876 was the first systematic study of the oceans*

the time they have finished collecting your data. Some of the events being studied are over in a matter of days or a few weeks. A space satellite's greater speed is often a considerable advantage, even though the measurements it makes may sometimes be less accurate. Most recently, NASA's Operation Icebridge has used satellite images of the oceans around the polar regions to monitor changes in thickness of sea ice, glaciers, and ice sheets to assess the impact that global warming is having on these areas.

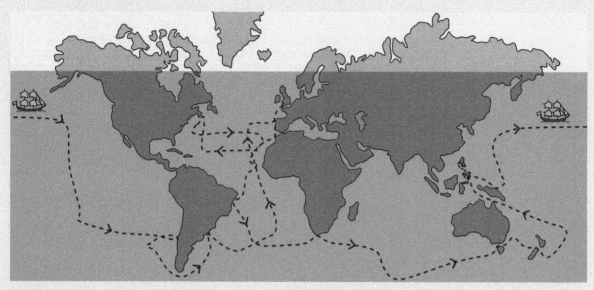

▲ **Figure 3** HMS Challenger's *voyage – it took three and a half years, compared with the area covered by a single satellite over a period of 10 days (shaded area)*

Surveys have shown that the depth of the oceans is far from constant. Mount Everest rises 8.85 km above sea level but there are several parts of the ocean that are deeper than 10 km. The average height of the land is only around 840 m above sea level, whereas the average depth of the oceans is 3700 m.

The oceans play an important role in controlling the Earth's climate. Together with the atmosphere, the oceans are at the centre of the system that controls global conditions – the conditions in which life exists and under which it has evolved for billions of years.

Salt of the Earth

The sea is salty. Salts are ionic compounds and over 99% of all the dissolved substances in sea water are ionic. The ions of the dissolved salts are free in solution.

Figure 4 shows the abundance of these ions. The proportion of one to another is remarkably constant, wherever a sample is taken from across the global seas.

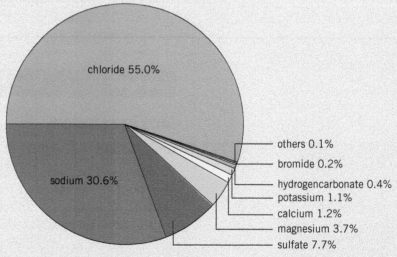

▲ **Figure 4** *Percentages by mass of different ions in sea water*

The composition of sea water has been known for over 100 years but it was only recently that scientists have begun to understand in any detail how the sea became salty.

Rainwater leaches salts from the soil and rivers on land and washes them into the sea. However, the seas contain, in abundance, some elements that are not found to any great extent in river water.

Underneath the sediments on the ocean floor are lavas, generated by long, thin, underwater volcanoes called mid-ocean ridges. The gases given off from these volcanoes are rich in compounds containing chlorine, bromine, and sulfur. Also, as molten lavas meet cold sea water they solidify and shatter. Water streams down through cracks, leaching out soluble minerals. The superheated solutions that re-emerge through hydrothermal vents are much richer sources of elements such as chlorine, bromine, and sulfur than crustal rock. The sources of the dissolved ions in sea water are shown in Figure 5.

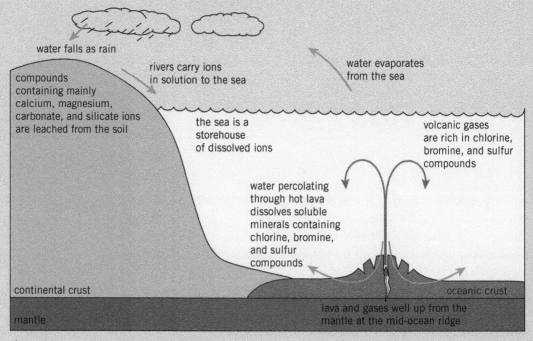

▲ **Figure 5** *Sources of dissolved ions in sea water*

Chemical ideas: Energy changes and chemical reactions 4.4

Ions in solution

Ionic solids

In ionic solids ions are held together by their opposite electrical charges. Each positive ion (called a **cation**) attracts several negative ions (called **anions**), and vice versa. The ions build up into a giant ionic lattice, in which very large numbers of ions are arranged in fixed positions.

One of the simplest examples is sodium chloride, $Na^+Cl^-(s)$ (Figure 6).

In the sodium chloride lattice, each Na^+ ion is surrounded by six Cl^- ions, and each Cl^- ion is surrounded by six Na^+ ions. Each Na^+ is attracted to the six Cl^- around it, but repelled by other Na^+ ions which are a bit further away (there are 12 of these), and attracted to the next lot of Cl^- ions (eight of them) which are further away still, and so on. It adds up to an infinite series of attractions and repulsions but overall the attractions are stronger than the repulsions, which is why the lattice holds together. The lattice holds together very strongly, which is why ionic solids are hard and have high melting and boiling points.

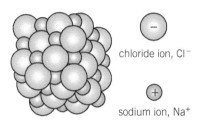

chloride ion, Cl^-

sodium ion, Na^+

▲ **Figure 6** *The sodium chloride lattice*

Ionic substances in solution

Many ionic substances dissolve readily in water. When they do the ions become surrounded by water molecules and spread out through the water (Figure 7). The dissolved ions, $Na^+(aq)$ and $Cl^-(aq)$, are no longer regularly arranged – they are scattered through the water at random. Once the Na^+ and Cl^- ions are separated they behave independently of each other.

This applies to all ionic substances. As soon as they are dissolved, the positive and negative ions separate and behave independently. So it is best to regard sea water, for example, as a mixture of positive and negative ions dissolved in water, rather than as a solution of ionic compounds.

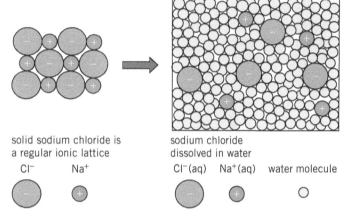

solid sodium chloride is a regular ionic lattice

Cl^- Na^+

sodium chloride dissolved in water

$Cl^-(aq)$ $Na^+(aq)$ water molecule

▲ **Figure 7** *What happens when an ionic substance such as sodium chloride dissolves in water*

Energy changes in solutions

However, not all ionic substances dissolve in water, so what decides whether an ionic substance will dissolve? One important factor is the energy changes that are involved.

Before an ionic solid such as sodium chloride can dissolve, the ions must be separated from the lattice so they can spread out in the solution. This is an endothermic process. The ions in solution are hydrated in an exothermic process.

Lattice enthalpy

Before an ionic solid can dissolve, the ions must be separated from their lattice so they can spread out in the solution. This means supplying energy to overcome the electrical attraction between the oppositely charged ions.

The strength of the ionic attractions is measure in a lattice by the **lattice enthalpy** of the solid. The lattice enthalpy $\Delta_{LE}H$ is the enthalpy change when one mole of solid is *formed* by the coming together of the separate ions. When ions are separated from one another, you can think of them as being in the gaseous state – when they are together in the lattice they are in the solid state.

> **Synoptic link**
>
> You found out about ionic bonding and ionic structures in Topic EL 7, Blood, sweat, and seas.

> **Synoptic link**
>
> In Topic DF 1, Getting energy from fuels, you found out about bond enthalpies.

Study tip

Forming an ionic lattice from gaseous ions is always an exothermic process.

$\Delta_{LE}H$ is always negative.

The *stronger* the ionic bonding in the lattice, the *more negative* is $\Delta_{LE}H$.

Synoptic link

You looked at bond polarity in Topic OZ 6, The CFC story.

So lattice enthalpy is defined as the enthalpy change involved in processes such as:

$Na^+(g) + Cl^-(g) \rightarrow NaCl(s)$ $\quad \Delta_{LE}H(NaCl) = -788\,kJ\,mol^{-1}$

$Mg^{2+}(g) + 2Br^-(g) \rightarrow MgBr_2(s)$ $\quad \Delta_{LE}H\,(MgBr_2) = -2434\,kJ\,mol^{-1}$

All lattice enthalpies are large *negative* quantities. If you want to break down a lattice, you have to put in energy equal to $-\Delta_{LE}H$ (which now becomes a *positive* quantity because you are putting energy *in*). This tends to stop substances dissolving unless the energy is 'paid back' later. Table 1 sets out some lattice enthalpy values for some simple ionic compounds.

▼ **Table 1** *Lattice enthalpies for some ionic compounds*

Compound	$\Delta_{LE}H$ / kJ mol^{-1}	Compound	$\Delta_{LE}H$ / kJ mol^{-1}
Li_2O	−2806	MgO	−3800
Na_2O	−2488	CaO	−3419
K_2O	−2245	SrO	−3222
LiF	−1047	MgF_2	−2961
NaF	−928	CaF_2	−2634
KF	−826	Al_2O_3	−15 916

▼ **Table 2** *Ionic radii for some cations*

ion	Radius / nm
Li^+	0.078
Na^+	0.098
K^+	0.133
Rb^+	0.149
Mg^{2+}	0.078
Ca^{2+}	0.106
Sr^{2+}	0.127
Al^{3+}	0.057

Table 2 lists values for the radii of the positive ions (cations) present in the compounds in Table 1. Lattice enthalpy depends on the *size* and *charge* of the ions. Lattice enthalpies become more negative (i.e., more energy is given out) when:

- the ionic charges increase
- the ionic radii decrease.

These two factors can be summarised by saying that the lattice enthalpy $\Delta_{LE}H$ becomes more negative for ions with greater charge density. This is because ions with a higher charge attract one another more strongly (electrostatic interactions increase). They also attract more strongly if they are closer together. Ions with a smaller radius can come closer together. Stronger attractions mean more negative lattice enthalpies. Substances with large negative lattice enthalpies, such as Al_2O_3, are usually insoluble.

Synoptic link

Charge density and stability of an ion was introduced in Topic EL 8, Spectacular metals.

Study tip

Water is not the only solvent that an ionic lattice can dissolve in.

Hydration and solvation

Despite the need to supply energy to break up the lattice, many ionic substances *do* dissolve. Something else must happen to supply the energy needed.

Most ionic substances can form aqueous solutions with water as the solvent. The covalent bonds in water molecules are polar because of the difference in electronegativity between oxygen and hydrogen. Because the water molecule has a bent shape, the whole molecule is polar and behaves as a tiny dipole (Figure 8).

▲ **Figure 8** *Water molecules are polar*

The tiny charges on the water molecules are attracted to the charges on the ions. These are known as ion–dipole interactions. This happens on the surface of an ionic solid which is placed in water, so the ions become separated from the lattice and surrounded by water

molecules – they are now in solution. In solution, the positive ions are surrounded by water molecules with the negative end of the dipole facing towards them. The negative ions are surrounded by water molecules with the positive end of the dipole facing towards them (Figure 9). The ions in solution are **hydrated** – they have water molecules bound to them.

Water molecules bind weakly to some ions, with the result that they are not extensively hydrated. Other ions are extensively hydrated and bind very strongly to the water molecules. Table 3 shows approximately how many water molecules are likely to be attached to particular positive ions in solution, though the ion will have an effect on all the water molecules surrounding it, not just the closest ones. The higher the charge density of the ion, the more water molecules it attracts, and the bigger the hydrated ion is. So an ion that is small in the absence of water can become large as a hydrated ion.

When bonds form between ions and water molecules, energy is released and this may supply enough energy for more ions to be removed from the lattice. It is not quite as simple as this. Water molecules are strongly attracted to each other by intermolecular bonds called **hydrogen bonds**. When ions dissolve in water, some of the water molecules must be pulled apart so that they can regroup around the ions. This process requires energy too.

The strength of the attractions between ions and water molecules is measured by the **enthalpy change of hydration** $\Delta_{hyd}H$. This is the enthalpy change for the formation of a solution of ions from one mole of gaseous ions. For example:

$$Na^+(g) + aq \rightarrow Na^+(aq) \qquad \Delta_{hyd}H = -406\,kJ\,mol^{-1}$$
$$Br^-(g) + aq \rightarrow Br^-(aq) \qquad \Delta_{hyd}H = -337\,kJ\,mol^{-1}$$

The symbol aq is used in an equation to represent water when it is acting as a solvent. Enthalpies of hydration depend on the concentration of the solution produced. Values quoted refer to a very dilute solution where you can assume interactions between the ions are negligible.

Enthalpies of hydration are always negative – hydration is *exothermic* and energy is given out. Some values are listed in Table 4. If you compare Table 4 with Table 2, you will see that the most exothermic values occur for the ions with:

- the greatest charge
- the smallest radii.

The reasons are very similar to those used earlier to explain the variation in $\Delta_{LE}H$. Small, highly charged ions can get close to water molecules and attract the water molecules strongly. This means the attraction between water molecules and the ions is stronger so the more negative $\Delta_{hyd}H$.

Molecules of some other solvents, such as ethanol, are also polar and can bind to ions. When dealing with solvents other than water, the more general term **enthalpy of solvation** $\Delta_{solv}H$ is used rather than enthalpy of hydration.

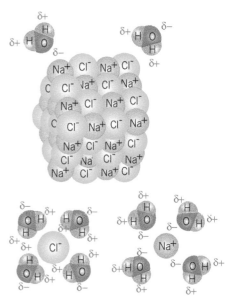

▲ **Figure 9** *Polar water molecules attract the ions in a solid lattice*

▼ **Table 3** *Approximate extent of hydration for some positive ions*

Ion	Average number of attached water molecules
Li^+	5
Na^+	5
K^+	4
Mg^{2+}	15
Ca^{2+}	13
Al^{3+}	26

Synoptic link

You met hydrogen bonding in Topic OZ 7, The ozone hole.

▼ **Table 4** *Enthalpies of hydration of some ions*

Ion	$\Delta_{hyd}H\,/\,kJ\,mol^{-1}$
Li^+	−520
Na^+	−406
K^+	−320
Rb^+	−296
Mg^{2+}	−1926
Ca^{2+}	−1579
Sr^{2+}	−1446
Al^{3+}	−4680

Enthalpy change of solution

The hydration of ions favours dissolving and helps to supply the energy needed to separate the ions from the lattice. The difference between the enthalpy changes of hydration of the ions and the lattice enthalpy gives the **enthalpy change of solution** $\Delta_{solution}H$. The enthalpy change of a solution is the enthalpy change when one mole of a solute dissolves to form a very dilute solution. $\Delta_{solution}H$ can be measured experimentally.

$$\Delta_{solution}H = \Delta_{hyd}H(\text{cation}) + \Delta_{hyd}H(\text{anion}) - \Delta_{LE}H$$

You can represent the enthalpy changes involved using an enthalpy cycle (Figure 10). The enthalpy change for breaking up the ionic lattice is *minus* $\Delta_{LE}H$, because $\Delta_{LE}H$ is defined as the enthalpy change when the lattice is *created* – the opposite of breaking it up. Hydration of ions is an exothermic process. $\Delta_{hyd}H$ is always negative.

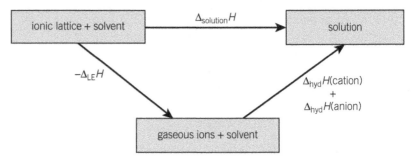

▲ **Figure 10** *An enthalpy cycle to show the dissolving of an ionic solid*

Figure 11, Figure 12, and Figure 13 show three different examples for this enthalpy cycle but in the form of an enthalpy level diagram. This makes it easier to compare the sizes of the different enthalpy changes involved.

Figure 11 represents a solute for which $\Delta_{solution}H$ is *negative*. The hydration of the ions provides slightly more energy than is needed to break up the lattice. This type of solute normally dissolves, giving out a little energy in the process. The process is energetically favourable.

Synoptic link

You learnt how to construct enthalpy level diagrams in Topic DF 2, How much energy?

Synoptic link 🧪

A general method for measuring the energy transferred in experiments involving enthalpy changes of solution is given in Techniques and procedures.

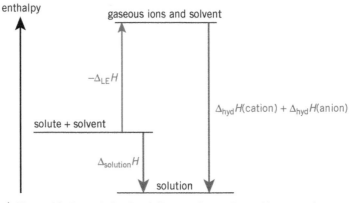

▲ **Figure 11** *An enthalpy level diagram for a solute with a negative $\Delta_{solution}H$. This type of solute will normally dissolve*

Figure 12 represents a solute for which $\Delta_{solution}H$ has a large *positive* value. The hydration of the ions does not provide as much energy as is needed to break up the lattice. This solute does not dissolve. The process is energetically unfavourable.

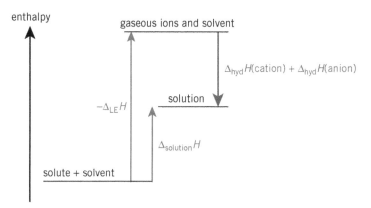

enthalpy

gaseous ions and solvent

$\Delta_{hyd}H(\text{cation}) + \Delta_{hyd}H(\text{anion})$

solution

$-\Delta_{LE}H$

$\Delta_{solution}H$

solute + solvent

▲ **Figure 12** *An enthalpy level diagram for a solute with a positive $\Delta_{solution}H$. The solute will not dissolve because too much energy is needed*

Figure 13 represents a solute for which $\Delta_{solution}H$ is slightly positive but where the ionic solute still dissolves. Many ionic solutes are like this – they dissolve even though it appears energetically unfavourable. Just why this can happen concerns entropy.

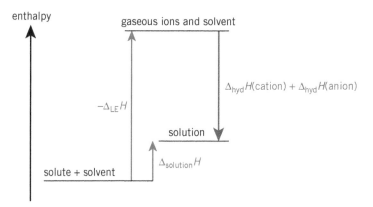

enthalpy

gaseous ions and solvent

$\Delta_{hyd}H(\text{cation}) + \Delta_{hyd}H(\text{anion})$

$-\Delta_{LE}H$

solution

$\Delta_{solution}H$

solute + solvent

▲ **Figure 13** *An enthalpy level diagram for a solute with a slightly positive $\Delta_{solution}H$. If the entropy increase is favourable, this solute may dissolve despite needing energy from the surroundings*

Non-polar solvents

Ionic solids such as sodium chloride, NaCl, are insoluble in non-polar solvents such as hexane. The molecules in non-polar solvents have no regions of slight positive and negative charge, so they are unable to interact strongly with ions. The enthalpy level diagram for dissolving an ionic solid in a non-polar solvent would look like Figure 14. The large positive value of $\Delta_{solution}H$ prevents the solid dissolving.

Activity O 1.1

In this activity you can practical determine enthalpy changes of solution and compare your values to existing data.

Synoptic link

You will find out more about entropy in Topic O 5, The global central heating system.

Activity O 1.2

In this activity you will observe solubilities of substances in different solvents and interpret these observations in terms of intermolecular bonding.

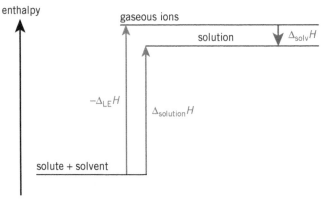

▲ **Figure 14** *The situation when you try to dissolve an ionic solid in a non-polar solvent such as hexane. The enthalpy change of solvation is so small that $\Delta_{solution}H$ is large and positive, and dissolving is unlikely to occur*

Summary questions

1 State the separate ions present in solutions of the following compounds. For example, in a solution of $Mg(NO_3)_2$ the ions present are $Mg^{2+}(aq)$ and $NO_3^-(aq)$:

a $Ca(OH)_2$ *(1 mark)*
b $MgSO_4$ *(1 mark)*
c KOH *(1 mark)*
d $AgNO_3$ *(1 mark)*
e $Al_2(SO_4)_3$ *(1 mark)*

2 Deduce the formulae for the following ionic compounds:

a sodium bromide *(1 mark)*
b magnesium hydroxide *(1 mark)*
c sodium sulfide *(1 mark)*
d barium oxide *(1 mark)*
e calcium carbonate *(1 mark)*
f calcium nitrate *(1 mark)*
g potassium carbonate *(1 mark)*

3 a Explain the statement that the lattice enthalpy of sodium fluoride is $-915\,kJ\,mol^{-1}$.
 b Describe how the sizes of the ions in an ionic compound affect its lattice enthalpy. *(4 marks)*

4 Suggest and explain which compound in each of the following pairs has the more negative (i.e., the more exothermic) lattice enthalpy:

a LiF and NaF *(2 marks)*
b Rb_2O and Na_2O *(2 marks)*
c MgO and Na_2O *(2 marks)*
d KF and KCl *(2 marks)*

5 The enthalpy of hydration of an ion is influenced by its size (small ions can attract water molecules more effectively than big ions) and the charge on the ion. Use these ideas to suggest which ion in each of the following pairs has the most negative (i.e., most exothermic) enthalpy of hydration:

a $Li^+(g)$ and $Na^+(g)$ *(2 marks)*
b $Mg^{2+}(g)$ and $Ca^{2+}(g)$ *(2 marks)*
c $Na^+(g)$ and $Ca^{2+}(g)$ *(2 marks)*

6 The lattice enthalpy of AgF is $-958\,kJ\,mol^{-1}$ and that of AgCl is $-905\,kJ\,mol^{-1}$. The enthalpy of hydration of $Ag^+(g)$ is $-446\,kJ\,mol^{-1}$, that of $F^-(g)$ is $-506\,kJ\,mol^{-1}$, and that of $Cl^-(g)$ is $-364\,kJ\,mol^{-1}$.

a Explain the difference in the lattice enthalpies of the two silver halides. *(2 marks)*
b Explain the difference in the enthalpies of hydration of the two halide ions. *(2 marks)*

02 The role of the oceans in climate control

Specification references: O(i), O(j), O(k), O(l), O(n)

The greenhouse effect

To understand the role that the oceans play in climate control you first need to understand the factors that affect the Earth's climate. In order to do this you need to think about the Earth's atmosphere and the role it plays in maintaining the moderate temperature range that makes life on Earth possible.

Atmospheric gases and global temperature

The Earth is surrounded by a mixture of gases which are called the atmosphere. The balance between the amount and type of radiation that reaches the Earth's atmosphere, is absorbed by the Earth, reradiated into the atmosphere, and subsequently absorbed by the atmosphere or allowed to escape back into space is crucial in determining global temperature.

Human influences are altering the composition of the Earth's atmosphere and the majority of scientists now believe that the increasing levels of greenhouse gases such as carbon dioxide and methane are the main cause of global warming.

Oceans soak up carbon dioxide

However, the oceans play an important role in absorbing carbon dioxide and preventing its build-up in the atmosphere. Oceans cover almost three-quarters of the Earth's surface and carbon dioxide is fairly soluble in water, so large amounts of atmospheric carbon dioxide, $CO_2(g)$, dissolve in the oceans.

When carbon dioxide dissolves in water, it forms hydrated CO_2 molecules:

$$CO_2(g) + aq \rightleftharpoons CO_2(aq)$$

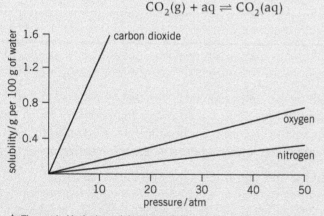

▲ **Figure 1** *Variation of the solubilities of some gases with pressure at 298 K*

Learning outcomes

Demonstrate and apply knowledge and understanding of:

→ the greenhouse effect

→ the Brønsted–Lowry theory of acids and bases

→ the difference between weak and strong acids and bases

→ performing pH calculations involving strong acids, strong bases, and weak acids.

Synoptic link

You need to understand equilibria before you can explain how carbon dioxide interacts with the sea. This was covered in Chapter 3, Elements from the sea.

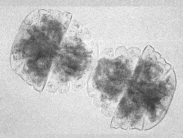

▲ **Figure 2** *Phytoplankton act as a biological pump, removing CO_2 from the atmosphere and transporting organic carbon compounds from surface waters to deeper layers as a rain of dead and decaying organisms. This is balanced by upward transport of carbon by deeper water, which is richer in CO_2 than surface water*

Study tip

Sometimes, an aqueous solution of carbon dioxide is represented by the formula $H_2CO_3(aq)$. It is as if the following reaction has occurred.

$$CO_2(aq) + H_2O(l) \rightarrow H_2CO_3(aq)$$

This is why some people use the name carbonic acid for a solution of carbon dioxide.

This is a reversible reaction, as you will know if you have watched what happens when you open a bottle of fizzy drink. However, it is a fairly slow reaction (which is a good thing from the point of view of fizzy drink consumers) and it takes quite a long time for equilibrium to be reached.

The uptake of carbon dioxide by the oceans is quicker than this. Small marine plants called phytoplankton (Figure 2) use up most of the carbon dioxide that goes into the sea. So the concentration of 'free' $CO_2(aq)$ is small, and gaseous carbon dioxide is encouraged to dissolve.

In addition, some of the carbon dioxide molecules react chemically with water and are removed from the equilibrium. More carbon dioxide therefore dissolves to maintain the equilibrium position:

$$CO_2(aq) + H_2O(l) \rightleftharpoons H^+(aq) + HCO_3^-(aq)$$
$$HCO_3^-(aq) \rightleftharpoons H^+(aq) + CO_3^{2-}(aq)$$

The reaction with water produces a mixture that contains mainly hydrogencarbonate ions, HCO_3^-, and H^+ ions, together with some carbonate ions, CO_3^{2-}. Since $H^+(aq)$ ions are formed, this reaction is responsible for the acidic nature of carbon dioxide.

Much of the excess carbon dioxide released into the atmosphere from the combustion of fuels is absorbed by the oceans. Estimates vary but it seems likely that 35–50% is removed in this way. The oceans continue to absorb carbon dioxide because surface water, rich in carbon dioxide, is constantly being removed and stored away for hundreds of years in the deep ocean. The maximum amount of carbon dioxide is removed in cold regions, where carbon dioxide is most soluble. The ocean currents then move the solution, so more carbon dioxide is absorbed.

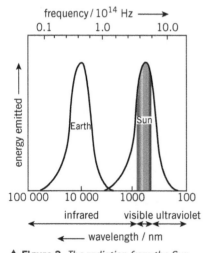

▲ **Figure 3** *The radiation from the Sun which reaches the outer limits of the atmosphere, and the radiation given off from the surface of the Earth (the frequencies and wavelengths are plotted on a logarithmic scale, so each division is a factor of 10 greater than the one before)*

Chemical ideas: Human impacts 15.3

Radiation in, radiation out

When things get hot, they emit electromagnetic radiation. The hotter the object, the higher the energy of the radiation.

The surface of the Sun has a temperature of about 6000 K. This means that it radiates energy in the ultraviolet, visible, and infrared regions.

The Earth is heated by the Sun's radiation. The Earth's average surface temperature is about 285 K (12 °C) – a lot cooler than the Sun but still hot enough to radiate electromagnetic radiation. At this lower temperature, the energy radiated is mainly in the infrared region.

The radiation from the Sun that reaches the outer limits of the atmosphere is mainly in the visible and ultraviolet regions. Part of this energy is absorbed by the Earth and its atmosphere, and part is reflected back into space. The part that is absorbed helps to heat the Earth, and the Earth in turn radiates energy back into space (Figure 3). A steady state is reached where the Earth is radiating energy as fast as it absorbs it, and the average temperature of the Earth remains constant.

As in all steady states, the delicate balance can be disturbed by changes to the system – in particular by changes to the quantities of various gases in the atmosphere.

Carbon dioxide and methane are examples of **greenhouse gases**. Carbon dioxide and methane in the troposphere absorb some of the infrared radiation emitted from the surface of the Earth and prevent the radiation from being re-radiated into space. The effect of this is to make the Earth warmer (Figure 4).

To understand how absorption of infrared radiation by carbon dioxide and methane molecules causes warming, you need to think about what happens to the energy once it is absorbed by the molecules. Two things can happen:

1 Absorption of infrared radiation increases the vibrational energy of the carbon dioxide and methane molecules and the bonds vibrate more vigorously. This vibrational energy can be transferred to other molecules in the air (i.e., nitrogen and oxygen molecules) by collisions. This increases their kinetic energy, raising the temperature of the air.

2 Some infrared radiation is re-emitted by the molecules – back towards the Earth and some is radiated out towards space.

The overall effect is to trap some of the radiation from the Earth that would otherwise be lost.

Carbon dioxide and methane are natural components of the atmosphere. However, since the Industrial Revolution, fossil fuels have been used to fuel industrial plants, and the level of carbon dioxide has been rising. Gradually the Earth's temperature has been rising too and the majority of scientists believe that the increasing level of carbon dioxide and methane in the atmosphere is the main cause of global warming.

IR window
Water vapour is the most abundant greenhouse gas in the atmosphere, however it only absorbs certain wavelengths of infrared radiation. The wavelengths of infrared radiation that water vapour does not absorb is called the **IR window**. Infrared radiation with the same wavelengths as the wavelengths that that water does not absorb escape through the atmosphere and back into space. The combined effects of absorption and the IR window means that water vapour helps to keep the temperature balance of the Earth.

However, carbon dioxide absorbs infrared radiation at the same wavelengths as the IR window. Whilst carbon dioxide is naturally a component of the atmosphere, the increasing levels mean that more radiation is being absorbed that would previously have escaped. Furthermore, whilst water is the most abundant greenhouse gas, its level in the atmosphere is dependent on temperature – the higher the temperature, the higher the level of water vapour in the atmosphere. As temperature increases due to the increasing levels of carbon dioxide in the atmosphere, the amount of water vapour in the atmosphere will also increases, exacerbating the issue.

Synoptic link

You first met the term steady state in Topic OZ 4, Ozone – here today and gone tomorrow.

Look back at topic OZ 1, What's in the air?, to revise the structure of the atmosphere and where the troposphere is.

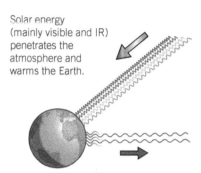

Solar energy (mainly visible and IR) penetrates the atmosphere and warms the Earth.

The Earth absorbs this energy, heats up and radiates IR'
'Greenhouse gases in the troposphere absorb some of this IR

▲ **Figure 4** *The greenhouse effect*

Synoptic link

You have already met the idea of IR radiation interacting with matter in Topic OZ 2, Screening the sun.

Activity O 2.1

This card sort activity helps you understand the sequence of events that make up the greenhouse effect.

All of the greenhouse gases absorb different wavelengths of infrared radiation, however as the levels of greenhouse gases are increased due to human activity, more radiation is being absorbed that would previously not have been.

Some scientists have predicted that the carbon dioxide concentration in the atmosphere could rise from the current levels around 400 ppm to over 650 ppm by 2100. To keep carbon dioxide levels stable, emissions of the gas worldwide will need to be reduced drastically.

Chemical ideas: Acids and bases 8.2

Strong and weak acids and pH
The Brønsted–Lowry theory of acids and bases

The general definition of an acid is that it is a substance which donates H^+ in a chemical reaction. The substance that accepts the H^+ is a base. The reaction in which this happens is called an acid–base reaction.

For example, the reaction of hydrogen chloride with ammonia forms a white salt, ammonium chloride (Figure 5):

$$HCl(g) + NH_3(g) \rightarrow NH_4^+Cl^-(s)$$

In this reaction, the hydrogen chloride transfers H^+ to ammonia. Hydrogen chloride is behaving as an acid and the ammonia is behaving as a base. Since a hydrogen atom consists of only a proton and an electron, an H^+ ion corresponds to just one proton, so you can refer to acids as proton donors and bases as proton acceptors.

This is known as the **Brønsted–Lowry theory** of acids and bases.

▲ **Figure 5** *The white of ammonium chloride is formed when acidic hydrogen chloride and the base ammonia react*

Study tip

The Brønsted–Lowry theory of acids and bases:

An acid is an H^+ donor.

A base is a H^+ acceptor.

Acid–base pairs

Once an acid has donated an H^+ ion, there is always the possibility that it will take it back again. For example, consider ethanoic acid – the acid in vinegar:

$$CH_3COOH(aq) \rightarrow CH_3COO^-(aq) + H^+(aq)$$

When you add a strong acid to a solution containing ethanoate ions, CH_3COO^-, the ethanoate ions accept H^+ from the stronger acid and go back to ethanoic acid:

$$CH_3COO^-(aq) + H^+(aq) \rightarrow CH_3COOH(aq)$$

In this reaction, the ethanoate ion is behaving as a *base* – it is called the **conjugate base** of ethanoic acid.

Every acid has a conjugate base, and every base has a **conjugate acid**. They are called a **conjugate acid–base pair**. If you represent a general acid as HA, then you have:

$$\underset{\text{conjugate acid}}{HA(aq)} \rightarrow H^+(aq) + \underset{\text{conjugate base}}{A^-(aq)}$$

For example, the conjugate base of HCl(aq) is Cl$^-$(aq). The conjugate acid of NH$_3$(aq) is NH$_4^+$(aq). Table 1 shows more examples.

Table 1 shows how water, H$_2$O, can be both an acid and a base – it all depends on what the water is reacting with. If it is reacting with a strong acid, such as HCl(aq), water acts as a base accepting H$^+$(aq) and forming H$_3$O$^+$(aq). This is called the oxonium ion:

$$HCl(aq) + H_2O(l) \rightarrow H_3O^+(aq) + Cl^-(aq)$$
$$\text{acid} \qquad \text{base}$$

Water may also act as an acid. For example:

$$H_2O + NH_3 \rightarrow OH^- + NH_4^+$$
$$\text{acid} \quad \text{base}$$

Here water is donating a proton to ammonia.

The pH scale

The **pH scale** was devised at the beginning of the 20th century by a Danish chemist called Søren Sørensen. He was studying the brewing of beer. Brewing requires careful control of acidity to produce conditions in which yeast will grow but unwanted bacteria will not, so he devised a simple way to indicate how much acid or alkali was present in the brewing solution. pH measures the power of H$^+$(aq) in a solution, that is, its concentration. pH is defined as $-\log_{10}[H^+(aq)]$.

Figure 6 shows how the pH of a solution is related to [H$^+$(aq)] in a solution. The pH value is the same as the negative power of 10 that relates to [H$^+$(aq)]. When [H$^+$(aq)] changes by a factor of 10, the pH changes by 1. This is known as a logarithmic scale. Solutions with a pH of 7 are neutral. Acids have a pH of less than 7 and alkalis have a pH of more than 7.

The pH scale runs in the opposite direction to the scale of [H$^+$(aq)] values – a *low* pH corresponds to a *high* concentration of H$^+$(aq).

Many solutions of acids or alkalis have a concentration that is less than 1 mol dm^{-3} and so the pH values of these solutions lie between 0 and 14.

An acid where [H$^+$(aq)] is more than 1 mol dm^{-3} has a pH of less than 0, that is, a negative pH. Similarly, an alkaline solution where [OH$^-$(aq)] is greater than 1 mol dm^{-3} has a pH greater than 14.

Strong and weak acids

Strong acids

Acids vary in strength because acids donate H$^+$ to differing extents. **Strong acids** have a strong tendency to donate H$^+$ and dissociate completely in aqueous solution. So if HA represents the strong acid, the equation for the reaction is:

$$HA(aq) + H_2O(l) \rightarrow H_3O^+(aq) + A^-(aq)$$

If water, which is present in excess, is left out this can be simplified to:

$$HA(aq) \rightarrow H^+(aq) + A^-(aq)$$

▼ **Table 1** *Conjugate acid–base pairs. State symbols have been omitted for clarity*

	Conjugate acid	Conjugate base
chloric(VII) acid	HClO$_4$	$\rightarrow$ H$^+$ + ClO$_4^-$
hydrochloric acid	HCl	$\rightarrow$ H$^+$ + Cl$^-$
sulfuric acid	H$_2$SO$_4$	$\rightarrow$ H$^+$ + HSO$_4^-$
oxonium ion	H$_3$O$^+$	$\rightarrow$ H$^+$ + H$_2$O
ethanoic acid	CH$_3$COOH	$\rightarrow$ H$^+$ + CH$_3$COO$^-$
hydrogen sulfide	H$_2$S	$\rightarrow$ H$^+$ + SH$^-$
ammonium ion	NH$_4^+$	$\rightarrow$ H$^+$ + NH$_3$
water	H$_2$O	$\rightarrow$ H$^+$ + OH$^-$
ethanol	C$_2$H$_5$OH	$\rightarrow$ H$^+$ + C$_2$H$_5$O$^-$

The pH scale

$$pH = -\log_{10}[H^+(aq)]$$

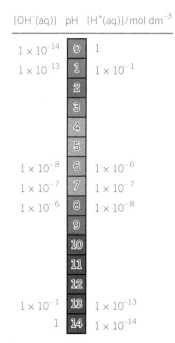

▲ **Figure 6** *The pH scale, including approximate colours of universal indicator solution*

Synoptic link

You first looked at acids in Topic EL 9, How salty?

Activity O 2.2

In this activity you can explore your understanding of strong and weak acids as well as dilute and concentrated acids.

Hydrochloric acid, HCl, and sulfuric acid, H_2SO_4, are examples of strong acids:

$$HCl(aq) \rightarrow H^+(aq) + Cl^-(aq)$$
$$H_2SO_4(aq) \rightarrow H^+(aq) + HSO_4^-(aq)$$

Weak acids

If a substance is a **weak acid**, its tendency to donate H^+ is weaker and it does not dissociate completely when dissolved in water. Some $H^+(aq)$ ions are formed but there is still some unreacted acid in solution. If HA represents a weak acid, its reaction with water can be represented as:

$$HA(aq) \rightleftharpoons H^+(aq) + A^-(aq)$$

The equilibrium sign shows that a significant concentration of undissociated HA(aq) is present along with the $A^-(aq)$ and $H^+(aq)$ ions formed from it.

Ethanoic acid is an example of a weak acid:

$$CH_3COOH(aq) \rightleftharpoons H^+(aq) + CH_3COO^-(aq)$$

The position of this equilibrium is well to the left, so ethanoic acid is a weak acid. The further the equilibrium lies to the right-hand side, the stronger the acid.

Carbon dioxide reacts with water to form a weakly acidic solution:

$$CO_2(aq) + H_2O(l) \rightleftharpoons H^+(aq) + HCO_3^-(aq)$$
$$HCO_3^-(aq) \rightleftharpoons H^+(aq) + CO_3^{2-}(aq)$$

Again, the position of equilibrium in this reaction is well to the left and so this is classified as a weak acid.

Calculating pH values

Calculating the pH of a strong acid

You can find the pH of a solution of a strong acid by a straightforward calculation. Since the reaction with water effectively goes to completion the amount in moles of $H^+(aq)$ ions is equal to the amount in moles of acid (HA) put into solution:

$$HA(aq) \rightarrow H^+(aq) + A^-(aq)$$

Calculating [H⁺(aq)] of a strong acid from pH

The expression for pH can be rearranged so that you can calculate $[H^+(aq)]$ from a pH value.

$$pH = -\log_{10}[H^+(aq)]$$
$$[H^+(aq)] = 10^{-pH}$$

Calculating the pH of a weak acid

Weak acids do not dissociate completely and so you cannot assume that the concentration of H^+ ions is equal to the initial concentration of the acid. Instead, the **equilibrium constant** is used to calculate $[H^+]$.

$$HA(aq) \rightleftharpoons H^+(aq) + A^-(aq)$$

 Worked example: Calculating pH

What is the pH of a $0.01\,mol\,dm^{-3}$ solution of a strong acid HA?

Step 1: It is a strong acid so it is fully dissociated.

$[H^+(aq)] = 0.01\,mol\,dm^{-3}$

Step 2: Calculate pH.

$pH = -\log_{10}(0.01) = 2$

 Worked example: Calculating [H⁺(aq)]

A sample of strong acid, HA, has a pH = 2.6.

What is the $[H^+(aq)]$?

$[H^+(aq)] = 10^{-pH}$

$= 10^{-2.6}$

$= 2.5 \times 10^{-3}$

You can write an equilibrium constant for this reaction in the usual way:

$$K_a = \frac{[H^+(aq)][A^-(aq)]}{[HA(aq)]}$$

This equilibrium constant is called the **acidity constant** K_a (or often the acid dissociation constant). Table 2 gives values for some weak acids.

To find $[H^+(aq)]$, two assumptions about weak acids are made that simplify the calculation.

Assumption 1

$$[H^+(aq)] = [A^-(aq)]$$

At first sight this might seem obvious because equal amounts of $H^+(aq)$ and $A^-(aq)$ are formed from HA. But there is another source of $H^+(aq)$, that is water itself through the equilibrium:

$$H_2O(l) \rightleftharpoons H^+(aq) + OH^-(aq)$$

Water produces far fewer $H^+(aq)$ ions than most weak acids, so you do not introduce a significant inaccuracy into the calculation by neglecting the ionisation of water.

Assumption 2

The amount of HA at equilibrium is equal to the amount of HA put into the solution.

In other words, when the concentration of HA at equilibrium is calculated, you can neglect the fraction of HA which has lost H^+. This can be done because you are dealing with weak acids and this fraction is very small. For example, in a solution of HA of concentration $0.1 \, mol \, dm^{-3}$, you can assume that when equilibrium has been established:

$$[HA(aq)] = 0.1 \, mol \, dm^{-3}$$

 Worked example: Finding pH of a weak acid

Find the pH of a solution of ethanoic acid with a concentration of $1 \, mol \, dm^{-3}$. (K_a for ethanoic acid is $1.7 \times 10^{-5} \, mol \, dm^{-3}$ at 298 K.)

Step 1: You can represent the reaction between ethanoic acid and water by the general equation:

$$HA(aq) \rightleftharpoons H^+(aq) + A^-(aq)$$

Therefore the acidity constant is given by:

$$K_a = \frac{[H^+(aq)][A^-(aq)]}{[HA(aq)]} = 1.7 \times 10^{-5} \, mol \, dm^{-3} \text{ (at 298 K)}$$

Step 2: Assumptions 1 and 2 then allow you to write:

$$1.7 \times 10^{-5} \, mol \, dm^{-3} = K_a = \frac{[H^+(aq)]^2}{1 \, mol \, dm^{-3}}$$

Therefore $[H^+(aq)]^2 = 1.7 \times 10^{-5} \, mol^2 \, dm^{-6}$

So $[H^+(aq)] = 4.12 \times 10^{-3} \, mol \, dm^{-3}$ and pH $= 2.38$ (at 298 K)

▼ **Table 2** *Acidity constants for some weak acids at 298 K*

Acid	$K_a / mol \, dm^{-3}$
methanoic acid, HCOOH	1.6×10^{-4}
benzoic acid, C_6H_5COOH	6.3×10^{-5}
ethanoic acid, CH_3COOH	1.7×10^{-5}
chloric(I) acid, HClO	3.7×10^{-8}
hydrocyanic acid, HCN	4.9×10^{-10}
nitric(III) acid, HNO_2	4.7×10^{-4}

Synoptic link

You need to be able to carry out equilibrium constant calculations in order to calculate the pH of a substance. These were covered in Topic CI 2, Manufacturing nitric acid.

Synoptic link

The procedure for measuring pH of a solution using a pH meter is given in Techniques and procedures.

Study tip

Measurement of $[H^+(aq)]$ or pH alone does not allow you to distinguish strong and weak acids.

Concentration is a measure of the amount of substance in a given volume of solution – typically measured in $mol \, dm^{-3}$. Strength is a measure of the extent to which an acid can donate H^+. The two terms have very different meanings in chemistry but are often interchangeable in everyday language.

Activity O 2.3

This activity supports your understanding of pH and how dilution of strong and weak acids impacts on pH.

Activity O 2.4

This practical helps check your understanding of pH and K_w.

Weak acids and pKₐ

Values for K_a can be very small for weak acids. As such, a logarithmic scale pK_a can be used.

$$pK_a = -\log_{10} K_a$$

The *weaker* the acid the *higher* the value of pK_a.

Calculating the pH of a strong alkali

In alkaline solutions it takes two steps to calculate $[H^+]$. The first involves using a special equilibrium constant called the **ionisation product of water** K_w.

Water is not thought of as an ionic substance but water does in fact ionise slightly. When water ionises, it behaves as both an acid and a base. In the equilibrium water is acting like other weak acids, because H—OH can be thought of as HA, and OH^- as A^-.

$$H_2O(l) \rightleftharpoons H^+(aq) + OH^-(aq)$$

You can write an expression for K_a for water as:

$$K_a = \frac{[H^+(aq)][OH^-(aq)]}{[H_2O(l)]}$$

However, you can leave out the $[H_2O(l)]$ term, which is effectively constant because water is present in excess. The expression becomes:

$$K_w = [H^+(aq)][OH^-(aq)]$$

At 298 K, $K_w = 1 \times 10^{-14}\,mol^2\,dm^{-6}$

An acidic solution is defined as one in which $[H^+(aq)] > [OH^-(aq)]$. An alkaline solution is defined as one in which $[H^+(aq)] < [OH^-(aq)]$.

You can then use K_w to calculate the pH of a solution of a strong base – a solution in which the production of hydroxide ions is complete.

 Worked example: Calculating the pH of pure water

In pure water
$[H^+(aq)] = [OH^-(aq)]$, so
$K_w = 1 \times 10^{-14}\,mol^2\,dm^{-6}$
$= [H^+(aq)]^2$

Therefore
$[H^+(aq)] = 1 \times 10^{-7}\,mol\,dm^{-3}$
and pH = 7

Worked example: Calculating pH of a strong base

Calculate the pH of $0.1\,mol\,dm^{-3}$ NaOH(aq).

Sodium hydroxide is completely ionised in solution – it is a **strong base** – so in $0.1\,mol\,dm^{-3}$ NaOH(aq), $[OH^-(aq)] = 0.1\,mol\,dm^{-3}$. You can neglect the small amount of OH^- formed from water, so:

$$K_w = 1 \times 10^{-14}\,mol^2\,dm^{-6}$$
$$= [H^+(aq)] \times 0.1\,mol\,dm^{-3}$$

Therefore:

$$[H^+(aq)] = \frac{1 \times 10^{-14}\,mol^2\,dm^{-6}}{0.1\,mol\,dm^3}$$

or

$$[H^+(aq)] = 1 \times 10^{-13}\,mol\,dm^{-3} \text{ and pH} = 13$$

Summary questions

1 Write out each of the following equations and state which of the reactants is the acid and which is the base.

  a $HNO_3 + H_2O \rightarrow H_3O^+ + NO_3^-$ (1 mark)
  b $NH_3 + H_2O \rightarrow NH_4^+ + OH^-$ (1 mark)
  c $NH_4^+ + OH^- \rightarrow NH_3 + H_2O$ (1 mark)
  d $SO_4^{2-} + H_3O^+ \rightarrow HSO_4^- + H_2O$ (1 mark)
  e $H_2O + H^- \rightarrow H_2 + OH^-$ (1 mark)
  f $H_3O^+ + OH^- \rightarrow 2H_2O$ (1 mark)
  g $NH_3 + HBr \rightarrow NH_4^+ + Br^-$ (1 mark)
  h $H_2SO_4 + HNO_3 \rightarrow HSO_4^- + H_2NO_3^+$ (1 mark)

2 Describe each of these reactions as either acid–base or redox.

  a $NH_4^+ + CO_3^{2-} \rightarrow NH_3 + HCO_3^-$ (1 mark)
  b $H_2S + 2OH^- \rightarrow S^{2-} + 2H_2O$ (1 mark)
  c $I_2 + 2OH^- \rightarrow I^- + IO^- + H_2O$ (1 mark)
  d $Mg + 2H^+ \rightarrow Mg^{2+} + H_2$ (1 mark)

3 Calculate the pH values of the following solutions of strong acids at 298 K.

  a $0.01 \, mol \, dm^{-3}$ hydrochloric acid (1 mark)
  b $0.2 \, mol \, dm^{-3}$ nitric acid (1 mark)
  c $0.2 \, mol \, dm^{-3}$ sulfuric acid (1 mark)
  d $0.1 \, mol$ of $HClO_4(l)$ in $250 \, cm^3$ of aqueous solution (1 mark)

4 Calculate the pH values of the following solutions of weak acids at 298 K. You will need to use the K_a values given in Table 2.

  a $0.1 \, mol \, dm^{-3}$ ethanoic acid (2 marks)
  b $0.05 \, mol \, dm^{-3}$ ethanoic acid (2 marks)
  c $0.001 \, mol \, dm^{-3}$ benzoic acid (2 marks)
  d $0.25 \, mol$ of methanoic acid in $100 \, cm^3$ of solution (2 marks)

5 $0.01 \, mol \, dm^{-3}$ hydrochloric acid solution and a $0.2 \, mol \, dm^{-3}$ nitric(III) acid (HNO_2) solution have a similar pH value of 2:

  a Classify each of these acids as strong or weak. (2 marks)
  b Explain the difference between a strong acid and a weak acid using HCl and HNO_2 as examples. (2 marks)

6 Calculate the pH of the following solutions of strong bases at 298 K.

  a $1 \, mol \, dm^{-3} \, KOH(aq)$ (1 mark)
  b $0.1 \, mol \, dm^{-3} \, Ba(OH)_2(aq)$ (1 mark)

7 Indicators are weak acids. The acidic form, HIn, of an indicator is one colour, with its conjugate base, In^-, being a different colour. At the end point for a titration, $[HIn(aq)] = [In^-(aq)]$.

For the indicator phenolphthalein, HIn is colourless and In is pink.

$$HIn(aq) \rightleftharpoons H^+(aq) + In^-(aq)$$

  a What is the colour of phenolphthalein in alkaline solution? Explain your answer. (1 mark)
  b Write an expression for the acidity constant K_a for phenolphthalein. (1 mark)
  c Use this expression to calculate $[H^+(aq)]$, and hence the pH at the end point of a titration when phenolphthalein is used as an indicator. The pK_a for phenolphthalein at 298 K is 9.3. (1 mark)

O 3   The oceans as a carbon store

Specification reference: O(m)

Learning outcomes

Demonstrate and apply knowledge
and understanding of:

→ the meaning of the term buffer

→ how buffers work (including
    everyday applications)

→ buffer solution calculations.

▲ **Figure 1** *The first land plants
appeared 400 million years ago*

Life on Earth

The Earth's early life forms evolved in the oceans and that is where
they stayed throughout most of the Earth's history. The planet was
nearly 4 billion years old and living things had existed for 3 billion
years before life moved out of the oceans onto the land.

The Earth's early atmosphere consisted mainly of carbon dioxide,
ammonia, methane, and hydrogen sulfide. Photosynthesis became
possible with the evolution of cyanobacteria. These bacteria produced
oxygen but it was used up by reducing agents dissolved in the sea
water before it could build up in the atmosphere. This was just as well
for the cyanobacteria because they cannot tolerate oxygen. Instead,
they use sulfate ions, SO_4^{2-}, and nitrate(V) ions, NO_3^-, as oxidising
agents in respiration. It was only later that organisms could use the
free oxygen that was dissolved in the sea water or that had built up in
the atmosphere.

The Earth's atmosphere has changed dramatically since the early days
of life. Reducing agents (e.g., methane) and acidic gases (e.g., carbon
dioxide) have been largely replaced by a neutral, oxidising mixture
of nitrogen and oxygen. If life had been forced to evolve on land in
contact with the air, the primitive organisms would have become
extinct. Instead they were protected by their watery environment,
which has altered remarkably little over billions of years.

Keeping things steady

The ability of the oceans to withstand external changes has been
essential for the unbroken evolution of life. For example, the pH of
the oceans has remained close to 8 for millions of years. Why were the
oceans not much more acidic when the atmosphere contained 35%
CO_2 – a thousand times greater than its present level?

One reason is that a solution of carbon dioxide in water is a weak acid
that reacts incompletely with water. If you represent the acid by the
formula HA, you can show the reaction with water by the equation:

$$HA + H_2O \rightleftharpoons H_3O^+ + A^-$$

H_3O^+ ions (called oxonium ions) make the solution acidic. The position
of equilibrium is well over to the left-hand side of the equation and only
a fraction of the acid added to the water reacts to produce oxonium ions.
So the solution is not as acidic as it would be if all the acid had reacted.

The equation can be simplified by leaving out the water, which is
present in excess:

$$HA(aq) \rightleftharpoons H^+(aq) + A^-(aq)$$

For carbon dioxide, only a small proportion of the $CO_2(aq)$ molecules
react to form hydrogen ions and hydrogencarbonate ions:

$$CO_2(aq) + H_2O(l) \rightleftharpoons H^+(aq) + HCO_3^-(aq)$$

If a small amount of alkali is added to a solution of a weak acid such as aqueous carbon dioxide, some of the $H^+(aq)$ ions are removed. But this causes more $CO_2(aq)$ to react with water to restore the position of equilibrium and so replace the $H^+(aq)$ ions. In this way a solution of carbon dioxide can maintain a constant pH, even when a small amount of alkali is added.

What happens if the ocean becomes more acidic? When the proportion of carbon dioxide in the atmosphere was much higher, the equilibrium would ensure that the concentration of $CO_2(aq)$ was also higher:

$$CO_2(g) \rightleftharpoons CO_2(aq)$$

This in turn would result in the equilibrium moving to the right and generating more $H^+(aq)$ ions:

$$CO_2(aq) + H_2O(l) \rightleftharpoons H^+(aq) + HCO_3^-(aq)$$

You would therefore predict that the pH of the ocean would have been lower (more acidic). But the ocean could resist even this change because it is an example of a **buffer solution** – one that remains within a narrow range of pH values, despite the addition of acid or alkali.

The most common type of buffer solution is made up of a weak acid with one of its salts dissolved in it. The weak acid acts as a reservoir of $H^+(aq)$ ions. These can react with OH^- ions that are added and so prevent the solution becoming more alkaline. The anions from the salt act as bases, reacting with additions of $H^+(aq)$ ions and keeping the solution from becoming acidic.

Simple buffers rely on the shifting back and forth of the equilibrium:

$$HA(aq) \rightleftharpoons H^+(aq) + A^-(aq)$$

The weak acid is represented by HA and the anions in the salt of this weak acid are represented by A^-.

For the ocean to resist an increase in acidity, there must be a supply of $HCO_3^-(aq)$ ions, corresponding to the $A^-(aq)$ ions from the salt of the weak acid in a normal buffer solution. One source of HCO_3^- ions is the material that dissolves in river water and then flows into the sea. But there are also other processes involving seashells, chalk, and limestone that can provide an almost limitless supply of hydrogencarbonate ions.

▲ **Figure 2** *Mitchell Falls in Western Australia. River waters provide a constant supply of HCO_3^- ions to the oceans from the weathering of limestone rocks*

Chemical ideas: Acids and bases 8.3

Buffers

Buffers are solutions that can resist changes in pH, despite the addition of small quantities acid or alkali. The pH of the buffer solution stays approximately constant because most of any acid or alkali that is added to the solution is removed. However, this is only the case for future quantities of added acid or alkali.

Study tip

Many processes in living systems must take place under fairly precise pH conditions. If the pH changes to a value outside a narrow range, the process will not occur at the correct rate, or it may not take place at all and the organism will die. The pH ranges for some fluids in our bodies are:

- stomach juices 1.6–1.8
- saliva 6.4–6.8
- blood 7.35–7.45

Study tip

Many reactions, especially those involving enzymes, are pH sensitive. For these reactions to be carried out in a laboratory a buffer solution of suitable pH can be added to the reaction mixture.

Buffer solutions are usually made from:

- a weak acid and one of its salts, for example, ethanoic acid and sodium ethanoate
- a weak base and one of its salts, for example, ammonia solution and ammonium chloride.

How do buffers work?

The action of a buffer solution depends on the weak acid equilibrium reaction:

$$HA(aq) \rightleftharpoons H^+(aq) + A^-(aq)$$

Two assumptions about the species present in the equilibrium are needed to explain how a buffer solution made from a weak acid and one of its salts would resist changes in pH on addition of a small amount of acid or alkali.

Assumption 1

All the A^- ions come from the salt.

The weak acid, HA, supplies very few A^- ions in comparison with the fully ionised salt.

Assumption 2

Almost all the HA molecules put into the buffer remain unchanged.

Adding acid, H^+ ions

When H^+ ions are added, some $A^-(aq)$ ions from the salt react with the extra $H^+(aq)$ ions to form HA(aq) and water. This removes the H^+ ions from solution and the pH is reestablished.

Adding alkali, OH^- ions

If alkali is added, $H^+(aq)$ ions are removed from the solution. The buffer solution counteracts this because $H^+(aq)$ can be regenerated from the acid HA. The pH is reestablished.

The presence of *both* a weak acid *and* its salt are necessary for a buffer to work. There must be plenty of HA to act as a source of extra $H^+(aq)$ ions when they are needed, and plenty of A^- to act as a sink for any extra $H^+(aq)$ ions which have been added (Figure 3).

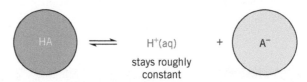

HA
H⁺(aq)
stays roughly constant
+
A⁻
plenty of HA to make more H⁺(aq) if some is used up by alkali that gets added
plenty of A⁻ to combine with any H⁺(aq) that gets added

▲ **Figure 3** *How a buffer solution keeps the pH constant*

Calculations with buffers

For calculations on buffer solutions, all that is needed is the K_a expression for the relevant weak acid:

$$K_a = \frac{[H^+(aq)][A^-(aq)]}{[HA(aq)]} = [H^+(aq)] \times \frac{[A^-(aq)]}{[HA(aq)]}$$

Using Assumptions 1 and 2 you get:

$$K_a = [H^+(aq)] \times \frac{[salt]}{[acid]}$$

The value of $[H^+(aq)]$, and therefore the pH of the buffer solution, depends on *two* factors – the value of K_a and the ratio of [salt] : [acid].

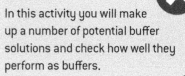

Activity O 3.1

In this activity you will make up a number of potential buffer solutions and check how well they perform as buffers.

The value of K_a

This provides the rough pH of the buffer solution. K_a values normally lie in the range $1 \times 10^{-4}\,mol\,dm^{-3}$ to $1 \times 10^{-10}\,mol\,dm^{-3}$. The choice of a particular weak acid determines which *region* of the pH range the buffer is in, from about pH = 4 to pH = 10.

The ratio of [salt] : [acid]

This provides the more precise pH of a buffer solution. Changing the ratio from about 3 : 1 to about 1 : 3 changes $[H^+(aq)]$ by a factor of 9, and alters the pH by approximately 1 unit. The ratio should not be too far outside this range, otherwise there will be insufficient HA or A^- for the buffer to be effective.

Activity O 3.2

This practical brings together ideas about acids, bases, and buffer solutions.

The expression for K_a shows that the pH of a buffer is not affected by dilution.

$$K_a = [H^+(aq)] \times \frac{[salt]}{[acid]}$$

When you add water, the concentrations of both the salt and the acid are reduced equally. Therefore the ratio of their concentrations remains the same, and the pH is unchanged.

 ## Worked example: Calculating the pH of a buffer solution

Calculate the pH of a buffer solution that contains $0.1\,mol\,dm^{-3}$ ethanoic acid and $0.2\,mol\,dm^{-3}$ sodium ethanoate. K_a for ethanoic acid is $1.7 \times 10^{-5}\,mol\,dm^{-3}$ at 298 K.

$$K_a = \frac{[H^+(aq)] \times [CH_3COO^-(aq)]}{[CH_3COOH(aq)]}$$

Using Assumptions 1 and 2, you can write:

$$1.7 \times 10^{-5}\,mol\,dm^{-3} = [H^+(aq)] \times \frac{0.2\,mol\,dm^{-3}}{0.1\,mol\,dm^{-3}}$$

Therefore $[H^+(aq)] = 1.7 \times 10^{-5}\,mol\,dm^{-3} \times \frac{0.1\,mol\,dm^{-3}}{0.2\,mol\,dm^{-3}}$

$$[H^+(aq)] = 8.5 \times 10^{-6}\,mol\,dm^{-3}$$

$$pH = 5.07 \text{ (at 298 K)}$$

Summary questions

1 Define the term buffer solution. *(1 mark)*

2 Calculate the pH values of the following buffer solutions at 298 K. You will find K_a values in Table 2, Topic O 2.
   a A solution in which the concentrations of methanoic acid and potassium methanoate are both 0.1 mol dm^{-3}. *(1 mark)*
   b A solution made by dissolving 0.01 mol benzoic acid and 0.03 mol sodium benzoate in 1 dm^3 of solution. *(1 mark)*
   c A solution made by mixing equal volumes of 0.1 mol dm^{-3} methanoic acid and 0.1 mol dm^{-3} potassium methanoate. *(1 mark)*

3 Calculate the pH values of the following buffer solutions at 298 K. You will find K_a values in Table 2, Topic O 2.
   a A solution which is 0.1 mol dm^{-3} with respect to propanoic acid and 0.005 mol dm^{-3} with respect to sodium propanoate. (K_a of propanoic acid = 1.3×10^{-5} mol dm^{-3}) *(1 mark)*
   b A solution made by dissolving 0.005 mol of methanoic acid and 0.015 mol of sodium methanoate in 500 cm^3 of solution. *(1 mark)*
   c A solution made by mixing 250 cm^3 of 0.1 mol dm^{-3} ethanoic acid and 500 cm^3 of 0.1 mol dm^{-3} sodium ethanoate. *(1 mark)*

4 Describe how a buffer solution of sodium ethanoate and ethanoic acid reacts in order to maintain a constant pH when:
   a a small amount of hydrochloric acid is added *(1 mark)*
   b a small amount of sodium hydroxide is added *(1 mark)*
   c a small amount of water is added. *(1 mark)*

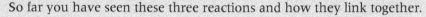
So far you have seen these three reactions and how they link together.

$$CO_2(g) \rightleftharpoons CO_2(aq) \qquad \text{Reaction 1}$$

$$CO_2(aq) + H_2O(l) \rightleftharpoons H^+(aq) + HCO_3^-(aq) \qquad \text{Reaction 2}$$

$$HCO_3^-(aq) \rightleftharpoons H^+(aq) + CO_3^{2-}(aq) \qquad \text{Reaction 3}$$

These three equations can be added together to produce just one equation (Reaction 4), which shows how the reactants in Reaction 1 lead to the products of Reaction 3:

$$CO_2(g) + H_2O(l) \rightleftharpoons 2H^+(aq) + CO_3^{2-}(aq) \qquad \text{Reaction 4}$$

The reaction does not happen as simply as this but this equation should make the next part of the story clearer.

Le Chatelier's principle tells you that any way of removing H^+ or CO_3^{2-} ions from solution (Reaction 4) will cause more carbon dioxide to dissolve. Removing H^+ ions by adding a base is one way of doing this. You may be familiar with this process, as carbon dioxide is an acidic gas and it dissolves well in alkaline solutions. That is why alkalis such as sodium hydroxide or calcium hydroxide are used to absorb carbon dioxide.

Making the sea alkaline is not a very feasible way of encouraging the oceans to take up carbon dioxide. However, many marine organisms build protective shells composed of insoluble calcium carbonate, using CO_3^{2-} ions in the sea water (Figure 1). The building of these shells provides a route for absorbing carbon dioxide and keeping the composition of our atmosphere constant.

Billions of years ago, the Earth's atmosphere contained much more carbon dioxide than it does now – probably about 35% carbon dioxide by volume. Once the process of photosynthesis had evolved, marine life had plenty of raw materials to work on in the form of carbon dioxide and water. Shell production flourished – limestone and chalk rocks (Figure 2) are the remains of the shells of marine organisms that lived at that time and changed carbon dioxide from the atmosphere into solid calcium carbonate.

Calcium carbonate is a good material for shellfish to use as protection at the surface of the oceans. It does not dissolve in sea water but it does dissolve, very slightly, in pure water. It is an example of a sparingly soluble solid – the dissolving of sparingly soluble solids is controlled by equilibria such as:

$$CaCO_3(s) \rightleftharpoons Ca^{2+}(aq) + CO_3^{2-}(aq) \qquad \text{Reaction 5}$$

The ions in the saturated solution are in dynamic equilibrium with the undissolved solid present.

Learning outcomes

Demonstrate and apply knowledge and understanding of:

→ the term solubility product for ionic compounds

→ solubility product calculations

→ techniques and procedures for determining solubility products.

▲ **Figure 1** Seashells are an important component in pH regulation in the oceans

▲ **Figure 2** The chalk cliffs of the Seven Sisters are the legacy of marine organisms that lived billions of years ago

The position of this equilibrium is determined by an equilibrium constant which, because it describes the solubility of a compound, is called a solubility product K_{sp}. The solubility product for Reaction 5 is:

$$K_{sp}(CaCO_3) = [Ca^{2+}(aq)][CO_3^{2-}(aq)]$$
$$= 5.0 \times 10^{-9} \, mol^2 dm^{-6} \text{ at } 298 \, K$$

One of two things can happen when Ca^{2+} ions and CO_3^{2-} ions are mixed together in a solution:

● If the dissolved calcium ion concentration multiplied by the dissolved carbonate concentration gives a value in *excess of* K_{sp} then calcium carbonate will precipitate out of solution.

● If the dissolved calcium ion concentration multiplied by the dissolved carbonate concentration gives a value *smaller than or equal to* K_{sp} then the ions stay in solution.

At the surface of the sea, calcium carbonate is an excellent material from which to build seashells because the concentrations of $Ca^{2+}(aq)$ ions and $CO_3^{2-}(aq)$ ions are already high enough for the calcium carbonate in the shells to be effectively insoluble (the equilibrium in Reaction 5 lies well over to the left). However, the shells are in equilibrium with the ions in sea water, so there will be a constant exchange of Ca^{2+} and CO_3^{2-} between the two.

Things are different deeper in the ocean, where the pressure is higher and the temperature is lower. Under these conditions calcium carbonate is more soluble. There is also a continuous downward drift of material from above. It is like a perpetual snowstorm, and the falling material is called marine snow. It contains the remains of dead organisms and the waste products from live creatures. Most of the organic material, such as tissue, is consumed or decomposed higher up but some reaches the deeper water where bacteria break it down to produce carbon dioxide. The shells fall intact but then react with the extra carbon dioxide and dissolve (Figure 3).

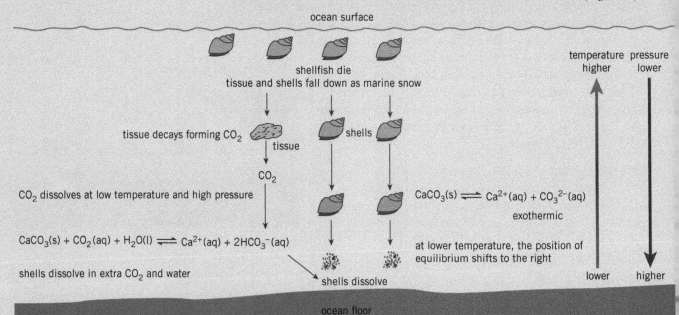

▲ **Figure 3** *Dissolving of shells on the deep ocean floor*

There are no shells on the deep ocean floor as they have all dissolved. Therefore, the creatures that live there cannot use calcium carbonate for a protective coating.

The calcium carbonate deposits that built up to form the limestone hills could not have formed in deep water. They must have been laid down when the landmass was in shallower seas. The abundance of life also suggests that it was warm, tropical water. Evidence like this helps scientists piece together the distant history of the Earth, and helps to explain how the continents have drifted and how the climate has changed throughout time.

Figure 4 summarises the processes that prevent the oceans from becoming acidic. The shells, chalk, and limestone in the seas provide the reservoir of anions needed to prevent changes in acidity.

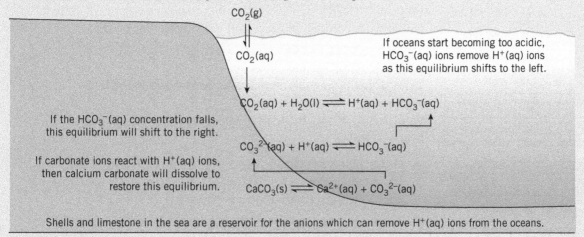

▲ **Figure 4** *Buffering action in the oceans – why the oceans do not become more acidic*

What would happen if the carbon dioxide in the atmosphere rose to the high level of several billion years ago? Limestone deposits could not form if the atmosphere contained 35% carbon dioxide. Solid calcium carbonate would dissolve to produce the carbonate and hydrogencarbonate ions needed to remove the extra H^+ ions. Few CO_3^{2-} ions would remain – most of the carbon would be in the form of HCO_3^- and dissolved carbon dioxide. The sea would be like a mixture of mineral water and bicarbonate of soda – the shells and white cliffs would disappear.

Chemical ideas: Equilibrium in chemistry 7.6

Solubility equilibria

What are solubility equilibria?

Substances are often described as being either soluble or insoluble but this is an oversimplification. No substance is totally insoluble – there will always be some quantity in solution, however small. Rainwater dissolves silicon(IV) oxide from the rocks and transports it to the sea, even though the sand on the beach appears insoluble. Calcium carbonate dissolves and re-precipitates in various ocean processes.

Synoptic link

K_{sp} is another type of equilibrium constant. You previously met equilibrium constants in Topic ES 4, From extracting bromine to making bleach, and Topic CI 2, Manufacturing nitric acid.

Activity O 4.1

In this activity you will determine practically the K_{sp} for potassium hydrogen tartrate.

Synoptic link

Determining K_{sp} is covered further in Techniques and procedures.

▼ **Table 1** *Some solubility products at 298 K*

Compound	K_{sp}
$CaCO_3$	$5.0 \times 10^{-9} \, mol^2 \, dm^{-6}$
$CaSO_4$	$2.0 \times 10^{-5} \, mol^2 \, dm^{-6}$
$BaSO_4$	$1.0 \times 10^{-10} \, mol^2 \, dm^{-6}$
PbI_2	$7.1 \times 10^{-9} \, mol^3 \, dm^{-9}$
$AgCl$	$2.0 \times 10^{-10} \, mol^2 \, dm^{-6}$
$AgBr$	$5.0 \times 10^{-13} \, mol^2 \, dm^{-6}$
AgI	$8.0 \times 10^{-17} \, mol^2 \, dm^{-6}$
PbS	$1.3 \times 10^{-28} \, mol^2 \, dm^{-6}$
$PbSO_4$	$1.6 \times 10^{-8} \, mol^2 \, dm^{-6}$

Summary questions

1 Write ionic equations for the equilibria set up when the following sparingly soluble compounds dissolve in water:
   a $AgI(s)$ (*1 mark*)
   b $BaSO_4(s)$ (*1 mark*)
   c $PbI_2(s)$ (*1 mark*)
   d $Fe(OH)_3(s)$ (*1 mark*)

2 Write an expression for K_{sp} for each of the equilibria in Question 1. Give the units for K_{sp} in each case.   (*8 marks*)

3 K_{sp} for silver bromate(V), $AgBrO_3$, is $6.0 \times 10^{-5} \, mol^2 \, dm^{-6}$ at 298 K. A student added $100 \, cm^3$ of $0.001 \, mol \, dm^{-3}$ silver nitrate(V) solution to $100 \, cm^3$ of $0.001 \, mol \, dm^{-3}$ potassium bromate(V) solution at 298 K. Calculate whether or not a precipitate of silver bromate(V) would form.   (*3 marks*)

A sparingly soluble ionic solid such as calcium carbonate, in contact with a saturated solution of its ions, is an example of a chemical equilibrium:

$$CaCO_3(s) \rightleftharpoons Ca^{2+}(aq) + CO_3^{2-}(aq)$$

$$K_c = \frac{[Ca^{2+}(aq)][CO_3^{2-}(aq)]}{[CaCO_3(aq)]}$$

Adding more solid will not cause the equilibrium to shift further to the product side because the solution is saturated at that temperature. In other words, the equilibrium is not affected by the amount of solid present, so you can rewrite the equilibrium expression as:

$$K_{sp}(CaCO_3) = [Ca^{2+}(aq)][CO_3^{2-}(aq)]$$

K_{sp} stands for solubility product. It represents the conditions for equilibrium between a sparingly soluble solid and its saturated solution. At 298 K, whenever you have solid calcium carbonate in equilibrium with its solution, you will always find that:

$$K_{sp}(CaCO_3) = [Ca^{2+}(aq)][CO_3^{2-}(aq)] = 5.0 \times 10^{-9} \, mol^2 \, dm^{-6}$$

Using solubility equilibria to predict whether a precipitate will form from a solution

You can use the K_{sp} value to predict whether a precipitate will form from a solution.

- If the dissolved calcium ion concentration multiplied by the dissolved carbonate concentration gives a value *in excess of K_{sp}*, then calcium carbonate will precipitate out of solution.

- If the dissolved calcium ion concentration multiplied by the dissolved carbonate concentration gives a value *smaller than or equal to K_{sp}*, then the ions stay in solution.

Worked example: Finding out whether a precipitate will form or not

Will a precipitate form from a solution containing calcium ions and carbonate ions at a concentration of $1.0 \times 10^{-5} \, mol \, dm^{-3}$? The solubility product of calcium carbonate is $5.0 \times 10^{-9} \, mol^2 \, dm^{-6}$. So:

$$[Ca^{2+}(aq)] = 1.0 \times 10^{-5} \, mol \, dm^{-3}$$
$$[CO_3^{2-}(aq)] = 1.0 \times 10^{-5} \, mol \, dm^{-3}$$

Therefore:

$$K_{sp} = [Ca^{2+}(aq)][CO_3^{2-}(aq)]$$
$$= 1.0 \times 10^{-5} \, mol \, dm^{-3} \times 1.0 \times 10^{-5} \, mol \, dm^{-3}$$
$$= 1.0 \times 10^{-10} \, mol^2 \, dm^{-6}$$

This is less than the solubility product of $5.0 \times 10^{-9} \, mol^2 \, dm^{-6}$, so no precipitate will form.

Some K_{sp} values are given in Table 1. Like all equilibrium constants, K_{sp} changes with temperature, so you must always quote the temperature to which a K_{sp} value relates.

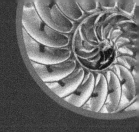

05 The global central heating system

Specification references: O(d), O(e), O(f), O(g)

So far you have considered the important role that the oceans play in climate control through their ability to absorb carbon dioxide. However, the oceans and the atmosphere also play an important role in distributing the energy that the Earth receives from the Sun more evenly around the globe.

If the Earth were a dry lump of rock with no atmosphere, each part of its surface would soon settle down to a situation in which the energy received from the Sun would, on average, be balanced by energy lost through radiation. The tropics would be much warmer and the poles even colder than they are – and the Earth would be far less hospitable to life.

But the Earth is surrounded by water and gas. Temperature differences set up currents in the oceans and the atmosphere that spread out the heating effect of the Sun more evenly. Just like warm air from a radiator spreads around a room, currents in the sea and air take thermal energy from the tropics to the colder regions of the Earth.

In fact, the ocean–atmosphere system is even more effective at spreading out energy. Warm water can do more than circulate, it can evaporate. Energy is taken in when water evaporates and is released when water condenses. The tropics are cooled by evaporation, and currents in the atmosphere carry the water vapour to colder, high-latitude regions. The water condenses, releasing energy and warming up these areas.

High-latitude regions receive more energy than is provided by the Sun alone – they are wetter but warmer. Figure 1 shows the balance of condensation and evaporation around the Earth.

Learning outcomes

Demonstrate and apply knowledge and understanding of:

→ entropy as a measure of the number of ways that molecules and their associated energy quanta can be arranged

→ qualitative predictions of the $\Delta_{sys}S$ for a reaction in terms of:

  • the differences in magnitude of the entropy of a solid, a liquid, and a gas

  • the difference in number of particles of gaseous reactants and products

→ the expressions:
  $\Delta_{tot}S = \Delta_{sys}S + \Delta_{surr}S$ and $\Delta_{surr}S = -\Delta H/T$; calculations using these expressions; the relation of the feasibility of a reaction to the sign of $\Delta_{tot}S$.

→ calculation of $\Delta_{sys}S$ for a reaction given the entropies of reactants and products

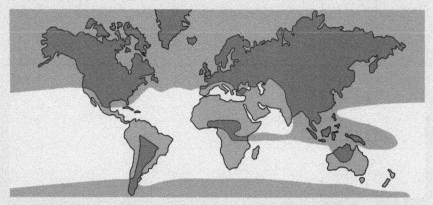

▲ **Figure 1** *The condensation–evaporation balance of the Earth. The dark shaded areas denote regions where condensation exceeds evaporation*

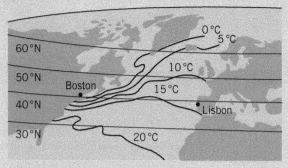

▲ **Figure 2** *Average surface temperatures in the North Atlantic in February*

In the North Atlantic region, the winds and warm water currents flow from south-west to north-east. Northern Europe, including the UK, is warmed by energy that has been transported from the tropics and the Caribbean. In winter, as much as 25% of our thermal energy may come this way. Eastern North America does not receive this energy, so winters are much more pleasant in Lisbon, Portugal (latitude 38°N) than in Boston, USA (latitude 42°N) (Figure 2).

Energy in the clouds

The molecules in liquid water and water vapour differ in one important aspect. In a liquid, attractive forces between the molecules (intermolecular bonds) keep the molecules quite close together. In vapour, the molecules are much further apart and move about freely.

▲ **Figure 3** *Energy is taken in from the surroundings when sea water evaporates, and is released when the water vapour condenses into clouds*

When water evaporates, changing from liquid to vapour, the intermolecular bonds must be overcome – a process that takes in energy. The enthalpy change of vaporisation $\Delta_{vap}H$ is a measure of this energy.

Evaporation is an endothermic process:

$$H_2O(l) \rightarrow H_2O(g) \qquad \Delta_{vap}H \text{ is positive}$$

In the reverse process – condensation – molecules come together again, intermolecular bonds reform, and an equal quantity of energy is released. In other words, condensation is exothermic and the enthalpy change is $-\Delta_{vap}H$.

Condensation is an exothermic process:

$$H_2O(g) \rightarrow H_2O(l) \qquad \Delta_{vap}H \text{ is negative}$$

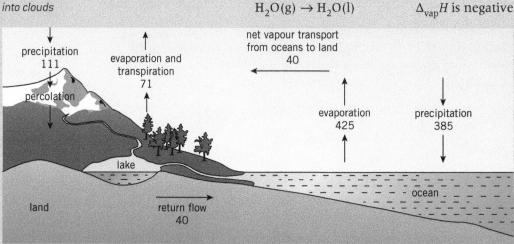

▲ **Figure 4** *The global water cycle, with the amount of water transferred in each process, in trillion tonnes. It summarises the main processes by which water circulates around the Earth*

More water evaporates from the oceans than is directly returned to them as precipitation (i.e., rain and snow). Each year, 40 trillion tonnes of water vapour produced from the sea falls as precipitation over the land. The process makes the land wetter and keeps the rivers flowing – it also makes the land warmer.

So evaporation and condensation of water affect the temperature of different parts of the Earth in two ways:

1 by transferring energy from low latitudes to high latitudes
2 by warming the land through condensation of water that comes from the oceans.

However, when a change of state occurs, it does not just lead to a change in enthalpy. Solids, liquids, and gases also have different entropies. Entropy is the measurement of disorder in a system and is an important factor in determining whether a chemical change is feasible.

Chemical ideas: Energy changes and chemical reactions 4.5

Enthalpy and entropy

Many of the reactions that occur of their own accord are exothermic (ΔH is negative). For example, if you add zinc to a copper sulfate solution, the reaction to form copper and zinc sulfate takes place and the solution gets hot. Negative ΔH is a factor in whether or not a reaction is spontaneous, but it does not explain why a number of endothermic reactions are spontaneous. For example, the following process, which occurs spontaneously, is endothermic (ΔH is positive):

$$NH_4NO_3(s) \quad + \quad (aq) \quad \rightarrow \quad NH_4NO_3(aq)$$
ammonium nitrate $\qquad\qquad\qquad$ aqueous ammonium nitrate

Entropy or disorder

Many processes which take place spontaneously involve mixing or spreading out, for example, liquids evaporating, solids dissolving to form solutions, or gases mixing.

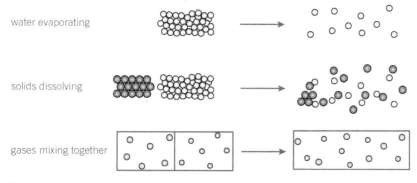

water evaporating

solids dissolving

gases mixing together

▲ **Figure 5** *Spontaneous processes*

This is the clue to the second factor which drives chemical processes – a tendency towards randomising or disordering. Gases are more random than liquids, which are more random than solids, because of the arrangement of their particles.

So, endothermic reactions may be spontaneous if they involve spreading out, randomising, or disordering. This is true of the reaction between zinc and copper sulfate and when dissolving ammonium nitrate in water – the arrangement of the particles in the products is more random than in the reactants.

The randomness of a system, expressed mathematically, is called the **entropy** of the system. Entropy is a measure of the number of ways that molecules and their associated energy can be arranged and is given the symbol S. Reactions such as that between zinc and copper sulfate in which the products are more disordered than the reactants will have positive values for the entropy change ΔS.

The amount of energy available to the molecules in a system is also important. Molecules do not all have the same energy. As molecules collide and exchange energy with each other, their energies change. Entropy is also a measure of the number of ways that energy can be arranged between particles. The greater the energy within a system, the higher the entropy will be.

Entropy values

Entropies have been determined for a vast range of substances and can be looked up in tables and databases. They are usually quoted for the standard conditions 298 K and 100 kPa pressure. Table 1 gives some examples.

▼ **Table 1** *Some values of entropy*

Substance	State at standard conditions	Entropy S / $J\,K^{-1}\,mol^{-1}$
carbon (diamond)	solid	2.4
carbon (graphite)	solid	5.7
copper	solid	33.0
iron	solid	27.0
ammonium chloride	solid	95.0
calcium carbonate	solid	93.0
calcium oxide	solid	40.0
iron(III) oxide	solid	88.0
water (ice)	solid	48.0
water (liquid)	liquid	70.0
mercury	liquid	76.0
water (steam)	gas	189.0
hydrogen chloride	gas	187.0
ammonia	gas	192.0
carbon dioxide	gas	214.0

Entropy

Entropy is a measure of the number of ways of arranging molecules and distributing energy.

Study tip

ΔS is positive when products are more disordered than reactants.

ΔS is negative when products are less disordered than reactants.

Study tip

In general, gases have higher entropies than liquids, and liquids have higher entropies than solids.

The units of entropy are $JK^{-1}mol^{-1}$, not $kJK^{-1}mol^{-1}$. So when you are dealing with enthalpy (the units of which are usually $kJmol^{-1}$) and entropy values together, you need to ensure that enthalpy values are converted into $Jmol^{-1}$ by multiplying by 1000.

A collection of molecules has higher entropy if:

- the molecules are spread out more
- the energy is shared among more molecules.

Calculating entropy changes

The entropy change for a reaction can be calculated by adding all the entropies of the products and subtracting the sum of the entropies of the reactants.

Will it or won't it?

When changes take place within a chemical system, such as during a chemical reaction or physical process such as ice melting, there are nearly always accompanying changes to the surroundings. To predict whether or not a change is feasible, you need to take account of the entropy changes to the system *and* its surroundings.

To find the total entropy change $\Delta_{total}S$ for a process, the entropy change for the chemical system $\Delta_{sys}S$ is combined with the entropy change in the surroundings $\Delta_{surr}S$:

$$\Delta_{total}S = \mathbf{\Delta}_{sys}S + \mathbf{\Delta}_{surr}S$$

For a reaction to be feasible, $\Delta_{total}S$ has a positive value.

Calculating the entropy change to the system is easy – it is simply a matter of subtracting the entropy of the reactants from the entropy of the products. Calculating the entropy change to the surroundings is a little bit more complicated. Look at freezing in more detail to see how this process affects the surroundings:

$$H_2O(l) \rightarrow H_2O(s)$$

$$\Delta S = -22.0\,JK^{-1}mol^{-1}$$

$$\Delta H = -6.01\,kJmol^{-1}$$

ΔS is the entropy change for this process. It is negative because the entropy *decreases* when liquid water becomes solid water. ΔH is also negative – the change is exothermic and energy is transferred from the water to the surroundings by heating. This change affects the entropy of the surroundings.

Freezing is an exothermic process. Energy is transferred from the chemical system to the surroundings by heating. When ice freezes on a window pane, energy is transferred to the glass but it soon spreads out – by conduction, convection, and radiation – into the things around the glass. Many things – such as the frame, the wall, the air, and the plant beside the window – get hotter, and therefore increase in entropy.

Activity O 5.1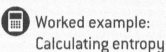

This activity gives you the opportunity to practise calculating $\Delta_{sys}S$ and $\Delta_{total}S$.

Worked example: Calculating entropy

$$CaCO_3(s) \rightarrow CaO(s) + CO_2(g)$$

Using the values from Table 1:

Entropy of products
$= 40 + 214 = 254\,JK^{-1}mol^{-1}$

Entropy of reactant
$= 93\,JK^{-1}mol^{-1}$

$\Delta S = 254 - 93$
$= +161\,JK^{-1}mol^{-1}$

This is a large positive value because a gas is formed from a solid.

Study tip

A chemical system is the products and reactants.

The surroundings are everything else.

Synoptic link

It may be useful to revise the concept of enthalpy change, which was covered in Chapter 2, Developing fuels.

You cannot work out how much each individual substance in the surroundings has increased in entropy because it is impossible to say exactly how the energy has been shared out. Fortunately, you do not need to know this. There is a very simple relationship which allows you to think just in terms *of surroundings*.

The relationship tells you that the entropy change in the surroundings is equal to the energy transferred (the enthalpy change in the surroundings) divided by the temperature. Since the energy *gained* by the *surroundings* is the same as the energy *lost* by the *chemical system*, and vice versa, the enthalpy change in the surroundings is equal to $-\Delta H$.

So the entropy change in the surroundings $\Delta_{surr}S$ is given by:

$$\Delta_{surr}S = -\frac{\Delta H}{T}$$

So you can now rewrite the expression $\Delta_{total}S = \Delta_{sys}S + \Delta_{surr}S$ as:

$$\Delta_{total}S = \Delta_{sys}S + \left(-\frac{\Delta H}{T}\right)$$

Remember that, for a spontaneous change, $\Delta_{total}S$ has a positive value.

 Worked example: Will water freeze at +10 °C?

$\Delta_{sys}S = -22.0\,J\,K^{-1}\,mol^{-1}$

$$\Delta_{surr}S = -\frac{\Delta H}{T} = -\frac{(-6010\,J\,mol^{-1})}{283\,K}$$
$$= +21.2\,J\,K^{-1}\,mol^{-1}$$

$$\Delta_{total}S = \Delta_{sys}S + \Delta_{surr}S$$
$$= (-22.0\,J\,K^{-1}\,mol^{-1}) + (+21.2\,J\,K^{-1}\,mol^{-1})$$
$$= -0.8\,J\,K^{-1}\,mol^{-1}$$

Overall, the process leads to a decrease in entropy, so this is not a spontaneous change. Water will not freeze at +10 °C.

 Worked example: Will water freeze at −10 °C?

$\Delta_{sys}S = -22.0\,J\,K^{-1}\,mol^{-1}$

$$\Delta_{surr}S = -\frac{\Delta H}{T} = -\frac{(-6010\,J\,mol^{-1})}{263\,K}$$
$$= +22.9\,J\,K^{-1}\,mol^{-1}$$

$$\Delta_{total}S = \Delta_{sys}S + \Delta_{surr}S$$
$$= (-22.0\,J\,K^{-1}\,mol^{-1}) + (+22.9\,J\,K^{-1}\,mol^{-1})$$
$$= +0.9\,J\,K^{-1}\,mol^{-1}$$

Overall, the process leads to an increase in entropy, so this is a spontaneous change. Water will freeze at −10 °C.

Melting and freezing do not involve changing one compound into another, but the $\Delta_{total}S$ rule also applies to the production of new chemicals. You can see this by looking at a chemical reaction which is responsible for putting the fizz into sparkling water – the decomposition of calcium carbonate.

Limestone rocks have existed for a long time. At the temperatures on the Earth's surface, the calcium carbonate has not broken down into calcium oxide and carbon dioxide but this is exactly what happens when limestone gets very hot, for example, when it is next to lava which has forced its way into the Earth's crust. The process you need to consider is:

$$CaCO_3(s) \rightarrow CaO(s) + CO_2(g)$$
$$\Delta S = +159\,J\,K^{-1}\,mol^{-1}$$
$$\Delta H = +179\,kJ\,mol^{-1}$$

Worked example: The decomposition of calcium carbonate

Will calcium carbonate decompose to produce carbon dioxide
at 298 K (25 °C) and 1273 K (1000 °C)?

At 298 K

$$\Delta_{sys}S = +159\,J\,K^{-1}\,mol^{-1}$$

$$\Delta_{surr}S = \frac{-\,(+179\,000\,J\,mol)^{-1}}{298\,K}$$
$$= -601\,J\,K^{-1}\,mol^{-1}$$

$$\Delta_{total}S = (+159\,J\,K^{-1}\,mol^{-1}) + (-601\,J\,K^{-}mol^{-1})$$
$$= -442\,J\,K^{-1}\,mol^{-1}$$

At 1273 K

$$\Delta_{sys}S = +\,159\,J\,K^{-1}\,mol^{-1}$$

$$\Delta_{surr}S = \frac{-\,(+179\,000\,J\,mol)^{-1}}{1273\,K}$$
$$= -141\,J\,K^{-1}\,mol^{-1}$$

$$\Delta_{total}S = (+159\,J\,K^{-1}\,mol^{-1}) + (-141\,J\,K^{-}mol^{-1})$$
$$= +18\,J\,K^{-1}\,mol^{-1}$$

At the Earth's surface at 298 K, the decomposition of limestone would
result in a large entropy *decrease*. Therefore it does not take place.
At higher temperatures, near hot lava, the total entropy change is
positive and the reaction can occur.

What about equilibrium?

A spontaneous change always takes place in the direction which
corresponds to an increase in $\Delta_{total}S$ and favourable kinetics. This is
the case when ice melts at 10 °C:

$$H_2O(s) \rightarrow H_2O(l)$$

However, at −10 °C, $\Delta_{total}S$ for the melting of ice is negative. Therefore
the reverse process occurs and water freezes.

What happens when $\Delta_{total}S$ is equal to zero? There is no net change in
either direction – liquid and solid are in equilibrium. You can check
this by looking at the figures again – this time at 0 °C (273 K).

 ## Worked example: Water at 273 K

$$H_2O(s) \rightleftharpoons H_2O(l) \qquad \Delta_{sys}S = +22.0\,J\,K^{-1}\,mol^{-1}$$

$$\Delta H = +6.01\,kJ\,mol^{-1}$$

$$\Delta_{sys}S = +22.0\,J\,K^{-1}\,mol^{-1}$$

$$\Delta_{surr}S = -\frac{\Delta H}{T} = -\frac{(+6010\,J\,mol^{-1})}{273\,K} = 22.0\,J\,K^{-1}\,mol^{-1}$$

$$\Delta_{total}S = (+22.0\,J\,K^{-1}\,mol^{-1}) + (-22.0\,J\,K^{-1}\,mol^{-1}) = 0\,J\,K^{-1}\,mol^{-1}$$

Liquid water and ice exist together at 0 °C, they are in
equilibrium, and at this temperature $\Delta_{total}S = 0$.

The requirement for equilibrium is that $\Delta_{total}S$ must be zero.

Kinetic factors

Neither enthalpy changes nor entropy changes tell you anything
about how quickly or slowly a reaction is likely to go. So you might
predict that a certain reaction should occur spontaneously because of
enthalpy and entropy changes but the reaction might take place so
slowly that for practical purposes it does not occur at all. This is why
changes where $\Delta_{total}S > 0$ are described as feasible. They may not be
spontaneous if they have a large activation enthalpy E_a value.

Activity O 5.2

This activity will help you
understanding of why chemical
reactions happen.

Synoptic link

You studied equilibria in Chapter 3,
Elements from the sea.

Activity O 5.3

This card sort activity
challenges you to think about
energy and entropy.

Synoptic link

Activation enthalpy was covered in
Topic OZ 4, Ozone – here today and
gone tomorrow.

Summary questions

1 For each of the following changes, state whether the entropy of the system described would increase, decrease, or stay the same:
   a petrol vaporises   (1 mark)
   b petrol vapour condenses   (1 mark)
   c sugar dissolves in water   (1 mark)
   d oil mixes with petrol   (1 mark)
   e a suspension of oil in water separates into two layers   (1 mark)
   f car exhaust gases are adsorbed onto the surface of a catalyst
      in a catalytic converter.   (1 mark)

2 State which substance would have the higher standard molar entropy for each of the following pairs of substances. Explain your reasons.
   a solid wax or molten wax   (2 marks)
   b $Br_2(l)$ or $Br_2(g)$ (2 marks)
   c separate samples of copper and zinc or a sample of brass
      (an alloy of copper and zinc)   (2 marks)
   d pentane $C_5H_{12}(l)$ or octane $C_8H_{18}(l)$.   (2 marks)

3 Explain the pattern in the entropies of the first five alkanes at 298 K, given in the following table.

Alkane	$CH_4(g)$	$C_2H_6(g)$	$C_3H_8(g)$	$C_4H_{10}(g)$	$C_5H_{12}(l)$
$S/\,J\,K^{-1}\,mol^{-1}$	186	230	270	310	261

(2 marks)

▼ **Table 2** *Entropy values for question 4*

Substance	$S/\,J\,K^{-1}\,mol^{-1}$
$Hg(l)$	76.0
$Hg(g)$	174.8
$C(s)$ (graphite)	5.7
$H_2O(g)$	188.7
$CO(g)$	197.6
$H_2(g)$	130.6
$NO_2(g)$	240.0
$N_2O_4(g)$	304.2
$N_2(g)$	191.6
$NH_3(g)$	192.3

4 Calculate the entropy change for each of the following reactions using the standard molar entropies provided (Table 2). In each case, discuss the values you have obtained.
   a $Hg(l) \rightarrow Hg(g)$ (2 marks)
   b $C(s) + H_2O(g) \rightarrow CO(g) + H_2(g)$ (2 marks)
   c $2NO_2(g) \rightarrow N_2O_4(g)$ (2 marks)
   d $N_2(g) + 3H_2(g) \rightarrow 2NH_3(g)$ (2 marks)

5 The values of ΔH and $\Delta_{sys}S$ that follow refer to changes at 298 K under standard conditions. Calculate whether or not the following changes will be spontaneous at 298 K. Explain your answer in each case.
   a $Ca^{2+}(aq) + CO_3^{2-}(aq) \rightarrow CaCO_3(s)$
      $\Delta_{sys}S = +203\ J\,K^{-1}\,mol^{-1}$ $\Delta H = +13\ kJ\,mol^{-1}$ (2 marks)
   b $H_2O_2(l) \rightarrow H_2O(l) + \frac{1}{2}O_2(g)$
      $\Delta_{sys}S = +63\ J\,K^{-1}\,mol^{-1}$ $\Delta H = -98\ kJ\,mol^{-1}$ (2 marks)
   c $N_2(g) + O_2(g) \rightarrow 2NO(g)$
      $\Delta_{sys}S = +25\ J\,K^{-1}\,mol^{-1}$ $\Delta H = +180\ kJ\,mol^{-1}$ (2 marks)
   d $NH_4NO_3(s) \rightarrow N_2O(g) + 2H_2O(l)$
      $\Delta_{sys}S = +209\ J\,K^{-1}\,mol^{-1}$ $\Delta H = -124\ kJ\,mol^{-1}$ (2 marks)
   e $C(graphite) \rightarrow C(diamond)$
      $\Delta_{sys}S = -4\ J\,K^{-1}\,mol^{-1}$ $\Delta H = +2\ kJ\,mol^{-1}$ (2 marks)

Practice questions

1 In the greenhouse effect molecules absorb IR radiation. The source of energy for this IR radiation comes from:

   A radiation bouncing off the Earth

   B the Earth absorbing IR and re-emitting it

   C the Earth absorbing UV and emitting IR

   D the Sun   (*1 mark*)

2 In order to absorb IR radiation a molecule must:

   A be a greenhouse gas

   B be present in the atmosphere

   C have bonds that vibrate more when hit by IR

   D be an organic molecule   (*1 mark*)

3 Which of the following reactions will only be feasible at high temperatures?

	$\Delta_{sys}S$	ΔH
A	positive	positive
B	positive	negative
C	negative	negative
D	negative	positive

(*1 mark*)

4 For which of the following processes is $\Delta_{sys}S$ positive?

   A $N_2(g) + 3H_2(g) \rightarrow 2NH_3(g)$

   B $H_2O(g) \rightarrow H_2O(l)$

   C $CaCO_3(s) + 2HCl(aq) \rightarrow CaCl_2(aq) + CO_2(g) + H_2O(l)$

   D $Ag^+(aq) + Cl^-(aq) \rightarrow AgCl(s)$ (*1 mark*)

5 A simple molecular solid is insoluble in water. This could be because:

   A its covalent bonds are stronger than the hydrogen bonds in water

   B the instantaneous dipole–induced dipole bonds between its molecules are weaker than the hydrogen bonds in water

   C the instantaneous dipole–induced dipole bonds between its molecules and water are weaker than the instantaneous dipole–induced dipole bond between its molecules

   D the instantaneous dipole–induced dipole bonds between its molecules and water are weaker than the hydrogen bonds in water   (*1 mark*)

6 Which of the following represents the lattice enthalpy of NaCl?

   A $Na^+(s) + Cl^-(s) \rightarrow NaCl(s)$

   B $Na(s) + \frac{1}{2}Cl_2(g) \rightarrow NaCl(s)$

   C $Na^+(g) + Cl^-(g) \rightarrow NaCl(s)$

   D $Na^+(aq) + Cl^-(aq) \rightarrow NaCl(s)$ (*1 mark*)

7 The ionic radius of metals increases down a group.

   Which of the following would have the most negative lattice enthalpy?

   A KCl

   B NaCl

   C $CaCl_2$

   D $MgCl_2$ (*1 mark*)

8 Which are correct statements about a solution of a weak acid, HA?

   1 it contains only HA, H^+, and A^- in solution

   2 it will neutralise less sodium hydroxide than the same volume of a strong acid of the same concentration.

   3 It has a pH between 3 and 7.

   A 1, 2, and 3 correct

   B 1 and 2 are correct

   C 2 and 3 are correct

   D only 1 is correct   (*1 mark*)

9 Which of the following can act as a buffer solution?

   1 a solution of a strong acid and its conjugate base

   2 a solution of a weak acid and its salt

   3 a partially neutralised solution of a weak acid

   A 1, 2, and 3 correct

   B 1 and 2 are correct

   C 2 and 3 are correct

   D only 1 is correct   (*1 mark*)

10 Which of the following describe ways in which molecules that have absorbed IR warm the Earth's atmosphere?

1 the energy is transferred to other molecules, increasing their kinetic energy

2 the molecules emit IR

3 the molecules form an insulating blanket around the Earth

A 1, 2, and 3 correct

B 1 and 2 are correct

C 2 and 3 are correct

D only 1 is correct *(1 mark)*

11 Ammonium nitrate, NH_4NO_3, is used as a fertiliser.

When heated, ammonium nitrate decomposes.

$$NH_4NO_3(s) \rightarrow N_2O(g) + 2H_2O(g)$$

	$\Delta_f H / kJ\,mol^{-1}$	$S / J\,mol^{-1}K^{-1}$
NH_4NO_3	−166	+151
N_2O	+82	+220
H_2O	−242	+189

a **(i)** Use the data to calculate ΔH, $\Delta_{sys}S$, and $\Delta_{total}S$ for the decomposition reaction at 298 K. *(4 marks)*

(ii) Explain what your answers from (i) tell you about the feasibility of the reaction at different temperatures. *(2 marks)*

(iii) How could you predict the sign of $\Delta_{sys}S$ from the equation for the decomposition? *(2 marks)*

b Ammonium nitrate is very soluble in water. Use the data given to work out a value for the enthalpy change of solution of ammonium nitrate.

Draw an enthalpy level diagram to illustrate your calculation.

	$kJ\,mol^{-1}$
lattice enthalpy of NH_4NO_3	−646
enthalpy change of hydration of NH_4^+	−307
enthalpy change of hydration of NO_3^-	−314

(2 marks)

c Use your answer to **b** to predict (with a reason) whether the solubility of ammonium nitrate will increase, decrease, or stay the same when the temperature is raised. *(1 mark)*

d The enthalpy change of hydration of Na^+ is −407 kJ mol^{-1}.

(i) Which bonds are formed when hydration occurs?

(ii) Explain why this value is more exothermic than the value for the ammonium ion. *(3 marks)*

12 The pH of our blood is very important to our health. The main reaction maintaining this pH is:

$$CO_2(aq) + H_2O(l) \rightleftharpoons H^+(aq) + HCO_3^-(aq)$$
Equation 1

For which $K_a = \dfrac{[H^+(aq)][HCO_3^-(aq)]}{[CO_2(aq)]}$

$$= 4.47 \times 10^{-7}\,mol\,dm^{-3}$$

a Write down the conjugate base of HCO_3^- and explain your answer. *(2 marks)*

b Explain, in terms of the equilibrium in **Equation 1**, how blood acts as a buffer solution when a small amount of acid is added. *(4 marks)*

c The solubility of carbon dioxide, CO_2, at 298 K and 1 atmosphere pressure is $3.29 \times 10^{-2}\,mol\,dm^{-3}$.

Calculate the pH of a saturated solution of carbon dioxide.

Give your answer to **two** decimal places. *(2 marks)*

d In healthy blood:

$$\frac{[CO_2(aq)]}{[HCO_3^-(aq)]} = 0.09$$

Calculate the pH of healthy blood to **one** decimal place. *(2 marks)* e An acceptable pH range for blood to remain healthy is ±0.05 either side of the value in **d**.

    (i) If the carbon dioxide level in blood rises, will the pH increase or decrease? Explain your answer.

    (ii) Calculate the ratio $\dfrac{[CO_2(aq)]}{[HCO_3^-(aq)]}$ at which blood becomes unhealthy in the presence of increased $[CO_2]$.

(3 marks)

13 Some students set out to measure the solubility of calcium hydroxide (in $mol\,dm^{-3}$) by titrating a saturated solution of calcium hydroxide with standard hydrochloric acid.

They start with solid calcium hydroxide and a $1.00\,mol\,dm^{-3}$ solution of acid.

The solubility of calcium hydroxide, $Ca(OH)_2$, at room temperature is around $0.015\,mol\,dm^{-3}$.

a Describe a method they could use to achieve an acceptable titration result.

(6 marks)

b Use the solubility of calcium hydroxide given in **a** to calculate the solubility product of calcium hydroxide with its units. *(3 marks)*

c A solution of sodium hydroxide has a pH of 13.1.

    (i) Calculate the concentration of NaOH in $mol\,dm^{-3}$.

    (ii) Calculate the solubility of calcium hydroxide in this NaOH solution. You may assume all the hydroxide ions come from the NaOH. *(4 marks)*

Chapter 8 Summary

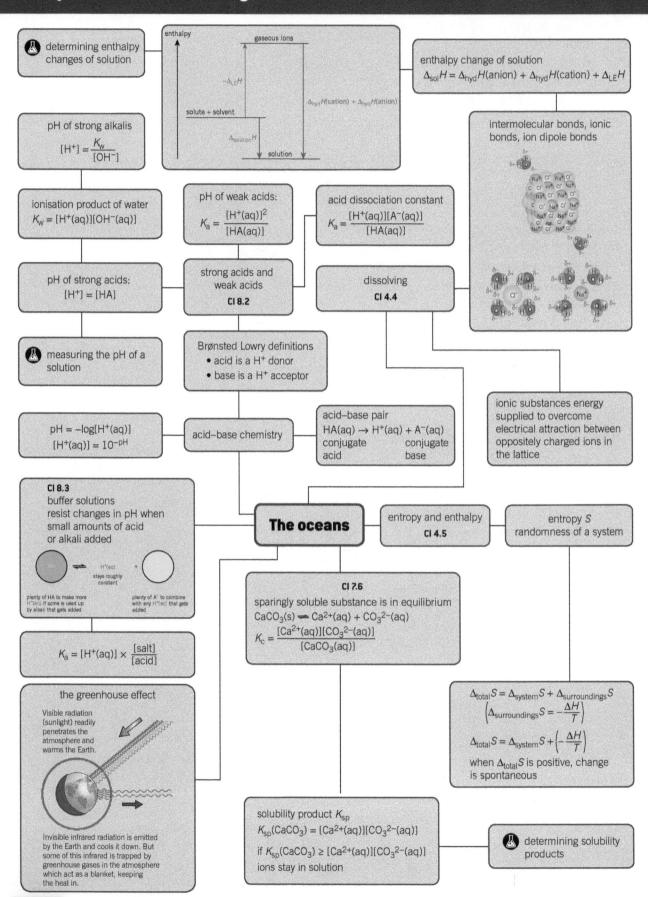

determining enthalpy changes of solution

enthalpy change of solution
$$\Delta_{sol}H = \Delta_{hyd}H(\text{anion}) + \Delta_{hyd}H(\text{cation}) + \Delta_{LE}H$$

enthalpy
gaseous ions
$-\Delta_{LE}H$
solute + solvent
$\Delta_{hyd}H(\text{cation}) + \Delta_{hyd}H(\text{anion})$
$\Delta_{solution}H$
solution

intermolecular bonds, ionic bonds, ion dipole bonds

pH of strong alkalis
$$[H^+] = \frac{K_w}{[OH^-]}$$

ionisation product of water
$$K_w = [H^+(aq)][OH^-(aq)]$$

pH of weak acids:
$$K_a = \frac{[H^+(aq)]^2}{[HA(aq)]}$$

acid dissociation constant
$$K_a = \frac{[H^+(aq)][A^-(aq)]}{[HA(aq)]}$$

pH of strong acids:
$$[H^+] = [HA]$$

strong acids and weak acids
CI 8.2

dissolving
CI 4.4

measuring the pH of a solution

Brønsted Lowry definitions
• acid is a H^+ donor
• base is a H^+ acceptor

$$pH = -\log[H^+(aq)]$$
$$[H^+(aq)] = 10^{-pH}$$

acid–base chemistry

acid–base pair
$$HA(aq) \rightarrow H^+(aq) + A^-(aq)$$
conjugate acid   conjugate base

ionic substances energy supplied to overcome electrical attraction between oppositely charged ions in the lattice

CI 8.3
buffer solutions resist changes in pH when small amounts of acid or alkali added

HA ⇌ H⁺(aq) + stays roughly constant

plenty of HA to make more H⁺(aq) if some is used up by alkali that gets added

plenty of A⁻ to combine with any H⁺(aq) that gets added

The oceans

entropy and enthalpy
CI 4.5

entropy S randomness of a system

$$K_a = [H^+(aq)] \times \frac{[salt]}{[acid]}$$

CI 7.6
sparingly soluble substance is in equilibrium
$$CaCO_3(s) \rightleftharpoons Ca^{2+}(aq) + CO_3^{2-}(aq)$$
$$K_c = \frac{[Ca^{2+}(aq)][CO_3^{2-}(aq)]}{[CaCO_3(aq)]}$$

the greenhouse effect

Visible radiation (sunlight) readily penetrates the atmosphere and warms the Earth.

Invisible infrared radiation is emitted by the Earth and cools it down. But some of this infrared is trapped by greenhouse gases in the atmosphere which act as a blanket, keeping the heat in.

$$\Delta_{total}S = \Delta_{system}S + \Delta_{surroundings}S$$
$$\left(\Delta_{surroundings}S = -\frac{\Delta H}{T}\right)$$
$$\Delta_{total}S = \Delta_{system}S + \left(-\frac{\Delta H}{T}\right)$$
when $\Delta_{total}S$ is positive, change is spontaneous

solubility product K_{sp}
$$K_{sp}(CaCO_3) = [Ca^{2+}(aq)][CO_3^{2-}(aq)]$$
if $K_{sp}(CaCO_3) \geq [Ca^{2+}(aq)][CO_3^{2-}(aq)]$ ions stay in solution

determining solubility products

Ionic liquids

Most ionic solids share similar properties – they have high melting points, they conduct in solution or when molten, and they tend to dissolve in water.

However there are some ionic substances which have melting points at or below room temperature. These are known as ionic liquids.

The substances $[1\text{-butylpyridinium}]^+[BF_4]^-$ and $[1\text{-butyl-3-methyl-}1H\text{-imidazolium}]^+[CF_3CO_2]^-$ are examples of ionic liquids. Their melting points are 25 °C and 20 °C respectively.

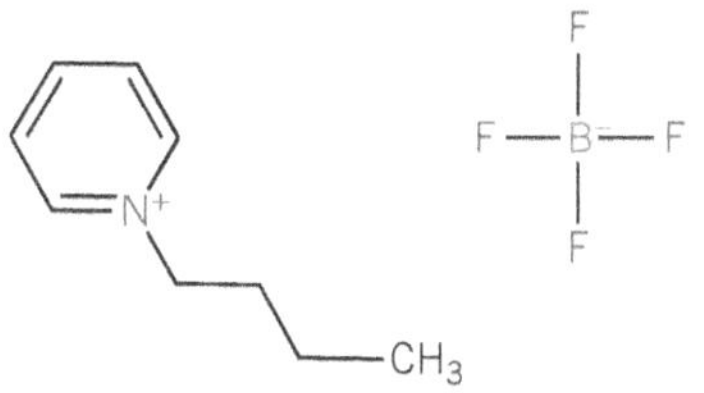

▲ **Figure 1** $[1\text{-butylpyridinium}]^+[BF_4]^-$

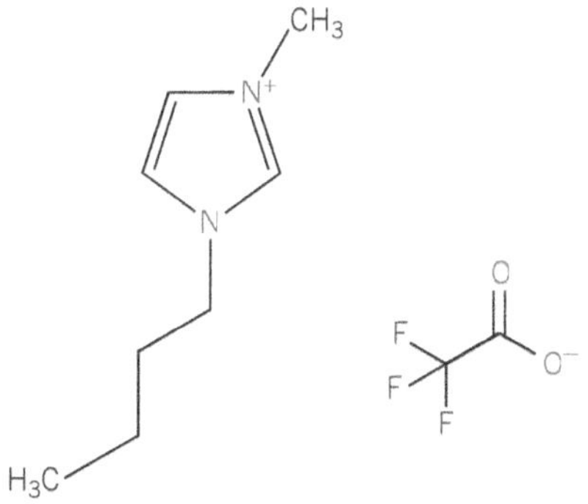

▲ **Figure 2** $[1\text{-butyl-3-methyl-}1H\text{-imidazolium}]^+[CF_3CO_2]^-$

They are increasingly used as solvents for a number of synthetic processes and can be considered an example of 'green' chemistry because they are neither volatile nor flammable. In addition, they are less toxic than organic alternatives, which makes them potentially safer to work with.

1　Using ideas about charge density and lattice enthalpy, explain why these ionic liquids have low melting points.

2　What factors are involved in an enthalpy-level diagram to calculate $\Delta_{soln}H$? Do you expect these ionic liquids would dissolve in water?

3　The $CF_3CO_2^-$ ion is formed from trifluoroethanoic acid, CF_3CO_2H, a weak acid. Write an expression for K_a and calculate the pH of a 0.1 mol dm^{-3} solution of trifluoroethanoic acid. $K_a = 5.89 \times 10^{-1}$ mol dm^{-3}

4　Calculate the pH of a buffer solution which contains 0.15 mol dm^{-3} CF_3CO_2H and 0.25 mol dm^{-3} $CF_3CO_2^-$.

5　An ionic substance A^+B^- is melted. The enthalpy change for this process, $\Delta H = +25.7$ kJ mol^{-1} and the entropy change, $\Delta_{sys}S = +35.0$ J K^{-1} mol^{-1}. Given that, at melting, the solid and liquid are in equilibrium and $\Delta_{tot}S = 0$, calculate the melting temperature T and explain whether A^+B^- is an ionic liquid.

Extension

1　Gibbs energy G is a quantity that relates enthalpy H, entropy S, and temperature T. Research Gibbs energy and how it relates to the spontaneity of a process.

2　Produce a series of instructions explaining how to calculate enthalpy change of solution, pH of a strong acid, pH of a weak acid, pH of a strong alkali, pH of a buffer, solubility product, entropy change of reaction, and whether a process will occur spontaneously.

3　The oceans are one of the defining features of planet Earth. Use your knowledge of the chemistry of the oceans to explain how the planet would be different if the oceans did not exist. Go on to explain how the planet would be different if the oceans were made of a liquid such as hexane, which is non-polar, has a much less endothermic enthalpy of vaporisation, and does not form ions.

CHAPTER 9
Developing metals

Topics in this chapter

→ **DM 1 Metals of antiquity**
introduction to transition metals and their oxidation states

→ **DM 2 Transition metals as catalysts**
catalytic activity

→ **DM 3 Colourful compounds**
colour in transition metal compounds and complexes

→ **DM 4 Electrochemistry**
electrochemical cells

→ **DM 5 Rusting**
rusting and methods of protection against rusting

→ **DM 6 Complexes and ligands**
structure and properties of complexes

Why a chapter on Developing Metals?

This chapter tells the story of the importance of metals to society, both ancient and modern, with emphasis on the d-block transition metals. These are the structural metals used by engineers to make the things needed in everyday life. The metals and their compounds are also of great importance, both in industry and in biological systems. The unique chemistry of transition metals is closely related to their electronic structure – variable oxidation states, catalytic activity, complex formation, and the formation of coloured compounds.

You will also study the development of the electrochemical cell and use standard electrode potentials to explain observations and make predictions about redox reactions, including those involved in rusting.

Knowledge and understanding checklist

In this chapter you will learn more about some ideas introduced in earlier chapters:

- [ ] electron energy levels in atoms (**Elements of life** and **Elements from the sea**)
- [ ] atomic absorption and emission spectra (**Elements of life**)
- [ ] redox reactions and oxidation numbers (**Elements from the sea**)
- [ ] chemical equilibrium (**The ozone story** and **The chemical industry**)
- [ ] catalysis (**Developing fuels, The ozone story,** and **Polymers and life**).

Maths skills checklists

In this chapter you will need to use the following maths skills. You can find support for these skills on Kerboodle and through MyMaths.

- [ ] Recognise and make use of appropriate units in calculations.
- [ ] Recognise and use expressions in decimal and ordinary form.
- [ ] Use calculators to find and use power and exponential functions.
- [ ] Use an appropriate number of significant figures.
- [ ] Understand and use the symbols $=, <, \ll, \gg, >, \propto, \sim, \rightleftharpoons$.
- [ ] Change the subject of an equation.
- [ ] Substitute numerical values into algebraic equations, using appropriate units for physical quantities.
- [ ] Solve algebraic equations.
- [ ] Use logarithms in relation to quantities that range over several orders of magnitude.

MyMaths.co.uk
Bringing Maths Alive

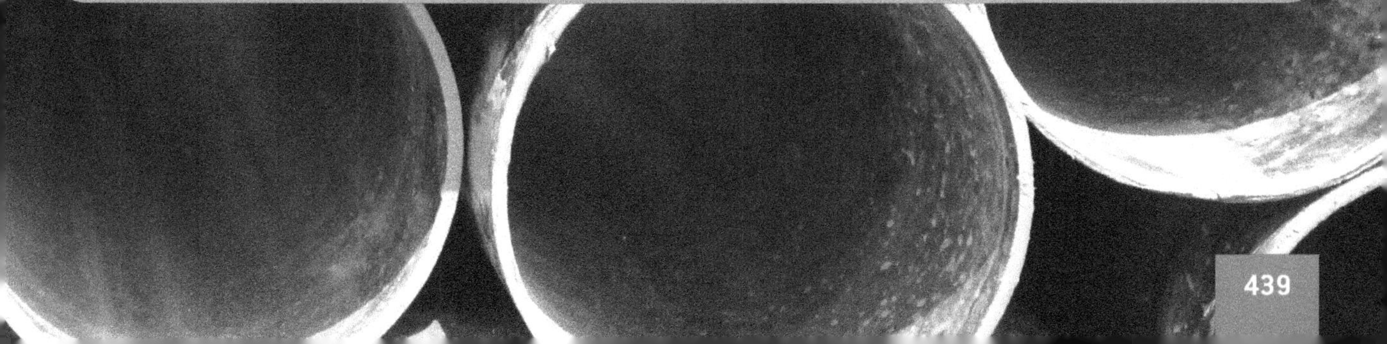

439

DM 1 Metals of antiquity

A history of metals

Gold in the hills

The earliest use of metals in ancient society was restricted to metals such as gold and silver.

Despite their relative rarity, gold and silver can actually be found native (as the element) in the Earth's crust. Gold and silver are some of the least reactive metals.

Copper is another metal that can be found in its native state, and new research suggests copper has been used for as long as gold.

Artistic gold

Native metals such as gold and silver were used in early art, particularly religious art (Figure 1). Gold was often used in altar pieces because it does not tarnish and it catches the light in churches when illuminated by candle.

Mixing it

Gold, silver, and copper are all soft metals and so they were mainly used as decorative artefacts by early civilisations but tin is relatively easy to produce from its ores, such as cassiterite, SnO_2. When tin is added to molten copper and the mixture allowed to solidify, a much harder substance is produced, capable of being shaped into tools. These discoveries led to the start of the Bronze Age.

▲ Figure 1 *Altar piece with liberal amounts of gold leaf*

▲ Figure 2 *Copper has been used for tools since at least 3500 BC but it may have been used even earlier*

Money makes the world go round

The lack of reactivity and relative ease of working meant that the metals of antiquity were used for a coinage bartering system. Arguably, the world's first metal coin was minted in Turkey (Figure 3).

▲ Figure 3 *The Lydian lion, minted sometime around 600 BC in Lydia, Asia Minor (current-day Turkey), made of electrum, an alloy of gold and silver*

Karats

The word karat refers to the amount of gold in a particular item. Karats are measured in units of 24, where 24 karat gold is pure gold. 18 karat gold is 18 parts gold and 6 parts alloys such as copper, nickel, silver, or zinc. 14 karat gold is 14 parts gold and 10 parts alloy.

Iron Age

The Iron Age goes back as far as 1500 BC or, some would argue, earlier. Nine small beads dated to 3200 BC from two burial sites in Gerzeh, Egypt, were not the result of chemical processing of iron ore but the mechanical working of meteorite iron.

The composition of the beads was analysed by non-invasive analytical techniques because of their inherent archaeological value. However, the iron content of materials can be found by redox titration.

A characteristic of transition metals is their ability to form more than one stable ion. The change of oxidation state from Fe^{2+} to Fe^{3+} forms the basis of the redox titration when analysing the iron content of materials.

Table 1 summarises the most common oxidation states of the d-block elements from scandium to zinc. The most important oxidation states are shaded.

▼ Table 1 *Oxidation states shown by elements in the first row of the d-block – the most important oxidation states are highlighted*

Element	Sc	Ti	V	Cr	Mn	Fe	Co	Ni	Cu	Zn
									+1	
			+2	+2	+2	+2	+2	+2	+2	+2
Common oxidation states	+3	+3	+3	+3	+3	+3	+3	+3	+3	
		+4	+4		+4					
			+5							
				+6	+6	+6				
					+7					

The d-block elements, including transition metals

The **d-block** consists of three horizontal series in Period 4, Period 5, and Period 6, each series containing 10 elements.

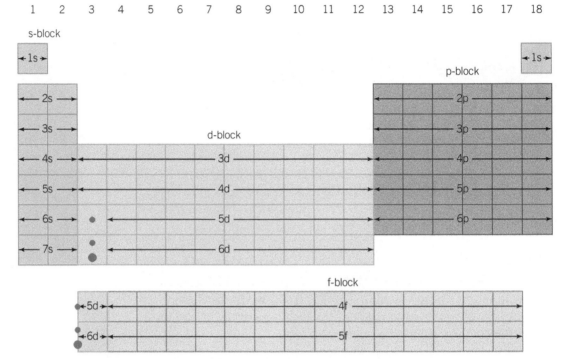

▲ Figure 4 The blocks of the periodic table

▼ Table 2 *Arrangement of electrons in the ground state of elements of the first row of the d-block, [Ar] represents the electronic configuration of argon*

Element	Electron configuration
Ar	$1s^2 2s^2 2p^6 3s^2 3p^6$
scandium	$[Ar]3d^1 4s^2$
titanium	$[Ar]3d^2 4s^2$
vanadium	$[Ar]3d^3 4s^2$
chromium	$[Ar]3d^5 4s^1$
manganese	$[Ar]3d^5 4s^2$
iron	$[Ar]3d^6 4s^2$
cobalt	$[Ar]3d^7 4s^2$
nickel	$[Ar]3d^8 4s^2$
copper	$[Ar]3d^{10} 4s^1$
zinc	$[Ar]3d^{10} 4s^2$

The chemistry of the d-block is different to that of the elements in other parts of the periodic table. This results from the special electronic configurations of d-block elements and the energy levels associated with their electrons. Differences between elements within a group in the d-block are less sharp than those in the s-block and p-block, and similarities across the period are greater. As such, you can discuss the d-block as a collection of elements with many features in common.

Electronic configuration of d-block elements

Across the first row of the d-block (the 10 elements from scandium to zinc), each element has one more proton in the nucleus and one more electron than the previous element. Each additional electron enters the 3d shell. Remember this is *not* the outermost shell because the outer 4s orbital has already been filled.

The electronic configurations of the atoms of the first row of the d-block are shown in Table 2. Only the two outer sub-shells are shown because the elements all have an identical core of electrons. The core is the electronic configuration of the noble gas argon, Ar, $1s^2 2s^2 2p^6 3s^2 3p^6$.

These elements have essentially the same *outer* electronic arrangement as each other, in the same way as the elements in a vertical group all have the same outer shell structure. Moreover, unlike the elements in a group, they do not differ by a complete electron shell but by having one more electron in the *inner*, incomplete 3d sub-shell.

Element	Atomic number	Electronic arrangement	3d					4s
Sc	21	[Ar]	↑					↑↓
Ti	22	[Ar]	↑	↑				↑↓
V	23	[Ar]	↑	↑	↑			↑↓
Cr	24	[Ar]	↑	↑	↑	↑	↑	↑
Mn	25	[Ar]	↑	↑	↑	↑	↑	↑↓
Fe	26	[Ar]	↑↓	↑	↑	↑	↑	↑↓
Co	27	[Ar]	↑↓	↑↓	↑	↑	↑	↑↓
Ni	28	[Ar]	↑↓	↑↓	↑↓	↑	↑	↑↓
Cu	29	[Ar]	↑↓	↑↓	↑↓	↑↓	↑↓	↑
Zn	30	[Ar]	↑↓	↑↓	↑↓	↑↓	↑↓	↑↓
			Building up of inner 3d sub-shell					Outer shell

▲ **Figure 5** *Arrangement of electrons in the ground state of elements in the first row of the d-block. [Ar] represents the electronic configuration of argon*

Why are chromium and copper different?

Look carefully at Table 2. The electronic configurations of chromium and copper do not fit the pattern of building up the 3d sub-shell.

In the ground state of an atom, electrons are always arranged to give the lowest total energy. Because of their negative charge, electrons repel one another so a lower total energy is obtained with electrons singly in orbitals than if they are paired in an orbital. The energies of the 3d and 4s orbitals are very close together in Period 4. In chromium, the orbital energies are such that putting one electron into each 3d and 4s orbital gives a lower energy than having two in the 4s orbital.

With copper, putting two electrons into the 4s orbital would give a higher energy than filling the 3d orbitals.

You might expect chromium to have the electronic arrangement:

[Ar]	↑	↑	↑	↑		↑↓

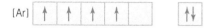

This would continue the pattern of the previous elements. However its electronic arrangement is actually:

[Ar]	↑	↑	↑	↑	↑	↑

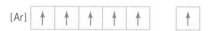

This has a lower energy by avoiding having a pair of electronic in the same orbital.

Synoptic link

You first met shells and sub-shells in Topic EL 3, Electrons, where would we be without them?

Study tip

The convention when writing electronic structures is to keep all the principal shell electrons together, for example, $3s^2\ 3p^6\ 3d^2\ 4s^2$, even though the 4s electronic sub-shell is filled before the 3d sub-shell.

▼ **Table 3** *Electronic configuration of the common oxidation states of iron and copper*

Atom or ion	Electronic configuration
Fe	$1s^2 2s^2 2p^6 3s^2 3p^6 3d^6 4s^2$
Fe^{2+}	$1s^2 2s^2 2p^6 3s^2 3p^6 3d^6$
Fe^{3+}	$1s^2 2s^2 2p^6 3s^2 3p^6 3d^5$
Cu	$1s^2 2s^2 2p^6 3s^2 3p^6 3d^{10} 4s^1$
Cu^+	$1s^2 2s^2 2p^6 3s^2 3p^6 3d^{10}$
Cu^{2+}	$1s^2 2s^2 2p^6 3s^2 3p^6 3d^9$

Ion formation

When the transition metals lose electrons to form simple ions such as Fe^{2+} or Cr^{3+}, it is *always* the 4s electrons that are lost first. As such, the electronic structure of Fe^{2+} is $[Ar]3d^6$ and Cr^{3+} is $[Ar]3d^3$.

Shells, sub-shells, and transition metal ions

Transition metals form a number of ions exhibiting a range of oxidation states. These depend on the ability to lose 4s and 3d electrons. For example, the most common oxidation states exhibited by iron are +2 and +3, that is, Fe(II) and Fe(III).

Transition metals form one or more stable ions with incompletely filled d-orbitals.

 Worked example: Electronic configuration of vanadium(III)

What is the electronic configuration of vanadium(III)?

Step 1: Write down the electronic configuration of vanadium.

$$[Ar]3d^3 4s^2$$

Step 2: The ion has a +3 charge so three electrons have been removed from the atom. Two will be removed from the 4s sub-shell and one from the 3d sub-shell.

The electronic configuration of vanadium(III) is $1s^2 2s^2 2p^6 3s^2 3p^6 3d^2$

Zinc and scandium are often not considered transition metals even though they are in the first row of the d-block. Zinc only forms the Zn^{2+} ion, with an electronic configuration of $1s^2 2s^2 2p^6 3s^2 3p^6 3d^{10}$. The Zn^{2+} ion has a full d-orbital. Scandium only forms the Sc^{3+} ion, which has an electronic configuration of $1s^2 2s^2 2p^6 3s^2 3p^6$. The Sc^{3+} ion has an empty d-orbital.

Zinc and scandium are not transition metals but they are classified as d-block elements.

Variable oxidation states and their relative stabilities

Transition metals exist in a number of oxidation states (Table 1). This is because there are several stable arrangements of the d-electrons and s-electrons. As you can see from the electronic configuration of chromium and copper, d^5 and d^{10} are particularly stable. This explains why only copper forms an ion with an oxidation state of +1 – the electronic configuration of Cu^{1+} is $[Ar]3d^{10}$, and why the Fe^{3+} ion ($[Ar]3d^5$) has greater stability than the Fe^{2+} ion ($[Ar]3d^6$).

Redox titration

Iron content can be analysed using redox titration. This type of analysis depends on the variable oxidation states of the transition metals involved in the reactions.

Iron(II) ions can be oxidised to iron(III) ions by potassium manganate(VII) in acidic solution. The two half equations are:

$Fe^{2+}(aq) \rightarrow Fe^{3+}(aq) + e^-$ oxidation

$MnO_4^-(aq) + 8H^+(aq) + 5e^- \rightarrow Mn^{2+}(aq) + 4H_2O(l)$ reduction
purple

The overall equation is:

$5Fe^{2+}(aq) + MnO_4^-(aq) + 8H^+ \rightarrow 5Fe^{3+}(aq) + Mn^{2+}(aq) + 4H_2O(l)$

If a known volume of the iron(II) solution is titrated with potassium manganate(VII) solution of known concentration, the end point is when the first permanent pink colour is observed – when $MnO_4^-(aq)$ is in excess. No additional indicator is required.

Synoptic link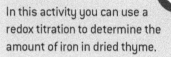

You met another type of titration, iodine–thiosulfate titrations, in Topic ES 5, The risks and benefits of chlorine. You were also introduced to acid-base titrations in Topic EL 9, How salty? You can find out more about these three types of titration in Techniques and procedures.

Activity DM 1.1

In this activity you can use a redox titration to determine the amount of iron in dried thyme.

Worked example: Determining iron content

$25\,cm^3$ of an iron(II) sulfate solution, in excess acid, requires an average concordant titre of $17.3\,cm^3$ $0.0200\,mol\,dm^{-3}$ potassium manganate(VII) solution to reach a permanent pink colour.

What is the concentration of $Fe^{2+}(aq)$ in the solution in $mol\,dm^{-3}$?

The original iron(II) sulfate solution was obtained by dissolving a sample of $27.50\,g$ in excess sulfuric acid and making it up to $2000\,cm^3$.

What is the percentage of iron in the iron ore?

Step 1: Write the equation for the reaction.

$5Fe^{2+}(aq) + MnO_4^-(aq) + 8H^+(aq) \rightarrow$
$5Fe^{3+}(aq) + Mn^{2+}(aq) + 4H_2O(l)$

Step 2: Calculate the number of moles of $MnO_4^-(aq)$ used and therefore the number of moles of $Fe^{2+}(aq)$.

$moles\ MnO_4^-(aq) = 0.02 \times \frac{17.3}{1000} = 3.46 \times 10^{-4}$

$moles\ Fe^{2+} = 3.46 \times 10^{-4} \times 5 = 1.73 \times 10^{-3}$

Step 3: Calculate the concentration of Fe^{2+} in the original solution.

$1.73 \times 10^{-3} \times \frac{1000}{25} = 0.069\,mol\,dm^{-3}$

Step 4: Calculate the percentage of iron in the ore.

$27.5\,g$ of ore was made up to $2000\,cm^3$. This would contain $3.86\,g \times 2 = 7.72\,g$ of Fe^{2+}

percentage of iron in the ore $= \frac{7.72}{27.50} \times 100 = 28.1\%$

Instead of acidified potassium manganate(VII) solution, the concentration of Fe^{2+} ions could be found using acidified potassium dichromate solution. The half equation for the reaction is:

$$Cr_2O_7{}^{2-}(aq) + 14H^+ + 6e^- \rightarrow 2Cr^{3+}(aq) + 7H_2O(l)$$

dichromate(VI) ion   chromium(III) ion

orange

Summary questions

1 Write electron configurations for the following ions.
   a Ti^{4+} (*1 mark*)   b Mn^{3+} (*1 mark*)   c Co^{2+} (*1 mark*)

2 Write electron configurations for the following ions.
   a Fe^{2+} (*1 mark*)   b V^{3+} (*1 mark*)   c Cr^{3+} (*1 mark*)

3 Scandium, nickel, and zinc only have one common oxidation state. Write electron configurations for Sc^{3+}, Ni^{2+}, and Zn^{2+} and use these to explain why nickel is classed as a transition metal but scandium and zinc are not.   (*4 marks*)

4 Suggest electron configurations for the following ions.
   a Pd^{2+} (*1 mark*)   b Ag^+ (*1 mark*)   c Cd^{2+} (*1 mark*)

5 Manganese(II) is more stable than manganese(III). Suggest why.

6 A student wants to determine the amount of iron in an iron(II) compound by a redox titration with a standard solution of acidified potassium dichromate solution.

The equation for the reaction is:

$$Cr_2O_7{}^{2-} + 6Fe^{2+} + 14H^+ \rightarrow 2Cr^{3+} + 6Fe^{3+} + 7H_2O$$

These are the student's results:

Mass of iron compound dissolved in 500 cm³ solution   = 11.05 g

Volume of iron(II) compound used in each titre   = 25.00 cm³

Average titre of 0.0150 mol dm⁻³ potassium dichromate solution   = 26.55 cm³

   a Calculate the number of moles of potassium dichromate in the average titre.   (*1 mark*)

   b Calculate the number of moles of iron(II) ions in 25 cm³ of its solution.   (*1 mark*)

   c Calculate the mass of iron in the iron(II) compound dissolved in 500 cm³ of solution.   (*2 marks*)

   d Calculate the percentage by mass of iron in the iron(II) compound.   (*1 mark*)

In addition to being used in jewellery, the element silver has many varied uses, for example:

- The Phonecians stored water and other liquids in silver coated bottles to keep the water safe for drinking.

- In the 1800s aqueous silver nitrate was put in the eyes of new born babies eyes to prevent the transmission gonorrhoea from infected mothers.

- Before refrigeration silver dollars used to be put into milk bottles to keep milk fresh. Now you can buy fridges coated with silver nanoparticle coatings on their inner surfaces.

- Wound dressings are often impregnated with silver.

Why has silver been used in all of these applications? Silver ions have antimicrobial properties. It is thought that these catalyse reactions in the cell walls of bacteria, resulting in the formation of disulfide bonds. These reactions change the shape of cellular enzymes, subsequently preventing the growth of more bacteria.

Once antibiotics were discovered, the use of silver as a bactericidal agent decreased. However, with the emergence of antibiotic-resistant strains of bacteria such as MRSA (Methicillin-resistant Staphylococcus aureus) there has been a renewed interest in using silver as an antibacterial agent.

Transition metals and their compounds have also been used extensively in the chemical industry to make processes more economical. For example:

- the hydrogenation of alkenes, catalysed by nickel or platinum

- the manufacture of ammonia by the Haber process, catalysed by iron

- the manufacture of sulfuric acid by the contact process, catalysed by vanadium(V) oxide

- alloys of platinum, rhodium, and palladium are used in car catalytic converters.

All these examples show how useful the catalytic properties of transition metals are in everyday life.

Learning outcomes

Demonstrate and apply knowledge and understanding of:

→ the catalytic activity of transition metals and their compounds.

Synoptic link

Enzymes were covered in Topic PL 6, Enzymes.

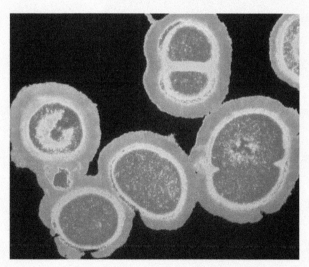

▲ Figure 1 *MRSA is often called a superbug because it is resistant to most anti-biotics and can be very difficult to treat*

Synoptic link

Heterogeneous catalysis was met in Topic DF 5, Getting the right sized molecules.

Chemical ideas: The periodic table 11.4b

Catalytic activity of transition metals

A **catalyst** alters the rate of a chemical reaction without being used up in the process. Transition metals and their compounds are effective and important catalysts, both in industry and in biological systems.

Chemists believe that catalysts offer an alternative reaction pathway that has a lower activation enthalpy than that of an uncatalysed reaction.

It is the availability of 3d as well as 4s electrons and the ability to change oxidation state that help make transition metals such good catalysts.

Heterogeneous catalysis

In heterogeneous catalysis, the catalyst is in a different phase from the reactants. In the case of transition metals, this usually means a solid metal catalyst with reactants in the gas phase or liquid phase.

Transition metals can use the 3d and 4s electrons of atoms on the metal surface to form weak bonds (**chemisorption**) to reactants. Once the reaction has occurred on the surface, these bonds can break to release the products. An example of heterogeneous catalysis is the contact process in which the manufacture of sulfuric acid is catalysed by vanadium(V) oxide.

Homogeneous catalysis

In homogeneous catalysis, the catalyst is in the same phase as the reactants. In the case of transition metals, this often means the reaction takes place in the aqueous phase – the catalyst being an aqueous transition metal ion.

Homogeneous catalysis usually involves the transition metal ion forming an intermediate compound with one or more of the reactants, which then breaks down to form the products.

An example is the reaction between 2,3-dihydroxybutanoate ions and hydrogen peroxide, which is catalysed by cobalt(II), Co^{2+}, ions.

Figure 2 shows a suggested mechanism for this catalysis. Figure 6 shows the corresponding enthalpy profile.

Synoptic link

Catalysts, heterogeneous catalysis, and homogeneous catalysis were first introduced in Topic DF 5, Getting the right sized molecules.

Synoptic link

Cracking in oil refining was introduced in Topic DF 5, Getting the right sized molecules. Catalytic converters in cars were introduced in Topic DF 10, The trouble with emissions.

Activity DM 2.1

In this activity you will investigate the catalytic activity of aqueous cobalt(II) ions in the reaction between hydrogen peroxide and 2,3-dihydroxybutanoate ions.

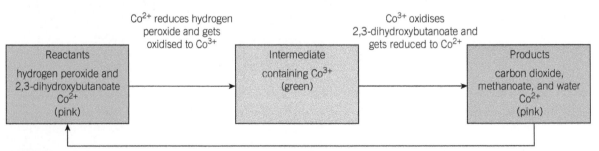

▲ Figure 2 *A suggested mechanism for the catalytic action of Co^{2+}. The reaction mixture turns from pink to green, then back to pink*

Transition metal ions are particularly effective catalysts in redox reactions. This is because they can readily move from one oxidation state to another, in the way that cobalt readily moves between the +2 and +3 states in the reaction above.

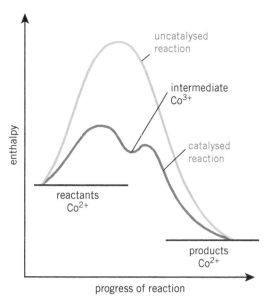

▲ **Figure 3** *The enthalpy profiles for the catalysed and uncatalysed reactions*

Summary questions

1 Solid, insoluble MnO_2 acts as a catalyst for the decomposition of hydrogen peroxide. Suggest how it does this. *(2 marks)*

2 When Fe^{2+} ions are added to a mixture containing $S_2O_8^{2-}$ ions and iodide ions the following two reactions occur sequentially.

Reaction 1 $2Fe^{2+}(aq) + S_2O_8^{2-}(aq) \rightarrow 2SO_4^{2-}(aq) + 2Fe^{3+}(aq)$

Reaction 2 $2Fe^{3+}(aq) + 2I^-(aq) \rightarrow 2Fe^{2+}(aq) + I_2(aq)$

a Write a balanced equation for the overall reaction. *(2 marks)*

b Using equations for the two reactions, what evidence is there to suggest that Fe^{2+} is acting as a catalyst? *(1 mark)*

c How could you test whether Fe^{2+} is a catalyst for this reaction? *(1 mark)*

d Draw an energy level diagram to represent what is happening in the two reactions given in the equation. *(2 marks)*

DM 3 Colourful compounds

Specification references: DM(g), DM(i), DM(j), DM(k), DM(m), DM(n)

▲ Figure 1 Coloured glass is made by adding metallic oxides to the molten glass

d-block elements in art

Coloured glass

Stained glass windows in churches are not the only example of a use of coloured glass. Splashbacks, worktops, table tops, doors, vases, and feature walls can all be made using beautiful coloured glass.

Glass is coloured by adding metallic oxides whilst it is still in a molten state. These additions are often oxides of transition metals. The exact colour will depend on the type of glass used, the concentration of metal compound added, and any other elements or compounds fused into the glass.

▼ Table 1 Some of the metal ions used to colour glass

Transition metal ion present	Colour of glass
Fe^{2+}	green
Cu^{2+}	blue
Au	red
Mn^{3+}	deep purple
Cd^{2+}	mustard yellow
Co^{2+}	blue
Pb^{2+}	yellow

Light shining through the glass can to a beautiful array of colour. How are all these colours formed?

▲ Figure 2 The Palais des congrès de Montréal is a convention centre located in Montreal, Canada. Completed in 1983, the Palais des congrès de Montréal has entire external walls made of coloured glass

Colours in pottery

Pottery can also be made in a variety of colours. Glazes are melted onto ceramic pottery then fired in a kiln in order to colour and waterproof the item. Oxides of transition metals such as titanium, iron, copper and cobalt are widely used in glazes. Some results are more predictable than others. Cobalt oxide usually gives rise to a deep blue glaze after firing whilst, depending on a number of variables, iron can give rise to red, yellow, brown, blue, or green. If the oxygen supply to a kiln is limited then the transition metals in the glaze might be reduced to a lower oxidation state.

How do these transition metal compounds give rise to colours in glass and pots? As light passes through coloured glass, certain frequencies of light are absorbed leaving the remaining frequencies to pass through. In the case of ceramic pots, when light falls onto the pots, certain frequencies are absorbed and the remaining frequencies are reflected from the surface into your eye.

▲ Figure 3 *These pots were fired using a copper containing glaze in a limited oxygen supply giving a characteristic red colour*

Raman spectroscopy – it's more than skin deep

Raman spectroscopy is a non-sampling, non-invasive, and non-damaging spectroscopic and analytical technique. It is therefore ideal in identifying the pigments used in fragments of potters.

In Raman spectroscopy, a single wavelength of coloured light from a laser source is shone upon a material, such as the artwork on a canvas. The majority of the light bounces off the sample with exactly the same energy and therefore, the same colour. However, a tiny fraction (about one in a million of the photons bouncing off the compound in the sample) come off at a slightly different frequency – they have lost some of their energy and therefore have a slightly different colour.

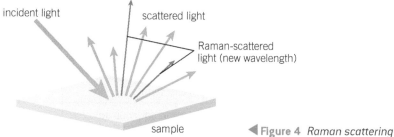

◀ Figure 4 *Raman scattering*

By examining the difference between the light that goes in and the light that comes back, it is possible to identify the absorbing substance, for example, a particular pigment used by an artist.

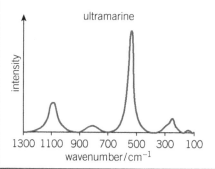
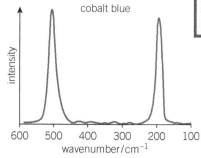

◀ Figure 5 *Raman spectra of ultramarine and cobalt blue. The spectra are unique with the peaks representing the difference between the energy/frequency of the scattered light and the incident light for different wavelengths of incident light*

A photon of laser light can excite a molecule. If the molecule relaxes back to a different state, the emitted photon will have a different frequency (energy) from the one that was absorbed and the shift in energy can be used to deduce the different vibrational (and rotational) states of a molecule. The difference in energy is used to produce the vibrational Raman spectra, and therefore give away the identity of the molecule itself.

A new type of Raman spectroscopy, spatially offset Raman spectroscopy (SORS), can even be used to reveal the chemical composition of substances beneath the surface of a solid.

1   Calculate the energy of a photon with frequency of $6.75 \times 10^{14}\, s^{-1}$.
2   What colour in the visible spectrum does the photon correspond to?

▲ **Figure 6** *Solutions of transition metals compounds come in a wide variety of colours*

▼ **Table 2** *Colours of some transition metal ions in aqueous solution*

Ion	Outer electrons	Colour
Ti^{3+}	$3d^1$	purple
V^{3+}	$3d^2$	green
Cr^{3+}	$3d^3$	violet
Mn^{3+}	$3d^4$	violet
Mn^{2+}	$3d^5$	pale pink
Co^{2+}	$3d^7$	pink
Ni^{2+}	$3d^8$	green

Study tip

You need to learn the colours for the copper and iron ions in solution, and be able to account for why they are coloured where appropriate.

Chemical ideas: The periodic table 11.4c

Coloured compounds of transition metals

Compounds of d-block transition metals are frequently coloured, both in the solid state and in solution (Table 2).

The intensity of colour varies greatly, for example, MnO_4^- is intensely deep purple but Mn^{2+} is very pale pink. The colour of transition metal compounds can often be related to the presence of unfilled or partly filled d-orbitals in the metal ion.

▼ **Table 3** *The colour of common iron and copper ions in aqueous solution*

Ion	Electronic configuration	Colour in aqueous solution
Fe^{2+}	$[Ar]3d^6$	green
Fe^{3+}	$[Ar]3d^5$	orange/brown
Cu^+	$[Ar]3d^{10}$	unstable in aqueous solution
Cu^{2+}	$[Ar]3d^9$	blue

Copper(I) ions are unstable in aqueous solution and disproportionate, for example:

$$Cu_2SO_4(s) + (aq) \rightarrow Cu(s) + CuSO_4(aq)$$

oxidation state
of copper   +1   0   +2

When white light falls on a substance, some may be absorbed, some transmitted, and some reflected. If light in the visible region of the spectrum is absorbed then the compound will appear coloured.

The fact that most transition metal compounds are coloured can also be used in qualitative analysis.

Qualitative analysis

The addition of drops of sodium hydroxide solution, or ammonia solution, to solutions containing transition metal ions such as Fe^{2+}, Fe^{3+}, and Cu^{2+}, results in characteristic coloured **precipitates** forming. These reactions can be represented by ionic equations:

$$Fe^{2+}(aq) + 2OH^-(aq) \rightarrow Fe(OH)_2(s)$$

green **gelatinous** precipitate

$$Fe^{3+}(aq) + 3OH^-(aq) \rightarrow Fe(OH)_3(s)$$

orange gelatinous precipitate

$$Cu^{2+}(aq) + 2OH^-(aq) \rightarrow Cu(OH)_2(s)$$

pale blue precipitate

If the source of the hydroxide ions is ammonia solution, the pale blue precipitate of $Cu(OH)_2$ will re-dissolve on addition of excess ammonia solution, to give a deep blue-purple solution. This solution contains a copper/ammonia complex ion (see Topic DM 6).

Neither of the iron hydroxides form complexes with ammonia and so do not re-dissolve.

Chemical ideas: Radiation and matter 6.7a

Coloured compounds of inorganic d-block elements

What happens when a d-block compound absorbs light?

Light is absorbed by an atom only if the energy of the light matches the energy gap between two energy states in the atom. If it does, then an electron is promoted from an orbital of lower energy to one of higher energy. So the atom or ion absorbing the radiation changes from its ground state to an excited state.

The frequency v of the light absorbed depends on the energy difference ΔE between these two levels. There are five d-orbitals in d-block metals. However, molecules or anions called **ligands** surrounding the metal atom or ion, for example, copper(II) ions in solution are surround by $6H_2O$ ligands, cause these orbitals to split in such a way that some are slightly higher in energy than others. The difference between the two levels ΔE is now such that the light absorbed falls in the visible part of the spectrum. The transmitted light is the light that is not absorbed, and is the colour seen (Figure 7). This is the case for coloured glass and coloured solutions.

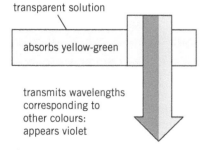

▲ Figure 7 *A visible representation of absorption and transmission of a coloured solution*

Synoptic link

Ionic equations were first introduced in Topic EL 7, Blood, sweat, and seas.

Activity DM 3.1

In this activity you will compare the reactions of copper(II), iron(II), and iron(III) ions with sodium hydroxide and ammonia

Activity DM 3.2

This activity checks you know the formulae and colours of some copper(II), iron(II), and iron(III) ions and compounds.

Synoptic link

The relationship between light and energy was first discussed in Topic EL 2, How do we know so much about outer space, and Topic OZ 2, Screening the Sun.

Study tip

$\Delta E = hv$

Where h is Planck's constant.

The absorption spectrum for the $[Ti(H_2O)_6]^{3+}$ complex ion is shown in Figure 8. It is called a complex ion because it is composed of both central ion and attached ligands. It absorbs mainly yellow-green light – therefore the solution looks violet.

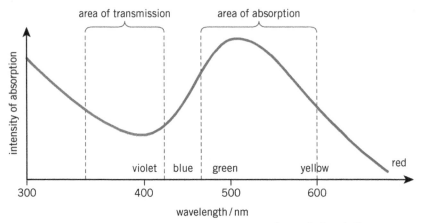

▲ **Figure 8** *The absorption spectrum of the hydrated Ti^{3+} ion, $[Ti(H_2O)_6]^{3+}$*

Green-yellow and violet are complementary colours – each is the colour that remains when the other is removed from white light.

Complementary colours are shown diagonally opposite each other in the colour wheel (Figure 9).

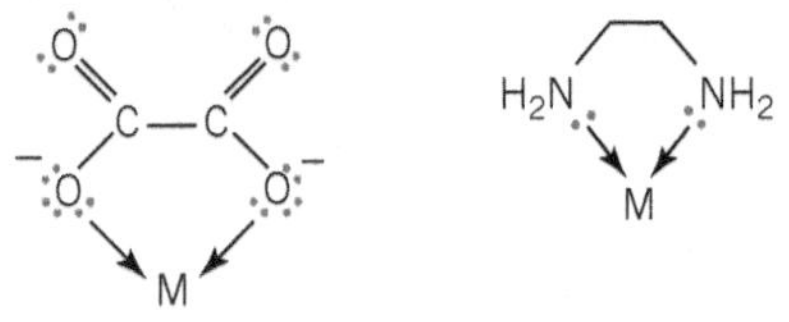

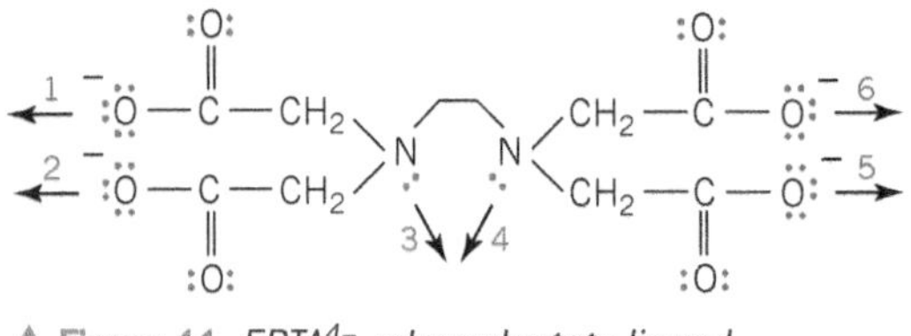

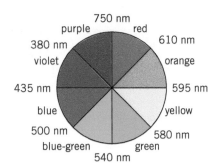

▲ **Figure 9** *Wavelength ranges for colours in a colour wheel*

$[Ti(H_2O)_6]^{3+}$ is a simple example because the ion contains only one d-electron. The situation is more complicated when there are several d-electrons, and many more transitions are possible between d-levels. However, the basic principles are the same.

In $[Ti(H_2O)_6]^{3+}$, the ligands are water molecules. Another common ligand is the ammonia molecule, NH_3. Ligands attach themselves to the central ion by using a lone pair of electrons to form a dative or **coordinate bond**.

Synoptic link

Dative covalent bonding has been covered in Topic CI 1, Nitrogen chemistry, and Topic PL 2, The nylon story.

Typical anions that can behave as ligands include the halide ions (e.g., chloride, Cl^-), cyanide, CN^-, and the hydroxide ion, OH^-. These ligands are described as **monodentate** because they can form only one dative bond to the central metal ion. Some ligands can form two dative bonds and are termed **bidentate** ligands. 1,2-diaminoethane or ethylenediamine (en) is a typical bidentate ligand (Figure 10).

▲ **Figure 10** *Bidentate ligands the ethanedioate ion (left) and the 1,2-diaminoethane molecule (right)*

Another well-known ligand, EDTA, can form six dative bonds from the one ligand (Figure 11). It is called a **hexadentate** ligand.

▲ **Figure 11** *$EDTA^{4-}$, a hexadentate ligand*

Study tip

A dative bond (also called coordinate bond) is represented by an arrow pointing from a lone pair of electrons towards the atom or ion they are bonding with.

Ligands affect the d-orbital electrons of the metal ion they are surrounding. **Orbitals** close to the ligands are pushed to slightly higher energy levels than those further away. As a result, the five d-orbitals are split into two groups at different energy levels (Figure 12).

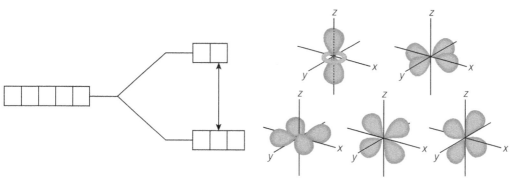

▲ Figure 12 *The shapes of the five d-orbitals and a typical splitting pattern*

Figure 13 shows why the Ti^{3+} ion in solution is purple. The size of the splitting means that light in the red, yellow, and green end of the spectrum is absorbed and promotes the one electron in the d-orbitals in a Ti^{3+} to an orbital in a higher energy state. The remaining visible light, blue and purple, passes through the solution and this is the colour it appears.

Synoptic link

Orbitals are first discussed in EL 3, Electrons, where would we be without them?

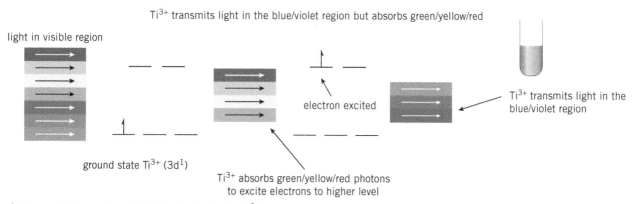

▲ Figure 13 *Absorption of light by the hydrated Ti^{3+} ion*

When light passes through a solution of $[Ti(H_2O)_6]^{3+}$ ions a photon of light may be absorbed. The energy of this photon corresponds to excitation of an electron from a low energy d-orbital to a high energy d-orbital.

Remember that the frequency v of the light absorbed depends on the energy difference ΔE between these two levels. For most d-block transition metals, the size of ΔE is such that the light absorbed falls in the visible part of the spectrum. The colour seen is white light minus the frequencies of absorbed light.

The energy needed to excite a d-electron to the higher energy level also depends on the oxidation state of the metal. This is why redox reactions of transition metal compounds can be accompanied by spectacular colour changes. For example, vanadium shows a different colour in each of its oxidation states in aqueous solution:

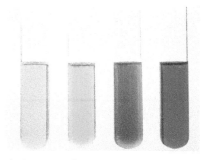

▲ Figure 14 *The colours of the four oxidation states of vanadium*

$$V(+5) \rightarrow V(+4) \rightarrow V(+3) \rightarrow V(+2)$$

yellow   blue   green   violet

Some ligands have a more powerful effect than others in splitting of the d-sub-shell. So changing the ligand complexed with a metal ion often results in a colour change:

$$[Ni(H_2O)_6]^{2+}(aq) + 6NH_3(aq) \rightleftharpoons [Ni(NH_3)_6]^{2+}(aq) + 6H_2O$$

light green   lilac/blue

The colour of a transition metal complex depends on:

- the number of d-electrons present in the transition metal ion
- the arrangement of ligands around the ion because this affects the splitting of the d-sub-shell
- the nature of the ligand because different ligands have a different effect on the relative energies of the d-orbitals in a particular ion.

For example, ammonia ligands cause a larger difference than water ligands in splitting d-orbital energies. The blue colour of hydrated Cu^{2+} ions, for example, changes to deep blue/violet when ammonia is added (Figure 15):

$$4NH_3(aq) + [Cu(H_2O)_6]^{2+} \rightarrow [Cu(NH_3)_4(H_2O)_2]^{2+} + 4H_2O(l)$$

blue   deep blue/violet

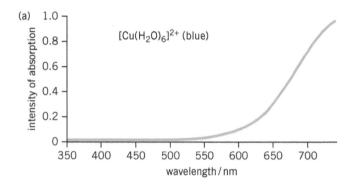

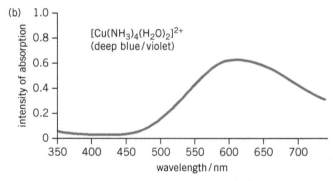

▲ **Figure 15** *Absorption spectrum for $[Cu(H_2O)_6]^{2+}$ and $[Cu(NH_3)_4(H_2O)_2]^{2+}$. The ammonia ligands cause a bigger splitting of d-orbital energies*

You can see from the following complexes of chromium that the colour also depends on the *number* of each kind of ligand present:

$$[Cr(H_2O)_6]^{3+} \qquad [Cr(H_2O)_5Cl]^{2+} \qquad [Cr(H_2O)_4Cl_2]^+$$

violet   green   dark green

Colorimetry

Whilst the colour of a solution depends on the colour of light it absorbs, the intensity of its colour depends on the concentration of the solution. The more concentrated the solution, the darker its colour, that is, the more light it absorbs.

Colorimetry is a technique that can be used to find the concentration of a coloured solution.

In colorimetry, a narrow beam of white light is passed through a coloured filter. Filters come in a range of colours and the filter colour must correspond to the colour of light that is *most strongly absorbed* by the solution being analysed.

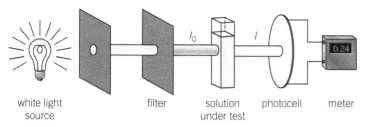

white light source   filter   solution under test   photocell   meter

▲ Figure 17 *Components in a colorimeter*

The absorbance A is proportional to the concentration of the solution c. For dilute solutions there is a direct relationship between the two:

$$A = kc$$

k is a constant. This means that as the concentration of a solution increases, its absorbance increases linearly.

A visible spectrophotometer can be used instead of a colorimeter.

Synoptic link

Colorimetry is covered further in Techniques and procedures.

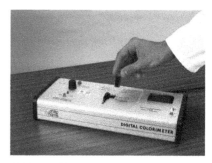

▲ Figure 16 *A typical colorimeter*

Activity DM 3.3

This activity helps you learn about the principles of colorimetry.

Activity DM 3.4

In this activity you will use colorimetry to find the concentration of thiocyanate ions in a sample of waste water.

Summary questions

1 a Explain why titanium(IV) oxide (the white pigment in white paint) is not coloured.   (*1 mark*)
 b Explain why compounds of Sc^{3+}, Zn^{2+}, and Cu^+ are not coloured.   (*2 marks*)

2 Zinc displaces copper from copper(II) sulfate to form zinc(II) sulfate. The solution turns from blue to colourless. Explain why the solution becomes colourless.   (*2 marks*)

3 Classify the following ligands as mono-, bi- or polydentate ligands.
 a H_2O (*1 mark*)
 b ethanedioate ion (oxalate ion)

ethanedioate ion   (*1 mark*)
 c thiocyanate ion SCN^- (*1 mark*)

DM 4 Electrochemistry

Specification references: DM(c), DM(d)

Frogs' legs

Luigi Galvani and Alessandro Volta were both scientists working at Italian universities in the late 18th century. Galvani, a biologist, was experimenting on frogs' legs. In one experiment, Galvani's steel scalpel touched a brass hook that was holding the leg in place. The leg twitched. Further experiments confirmed this effect and Galvani was convinced that he was seeing the effects of what he called animal electricity – the life force within the muscles of the frog.

Alessandro Volta was able to reproduce the results but was sceptical of Galvani's explanation. By experiment Volta found that it was the two dissimilar metals, not the frog's leg, that produced the electricity. The frog's leg was just an indicator of presence of the electricity.

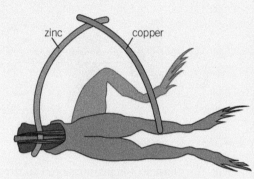

▲ **Figure 1** *Volta's frog experiment*

Alessandro Volta was a physicist, chemist, and a pioneer of electrical science. He is widely given credit for producing the first electrical battery – which people called the voltaic pile (Figure 2). His voltaic pile was a stack of zinc and silver discs separated by a wet cloth containing a salt or a weak acid solution. With this invention, scientists could produce steady flows of electric current, unleashing a wave of new discoveries and technologies.

Michael Faraday and John Daniell

Michael Faraday was a British physicist and chemist. His work on electrochemistry and electromagnetism, again in the late 18th century, laid the foundation for many areas of science. He played a key

▲ **Figure 2** *An original voltaic pile from 1799*

role in the development of electricity for use in technology and coined such terms as electrode, cathode, anode, and ion as well as producing his laws of electrolysis.

John Daniell was a British chemist and meteorologist who developed the first modern storage cell based on Faraday's principles. These consisted of a large glass jar with a copper star-shaped electrode in the bottom and a zinc electrode suspended near the top. The bottom of the jar was filled with a concentrated copper sulfate solution. On top of this was poured dilute sulfuric acid. The lower density of sulfuric acid meant it remained on top. This was the first practical battery to find wide use, powering telegraphs, railway signalling systems, and home doorbells.

Faraday and Daniell realised the basis of generating electricity was redox, and the d-block elements and their compounds were ideal as a rich area of redox chemistry.

Look after your smile

Dental amalgam

Dental amalgam is a dental filling material used to fill cavities caused by tooth decay (Figure 3). It has been used for more than 150 years in hundreds of millions of patients.

Dental amalgam is a mixture of metals, consisting of liquid mercury and a powdered alloy composed of silver, tin, and copper. Approximately 50% of dental amalgam is elemental mercury by weight.

There have been some concerns about the presence of the mercury, however dental amalgam fillings are strong and long-lasting so they are less likely to break than some other types of fillings.

Dental amalgam is also the least expensive type of filling material.

Palladium alloys

Since the late 1970s, palladium has been a key metal used worldwide by the dental industry in the development of alloys for the manufacture of crown and bridge restorations.

Gold dental alloys typically also contain palladium because it improves resistance to tarnishing and corrosion without affecting colour (Figure 4).

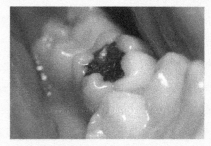
▲ Figure 3 *A dental amalgam filling*

▲ Figure 4 *Teeth with a gold crown*

Mouth tingling?

Dissimilar metals in your mouth, with saliva serving as the electrolyte, can cause a shock to your teeth. It can be caused by a piece of aluminium foil or a spoon that touches a silver or mercury filling in a mouth – it is a galvanic event. There are several different types of galvanism.

- A silver/mercury filling is placed in opposition or adjacent to a tooth restored with gold. These dissimilar metals in conjunction with saliva and body fluids constitute an electric cell. When brought into contact, the circuit is shorted, the flow of electrical current passes through the pulp, and the patient experiences pain.

- Dissimilar metals coming into contact when the upper and lower teeth come together and touch each other.

- Two adjacent teeth are restored with dissimilar metals. The current flows from metal to metal through the dentine, bone, and tissue fluids of both teeth, resulting in discomfort and tooth sensitivity.

Of course, the electrolyte in cell design does not have to be saliva, as lemon juice is fine (Figure 5).

▲ Figure 5 *A lemon battery used to power a clock*

Modern cells

There are lots of different types of electrochemical cells but they all work on the principle that at one side of the cell (the anode/negative terminal) electrons must be released and at the other (the cathode/positive terminal) electrons must be gained. An electrolyte must also be present to allow for the movement of ions.

In the original Daniell cell, zinc metal lost electrons (and formed zinc ions in the process) and copper ions gained them to form copper metal. This type of cell is not rechargeable and is called a primary cell. One type of modern cell found in the lithium ion cell also uses d-block compounds.

Chemical ideas: Redox 9.3

Redox reactions, cells, and electrode potentials

Redox reactions

Redox reactions involve electron transfer. Redox reactions can be split into two half-reactions – one producing electrons and one accepting them.

For example, when zinc is added to copper(II) sulfate solution, a redox reaction takes place. The blue colour of the solution becomes paler, and copper metal deposits on the zinc. The temperature rises – it is an exothermic reaction. The overall equation is:

$$Zn(s) + CuSO_4(aq) \rightarrow ZnSO_4(aq) + Cu(s)$$

This is an example of a **displacement reaction** as well as a redox reaction.

> **Synoptic link**
>
> You first met displacement reactions in Topic ES 1, The lowest point on Earth.

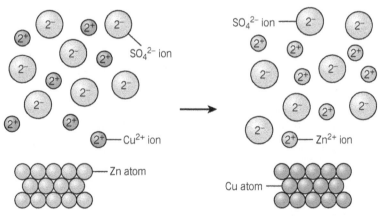

▲ Figure 6 *The reaction of zinc with copper(II) sulfate solution*

The sulfate ions play no part in the reaction and are **spectator ions**. By removing spectator ions from the overall equation, you obtain the ionic equation:

$$Zn(s) + Cu^{2+}(aq) \rightarrow Cu(s) + Zn^{2+}(aq)$$

| grey solid | blue solution | orange solid | colourless solution |

In the reaction, zinc atoms transfer electrons to copper(II) ions.

The ionic equation can be written as two half-equations:

$$Zn(s) \rightarrow Zn^{2+}(aq) + 2e^- \qquad \text{oxidation}$$
electrons produced in this half-reaction

$$Cu^{2+}(aq) + 2e^- \rightarrow Cu(s) \qquad \text{reduction}$$
electrons accepted in this half-reaction

Zinc provides the electrons which reduce Cu^{2+} to Cu, so zinc is the **reducing agent**, whilst Cu^{2+} is the **oxidising agent**.

If copper is added to zinc sulfate solution no change is observed – a reaction does not occur in the reverse direction. Zinc reacts with copper ions but copper ions do not react with zinc.

However, if copper is added to silver nitrate(V) solution the copper does react. A grey precipitate forms and the solution turns from colourless to blue. The overall reaction is:

$$Cu(s) + 2Ag^+(aq) \rightarrow Cu^{2+}(aq) + 2Ag(s)$$

| orange solid | colourless solution | blue solution | grey solid |

The half-reactions are:

$$Cu(s) \rightarrow Cu^{2+}(aq) + 2e^- \qquad \text{oxidation}$$

$$2Ag^+(aq) + 2e^- \rightarrow 2Ag(s) \qquad \text{reduction}$$

No reaction is observed when silver is added to copper(II) sulfate solution.

Individual half-reactions are reversible – they can go either way:

$$Cu^{2+}(aq) + 2e^- \rightleftharpoons Cu(s)$$

The actual direction they take depends on what they are reacting with. For example, zinc atoms can supply electrons to copper ions, so that the copper half-reaction is:

$$Cu^{2+}(aq) + 2e^- \rightarrow Cu(s)$$

But copper atoms supply electrons to silver ions, so in this case the copper half-reaction is:

$$Cu(s) \rightarrow Cu^{2+}(aq) + 2e^-$$

Synoptic link

Oxidation states were first introduced in Topic ES 2, Bromine from the sea.

Study tip

Remember that in a redox reaction the oxidising agent is reduced by the end of the reaction and the reducing agent is oxidised by the end of the reaction.

Activity DM 4.1

This activity helps you think about redox reactions and balance half equations.

Synoptic link

You first met the principle of equilibrium in Topic ES 4, From extracting bromine to making bleach.

Combining half-equations

Once you know the direction in which each half-reaction will go, you can add the half-equations together to get an equation for the overall reaction. For example, if you add zinc to silver ions, the zinc atoms supply electrons to the silver ions. The half-equations are:

$$Zn(s) \rightarrow Zn^{2+}(aq) + 2e^-$$

$$Ag^+(aq) + e^- \rightarrow Ag(s)$$

To combine the two half-equations together, the number of electrons has to be the same in each half-equation because every electron released by a zinc atom must be accepted by a silver ion.

This means the silver half-equation should be multiplied by two so there are $2e^-$ in each half-equation:

$$Zn(s) \rightarrow Zn^{2+}(aq) + 2e^-$$

$$2Ag^+(aq) + 2e^- \rightarrow 2Ag(s)$$

Now you can add the two half-equations together to give the overall equation:

$$Zn(s) + 2Ag^+(aq) \rightarrow 2Ag(s) + Zn^{2+}(aq)$$

The electrons cancel out because they are on both sides of the equation.

Balancing redox equations

Redox equations can be balanced using oxidation states, for example:

$$Ca + Al^{3+} \rightarrow Ca^{2+} + Al$$

oxidation states of Ca	0	+2	change +2
oxidation states of Al	+3	0	change −3

The changes in oxidation state must be equal. The balanced equation for the reaction of calcium with aluminium(III) is:

$$3Ca + 2Al^{3+} \rightarrow 3Ca^{2+} + 2Al$$

oxidation states of Ca	0	+6	change +6
oxidation states of Al	+6	0	change −6

In this example it is easy to relate changes in oxidation states to electrons transferred but this is not so obvious in more complex examples. For example:

$$Sn + HNO_3 \rightarrow SnO_2 + NO_2 + H_2O$$

Tin and nitrogen are the only elements that change oxidation state in this reactions, so:

$$Sn + HNO_3 \rightarrow SnO_2 + NO_2 + H_2O$$

oxidation states of Sn	0	+4	change +4
oxidation states of N	+5	+4	change −1

To balance the changes in oxidation state, four NHO_3 and one Sn is needed:

$$Sn + 4HNO_3 \rightarrow SnO_2 + 4NO_2 + H_2O$$

Synoptic link

Balancing equations with oxidation states was covered in Topic ES 2, Bromine from sea water.

To balance the hydrogen and oxygen, two H_2O are required.

$$Sn + 4HNO_3 \rightarrow SnO_2 + 4NO_2 + 2H_2O$$

For ionic equations, the charges must be balanced as well, for example:

$$MnO_4^- + Fe^{2+} + H^+ \rightarrow Mn^{2+} + Fe^{3+} + H_2O$$

Again, the oxidation states of hydrogen and oxygen do not change, so:

$$MnO_4^- + Fe^{2+} + H^+ \rightarrow Mn^{2+} + Fe^{3+} + H_2O$$

oxidation states of Mn +7		+2	change −5
oxidation states of Fe	+2	+3	change +1

To balance the oxidation states, five Fe^{2+} need to react with one MnO_4^-. As such, the total charge on the left hand side (excluding H^+ is now +9 and the total charge on the right hand side is 17+. Eight H^+ ions on the left hand side will balance the charges.

To balance the hydrogens and oxygens, four H_2O are needed on the right hand side. The equation is now fully balanced.

$$MnO_4^- + 5Fe^{2+} + 8H^+ \rightarrow Mn^{2+} + 5Fe^{3+} + 4H_2O$$

It is possible to balance complex half-equations by this method too, for example:

$$MnO_4^- + H^+ \rightarrow Mn^{2+} + H_2O$$

oxidation state of Mn +7   +2   change −5

As this is a reduction reaction, electrons are gained, which go on the right hand side of the half-equation. The number of electrons equals the change of oxidation state:

$$MnO_4^- + H^+ + 5e^- \rightarrow Mn^{2+} + H_2O$$

Then balance charges and hydrogens:

$$MnO_4^- + 8H^+ + 5e^- \rightarrow Mn^{2+} + 4H_2O$$

Electrochemical cells

Something must control the direction of electron transfer in a redox reaction. To find out more about redox reactions, and what makes them go in a particular direction, you need to study the half-reactions.

You can arrange for the two half-reactions to occur separately with electrons flowing through an external wire from one half-reaction to the other.

A system like this is used in all batteries and dry cells. In one part of the cell an oxidation reaction occurs. Electrons are produced and transferred through an external circuit to the other part of the cell, where a reduction reaction takes place accepting the electrons. The two parts are called half-cells which, when combined, make an electrochemical cell. Figure 7 shows the general arrangement, and Figure 8 shows a familiar example.

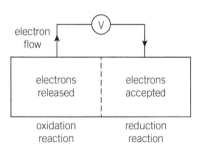

▲ Figure 7 *The general arrangement for an electrochemical cell*

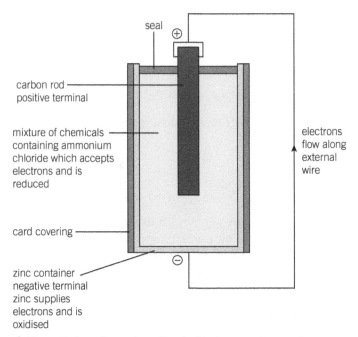

▲ **Figure 8** *An ordinary dry cell – the kind you use in a torch*

Electrical units

- Electric charge is measured in coulombs (C).
- Electric current is a flow of charge and is measured in amperes (A).
- One ampere is a flow of charge of one coulomb per second.

The **potential difference** between the terminals of the cell is measured in volts (V). The voltage of the cell tells you the number of joules of energy transferred whenever one coulomb of charge flows round the circuit:

$$1\,V = 1\,J\,C^{-1}$$

The energy given out, instead of heating the surroundings, becomes available as electrical energy which is used to do work.

Cells are labelled with positive and negative terminals and a voltage. The voltage measures the potential difference between the two terminals. As current flows in a circuit, the voltage can drop – the higher the current drawn, the lower the voltage the cell may give.

If you want to compare cells and half-cells by measuring voltages, you can measure the potential difference between the terminals of the cell when no current flows. This potential difference is given the symbol E_{cell}.

To measure E_{cell} a high-resistance voltmeter is used so that almost zero current flows. The maximum potential difference between the electrodes of the two half-cells is recorded.

Metal ion to metal half-cells

You can set up a simple half-cell by using a strip of metal dipping into a solution of metal ions. For example, the copper–zinc cell consists of the two half-cells (Figure 9).

Potential difference

The potential difference is a measure of how much each electrode is tending to release or accept electrons.

Activity DM 4.2

This activity allows you to set up some more complex electrochemical cells.

Activity DM 4.3

This practical allows you to measure the potential difference of electrochemical cells.

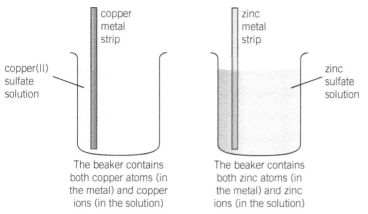

The beaker contains both copper atoms (in the metal) and copper ions (in the solution)

The beaker contains both zinc atoms (in the metal) and zinc ions (in the solution)

▲ Figure 9 *The copper and zinc half-cells*

Each of these half-cells has its own electrode potential. For example, take the zinc half-cell in Figure 10. The zinc atoms in the zinc strip form Zn^{2+} ions by releasing electrons:

$$Zn(s) \rightarrow Zn^{2+}(aq) + 2e^-$$

The electrons released make the zinc strip negatively charged relative to the solution, so there is a potential difference between the zinc strip and the solution. The Zn^{2+} ions in the solution accept electrons, reforming zinc atoms:

$$Zn^{2+}(aq) + 2e^- \rightarrow Zn(s)$$

When Zn^{2+} ions are turning back to zinc atoms as fast as they are being formed, an equilibrium is set up:

$$Zn^{2+}(aq) + 2e^- \rightleftharpoons Zn(s)$$

For a general metal, M:

$$M^{2+}(aq) + 2e^- \rightleftharpoons M(s)$$

The position of this equilibrium determines the size of the potential difference (the electrode potential) between the metal strip and the solution of metal ions. The further to the right the equilibrium lies, the greater the tendency of the electrode to accept electrons, and the more positive the electrode potential.

When two half-cells are put together the one with the more positive potential will become the positive terminal of the cell and the other one will become the negative terminal.

Making a cell from two half-cells

A connection is needed between the two solutions but the solutions should not mix together. A strip of filter paper soaked in saturated potassium nitrate(V) solution can be used as a junction, or **salt bridge**, between the half-cells. Sometimes this is called an ion bridge because the circuit is completed by the movement of ions, not

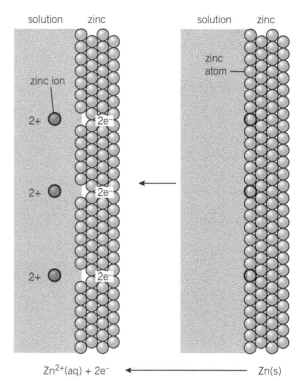

▲ Figure 10 *Zinc atoms form zinc ions, releasing electrons and setting up a potential difference between the metal strip and the solution of metal ions*

Synoptic link

You can find out how to set up an electrochemical cell in Techniques and procedures.

electrons. The potassium ions and nitrate(V) ions carry the current in the salt bridge so that there is electrical contact between the solutions but no mixing. The complete set-up is shown in Figure 11.

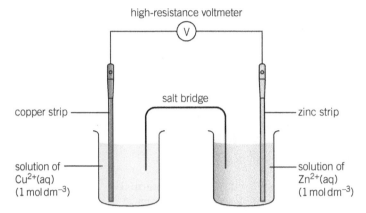

▲ Figure 11 *A copper–zinc cell*

The circuit is completed by a metal wire connecting the copper and zinc strips. A high-resistance voltmeter can be included in the circuit to measure the maximum voltage E_{cell} produced by the cell.

An electrochemical cell consists of two half-cells connected by a salt bridge. You measure the maximum potential difference between the two electrodes of the half-cells with a high-resistance voltmeter, so that negligible current flows. Table 1 shows data obtained from a number of different cells.

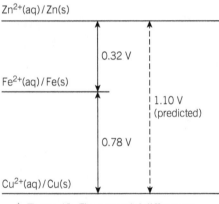

▲ Figure 12 *The potential differences between three half-cells*

▼ Table 1 *The potential differences E_{cell} generated by some cells*

Positive half-cell	Negative half-cell	E_{cell}/V
$Cu^{2+}(aq)/Cu(s)$	$Zn^{2+}(aq)/Zn(s)$	1.10
$Cu^{2+}(aq)/Cu(s)$	$Fe^{2+}(aq)/Fe(s)$	0.78
$Zn^{2+}(aq)/Zn(s)$	$Mg^{2+}(aq)/Mg(s)$	1.60
$Au^{3+}(aq)/Au(s)$	$Zn^{2+}(aq)/Zn(s)$	2.26
$Cu^{2+}(aq)/Cu(s)$	$Mg^{2+}(aq)/Mg(s)$	2.70
$Ag^{+}(aq)/Ag(s)$	$Zn^{2+}(aq)/Zn(s)$	1.56
$Fe^{2+}(aq)/Fe(s)$	$Zn^{2+}(aq)/Zn(s)$	0.32

It is difficult to see any patterns because differences between the electrode potentials of the various half-cells are being measured.

Figure 12 shows the potential differences between three half-cells – copper, iron, and zinc. The potential difference measured between the electrodes of the copper and iron half-cells is 0.78 V, with the copper positive. The potential difference measured between the iron and zinc half-cells is 0.32 V, with the iron positive.

Using Figure 12, you can predict that the potential difference between copper and zinc will be 1.10 V, which is the value obtained when measured.

Figure 13 shows the **standard hydrogen half-cell** or standard hydrogen electrode.

The standard conditions for the hydrogen half-cell are concentration of hydrogen [$H^+(aq)$] $1.00\,mol\,dm^{-3}$, pressure of hydrogen gas $10^5\,Pa$, and temperature 298 K.

Standard electrode potentials

The standard hydrogen half-cell is used as a reference half-cell and all other half-cells are measured against it. A list of electrode potentials has been generated relative to the standard hydrogen half-cell. The half-reaction in this half-cell is:

$$2H^+(aq) + 2e^- \rightleftharpoons H_2(g)$$

Standard conditions

Electrode potentials vary with temperature and so a standard temperature is defined. This is 298 K. Altering the concentrations of any ions appearing in the half-reactions also affects the voltages, so a standard concentration of $1.00\,mol\,dm^{-3}$ is chosen. Standard pressure is $10^5\,Pa$ (1 atmosphere).

The potential of the standard hydrogen half-cell is defined as 0.0 V, a value chosen for convenience.

The **standard electrode potential** of a half-cell $E^\ominus$ is defined as the potential difference between the half-cell and a standard hydrogen half-cell.

$E^\ominus$ values have a sign depending on whether the half-cell is at a higher or lower positive potential than the standard hydrogen half-cell. Measurements are made at 298 K, with the metal dipping into a $1.00\,mol\,dm^{-3}$ solution of a salt of the metal. Some values are shown in Table 2.

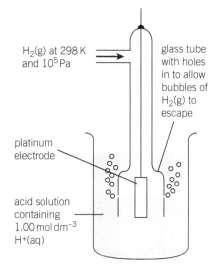

H₂(g) at 298 K and 10^5 Pa

glass tube with holes in to allow bubbles of $H_2(g)$ to escape

platinum electrode

acid solution containing $1.00\,mol\,dm^{-3}$ $H^+(aq)$

▲ **Figure 13** *The standard hydrogen half-cell*

▼ Table 2 *The standard electrode potentials of some half-cells*

Half-cell	Half-reaction	$E^\ominus/V$
$Mg^{2+}(aq)/Mg(s)$	$Mg^{2+}(aq) + 2e^- \rightleftharpoons Mg(s)$	-2.36
$Zn^{2+}(aq)/Zn(s)$	$Zn^{2+}(aq) + 2e^- \rightleftharpoons Zn(s)$	-0.76
$2H^+(aq)/H_2(g)$	$2H^+(aq) + 2e^- \rightleftharpoons H_2(g)$	0 (by definition)
$Cu^{2+}(aq)/Cu(s)$	$Cu^{2+}(aq) + 2e^- \rightleftharpoons Cu(s)$	$+0.34$
$Ag^+(aq)/Ag(s)$	$Ag^+(aq) + e^- \rightleftharpoons Ag(s)$	$+0.80$

This is the **electrochemical series.**

The half-cell at the bottom of the series has the greatest tendency to accept electrons and the most positive $E^{\ominus}$ value. The half-cell at the top has the least tendency to accept electrons and the most negative $E^{\ominus}$ value. In fact, the half-cell at the top has the greatest tendency to go in the reverse direction and release electrons. For this reason, the most reactive metals are at the top of the electrochemical series.

Other half-cell reactions

Metal ion/metal reactions are only one type of redox reaction. There are many others, for example, between ions:

$$Fe^{3+}(aq) + e^- \rightarrow Fe^{2+}(aq)$$

$$Cr_2O_7^{2-}(aq) + 14H^+(aq) + 6e^- \rightarrow 2Cr^{3+}(aq) + 7H_2O(l)$$

$$MnO_4^-(aq) + 8H^+(aq) + 5e^- \rightarrow Mn^{2+}(aq) + 4H_2O(l)$$

and between molecules and ions:

$$Cl_2(aq) + 2e^- \rightarrow 2Cl^-(aq)$$

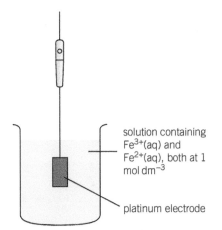

solution containing $Fe^{3+}(aq)$ and $Fe^{2+}(aq)$, both at 1 $mol\,dm^{-3}$

platinum electrode

▲ **Figure 14** *A standard half-cell for the $Fe^{3+}(aq)/Fe^{2+}(aq)$ half-reaction*

All these half-reactions can be set up as half-cells. However, there is no metal in the half-reaction to make electrical contact, so an electrode made of an unreactive metal such as platinum needs to be used. It dips into a solution containing all the ions and molecules involved in the half-reaction. Figure 14 shows the set-up for a $Fe^{3+}(aq)/Fe^{2+}(aq)$ half-cell.

Table 3 shows values of standard electrode potentials for a selection of half-cells of this type.

▼ **Table 3** *Some standard electrode potentials*

Half-reaction	$E^{\ominus}/V$
$I_2(aq) + 2e^- \rightleftharpoons 2I^-(aq)$	+0.54
$Br_2(aq) + 2e^- \rightleftharpoons 2Br^-(aq)$	+1.07
$Cl_2(g) + 2e^- \rightleftharpoons 2Cl^-(aq)$	+1.36
$MnO_4^-(aq) + 8H^+(aq) + 5e^- \rightleftharpoons Mn^{2+}(aq) + 4H_2O(l)$	+1.51

Working out E_{cell} from standard electrode potentials

An electrochemical cell consists of two half-cells. If you know the electrode potential of each half-cell then you can work out the potential difference E_{cell} of the cell as a whole.

For example, suppose you set up a cell using the two half-reactions:

$$Zn^{2+}(aq) + 2e^- \rightarrow Zn(s) \qquad E^{\ominus} = -0.76\,V$$

$$Cu^{2+}(aq) + 2e^- \rightarrow Cu(s) \qquad E^{\ominus} = +0.34\,V$$

E_{cell} is the voltage difference between the standard electrode potentials of the two half-cells. It is easiest to see this if you draw up an electrode potential chart (Figure 15).

From Figure 15:

$$E_{cell} = (+0.34\,V) - (-0.76\,V) = +1.10\,V$$

E_{cell} is an experimentally measured potential difference and is *always* positive, so always subtract the less positive potential from the more positive one.

To find the overall reaction occurring in the cell as a whole, you can add together the two half-equations.

$$Zn(s) \rightarrow Zn^{2+}(ac) + 2e^- \qquad \text{oxidation}$$

$$\underline{Cu^{2+}(aq) + 2e^- \rightarrow Cu(s) \qquad \text{reduction}}$$

$$Zn(s) + Cu^{2+}(aq) \rightarrow Cu(s) + Zn^{2+}(aq) \qquad \text{overall}$$

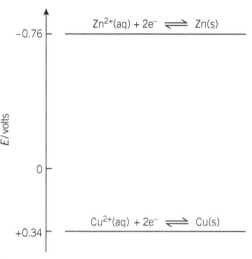

▲ **Figure 15** *Electrode potential chart for the Daniell cell*

A table of standard electrode potentials allows you to calculate the maximum voltage obtainable from any cell under standard conditions. However electrode potentials are also useful in other ways, for example, they can be used to predict the reactions which can take place when two half-cells are connected so that electrons can flow. In fact, they can be used to make predictions about any redox reaction, whether or not it occurs in a cell.

Predicting the direction of redox reactions

Electrode potentials measure the tendency of a half-reaction to accept electrons. The more positive the electrode potential, the greater is this tendency. Figure 16 illustrates this for three half-reactions. If a half-reaction has a large tendency to *accept* electrons, it will have a small tendency to *supply* them, and vice versa.

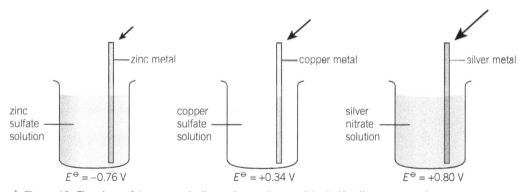

▲ **Figure 16** *The sizes of the arrows indicate the tendency of the half-cells to accept electrons*

Copper ions will accept electrons supplied by zinc. Silver ions will accept electrons supplied by copper.

For zinc to supply electrons the reaction must be:

$$Zn(s) \rightarrow Zn^{2+}(aq) + 2e^-$$

For the copper ions to accept electrons, the reaction must be:

$$Cu^{2+}(aq) + 2e^- \rightarrow Cu(s)$$

The prediction for the overall reaction agrees with the observed changes:

$$Zn(s) + Cu^{2+}(aq) \rightarrow Zn^{2+}(aq) + Cu(s)$$

When the copper and silver half-cells are connected, the predicted changes are:

$$2Ag^+(aq) + 2e^- \rightarrow 2Ag(s)$$

$$Cu(s) \rightarrow Cu^{2+}(aq) + 2e^-$$

The overall reaction predicted is:

$$Cu(s) + 2Ag^+(aq) \rightarrow Cu^{2+}(aq) + 2Ag(s)$$

This again agrees with the observed changes.

Electrode potential charts

Electrode potential charts provide a useful way of displaying and using the data. You can use them to make predictions about the direction a particular redox reaction will take. Figure 17 shows an electrode potential chart for the three half-reactions that have just been discussed.

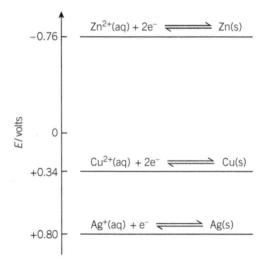

Half-reactions are all written with electrons on the left-hand side.

The most positive potential is placed at the bottom of the chart.

If two half-cells are connected, electrons will flow through the external circuit to the half-cell with the more positive electrode potential.

▲ Figure 17 *An electrode potential chart*

Table 4 gives standard electrode potentials of some half-reactions.

▼ Table 4 *Standard electrode potentials for a number of half-cells*

Half-cell	Half-reaction	$E^{\ominus}$/V
$I_2(aq)/2I^-(aq)$	$I_2(aq) + 2e^- \rightleftharpoons 2I^-(aq)$	+0.54
$Fe^{3+}(aq)/Fe^{2+}(aq)$	$Fe^{3+}(aq) + e^- \rightleftharpoons Fe^{2+}(aq)$	+0.77
$Br_2(aq)/2Br^-(aq)$	$Br_2(aq) + 2e^- \rightleftharpoons 2Br^-(aq)$	+1.07
$Cl_2(g)/2Cl^-(aq)$	$Cl_2(aq) + 2e^- \rightleftharpoons 2Cl^-(aq)$	+1.36
$MnO_4^-(aq)/Mn^{2+}(aq)$	$MnO_4^-(aq) + 8H^+(aq) + 5e^- \rightleftharpoons Mn^{2+}(aq) + 4H_2O(l)$	+1.51

Worked example: Predicting the reaction

What reaction could occur if you connected the $MnO_4^-(aq)/Mn^{2+}(aq)$ and $Fe^{3+}(aq)/Fe^{2+}(aq)$ half-cells so that electrons can flow? Write an equation for the overall reaction.

Step 1: Construct an electrode potential chart.

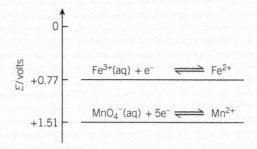

Step 2: Use the electrode potential chart to predict whether a reaction could occur.

Electrons flow to the positive terminal of a cell. This will be the $MnO_4^-(aq)/Mn^{2+}(aq)$ half-cell, which has an electrode potential of +1.51 V.

A reduction reaction will occur in the positive half-cell:

$$MnO_4^-(aq) + 8H^+(aq) + 5e^- \rightarrow Mn^{2+}(aq) + 4H_2O(l)$$

The other half-cell must supply electrons and is the negative terminal of the cell. Oxidation occurs.

$$Fe^{2+}(aq) \rightarrow Fe^{3+}(aq) + e^-$$

Step 3: Use the half-equations to give an overall equation.

The number of electrons supplied and accepted must be equal. Therefore the reaction in the $Fe^{3+}(aq)/Fe^{2+}(aq)$ half-cell must occur five times each time one $MnO_4^-(aq)$ ion is reduced:

$$5Fe^{2+}(aq) \rightarrow 5Fe^{3+}(aq) + 5e^-$$

The overall equation is:

$$MnO_4^-(aq) + 8H^+(aq) + 5Fe^{2+}(aq) \rightarrow Mn^{2+}(aq) + 4H_2O(l) + 5Fe^{3+}(aq)$$

What exactly can be predicted?

You can use the electrode potentials to make predictions about the feasibility of redox reactions, whether or not you have physically arranged the reagents into two half-cells with electrodes. This makes electrode potentials useful for making predictions about *any* redox reactions.

Synoptic link

You met the idea of feasibility of reactions when studying entropy in Chapter 8, Oceans.

Study tip

Remember electrons flow from the more negative half-cell to the more positive half-cell. Once you have decided which half-cell supplies electrons, and which half-cell receives them, it is easy to predict the direction of the half-reactions.

Synoptic link

Rates of reactions were introduced in Chapter 4, The ozone story.

Study tip

It is useful to picture an electrode potential chart going from high negative $E^\ominus$ values at the top to high positive $E^\ominus$ values at the bottom when answering question about the likelihood of particular redox reactions – reactions will follow an anticlockwise direction from top right to bottom right.

Activity DM 4.5

In this activity you can use electrode potentials to predict the feasibility of redox reactions.

You can use electrode potentials to say whether it is possible for a reaction to happen. But they tell you nothing about the rate of the reaction. So a reaction which is predicted as possible may not actually occur in practice because it is too slow.

You may predict that a change is feasible, mix the reagents, and find that nothing in fact happens. The rate of the reaction may be so slow that no change is observable. But reaction rates can sometimes be altered.

If a reaction is slow it means that the activation enthalpy for the reaction must be very high. If you want the reaction to happen faster, you could look for a catalyst to provide an alternative route with a lower activation enthalpy, and so increase the rate.

However, if you predict from the electrode potentials that the reaction is not possible, no catalyst is going to make it happen. Electrons will not flow spontaneously from a positive potential to a less positive one.

Changing the conditions

Standard electrode potentials refer to reactions occurring in aqueous solution under standard conditions – at 298 K and 1 atm pressure. Under different conditions, the electrode potentials will be different.

If it is predicted that a reaction is not feasible under standard conditions, changing the conditions in order to alter the values of the electrode potentials could make the reaction feasible.

Electrode potentials may vary with the concentration of the ions and molecules involved in the cell reaction. For example, if hydrogen or hydroxide ions are involved then pH changes will change electrode potentials and reactions may become possible.

Summary questions

1 a Calculate the values for the cells made from the following pairs of standard half-cells.
    i $Ag^+(aq)/Ag(s)$ and $Mg^{2+}(aq)/Mg(s)$ (1 mark)
    ii $Zn^{2+}(aq)/Zn(s)$ and $Cu^{2+}(aq)/Cu(s)$ (1 mark)
    iii $Zn^{2+}(aq)/Zn(s)$ and $Fe^{2+}(aq)/Fe(s)$ (1 mark)
   b For each cell in part a identify which electrode will be the positive terminal of the cell.   (5 marks)
   c Write equations for the two half-reactions and balanced equations for the overall reactions occurring in each cell in part a.   (5 marks)

2 The standard electrode potentials for some half-reactions are:

$$Sn^{2+}(aq) + 2e^- \rightleftharpoons Sn(s) \qquad -0.14\,V$$

$$Fe^{3+}(aq) + e^- \rightleftharpoons Fe^{2+}(aq) \qquad +0.77\,V$$

$$2Hg^{2+}(aq) + 2e^- \rightleftharpoons Hg_2^{2+}(aq) \qquad +0.92\,V$$

$$Cl_2 + 2e^- \rightleftharpoons 2Cl^-(aq) \qquad +1.36\,V$$

Construct an electrode potential chart and use it to predict which of the following reactions can occur:

a $2Fe^{2+}(aq) + 2Hg^{2+}(aq) \rightarrow 2Fe^{3+}(aq) + Hg_2^{2+}(aq)$ (1 mark)
b $Sn(s) + 2Hg^{2+}(aq) \rightarrow Sn^{2+}(aq) + Hg_2^{2+}(aq)$ (1 mark)
c $2Cl^-(aq) + Sn^{2+}(aq) \rightarrow Cl_2(g) + Sn(s)$ (1 mark)
d $2Fe^{2+}(aq) + Cl_2(aq) \rightarrow 2Fe^{3+}(aq) + 2Cl^-(aq)$ (1 mark)

3 Balance the following equations.
a $I_2 + S_2O_3^{2-} \rightarrow I^- + S_4O_6^{2-}$ (1 mark)
b $Br_2 + S_2O_3^{2-} \rightarrow Br^- + SO_4^{2-}$ (1 mark)
c $Cl_2 + OH^- \rightarrow ClO^- + Cl^- + H_2O$ (1 mark)
d $H^+ + VO_2^+ \rightarrow VO^{2+} + H_2O$ (1 mark)
e $Cr_2O_7^{2-} + H^+ + I^- \rightarrow I_2 + Cr^{3+} + H_2O$ (1 mark)
f $H^+ + H_2O_2 + Fe^{2+} \rightarrow H_2O + Fe^{3+}$ (1 mark)

4 Electrode potentials can be useful when considering redox reactions of organic compounds.

a Use the data given below to decide which reactions are feasible, under standard conditions, between oxygen and:

Half-reaction	$E^\ominus$ / V
$HCOOH(aq) + 2H^+(aq) + 2e^- \rightleftharpoons HCHO(aq) + H_2O(l)$	+0.06
$O_2(g) + 4H^+(aq) + 4e^- \rightleftharpoons 2H_2O(l)$	+1.23
$CO_2(g) + 8H^+(aq) + 8e^- \rightleftharpoons CH_4(g) + 2H_2O(l)$	+0.17
$HCHO(aq) + 2H^+(aq) + 2e^- \rightleftharpoons CH_3OH(aq)$	+0.23
$CH_3OH(aq) + 2H^+(aq) + 2e^- \rightleftharpoons CH_4(g) + H_2O(l)$	+0.59

i methane
ii methanol
iii methanal (HCHO) (3 marks)
b Give equations for those reactions which are feasible. (3 marks)

DM 5 Rusting

The addition of high levels of chromium to produce stainless steels is one way of stopping a perfectly natural and spontaneous process – the tendency of elemental iron to return to a compound state.

Many metals, including iron, occur in the Earth's crust as oxides. This is because the change from a metal to its oxide is an energetically favourable process – in other words, the oxide is more stable than the metal. Indeed, to reverse the process and extract the metal from its ore requires a great deal of energy. Just think about the high temperatures needed in the blast furnace. No wonder then that iron tends to reform its oxide – in other words, it *rusts*. **Rusting** is a common name for the **corrosion** of iron.

The rusting of steel is a familiar problem for many car owners – the iron simply returns to its oxide. Cars rust because the steel they are made from reacts with oxygen and water in the atmosphere. When iron or steel rusts, a hydrated form of iron(III) oxide with variable composition ($Fe_2O_3 \cdot xH_2O$) is produced. This oxide is permeable to air and water and does not form a protective layer on the metal surface – so the metal continues to corrode under the layers of rust.

▲ Figure 1 *A rusty car*

Chemical ideas: Redox 9.4

Rusting and its prevention

Rusting is an electrochemical process. Electrochemical cells are set up in the metal surface, where different areas act as sites of oxidation and reduction. The two half-reactions involved in rusting are:

$$Fe^{2+}(aq) + 2e^- \rightleftharpoons Fe(s) \qquad\qquad E^\ominus = -0.44\,V$$

$$\tfrac{1}{2}O_2(g) + H_2O(l) + 2e^- \rightleftharpoons 2OH^-(aq) \qquad E^\ominus = +0.40\,V$$

The reduction of oxygen to hydroxide ions occurs at the more positive potential, and so electrons flow from the half-cell in which the iron is oxidised to iron(II) ions.

Figure 2 shows what happens when a drop of water is left in contact with iron or steel.

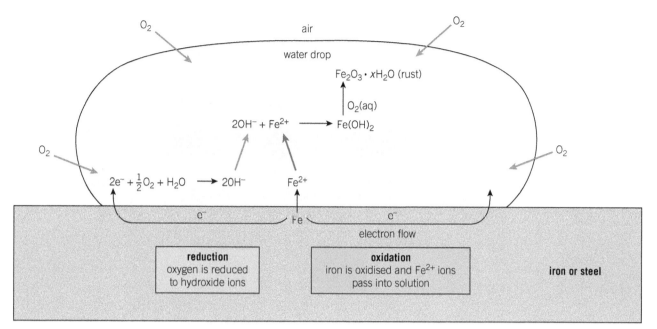

▲ Figure 2 *Rusting is an electrochemical process*

The concentration of dissolved oxygen in the water droplet determines which regions of the metal surface are sites of reduction and which are sites of oxidation. At the edges of the droplet, where the concentration of dissolved oxygen is higher, oxygen is reduced to hydroxide ions:

$$\frac{1}{2}O_2(g) + H_2O(l) + 2e^- \rightarrow 2OH^-(aq)$$

The electrons needed to reduce the oxygen come from the oxidation of iron at the centre of the water droplet, where the concentration of dissolved oxygen is low. The $Fe^{2+}(aq)$ pass into solution:

$$Fe(s) \rightarrow Fe^{2+}(aq) + 2e^-$$

The electrons that are released flow in the metal surface to the edges of the droplet.

This explains why corrosion is always greatest at the centre of a droplet of water or under a layer of paint – these are the regions where the oxygen supply is limited. Pits are formed here where the iron has dissolved away.

Rust forms in a series of secondary processes within the solution as Fe^{2+} and OH^- ions diffuse away from the metal surface. It does not form as a protective layer in contact with the iron surface:

$$Fe^{2+}(aq) + 2OH^-(aq) \rightarrow Fe(OH)_2(s)$$
$$Fe(OH)_2(s) \xrightarrow{O_2(aq)} Fe_2O_3 \cdot xH_2O(s)$$

Activity DM 5.1

This practical helps you explain rusting and its prevention.

You can see the reason for using zinc as a sacrificial metal by comparing the standard electrode potentials of the iron and zinc half-cells (Table 1). Any metal with a more negative value than iron could be used as a sacrificial metal.

▼ Table 1 *Standard electrode potentials for a number of half-cells*

Half-reaction	$E^{\ominus}/V$
$Mg^{2+}(aq) + 2e^- \rightleftharpoons Mg(s)$	−2.36
$Zn^{2+}(aq) + 2e^- \rightleftharpoons Zn(s)$	−0.76
$Cr^{3+}(aq) + 3e^- \rightleftharpoons Cr(s)$	−0.74
$Fe^{2+}(aq) + 2e^- \rightleftharpoons Fe(s)$	−0.44
$Sn^{2+}(aq) + 2e^- \rightleftharpoons Sn(s)$	−0.14

The zinc half-reaction takes place from right to left, forming $Zn^{2+}(aq)$ and $2e^-$. This shifts the position of the iron half-reaction equilibrium to the right, effectively preventing the formation of $Fe^{2+}(aq)$.

Tin cannot be used as a sacrificial metal to protect steel because the less negative standard electrode potential of the tin half-reaction would mean that the Fe/Fe^{2+} half-reaction would release electrons, that is, go from right to left.

Notice the standard electrode potential of the chromium half-reaction also explains why chromium atoms, in contact with oxygen at the surface of stainless steel, form the protective layer of chromium(III) oxide.

Protecting against rust

The simplest way of protecting steel against rust is to provide a barrier between the metal and the atmosphere. The barrier may be oil, grease, or a coat of paint.

Barriers made from organic polymers are increasingly used. The steel is coated with a plastic film – a colourful and flexible answer to the rusting problem. A quick look around at home will provide many examples – sink drainers, refrigerator and dishwasher shelves, and even car bodies.

Sometimes iron is covered with a thin layer of another metal. Many car manufacturers make their car bodies from **galvanised** steel – this has a protective coating of zinc. As long as the galvanised surface remains undamaged, the zinc layer is protected from corrosion by a firmly adherent layer of zinc oxide.

Even if the coating is scratched, the protection is still maintained, because the zinc corrodes in preference to the iron – the zinc is being used as a **sacrificial metal**. One of the earliest examples of using a sacrificial metal was suggested by Humphry Davy in 1824, to protect the metal sheathing on sailing ships from corrosion.

Great improvements in protecting cars from rusting have been made over the last few years. Indeed, one manufacturer offers a 12-year anti-corrosion warranty on all new car models. Even so, if steel is used in car bodies then the rusting problem is only postponed, not eliminated.

Impressed current (applied emf)

Destructive oxidation of a metal occurs at an anode site.

Any technique that makes a metal a cathode site will ensure that reduction occurs on that cathode surface and therefore ensures the metal is protected from corrosion.

Apart from attaching sacrificial anode metals, another way of protecting a metal is to make it a cathode by supplying electrons to it by application of an emf from an external electrical source. This technique is used for protecting bridges, piers, pipelines, underground fuel tanks, etc.

▲ Figure 3 *The 'Angel of the North' sculpture near the A1 road near Gateshead is made from weather-resistant, high-strength COR-TEN steel. The steel contains 0.25–0.40% copper and weathers to a rich brown colour. Unlike most steels, in this 'weathering steel' (more correctly called 'atmospheric corrosion-resistant steel') the oxide matrix on the surface forms a protective layer, preventing further rusting.*

Using an impressed current to protect metal structures is another form of cathodic protection.

The negative terminal of a battery or DC power supply is connected to the metal to be protected. Remember, electrons flow from anode to the cathode.

This makes the metal a protected cathode.

The anode is usually an inert electrode which is corrosion-resistant and rather than the anode being oxidised, water is oxidised:

$$\text{Anode half-equation: } 2H_2O(l) \rightarrow O_2(g) + 4H^+(aq) + 4e^-$$

Consider an underground steel pipeline passing through moist soil where sufficient oxygen is present to result in severe corrosion to the pipe. An impressed current (or applied emf) technique would prove an ideal method of protecting the pipe (Figure 4).

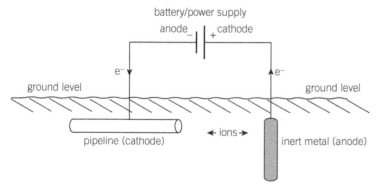

▲ Figure 4 *Protecting a pipeline using impressed current*

Summary questions

1   Figure 5 shows a piece of iron or steel which has a small depression in its surface that has partly filled with water. 'Pitting' of the metal has occurred.

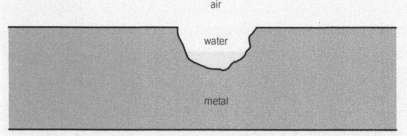

▲ Figure 5

   a   Where will the reduction of oxygen occur? Is this an anode or cathode site?   (2 marks)
   b   Where will the oxidation of iron occur? Is this an anode or cathode site?   (2 marks)
   c   Copy the diagram and label with all relevant features and show the direction of electron flow as well as the movement of ions.   (2 marks)

2   A coach screw, made of steel, was used to secure a wooden section of a jetty in a marine (seawater) environment. When the bolt was removed for inspection, the internal section revealed severe corrosion, although the head of the bolt appears in good order.
   a   Explain the reason for this severe deterioration of the bolt in parts that were not so directly exposed to the air.   (2 marks)
   b   Why does an occurrence like this present potential safety hazards for structures subject to possible corrosion?   (1 mark)
   c   Suggest why the head of the bolt was not so significantly affected by corrosion.   (2 marks)

3   Why does the presence of salt water, as opposed to fresh water, accelerate the rate of metallic corrosion?   (2 marks)

4   A marine biologist drops and breaks a mercury thermometer in the bottom of an aluminium dinghy. The nature of the spillage made it extremely difficult to retrieve all the droplets of mercury metal from the bottom of the boat. Quite soon after this accident, the boat developed small holes in its hull. Why did the boat corrode?   (2 marks)

DM 6 Complexes and ligands

Specification references: DM(b), DM(j)

Alfred Werner

Alfred Werner (1866–1919) was a Swiss chemist who was a professor at the University of Zurich. He won the Nobel Prize in Chemistry in 1913 for proposing the octahedral configuration of transition complexes. Werner developed the basis for modern coordination chemistry.

In 1893, Werner was the first to propose correct structures for coordination compounds containing **complex ions**, in which a central transition metal atom or ion is surrounded by neutral or anionic **ligands**.

The ligands bind to the central metal atom or ion by donating a pair of electrons forming dative or **coordinate** bonds.

One of the complex ions Werner described the structure of is shown in Figure 1.

In this complex the central metal ion is Co^{3+} and six ammonia molecules act as ligands, each using the lone pair on the nitrogen atom to form a coordinate bond to the cobalt ion. This arrangement produces the octahedral complex, $[Co(NH_3)_6]^{3+}$.

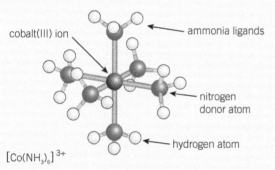

cobalt(III) ion
ammonia ligands
nitrogen donor atom
hydrogen atom

$[Co(NH_3)_6]^{3+}$

▲ **Figure 1** *Cobalt complex ion*

Complex ions in medicine

Many complex ions have an increasingly important role in the fight against disease and ill-health.

One of the earliest anti-cancer drugs, and still one of the most widely used, is the platinum complex *cis*-platin (Figure 2).

Cis-platin is a neutral complex overall, with a Pt^{2+} ion at the centre and two ammonia molecules and two chloride ions as ligands.

It has a trans isomer, *trans*-platin, which does not show the same anti-cancer activity.

cis-platin   *trans*-platin

▲ **Figure 2** *The two stereoisomers, cis-platin and trans-platin*

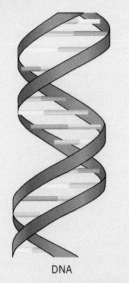

DNA

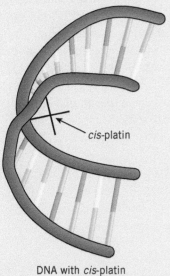

cis-platin

DNA with *cis*-platin

▲ **Figure 3** *structural change to DNA when cis-platin binds*

Cis-platin binds to cell DNA and causes critical structural changes in the DNA which cancer cells cannot repair and therefore the cancer cells die (Figure 3).

Healthy cells, however, are able to repair the structural changes brought about by the attachment of *cis*-platin.

Table 1 shows several transition complexes with medicinal roles.

▼ Table 1 *Some transition complexes with medicinal roles*

Complex	Central ion	Ligand(s)	Medicinal roles
	Pt^{2+}	dicarboxylate anion	anti-cancer drug
	Fe^{3+}	CN^- and NO	lowers blood pressure during surgery
	Ru^{3+}	Cl^- thioketone heterocyclic ring	anti-cancer drug
	Au^+	sulfide containing dicarboxylate anion	treatment of rheumatoid arthritis
	V^{4+}	ring system	treatment of diabetes

Finally, one biological complex that you really cannot do without is haemoglobin (Figure 4).

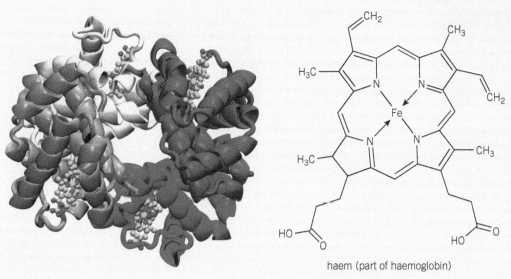

haem (part of haemoglobin)

▲ **Figure 4** *The haem complex (right) and the four haem groups in haemoglobin (left), where each haem can form one coordinate bond with an oxygen molecule acting as a ligand*

Chemical ideas: The periodic table 11.4d

Chemistry of complexes

A complex consists of a central metal atom or ion surrounded by a number of negatively charged ions or neutral molecules possessing a lone pair of electrons. These surrounding anions or molecules are called **ligands**.

A complex may have an overall positive charge, a negative charge, or no charge at all. For example:

$$[Fe(H_2O)_6]^{3+} \qquad [NiCl_4]^{2-} \qquad Ni(CO)_4$$

If a complex is charged, it is called a complex ion – the overall charge is the sum of the charge on the central metal ion and the charges on the ligands.

For $[NiCl_4]^{2-}$ the charge is $(2+) + 4(1-) = 2-$

In reality, the charges on a complex ion are delocalised over the whole ion.

Bonding in complexes is complicated – it usually involves electron pairs from the ligand being shared with the central ion. This means that ligands are electron donors. They form dative bonds (also called coordinate bonds).

The number of bonds from the central ion to ligands is known as the **coordination number** of the central ion. The most common coordination numbers are six and four, but two does occur, for example, in complexes of Ag(I) and Cu(I).

Synoptic link

Dative covalent bonds were introduced in Topic EL 5, The molecules of life.

Synoptic link

Shapes of molecules and ions were first discussed in Topic EI 5, The molecules of life.

Shapes of complexes

The shape of a complex depends on its coordination number.

Complexes with coordination number of six usually have an octahedral arrangement of ligands around the central metal ion (Figure 5) – this is the most common coordination number.

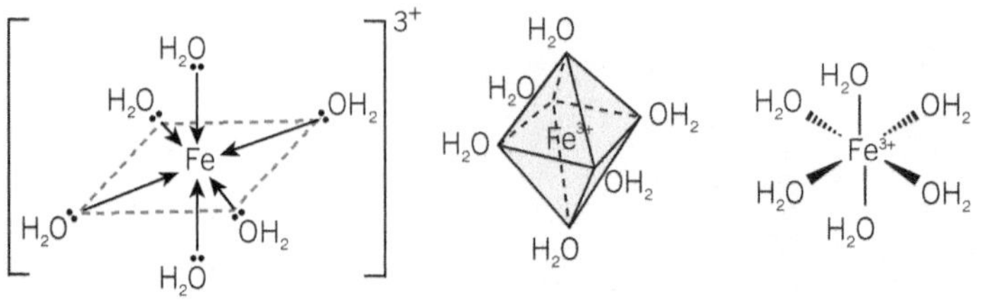

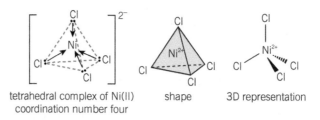

octahedral complex of Fe(III) coordination number six

shape

3D representation

▲ **Figure 5** *An octahedral complex of Fe(III) – coordination number six*

Complexes with coordination number four usually have a tetrahedral arrangement of ligands around the central metal ion (Figure 6).

tetrahedral complex of Ni(II) coordination number four

shape

3D representation

▲ **Figure 6** *A tetrahedral complex of Ni(II) – coordination number four*

Some four-coordinate complexes have a square planar structure (Figure 7).

square planar complex of Ni(II) coordination number four

shape

3D representation

▲ **Figure 7** *A square planar complex of Ni(II) – coordination number four*

▼ Table 2 *A summary of the shapes of complexes*

Coordination number	Shape of complex	Example
6	octahedral	$[Fe(CN)_6]^{3-}$
4	tetrahedral	$[NiCl_4]^{2-}$
4	square planar	$[Ni(CN)_4]^{2-}$
2	linear	$[Ag(NH_3)_2]^{+}$

Complexes with coordination number two usually have a linear arrangement of ligands (Figure 8).

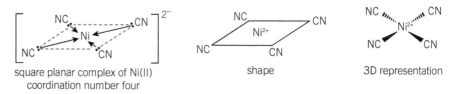

▲ **Figure 8** *A linear complex of Ag(I) – coordination number two*

Some complexes can have irregular, or distorted, shapes. For example, in the $[Cu(H_2O)_6]^{2+}$ ion, four of the water ligands are held more strongly than the remaining two, so that four of the copper–oxygen bonds are shorter. This produces a distorted octahedral arrangement.

Polydentate ligands

As first discussed in DM 3, some ligand molecules, such as ammonia, NH_3, and water, H_2O, can only bond to a metal ion through a single atom or ion and are called monodentate ligands.

Others can bond through more than one atom and are polydentate ligands. For example, the ethanedioate ion and the 1,2-diaminoethane molecule are bidentate ligands, because they can form *two* bonds with a metal ion by using pairs of electrons from two oxygen or nitrogen atoms. The metal ion is held in a five-membered ring (Figure 9). The ring is called a **chelate ring**.

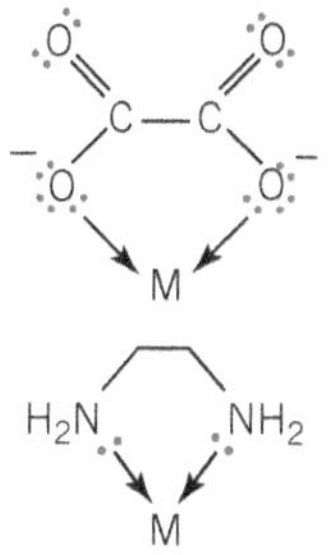

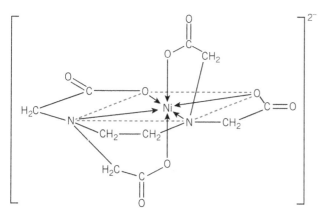

▲ Figure 9 *Bidentate ligands the ethanedioate ion (top) and the 1,2-diaminoethane molecule (bottom)*

$EDTA^{4-}$ forms six bonds to metal ions and so is a hexadentate ligand. $EDTA^{4-}$ forms complexes using the two nitrogen atoms and the four oxygen atoms of the COO^- groups. $EDTA^{4-}$ holds the metal atom inside (Figure 10).

Study tip

You only need to know the structure of the ethanedioate ion.

▲ Figure 10 *The nickel–EDTA complex ion $[Ni(EDTA)]^{2-}$*

Ligand substitution reactions

When copper sulfate is dissolved in water, the copper ions form a complex ion with a coordination number of six (Figure 11).

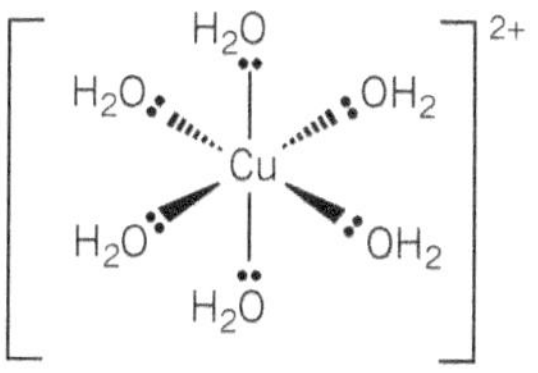

$[Cu(H_2O)_6]^{2+}$

▲ Figure 11 *The Cu^{2+} ion forms a complex ion with a coordination number of 6*

However, reactions can occur where one ligand displaces another. For example, if you add concentrated hydrochloric acid to $[Cu(H_2O)_6]^{2+}$, then chloride ions will displace the water ligands and the $[CuCl_4]^{2-}$ complex forms. This is an example of a **ligand substitution reaction**:

$$[Cu(H_2O)_6]^{2+}(aq) + 4Cl^- \rightarrow [CuCl_4]^{2-} + 6H_2O$$

blue	yellow
octahedral	tetrahedral

If concentrated aqueous ammonia is added to the same complex ion, ammonia molecules will displace the water ligands:

$$[Cu(H_2O)_6]^{2+}(aq) + 4NH_3 \rightarrow [Cu(NH_3)_4(H_2O)_2]^{2+} + 4H_2O$$

blue	deep blue/violet
octahedral	octahedral

Ligand substitution occurs if the new complex formed is more stable than the previous complex. Therefore the stability of a complex depends on its ligands. For example, the complex ion of copper(II) with ammonia ligands is more stable than with water ligands.

Summary questions

1 Give the coordination number for each of the following complex ions:
   a $[Ag(H_2O)_2]^+$ (1 mark)   b $[CoCl_4]^{2-}$ (1 mark)
   c $[Co(CN)_6]^{3-}$ (1 mark)   d $[Cr(H_2O)_5OH]^{2+}$ (1 mark)

2 What is the oxidation state of the central metal ion of each of the complexes a–d in question 1? (4 marks)

3 Titanium(IV) chloride dissolves in concentrated hydrochloric acid to give the ion $[TiCl_6]^{2-}$.
   a What is the coordination number of the titanium ion? (1 mark)
   b What is the oxidation number of the titanium ion? (1 mark)
   c Suggest a name for $[TiCl_6]^{2-}$. (1 mark)
   d Draw a likely structure for the ion. (2 marks)

Practice questions

1 Which row is correct:

	Fe^{3+}(aq)	Fe^{2+}(aq)	Cu^{2+}(aq)
A	brown	green	blue
B	green	brown	blue
C	brown	green	green
D	green	brown	green

(1 mark)

2 Which of the following is correct about transition metals as catalysts?

A Homogeneous catalysis works by absorbing ions onto the surface.

B Heterogeneous catalysis works because ions can change oxidation states.

C Heterogeneous catalysts form weak bonds to reactants on their surface.

D Homogeneous catalysts have a large surface area. *(1 mark)*

3 Many transition metal ions are coloured because:

A they absorb UV radiation

B electrons are promoted within the d-sub-shell

C electrons fall within the d-sub-shell

D ligands split the p-sub-shell. *(1 mark)*

4 For a colorimeter, the transmittance:

A increases with the concentration of a coloured substance

B is used in a calibration curve to measure the concentration of a coloured substance.

C is inversely proportional to the concentration of a coloured substance

D is set to zero using a tube of water. *(1 mark)*

5 Which of the following rows is correct about the possible shapes of complexes?

	Four-coordinate complex	Six-coordinate complex
A	square planar	hexagonal
B	tetrahedral	hexagonal
C	tetrahedral	octahedral
D	quadrilateral	octahedral

(1 mark)

6 In which of these groups does vanadium have the same oxidation state in both the compounds?

A VCl_2 and $V(SO_4)_2$

B V_2O_5 and VO^{2+}

C VO_3^- and VO^{2+}

D VCl_3 and VO_4^{3-} *(1 mark)*

7 Which row correctly describes the half-cell to measure the $E^{\ominus}$ for the half-equation:
$MnO_4^- + 5e^- + 8H^+ \rightleftharpoons Mn^{2+} + 4H_2O$?

	Ions in half-cell	Electrode
A	1.0 mol dm^{-3} MnO$_4^-$ 1.0 mol dm^{-3} Mn^{2+}	platinum
B	1.0 mol dm^{-3} MnO$_4^-$ 1.0 mol dm^{-3} H$_2$SO$_4$	manganese
C	1.0 mol dm^{-3} MnO$_4^-$	manganese
D	1.0 mol dm^{-3} MnO$_4^-$ 1.0 mol dm^{-3} Mn^{2+} 1.0 mol dm^{-3} H$^+$	platinum

(1 mark)

8 A reaction that is predicted by standard electrode potentials may not occur because:

1 the activation enthalpy is large

2 the conditions are not standard

3 the reaction is reversible

A 1, 2, and 3 correct

B 1 and 2 are correct

C 2 and 3 are correct

D only 1 is correct *(1 mark)*

9 A complex of iron is:

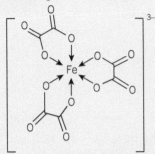

Which of the following is/are correct?

1 the coordination number of the complex is three

2 the charge on the ligand is 2−

3 the charge on the iron ion is 3+

A 1, 2, and 3 correct

B 1 and 2 are correct

C 2 and 3 are correct

D only 1 is correct *(1 mark)*

10 Which of the following could be a bidentate ligand?

1 −OOCCOO−

2 $H_2NCH_2CH_2NH_2$

3 $H_2NCH_2CH_2COOH$

A 1, 2, and 3 correct

B 1 and 2 are correct

C 2 and 3 are correct

D only 1 is correct *(1 mark)*

11 Gold occurs native (as the unreacted element) in very small amounts in gold ores. The gold can be extracted as $Na[Au(CN)_2]$.

$$4Au + 8NaCN + O_2 + 2H_2O \rightarrow 4Na[Au(CN)_2] + 4NaOH$$

Equation 1

a Explain how you can tell from a formula in the equation that the charge on the cyanide ion is CN^-. *(1 mark)*

b The cyanide ion has a triple bond. Draw a dot-and-cross diagram to show the distribution of electrons in this ion and hence which atom has the negative charge. *(2 marks)*

c Suggest the name of the shape of the $[Au(CN)_2]^-$ ion. *(1 mark)*

d What terms are given to the following in this context?

(i) CN^-

(ii) $[Au(CN)_2]^-$ *(2 marks)*

e Write the oxidation numbers under all the atoms that change oxidation state in the reaction in **Equation 1**. *(2 marks)*

f $[Au(CN)_2]^-$ is reacted with zinc to release gold metal. Some standard redox potential data are given in the table.

	$E^⦵/V$
$Zn^{2+}(aq) + 2e^- \rightleftharpoons Zn(s)$	−0.76
$[Au(CN)_2]^-(aq) + e^- \rightleftharpoons Au(s) + 2CN^-(aq)$	−0.60

Use these data to explain why zinc reacts with $[Au(CN)_2]^-$, naming the reducing agent and writing an equation for the reaction. *(4 marks)*

g Gold also forms the complexes $[AuBr_4]^-$ and $[AuBr_2]^-$. $[AuBr_2]^-$ is less stable than $[Au(CN)_2^-]$. Based on this information:

(i) Give another oxidation state in which gold exists apart from that in **d**.

(ii) Write the equation for a ligand exchange reaction. *(2 marks)*

12 Some students have a sample of solid ethanedioic acid, $H_2C_2O_4 \bullet xH_2O$, and wish to determine the value of x.

They discover that the acid can be titrated with acidified potassium manganate(VII), $KMnO_4$. The reaction goes slowly at first and needs warming.

The students make up a solution of 1.26 g of $H_2C_2O_4 \bullet xH_2O$ in $1.00\,dm^3$ of solution. They found that $25.0\,cm^3$ of this solution reacts exactly under acid conditions with $8.00\,cm^3$ (average titre) of a $0.0125\,mol\,dm^{-3}$ potassium manganate(VII) solution provided for them.

a Describe how the students carry out the titration to get consistent results once they have made the solution of ethanedioic acid. *(6 marks)*

b Under acid conditions, manganate(VII) ions oxidise $H_2C_2O_4$ to CO_2 and the manganese is reduced to Mn^{2+}.

Write the equation for the reaction of manganate(VII) ions with $H_2C_2O_4$ under acid conditions. *(2 marks)*

c Calculate the value of x. *(4 marks)*

d The students' teacher says that their titration result could be made more precise. How could the students do this without changing the apparatus they used? *(1 mark)*

13 Much food is still sold in tins which are made of tin-plated steel. The layer of tin protects the steel from the oxygen and water that causes rusting. However, tin causes faster rusting if the layer is broken.

		$E^{\ominus}$/V
A	$Fe^{2+} + 2e^- \rightleftharpoons Fe(s)$	-0.44
B	$Sn^{2+} + 2e^- \rightleftharpoons Sn(s)$	-0.14
C	$Fe^{3+} + e^- \rightleftharpoons Fe^{2+}$	$+0.77$

a Write the two half-equations by which iron begins to rust and give the overall equation. *(2 marks)*

b Write the equation that shows that tin speeds up rusting. *(1 mark)*

c Rust contains Fe^{3+} ions. Give a test for these ions in solution and write an ionic equation (with state symbols) for the reaction. *(3 marks)*

d (i) Write the electron configurations for Fe^{3+} and Fe^{2+} ions.

(ii) Explain how these show that Fe^{3+} is more stable than Fe^{2+}. *(2 marks)*

e (i) A cell is set up between standard electrode systems B and C in the table. Draw a labelled diagram of this cell.

(ii) Give the value of $E^{\ominus}_{cell}$. *(5 marks)*

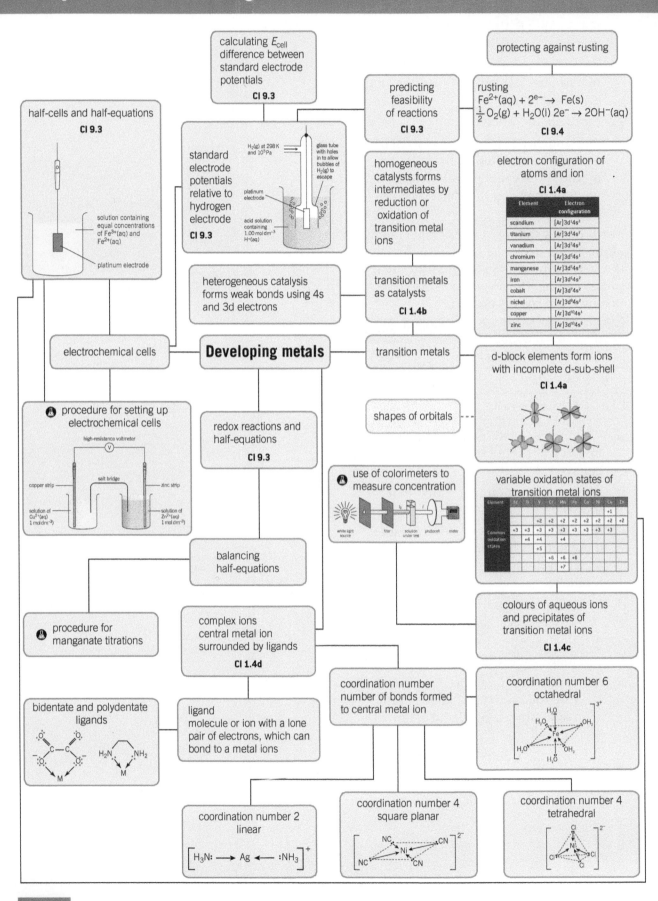

calculating E_{cell} difference between standard electrode potentials
Cl 9.3

protecting against rusting

half-cells and half-equations
Cl 9.3

solution containing equal concentrations of $Fe^{3+}(aq)$ and $Fe^{2+}(aq)$

platinum electrode

predicting feasibility of reactions
Cl 9.3

rusting
$Fe^{2+}(aq) + 2e^- \rightarrow Fe(s)$
$\frac{1}{2}O_2(g) + H_2O(l)\ 2e^- \rightarrow 2OH^-(aq)$
Cl 9.4

standard electrode potentials relative to hydrogen electrode
Cl 9.3

$H_2(g)$ at 298 K and 10^5 Pa
glass tube with holes in to allow bubbles of $H_2(g)$ to escape
platinum electrode
acid solution containing 1.00 mol dm^{-3} $H^+(aq)$

homogeneous catalysts forms intermediates by reduction or oxidation of transition metal ions

electron configuration of atoms and ion
Cl 1.4a

Element	Electron configuration
scandium	[Ar]$3d^14s^2$
titanium	[Ar]$3d^24s^2$
vanadium	[Ar]$3d^34s^2$
chromium	[Ar]$3d^54s^1$
manganese	[Ar]$3d^54s^2$
iron	[Ar]$3d^64s^2$
cobalt	[Ar]$3d^74s^2$
nickel	[Ar]$3d^84s^2$
copper	[Ar]$3d^{10}4s^1$
zinc	[Ar]$3d^{10}4s^2$

heterogeneous catalysis forms weak bonds using 4s and 3d electrons

transition metals as catalysts
Cl 1.4b

electrochemical cells

Developing metals

transition metals

d-block elements form ions with incomplete d-sub-shell
Cl 1.4a

procedure for setting up electrochemical cells

high-resistance voltmeter
copper strip
salt bridge
zinc strip
solution of $Cu^{2+}(aq)$ 1 mol dm^{-3}
solution of $Zn^{2+}(aq)$ 1 mol dm^{-3}

redox reactions and half-equations
Cl 9.3

shapes of orbitals

use of colorimeters to measure concentration

white light source
filter
solution under test
photocell
meter

variable oxidation states of transition metal ions

Element	Sc	Ti	V	Cr	Mn	Fe	Co	Ni	Cu	Zn
									+1	
			+2	+2	+2	+2	+2	+2	+2	+2
Common oxidation states	+3	+3	+3	+3	+3	+3	+3	+3		
		+4	+4	+4						
			+5							
				+6	+6	+6				
					+7					

balancing half-equations

colours of aqueous ions and precipitates of transition metal ions
Cl 1.4c

procedure for manganate titrations

complex ions central metal ion surrounded by ligands
Cl 1.4d

coordination number number of bonds formed to central metal ion

coordination number 6 octahedral

$\begin{bmatrix} & H_2O \\ H_2O & OH_2 \\ & Fe \\ H_2O & OH_2 \\ & H_2O \end{bmatrix}^{3+}$

bidentate and polydentate ligands

$O-C-C-O$ ... M
$H_2N\quad NH_2$ ... M

ligand molecule or ion with a lone pair of electrons, which can bond to a metal ions

coordination number 2 linear

$[H_3N: \longrightarrow Ag \longleftarrow :NH_3]^+$

coordination number 4 square planar

$\begin{bmatrix} NC & CN \\ & Ni \\ NC & CN \end{bmatrix}^{2-}$

coordination number 4 tetrahedral

$\begin{bmatrix} & Cl \\ & Ni \\ Cl & Cl \\ & Cl \end{bmatrix}^{2-}$

Coinage metals

The composition of coins has changed over time as the cost and availability of metals has fluctuated. The earliest coins were made of precious metals such as gold, and the value of the coin related exactly to the value of the metal it was made from.

However modern coins are made from cheaper metals. The alloys used in some British coins are shown in Table 1.

▼ Table 1 *Composition of British coins*

1p coin 2p coin	Pre-1992: 97% Cu, 2.5% Zn, 0.5% Sn	Post-1992: Mild steel (94%) coated with Cu (6%)
5p coin 10p coin	Pre-2011: 75% Cu, 25% Ni	Post-2011 Mild steel (94%) coated with Ni (6%)
20p coin	84% Cu, 16% Ni	
50p coin	75% Cu, 25% Ni	
£1 coin	70% Cu, 24.5% Zn, 5.5% Ni	

▲ Figure 1 *British coins*

The 1p, 2p, 5p and 10p coins changed composition because the cost of the metals, particularly copper, increased significantly, making these metals unviable to use in low-value coins. Mild steel is an alloy of iron, carbon, and manganese.

20p coins have a higher proportion of copper in the alloy than 50p coins, which explains their slightly pinker colour compared with 50p coins.

1   A 50p coin has a mass of 8g. Calculate the mass of Cu and Ni in the coin, and hence calculate the mole ratio of Cu : Ni. Explain why this is different from the ratio in the table.

2   Give the electron configurations of a Cu atom, a Cu^{2+} ion, a Cu^+ ion, a Ni atom and a Ni^{2+} ion.

3   By referring to variable oxidation states, explain why transition metals can act as catalysts.

4   A student dissolved some nickel and some copper in nitric acid, to give a green and a blue solution respectively. Explain why the solutions are coloured, and why the colours are different.

5   When some ammonia was added to the blue solution in question 4, the solution became a dark blue colour. When some hydrochloric acid was added to a separate sample of the blue solution, it became yellow. Explain these observations giving chemical equations.

6   A student set up an electrochemical cell using the copper and nickel solutions from the dissolved metals in question 4. Draw a labelled diagram of the arrangement of this cell, stating what other materials are needed. Given $E_{cell} = 0.57$ V and $E^{\ominus}$ $(Cu^{2+}(aq)/Cu(s)) = +0.34$V, calculate the standard electrode potential of the nickel half-cell.

Extension

1   This topic has considered metals from the first row of the d-block, from Sc to Zn. Research some properties of second row d-block metals from Y to Cd and/or the properties of f-block metals such as Ce, Sm and Gd. What similarities do they have with first row d-block metals?

2   Produce a revision resource about the chemistry of transition metals. Include oxidation states and redox characteristics, and explanation of their colour and the use of colorimetry, electrochemical cells and electrode potentials, and complex formation.

3   The cost of many metals has increased significantly in recent years due to their scarcity. How are chemists involved in the extraction of metals from previously unviable ores, and in the development of new recycling techniques?

CHAPTER 10
Colour by design

Topics in this chapter

Why a chapter on Colour by design?

This chapter begins by looking at coloured molecules that are found in nature. You will discover how the colour of molecules is related to the presence of extended delocalised systems in their structures and how this idea of delocalisation was used to develop the currently accepted model of the bonding in benzene rings.

The electrophilic substitution reactions of benzene rings are then explored and you will see how benzene rings have been used in creating synthetic dye molecules. The structures of these dye molecules can be modified to improve their colour, solubility, or ability to bond to fabrics.

A second naturally occurring class of molecules is then discussed – the triglyceride structures found in fats and oils; you will also find out more about the technique of gas-liquid chromatography, used to analyse mixtures of oils and other organic compounds

Carbonyl compounds (aldehydes and ketones) are re-introduced, and a new reaction type – nucleophilic addition is described. This reaction allows carbon-carbon bonds to be formed in organic synthesis and the chapter ends with a discussion of the strategies used to design these synthetic routes and some of the problems that need to be overcome. In the course of this discussion, you get the chance to review idea about functional groups and their reactions from the whole of your A-level course.

Knowledge and understanding checklist

In this chapter you will learn more about some ideas introduced in earlier chapters:

- [ ] origin of colour in coloured complexes (**Developing metals**)
- [ ] σ- and π-bonds (**Developing fuels**)
- [ ] electron-pair repulsion theory and shapes of molecules (**Elements of life**)
- [ ] enthalpy changes (**Developing fuels**)
- [ ] organic functional groups and nomenclature of organic molecules (**Developing fuels, Elements from the sea, What's in a medicine, and Polymers of life**)
- [ ] esters and ester formation (**What's in a medicine?** and **Polymers of life**)
- [ ] electrophiles (**Developing fuels**)
- [ ] nucleophiles (**Elements from the sea**)
- [ ] organic mechanisms using curly arrows (**Developing fuels** and **Elements from the sea**)
- [ ] intermolecular bonding (**The ozone story** and **Polymers of life**)
- [ ] solubility of a solute in aqueous solvents (**Oceans**)
- [ ] chromatography (**What's in a medicine?**)
- [ ] organic synthesis (**What's in a medicine?**)
- [ ] carbonyl compounds (**What's in a medicine?**)
- [ ] addition, elimination, and substitution reactions (**Developing fuels, What's in a medicine?, and The ozone story**).

Maths skills checklists

In this chapter you will need to use the following maths skills. You can find support for these skills on Kerboodle and through MyMaths:

- [ ] Use ratios, fractions and percentages.
- [ ] Use angles and shapes in regular 2-D and 3-D structures.
- [ ] Visualise and represent 2-D and 3-D forms.
- [ ] Understand the symmetry of 2-D and 3-D shapes.

MyMaths.co.uk
Bringing Maths Alive

CD 1 Coloured molecules

▲ **Figure 1** *The pink colour of flamingos is caused by derivatives of the carotene molecules they absorb from algae in their diet*

Colour in the world

The natural world is full of colour. Some colours, such as the blue of the sky or the colours in a rainbow, are produced by the scattering or refraction of light. But in most cases colour is due to the presence of certain chemical compounds and arises from the way these compounds interact with light.

Many rocks and minerals are coloured. Very often this colour is due to the presence of transition metal ions within the compounds in the rock or mineral.

The biological world is even more colourful – from the vibrant greens of plant leaves to the dramatic plumage of birds' feathers. Such structures contain biological pigments – coloured organic molecules produced naturally by specific organisms. A good example of such a pigment is the orange-red β-carotene molecule present in many plants and algae – this even provides the vivid pink feathers of flamingos (Figure 1).

Biological pigments and conjugated systems

Carotenoids

The β-carotene molecule, which gives vegetables such as carrots and sweet potatoes their characteristic orange colour, is just one member of a series of biologically important molecules known as the carotenoids.

β-carotene

lycopene

▲ **Figure 2** *β-carotene is responsible for the orange pigment present in carrots and sweet potatoes and lycopene is responsible for the red pigment present in tomatoes and red peppers*

Although the colour of these carotenoids is most obvious in the roots and fruits of plants, carotenoids are also present in the leaves of all plants. This gives a clue to one of their functions in plants – they have the ability to absorb energy from sunlight, and they can transfer this energy to chlorophyll, which increases the efficiency of photosynthesis in the leaves.

This ability to absorb energy from sunlight also explains why you see plant material containing carotenes as coloured. When sunlight (or any other source of white light) hits the surface of the plant, the carotenes absorb some of the wavelengths of the sunlight, but reflect the remainder.

Conjugated systems

So what is it about carotenoids that enables them to absorb light in the visible part of the spectrum? The clue comes from the structural feature which all carotenoids share. The cartenoids contain long hydrocarbon chains consisting of alternating double and single bonds (Figure 2). This type of arrangement is called a **conjugated system** of bonds. For each double bond, one pair of electrons is not confined to linking two particular atoms, as in a single bond, but is spread out or delocalised over the whole conjugated system. This extensive **delocalisation** affects the size of the energy gap between the electron energy levels in the molecule. The greater the number of delocalised electrons, the smaller the energy gap. In β-carotene, the 22 electrons are delocalised over the conjugated system, from the 11 double bonds which it contains.

> **Conjugated system**
>
> A conjugated system has alternating double and single bonds, allowing the overlap of p-orbitals.

> **Delocalised**
>
> Delocalised electrons are not associated with a particular pair of atoms, but are able to spread out over several atoms.

Chemical ideas: Radiation and matter 6.7b

Coloured organic molecules

Absorption of light

Colourless objects

White opaque solids appear white in sunlight (or white light) because none of the wavelengths of incident light are absorbed by the surface of the object – they are all reflected. Our brain perceives a mixture of all the wavelengths of visible light as white.

Similarly, when light shines through colourless transparent substances, none of the wavelengths of incident light are absorbed – they are all transmitted.

> **Synoptic link**
>
> You have already seen how colour arises in solutions of transition metals in Topic DM 3, Colourful compounds.

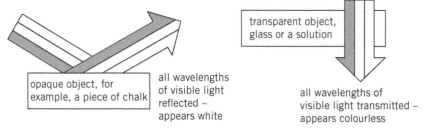

opaque object, for example, a piece of chalk

all wavelengths of visible light reflected – appears white

transparent object, glass or a solution

all wavelengths of visible light transmitted – appears colourless

▲ Figure 3 *How light behaves with opaque and transparent objects*

Coloured objects

Objects appear coloured because wavelengths corresponding to particular colours are absorbed by substances in the object.

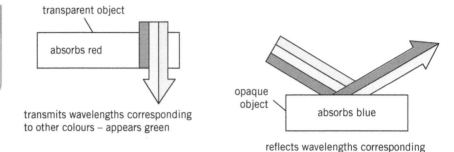

transparent object

absorbs red

transmits wavelengths corresponding to other colours – appears green

opaque object

absorbs blue

reflects wavelengths corresponding to other colours – appears orange

▲ Figure 4 *How colours arise from the absorption of light*

Complementary colours

When absorption occurs, wavelengths corresponding to one colour are removed from the white light and you see the **complementary colour**. The relationship is sometimes shown in a colour wheel, in which complementary colours are opposite one another (Figure 5). So an object absorbing violet appears yellow.

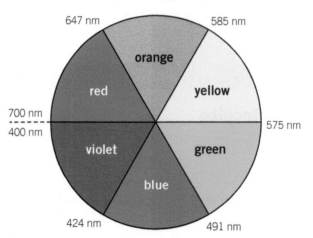

▲ Figure 5 *A colour wheel shows the relationship between a colour and its complementary colour*

Electronic transitions and colour

Colour in organic molecules

When substances absorb radiation in the visible region of the spectrum, the energy absorbed causes changes in electronic energy, and electrons are excited from their normal state (ground state) to an **excited state**.

Colourless compounds

Not all electronic transitions involve the absorption of visible light. Many require a greater energy. In these cases, the compound absorbs ultraviolet radiation but not visible light – it appears colourless. Figure 6 shows the energy needed to excite an electron in a coloured compound and in a colourless compound.

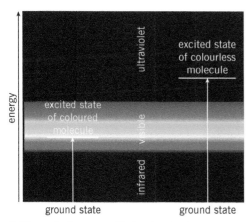

▲ Figure 6 *The type of light absorbed by a compound depends on the energy gap between the ground state and the excited state*

Conjugated systems and delocalisation

Some organic molecules consist of a system of alternating double and single bonds. In this system, known as a **conjugated system**, the electrons in the p-orbitals that make up the π-bond of each double bond are able to spread out over all of the atoms in the system. The electrons are said to be **delocalised**. The p-electrons from C=C, C≡C, and other multiple bonds (i.e., C=O or N=N bonds) can all become delocalised in these conjugated systems.

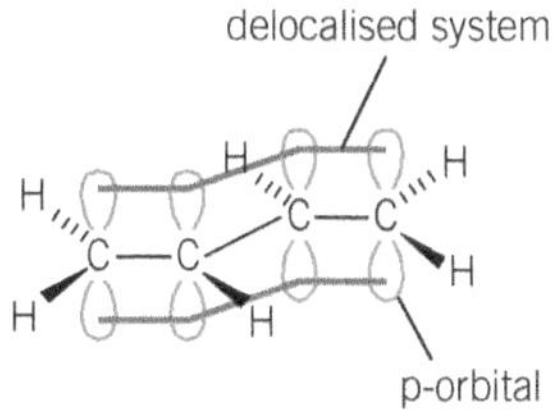

delocalised system

p-orbital

▲ Figure 7 *The p-electrons from the C=C bonds in buta-1,3-diene are able to spread out to form a delocalised system*

Conjugated systems can even include lone pairs from oxygen and nitrogen atoms, if they are aligned in the correct direction to allow them to overlap with other delocalised electrons in the system.

One of the most important examples of a delocalised system is the benzene ring.

Factors affecting colour

Colour in organic molecules is due to the presence of a conjugated system in a molecule, which decreases the gap between the ground state and the excited state:

* The more electrons that are delocalised in the conjugated system, the smaller the energy gap.

* Smaller energy gaps result in the absorption of longer wavelengths of the light absorbed.

As a useful rule of thumb, delocalised systems with five π-bonds in the conjugated system are likely to absorb light in the visible range of the spectrum (Figure 8).

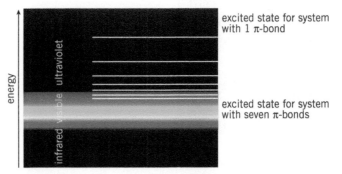

▲ **Figure 8** *The effect of increasing the number of π-bonds in a conjugated system. The energy gap between the ground state and the excited state gets smaller as the number of π-bonds increases*

Summary questions

1 Molecules appear coloured if they absorb light in the visible range of the spectrum. What process occurs in the molecule to allow the absorption of light? *(2 marks)*

2 The molecule retinol is used in some skincare creams and gives them a yellow appearance. Use ideas about the absorption of light to explain why creams containing retinol are yellow. *(2 marks)*

3 Figure 9 shows the structure of retinol. It contains an extended conjugated system.

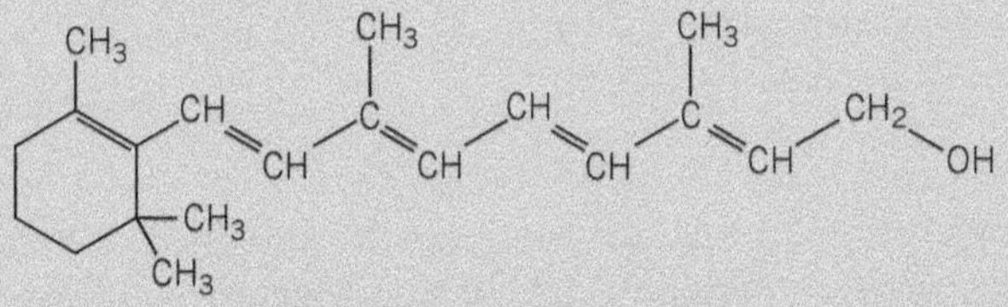

▲ **Figure 9** *The structure of retinol*

a Describe the bonding in the conjugated system. *(3 marks)*

b Explain how the presence of an extended conjugated system causes the retinol molecule to be coloured. *(2 marks)*

4 Eicosane, $C_{20}H_{42}$, is a white solid, and is a saturated hydrocarbon. Use ideas about the structure of eicosane and electronic energy levels to explain why eicosane is white. *(3 marks)*

CD 2 Benzene and delocalisation

Specification reference: CD(e)

Introduction

Until about 150 years ago, chemists had very little idea about the way in which atoms were arranged in organic molecules, and the synthesis of organic molecules was a haphazard process. In this topic, you will learn about a key event in the history of chemistry – the development of a model for the structure and bonding in benzene, and how modern techniques have allowed chemists to test and adapt this model. In the process, the concept of delocalisation was developed which is so critical to our understanding of colour.

Deducing structure

The discovery of benzene

One of the most puzzling of molecules known to chemists in the 19th century was benzene. First isolated by Michael Faraday in 1825 from gas oil, it was found to have an empirical formula of C_1H_1. Calculations of its molecular mass later showed it to have the molecular formula C_6H_6, but there was no obvious way in which carbon and hydrogen atoms could bond together to produce a molecule with this formula.

Imagining the structure

A crucial breakthrough in deducing molecular structures came in 1855 when it was first proposed that carbon was tetravalent (could form four bonds to other atoms). Chemists came up with a range of suggestions for the structure of benzene (Figure 1).

$$H_2C = C = CH - CH = C = CH_2$$

▲ **Figure 1** *Some of the 19th-century suggestions for the structure of benzene*

Yet the more that was discovered about benzene, the more enigmatic it became. Benzene seemed to have an extraordinarily robust and stable structure – it was unusually unreactive compared to other unsaturated hydrocarbons – and it was found to be a structural component of a large number of other compounds.

Kekulé's model

In 1865, the German chemist Friedrich August Kekulé published a paper suggesting a structure consisting of a six-membered ring of carbon atoms with alternating single and double bonds (Figure 2).

The simplicity and geometric beauty of the structure appealed to the scientific community, and the model rapidly became accepted.

Learning outcomes

Demonstrate and apply knowledge and understanding of:

→ the two common representations of the benzene molecule and their relation to

- the shape of the molecule

- bonding in the molecule (including a treatment of enthalpy change of hydrogenation).

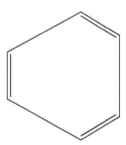

▲ **Figure 2** *Kekulé's suggestion for the structure of benzene*

Kekulé's dream?

Kekulé described how he first visualised this structure in a day-dream involving a snake biting its own tail. But what might seem on the surface to be a charming story about the origins of scientific creativity may have had a more cynical purpose. Kekulé was not in fact the first to suggest this structure. Four years before Kekulé's paper, an Austrian schoolteacher, Josef Loschmidt, had published a book in which he had shown a six-membered ring as the basis for the structure of benzene and other aromatic molecules. It seems certain that Kekulé had seen this book before producing his own paper. So was his dream story an attempt to convince the scientific world that his inspiration for his benzene structure was his own imagination, rather than the plagiarised work of a fellow scientist?

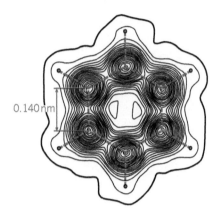

▲ Figure 3 *The electron density contour map of the benzene molecule*

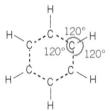

▲ Figure 4 *The geometry of the benzene ring*

Chemical ideas: Organic chemistry: frameworks 12.3a

Arenes

Structure and bonding in benzene

The geometry of the benzene molecule

The modern model of the structure and bonding of benzene is based on evidence from work by British crystallogtapher Kathleen Lonsdale. This work used X-ray crystallography to produce a contour map of the electron density in a benzene molecule (Figure 3).

Analysis of this contour map reveals the following features of the geometry of the benzene molecule:

- The benzene ring is a regular, planar hexagon.
- All the bond angles are 120°.
- All the carbon–carbon bonds are the same length – less than for a carbon–carbon single bond but greater than for a carbon–carbon double bond (Table 1).

▼ Table 1 *Some carbon–carbon bond lengths*

Molecule	Bond	Bond length / nm
alkane	C—C	0.154
alkene	C=C	0.134
benzene	all bonds	0.140

Bonding in benzene – the delocalised model

The model now accepted as being correct for the bonding in the benzene molecule was proposed in 1931, using mathematical ideas derived from quantum theory.

The delocalised model is shown in Figure 5.

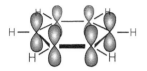 

▲ Figure 5 *The p-electrons in the benzene ring form a delocalised charge cloud above and below the ring*

- Each carbon atom has four outer electrons, which can be used to form bonds.

- Three of these electrons are used to form single σ-bonds to either carbon or hydrogen, leaving one p-electron on each carbon atom.

- Instead of overlapping in pairs to form three separate π-bonds, these six p-electrons delocalise and spread out evenly so that they are shared by all six carbon atoms in the ring.

- These electrons form a delocalised charge cloud, above and below the plane of the molecule.

To represent this in a structural formula, a circle is usually drawn inside the hexagon (Figure 6).

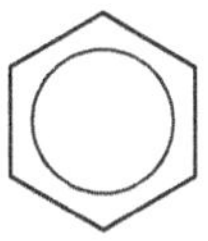

In this model, the electron density in all the carbon–carbon bonds would be uniform all the way around the ring, as shown in the contour map in Figure 3.

The stability of the benzene ring

An important point about delocalisation is that it stabilises structures in which it is present. As a general principle, the *more delocalised* a structure is, the *more stable* the molecule becomes.

Because of its planar cyclic structure and the number of electrons involved, the delocalisation in benzene is particularly strong and the molecule is extremely stable. As a result, it does not readily undergo reactions that would disrupt the delocalisation of the electrons, such as addition reactions.

Comparing models of benzene

The delocalised model of the bonding in benzene has developed from what is known as the Kekulé model. In this model, the six-membered ring consists of alternating double and single bonds.

Testing the models

The Kekulé model explained some of the observed properties of benzene, but even in Kekulé's lifetime experimental evidence was obtained which cast doubt on the validity of the model. Measurements of the energy released during reactions of benzene produced results which could not be explained using this model.

Synoptic link

You met π-bonds in Topic DF 6, Alkenes – versatile compounds.

▲ Figure 6 *The usual way of representing the benzene molecule, clearly indicating the delocalisation of the π-electrons*

Thermochemical data

Benzene reacts (under rather special conditions) with hydrogen in an addition reaction to form cyclohexane.

$$C_6H_6 + 3H_2 \xrightarrow[\text{300°C, 30 atm}]{\text{finely divided Ni}} C_6H_{12}$$

When one $C\!=\!C$ bond is present in a six-membered ring (in the molecule cyclohexene), the enthalpy of hydrogenation is $-120\,\text{kJ mol}^{-1}$ (Figure 7).

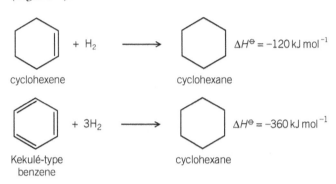

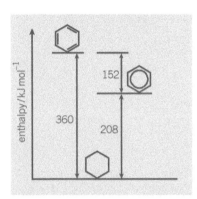

▲ **Figure 8** *The delocalised structure of benzene means that it is $152\,\text{kJ mol}^{-1}$ more stable than would be expected from the Kekulé model*

▲ **Figure 7** *The enthalpy changes for the hydrogenation of cyclohexene and the prediction for benzene using the Kekulé model*

Testing the model

Thermochemical experiments allowed chemists to test the model.

When using the Kekulé model, the enthalpy change of hydrogenation of benzene is predicted to be three times the value for that obtained from one $C\!=\!C$ bond.

In fact, when one mole of benzene molecules are hydrogenated, the enthalpy change is $-208\,\text{kJ mol}^{-1}$. The amount of energy given out is $152\,\text{kJ mol}^{-1}$ *less* than would be predicted from the Kekulé model.

This is exactly what would be predicted from the delocalised model. Delocalisation makes the benzene molecule more stable than the Kekulé structure would suggest, and so less energy is released when it is converted into cyclohexane (Figure 8).

So the thermochemical data provides good evidence that the delocalised model is a much better description of the bonding in benzene than the Kekulé model.

Molecular geometry

Benzene is a planar molecule (all the atoms lie in one plane), and that would also be true of the Kekulé structure. Kekulé's structure would also have bond angles of 120°.

The problem is that the single and double bonds between carbon atoms in the Kekulé model would have different lengths which would mean that the Kekulé structure would not actually be a regular hexagon, but the shape shown in Figure 9.

Synoptic link

You can remind yourself how to use electron pair repulsion theory to predict bond angles and molecular shapes in molecules containing $C\!=\!C$ and $C\!-\!C$ bonds in Topic EL 5, The molecules of life.

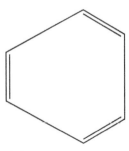

▲ **Figure 9** *The shape of benzene, according to the Kekulé model*

Comparing the models

Table 2 shows how the predictions from the Kekulé model compare with those from the delocalised model.

▼ Table 2 *Comparing the Kekulé model and the delocalised model*

Property	Predictions from the Kekulé model	Predictions from the delocalised model
shape	asymmetrical hexagonal planar	symmetrical hexagonal planar
bond angles	120°	120°
bond lengths	alternating short and long	equal lengths
enthalpy change of hydrogenation	$-360\,kJ\,mol^{-1}$	much less negative than $-360\,kJ\,mol^{-1}$

The way in which benzene reacts with electrophiles provides further ways to test the Kekulé and delocalised models. These will be discussed in Topic CD 4.

Summary questions

1 The delocalised model of the bonding in benzene is now accepted as correct. Describe the bonding in benzene according to this model.

(*4 marks*)

2 Kekulé's model of the bonding in benzene can explain some of the features of the structure of benzene, but not others. Describe and explain:
   a one feature of the structure of benzene which can be explained using the Kekulé model (*3 marks*)
   b one feature that cannot be explained using the Kekulé model.

(*2 marks*)

3 The enthalpy change of hydrogenation of cyclohexene is $-120\,kJ\,mol^{-1}$, whereas the enthalpy change of hydrogenation of benzene is $-208\,kJ\,mol^{-1}$. Discuss the likely value for the enthalpy change of hydrogenation of cyclohexa-1,3-diene. (*5 marks*)

4 Kekulé's model for the structure of benzene suggested that several isomers would exist for di-substituted benzene rings. However, chemists found fewer isomers than Kekulé predicted.
   a According to the Kekulé model, how many different isomers would you expect for the molecule dichlorobenzene, $C_6H_4Cl_2$, in which two chlorine atoms have replaced two hydrogen atoms in the benzene ring? Draw structures for the isomers you predict. (*5 marks*)
   b How many isomers would be predicted using the delocalised model? Explain why the two models predict different numbers of isomers.

(*2 marks*)

Dye

A dye is a soluble, coloured organic molecule that is able to bind to a substrate such as a fibre, and impart colour to it.

▲ Figure 1 *The root of the madder plant*

▲ Figure 2 *Carpets manufactured using the bright red colour of the alizarin dye*

Introduction

Following Kekulé's announcement of the structure and bonding in benzene, chemists realised that many of the molecules found both in natural products and in crude oil contained one or more benzene rings. Like benzene, many of them proved to be unreactive and new methods were developed which enabled derivatives of these aromatic compounds to be created, paving the way for the design and manufacture of synthetic dyes and other important industrial chemicals.

Mimicking nature's rainbow

Until the middle of the 19th century, all coloured dyes were extracted from plant or animal material. Although nature is rich in colours, only a small number of the coloured substances found in living organisms were suitable for use as **dyes**. Others proved too difficult to extract, did not bind strongly enough to cloth, or faded too quickly when exposed to light.

Alizarin

In the early years of the 19th century, one of the most commercially important dyes was the bright red compound known as alizarin. This was extracted from the root of the madder plant, which was then widely cultivated across the world (Figure 1). The bright red colour of the alizarin molecule was responsible for products as varied as cricket ball casings and Turkish carpets (Figure 2).

Probing the structure of alizarin

Alizarin had been used as a dye for thousands of years, but no one had any idea of its chemical nature. However, Kekulé's breakthrough in deducing the structure of benzene gave chemists the key to unlocking structural problems in organic chemistry. In 1868, Carl Graebe and Carl Liebermann attempted to solve the mystery of the alizarin structure. Alizarin had been found to be a mixture of two similar molecules, and they knew that the molecular formula of the main component of the mixture was $C_{14}H_8O_4$.

The very low ratio of H : C suggested that the structure might be based on molecules such as naphthalene, $C_{10}H_8$, or anthracene, $C_{14}H_{10}$. Previously it had been suggested that these molecules might contain fused benzene rings (Figure 3).

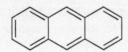

naphthalene   anthracene

▲ Figure 3 *Structures of naphthalene and anthracene. When showing the structure of fused benzene rings, a Kekulé structure is normally used to avoid any confusion about the number of double bonds that are part of the delocalised structure*

Graebe and Liebermann's strategy was to use a newly developed method which could convert **aromatic compounds** to their parent **arene** by reducing them with zinc dust (Figure 4).

A synthesis for alizarin

Reduction of alizarin by this method produced anthracene, suggesting that there was some structural similarity between them. Importantly, it also suggested that it might be possible to synthesise alizarin from anthracene. Graebe and Liebermann were unable to deduce any further structural details, but they were able to find a way of converting anthracene back into alizarin. However, the method they used was expensive and produced alizarin in very low yields.

But now the race was on to find a cheaper commercial route to synthetic alizarin. Even though some of the structural features of the molecule were known, the full structure was not – and yet only a year later two groups of chemists had independently solved the problem of the synthesis.

The first synthetic route to a natural dye created a dramatic effect in the dye industry. The madder crop fields, which had occupied hundreds of thousands of hectares of agricultural land in Europe and Asia, were replaced as a source of alizarin by the rapidly growing chemical industry.

The structure of alizarin

Alizarin was eventually found to be a mixture of two very similar molecules, each based on two benzene rings – but rather than being fused directly they are joined together by a quinone group (Figure 5).

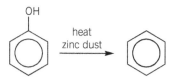

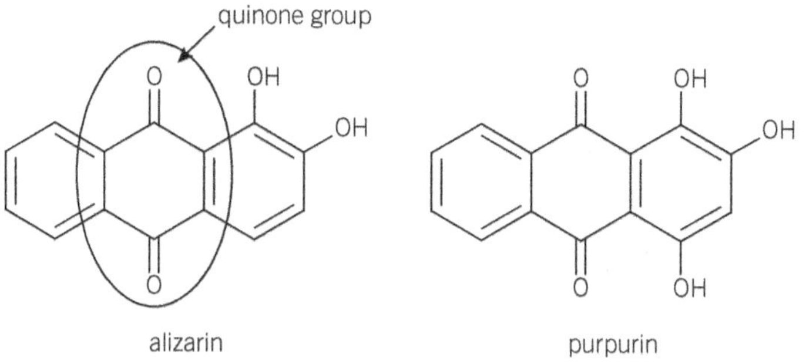

▲ Figure 4 *Reduction of phenol with zinc*

![Structures of alizarin and purpurin showing the quinone group]

alizarin

purpurin

▲ Figure 5 *The structure of alizarin and purpurin, showing the quinone group*

The fact that natural dyes are often a mixture of several closely related molecules is one reason why even the modern chemical industry can find it hard to precisely mimic the colour of naturally occurring dyes.

Chemical ideas: Organic chemistry: frameworks 12.3b

Naming arenes

Arenes

Hydrocarbons like benzene, which contain rings stabilised by electron delocalisation, are called **arenes**. The ending -ene tells you they are unsaturated, like alkenes.

> **Arene**
>
> Arenes are aromatic compounds containing C and H only, and are also known as aromatic hydrocarbons.

Simple arenes

There are many arenes. Figure 6 shows some examples in which hydrogen atoms on the benzene ring have been replaced by alkyl groups.

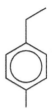

methyl benzene 1,3-dimethyl benzene 1-ethyl-4-methylbenzene

▲ **Figure 6** *Naming simple arenes*

> **Study tip**
>
> As with other compounds you have met, the numbers used are as low as possible and the groups are listed in alphabetical order.

More complex arenes

A different strategy is used to name arenes in which the group attached to the benzene ring contains an alkene functional group. For example, the molecule shown in Figure 7 is called phenylethene.

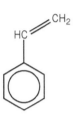

▲ **Figure 7** *Phenylethene*

Phenylethene is named from the parent compound ethene, rather than from benzene. A benzene ring in which one hydrogen atom has been substituted by another group is known as a **phenyl group**. Benzene rings can be found in fused ring systems, for example, napthalene and anthracene (Figure 3). In these molecules there10 and 14 delocalised electrons respectively.

> **Phenyl group**
>
> A phenyl group is a group of atoms with the formula C_6H_5. It is related to benzene in the same way that a methyl group is related to methane.

Aromatic compounds

The term aromatic was originally used to describe the characteristic aroma of some naturally occurring oils that contain benzene rings. It is now used in a much wider sense, to mean molecules that contain benzene rings.

> **Aromatic compound**
>
> Aromatic compounds contain a benzene ring.

Aromaticity

The word aromatic is used to describe molecules that contain a benzene ring because all such molecules possess the special stability that results from the delocalisation of the electrons in the benzene ring.

Other conjugated systems also have the extra stability that exists in a benzene ring. For example, naphthalene and anthracene (Figure 3) also have this stability and are also described as aromatic.

Other molecules such as cyclooctatetrene (Figure 8) look as though they are aromatic but in fact are not.

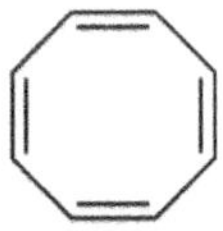

▲ **Figure 8** *The structure of cyclooctatetrene, a non-aromatic hydrocarbon*

To decide whether a conjugated system is aromatic, you need to look and see whether it has the following structural features:

- planar
- cyclic
- possessing $4n + 2$ delocalised π-electrons (i.e., 6, 10, 14, etc.)

> **Study tip**
>
> You may be asked to write the molecular formula of molecules containing benzene rings. If the benzene ring has several groups attached to it, you need to remember that each group will have replaced an H atom, so count the H atoms carefully.

> **Synoptic link**
>
> You first met aromatic compounds in Topic DF 3, What's in your tank?

Aromatic systems can include rings where one or more of the atoms is not a carbon – such as the nitrogen atoms found in the bases which make up DNA.

1 Cyclooctatetrene is not aromatic. Use information from Topic CD 2 to predict:
   a the enthalpy change of hydrogenation when it is converted to cyclooctane
   b the carbon–carbon bond lengths in the cyclooctatetrene molecule.
2 Which of the molecules shown in Figure 9 would be called aromatic?

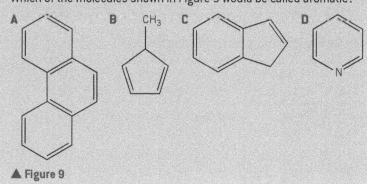

▲ Figure 9

Compounds derived from arenes

Names based on benzene

Hydrogen atoms on the benzene ring can be replaced by different functional groups (Figure 10).

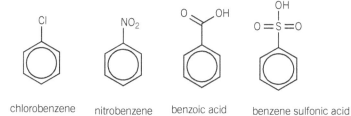

chlorobenzene   nitrobenzene   benzoic acid   benzene sulfonic acid

▲ Figure 10 *Aromatic compounds derived from benzene*

Two of these groups, the nitro group, $-NO_2$, and the sulfonic acid group, $-SO_3H$, may be new to you – these groups have important roles in dye molecules which are derived from benzene.

Names based on the phenyl group

Figure 11 shows two important aromatic compounds whose names are based on the phenyl group rather than on benzene – they are phenol, C_6H_5OH, and phenylamine, $C_6H_5NH_2$. Both of these are important **feedstocks** for the chemical industry, and in particular for the manufacture of dyes.

Although you will be expected to know and use just these systematic names, you should be aware that some aromatic molecules also have traditional names that you may encounter in books and websites. For example, phenylamine is often known by its traditional names of aniline.

Feedstocks

Feedstocks are the reactants that go into a chemical process.

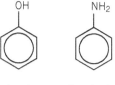

phenol   phenylamine

▲ Figure 11 *These are important feedstocks*

Study tip

Remember that you can use the word hydroxy to represent the $-OH$ group and amino to represent the $-NH_2$ group, if you are trying to name aromatic molecules with several groups attached.

Synoptic link

You need to be able to recognise and name esters which contain benzene rings. See Topic WM 2, Identifying the active chemical in willow bark, and Topic PL 1, The polyester story.

Non-systematic names

You may also come across the names in Figure 12, although you are not expected to be able to remember them.

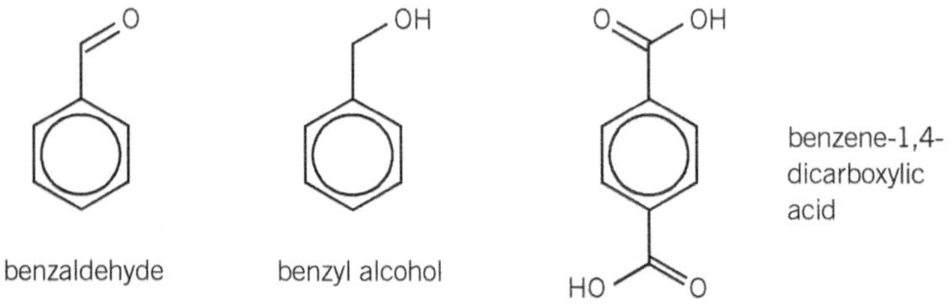

O
OH
O
OH
benzene-1,4-dicarboxylic acid
benzaldehyde
benzyl alcohol
HO
O

▲ **Figure 12** *Some aromatic compounds are named in unpredictable ways*

Esters derived from phenol and benzoic acid

Figure 13 shows some examples of esters containing benzene rings.

O
O
CH3
O
OCH3
phenyl ethanoate
methyl benzoate

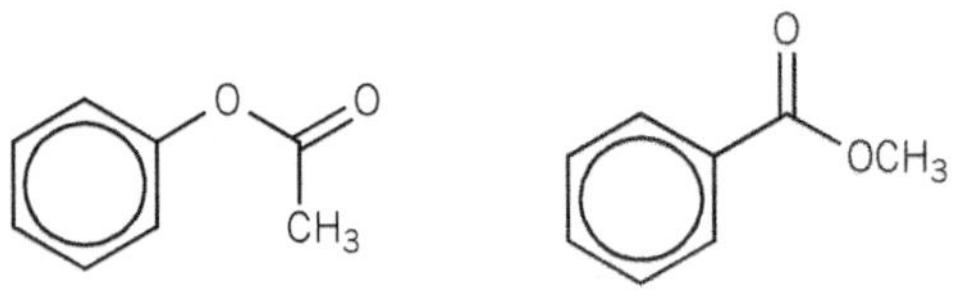

▲ **Figure 13** *Esters containing benzene rings*

Summary questions

1 Give the systematic name of these arene molecules: *(3 marks)*

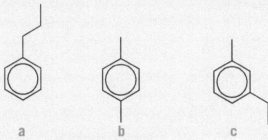

a
b
c

2 Draw out the structure of:
 a 3-nitrophenol *(1 mark)*
 b 3-hydroxybenzenesulfonic acid *(1 mark)*
 c pentyl benzoate. *(1 mark)*

3 Write down the molecular formula of:
 a 1,2-dihydroxybenzene *(1 mark)* b purpurin (Figure 5). *(1 mark)*

4 The bonding in the molecule naphthalene (Figure 3) has some similarities to the bonding in benzene. Suggest and explain the key features of the bonding in naphthalene. *(4 marks)*

CD 4 Reactions of aromatic compounds

Specification references: CD(d), CD(g)

Introduction

The delocalised electrons in the benzene ring mean that the reactions of benzene and benzene derivatives are completely different from those of other unsaturated compounds, such as alkenes, that you have studied previously in this course.

Chemists in the 19th century found ways in which they could insert a range of useful side groups onto benzene rings. These reactions provide a way for chemists to alter the physical and chemical properties of the benzene ring. Some of these reactions involve joining carbon chains to benzene rings, and these are the starting point for the synthesis of an even wider range of organic molecules.

The first synthetic dye

Natural dyes and synthetic dyes

Indigo

An important natural dye was indigo (Figure 1). This intense blue substance formed when the colourless molecules obtained from plants such as woad were fermented in alkaline conditions (by soaking the plants in animal urine) and then allowed to oxidise.

▲ Figure 1 *The structure of indigo, a valuable natural dye*

The extended delocalised system in indigo means that it strongly absorbs light in the yellow part of the spectrum, giving the molecule an intense blue colour.

Picric acid

In 1779, probably in an attempt to produce variations on the colour of indigo, an Irish chemist called Peter Woufle tried treating indigo with nitric acid. The product of the reaction was picric acid, or 2,4,6-trinitrophenol, a bright yellow substance, which was able to dye wool and other fabrics.

When, in the 1840s, it was found that it could be synthesised by nitrating phenol molecules found in coal tar, picric acid became the first fully **synthetic** dye.

Learning outcomes

Demonstrate and apply knowledge and understanding of:

→ electrophilic substitution reactions of arenes and the names of the benzene derivatives formed from:
  • halogenation of the ring
  • Friedel–Crafts alkylation and acylation

→ how delocalisation accounts for the characteristic properties.

▲ Figure 1 *Indigo dye*

507

Synthetic

A synthetic compound is one made artificially by chemical reactions.

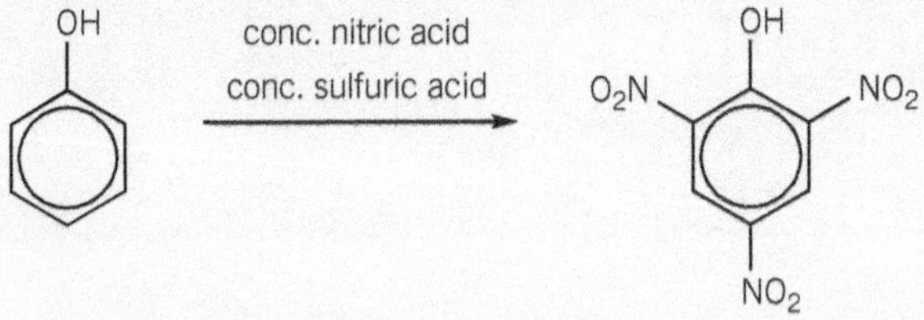

▲ Figure 2 *Picric acid can be synthesised by nitrating phenol*

Although benzene and phenol are both colourless molecules, the presence of the nitro groups in picric acid alters the wavelength at which the delocalised system absorbs light. This results in picric acid absorbing in the violet part of the spectrum, meaning that it appears yellow.

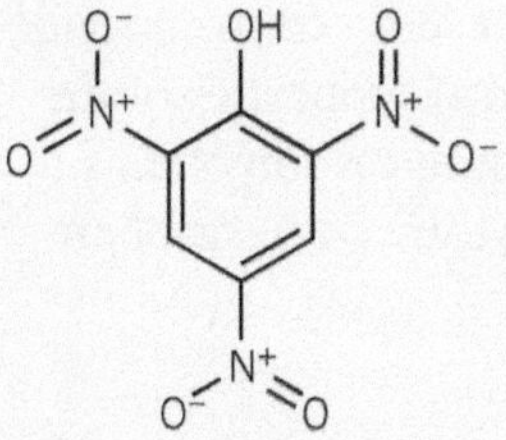

▲ Figure 3 *The structure of the nitro groups in picric acid shows how they can extend the delocalised system of the benzene ring*

You will not hear much about the use of picric acid as a dye nowadays. In the 1880s it was discovered to be highly explosive. When solid, it is extremely unstable and since then its use as a dye has been limited for use in solution to stain biological samples. The nitration of phenol is an example of an electrophilic substitution reaction.

Electrophile

A positive ion, or a molecule with a full or partial positive charge, that will be attracted to a negatively charged region and react by accepting a covalent bond.

Synoptic link

Electrophiles and electrophilic addition were covered in Topic DF 6, Alkenes – versatile compounds.

Study tip

You need to be able to recognise reactions as substitution or addition and to classify reagents as electrophiles or nucleophiles.

Chemical ideas: Organic chemistry: framework 12.3c

Reactions of arenes

Classifying the reactions of arenes

The benzene ring in arenes is an area of high electron density. So benzene, like alkenes, reacts with **electrophiles**.

The effect of delocalisation

There is a difference in the behaviour of an alkene and an arene towards electrophiles, which is due to the presence of the delocalised system of electrons:

- Alkenes react with electrophiles in addition reactions.
- The product is a saturated molecule.

For example, cyclohexene reacts with bromine to form 1,2-dibromocyclohexane:

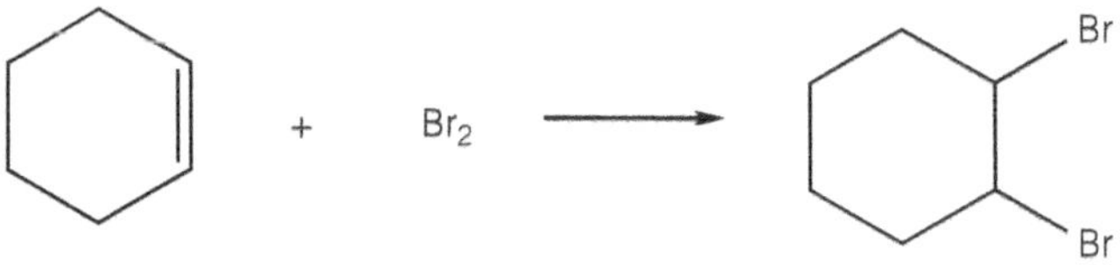

▲ Figure 4 *Alkenes take part in addition reactions*

- Benzene undergoes substitution rather than addition reactions with electrophiles.
- The product still contains an unsaturated benzene ring.
- The overall reaction is described as electrophilic substitution.

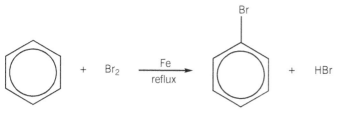

▲ **Figure 5** *Arenes take part in substitution reactions*

Study tip

Remember in skeletal structures hydrogen is not shown explicitly. In Figure 5, bromine substitutes for a hydrogen on the benzene ring.

By reacting in this way, the stable benzene ring system is kept intact. The reactions are relatively slow because the first step in the reaction mechanism disrupts the delocalised electron system, and this requires a substantial input of energy. Often the presence of a catalyst is required to create an electrophile reactive enough to attack the benzene ring.

Electrophilic substitution reactions of benzene

Benzene undergoes many substitution reactions. In every case, there is an electrophile which is attracted to the delocalised electron system in the benzene ring.

Bromination and chlorination of benzene

Bromination
Benzene reacts with liquid bromine in the presence of a catalyst, such as iron filings or iron(III) bromide. Heat is necessary. The bromine is decolorised and fumes of HBr are given off. The equation for the reaction is shown in Figure 5.

The role of the iron catalyst
The first step involves reaction of the benzene ring with the electrophile Br^+. The Br^+ is generated by the action of the catalyst.

If iron is used as the catalyst, the steps in the mechanism are:

- Iron reacts with bromine to form iron(III) bromide, $FeBr_3$:

$$2Fe + 3Br_2 \rightarrow 2FeBr_3$$

- The $FeBr_3$ then interacts with a neighbouring bromine molecule – a Br^- ion is removed from it to form the Br^+ electrophile.

$$Br-Br \quad + \quad \begin{array}{c} Br \\ | \\ Fe-Br \\ | \\ Br \end{array} \quad \longrightarrow \quad Br^+ \quad + \quad \left[\begin{array}{c} Br \quad Br \\ \diagdown \diagup \\ Br-Fe \\ | \\ Br \end{array} \right]^-$$

▲ **Figure 6** *The formation of Br^+ from Br_2 and $FeBr_3$*

- Br^+ then bonds to a carbon atom in the benzene ring and an H^+ ion is lost from the ring.

Study tip

You are not required to be able to draw the mechanism for this process using curly arrows, but you need to be able to explain the role of the $FeBr_3$ catalyst in forming the Br^+ electrophile.

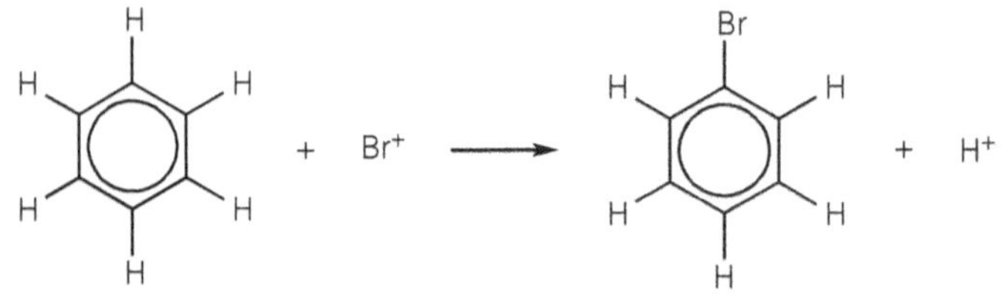

▲ **Figure 7** *Br^+ takes the place of an H^+ ion in the benzene ring*

- The H^+ reacts with the $FeBr_4^-$ to produce HBr and regenerate the $FeBr_3$ catalyst:

$$H^+ + FeBr_4^- \rightarrow HBr + FeBr_3$$

Substances that are able to remove halogen atoms from molecules in this way are known as halogen carriers and play an important part in several other reactions of benzene.

Chlorination

A chlorine atom may be substituted into a benzene ring, in much the same way as a bromine atom. An aluminium chloride catalyst is often used as the halogen carrier in this case:

$$Cl_2 + AlCl_3 \rightarrow AlCl_4^- + Cl^+$$

The combination of reactive chlorine and the powerful halogen carrier aluminium chloride means that the reaction will occur at room temperature. Aluminium chloride reacts violently with water, so the reaction must be carried out under anhydrous conditions.

The importance of halogenated benzene derivatives

Benzene rings with halogens attached have been found to be important intermediates in synthetic routes. The carbon atom attached to the halogen can be made to attack other carbon atoms and this ability to form carbon–carbon bonds is crucial if chemists wish to make large complex molecules from simpler starting materials.

The Friedel–Crafts reaction is an example of a carbon–carbon bond forming reaction.

Alkylation

Alkylation is a reaction that introduces an alkyl group into a molecule.

Friedel–Crafts reactions

Aluminium chloride, $AlCl_3$, has similar properties to $FeCl_3$ and can also be used as a halogen carrier catalyst to help polarise organic molecules that contain halogens and cause them to substitute in a benzene ring. This type of reaction is called a Friedel–Crafts reaction, after its discoverers.

Alkylation

If benzene is heated under reflux with chloroethane and anhydrous aluminium chloride, a substitution reaction called an **alkylation** occurs and ethylbenzene is formed:

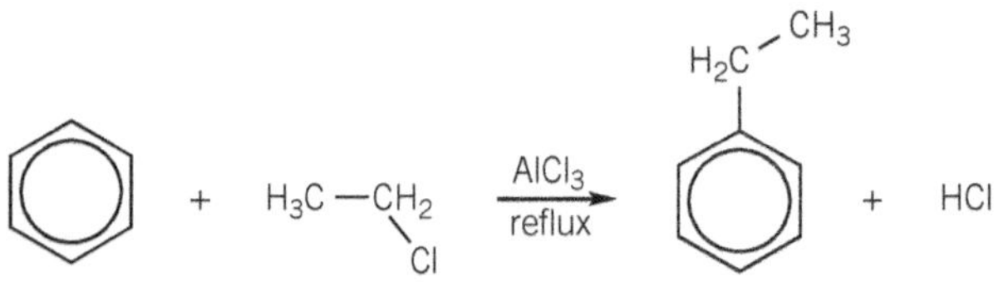

▲ **Figure 8** *Friedel-Crafts alkylation*

Study tip

When drawing the structure of products of alkylations, it is very important to think carefully about which atom of the alkyl group becomes attached to the benzene ring. It will be the carbon attached to the halogen in the alkyl halide.

The aluminium chloride helps to form the reactive electrophile that attacks the benzene ring:

$$CH_3CH_2Cl + AlCl_3 \rightarrow CH_3CH_2^+ + AlCl_4^-$$

The positively charged carbon atom then reacts with the benzene ring to form ethylbenzene.

Acylation

A similar reaction takes place when benzene is treated with an acyl chloride (or an acid anhydride) and aluminium chloride under anhydrous conditions. Here the reaction is an **acylation**, because an **acyl group** becomes attached to the benzene ring.

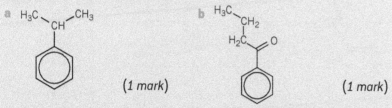

▲ **Figure 9** *Friedel-Crafts acylation*

The importance of Friedel–Crafts reactions

Friedel–Crafts alkylations and acylations are particularly useful to synthetic chemists, because they provide a way of adding carbon atoms to the benzene ring and building up side chains.

Summary questions

1   Benzene reacts with bromine under suitable conditions to form bromobenzene.
    a   State the type and mechanism of this reaction.   (*2 marks*)
    b   i   What other substance must be present in order for the reaction to occur?
        ii State one other essential condition for the reaction.   (*2 marks*)

2   Draw the structural formulae of products formed when benzene reacts with:
    a   chloromethane in the presence of aluminium chloride   (*1 mark*)
    b   ethanoyl chloride in the presence of aluminium chloride. (*1 mark*)

3   Name the organic reagent that would be used in a reaction with benzene to form these molecules.
    a (*1 mark*)   b (*1 mark*)

4   Predict, with reasons, the main product that will be formed when the molecule below reacts with chlorine at room temperature.

(*2 marks*)

Acyl group

An acyl group of atoms has the structure R—C=O, usually formed from a carboxylic acid RCOOH or an acyl chloride RCOCl.

Acylation

Acylation is a reaction in which an acyl group is introduced into a molecule.

Study tip

You may be asked to predict the structures formed when derivatives of benzene react with electrophiles. In these cases you can expect to be given information about where on the molecule substitution occurs, since predicting where this happens is beyond the scope of this course.

CD 5 Benzene rings and dye molecules

Introduction

You have seen how chemists discovered ways of substituting halogens, alkyl groups, and acyl groups into benzene rings. In this topic you will learn about several more substitution reactions. These reactions are of particular importance to chemists designing dyes – they affect some key properties of the dyes and one of them also provides a way of making an important class of synthetic dyes.

Azo dyes

The revolution in the development of the chemical dyestuff industry in the second half of the 19th century saw a host of new molecules produced by chemists eager to create the next fashionable shade of colour. Although chemists were becoming increasingly confident about the structures of the molecules they were creating, there was much less understanding of the relationship between structure and properties.

Designing colours

However, the work of Otto Witt in 1875 changed all that. He was working on a newly discovered class of synthetic molecules, the azo dyes. These were produced from phenylamine by a two-stage process involving a diazotisation and a coupling reaction.

The resulting dyes contained a functional group that was new to the chemists of the time – the azo group.

▲ Figure 1 *The structures of yellow and brown azo dyes that Witt investigated*

Witt thought that the colour of an azo compound was related to its structure. Knowing the structures of two of these azo dyes (Figure 1) he suggested that it should be possible to synthesise an azo compound in the series with *two* amine groups on the benzene ring. He predicted that this would be an orange colour – an intermediate between the yellow and brown colour of the dye molecules. Then when he made this compound he found it had the predicted colour. It proved to be a successful dye for cotton and was marketed as *chrysoidine* – it was the first commercially useful azo dye (Figure 2).

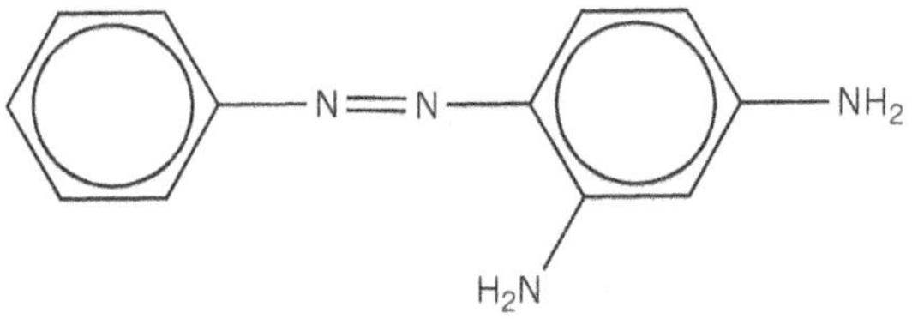

▲ **Figure 2** *The orange dye chrysoidine*

More modifications

Modifying colour

There is a vast range of azo dyes in use today, in an array of colours.

All contain at least one azo group, but also have groups attached to the benzene rings which modify the colour of the chromophore. Groups that can do this include phenol groups and nitro groups, as well as the amine group used by Witt.

Modifying other properties

Another group often seen as part of the structure of azo dyes is the sulfonate group, SO_3^-, derived from the sulfonic acid group, SO_3H. This ionic group helps to make the dye molecule water soluble, as well as providing a way of attaching the dye to fabric by bonding to positively charged groups in the molecules of the fabric.

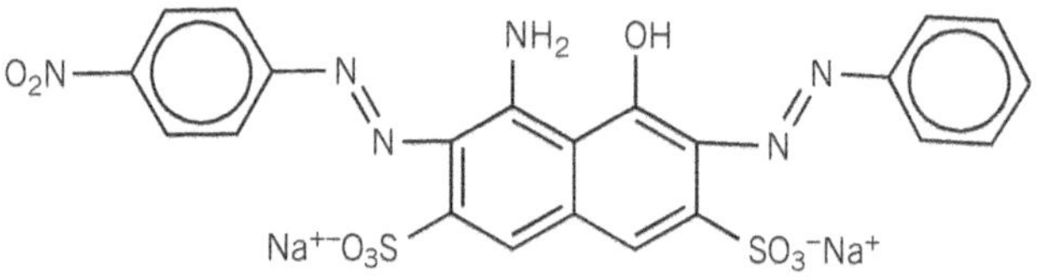

▲ **Figure 3** *The vivid colours of these textiles are due to the presence of different azo dyes which bind strongly to the fibres of the fabric*

▲ **Figure 4** *The structure of the azo dye Acid Black 1. As well as an extended delocalised system involving two azo groups, there are nitro, amino, and phenol groups that modify the colour of the chromophore. The ionic SO_3Na group helps to increase the solubility of the dye*

Chemical ideas: Organic chemistry: frameworks 12.4

More reactions of benzene rings

Nitration

Benzene reacts with a mixture of concentrated nitric acid and concentrated sulfuric acid – this is called a nitrating mixture. If the temperature is kept below 55 °C, then the product is nitrobenzene (Figure 5).

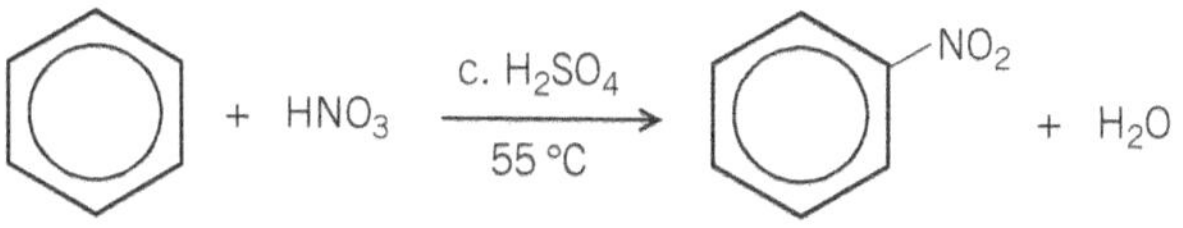

Study tip

Concentrated is sometimes shortened to c. in this type of equation.

▲ **Figure 5** *The reaction between benzene and nitric acid*

At higher temperatures, further substitution of the ring takes place to give the di- and tri-substituted compounds (Figure 6).

▲ **Figure 6** *1,3-dinitrobenzene and 1,3,5-trinitrobenzene, formed when benzene is nitrated at temperatures above 55 °C. Molecules with multiple nitro groups on a benzene ring may be dangerously explosive*

The electrophile which reacts with the benzene ring is the NO_2^+ cation. This is formed in the nitrating mixture by the reaction of sulfuric acid with nitric acid:

$$HNO_3 + 2H_2SO_4 \rightarrow \underbrace{NO_2^+ + 2HSO_4^- + H_3O^+}_{\text{nitrating mixture}}$$

The importance of nitration reactions

Inserting nitro groups into a benzene ring that is part of a chromophore will modify the properties of the chromophore. It will change the wavelength of light that it absorbs and the molecule will have a different colour.

Sulfonation

When benzene and concentrated sulfuric acid are heated together under reflux for several hours, benzenesulfonic acid is formed (Figure 7).

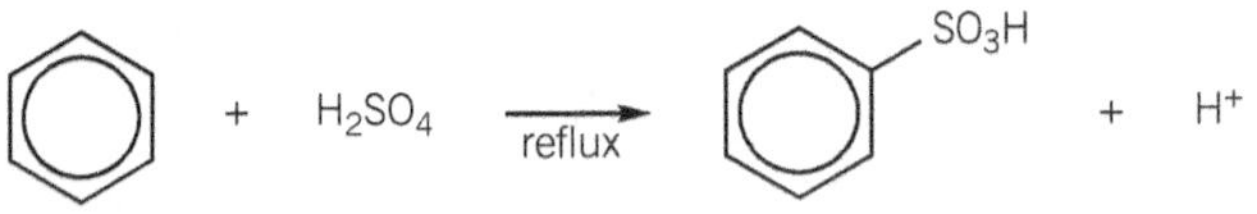

▲ **Figure 7** *The reaction between benzene and sulfuric acid*

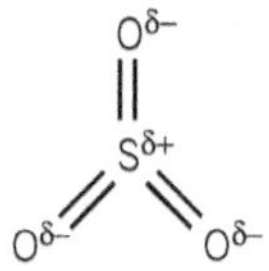

▲ **Figure 8** *One way of drawing the structure of the SO_3 molecule, showing that the sulfur atom is the electrophilic part of the molecule*

The electrophile in this case is SO_3, which is present in the concentrated sulfuric acid. SO_3 carries a large partial positive charge on the sulfur atom and it is this atom which becomes bound to the benzene ring (Figure 8). The full structural formula of benzenesulfonic acid is shown in Figure 9.

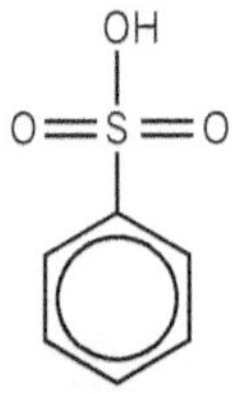

Benzenesulfonic acid is a strong acid and forms salts in alkaline solution (Figure 10).

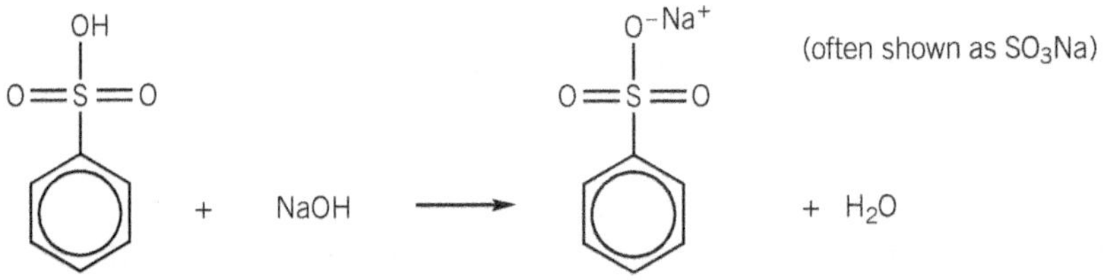

▲ **Figure 9** *The structure of benzenesulfonic acid*

(often shown as SO_3Na)

▲ **Figure 10** *The formation of sodium sulfonate from benzenesulfonic acid*

The importance of sulfonation reactions

Because the sulfonic acid group is able to form ionic salts, sulfonation provides a way of forming more soluble derivatives of aromatic compounds. Dyes are often made more soluble by adding sulfonic acid groups.

Azo compounds

Azo dyes

Azo compounds contain the azo functional group, $R_1–N{=}N–R_2$. Azo compounds are formed as a result of a coupling reaction between a diazonium salt and a coupling agent.

If R_1 and R_2 are arene groups, then the azo compound is much more stable than if R_1 and R_2 are alkyl groups. This is because the $–N{=}N–$ group is stabilised by becoming part of an extended delocalised system involving the arene groups. This stability enables the highly coloured aromatic azo compounds to be used as dyes.

> **Coupling**
>
> Coupling is a reaction between a diazonium compound and a phenol or aromatic amine to produce an azo dye.

Diazonium compounds

Diazonium compounds contain the diazonium group (Figure 11), sometimes known as the diazo group.

R ——— N⁺≡≡≡N

▲ Figure 11 *The diazonium functional group*

Stability of diazonium compounds

Diazonium compounds (also known as diazonium salts) are usually unstable because the diaozonium group is very easily lost, forming nitrogen gas. However, if the group is attached to a benzene ring, the electrons in the $N{\equiv}N$ bond will become part of the delocalised system in the benzene ring. This stabilises it enough to allow diazonium compounds to be formed as intermediates in synthesis. The compounds are prepared in ice-cold solutions (the solids are explosive), and are used immediately.

An example of a relatively stable diazonium compound is benzenediazonium chloride (Figure 12).

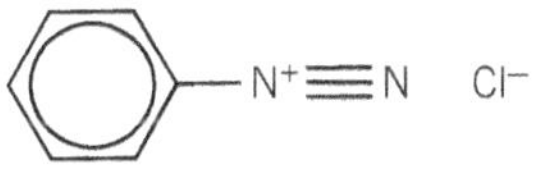

▲ Figure 12 *Benzenediazonium chloride is a relatively stable diazonium compound*

Diazotisation

Aromatic diazonium compounds are prepared from aromatic amines (such as phenylamine):

- The aromatic amine is dissolved in dilute hydrochloric acid.
- A cold solution of sodium nitrite (sodium nitrate(III)), $NaNO_2$, is added.
- The temperature is kept below 5 °C by performing the reaction in an ice-bath.

The reaction that occurs is known as diazotisation.

> **Diazotisation**
>
> Diazotisation is a reaction in which an amine group is converted into a diazonium salt.

The actual reactant in this process is the unstable acid HNO_2 (nitrous acid), which reacts with the amine functional group (Figure 13).

▲ Figure 13 *A diazotisation reaction*

Coupling reactions

Activity CD 5.3

Using different dizonium ions and coupling agents will allow you to make a number of different azo dyes.

In a coupling reaction, a diazonium compound reacts with a coupling agent – a compound containing a relatively reactive benzene ring.

The diazonium ion acts as an electrophile and reacts with the benzene ring of the coupling agent. Generally, coupling agents contain phenol or amino groups attached to the benzene ring, as the lone pairs of electrons from these groups increase the electron density on the benzene ring and increase its reactivity towards electrophiles.

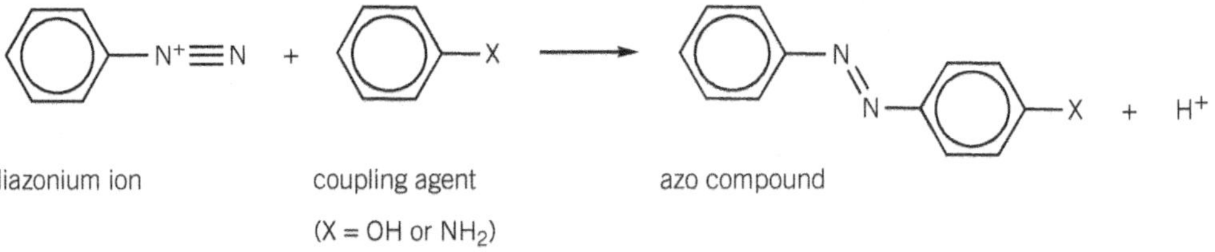

diazonium ion   coupling agent   azo compound

$(X = OH$ or $NH_2)$

▲ Figure 14 *A generalised coupling reaction*

In a coupling reaction:

- A solution of a coupling agent is made up.
- An ice-cold solution of the diazonium salt is added to it.
- A coloured precipitate of the azo compound immediately forms.

Coupling with phenols

With phenols, the coupling agent is usually dissolved in alkali to enable the reaction to occur.

When a solution of benzenediazonium salt is added to an alkaline solution of naphthalen-2-ol, a red azo compound is precipitated (Figure 15).

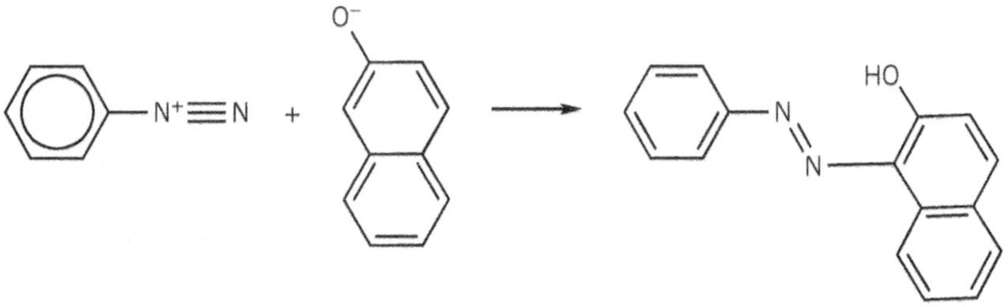

Study tip

Predicting the position on the coupling agent, where the diazonium ion joins, is beyond the scope of this course.

You will not be required to name the azo compounds that are formed.

▲ Figure 15 *Coupling using naphthalene*

Coupling with amines

Diazonium salts also couple with aromatic amines such as phenylamine.

The range of azo compounds

By altering the structure of the diaozonium compound and coupling agents used in the coupling reaction, a wide range of coloured azo compounds can be formed, differing in their colour, water solubility, and ability to bind to fabrics. An example is methyl orange (Figure 16).

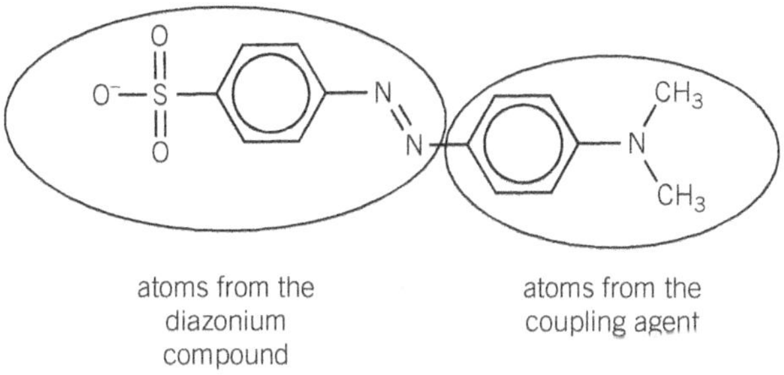

atoms from the diazonium compound

atoms from the coupling agent

▲ Figure 16 *The structure of methyl orange, showing the parts of the molecules which originated in the diazonium compound and the coupling agent*

Introducing –OH groups, NH_2, and ionic groups such as SO_3^- into a dye molecule will impact on the ability of a dye molecule to bond to fibres.

Study tip

You may be asked to identify the diazonium compound and coupling agent used to form a particular azo dye structure. The benzene ring which came from the coupling agent will have an amino or phenol group still attached to it, making the coupling agent easy to spot.

Summary questions

1 What reagents and conditions are needed to convert benzene into:
   a nitrobenzene
   b benzene sulfonic acid? *(4 marks)*

2 Dye molecules often contain functional groups such as nitro, amine, and sulfonate groups. For each of these three functional groups, state one role of the group in a dye molecule. *(2 marks)*

3 a Draw the structure of the diazonium salt that would form when 4-aminophenol is treated with sodium nitrite in the presence of ice-cold hydrochloric acid. *(2 marks)*
   b Write equations for the coupling reaction of this diazonium salt with:
      i phenol (coupling takes place at the 4-position)
      ii naphthalen-2-ol (coupling takes place at the carbon atom adjacent to the OH). *(2 marks)*

4 Give the structures of the diazonium compound(s) and coupling agent that you would need to make each of the azo compounds shown in Figure 2 (chrysoidine) and Figure 4 (Acid Black 1). (Hint: The coupling agent usually contains a phenol group or an amine group attached to an arene ring system.) *(4 marks)*

Introduction

You have seen how chemists have been able to design and synthesise coloured molecules. But being coloured alone is not sufficient for a molecule to be classified as a dye, as there must also be some kind of interaction between the dye molecule and the substance which is to be coloured by the dye. In this topic you will look at the type of interactions which are important in enabling dyes to attach to materials such as the fibres of polymers, and how these interactions are related to the structures of the dyes and the fabrics.

From accident to design – attaching synthetic dyes to fibres

Mauveine

In the previous three topics you have seen how developments in the understanding of structure enabled 19th-century chemists to develop two synthetic dyes:

- alizarin – produced by chemists using ideas about the structure of aromatic compounds to find a way of mimicking a molecule found in nature
- azo dyes – developed by Witt using his intuition about the link between colour and structure.

But many chemical discoveries, even today, owe more to chance than to any reasoned scientific planning. The discovery of mauveine, the molecule that truly launched the dyestuffs industry, happened in spite of some very flawed chemical reasoning.

Perkin's mauve

An 18-year-old English chemist, William Henry Perkin, was trying to synthesise the molecule quinine using aromatic amines as starting materials. His attempts ended in failure but in one experiment, using the molecule aniline, he discovered that the black sludge which he had produced dissolved in ethanol to produce a deep purple solution. It certainly was not quinine, but Perkin realised that it might have a use as a synthetic alternative to the expensive purple dyes obtained from natural products. He found that it was able to dye silk, attaching itself strongly enough to the fabric that the colour did not wash out or fade with use (Figure 1).

By accident he had stumbled upon the first important industrial synthesis of a dye, in 1856.

He called the new dye mauveine, although he still had no idea about the structure of his new discovery. In fact, because mauveine turned out to be a mixture of several closely related molecules, the actual structure of the main component was not confirmed until 1994 (Figure 2).

▲ **Figure 1** *A silk dress dyed with Perkin's mauve dye*

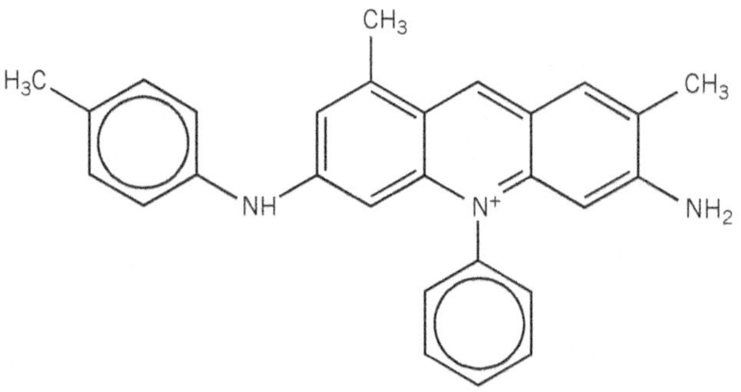

▲ Figure 2 *The structure of the main component of mauveine*

Bonding to silk

The structural formula of mauveine illustrates another key link between the properties of the dye and the structure of the molecule. You can see that the molecule has an extensive delocalised system which provides the chromophore for the dye. However, the presence of the positive nitrogen atom is also crucial. This charged group enables the molecule to bond ionically to negatively charged $-COO^-$ groups on the amino acid side chains of the protein fibres that make up silk. This ensures that the dye remains attached firmly to the silk fibres. Dye molecules like mauveine are described as cationic or basic dyes, as they are applied in alkaline conditions, ensuring that the $-COOH$ groups are found in their ionised form.

Designing colourfast dyes

In order to ensure that dyed fabric retains its colour (is 'colourfast'), the bond between the dye and the fabric needs to be as strong as possible. The dream of dye chemists in the 20th century was to develop dyes that were held to fibres by strong covalent bonds.

Dyes that had been developed for cotton and cellulose relied on hydrogen bonding to attach the dye to the fabric, so it was difficult to ensure that they remained colourfast.

In the early 1950s, a group of chemists working at ICI's research laboratories in Blackley started to modify some azo dye molecules by adding a group of atoms known as trichlorotriazine (Figure 3).

Trichlorotriazine is readily attacked by nucleophilic groups such as amine or hydroxyl groups, so it can attach easily to azo dye molecules containing amine groups. The resulting modified dye molecule is then able to react with the hydroxyl groups in **cellulose**-based fibres such as cotton. The overall effect is that it forms a strong covalent bridge between a dye molecule and a fabric (Figure 4).

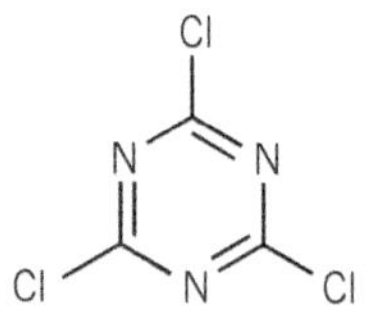

▲ Figure 3 *Trichlorotriazine*

Cellulose

Cellulose is a natural polymer found in plant cells, consisting of chains of glucose molecules joined together.

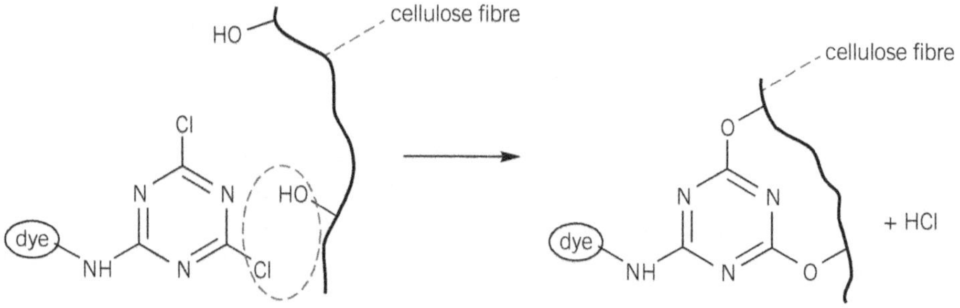

◀ Figure 4 *The formation of bonds between a fibre-reactive dye and cellulose*

Fibre-reactive dye

A fibre-reactive dye molecule contains a reactive group of atoms that can bond covalently to molecules of the fibre.

The same principle can be used to design **fibre-reactive dyes** to attach to proteins such as silk and wool, which contain amine groups. However, it is more difficult to dye non-polar fabrics such as polyester. The majority of dyes developed in the past 20 years make use of fibre-reactive technology to ensure that they are colourfast.

Synoptic link

In order to understand this section, you will need to remind yourself about the different types of intermolecular bonds (Topic OZ 6, The CFC story), and the structure of polymers such as nylon, polyester, and proteins (which make up fibres such as wool and silk) in Chapter 7, Polymers and life.

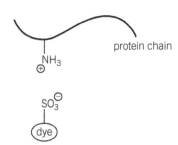

▲ Figure 6 *Aluminium chelates with cotton fibres and alizarin dye under alkaline conditions*

Chelate

A chelate is a complex ion in which the metal ion is bonded to two or more atoms in the same molecule.

Chemical ideas: Structure and properties 5.7

Bonding dyes to fibres

Ionic bonds

Dyes described as acid dyes have a negative charge when dissolved in water, because of the presence of the sulfonate groups. They are applied in acid conditions to fibres, and under these conditions, polyamide and protein fibres will have positively charged NH_3^+ groups present at the end of polyamide chains or on side groups in proteins.

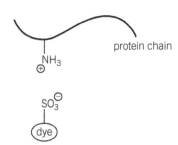

▲ Figure 5 *Ionic bonding between an acid dye and a nylon molecule*

Covalent bonding

There are two strategies for attaching dyes to fabric using covalent bonds:

- Mordanting – this has been in use for thousands of years.
- Adding fibre – reactive groups to the dye molecule – this was developed in the middle of the last century.

Mordanting

This makes use of a metal ion to join the dye to the fabric. Groups on the fabric and the dye form dative covalent bonds to the metal, forming **chelate** complex ions. Metal ions used as mordants include Al^{3+} and Cr^{3+} ions.

Fibre-reactive dyes

These make use of a reactive group attached to the dye molecule to form a bridge between the dye and the fibre.

Intermolecular bonds

Hydrogen bonds

Fibres such as cotton and cellulose (which makes up paper) consist of molecules that contain large numbers of hydroxyl groups. If the dye molecule contains amine groups, or alcohol groups, then it will be able to hydrogen bond to the fibre.

However, it is important that this hydrogen bonding is strong enough so it will not to be broken in the presence of water. This means that dye molecules that attach to fibres in this way are usually linear molecules. They align themselves closely to the fibre molecule and attach in several places (Figure 7). Such dye molecules are often known as direct dyes, to contrast them with dyes that bond to cotton via mordants.

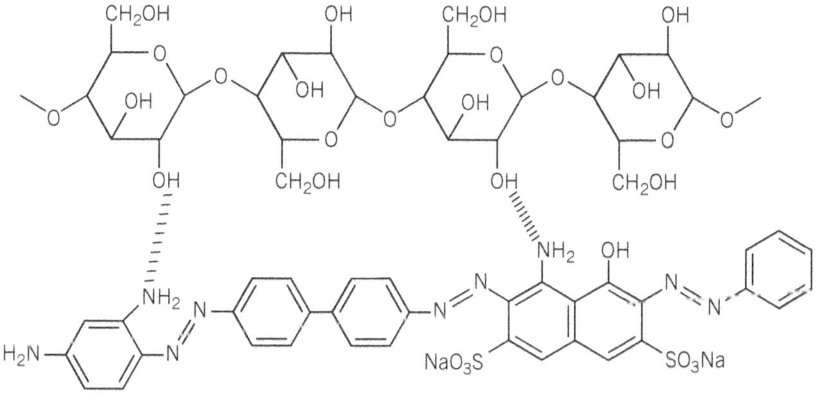

▲ **Figure 7** *This direct dye has a linear shape, enabling it to line up close to a cellulose fibre*

Study tip

You do not need to learn the structures of these molecules. You will be expected to predict the types of intermolecular bonds between molecules when given structures of dyed and fibres.

Instantaneous dipole – induced dipole bonds

Some polymers, such as polyesters, have very few polar groups and can only bond to dyes using instantaneous dipole–induced dipole bonds.

These dyes are known as disperse dyes (disperse forces is an alternative description of instantaneous dipole–induced dipole forces), and typically have very few polar groups, so they will not be soluble in water. So, whereas most dyes are dissolved in water in order to carry out the dyeing process, these dyes are

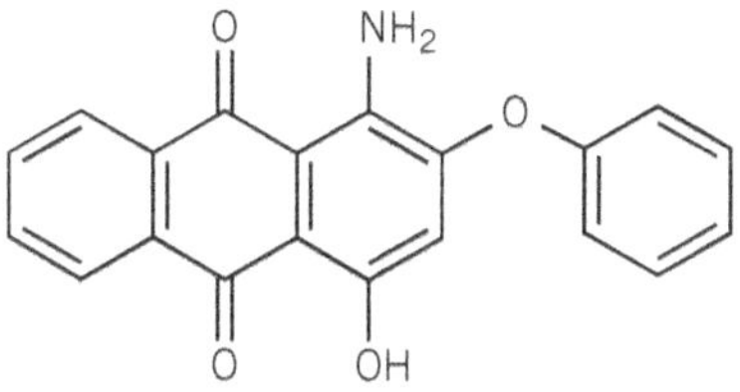

▲ **Figure 8** *The structure of Disperse Red 60 dye, showing its relatively small size and non-polar structure*

suspended (dispersed) in water. When the mixture of dye and water is added to a fabric, the small dye molecules are able to diffuse into the fabric and are held there by weak instantaneous dipole–induced dipole forces between dye and fabric.

Summarising dye – fibre interactions

Table 1 shows the features of dye molecules and polymer molecules which create the different types of interaction that attach dyes to fibres.

Study tip

To decide what type of bonding is likely to exist between a particular pairing of dye molecule and fabric, look for specific structural features on the molecules.

▼ **Table 1** *Features of dye molecules and polymer molecules*

Type of interaction	Features of dye molecule	Feature of polymer molecule	Example of fabric dyed by this interaction
ionic bonds	SO_3^- groups	NH_3^+ groups in acidic solution	nylon, wool, silk
covalent bonds (fibre-reactive dyes)	presence of reactive group, e.g., triazine derivatives	OH or NH_2 groups	cotton, cellulose

▼ Table 1 (continued)

Type of interaction	Features of dye molecule	Feature of polymer molecule	Example of fabric dyed by this interaction
hydrogen bond	several NH$_2$ groups linear molecule	frequent OH groups	cotton
instantaneous dipole–induced dipole	few polar groups small molecule	no OH or NH groups	polyester

Activity CD 6.1

This practical will help develop your understanding of the ways in which dyes attach to fabrics.

Summary questions

1 Name a group present on dye molecules which enables them to form ionic bonds to positively charged fibres. *(1 mark)*

2 Which type of fibre is most likely to be attached to dye molecules using instantaneous dipole–induced dipoles? *(1 mark)*

3 Look at the following structures A–C which show three dye molecules, and D–F which show three polymer molecules. Match each dye molecule with the polymer to which it is most likely to attach, giving reasons for your answers.

A

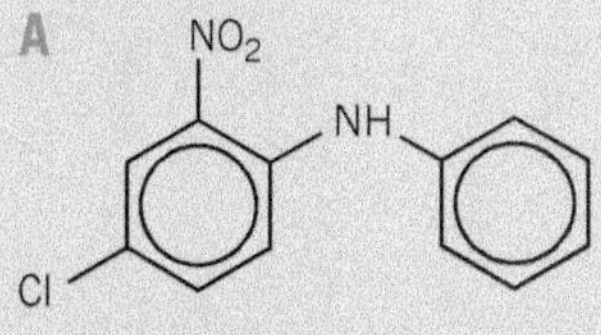

B

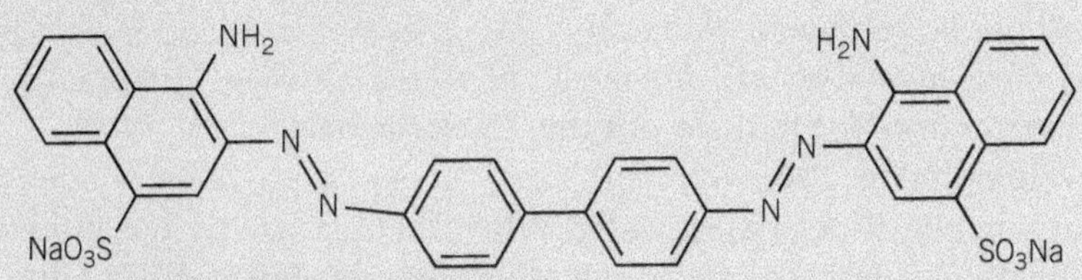

C

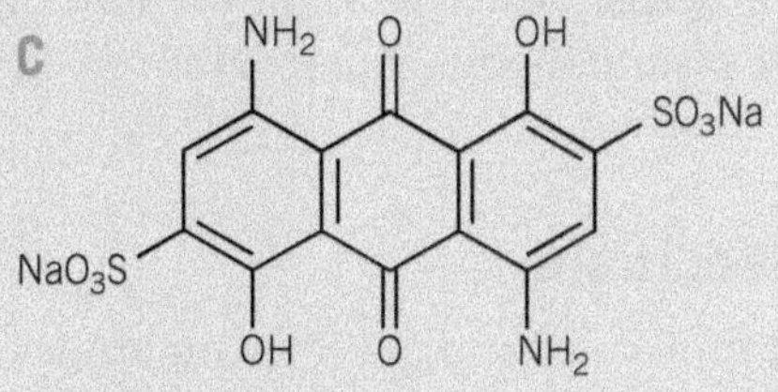

D

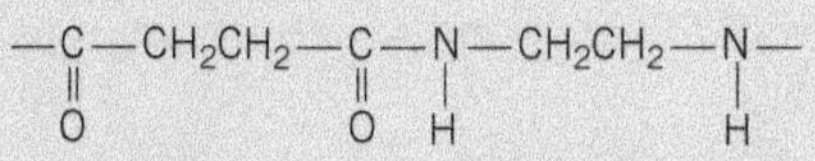

E

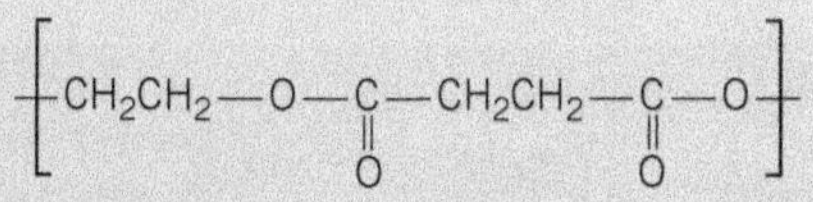

F

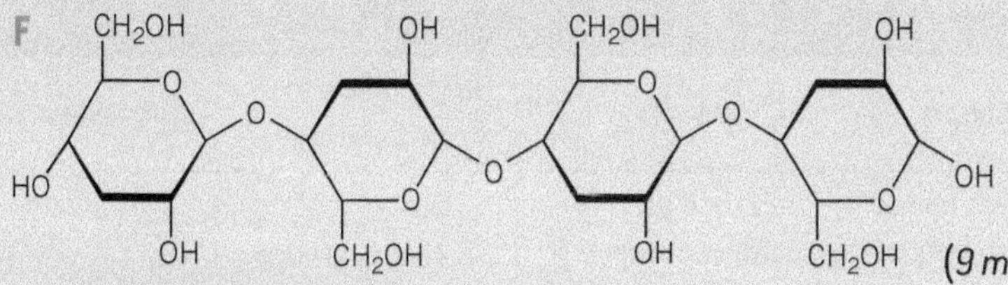

(9 marks)

4 Explain, using ideas about intermolecular forces, why disperse dyes such as Disperse Red 60 (Figure 8) are not soluble in water. *(4 marks)*

CD 7 Fats and oils

Specification reference: CD(c)

Introduction

Molecules containing chromophores are present in many spices, such as turmeric and paprika, and using these spices in cooking imparts colour as well as flavour. These spices often contain non-polar molecules, and these dissolve well in the oils and fats that are often present in food.

What is in your curry?

The attractive appearance of dishes from an Indian restaurant owe their vibrant colours to the presence of coloured molecules from spices such as turmeric (Figure 1).

Curcumin

Turmeric contains the coloured molecule curcumin, and its structure is shown in Figure 2.

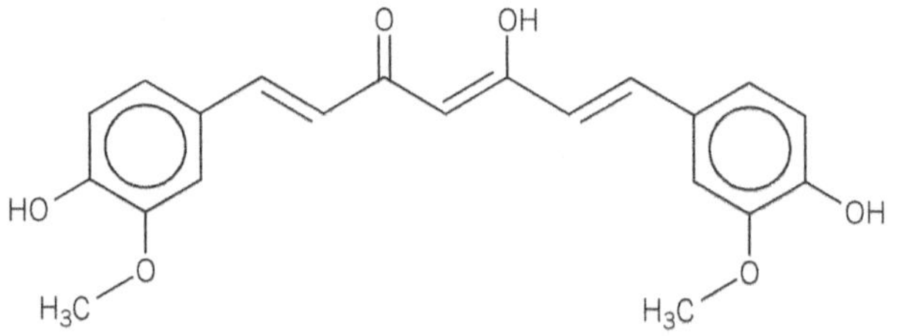

▲ **Figure 2** *Curcumin, the coloured molecule present in turmeric*

The relatively large non-polar region of the molecule means that the interactions between it and water molecules are not strong enough for it to dissolve well in water. However, it dissolves well in oils and fats, such as the coconut oil traditionally used in Indian cooking. These consist of **triglyceride** molecules, which possess long non-polar chains (Figure 3).

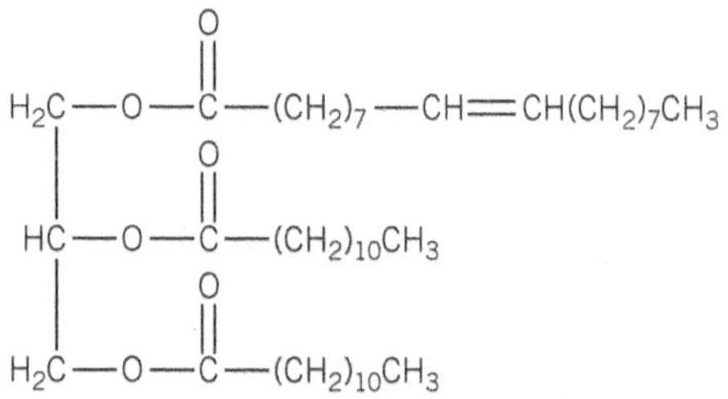

▲ **Figure 3** *Coconut oil is a triglyceride, typically containing two fatty acid chains derived from lauric acid and a fatty acid chain derived from oleic acid*

The instantaneous dipole–induced dipole bonds formed between the triglyceride molecule and curcumin molecules are similar in strength to those in the separate substances, and so the curcumin molecule dissolves in the triglyceride.

Learning outcomes

Demonstrate and apply knowledge and understanding of:

→ the structure of fats and oils, consisting of mixed esters of propane-1,2,3-triol with varying degrees of nonsaturation.

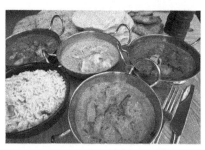

▲ **Figure 1** *Colourful dishes from an Indian restaurant*

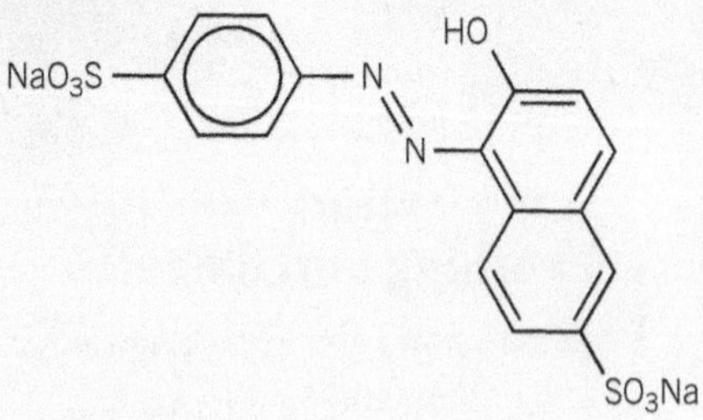

Sunset yellow

Spices are expensive, so some food manufacturers have attempted to reduce costs by replacing the natural colour produced by spices with similar colours that are produced by azo dyes. One such molecule is known as Sunset Yellow (Figure 4).

The use of azo dyes in food has been linked to a number of health problems, including allergic reactions and hyperactivity in children.

▲ **Figure 4** *The azo dye known as Sunset Yellow*

Chemical ideas: Organic chemistry: modifiers 13.10

Oils and fats

Oils and fats: examples of esters

Oils and fats provide an important way of storing chemical energy in living systems. Most of them have the same basic structure. The only difference is that oils are liquid at room temperature, whereas fats are solid.

Most oils and fats are esters of propane-1,2,3-triol (commonly called glycerol) with long-chain carboxylic acids, R—COOH. In lauric acid, for example, R is a $CH_3(CH_2)_{10}$ group.

Because glycerol has *three* alcohol groups in each molecule, three carboxylic acid molecules can form ester linkages with each glycerol molecule to form a triester (often called a triglyceride). The structure of a two triesters are shown in Figure 5.

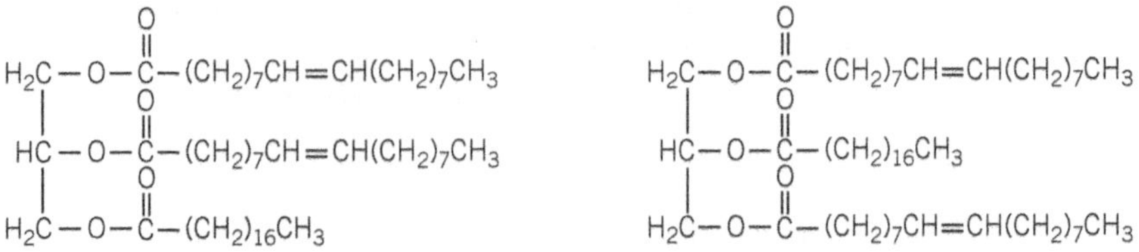

> **Triglyceride (also known as triester)**
>
> A triglyceride is an ester molecule formed by the reaction of one molecule of glycerol (propan-1,2,3-triol) and three fatty acid molecules.

> **Synoptic link**
>
> You can remind yourself of the structure and chemistry of esters by reading Chapter 5, What's in a medicine?

> **Study tip**
>
> There is no need for you to learn how t name these triesters.

1,2-dioleoyl-3-stearoylglycerol   1,3-dioleoyl-2-stearoylglycerol

▲ **Figure 5** *The structure of a triester, formed from glycerol and fatty acids*

The triesters found in natural oils and fats are often mixed triesters in which the three acid groups are not all the same. You saw this in the structure of the triester present in coconut oil (Figure 3).

The presence of the long, non-polar fatty acid chains in fats and oils means that these substances do not mix with water.

> **Activity CD 7.1**
>
> In this activity you can use molecular modelling to check your understanding of fats and oils.

Fats and fatty acids

The carboxylic acids in fats and oils usually have unbranched hydrocarbon chains. They contain an even number of carbon atoms, ranging to C_{24}, but often contain 16 or 18 carbon atoms. They are sometimes called fatty acids because of their origin. The alkyl groups, R^1, R^2, and R^3, are either fully saturated or contain one or more double bonds. Table 1 shows some common fatty acids. There are many others, some of which contain three or four double bonds.

▼ **Table 1** *Common fatty acids*

Structure	Traditional name	Origin of name
$CH_3(CH_2)_{14}COOH$	palmitic acid	palm oil
$CH_3(CH_2)_{16}COOH$	stearic acid	suet
$CH_3(CH_2)_7CH=CH(CH_2)_7COOH$	oleic acid	olive oil
$CH_3(CH_2)_4CH=CHCH_2CH=CH(CH_2)_7COOH$	linoleic acid	oil of flax

Study tip

There is no need to remember any of these structures.

Hydrolysis

An oil or fat can be identified by breaking it down into glycerol and its fatty acids, and then measuring the amount of each acid present.

Like all esters, oils and fats can be split up by hydrolysis. This is usually done by heating the oil or fat with concentrated sodium hydroxide solution to give glycerol and the sodium salts of the (Figure 6).

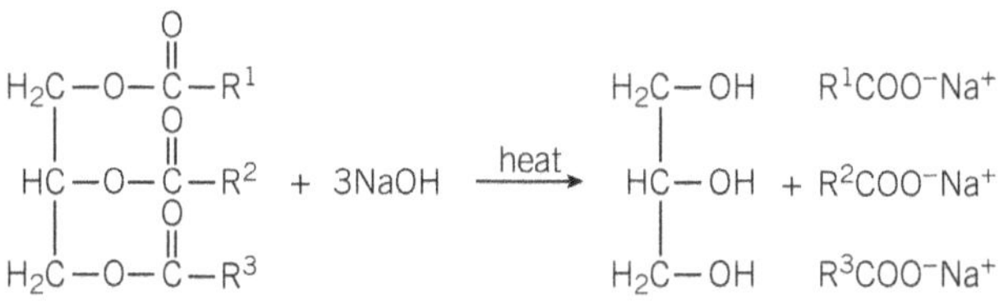

▲ **Figure 6** *The hydrolysis of a triester using sodium hydroxide*

Synoptic link

You first met salts in Chapter 1, Elements of life.

The free fatty acids can be released from the sodium salts by adding a dilute mineral acid, such as hydrochloric acid, for example:

$$RCOO^-Na^+ + HCl \rightarrow RCOOH + NaCl$$

Summary questions

1 Explain why fats and oils are described as triesters. *(2 marks)*

2 Draw out the structure of the triester formed when glycerol (propane-1,2,3-triol) combines with one molecule of linoleic acid and two molecules of palmitic acid. *(2 marks)*

3 Look at the structure of the triglyceride present in coconut oil (Figure 3). Write an equation for the hydrolysis of this molecule, using sodium hydroxide. *(2 marks)*

4 Oils and fats do not mix with water. Use ideas about the structure of these molecules to explain this fact. *(4 marks)*

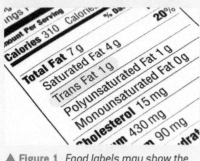

▲ **Figure 1** *Food labels may show the proportion of trans fats present in food products*

Introduction

Gas–liquid chromatography is an important analytical technique used in a wide range of situations. One important way in which it is used is in the testing of foods. In the process of finding out how the technique works, you will look at how it may be used to detect and measure the *trans* fats which are a controversial ingredient of some foods.

Trans fats and food testing

Food manufacturers are now required by law to include details of the proportion of fats, sugars, and other food components on the labels of the products that are sold to the public. Some manufacturers go further and give details of the different types of fat present in the food. You may see labels like the one shown in Figure 1. One of the most controversial components shown on this label is *trans* fat.

Forming *trans* fats

Solid or liquid?

Natural oil or fat contains a mixture of triglycerides. The nature of the fatty acids present affects the properties of the triglyceride and determines whether the substance is a liquid oil or a solid fat. The shape of saturated and unsaturated triglyceride molecules can be represented as in Figure 2.

saturated triglyceride

unsaturated triglyceride

▲ **Figure 2** *The shapes of triglycerides containing saturated fatty acids and some unsaturated fatty acids*

The saturated triglyceride molecules can pack together closely, with intermolecular bonds holding the molecules together. The unsaturated triglyceride molecules cannot pack so closely because the presence of Z (or *cis*) double bonds causes the molecules to 'kink'. This means that the attractive bonds between molecules will be weaker than between the saturated molecules. As a result, the more unsaturated fatty acid molecules there are in a triglyceride, the more likely it is to be a liquid at room temperature, such as an oil.

Converting oils to fats

Although many natural oils and fats can be eaten directly, most of them require processing in order to make them fit a number of requirements regarding taste, texture, and so on. One of the commonest processes is the hydrogenation of unsaturated oils to make margarine. Because the intermolecular bonds are stronger between these hydrogenated molecules, the margarine which contains these hydrogenated fats becomes solid.

The addition of hydrogen to unsaturated molecules reduces the number of double bonds, that is, the molecules become less unsaturated and more saturated. Fats which still contain one or more double bonds after hydrogenation are described as partially hydrogenated.

Forming *trans* fats

The partially hydrogenated fats produced artificially are not quite identical to naturally occurring unsaturated fats. One key difference is that any double bonds in the artificial molecules will be the E (*trans*) isomer, rather than the Z (*cis*) isomer which occurs in natural fats.

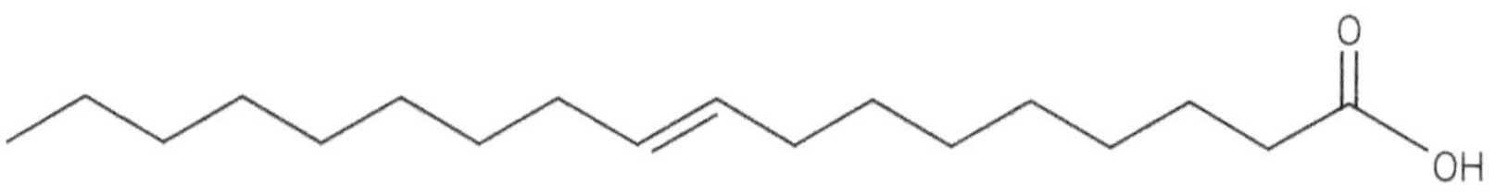

Synoptic link

Hydrogenation is an example of an addition reaction. You can read about the addition reactions of alkenes in Topic DF 6, Alkenes — versatile compounds.

E (*trans*) isomer (elaidic acid) formed during hydrogenation

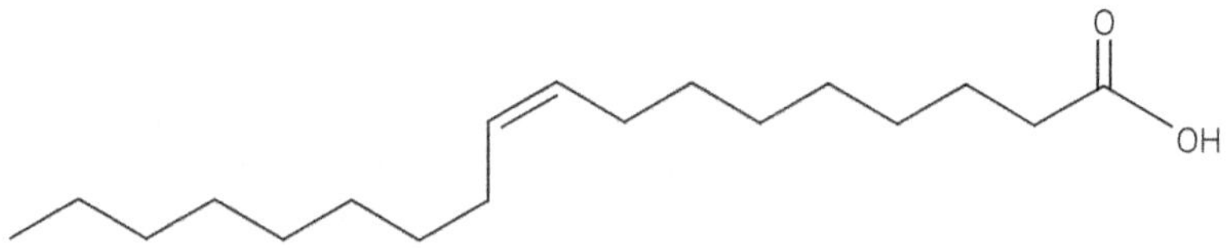

Z (*cis*) isomer (oleic acid) occurring in natural fats

▲ **Figure 3** *E/Z isomers of the fatty acid with molecular formula* $C_{18}H_{34}O_2$

There is also evidence to suggest that naturally occurring Z (*cis*) fatty acids may be converted into E (*trans*) fatty acids when they are repeatedly heated to high temperatures. This may happen if cooking oil is continually reused or recycled.

Health worries about *trans* fats

Hydrogenation provides an economic way of producing fats with suitable properties for use in cooking. It used to be thought that reducing the amount of saturated fats present in food would also mean that the health risks associated with eating fatty food would be reduced. However, over the last 20 years, a large number of studies concluded that the *trans* fatty acids present in partially hydrogenated fats were themselves a serious health risk, since they contribute to the build up of LDL cholesterol in blood, linked to the development of heart disease and strokes.

Food testing

Trans fats

Food testing agencies carry out testing and research to monitor levels of *trans* fats in food. One method that can be used is **gas–liquid chromatography (GLC)**, sometimes referred to as just gas chromatography. In order to be analysed using this technique, fats are hydrolysed into their component fatty acids, and these fatty acids are converted into their methyl esters. The resulting compounds are relatively volatile, so that they vaporise in the conditions used in the gas–liquid chromatograph. Esters of *trans* and *cis* fatty acids will have different boiling points, so they are detected at different times.

Food colours

As you saw in the previous topics, some food dyes may be associated with health issues as well. These dye molecules are often not volatile enough to detect using GLC, so a related technique, HPLC (high-performance liquid chromatography) is used instead. This uses high pressure but lower temperatures, and relies more on solubility differences rather than boiling point differences to separate the molecules.

Gas–liquid chromatography, GLC

This is an analytical technique used to separate and identify the components of a mixture. Compounds are separated because they distribute themselves differently between a gas phase and a liquid phase.

Synoptic link

The principles of thin-layer chromatography were explained in Topic WM 5, The synthesis of salicylic acid and aspirin, and in Techniques and procedures.

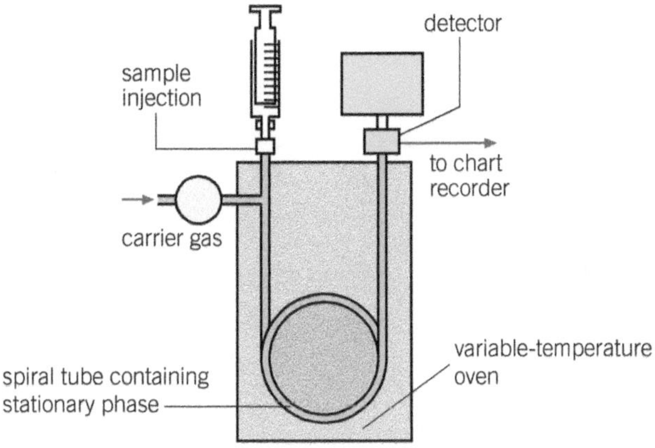

Chemical ideas: Equilibrium in chemistry 7.7

Gas–liquid chromatography

Principles

The principle in gas–liquid chromatography (GLC) is the same as in thin-layer chromatography. However, the stationary and mobile phases are very different to those used in thin-layer chromatography:

- The mobile phase is an unreactive gas, such as nitrogen, called the carrier gas.
- The stationary phase is a small amount of liquid with a high boiling point, held on a finely divided inert porous solid support.

This material is packed into a long thin tube called a column. The column is coiled inside an oven.

The complete apparatus used to carry out gas–liquid chromatography is known as a gas–liquid chromatograph and is shown in Figure 4.

How it works

The sample to be analysed is injected into the gas stream just before it enters the column. The components of the mixture are carried through the column in a stream of gas.

▲ **Figure 4** *The design of the apparatus used to carry out gas–liquid chromatography (gas–liquid chromatograph)*

Separating the components

Each component has a different affinity for the stationary phase compared with the mobile phase. Components will have different solubilities in the liquid stationary phase. Some components will dissolve in the liquid stationary phase (to different amounts) and some components will remain in the gaseous mobile phase. Each component distributes itself to different extents between the two phases, and so emerges from the column at different times.

The compounds that favour the mobile phase (the carrier gas) are carried along more quickly than components that dissolve in the stationary phase. The most volatile compounds usually emerge first. The compounds that favour the stationary phase (the liquid) and dissolve in it get held up in the column and come out last. The greater the affinity a compound has in the stationary phase, the more slowly it will be carried through the chromatograph.

The distribution of the components of a sample between the liquid stationary phase and the gaseous mobile phase is described using the partition coefficient K_{pc}.

$$K_{pc} = \frac{[\text{component in stationary phase}]}{[\text{component in mobile phase}]}$$

The greater the partition coefficient, the greater affinity the component of the sample has for the stationary phase.

Detecting the components

A detector on the outlet tube monitors the compounds coming out of the column. Signals from the detector are plotted out by a recorder as a **chromatogram**. This shows the recorder response against the time which has elapsed since the sample was injected onto the column. Each component of the mixture gives rise to a peak. The technique is very sensitive and very small quantities can be detected, such as traces of explosives or drugs in forensic tests. With larger instruments, a pure sample of each compound can be collected as it emerges from the outlet tube. In more sophisticated instruments, the outlet tube is connected to a mass spectrometer so that each compound can be identified directly.

Figure 5 shows a typical gas chromatogram.

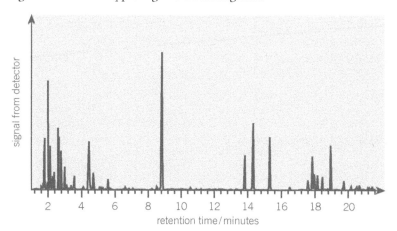

▲ **Figure 5** *A typical gas chromatography trace (gas chromatogram), where each peak represents a different compound*

Synoptic link

You have come across other equilibrium constant K values such as K_c in Chapter 3, Elements from the sea, and K_a in Chapter 8, Oceans.

Activity CD 8.1

This activity checks your understanding of gas–liquid chromatography through looking at the analysis of oils used in paints.

Synoptic link

You covered mass spectrometry in Topic WM 4, Mass spectrometry.

Retention time

The time that a compound is held in a column, under given conditions, is characteristic of the compound and is called its **retention time**.

It is important to calibrate a GLC instrument using samples of known compounds and to keep the conditions constant throughout the analysis, since small changes in conditions such as temperature will affect the observed retention time.

Quantitative analysis

The area under each peak depends on the amount of compound present, so you can use a gas chromatogram to work out the *relative amount* of each component in the mixture. If the peaks are very sharp, their relative heights can be used.

Summary questions

1 Describe the main features of the stationary phase used in GLC.

(*3 marks*)

2 Explain why it is important to enclose the column of a gas–liquid chromatograph inside an oven kept at a constant, known temperature. (*2 marks*)

3 The chromatogram in Figure 6 was obtained from a mixture containing a pair of *Z* and *E* isomers.

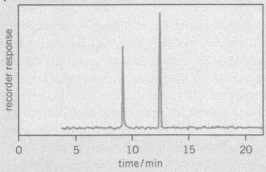

▲ **Figure 6** *A gas chromatogram for a mixture containing a pair of E/Z isomers*

a The pure *Z* isomer had a retention time of 9.0 min under the same conditions. The pure *E* isomer had a retention time of 12.5 min. Use the heights of the peaks to find the percentages of the *Z* and *E* isomers in the mixture. (*2 marks*)

b Explain how mass spectrometry could be used to confirm that these two compounds are isomers of each other. (*2 marks*)

4 Look at the structures of the *E* and *Z* fatty acid isomers shown in Figure 3.

a Draw out the structures of the methyl esters of these isomers. (*2 marks*)

b Explain why the methyl esters are used in the GLC analysis of these compounds, rather than the fatty acids themselves. (*2 marks*)

c Suggest, with a reason, which isomer will give rise to the GLC peak with the longest retention time. (*2 marks*)

CD 9 Carbonyl compounds and organic synthesis

Specification references: CD(i), CD(k)

Introduction

In this topic you will start to think about how modern chemists design synthetic routes to complex molecules, such as dyes and pharmaceuticals. One of the most important types of synthetic reaction is the formation of carbon–carbon bonds, and you will learn about an important reaction of aldehydes and ketones that enables this to happen.

Organic synthesis

The revolution in synthetic organic chemistry that began with the development of synthetic dyes soon extended to the synthesis of other important molecules, such as aspirin.

Forming carbon–carbon bonds

The synthesis of aspirin and dyes demonstrates one important feature of synthetic routes – the modification of functional groups.

A second important feature is the modification of the carbon skeleton. This will involve the formation of carbon–carbon bonds, which is not easy to do. One method of extending the carbon chain length is the Friedel–Crafts reaction, which you covered in Topic CD 4.

Using carbonyl groups

Another way of forming carbon–carbon bonds involves the reaction of a carbonyl group – an aldehyde or ketone. Figure 1 shows the synthesis of mandelic acid, which has some useful medical applications including treating skin complaints. It can be produced using oils extracted from almonds, but there is a simple two-step synthesis which enables it to be formed from benzaldehyde, a common industrial reagent.

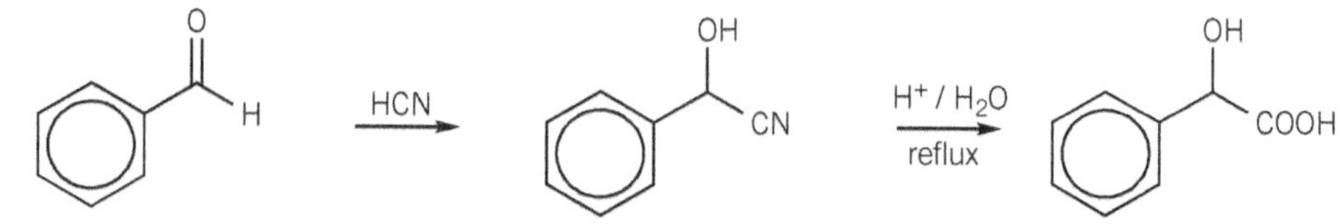

benzaldehyde

mandelic acid

▲ **Figure 1** *The synthesis of mandelic acid from benzaldehyde*

Cyanide ions in synthesis

The cyanide group –CN in the HCN molecule may not be familiar to you. It has the structure C≡N, and is known as the **nitrile** functional group when present in organic molecules.

Learning outcomes

Demonstrate and apply knowledge and understanding of:

→ reactions involving carbonyl compounds:
- oxidation of aldehydes to carboxylic acids
- reaction with Fehling's solution and Tollens' reagent
- reaction with hydrogen cyanide to form cyanohydrin

→ the mechanism of the nucleophilic addition reaction between a carbonyl compound and HCN, using curly arrows and bond polarities.

Synoptic link

You can remind yourself of the synthesis and preparation of aspirin in Topic WM 5, The synthesis of salicylic acid and aspirin.

The great advantage for the organic chemist of introducing nitrile groups into molecules is that it provides a way of lengthening the carbon chain in a molecule. The nitrile group can then be hydrolysed into a carboxylic acid group (as in the synthesis of mandelic acid) or reduced to an amine group by powerful reducing agents:

$$R{-}C{\equiv}N \rightarrow R{-}CH_2{-}NH_2$$

Chemical ideas: Organic chemistry: modifiers 13.11

Aldehydes and ketones

The carbonyl group

Synoptic link

You first met the structure of the carbonyl group in Topic WM 1, The development of modern ideas about medicines.

Aldehydes and ketones are two types of organic compounds that contain the carbonyl group, $C{=}O$.

Reactions of aldehydes and ketones

Oxidation

Synoptic link

You need to be able to name aldehydes and ketones. This was covered in Topic PL 1, The polyester story.

Many of the reactions of aldehydes and ketones are similar because they both contain a carbonyl functional group. However, there are key differences because aldehydes contain a hydrogen atom attached to the carbonyl group. This means that aldehydes will react with a range of oxidising agents, whereas ketones will not.

Distinguishing aldehydes and ketones

It is possible to tell a sample of an aldehyde apart from a sample of a ketone by heating the sample under reflux with acidified potassium dichromate. The aldehyde sample will cause the acidified dichromate solution to turn from orange to green. However, because acidified dichromate also oxidises primary and secondary alcohols, alternative tests as discussed below are used to positively identify aldehydes.

Fehling's test

Aldehydes are oxidised to carboxylic acids by Fehling's solution. A mixture of Fehling's A and B is warmed (in a water-bath) with an aldehyde and a colour change from a blue solution to a brick-red precipitate is observed.

▲ Figure 2 *When an aldehyde is warmed with Fehling's solution, the blue colour will gradually turn to a brick-red colour as a precipitate of copper(I) oxide forms*

Fehling's solution is made from a mixture of two solutions – Fehling's A, which contains the weak oxidising agent $Cu^{2+}(aq)$, and Fehling's B, which is an alkaline solution containing a bidentate ligand. The blue $Cu^{2+}(aq)$ is reduced to copper(I) ions, which are precipitated as copper(I) oxide, Cu_2O, forming the brick-red precipitate.

Activity CD 9.1

This practical helps check your understanding of the reactions of carboyl compounds.

Fehling's solution is always made freshly in the laboratory by mixing Fehling's A and B together.

Tollens' reagent

Aldehydes are oxidised to carboxylate ions by Tollens' reagent.

- Tollens' reagent is warmed with an aldehyde in a test tube (using a water-bath).
- A silver mirror forms on the inside of the test tube.

Tollens' reagent is formed by dissolving silver(I) oxide, Ag_2O, in ammonia solution to form the $[Ag(NH_3)_2]^+(aq)$ complex ion. When Tollens' reagent is added to an aldehyde, the silver(I) ions are reduced to elemental silver, which appears as a silvery layer on the inner surface of the test tube (Figure 3).

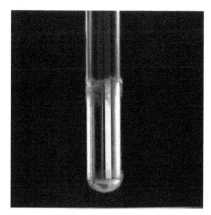

▲ Figure 3 *The formation of a 'silver mirror' when Tollen's reagent is warmed with an aldehyde*

Reduction

Aldehydes and ketones can both be formed by oxidising alcohols. Sometimes, when devising organic synthetic routes, it is necessary to reduce aldehydes and ketones back to alcohols. This does not take place readily and requires a powerful reducing agent. A metal hydride called sodium tetrahydridoborate(III), $NaBH_4$, is often used.

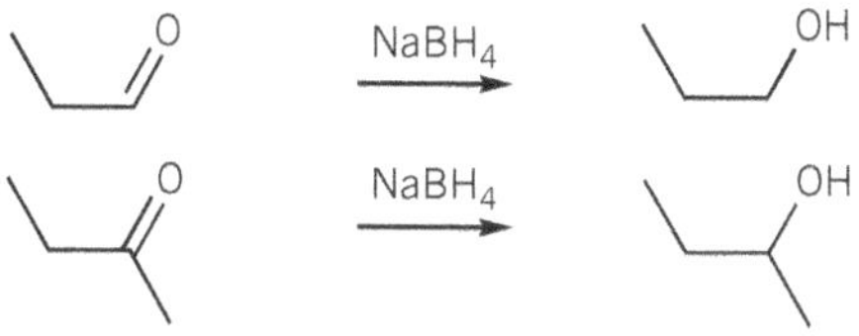

▲ Figure 4 *The reduction of aldehydes and ketones using $NaBH_4$*

Aldehydes are reduced to primary alcohols, and ketones are reduced to secondary alcohols.

Addition reactions

The carbonyl groups in both aldehydes and ketones undergo addition reactions. For example, hydrogen cyanide, HCN, in acidic conditions adds across the C=O bond to form **cyanohydrins** (sometimes called 2-hydroxynitriles).

For ethanal, the overall reaction is:

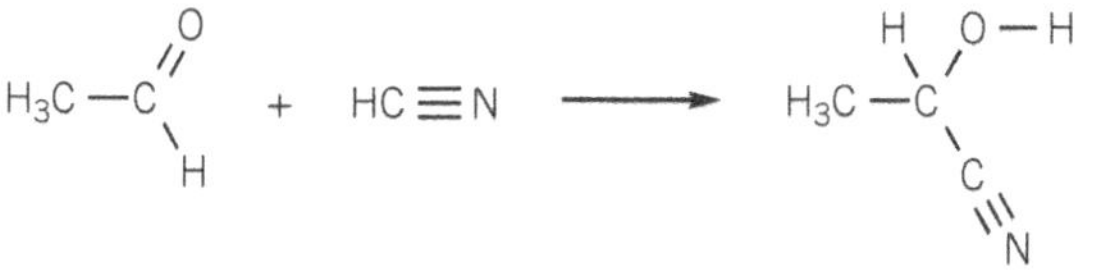

2-hydroxypropanenitrile

For propanone, the overall reaction is:

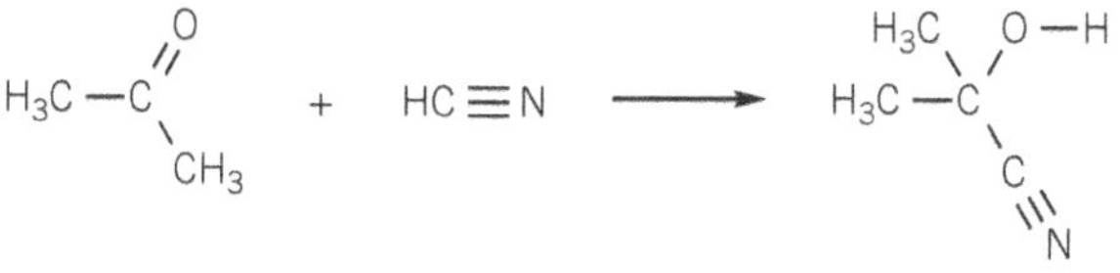

2-hydroxy-2-methylpropanenitrile

Study tip

The reduction of aldehydes and ketones and reactions involving cyanide ions and nitrile groups appear on the data sheet. You do not need to learn these details but you may need to use the information in questions about organic synthesis.

Cyanohydrin

Cyanohydrin is a functional group in which a hydroxyl group and a nitrile group, CN, are attached to the same carbon atom.

Study tip

You need to be able to draw out the structures of these products, and to recognise them as cyanohydrins, but you will not be required to know the system for naming the molecules.

Nitrile

Nitrile is a functional group in which a carbon atom is bonded by a triple bond to a nitrogen, C≡N.

Activity CD 9.2

In this activity you will make a stop-motion animation to explain the mechanism of nucleophilic addition reactions.

Study tip

When drawing out this mechanism, it is very important to make it clear that it is the lone pair of electrons on the carbon atom of the CN^- ion that attacks the $^{\delta+}C$ of the carbonyl. Make sure that the curly arrow starts at this lone pair.

Synoptic link

You may be required to compare the mechanism of this reaction with that of nucleophilic substitution (Topic OZ 8, What is the state of the ozone layer now?) or electrophilic addition (Topic DF 6, Alkenes – versatile compounds).

In practice, hydrogen cyanide is too hazardous to use in the laboratory, so acidified potassium cyanide is often used as a source of CN^- and H^+ ions.

Mechanism of the reaction

The reaction occurs in two stages (Figure 5). The cyanide ion is a nucleophile and it is attracted to the carbon atom in the carbonyl group which carries a partial positive charge. A new carbon–carbon bond forms, and a pair of electrons in the C═O bond moves onto the oxygen atom, which then carries a negative charge. The new negatively charged ion then takes up a proton, H^+, from the solvent – water.

The overall reaction is addition of H—CN across the C═O bond.

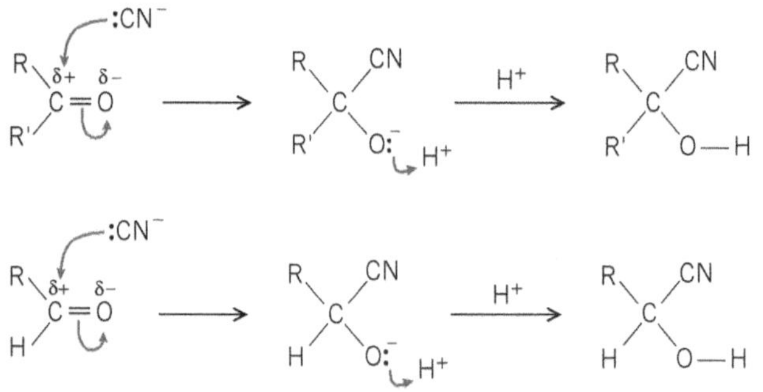

▲ **Figure 5** *The mechanism of nucleophilic addition for a ketone (top) and an aldehyde (bottom)*

The mechanism is described as **nucleophilic addition**, because the first step involves attack by a nucleophile.

The reaction of aldehydes and ketones with HCN is useful in organic synthesis, because a new carbon–carbon bond is formed – it is one way of increasing the length of a carbon chain.

Summary questions

1 A student suspects that a sample of an unknown organic compound is an aldehyde. She decides to use a test using Fehling's solutions A and B to check this.

  a Describe how she should carry out the test. *(2 marks)*

  b What observation would identify the unknown compound as an aldehyde? *(2 marks)*

2 Aldehydes and ketones undergo nucleophilic addition reactions with hydrogen cyanide in the presence of acid. Draw the structural formula for the product of the reaction between hydrogen cyanide and each of the following compounds:

  a $CH_3CH_2CH_2CHO$ *(1 mark)*

  b $CH_3CH_2COCH_2CH_3$ *(1 mark)*

3 Name the class of compounds formed by the reactions in question 2. *(1 mark)*

4 a Draw the mechanism of the reaction between benzaldehdye (Figure 1) and hydrogen cyanide *(5 marks)*

  b Explain why the mechanism is classified as nucleophilic addition. *(3 marks)*

5 Explain why the reaction of carbonyl compounds with hydrogen cyanide is important to chemists designing synthetic routes to form organic molecules. *(3 marks)*

CD 10 Planning an organic synthesis

Specification references: CD(f), CD(j)

Putting it all together – planning a synthesis

Why make organic compounds?

About 7 million organic compounds are known to chemists, and more are constantly being made or discovered. Fortunately, their behaviour can be understood in terms of the functional groups they contain, and the way these functional groups are arranged in space.

New compounds are made by a process of synthesis, a series of reactions that produce a particular molecule (usually called the target molecule). The synthetic chemist is a sort of molecular architect, who plans and carries out strategies for making new and useful substances. There may be many steps in the synthesis, each involving the preparation of a new compound from the previous compound.

Nowadays, most new substances are made in the hope that they will be useful in everyday life. In the pharmaceutical industry, for every medicine that becomes commercially available, many thousands of new compounds are prepared and tested. Minor variations in the chemical structure of molecules, such as in the penicillin range of antibiotics, can result in significant changes in their biological activity.

The principles of synthesis

Starting molecules

The starting molecule for synthesis needs to be one that is available in suitable quantities and at a cost that is low enough to make the synthesis economic.

Starting materials may be derived from crude oil, or may be natural products that can be harvested from living organisms. A synthesis that starts from natural products is called a semi-synthesis. If no products from living organisms are used, it is called a total synthesis.

Synthetic reactions

These may be of two types:

1 functional group interconversion – reactions that convert one functional group into another
2 reactions that alter the carbon skeleton of the molecule (reactions that form or break carbon–carbon bonds).

Case study of synthesis – making penicillins

Penicillin was the first antibiotic, discovered by chance in 1928. Antibiotics are substances which kill bacteria and so are used to prevent or treat infections. In fact, there are a range of naturally occurring penicillin molecules that have antibiotic properties and which all share the same general structure (Figure 1). This is based around a group of atoms known as a β-lactam ring, but with different side chains attached.

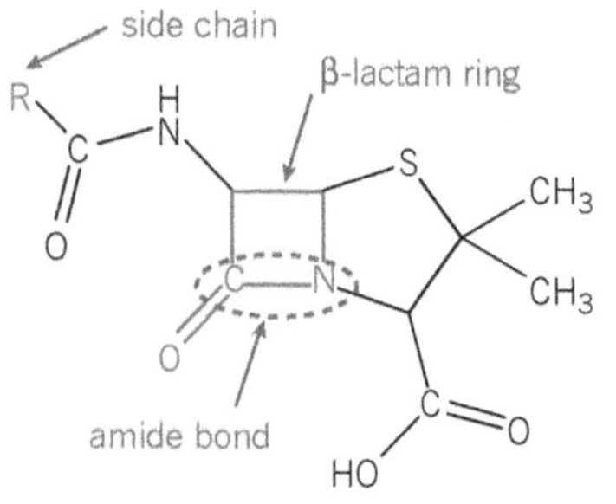

> ### Learning outcomes
>
> Demonstrate and apply knowledge and understanding of:
>
> → naming individual functional groups within a polyfunctional molecule
>
> → making predictions about properties of a polyfunctional molecule
>
> → using organic reactions and reaction conditions from across the whole course to suggest and explain synthetic routes for preparing organic compounds.

▲ Figure 1 *The general structure of penicillins, showing the β-lactam ring and the R side chains*

Since they were first used in the 1940s, millions of lives have been saved by the use of antibiotics. But their widespread use has resulted in the rise of resistant bacteria.

Chemists have attempted to overcome the problem of antibiotic resistance in bacteria by developing a range of new, semi-synthetic penicillins. Starting from the natural product 6-APA, shown in Figure 2, they have modified its structure by attaching a range of different side chains. One such semi-synthetic penicillin is ampicillin (Figure 3).

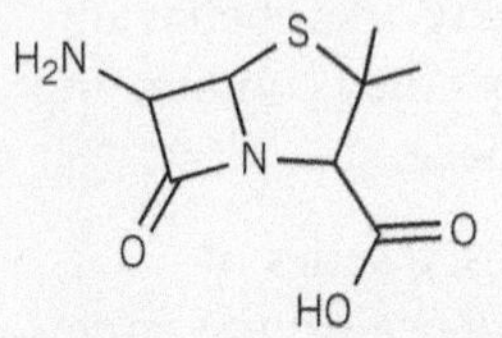

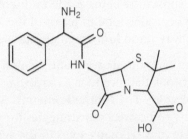

▲ Figure 2 The structure of 6-APA (6-aminopenicillanic acid), the natural product that is the starting material for the production of semi-synthetic penicillins

▲ Figure 3 The structure of ampicillin

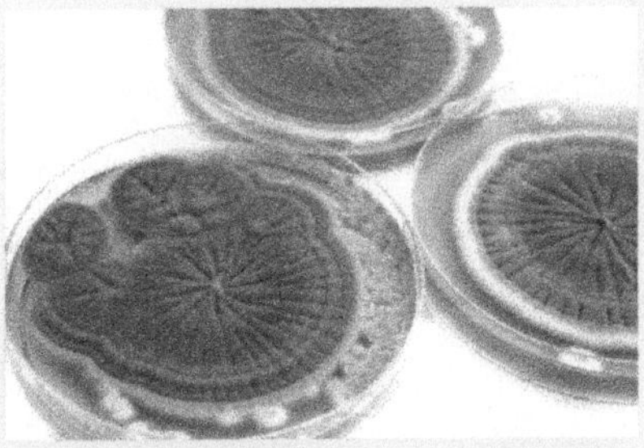

You can see that in order to produce ampicillin, the amine group on the 6-APA molecule must undergo a functional group interconversion to make it into an amide group. From the chemistry that you have encountered, this could be done by reacting it with a suitable acyl chloride.

Acyl chlorides may in turn need to be made from other, more readily available molecules. Your data sheet shows you the details of a reaction that can form an acyl chloride from a carboxylic acid.

Here is a possible two-step synthesis that could be tested in the laboratory to see whether it is a possible route to forming ampicillin.

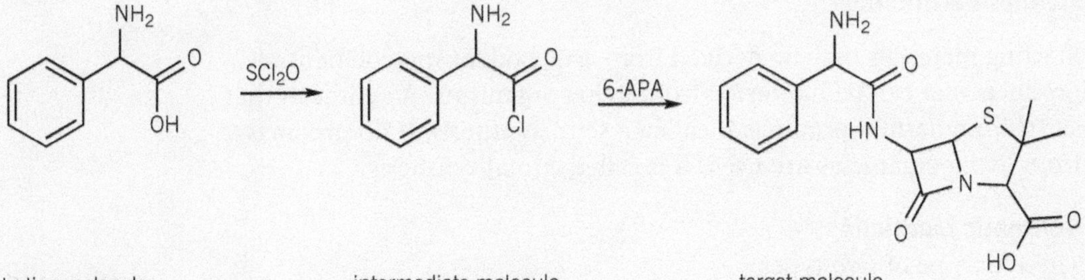

▲ Figure 4 The antibiotics bases on penicillin are all based on a compound originally found in penicillin fungi

starting molecule   intermediate molecule   target molecule

▲ Figure 5 A possible route for the semi-synthesis of ampicillin

The organic starting materials for this two-step synthesis are the natural product 6-APA and 2-amino-2-phenylethanoic acid, which can easily be produced from the amino acid glycine.

Polyfunctional molecules

One of the problems in synthesis is that many of the starting materials or intermediates may have several functional groups – they are **polyfunctional** molecules. It is important to check in each step that the reagents do not react with the other groups present. This may influence the order in which the steps are carried out.

For example, in the first step of the synthesis in Figure 4, the amine group in the intermediate might react with the acyl chloride group in the starting material. This may mean that apparently simple synthetic routes may not work in practice, and longer routes involving many more steps may need to be used.

Another approach is to use enzymes, which are much more selective in the reactions that they catalyse. This is the approach used in the actual industrial synthesis of ampicillin.

Chemical ideas: Organic synthesis 17.1

Functional group reactions

Throughout the course you have encountered a wide range of functional groups.

Table 1 below is a summary of the chemical behaviour of various functional groups.

▼ **Table 1** *A summary of the chemical behaviour of various functional groups and topic references to find out more*

Functional group	Structure	Typical properties of the functional group	Topic reference
alkene	–C=C–	Addition reactions with electrophiles, including halogens, hydrogen halides, hydrogen, and water.	DF 6
haloalkanes	R—X (X = Cl, Br, etc.)	Substitution reactions with nucleophiles, including CN^-.	OZ 8 Data sheet
alcohols	R—OH	Oxidised to carbonyl compounds and/or carboxylic acids.	WM 1
		Esterification reactions with carboxylic acids and acyl chlorides.	WM 1
		Substitution reactions with nucleophiles (halides).	WM 1
		Dehydrated to alkenes.	WM 1
carboxylic acids	R—COOH	Weak acids.	WM 2
		Esterification reactions with alcohols.	WM 1
		Converted into acyl chlorides by SCl_2O.	Data sheet
phenols	C_6H_5—OH	Weak acids.	WM 1
		Esterification reaction with acyl chlorides and acid anhydrides.	WM 2

Study tip

You may need to predict the properties of a polyfunctional molecule. To do this you need to be able to identify the functional groups present in a molecule and predict how each functional group will behave.

Study tip

Use the topic references in Table 1 to learn the properties of each of these functional groups. Some of the reactions are provided on the data sheet. You do not need to learn these, but you may need to use this information in questions.

Study tip

For any given reaction you will need to know reagents, conditions, and products.

▼ Table 1 (continued)

Study tip

It is worth familiarising yourself with the data sheet you will be provided during your assessments.

Functional group	Structure	Typical properties of the functional group	Topic reference
esters	R—COOR	Hydrolysed to carboxylic acids and alcohols.	PL 3
amides	R—CONH$_2$, R—CONHR, etc.	Hydrolysed to carboxylic acid and ammonia.	PL 3
acyl chlorides and acid anhydrides	R—COCl (RCO)$_2$O	Esterification with alcohols and phenols.	WM 2, PL 1
		Amide formation with amines.	PL 2
amines	R—NH$_2$, R—NH—R, etc.	Act as nucleophiles in reactions with acyl chlorides.	PL 2
		Act as bases.	
aldehydes and ketones	R—CHO R—CO—R	Oxidised to carboxylic acids (aldehydes only).	CD 9
		Addition reactions with HCN.	
		Reduced to alcohols by NaBH$_4$.	Data sheet
nitro groups	R—NO$_2$	Reduced to amines by Sn and conc HCl.	Data sheet
nitriles	R—C≡N	Hydrolysed to carboxylic acids by reflux with acids.	Data sheet

Activity CD 10.1

Using a range of qualitative tests will check your understanding of tests for functional groups.

Tests for functional groups

Predicting the behaviour of a polyfunctional molecule may also require you to know some specific tests for the functional groups present. A test should always give a clearly observable change, such as a colour change, or formation of a precipitate or evolution of a gas (Table 2).

▼ Table 2 *Tests for various functional groups*

Functional group	Test	Observation	Notes
alkene	Shake with bromine water or bromine water.	Red-brown colour of bromine decolourises.	
haloalkane	Warm with NaOH(aq), then acidify and add AgNO$_3$(aq).	Precipitate forms.	A white precipitate indicates chlorine. A cream precipitate indicates bromine. A yellow precipitate indicates iodine.
alcohol	Warm with acidified potassium dichromate(VI).	Colour change from orange to green.	Not for tertiary alcohols. Aldehydes also give this reaction.
aldehydes	Add Fehling's solution and warm. or Add Tollens' reagent and warm.	Change from blue solution to red precipitate. Silver mirror forms.	
phenol	Add iron(III) chloride.	Purple colour forms.	Also for any phenolic OH.
carboxylic acid	Add sodium carbonate solution.	Bubbles of gas formed.	

Synthetic reactions

Functional group interconversion

In Table 1 you were reminded of the reactions of the various functional groups that you have encountered.

Many of these reactions are functional group interconversions, which are important in synthetic routes.

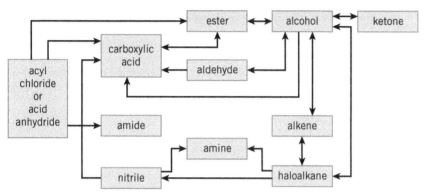

▲ **Figure 6** *Some of the important functional group interconversions*

> **Activity CD 10.2**
>
> In this activity you can add detail to flow charts, like the ones in Figures 5 and 6, to build up a toolkit of useful synthetic reactions.

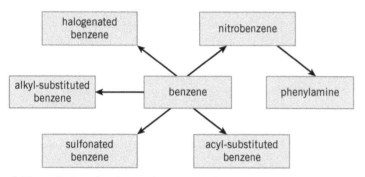

▲ **Figure 7** *Functional group interconversions involving aromatic compounds*

Carbon–carbon bond formation

Some of the functional group interconversions in Figures 6 and 7 result in the formation of carbon–carbon bonds. These are particularly important if the synthesis involves the alteration of the carbon skeleton of the molecule in:

- conversion of aldehydes and ketones into nitriles
- acylation and alkylation of benzene rings using the Friedel–Crafts reaction.

Summary questions

1 The table below shows the results from tests done on two molecules, A and B:

Molecule	Observation on adding iron (III) chloride	Observation on adding sodium carbonate	Observation on warming with Fehling's solution
A	purple colour developed	no reaction	brick-red precipitate
B	no change	bubbles of gas	brick-red precipitate

List the functional group(s) present in these two molecules, according to the results of these tests. (*3 marks*)

2 Look at the molecules shown in Figure 8 which are used as medicines to relieve pain and reduce inflammation.

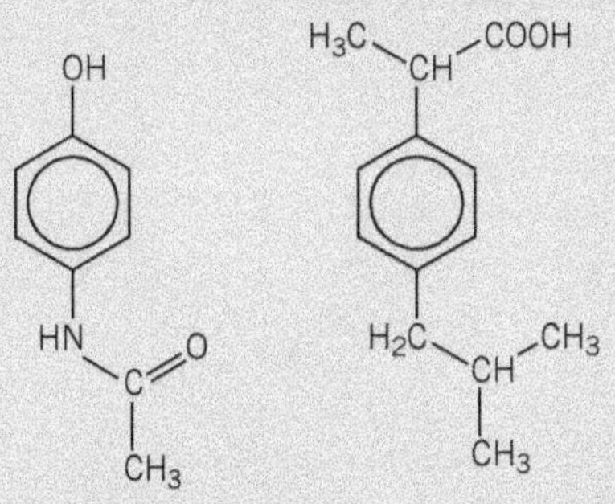

▲ **Figure 8** *The structures of paracetamol (left) and ibuprofen (right)*

a Both these molecules contain a benzene ring. List the other functional groups contained in each molecule. (*3 marks*)
b Describe one way in which the behaviour of these molecules will differ from one another. (*2 marks*)

3 Propan-2-ol was used as the starting material for a three-step synthesis. The details are shown below.

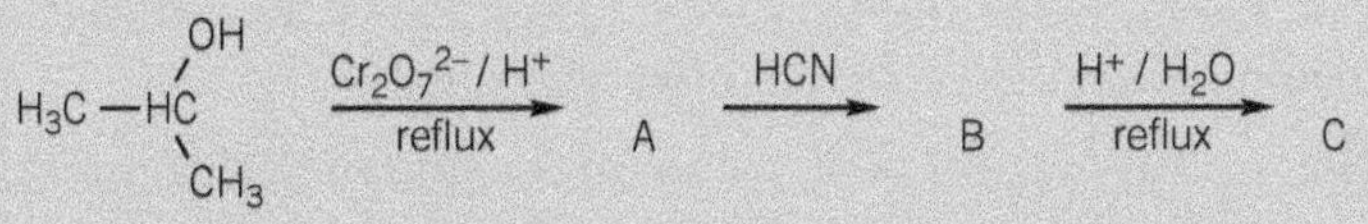

Draw out the structures of molecules A, B and C. (*3 marks*)

4 List the reagents and conditions necessary for each step in the two-step synthetic pathway below. (*5 marks*)

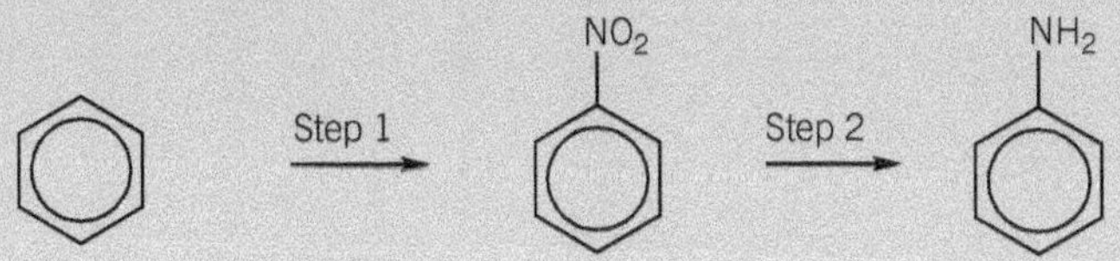

CD 11 Solving synthetic problems

Specification references: CD(j), CD(l)

Introduction

By thinking about the mechanisms of the reactions involved, you will start to appreciate some of the problems that organic chemists might have to solve in designing a synthesis. In the process, you will be reminded of how to classify the different types and mechanisms of organic reactions that you have encountered in this course.

Three synthetic routes

3-aminobenzoic acid from benzoic acid

3-aminobenzoic acid is used as the starting material for the formation of some azo dyes, via a diazotisation reaction. Benzoic acid is a convenient and cheap starting material.

Suggesting a synthesis

From Figure 1, you can deduce that:

- there is no change to the carbon skeleton, so no carbon–carbon bond forming reactions are needed

- an amine group needs to be placed on the benzene ring in a functional group interconversion.

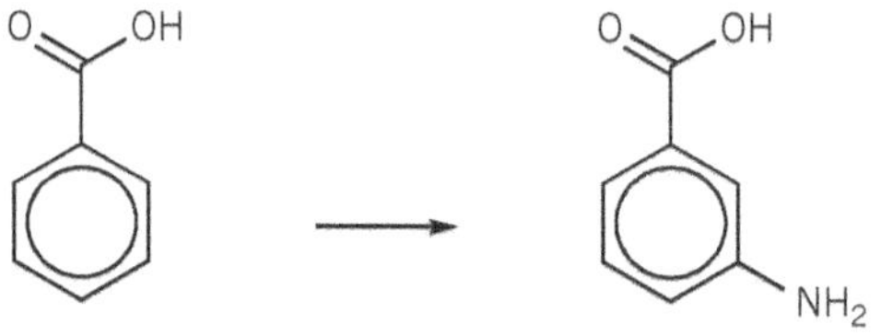

▲ **Figure 1** *Conversion of benzoic acid into 3-aminobenzoic acid*

However, there is no direct way of placing an amine group into a benzene ring.

To work out a synthetic route to achieve this conversion, you can consult the list of functional group interconversions (see Figure 5 in Topic CD 10). Here you will see that it is possible to form an amine group by reducing a nitro group. Nitro groups can easily be substituted into a benzene ring. Figure 2 shows a possible synthesis.

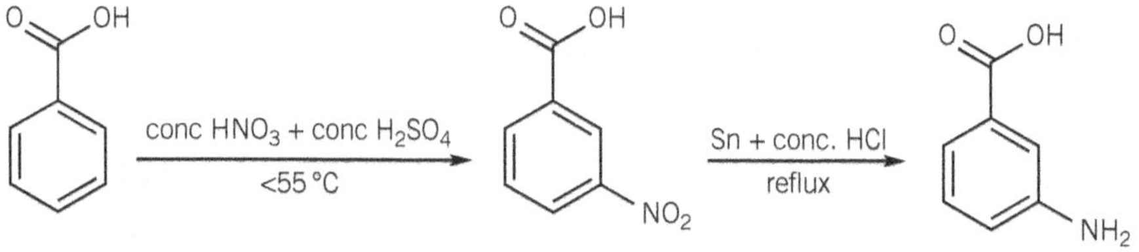

▲ **Figure 2** *The synthesis of 3-aminobenzoic acid*

- Step 1 is a substitution reaction (a nitro group replaces a hydrogen atom).
- Step 2 is a reduction (oxygen atoms are lost and hydrogen atoms are gained).

Substitution reactions and position isomers

Step 1 is a substitution reaction. However, there are several hydrogen atoms on the benzene rings which can be substituted by nitro groups. This means that there are three possible isomers that can be formed in Step 1 (Figure 3).

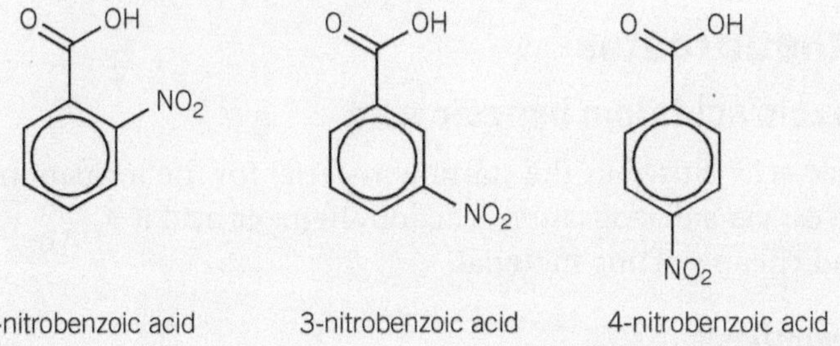

2-nitrobenzoic acid   3-nitrobenzoic acid   4-nitrobenzoic acid

▲ **Figure 3** *The three isomers possible when benzoic acid is nitrated*

About 75% of the product will be the 3-nitrobenzoic acid isomer – the other isomers must be removed by purification.

The synthesis of mandelic acid

You have already seen how mandelic acid is synthesised in Topic CD 9 (Figure 4).

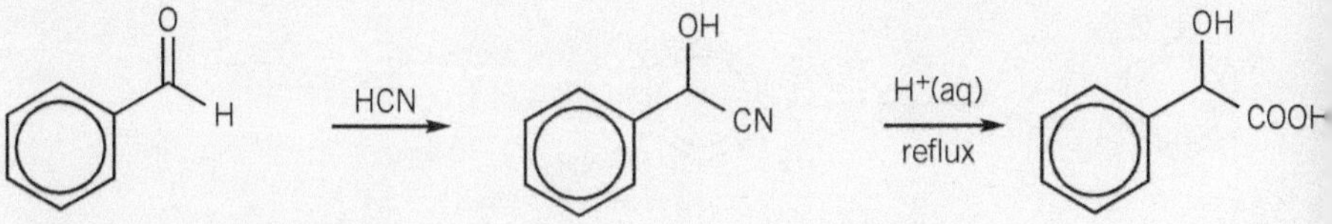

▲ **Figure 4** *The synthesis of mandelic acid*

The synthesis involves:

- a carbon–carbon bond forming reaction in Step 1
- a functional group interconversion in Step 2.

Step 1 is an addition reaction – a molecule of HCN combines with the benzaldehye to form a single product molecule.

Step 2 is a **hydrolysis** – the C≡N bond is broken by the action of water under acidic conditions.

Hydrolysis

Hydrolysis is a reaction in which bonds are broken by the action of water.

Nucleophilic addition reactions and optical isomers

Step 1 in the synthesis of mandelic acid involves addition of a carbonyl group, and the cyanohydrin that forms has a chiral carbon atom. The problem arises because the carbonyl group is planar. This means that the CN⁻ ion can attack from two different sides of the group, giving rise to a mixture of the two different optical isomers, known as a racemic mix:

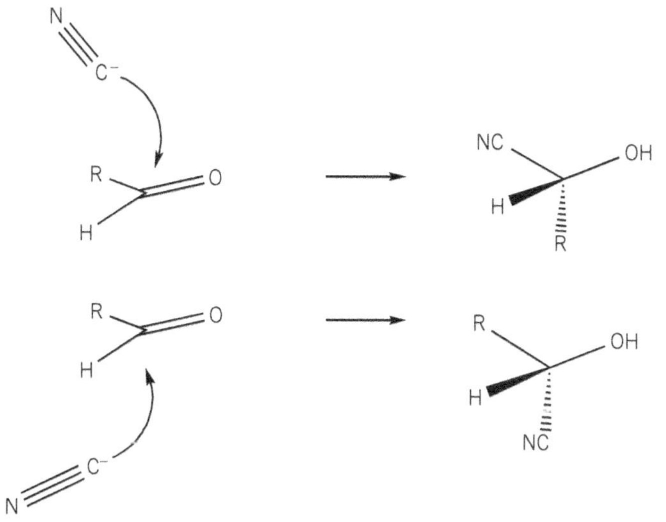

▲ Figure 5 *The formation of a racemic mix from nucleophilic attack on a carbonyl group*

If the racemic mix is then used for Step 2, the product will also consist of a racemic mix. The different isomers may have different biological activities and so may need to be separated, adding to the cost of a synthesis.

Draw the full skeletal formula of the two racemic products formed in Step 2.

Methyl phenylethanoate

Methyl phenylethanoate (Figure 6) is an ester used in the food and perfume industry to produce honey-like smells and flavours. A convenient starting material could be phenylethene (Figure 7).

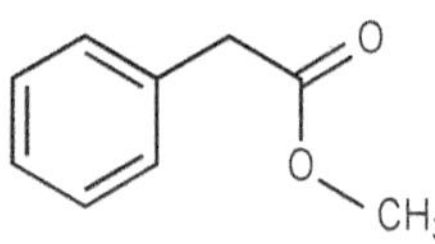

▲ Figure 6 *Methyl phenylethanoate*

▲ Figure 7 *Conversion of phenylethene into methyl phenylethanoate*

- There is no change to the carbon skeleton, so no carbon–carbon bond forming reactions are needed.
- The final product contains an ester. This can be formed in a condensation reaction between an alcohol and a carboxylic acid.

The alkene groups therefore need to be converted into a carboxylic acid. This cannot be done directly. However, looking at Figure 5 in Topic CD 10, carboxylic acids can be formed from alcohols, and alkenes can be converted into alcohols.

You could propose this laboratory synthesis:

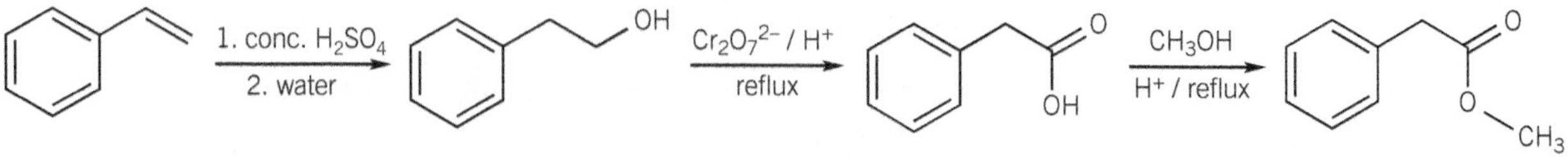

▲ Figure 8 *Synthesis of methyl phenylethanoate*

- Step 1 is an addition reaction.
- Step 2 is an oxidation – oxygen atoms are added and hydrogen atoms are removed.
- Step 3 is a condensation – two molecules bond together to make a larger product molecule, and a small molecule (water) is removed.

Condensation can be thought of as addition followed by elimination – it is sometimes referred to as addition – elimination.

Addition reactions and position isomers

Unfortunately, when this synthesis is tried it will fail. The problem is that the first step involves addition of water to the carbon–carbon double bond. Two possible products could form (Figure 9).

You might expect that these products are formed in approximately equal ratios, but in fact the first isomer is formed almost exclusively. The yield of the isomer that has the –OH group on the correct carbon atom would be too low to make this an economic synthesis. Instead, phenylethanoic acid is made using routes that start from molecules such as ethylbenzene, or even mandelic acid.

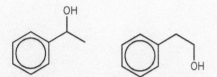

▲ **Figure 9** *The two possible isomers formed from addition of water to phenylethene*

Overall yield

As you have seen in these examples, yields of the steps in an organic synthesis may vary greatly. When evaluating whether a particular synthesis is economic, chemists will often calculate the overall yield of the reaction. Even if the individual yields of each step are quite high, the overall yield will be considerably lower. In general, having more steps in a synthesis means a lower overall yield, as well as increased purification and separation costs, so the preferred route to a compound is *usually* the one with the fewest steps, but this may not always be the case. Sometimes other factors are more important, such as the cost of the starting materials and the reagents, the time involved, atom economy, disposal of waste materials, and possible safety and health hazards.

> 🖩 **Worked example: Calculating an overall yield**
>
> Molecule D is made in a three-step synthesis, as shown below:
>
> $$A \xrightarrow[80\%]{\text{Step 1}} B \xrightarrow[45\%]{\text{Step 2}} C \xrightarrow[70\%]{\text{Step 3}} D$$
>
> Use the yields of each step to calculate the overall yield of the synthesis.
>
> 1 Express each of these yields as a fraction of 100.
>
> $\dfrac{80}{100}$ $\dfrac{45}{100}$ $\dfrac{70}{100}$
>
> 2 Multiply these numbers together.
>
> $\dfrac{80}{100} \times \dfrac{45}{100} \times \dfrac{70}{100} = 0.252$
>
> 3 Express this number as a percentage to find the overall percentage.
>
> yield $= 0.252 \times 100 = 25.2\%$

Classification of organic reactions

You need to be able to classify reactions by looking at an equation or description of the reaction. Table 1 reminds you of what happens in each of these reaction types and also lists the types of functional group that take part in each reaction type.

> **Activity CD 11.1**
> This card sort activity is aimed at helping you to classify organic reactions.

▼ **Table 1** *Summary of reaction types – what happens and the functional groups that take part*

Type of reaction	Definition of reaction	Examples of functional groups that take part in this type of reaction	Topic reference
addition	Two molecules react together to form a single product.	alkenes	DF 6
		aldehydes and ketones	CD 9
elimination	A small molecule (such as water or HCl) is removed from a larger one, leaving an unsaturated molecule.	alcohols	WM 1
		haloalkanes	
condensation	Two molecules react together to form a larger molecule, and a small molecule (such as water or HCl) is removed.	carboxylic acids and alcohols	WM, PL 1
		acyl chlorides, alcohols and amines	PL 1, PL 2
substitution	One group of atoms takes the place of another.	haloalkanes	OZ 8
		alcohols (reacting with hydrogen halides)	OZ 8, WM 1
		alkanes and alkyl groups	OZ 3
oxidation	Oxygen atoms are gained and/or hydrogen atoms are lost.	alcohols	WM 1
		aldehydes	WM 1, CD 9
reduction	Oxygen atoms are lost and/or hydrogen atoms are gained.	aldehydes and ketones	CD 9
		nitro groups	CD 11
hydrolysis	Bonds are broken by the action of water (although OH⁻ may appear in the equation in base-catalysed reactions).	esters	PL 1
		amides/peptides	PL 2
		nitriles	CD 9
		haloalkanes	OZ 8

Reaction mechanisms

As part of thinking about synthesis you may be asked to write mechanisms for reactions using curly arrows and partial charges. These may be for familiar or unfamiliar mechanisms.

Summary questions

1 State the type of reaction shown in each of these reactions:

a $CH_3Br + CN^- \rightarrow CH_3CN + Br^-$ (1 mark)

b $CH_3CHBrCH_3 \rightarrow CH_3CH\!\!=\!\!CH_2 + HBr$ (1 mark)

c $CH_3COCl + C_6H_5OH \rightarrow CH_3COOC_6H_5 + HCl$ (1 mark)

d $CH_3CHO + [O] \rightarrow CH_3COOH$ (1 mark)

e $CH_3CONH_2 + OH^- \rightarrow CH_3COO^- + NH_3$ (1 mark)

2 Compound Z can be prepared from the starting material, compound W, by two alternative routes. Route 1 (W → X → Y → Z) involves three steps. Route 2 (W → V → Z) involves two steps. The yields for each step are shown below.

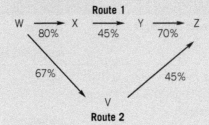

Route 1

W →(80%) X →(45%) Y →(70%) Z

67%

45%

V

Route 2

a Which route has the highest overall yield? Show your working. (2 marks)

b State two other factors which might need to be taken into account in selecting the best route for synthesis. (2 marks)

3 The reaction scheme below shows an incomplete three-step synthesis:

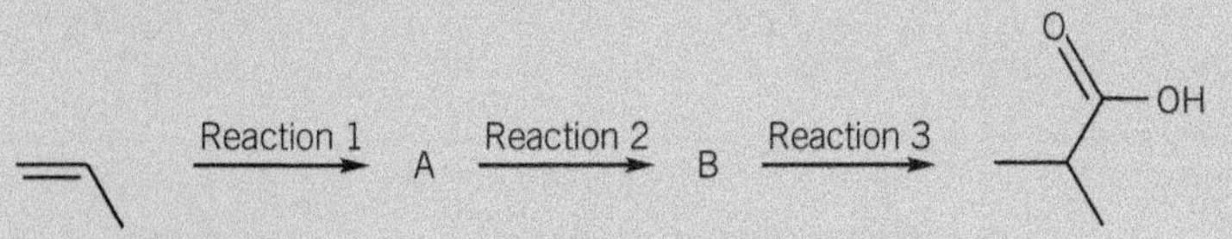

a Suggest structures for intermediates A and B. (2 marks)

b Classify each of the reactions in this scheme using names from Table 1. (3 marks)

Practice questions

1 Which of the following cannot be made in one step from benzene?

A $C_6H_5NH_2$

B $C_6H_5NO_2$

C C_6H_5Br

D $C_6H_5COCH_3$ (*1 mark*)

2 Which of the following tests would distinguish between a primary and a secondary alcohol?

A reaction with bromine

B reaction with iron(III) chloride

C oxidation followed by analysis of the product

D reaction with Fehling's solution   (*1 mark*)

3 Which of the following groups, when present in a chromophore, would not increase its solubility in water?

A $-CH_3$

B $-NH_2$

C $-OH$

D $-SO_3Na$ (*1 mark*)

4 A correct equation for the chlorination of benzene is:

A $C_6H_6 + Cl^- \rightarrow C_6H_6Cl$

B $C_6H_6 + \frac{1}{2}Cl_2 \rightarrow C_6H_6Cl$

C $C_6H_6 + Cl_2 \rightarrow C_6H_5Cl + HCl$

D $C_6H_6 + Cl_2 \rightarrow C_6H_6Cl_2$ (*1 mark*)

5 Which of the following is correct about the compound shown?

A It will be oxidised by acidified dichromate.

B It will react with ethanoic acid.

C It will react with ethanol and conc. sulfuric acid.

D One mole will react with one mole of sodium hydroxide.   (*1 mark*)

6 Which row of the table is correct for the sequence given below?

$CH_2\text{=}CHCH_3$

 ↓1

$CH_3CHBrCH_3$

 ↓2

$CH_3CH(NH_2)CH_3$

 ↓3

$CH_3CH(CH_3)NHCOCH_3$

	1	2	3
A	addition	substitution	condensation
B	elimination	substitution	hydrolysis
C	addition	oxidation	hydrolysis
D	elimination	oxidation	condensation

(*1 mark*)

7 The benzenediazonium ion can decompose to give nitrogen and an aromatic cation. This cation can be attacked by anions present.

Which of the following is correct about this process?

A The initial electron movement can be represented:

B The cation is $C_6H_6^+$.

C The attacking anion is an electrophile.

D A di-substituted benzene compound is formed by the anion attack.   (*1 mark*)

8 When bromine reacts with benzene:

1 electrophilic substitution occurs

2 HBr is formed

3 the reaction is rapid at room temperature and pressure.

A 1, 2, and 3 correct

B 1 and 2 are correct

C 2 and 3 are correct

D Only 1 is correct   (*1 mark*)

9 A compound has the structure:

Which of the following are correct about this compound?

1 It could be oxidised to a carboxylic acid with $K_2Cr_2O_7$ and acid?

2 It would give a purple colour with iron(III) chloride.

3 It would undergo electrophilic substitution with bromine.

A 1, 2, and 3 correct

B 1 and 2 are correct

C 2 and 3 are correct

D Only 1 is correct   (*1 mark*)

10 The following formula is sometimes used to represent benzene:

This formula implies that:

1 the carbon–carbon bond lengths are different

2 addition is more likely to occur than substitution

3 the enthalpy change of hydrogenation of benzene will not be the same as three times the enthalpy change of hydrogenation of cyclohexene.

A 1, 2, and 3 correct

B 1 and 2 are correct

C 2 and 3 are correct

D Only 1 is correct   (*1 mark*)

11 Acid orange 7 is used to dye wool and other protein fibres.

acid orange 7

a Name **two** functional groups in this molecule (not the arene rings)   (*2 marks*)

b (i) Complete the flow diagram for a possible route by which this dye could be synthesised. Give the initial reactant and the reagents at each stage.

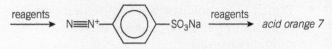

(ii) Name the type of reaction that is occurring in (i), in terms of dye formation.   (*5 marks*)

c A chemist wishes to substitute a $(CH_3)_2CH-$ group onto the benzene ring in acid orange 7.
Name the reagent, the catalyst, and the conditions that would be used.   (*3 marks*)

d A protein fibre has free $-NH_2$ groups. Acid orange 7 will bind to these by various bonds. Name the strongest bond by which the dye binds in neutral solution and in acidic solution.
Illustrate your answer by diagrams showing how these bonds form from different parts of the structure of acid orange 7.   (*4 marks*)

e Both the organic starting materials in the synthesis of acid orange 7 in **b** are colourless. Explain why acid orange 7 is coloured whilst the two starting materials are colourless.   (*6 marks*)

12 Oils used by oil painters can be analysed by gas–liquid chromatography.

   a First the paint sample is warmed in KOH solution to hydrolyse the ester links in the oil and form the potassium salts of the carboxylic acids.
Complete and balance the equation below to illustrate this and name the product shown. Use 'R' to represent all the carboxylic acid side chains.

$$\text{.........} + 3KOH \longrightarrow \begin{array}{c} CH_2OH \\ | \\ CHOH \\ | \\ CH_2OH \end{array} + \text{.........}$$

                                                        (3 marks)

   b The acids are then converted into their methyl esters. Write the structural formula of the methyl ester of a saturated straight-chain carboxylic acid. The acid has 16 carbon atoms.   *(1 mark)*

   c The methyl esters are then separated by gas–liquid chromatography.

      (i) Describe the interior of the column through which the gases pass.

     (ii) Suggest a method by which the emerging compounds are detected that enables their M_r values to be determined.   *(2 marks)*

13 Propanal is a useful intermediate in manufacturing processes.

   a Write the skeletal formula for propanal.   *(1 mark)*

   b Propanal has an isomer which is also a carbonyl compound but does not react with Tollens' reagent.

      (i) What is seen when propanal reacts with Tollens' reagent?

     (ii) Give the structural formula of the organic product of the reaction of propanal with Tollens' reagent.

    (iii) Give the name of the isomer of propanal.   *(3 marks)*

   c Propanal can be reduced to a compound that is used as a solvent in the pharmaceutical industry.

      (i) Give a reagent for this reaction.

     (ii) Name the product.   *(2 marks)*

   d Propanal reacts with cyanide ions to form a cyanohydrin. Give the mechanism of this reaction and show the product.

                                                (4 marks)

   e Propose a synthesis of 2-ethylpropanedioic acid from the cyanohydrin in **d**. (Write the structural formulae of the organic compounds connected by arrows, with reagents over the arrows.)   *(4 marks)*

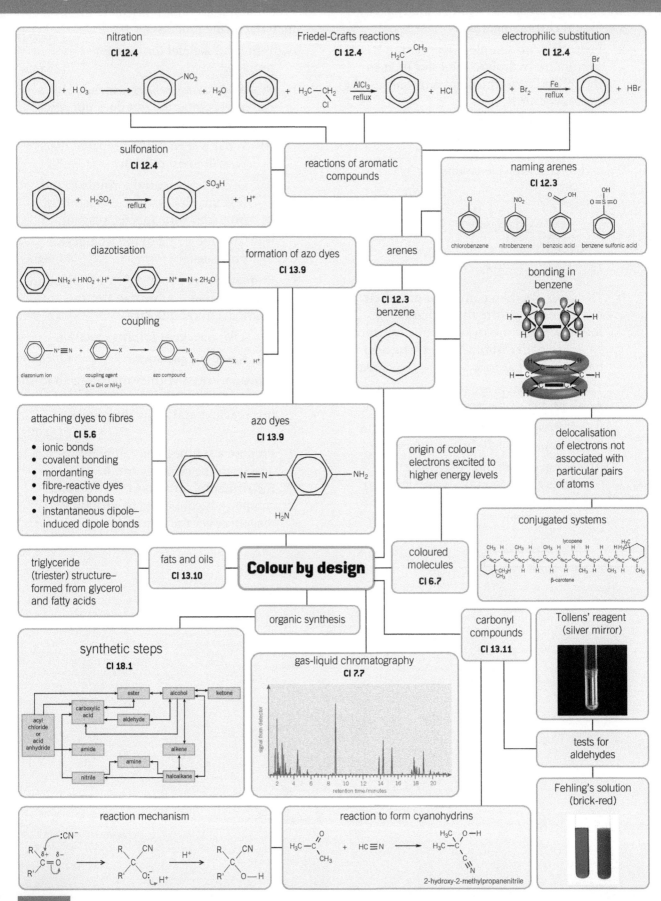

nitration
Cl 12.4

Friedel-Crafts reactions
Cl 12.4

electrophilic substitution
Cl 12.4

sulfonation
Cl 12.4

reactions of aromatic compounds

naming arenes
Cl 12.3

chlorobenzene   nitrobenzene   benzoic acid   benzene sulfonic acid

diazotisation

formation of azo dyes
Cl 13.9

arenes

bonding in benzene

Cl 12.3
benzene

coupling

diazonium ion   coupling agent (X = OH or NH₂)   azo compound

attaching dyes to fibres
Cl 5.6
• ionic bonds
• covalent bonding
• mordanting
• fibre-reactive dyes
• hydrogen bonds
• instantaneous dipole–induced dipole bonds

azo dyes
Cl 13.9

origin of colour electrons excited to higher energy levels

delocalisation of electrons not associated with particular pairs of atoms

conjugated systems

lycopene

β-carotene

triglyceride (triester) structure– formed from glycerol and fatty acids

fats and oils
Cl 13.10

Colour by design

coloured molecules
Cl 6.7

organic synthesis

carbonyl compounds
Cl 13.11

Tollens' reagent (silver mirror)

synthetic steps
Cl 18.1

ester   alcohol   ketone

carboxylic acid

aldehyde

acyl chloride or acid anhydride

amide

alkene

amine

nitrile   haloalkane

gas-liquid chromatography
Cl 7.7

signal from detector

retention time/minutes

tests for aldehydes

Fehling's solution (brick-red)

reaction mechanism

reaction to form cyanohydrins

2-hydroxy-2-methylpropanenitrile

Hair dyes

Hair is made of the protein keratin, which contains approximately 14% cysteine amino acid units. The natural colour of hair depends on two proteins called eumelanin and phaeomelanin, but many people wish to alter the colour of their hair using temporary or permanent hair dyes.

▲ Figure 1

Some dyes contain lead (II) ethanoate, $Pb(CH_3COO)_2$. When the dye is rubbed into the hair, the Pb^{2+} ions react with sulfur atoms in the cysteine units to form lead (II) sulfide, PbS, which is dark in colour.

Other dyes are based on organic chemicals, using principles developed from the work of the French chemist Eugene Schuller who created the first safe hair

dye in 1909. This involves oxidising a diamine such as 1,4-diaminobenzene under alkaline conditions using a mixture of hydrogen peroxide and ammonia. The product reacts with a coupling agent such as benzene-1,3-diol to produce a coloured compound.

▲ Figure 2 1,4-diaminobenzene

▲ Figure 3 benzene-1,3-diol

▲ Figure 4 Green dye molecule produced from 1,4-diaminobenzene and benzene-1,3-diol

By changing the starting diamine or the coupling agent, a wide range of different coloured dyes can be produced.

1 Identify the functional groups in the green dye molecule.
2 Explain why the green dye molecule is coloured whilst the starting materials are colourless.
3 Why are different colours formed when different diamines or coupling agents are used?
4 Suggest the type of bonding would you expect between the green dye molecule and the protein keratin.

 Extension

1 Research the different starting diamines and coupling agents used in hair dyes and find out about the reactions that produce the dye molecules.
2 Produce a flowchart showing all the aspects of arene chemistry that have been covered in this chapter.
3 Colour chemistry is an important branch of chemistry, linking organic and inorganic chemistry and finding a wide range of applications. Find out about the various industries, technologies, and manufacturing processes that colour chemists contribute to.

SCIENTIFIC LITERACY IN CHEMISTRY

Introduction

The ability to understand and communicate science is integral part of your chemistry course. In your scientific literacy in chemistry exam paper, you will be given a scientific passage. You will then be asked questions which test your comprehension of the scientific passage, as well as your knowledge and understanding of the chemistry concepts you have studied throughout your A Level qualification.

Throughout the course you will have extracted and manipulated data, interpreted and used information, and written logical accounts using appropriate technical terms. These are all examples of scientific literacy.

In all of your A Level written assessments, you will demonstrate these skills. This may be via questions set in unfamiliar contexts and might involve extended response answers.

In addition, you will be given a piece of scientific literacy in advance of the scientific literacy in chemistry exam paper. The examination will have associated questions with a particular emphasis on scientific literacy. There will also be questions assessing comprehension of and use of data from the practical insert of the Practical skills in chemistry exam paper.

Analysing and answering chemical literacy questions

For your Scientific literacy in chemistry exam, you will be provided with a pre-release Advance notice article, which you should read carefully before the examination date. In the scientific literacy in chemistry exam, there will be at least one question relating to the pre-release article.

To help you practise extracting information from a scientific passage, such as the pre-release advanced notice article, an example article is provided below, based on the Storylines content of Chapter 8, Oceans. The Scientific literacy in chemistry exam paper could ask questions on **any** of the content you have studied in your course.

This activity will enable you:

- extract and manipulate data
- interpret and use information
- show comprehension by written communication with regard to logical presentation and the correct use of appropriate technical terms
- check your understanding of entropy, pH, solubility product, and equilibria.

> **Chemical literacy**
>
> The practice chemical literacy activities give a piece of scientific literacy and a series of questions to test your understanding of chemistry and ability to extract information.

Dissolving Seashells

Seashells are involved in the reactions that influence the solubility of carbon dioxide in the oceans and in this section you are going to look at the role that they play. The three reactions below summarise what happens:

Reaction 1 $CO_2(g) \rightleftharpoons CO_2(aq)$

Reaction 2 $CO_2(aq) + H_2O(l) \rightleftharpoons H^+(aq) + HCO_3^-(aq)$ $pK_a = 6.34$

Reaction 3 $HCO_3^-(aq) \rightleftharpoons H^+(aq) + CO_3^{2-}(aq)$ $pK_a = 10.3$

▼ **Table 1** *Average concentrations in surface seawater*

Ion	Concentration / $mol\,dm^{-3}$
Cl	5.46×10^{-1}
Na^+	4.68×10^{-1}
SO_4^{2-}	2.81×10^{-2}
Mg^{2+}	5.33×10^{-2}
Ca^{2+}	1.04×10^{-2}
K^+	9.97×10^{-3}
HCO_3^-	2.34×10^{-3}
H^+	pH = 8.14

From Reactions 1 to 3, it can be deduced that decreasing [H^+] will cause more CO_2 to dissolve. Removing H^+ ions by adding a base is one way of doing this. You should be familiar with this process – carbon dioxide is an acidic gas and it dissolves well in alkaline solutions. That's why alkalis such as sodium hydroxide or calcium hydroxide are used to absorb CO_2.

Making the sea alkaline is not a very feasible way of encouraging the oceans to take up carbon dioxide! However, many marine organisms build protective shells composed of insoluble calcium carbonate, using CO_3^{2-} ions in the sea water. The building of these shells provides a route for "mopping up" carbon dioxide and keeping the composition of the Earth's atmosphere constant.

Billions of years ago, the Earth's atmosphere contained very much more carbon dioxide than it does now – probably about 35% CO_2 by volume. Once the process of photosynthesis had evolved, marine life had plenty of raw materials to work on in the form of carbon dioxide and water. Shell production flourished – limestone and chalk rocks are the remains of the shells of marine organisms that lived at that time and changed carbon dioxide from the atmosphere into solid calcium carbonate.

Calcium carbonate is a good material for shellfish to use for protection at the surface of the oceans. It does not dissolve in sea water but it does dissolve, very slightly, in pure water. It is an example of a sparingly soluble solid – the dissolving of sparingly soluble solids is controlled by equilibria such as Reaction 4 in which the ions in the saturated solution are in dynamic equilibrium with the undissolved solid present.

Reaction 4 $CaCO_3(s) \rightleftharpoons Ca^{2+}(aq) + CO_3^{2-}(aq)$

The position of this equilibrium is determined by an equilibrium constant called a solubility product K_{sp}, because it describes the solubility of a compound. The solubility product for Reaction 4 is:

$$K_{sp}(CaCO_3) = [Ca^{2+}(aq)][CO_3^{2-}(aq)] = 5.0 \times 10^{-9} \text{ mol}^2 \text{dm}^{-6} \text{ at } 298 \text{ K}$$

Other data for Reaction 4 are in Table 2.

▼ **Table 2** *Data for Reaction 4*

	$\Delta_f H / \text{kJ mol}^{-1}$	$S / \text{J mol}^{-1} \text{K}^{-1}$
$CaCO_3$	−1207	+93
$Ca^{2+}(aq)$	−543	−53
$CO_3^{2-}(aq)$	−677	−57

One of two things can happen when Ca^{2+} ions and CO_3^{2-} ions are mixed together in a solution.

1   If the dissolved calcium ion concentration multiplied by the dissolved carbonate concentration gives a value in excess of K_{sp} then calcium carbonate will precipitate out of solution.

2   If the dissolved calcium ion concentration multiplied by the dissolved carbonate concentration gives a value smaller than or equal to K_{sp} then the ions stay in solution.

At the surface of the sea calcium carbonate is an excellent material from which to build seashells because the concentrations of $Ca^{2+}(aq)$ ions and $CO_3^{2-}(aq)$ ions are already high enough for the calcium carbonate in the shells to be effectively insoluble (the equilibrium position in Reaction 4 lies well over to the left). Remember that the shells are in equilibrium with the ions in sea water, and there will be a constant exchange of Ca^{2+} and CO_3^{2-} between the two.

Things are different deeper in the ocean, where the pressure is higher and carbon dioxide is more soluble. The temperature is also lower and both carbon dioxide and calcium carbonate are more soluble. There is also a continuous downward drift of material from above. It's like a perpetual snowstorm – in fact the falling material is called marine snow. It contains the remains of dead organisms and the waste products from live creatures. Most of the organic material, such as tissue, is consumed or decomposed higher up but some reaches the deeper water where bacteria break it down to produce carbon dioxide. The shells fall intact but then react with the extra carbon dioxide and dissolve.

There are no shells on the deep ocean floor – they've all dissolved. The creatures that live there cannot use calcium carbonate for a protective coating.

The calcium carbonate deposits that built up to form our limestone hills could therefore not have formed in deep water. They must have been laid down when our landmass was in shallower seas.

The abundance of life also suggests that it was warm, tropical water. Evidence like this helps scientists piece together the distant history of the Earth, and helps to explain how the continents have drifted and how the climate has changed throughout time.

Questions

1 (i)   Explain why the lowering of $[CO_3^{2-}]$ by making seashells causes more CO_2 to dissolve from the air.

> Reactions 1 to 3 in the article show the equilibrium reactions involved. Equilibrium was covered in Topic ES 4, From extracting bromine to making bleach.

> The equilibrium position of Reaction 3 moves to the right, lowering $[HCO_3^-]$, which moves the equilibrium position of Reaction 2 to the right, lowering $[CO_2(aq)]$, which in turn moves the equilibrium position of Reaction 1 to the right allowing more CO_2 to dissolve.

   (ii) Why is this a good thing?

> The oceans reduce the amount of extra CO_2 in the atmosphere which lowers the effect of the extra CO_2 on the climate (i.e., reduces global warming).

> In Topic OZ 1, What's in the air?, and Topic O 2, The role of the oceans in climate control, you saw how human activities release more carbon dioxide into the air.
> The article mentions how this the production of seashells helps "mop up" carbon dioxide to keep the atmosphere constant.

2 *Use data from the Article to answer these parts:*
   (i)   Write an expression for the K_a of Reaction 3

> $$K_a = \frac{[H^+(aq)]\,[CO_3^{2-}(aq)]}{[HCO_3^-(aq)]}\,mol\,dm^{-3}$$

> The acid dissociation constant K_a is just an equilibrium constant. You learnt about writing an expression of an equilibrium constant in Topic CI 2, Manufacturing nitric acid.

   (ii) Calculate the value of this K_a, giving units.

> $$K_a = 10^{-pKa} = 5.01 \times 10^{-11}$$

> pK_a is a logarithmic form of K_a. You covered pK_a in Topic O 2, The role of the oceans in climate control. The value of pK_a came from the article.

   (iii) Calculate the ratio of $[CO_3^{2-}(aq)]/[HCO_3^-(aq)]$ at the pH of surface seawater.

> $$pH = -\log_{10}[H^+(aq)]$$
> $$[H^+] = 10^{-pH} = 10^{-8.14} = 7.24 \times 10^{-9}\,mol\,dm^{-3}$$
> $$\frac{[CO_3^{2-}(aq)]}{[HCO_3^-(aq)]} = \frac{5.01 \times 10^{-11}}{7.24 \times 10^{-9}} = 6.9 \times 10^{-3}\ \text{(not units)}$$

> The pH of surface seawater is given in Table 1. If you know the pH, you can calculate $[H^+]$, see Topic O 2, The role of oceans in climate control.

(iv) Calculate the corresponding $[CO_3^{2-}(aq)]$.

$[CO_3^{2-}(aq)] = 6.9 \times 10^{-3} \times 2.34 \times 10^{-3}$

$\qquad\qquad = 1.6 \times 10^{-5}\,mol\,dm^{-3}$

> Multiply the ratio by the $[HCO_3^-]$ value from Table 1 to cancel out $[HCO_3^-]$.

3 Using data from the article, calculate the lowest $[CO_3^{2-}(aq)]$ concentration that seashells could tolerate before they started to dissolve in surface seawater.

$[CO_3^{2-}(aq)] = \dfrac{K_{sp}}{[Ca^{2+}\,(aq)]} = \dfrac{5.0 \times 10^{-9}}{1.04 \times 10^{-2}}\,\dfrac{(mol\,dm^{-3})(mol\,dm^{-3})}{(mol\,dm^{-3})}$

$\qquad\qquad = 4.8 \times 10^{-7}\,mol\,dm^{-3}$

The lowest value of $[CO_3^{2-}(aq)]$ before

seashells dissolve is $4.8 \times 10^{-7}\,mol\,dm^{-3}$.

> K_{sp} is the equilibrium constant for solubility. You covered K_{sp} in Topic 0 4, Seashells as a carbon store. Both values are given in the article $-[Ca^{2+}(aq)]$ in Table 1 and K_{sp} in the main text.

4 Explain the meaning of:
'Remember that the shells are in equilibrium with the ions in sea water, and there will be a constant exchange of Ca^{2+} and CO_3^{2-} between the two.'

In a dynamic equilibrium, the forward and back

reactions occur at the same rate and therefore ions in

solution will be incorporated into seashells at the same

rate that other ions dissolve back into the seawater.

> This is a small extract of text from the article. You need to identify that the article is describing dynamic equilibrium, and therefore explain what is happening in a dynamic equilibrium. Dynamic equilibrium was covered in Topic ES 4, From extracting bromine to making bleach.

5 (i) Using data in Table 2, calculate the temperature below which $\Delta_{tot}S$ for Reaction 4 becomes positive.

$\Delta H = \Delta_{products}H - \Delta_{reactants}H$

$\qquad = (-677) + (-543) - (-1207) = -13\,kJ\,mol^{-1}$

$\Delta_{sys}S = \Delta_{products}S - \Delta_{reactants}S$

$\qquad = (-53) + (-57) - 93 = -203\,J\,mol^{-1}\,K^{-1}$

> $\Delta_{tot}S$ is the total enthalpy change of a reaction. A positive value of $\Delta_{tot}S$ means the change occurs spontaneously. You covered $\Delta_{tot}S$ in Topic 0 5, The global central heating system. Remember to convert 13 kJ into 13000 J.

substitute values into the equation

$\Delta_{tot}S = \Delta_{sys}S + \left(-\dfrac{\Delta H}{T}\right)$

$\Delta_{tot}S = -203 + \left(\dfrac{13000}{T}\right)$

$\Delta_{tot}S = 0$ when $\dfrac{13000}{T} = 203$

$T = \dfrac{13000}{203} = 64$ K (so $\Delta_{tot}S$ becomes positive below this temperature.)

(ii)　Explain the significance of the variation of the solubility of calcium carbonate with temperature.

The solubility of calcium carbonate increases with decreasing temperature, therefore it is more soluble (K_{sp} is larger) in deeper ocean water where the temperature is lower. This means shells are more likely to dissolve in deep oceans than at the surface.

6　In 5(i) you should have shown that $\Delta_{sys}S$ for the dissolving of calcium carbonate is negative. Suggest why this is so.

On dissolving, the high organisation of the lattice is lost, so this would contribute a positive term to $\Delta_{sys}S$. However, Ca^{2+} and CO_3^{2-} attract a lot of water molecules for hydration as they are doubly charged. This organisation of the water molecules is greater than the loss of organisation of the lattice, so overall $\Delta_{sys}S$ is negative.

In question 5 to 7 you are extracting and manipulating data in order and interpreting your results. This is demonstrating comprehension in your written answers, using correct technical terms.

7　Give reasons from the Article why there is more CO_2 deep in the oceans and $[CO_2(aq)]$ increases there.

There is more CO_2 because organic matter (marine snow) is decomposed by bacteria to give CO_2. Also increased pressure and decreased temperature both increase the solubility of CO_2. The CO_2 produced dissolves to give $[CO_2(aq)]$:

$CO_2(g) \rightleftharpoons CO_2(aq)$

Understanding and being familiar with practical techniques and procedures is an important part of being an effective chemist. This section outlines the techniques and procedures you need to know about as part of your course.

Measurement

Weighing a solid

Typically an accurate weighing will use a balance that records to two or three decimal places.

1 Zero the balance (sometimes called tare).
2 Place a weighing bottle or similar container onto the balance and add in approximately the required mass of solid.
3 Accurately weigh the mass of solid plus weighing bottle and record this information.
4 Empty the solid into the glassware where you will be using it.
5 Accurately reweigh the empty weighing bottle.
6 Subtract the recorded mass for the empty weighing bottle from the mass recorded for the solid and the weighing bottle.

Measuring volumes of liquids

Beakers or measuring cylinders can be used to give rough measurements of liquids. In order to measure volumes of liquids *accurately* two methods can be used – a pipette or a burette.

Pipette

A pipette is used for accurately dispensing a *fixed* volume of a liquid (typically $1.0 \, cm^3$ to $50 \, cm^3$ or $25 \, cm^3$).

1 Ensure the pipette is completely clean by rinsing out with water and then a small volume of the solution to be pipetted.
2 Dip the pipette into the solution to be pipetted and, using a pipette filler, draw enough liquid into the pipette until is it exactly the right volume – when the bottom of the meniscus is level with the line on the neck of the pipette when viewed at eye level.
3 Run the liquid out of the pipette into the piece of glassware the solution is being transferred to.
4 Allow the liquid to run out of the pipette until it stops. Touch the end of the pipette on the side of the conical flask and remove. There will still be a drop in the pipette – this is how it should be. The precise volume you require will have been dispensed.

Burette

1 Clean the burette by rinsing out with water and then a small volume of liquid of the solution to be used.

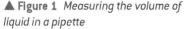

▲ **Figure 1** *Measuring the volume of liquid in a pipette*

2 Make sure the burette tap is closed. Pour the solution into the burette using a small funnel. Fill the burette above the zero line.

3 Use a clamp to hold your burette in place and allow some of the solution to run into a beaker until there are no air bubbles in the jet of the burette. Record the burette reading to the nearest $0.05\,cm^3$.

4 Carry out the titration to the end point.

5 Record the reading on the burette to the nearest $0.05\,cm^3$. Subtract the reading taken at the beginning of the titration from this reading taken at the end. This is known as the titre.

Measuring volumes of gases

The volume of a gas produced in a chemical reaction can be measured using either a gas syringe or an inverted buratte/measuring cyliner. The latter is called 'collecting gas over water'.

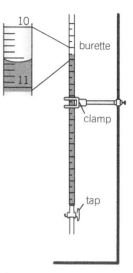

▲ **Figure 2** *Measuring the volume of liquid using a burette*

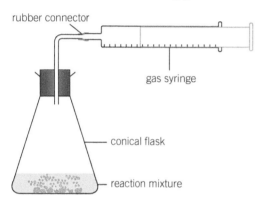

▲ **Figure 3** *Collecting gas using a gas syringe*

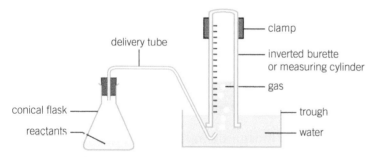

▲ **Figure 4** *Collecting a gas using a measuring cylinder or inverted burette*

In order for as much as possible of the gas to be collected the system needs to be gas tight.

The volume of gas collected in an inverted burette is the initial volume minus the final volume of gas.

Synthesis

Heating under reflux

Heating under reflux is used for reactions involving volatile liquids. It ensures that reactants and/or products do not escape whilst the reaction is in progress.

> **Study tip**
>
> If the gas is soluble in water you need to use the gas syringe method, as some of the gas would dissolve before reaching the burette in the second method.

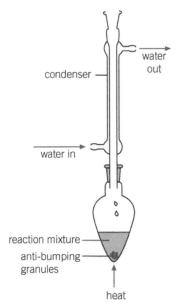

▲ **Figure 5** *Reflux apparatus*

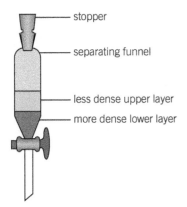

▲ **Figure 6** *Separating funnel*

Study tip

Take care to keep the correct layer. An easy way to test which layer is the aqueous layer is to add water to the separating funnel and wait for the layers to separate again. The layer that is bigger will be the aqueous layer.

1 Put the reactants into a pear-shaped or round-bottomed flask and add a few anti-bumping granules – these granules burst the bubbles in the boiling mixture and reduce the chance of boiling over.

2 Do not stopper the flask – doing this would cause pressure to build up and the glassware could crack or the stopper could fly out. In either case, a serious accident could result.

3 Attach a condenser vertically to the flask so that water flows into the condenser at the bottom and out of the condenser at the top. This ensures that the condenser is always full of cold water.

4 Heat so that the reaction mixture boils gently, using a Bunsen flame or heating mantle. When refluxing correctly, any vapours should reach no more than half way up the condenser before condensing back into a liquid. The liquid should drip back into the reaction flask steadily.

Purifying an organic liquid product

Once organic liquid products have been synthesised, they must then be purified to remove any solvents or impurities present.

1 When the organic product is mixed with another immiscible liquid (often an aqueous liquid) the two layers can be separated using a separating funnel. The layers separate, with the denser liquid forming the lower layer. Allow the layers to settle and then run off and dispose of the aqueous layer. Run the organic (product) layer into a clean conical flask.

2 If acidic impurities are present, add sodium hydrogen carbonate solution and shake well to remove them. If the crude product is alkaline and needs neutralising then add a dilute acid until the mixture is neutral.

3 Dry the crude product by adding anhydrous sodium sulfate and swirling the mixture. It is possible to use other anhydrous salts, such as calcium chloride, to dry organic compounds.

4 The pure product can then be separated by distillation.

Making water-soluble inorganic salts

Soluble salts can be made by two techniques – reacting an acid with a soluble base (alkali) or by reacting an acid with an insoluble base.

Reacting an acid and a soluble base (alkali)

Acid–base titrations can be used to produce a soluble salt.

1 Carry out an acid–base titration to find out how much acid solution is needed to neutralise 25 cm³ of the alkaline solution.

2 Transfer 25.0 cm³ of the alkaline solution to a clean conical flask.

3 Using a burette, add the correct amount of acid to neutralise the alkali. *Do not* add any indicator.

4 Transfer the neutralised solution to a clean evaporating basin and heat over a Bunsen flame to evaporate the water. Take care not to

heat too strongly, in order to avoid spitting. Once crystals of solid start to appear stop heating.

5   Leave the mixture to cool in the evaporating basin.

6   Filter the mixture.

7   Wash the solid residue with cold distilled water.

8   Transfer the residue to a watch glass and heat in an oven to dry the solid. Ensure the oven is set at a temperature below the melting point of the salt you have prepared.

9   At regular intervals remove the watch glass and solid, cool in a desiccator, and weigh. Once the solid has dried to a constant mass (the mass between two readings does not change) stop heating the solid salt and leave to cool in a desiccator.

> **Study tip**
>
> A desiccator allows materials to cool in a dry atmosphere, so preventing the reabsorption of moisture.

Reacting an acid and an insoluble base

1   In a beaker, warm excess insoluble base in dilute acid.

2   Continue to warm (but not boil) until the solution is neutral (use universal indicator paper for this step), adding more solid base if needed. Leave to cool.

3   Filter off the excess base and transfer the filtrate to a clean, dry evaporating basin.

4   Heat the evaporating basin until salt crystals begin to appear on the sides of the basin.

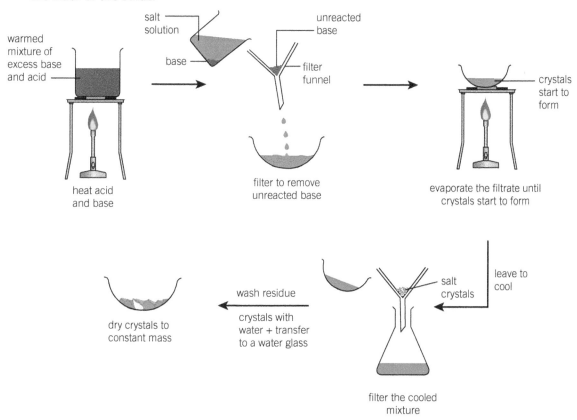

summary of the preparation of an inorganic salt by reacting an molecule base cool an acid

▲ **Figure 7** *Summary of the preparation of an inorganic salt by reaction of an insoluble base with an acid*

5   Cool the basin and contents.

6   Filter the mixture, and discard the filtrate.

7   Wash the solid residue with cold distilled water.

8   Transfer the residue to a watch glass and heat in an oven to dry the solid. Ensure the oven is set at a temperature below the melting point of the salt you have prepared.

9   At regular intervals remove the watch glass and solid, cool in a desiccator, and weigh. Once the solid has dried to a constant mass (the mass between two readings does not change) stop heating the solid salt and leave to cool in a desiccator.

Making water-insoluble inorganic salts

Insoluble salts can be prepared by the reaction of two soluble salts in solution.

1   Add equal volumes of the desired salt solutions in a beaker to form a precipitate of the insoluble salt.

2   Filter the precipitate

3   Wash the precipitate several times with cold deionized water.

4   Transfer the filtered, washed precipitate to a clean watch glass and place in a drying oven. Ensure the oven is set at a temperature below the melting point of the salt you have prepared

5   At regular intervals remove the watch glass and solid, cool in a desiccator, and weigh. Once the solid has dried to a constant mass (the mass between two readings does not change) stop heating the solid salt and leave to cool in a desiccator.

Purification

Simple distillation

Distillation can be used to separate a mixture of miscible liquids with unique boiling points. By heating the mixture, each pure component is vaporised, condensed, and collected. The components will evaporate in the order of their boiling points – the one with the lowest boiling point will evaporate first. Quickfit glassware is commonly used for distillation. It has ground glass joints that can be sealed using grease to prevent the loss of reagents. Other small scale systems are also available.

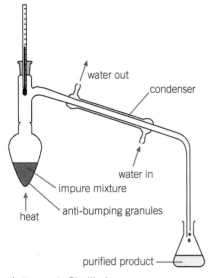

▲ **Figure 8** *Distillation apparatus*

1   Put the mixture into a pear-shaped flask and add a few anti-bumping granules. Set up the distillation apparatus as shown in Figure 8. The position of the thermometer is important – it gives an accurate reading of the vapour temperature.

2   Heat the mixture until it boils gently, using a Bunsen flame or heating mantle. Heating mantles are safer to use when heating flammable liquids as they do not have a naked flame.

3   When the vapour temperature is approximately two degrees below the boiling point of the liquid you are about to collect, put the collecting beaker in place. Collecting the distilled liquid until

the temperature of the vapour rises above the boiling point of the liquid you are collecting. Stop heating.

4 If another compound is required of a higher boiling point, repeat step **3** using a clean collecting beaker.

Thin layer or paper chromatography

Thin layer chromatography (TLC) and paper chromatography are used to separate small quantities of organic compounds, purify organic substances, and follow the progress of a reaction over time. Chromatography relies on the fact that different organic compounds have different affinities for a particular solvent, and so will be carried through the chromatography medium at different rates. Paper chromatography is carried out using chromatography paper as the stationary phase. Thin layer chromatography is carried out using a silica plate as the stationary phase.

1 Spot the test mixture and reference samples on a pencil line 1 cm from the base of the chromatography paper or silica plate. Pencil is used because it will not run into the solvent.

2 Suspend the plate in a beaker containing the solvent (Figure 9) and cover the beaker with a watch glass to prevent the solvent from evaporating.

3 Remove the plate when the solvent front is near the top. Mark how far the solvent has reached. Allow the plate to dry.

4 Locate any spots with iodine, ninhydrin, or under an ultraviolet lamp.

5 Match the heights reached, or R_f values, with those of known compounds musing the same chromatography solvent mix.

Recrystallisation

Recrystallisation is used to purify solid crude organic products with small amounts of impurities. A suitable hot solvent is chosen that only the desired compound dissolves in to an appreciable extent. When cooled, the pure organic compound will drop out of solution (recrystallise) – any other soluble impurities stay in solution.

1 Select a solvent in which the desired substance is very soluble at higher temperatures and insoluble, or nearly so, at lower temperatures.

2 Dissolve the mixture in the minimum quantity of hot solvent – the smaller the amount of solvent used the better the yield of the desired substance.

3 Filter to remove any insoluble impurities and retain the filtrate. It is best to preheat the filter funnel and conical flask to prevent any solid crystallising out at this stage.

4 Leave the filtrate to cool until crystals form.

5 Collect the crystals by vacuum filtration.

6 Dry the crystals in an oven or by leaving them in the open, covered by an inverted filter funnel.

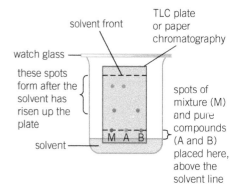

▲ **Figure 9** *Thin layer chromatography*

> ### Study tip
>
> R_f value is the distance travelled by the spot you are interested in divided by the distance moved by the solvent. It is always a value of 1 or less.

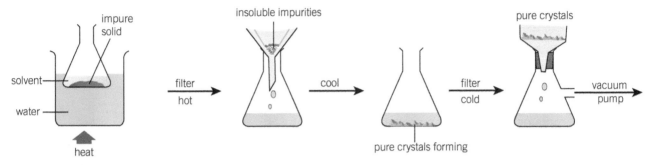

▲ **Figure 10** *Recrystallisation of an impure solid*

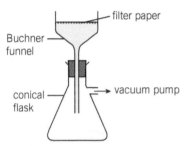

▲ **Figure 11** *Vacuum filtration*

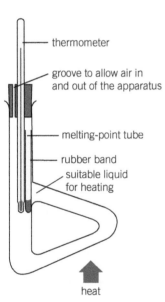

▲ **Figure 12** *Melting point apparatus*

Vacuum filtration

Vacuum filtration is used to separate a solid from a filtrate rapidly.

1 Connect a conical flask to a vacuum pump via the side arm (Figure 11). Do not switch the pump on yet.

2 Dampen a piece of filter paper and place it flat in the Buchner funnel.

3 Switch the vacuum pump on and then carefully pour in the mixture to be filtered. The pump creates a partial vacuum so that the filtrate gets 'pulled through' quickly.

4 Disconnect the flask from the vacuum pump before turning the pump off – this avoids 'suck back'.

Analysis

Determining melting points

This technique is used to determine the melting point of organic solids. The melting point can then be used as evidence of a product's identity and purity.

1 Seal the end of a glass melting point tube by heating it to melting in a Bunsen flame.

2 Tap the open end of the tube into the solid so a small amount goes into the tube. Tab the tube so that the solid falls to the bottom of the sealed end.

3 Fix the tube in the melting point apparatus and heat the surrounding liquid gently, stirring to ensure even heating throughout. The temperature should rise very slowly.

4 Note the temperature at which the solid starts and finishes melting. The difference between the highest and lowest temperatures recorded is known as the melting range.

5 Compare the experimental value to the published value for the melting point. The wider the melting range, the more impure the substance. A pure compound will melt within 0.5 °C of the true melting point.

Making a standard solution

A standard solution is one where its concentration is accurately known. It can be used to determine the concentration of another unknown solution or the purity of a solid. A standard solution can be made up either from a solid or by accurate dilution of another standard solution. A standard solution is made up fresh whenever a concentration has to be accurate. Here it is assumed that $250\,cm^3$ of solution is being made up.

Making a standards solution from a solid

1 Calculate the mass of solute required. In a weighing bottle, weigh out this amount accurately, to the nearest $0.01\,g$. Make a note of the mass of the weighing bottle and solute.

2 Pour $100\,cm^3$ of deionised water into a $250\,cm^3$ beaker. Carefully transfer the weighed solute into the water from the weighing bottle.

3 Reweigh the weighing bottle. The difference between the mass of the weighing bottle and solute and the mass of just the weighing bottle is the mass of solute transferred.

4 Stir the mixture in the beaker to ensure complete dissolving of the solute.

5 Transfer the solution to a clean $250\,cm^3$ volumetric flask. Rinse the beaker and stirring rod well with deionised water, making sure that all the washings go into the volumetric flask.

6 Add deionised water to the solution, swirling at intervals to mix the contents, until the level is within about $1\,cm$ of the $250\,cm^3$ mark on the neck of the volumetric flask.

7 Using a dropping pipette, add deionised water so that the bottom of the meniscus is level with the mark on the neck of the flask when looking at it at eye level.

8 Insert the stopper in the flask and invert it, shaking thoroughly to ensure complete mixing.

Making a standard solution by dilution

If the concentration of an existing solution is too high it can be adjusted. The existing solution is called the stock solution. In this example, the concentration is diluted by a factor of 10 to produce $250\,cm^3$ of a $0.1\,mol\,dm^{-3}$ solution from a standard solution of $1.0\,mol\,dm^{-3}$

1 Rinse a clean dry beaker with the stock solution, and then half fill it.

2 Use a pipette filler to rinse a clean $25.0\,cm^3$ pipette with some of the stock solution. Fill the pipette to the $25.0\,cm^3$ mark – with the bottom of the meniscus exactly on the mark.

3 Run the solution into a $250\,cm^3$ volumetric flask.

4 Add deionised water to the solution, swirling at intervals to mix the contents, until the level is within about $1\,cm$ of the mark on the neck of the volumetric flask.

5 Using a dropping pipette, add deionised water so that the bottom of the meniscus is level with the mark on the neck of the flask.

6 Insert the stopper in the flask and invert it, shaking thoroughly to ensure complete mixing. You now have your new diluted standard solution.

Study tip

Impurities in a crude product will cause the melting point to be lowered. Following recrystallization the melting point will be higher.

Study tip

The solute should be pure. Pure reagents can be supplied, and are known as Analar reagents. For standard solutions use Analar reagents.

Calculating the volumes needed

In order to calculate the volumes needed you need to use the equation:

concentration of diluted solution being made C_1 (mol dm⁻³)
× volume of diluted solution being made V_1 (dm³)
= concentration of stock solution C_2 (mol dm⁻³)
× volume of stock solution V_1 (dm³)

 Worked example: Standard solution by dilution

To prepare 50 m³ of 1.0 mol dm⁻³ solution from a stock solution of concentration 2.0 mol dm⁻³, what volume of stock solution is required?

Step 1: Use the equation $C_1V_1 = C_2V_2$ and insert known values.
$$1.0 \, \text{mol dm}^{-3} \times 0.05 \, \text{dm}^3 = 2.0 \, \text{mol dm}^{-3} \times Z \, \text{dm}^3$$

Step 2: Rearrange the equation to calculate the required volume of stock solution.
$$Z = \frac{1.0 \, \text{mol dm}^{-3} \times 0.05 \, \text{dm}^3}{2.0 \, \text{mol dm}^{-3}} = 0.025 \, \text{dm}^3$$

Step 3: Convert dm³ reading to cm³ reading −25 cm³ of stock solution needed.

Acid–base titration

Acid–base titration is used to determine the concentration of an acid or an alkali accurately. The method described below assumes that you are titrating an alkali of known concentration against an acid to calculate the concentration of the acid.

1   Rinse a burette with some of the acid solution then fill it with the acid. Run a little of the acid through the burette into a waste beaker to fill the tip. Record the initial burette reading to the nearest 0.05 cm³ (Figure 13).

2   Fill a clean 25.0 cm³ pipette with some of the alkaline solution.

3   Run the alkaline solution into a clean 250 cm³ conical flask.

4   Add two or three drops of a suitable indicator and swirl to mix

5   Run the acid from the burette into the flask. Swirl the flask continually and watch for the first hint of the solution changing colour. This first titration should be used as a trial run to give a rough indication of the amount of acid required. Record the final burette reading – the volume of acid used is called the titre.

6   Refill the burette and record the initial burette reading.

7   Using the pipette, transfer 25.0 cm³ of the alkaline solution to a clean conical flask. Add two to three drops of the indicator and swirl to mix.

8   Run in the acid solution to 1 cm³ below the rough titre. Then add the acid dropwise, swirling after each drop, until the colour of the indicator changes.

5   Repeat steps **6**, **7**, and **8** until there are three concordant results, that is, three results within 0.10 cm³ of each other.

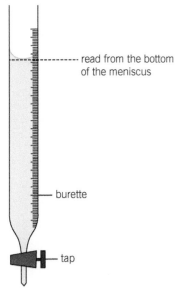

read from the bottom of the meniscus

burette

tap

▲ **Figure 13** *Taking a reading from a burette*

The volume and concentration of the standard alkaline solution required for neutralisation can be used to calculate the concentration of the hydrochloric acid.

Iodine–thiosulfate titration

Iodine-thiosulfate titrations involve redox reactions. They are used to find the concentration of a chemical that is a strong enough oxidizing agent to oxidise iodide ions to iodine. The liberated iodine is titrated with thiosulfate ions, with starch as an indicator. Once the amount of thiosulfate ions are calculated the amount of the chemical being analysed (in the example below chlorate(I) found in domestic bleach solutions) can be calculated.

Step 1

Chlorate(I) ions are strong enough oxidising agents to reduce iodide ions to iodine. An iodine-thiosulfate titration can be used to find the concentration of a solution of chlorate(I), for example, sodium chlorate(I).

$$ClO^- + 2I- + 2H^+ \rightarrow I_2 + Cl^- + H_2O \qquad \textbf{Equation 1}$$

| Cl | +1 | | −1 | reduced |
| I | | −1 | 0 | oxidised |

1 Pour some of the chlorate(I) solution into a clean and dry beaker.

2 Rinse a $25.0\,cm^3$ volumetric pipette with water and then the chlorate(I) solution. This ensures that the pipette is clean and that the chlorate(I) solution hasn't been diluted.

3 Transfer a carefully measured $25.0\,cm^3$ **aliquot** of the chlorate(I) solution to a conical flask using a volumetric pipette and filler. This piece of equipment has a low uncertainty. Ensure that the bottom of the meniscus is exactly on the mark when filling. Before emptying the solution into the flask, dry the outside with a paper towel and after emptying, touch the tip of the pipette to the inside of the conical flask. These procedures ensure that the $25.0\,cm^3$ are measured as accurately as possible.

4 Add excess iodide ions using a measuring cylinder to transfer $15\,cm^3$ of $0.5\,mol\,dm^{-3}$ potassium iodide to the conical flask. Since the amount of I_2 liberated is determined by the amount of chlorate(I) used, the volume of potassium iodide does not need to be precise. Add excess hydrogen ions using a measuring cylinder to transfer $20\,cm^3$ of $1\,mol\,dm^{-3}$ sulfuric acid.

5 The contents of the flask will be brown due to the iodine produced. (Figure 14a).

Step 2

Looking at Equation 1, you can see that 1 mole of chlorate(I) ions make 1 mole of iodine molecules.

$$I_2 + 2S_2O_3^{2-} \rightarrow 2I^- + S_4O_6^{2-} \qquad \textbf{Equation 2}$$

| I | 0 | | −1 | Reduced |
| S | | +2 | +2.5 | Oxidised |

Study tip

If you add too much indicator it could react with acids and alkalis and therefore produce less accurate results.

▲ Figure 14 *Stages in an iodine–thiosulfate titration*

You can titrate the iodine produced with sodium thiosulfate solution to work out the number of moles of iodine that are produced in Equation 1. Equation 2 is also redox. (Don't worry about the +2.5 oxidation state for the sulfur.)

1 Wash a burette is with water followed by a standard solution of $0.100\,mol\,dm^{-3}$ sodium thiosulfate solution. This ensures that the burette is clean and that the sodium thiosulfate solution hasn't been diluted.

2 Fill the burette with the sodium thiosulfate solution, making sure that the jet is full. The titre will be inaccurate if you start with an empty jet.

3 Put the conical flask on a white tile to make the end point easier to see.

4 Record the initial burette reading to the nearest $0.05\,cm^3$. It will be easier to read the burette if you put a white tile behind the graduations to see where the bottom of the meniscus is.

5 Start the rough titration. This gives you a rough idea of how much sodium thiosulfate is needed. You can add $1\,cm^3$ of sodium thiosulfate at a time. Near the end point the contents of the conical flask will be a pale straw colour (Figure 14b). When you see this it is the best time to add a few drops of starch solution. Now the contents of the flask will be blue black (Figure 14c). It will be easier to tell when you have reached the end point of colourless now (Figure 14d).

6 Record the final burette reading and calculate your rough **titre** by subtraction.

7 Wash the conical flask with distilled water. You don't need to dry it. Add the contents described in Step 1 to the conical flask.

8 Now carry out some accurate titrations. You can run the sodium thiosulfate into the conical flask until it is $1\,cm^3$ below the rough titre.

9 Now add the sodium thiosulfate drop wise so that you don't overshoot the end point. Also wash the inside of the flask and the end of the burette down regularly with distilled water from a wash bottle to ensure that all the reactants are together in the bottom of the conical flask.

10 Continue with the accurate titres until they are **concordant** – within $0.1\,cm^3$ of each other.

Redox titration

The procedure for a redox titration is similar to that used for acid–base titrations. The difference is in the type of reaction occurring, which is a redox reaction – electrons are transferred from one species to another. Often, there is no need for an indicator since one of the reactants or products is coloured, for example, when manganate(VII) ions are in the conical flask and are being reduced, the reaction is over when the last of the purple colour disappears.

Alternatively, if the manganate(VII) ions are being added from a burette to a colourless solution, the end point is when the first pale pink colour appears.

Using a colorimeter or visible spectrophotometer

Colorimeter

A colorimeter or visible spectrophotometer is used to determine the concentration of a coloured solution. Coloured solutions can absorb certain wavelengths of light. The amount of this light that is either absorbed or transmitted can be measured. This is proportional to the concentration of the solution.

Colorimeter

1   Select a filter with the complementary colour to the solution being tested. For example, a purple solution absorbs yellow light so you would choose a yellow filter. This allows only those wavelengths absorbed most strongly by the solution to pass through to the sample.

2   Make up a range of standard solutions of the test solution. There should be solutions both above and below the concentration of the unknown solution.

3   Zero the colorimeter using a tube/cuvette of pure solvent – this will be water in most cases.

4   Measure the absorbance of each of the standard solutions and plot a calibration curve of concentration against absorbance.

5   Measure the absorbance of the unknown sample and use the calibration curve to determine the concentration of the unknown solution.

By determining the concentrations of a reactant at different time intervals in a reaction you can follow the progress of that reaction.

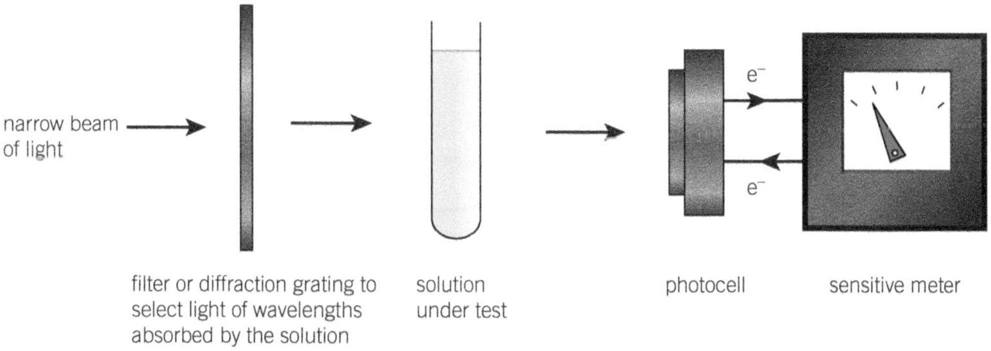

> **Synoptic link**
>
> See Topic EL 9, How Salty? for how to calculate the concentration of the acid.

> **Synoptic link**
>
> Making a standard solutions was covered earlier in Techniques and procedures.

narrow beam of light

filter or diffraction grating to select light of wavelengths absorbed by the solution

solution under test

photocell

sensitive meter

▲ **Figure 15** *A simple calorimeter*

Visible spectrophotometer

A visible spectrophotometer works in a similar way to a colorimeter in that a light source is passed through a sample and the amount of that light absorbed or transmitted is measured.

In the case of a colorimeter, the coloured light used as a source can only be selected from a specific number of different wavelengths (depending on the filter selected. A visible spectrophotometer can give data for absorption or transmission for any given value of the visible spectrum.

Measuring the energy transferred in a chemical reaction

Measuring the energy transfer when a fuel burns

This technique is used to determine the enthalpy change of combustion when a fuel is burnt. The values obtained experimentally can then be compared with theoretical values.

1 Using a measuring cylinder, pour a known volume of water into a copper calorimeter. Record its temperature.

2 Weigh a spirit burner – keep the cap on the burner to reduce loss of the fuel by evaporation.

3 Support the calorimeter over a spirit burner containing the fuel to be tested. Surround it with a draught excluder to help to reduce energy losses (Figure 16).

4 Remove the cap of the spirit burner and light the wick.

5 Use the thermometer to stir the water all the time it is being heated – carry on heating until the temperature has risen by 15 to 20 °C.

6 Extinguish the spirit burner and put the cap back in place. Keep stirring the water and make a note of the highest temperature reached.

7 Weigh the burner again.

The results can be used to calculate the enthalpy of combustion for the fuel under test.

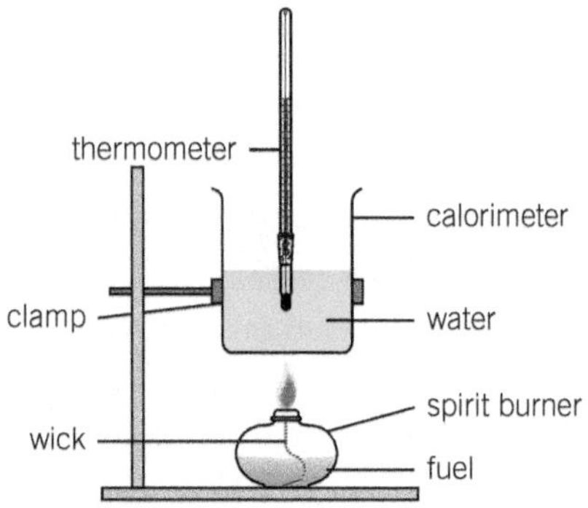

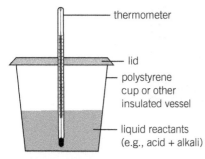

▲ **Figure 16** *Apparatus to determine the enthalpy of combustion*

Synoptic link

Look at Topic DF2, How much energy? For how to calculate the enthalpy of combustion.

Measuring the energy transferred when reactions occur in solution

This technique allows the enthalpy change of a reaction to be calculated through the measurement of changes in temperature when known quantities of reactants react together, for example, the enthalpy of neutralisation.

1 Using a measuring cylinder, add a known volume of a known concentration of acid to an insulated vessel and take the temperature.

2 Using a measuring cylinder, add a known volume of a known concentration of the alkali. Stir well to mix the reactants.

3 Top the vessel with a lid with a hole in (Figure 17).

4 Place the thermometer through the hole in the lid and record changes in temperature every 30 seconds until there are no further changes in temperature.

5 Calculate the maximum increase in temperature.

The reactions take place in insulated vessels to minimize the transfer of thermal energy to the surroundings. Use this information to calculate the enthalpy change of neutralisation.

These reactions may involve two solutions or solids reacting with solutions, for example, dissolving solids in solution.

Reactions of solids and solutions

1 Using a measuring cylinder add a known volume of a known concentration of the reactant solution into an insulated vessel. Take the temperature.

2 Add a known mass of solid reactant. This should be in excess.

▲ **Figure 17** *Apparatus to determine the enthalpy of neutralisation*

Study tip

Remember a rise in temperature indicates an exothermic reaction while a fall in temperature indicates an endothermic reaction.

3 Top the vessel with a lid with a hole in.

4 Place the thermometer through the hole in the lid and record changes in temperature every 30 seconds until there are no further changes in temperature.

In order to calculate the maximum change in temperature, you will need to plot a graph of temperature against time (Figure 18).

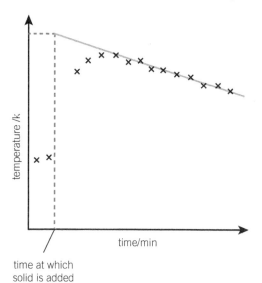

time at which
solid is added

▲ **Figure 18** *Graph of temperature against time showing a line of best fit and extrapolation*

Once a line of best fit has been drawn you need to be able to estimate the temperature rise immediately following the mixing of reagents.

The values recorded experimentally are too low because the reaction does not occur instantaneously but over a period of time. Throughout the reaction, heat is released and some is lost from the reaction vessel to the wider surroundings. The theoretical maximum temperature change is never achieved experimentally and must be found by extending the line of best fit back to the time at which the reactants were mixed. Extending data points beyond those actually measured is called **extrapolation**.

Other techniques

Carrying out the electrolysis of aqueous solutions

During electrolysis of an aqueous solution, an electric current is passed through the electrolyte. For electrolysis to occur in an aqueous solution an electrical circuit needs to be set up (Figure 19)

The dc power supply may be a powerpack or batteries. The most common material for electrodes (anode and cathode) is graphite. This conducts electricity, is cheap, and is relatively inert (unreactive). Platinum may also be used but it is expensive.

If the products of electrolysis are gaseous they can be collected using the apparatus in Figure 20.

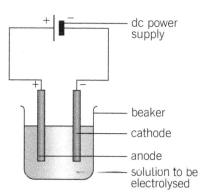

▲ **Figure 19** *Apparatus for electrolysis*

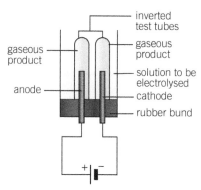

▲ **Figure 20** *Apparatus to collect gaseous products in electrolysis*

The test tubes are filled with water at the start of the reaction. The gaseous products displace the water and are collected in the test tubes. Once the powerpack is switched on the electrolys is will proceed.

If electrolysis is to be carried out for the purpose of purification of a metal, the anode will need to be made of the impure metal, the electrolyte must contain ions of that metal, and the cathode should be made of the pure metal.

Setting up and using electrochemical cells

This technique is used to determine the potential of an electrochemical cell. Standard electrode potentials can be measured by connecting any half-cell to a standard hydrogen half-cell, or a calibrated reference half-cell.

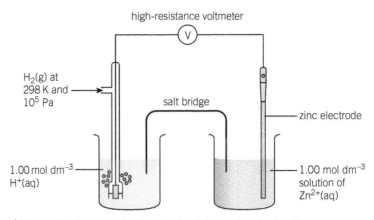

▲ **Figure 21** *An example of a standard electrochemical cell.*

1   Construct the half-cell whose electrode potential is to be measured. For a metal ion/metal half-cell (e.g., the Cu^{2+}/Cu half-cell), the electrode will be made from the metal which is in its ionic form in the solution. If the reaction involves two ions of the same element in different oxidation states (e.g., Fe^{3+}/Fe^{2+}), the electrode should be either platinum or carbon (graphite) and the solution will contain a mixture of the two ions. Ensure in all cases all the solutions have a concentration of $1.0\,mol\,dm^{-3}$ and are at $298\,K$.

2   Connect the half-cell to a standard hydrogen half-cell or other reference cell using a high-resistance voltmeter and salt bridge (Figure 21).

3   Connect the two electrodes to a high resistance voltmeter. Check that the reading on the voltmeter is positive – if it is then the half-cell connected to the positive terminal of the voltmeter is the positive electrode. If the reading is negative, change the connections round on the voltmeter to give a positive reading.

4   Record the voltmeter reading – in the case of a measurement taken using a hydrogen half-cell this value is the required cell e.m.f. $E^{\ominus}_{cell}$.

Measuring the pH of a solution

Before taking the pH of a solution, the electrode (sometimes called pH probe) will need to be calibrated using at least two solutions of known pH.

Calibrating the pH electrode

Since pH is temperature dependent you will need to calibrate for temperature once the electrode is connected to the meter (Figure 22). Sometimes this is automatically calibrated for by the machine.

Wash the electrode with distilled water. Transfer the electrode into a buffer solution of pH 7.00. Check the bulb of the electrode is completely immersed in the buffer solution. Wait for the reading to stabilize. Ensure the meter reads 7.00, adjusting to that value if necessary.

To measure the pH of acidic solutions, calibrate the electrode using an acidic buffer solution, typically with a pH 4.00. To measure alkaline solutions, calibrate the electrode using an alkaline buffer solution with pH 10.00.

To measure both acidic and alkaline solutions with a wide range of values, calibrate with acidic and alkaline buffer solutions, as well as a pH 7.00 buffer solution.

Once calibrated you can measure the pH of any solution by rinsing the electrode then immersing it in the solution to be measured.

Study tip

Buffers with a high pH tend to absorb carbon dioxide, so they need to be fresh when used.

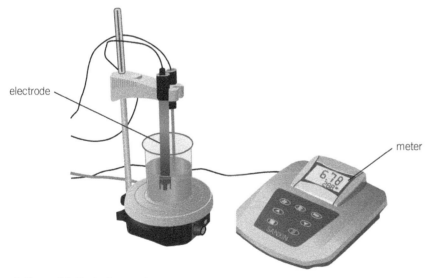

▲ **Figure 22** *Typical setup for a pH meter*

Cracking a hydrocarbon vapour over a heated catalyst

Cracking is used to break longer molecules down into smaller molecules.

1 Set up the apparatus below, ensuring there is space above the catalyst (Al_2O_3) to allow gases to pass freely over over it.

Study tip

Be aware of what the products of the electrolysis will be. For example, if chlorine gas is produced the activity will need to be carried out in a well-ventilated space such as a fume cupboard.

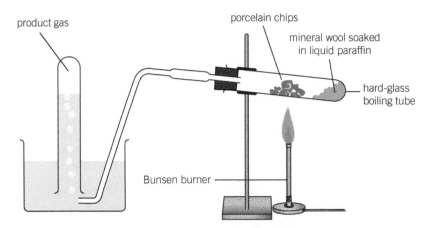

▲ **Figure 23** *Apparatus for cracking a mixture of alkanes*

2   Place several test tubes in the water in the collection trough.

3   Heat the catalyst strongly. This ensures that when the alkane vapour passes over it the temperature is high enough for the cracking reactions to take place.

4   Heat the alkane gently, collecting any gases that pass into the collection tubes, changing and corking full tubes. Continue to heat whilst changing the collection tubes. This prevents 'suck-back'.

5   Discard the first tube of gas. This will just be displaced air, rather than product.

6   Continue heating both the catalyst and the alkane mixture to be cracked until you have collected several tubes of gas or until no more gas is produced.

7   Remove the delivery tube from the collection trough before stopping heating the catalyst and alkane mixture. This prevents suck back.

8   Leave to cool then dismantle the apparatus.

9   Test any liquid product from the middle collection tube with bromine water. The bromine water should remain yellow/brown.

10 Test the gas collected in tubes by shaking with bromine water. The bromine water should decolourise.

Testing for the presence of unsaturation in alkenes

The addition of bromine water to an alkene is a reliable test for the presence of carbon-carbon double bonds.

1   To a few drops of the unknown liquid or gas add a few drops of bromine water and shake well.

2   If the unknown is an alkene the bromine water will change from orange/yellow to colourless (it is decolourised). If the sample is not unsaturated it will not decolourise the bromine water and it will remain orange/yellow.

Determining K_{sp}

The K_{sp} of any sparingly soluble salt is determined from a saturated solution of that salt. The first stage is to prepare a saturated solution in deionised water.

Study tip

Tap water contains ions in solution and would therefore alter the concentration of ions in solution.

Making a saturated solution.

1   Warm distilled water in a small conical flask and add the salt you are looking to determine the K_{sp} of, shaking frequently.

2   Keep adding solute until no more dissolves (you will know this is the case because there will be undissolved solid).

3   Leave the mixture to cool to room temperature.

4   There should still be solid at the bottom of the flask. If not repeat Steps 1–3 until this is the case. Filter the mixture through a filter paper and discard the residue.

Determining the concentration of ions in solution

K_{sp} is temperature dependent so take the temperature of the solutions you will be working with.

The second stage is to determine the concentration of one of the ions in solution. This will allow K_{sp} to be calculated. The concentration of one ion is sufficient since the concentration of one ion is proportional to the concentration of the other ion in solution, for example, for AgCl:

$$[Ag^+(aq)] = [Cl^-(aq)].$$

The method of determination for one of the ions will depend on the ions in solution.

●   For a basic solution, for example, calcium hydroxide, the concentration of the hydroxide ions can be determined by titration with hydrochloric acid, or potassium hydrogen tartrate can be titrated against standardised sodium hydroxide solution.

●   For a coloured solution, the concentration of the coloured ion (e.g., a metal cation such as Cu^{2+} or an anion such as MnO_4^{2-}) can be determined by using colorimetry.

Determining an equilibrium constant

In order to determine an equilibrium constant, the following stages are needed.

1   A mixture of reactants is allowed to reach equilibrium at a known temperature. This may happen rapidly or may take a number of hours.

2   Once a mixture has reached equilibrium, determine the concentration of one of the components in the equilibrium mixtures. The concentrations of the other components of the mixture can be calculated from this value and the equation for the reaction. The method you use will depend on the substances in the equilibrium mixture. Quantitative analysis methods might include redox titrations, acid–base titrations, colorimetric analysis, or pH measurements.
You will also need to know the concentrations of the substances in the initial mixture before the reaction starts.

3   Use a balanced equation to write K_c and insert the data from step 2.

4   Calculate K_c, including the units and the temperature at which the value of K_c is valid.

Study tip

When calculating K_{sp} the ratio of ion concentration is not always 1:1, so ensure you know the formula of the sparingly soluble salt before carrying out any calculation.

Study tip

Remember when you report K_{sp} to use the correct units and report the temperature for which this value is valid.

Synoptic link

Acid–base titrations were covered earlier in Techniques and procedures.

Synoptic link

Colorimetry was covered earlier in Techniques and procedures.

Synoptic link

Details on these titrations, colorimetry, and using a pH meter and covered earlier in Techniques and procedures.

For example, the hydrolysis of ethyl ethanoate forms an equilibrium system.

$$CH_3COOCH_2CH_3(l) + H_2O(l) \rightleftharpoons CH_3COOH(l) + CH_3CH_2OH$$

ethyl ethanoate $\qquad$ ethanoic acid $\quad$ ethanol

The concentration of ethanoic acid in the equilibrium mixture at 298 K can be found by titrating a known volume of the equilibrium mixture with a standardised sodium hydroxide solution. The following assumptions can be made:

1 $[CH_3COOH(l)]_{eq} = [CH_3CH_2OH(l)]_{eq}$

2 If the concentration of ethanoic acid in the equilibrium mixture is $x\,mol\,dm^{-3}$, the initial concentration of ethylethanoate was $y\,mol\,dm^{-3}$ and the initial concentration of water was $z\,mol\,dm^{-3}$.

$$K_c = \frac{[CH_3COOH(l)][CH_3CH_2OH(l)]}{[CH_3COOCH_2CH_3(l)][H_2O(l)]} = \frac{x^2}{(y-x)(z-x)} \text{ at } 298\,K$$

Uncertainties of measurements

There are several words that are used in connection with measurements that often get mixed up.

Accuracy is a measure of the closeness of agreement between an individual test result and the accepted reference value. If a test result is accurate, it is in close agreement with the accepted reference value.

Precision is the closeness of agreement between independent measurements obtained under the same conditions. It depends only on the distribution of random errors (i.e. the spread of measurements) and does not relate to the true value.

Error (of measurement) is the difference between an individual measurement and the true value (or accepted reference value) of the quantity being measured.

Uncertainty is an estimate attached to a measurement which characterises the range of values within which the true value is asserted to lie. This is normally expressed as a range of values such as 44.0 ± 0.4.

Reliability is the opposite of uncertainty, that is, if the uncertainty is great, the measurement is not very reliable.

not accurate
not precise

accurate
not precise

not accurate
precise

accurate
precise

▲ **Figure 24** *The different between accuracy and precision*

Uncertainties

When you carry out practical work, uncertainties are always present in any measurement. All measuring apparatus you use have an uncertainty associated with them.

As a general rule, the uncertainty is usually taken to be half a division on either side of the smallest unit on the scale you are using. However, the accuracy of measurements also depends on the quality of the apparatus used, such as a balance, thermometer, or glassware. For example, a $100\,cm^3$ measuring cylinder is graduated in divisions every $1\,cm^3$ – it has an uncertainty of half a division or $0.5\,cm^3$

The uncertainty of a piece of glassware is usually marked on it.

You will normally be told the absolute uncertainty in any measurement that you make so that you can use this to calculate the percentage uncertainty in a given measurement. If you are using a measuring cylinder, you should choose the smallest measuring cylinder for the volume to be measured because this will offer the lowest uncertainty. Measuring cylinders themselves have higher uncertainty than equipment such as burettes, volumetric pipettes, and volumetric flasks.

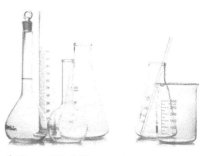

▲ Figure 25 *Different laboratory glassware will have different uncertainties in their measurements*

Examples of uncertainties

Some examples are shown below, although the actual uncertainty on a particular item of glassware you use may be different from the values given below.

Glassware	Uncertainty
volumetric or standard flask (Class B)	0.2 cm^3
pipette (Class B)	0.06 cm^3
burette (Class B)	0.05 cm^3 in each measurement

Calculating percentage uncertainties

Calculating the percentage uncertainties of measurements allows you to quantify uncertainties and enables you to compare them. The significance of the uncertainty in a measurement depends on how large a quantity is being measured.

$$\text{percentage uncertainty} = \frac{\text{uncertainty}}{\text{measured quantity}} \times 100$$

For example, a two-decimal place balance may have an uncertainty of 0.005 g. For a mass measurement of 2.56 g:

$$\text{percentage uncertainty} = \frac{0.005}{2.56} \times 100 = 0.2\%$$

For a mass measurement of 0.12 g, the percentage uncertainty is much greater:

$$\text{percentage uncertainty} = \frac{0.005}{0.12} \times 100 = 4.2\%$$

Multiple measurements

Where you measure quantities by difference, there will be an uncertainty in each measurement, which you should combine to give the uncertainty in the final value. For multiple measurements using the same two-decimal place balance, there will be an uncertainty of 0.005 g for each measurement.

For two mass measurements that give a resultant mass by different, there are two uncertainties. These uncertainties are combined to give the uncertainty in the resultant mass.

The formula for the percentage uncertainty is therefore:

$$\text{percentage uncertainty} = \frac{2 \times \text{uncertainty}}{\text{measured quantity}} \times 100$$

For example, using the same two-decimal place balance in an experiment to find the water of crystallisation of a salt:

mass of crucible + crystals before heat = 23.45 g   uncertainty = 0.005 g

mass of crucible + crystals after heat = 23.21 g   uncertainty = 0.005 g

mass lost = 0.23 g   overall uncertainty = 2×0.005 g

The uncertainty in each mass measurement is small but the overall percentage uncertainty in the mass loss is much greater:

$$\text{percentage uncertainty in mass loss} = \frac{2 \times 0.005}{0.23} \times 100 = 4\%$$

The same principle can be applied to other quantities measured by difference, such as temperature difference and the titre from a titration.

For example, in a titration:

initial burette reading = 0.50 cm^3 uncertainty = 0.05 cm^3

final burette reading = 23.00 cm^3 uncertainty = 0.05 cm^3

titre = 22.50 cm^3 overall uncertainty = 2×0.05 cm^3

$$\text{percentage uncertainty} = \frac{2 \times 0.005}{22.50} \times 100 = 0.44\%$$

Recording measurements

When you use a digital measuring device (such as a top pan balance or ammeter) you should record all the digits shown on the instrument. When using a non-digital device (such as a ruler or a burette) however, you should record all the figures that are known for certain plus one that is estimated.

A burette is graduated in divisions every 0.1 cm^3. A burette is a non-digital device, so you record all figures that are known for certain plus one that is estimated. Using the half-division rule, the estimation is one of 0.05 cm^3. You therefore record burette measurements to two decimal places with the last figure either 0 or 5.

Periodic table

Key
atomic number
Symbol
name
relative atomic mass

(1)	(2)												(3)	(4)	(5)	(6)	(7)	(0)
1																		**18**
1 **H** hydrogen 1.0																		2 **He** helium 4.0
	2											**13**	**14**	**15**	**16**	**17**	
3 **Li** lithium 6.9	4 **Be** beryllium 9.0												5 **B** boron 10.8	6 **C** carbon 12.0	7 **N** nitrogen 14.0	8 **O** oxygen 16.0	9 **F** fluorine 19.0	10 **Ne** neon 20.2
11 **Na** sodium 23.0	12 **Mg** magnesium 24.3	**3**	**4**	**5**	**6**	**7**	**8**	**9**	**10**	**11**	**12**	13 **Al** aluminium 27.0	14 **Si** silicon 28.1	15 **P** phosphorus 31.0	16 **S** sulfur 32.1	17 **Cl** chlorine 35.5	18 **Ar** argon 39.9
19 **K** potassium 39.1	20 **Ca** calcium 40.1	21 **Sc** scandium 45.0	22 **Ti** titanium 47.9	23 **V** vanadium 50.9	24 **Cr** chromium 52.0	25 **Mn** manganese 54.9	26 **Fe** iron 55.8	27 **Co** cobalt 58.9	28 **Ni** nickel 58.7	29 **Cu** copper 63.5	30 **Zn** zinc 65.4	31 **Ga** gallium 69.7	32 **Ge** germanium 72.6	33 **As** arsenic 74.9	34 **Se** selenium 79.0	35 **Br** bromine 79.9	36 **Kr** krypton 83.8
37 **Rb** rubidium 85.5	38 **Sr** strontium 87.6	39 **Y** yttrium 88.9	40 **Zr** zirconium 91.2	41 **Nb** niobium 92.9	42 **Mo** molybdenum 95.9	43 **Tc** technetium	44 **Ru** ruthenium 101.1	45 **Rh** rhodium 102.9	46 **Pd** palladium 106.4	47 **Ag** silver 107.9	48 **Cd** cadmium 112.4	49 **In** indium 114.8	50 **Sn** tin 118.7	51 **Sb** antimony 121.8	52 **Te** tellurium 127.6	53 **I** iodine 126.9	54 **Xe** xenon 131.3
55 **Cs** caesium 132.9	56 **Ba** barium 137.3	57–71 lanthanoids	72 **Hf** hafnium 178.5	73 **Ta** tantalum 180.9	74 **W** tungsten 183.8	75 **Re** rhenium 186.2	76 **Os** osmium 190.2	77 **Ir** iridium 192.2	78 **Pt** platinum 195.1	79 **Au** gold 197.0	80 **Hg** mercury 200.6	81 **Tl** thallium 204.4	82 **Pb** lead 207.2	83 **Bi** bismuth 209.0	84 **Po** polonium	85 **At** astatine	86 **Rn** radon
87 **Fr** francium	88 **Ra** radium	89–103 actinoids	104 **Rf** rutherfordium	105 **Db** dubnium	106 **Sg** seaborgium	107 **Bh** bohrium	108 **Hs** hassium	109 **Mt** meitnerium	110 **Ds** darmstadtium	111 **Rg** roentgenium	112 **Cn** copernicium		114 **Fl** flerovium		116 **Lv** livermorium		

57 **La** lanthanum 138.9	58 **Ce** cerium 140.1	59 **Pr** praseodymium 140.9	60 **Nd** neodymium 144.2	61 **Pm** promethium 144.9	62 **Sm** samarium 150.4	63 **Eu** europium 152.0	64 **Gd** gadolinium 157.2	65 **Tb** terbium 158.9	66 **Dy** dysprosium 162.5	67 **Ho** holmium 164.9	68 **Er** erbium 167.3	69 **Tm** thulium 168.9	70 **Yb** ytterbium 173.0	71 **Lu** lutetium 175.0
89 **Ac** actinium	90 **Th** thorium 232.0	91 **Pa** protactinium	92 **U** uranium 238 1	93 **Np** neptunium	94 **Pu** plutonium	95 **Am** americium	96 **Cm** curium	97 **Bk** berkelium	98 **Cf** californium	99 **Es** einsteinium	100 **Fm** fermium	101 **Md** mendelevium	102 **No** nobelium	103 **Lr** lawrencium

Answers

EL 1

1

Isotope	Symbol	Atomic Number	Mass Number	Number of neutrons
carbon-12	$^{12}_{6}C$	6	12.0	6
carbon-13	$^{13}_{6}C$	6	13.0	7
oxygen-16	$^{16}_{8}O$	8	16.0	8
strontium-90	$^{90}_{38}Sr$	38	90.0	52
iodine-131	$^{131}_{53}I$	53	131.0	78
iodine-121	$^{121}_{53}I$	53	121.0	68

2 a 35 protons, 44 neutrons, 35 electrons

  b 35 protons, 46 neutrons, 35 electrons

  c 17 protons, 18 neutrons, 17 electrons

  d 17 protons, 20 neutrons, 17 electrons

3 a $A_r(Br) = 80.0$

  b $A_r(Ca) = 40.1$

4 a $100 - x$

  b $193x$

  c $191(100 - x)$

  d $193x + 191(100 - x)$

  e $[193x + 191(100 - x)] \div 100$

  f 60% iridium-193, 40% iridium-191

5 a $^{7}_{3}Li + ^{1}_{1}p \rightarrow 2^{4}_{2}He$

  b $^{14}_{7}N + ^{1}_{0}n \rightarrow ^{14}_{6}C + ^{1}_{1}p$

6 40% antimony-123, 60% antimony-121

EL 2

1 green

2 a bright red   **d** brick red

  b yellow   **e** apple green

  c lilac   **f** green-blue

3 similarities – line spectrum; lines in same place (same frequency); lines get closer up; frequency

differences – black lines on a bright background for absorption spectrum

4 a Arrow points down. Shorter arrow represents red line.

  b Arrow points down. Longer arrow represents green line.

(Both arrows must start and finish on lines; does not matter which levels they go between.)

5 $5.5 \times 10^{14} s^{-1}$

EL 3

1 a The electron is in the first electron shell.

  b The electrons are in an s type orbital.

  c There are two electrons in this orbital.

2 a $1s^2 2s^2 2p^1$

  b $1s^2 2s^2 2p^6 3s^2 3p^3$

3 $Z = 16$. The element is sulfur.

4 a chlorine   **c** titanium

  b potassium   **d** tin

EL 4

1 a 2,1

  b 2,8,5

  c 2,8,8,2

2 A, C, and E (Group 2)

3

Electronic shell configuration	Group	Period
2,8,7	7	3
2,3	3	2
2,8,6	6	3
2,1	1	2
2,8,1	1	3
2,8,8,1	1	4

4 a s-block   **c** p-block

  b p-block   **d** s-block

5 a d-block   **c** f-block

  b p-block   **d** s-block

EL 5

1 a   **d**   **b**   **e**   **c**   **f**

2 a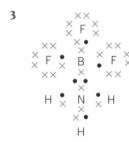

b

H×C×C×H

c

H×C×O×H
  H

3

×F×
×F× B ×F×
H×N×H
  H

4 a   H   109.5°

  H   Si   H

  H

b H×S×H   104.5°

c H×P×H   107°
     H

d O×C×O   180°

e F×S×F   104.5°

f   Cl   120°
     Cl× B ×Cl

g H×C×C×H   180°

5 a   H H   109.5°
   H—C—C—H   is written CH₃CH₃
      H H

b 109.5° H 104.5°
   H—C—O—H
      H

c 109.5° H   H
   H—C—N—H
      H H 107°

d H   H 120°
      C=C
     H   H

e   H 180°
   H—C—C≡N
   109.5° H

6 a pyramidal

b trigonal planar

c octahedral

EL 6

1 a 144

b neodymium, Nd

2 a $2Mg + O_2 \rightarrow 2MgO$

b $2H_2 + O_2 \rightarrow 2H_2O$

c $CaCO_3 + 2HCl \rightarrow CaCl_2 + CO_2 + H_2O$

d $2HCl + Ca(OH)_2 \rightarrow CaCl_2 + 2H_2O$

e $2CH_3OH + 3O_2 \rightarrow 2CO_2 + 4H_2O$

3 a $Zn(s) + H_2SO_4(aq) \rightarrow ZnSO_4(aq) + H_2(g)$

b $MgCO_3(s) \rightarrow MgO(s) + CO_2(g)$

c $BaO(s) + 2HCl(aq) \rightarrow BaCl_2(aq) + H_2O(l)$

4 a The mass of the sample is needed to be sure that iodine and oxygen are the only elements in the compound.

b The relative number of moles of iodine and oxygen.

c To change the relative number of moles into the ratio of moles of oxygen relative to 1 mole of iodine.

d In order to produce a ratio involving whole numbers.

e I_2O_5, I_4O_{10}, I_6O_{15}, etc.

f The molar mass is needed.

5 a 1 **b** 0.5

c Atoms of copper are approximately twice as heavy as atoms of sulfur. Thus the same mass contains only half as many moles of copper as it does of sulfur.

6 a 2 **b** 0.02 **c** 5 **d** 1.0×10^6

EL 7

1 a $\left[Li \right]^+$ $\left[H× \right]^-$ **b** $\left[K \right]^+$ $\left[×F× \right]^-$

c $\left[\text{Mg} \right]^{2+} \left[\overset{\bullet\times}{\underset{\times\times}{\overset{\bullet}{\text{O}}\times}} \right]^{2-}$ **d** $\left[\text{Ca} \right]^{2+} \left[\overset{\bullet\times}{\underset{\times\times}{\times\text{S}\times}} \right]^{2-}$

2 a macromolecular (covalent molecular)

   b simple molecular (covalent molecular)

   c ionic (giant lattice)

3 a $\left[\text{Ca} \right]^{2+} \left[\overset{\times\times}{\underset{\times\times}{\bullet\text{ Cl}\times}} \right]^{-} \left[\overset{\times\times}{\underset{\times\times}{\bullet\text{ Cl}\times}} \right]^{-}$

   b $\left[\text{Na} \right]^{+} \left[\text{Na} \right]^{+} \left[\overset{\bullet\times}{\underset{\times\times}{\bullet\text{ S}\times}} \right]^{2-}$

4 a In a normal covalent bond, each atom supplies a single electron to make up the pair of electrons involved in the bond. In a dative covalent bond one atom supplies both electrons.

   b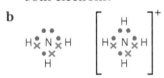

5 a ionic (giant lattice)

   b metallic (giant lattice)

6 a $\left[\text{Na} \right]^{+} \left[\text{Na} \right]^{+} \left[\text{Na} \right]^{+} \left[\overset{\bullet\times}{\underset{\bullet\times}{\times\text{ N}\times}} \right]^{3-}$

   b $\left[\text{Al} \right]^{3+} \left[\overset{\bullet\times}{\underset{\times\times}{\times\text{ F}\times}} \right]^{-} \left[\overset{\bullet\times}{\underset{\times\times}{\times\text{ F}\times}} \right]^{-} \left[\overset{\bullet\times}{\underset{\times\times}{\times\text{ F}\times}} \right]^{-}$

EL 8

1 charge on the ion and radius of the ion

2 a magnesium + steam →
   magnesium hydroxide + hydrogen
   $Mg(s) + 2H_2O(g) \rightarrow Mg(OH)_2(s) + H_2(g)$

   b calcium oxide + hydrochloric acid →
   calcium chloride + water
   $CaO(s) + 2HCl(aq) \rightarrow CaCl_2(aq) + H_2O(l)$

   c beryllium carbonate →
   beryllium oxide + carbon dioxide
   $BeCO_3(s) \rightarrow BeO(s) + CO_2(g)$

   d barium hydroxide + sulfuric acid →
   barium sulfate + water
   $Ba(OH)_2(aq \text{ or } s) + H_2SO_4(aq) \rightarrow$
   $BaSO_4(s) + 2H_2O(l)$

3 a 1st ionisation $Ca(g) \rightarrow Ca^+(g) + e^-$
   2nd ionisation $Ca^+(g) \rightarrow Ca^{2+}(g) + e^-$
   3rd ionisation $Ca^{2+}(g) \rightarrow Ca^{3+}(g) + e^-$

   b Once an electron has been removed the remaining electrons are held more tightly. Hence it is more difficult to remove a second electron.

   c The second ionisation enthalpy involves removal of an electron from shell 4 but the third involves removal of an electron from shell 3 which is closer to the nucleus.

EL 9

1

Ion	Concentration / g dm^{-3}	Concentration / mol dm^{-3}
Cl$^-$	183.0	5.15
Mg^{2+}	36.2	1.51
Na$^+$	31.5	1.37
Ca^{2+}	13.4	0.355
K$^+$	6.8	0.174
Br$^-$	5.2	0.065
SO$_4{}^{2-}$	0.6	0.006 25

2 a 117 g   **b** 3.95 g   **c** 1.4 g
   d 9930 g   **e** 0.0024 g

3 a 4.4×10^{-3} moles of NaOH(aq) in 25 cm^3 gives concentration of 0.176 mol dm^{-3}

   b No effect on next titre because pipette delivers same amount of NaOH(aq) as before.

4 a A solution of accurately known concentration.

   b Solid NaOH has the property of absorbing water (and carbon dioxide) from the air so it is not possible to accurately weigh NaOH.

DF 1

1 a Enthalpy change when one mole of the compound is burnt completely in oxygen, under standard conditions.

   b Enthalpy change when one mole of a compound is formed from its elements, with both the compound and its elements being in their standard states.

2 Formation of a compound from its elements may be an exothermic reaction ($\Delta_f H$ negative) or an endothermic reaction ($\Delta_f H$ positive). However, energy is liberated whenever a substance burns, so combustion reactions are always exothermic ($\Delta_c H$ negative).

3

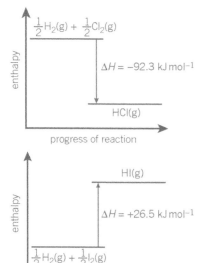

4 $\Delta_c H^{\ominus}(C)$ is enthalpy change when one mole of carbon is burnt completely under standard conditions.

$\Delta_f H^{\ominus}(CO_2)$ is enthalpy change when one mole of carbon dioxide is formed from its elements with both the carbon dioxide and its constituent elements in their standard states.

same equation for both:
$C(s) + O_2(g) \rightarrow CO_2(g)$

5 **a** $H_2(g) + \frac{1}{2}O_2(g) \rightarrow H_2O(l)$

b $\Delta_f H(H_2O) = -286\,kJ\,mol^{-1}$

c $1g\,H_2 = \frac{1}{2}\,mol = -143\,kJ$

Assumptions: complete reaction occurred and no heat is lost.

d $\Delta H = +286\,kJ\,mol^{-1}$

6 **a** $n(NaOH) = n(HCl) = 2 \times 10^{-3}\,mol$

$58\,000 \times 2 \times 10^{-3}\,(=116)\,J$

$\Delta T = \dfrac{116}{40 \times 4.18} - 0.69\,^{\circ}C$

b amount (mol) doubles in same volume, temp doubles = $1.4\,^{\circ}C$

c $n(NaOH) = 1 \times 10^{-3}\,mol$ (limiting)

$\Delta T = \dfrac{58}{30 \times 4.18} = 0.46\,^{\circ}C$

d same reaction $H^+ + OH^- \rightarrow H_2O$

same temp rise as **a** = $0.69\,^{\circ}C$

e $n(NaOH) = 2 \times 10^{-3}$ (limiting)

volume same as **a**, so temp rise same = $0.69\,^{\circ}C$

f $n(NaOH) = n(H^+)$

$n(H_2O)$ formed $= 4 \times 10^{-3}\,mol$

$58\,000 \times 4 \times 10^{-3} = 232\,J$

$\Delta T = \dfrac{232}{60 \times 4.18} = 0.93\,^{\circ}C$

DF 2

> ### Using a bomb calorimeter to accurately measure energy changes
>
> **1** Increased number of successful collisions leading to complete combustion of the whole sample.
> **2** Much lower heat losses in a bomb calorimeter.
> **3** Energy requirements to keep the volume constant are different to the energy requirements to keep the pressure constant.

1 **a** thermometer, measuring cylinder, gas meter

b volume of water used, temperature rise of water, volume of gas used

c cooling losses, impurities in gas, incomplete combustion, non-standard conditions

2 **a** $4C(s) + 5H_2(g) \rightarrow C_4H_{10}(g)$

b

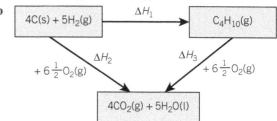

c $\Delta H_1 = \Delta H_2 - \Delta H_3$
$= 4(-393) + 5(-286) - (-2877)$
$= -125\,kJ\,mol^{-1}$

3 **a** The water starts dissolving products or reactants before completing the reaction.

b $n(H_2O) = 5 \times n(CuSO_4) = 5 \times 0.025$

mass H_2O = vol $H_2O = 5 \times 0.025 \times 18 = 2.25\,cm^3$

This is $(50 - 2)$ to the precision with which the volume is measured.

c first expt: $-65.9\,kJ\,mol^{-1}$

second expt: $+10.7\,kJ\,mol^{-1}$

d no heat losses

(specific heat capacity of water × vol of water) = (specific heat capacity of solution × mass of solution)

calorimeter absorbs negligible thermal energy

e

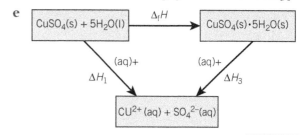

f $\Delta_r H = \Delta H_1 - \Delta H_2 = -65.9 - 10.7 = -76.6$

g $H = -77 \pm 1 \text{ kJ mol}^{-1}$

h the measuring cylinder used to measure the volume of the water

I −77.4 lies within −77±1.

In this experiment systematic errors such as 'heat losses' have been reduced to below the level of apparatus uncertainty. So to improve the acuracy of the measurement, these uncertainties would be reduced.

DF 3

1

Empirical formula	Molecular formula	M_r
C_3H_8	C_3H_8	44.0
CH_2	$C_{12}H_{22}$	168.0
CH	C_6H_6	78
$C_{10}H_{21}$	$C_{20}H_{42}$	282.0
CH_2	C_5H_{10}	70
CH	C_2H_2	26.0
C_5H_4	$C_{10}H_8$	128

2 a C_7H_{16} **b** $C_{16}H_{34}$ **c** $C_{20}H_{42}$

3 a

$$H \overset{\bullet}{\underset{\bullet}{\times}} \overset{H}{\underset{H}{C}} \overset{\times}{\underset{\times}{\bullet}} \overset{H}{\underset{H}{C}} \overset{\times}{\underset{\bullet}{\times}} H$$

b

$$\begin{matrix} H & & & H \\ & C & \times & C \\ H & & & H \end{matrix}$$

c

$$H \overset{\bullet}{\underset{\bullet}{\times}} \overset{H}{\underset{H}{C}} \overset{\times}{\underset{\bullet}{\times}} \overset{H}{\underset{H}{C}} \overset{\times}{\underset{\bullet}{\times}} \overset{H}{\underset{H}{C}} \overset{\times}{\underset{\bullet}{\times}} H$$

4 a CH_2 **b** C_2H_4

5 a C_2H_5

b Mathematically, the molecular formula could be any multiple of C_2H_5 but the only one that is chemically feasible is C_4H_{10}.

DF 4

1 a $C_3H_8(g) + 5O_2(g) \rightarrow 3CO_2(g) + 4H_2O(g)$

b

$$H-\overset{H}{\underset{H}{C}}-\overset{H}{\underset{H}{C}}-\overset{H}{\underset{H}{C}}-H + 5O=O \longrightarrow 3O=C=O + 4H-O-H$$

c 2 C—C, 8 C—H, 5 O=O

d 6 C=O, 8 O—H **f** −8542 kJ mol^{-1}

e +6488 kJ mol^{-1} **g** −2054 kJ mol^{-1}

2 $\Delta H = -122 \text{ kJ mol}^{-1}$

3 $\Delta H = -131 \text{ kJ mol}^{-1}$

4 a $-CH_2 + 1\frac{1}{2}O_2 \rightarrow CO_2 + H_2O$

b $\Delta H = -618 \text{ kJ mol}^{-1}$

c bond enthalpies are averages
alcohols and water are in the liquid state (which has not been allowed for)

DF 5

1 a zeolite – heterogeneous

b platinum on aluminium oxide – heterogeneous

2 a $2CO + 2NO \rightarrow 2CO_2 + N_2$

b both are toxic/form photochemical smog

c chemically bound to the surface of another substance

d the reaction they catalyse is slow at low temperatures **or** do not work properly until at high temperatures

3 The mechanism should show CO and NO being adsorbed to the catalyst surface.

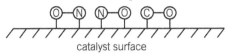

The bonds in the molecules are weakened and new bonds form between the atoms.

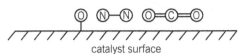

Products are released from the surface.

DF 6

Name	Structure	Skeletal formula
pent-2-ene	$CH_3CHCHCH_2CH_3$	
hepta-1,4-diene	$CH_2CHCH_2CHCHCH_2CH_3$	
4-methylpent-2-ene		
3-ethylhept-1-ene	$CH_2CHCH(CH_2CH_3)$ $CH_2CH_2CH_2CH_3$	
cyclopentene		
cyclopenta-1,3-diene		

2 a

H—C—C=C + Br₂ ⟶ H—C—C—C—H

b

H—C—C=C + H₂ ⟶ H—C—C—C—H

3

H—C—C—C—H   H—C—C—C—H

4 a bromine, room temperature

b steam with phosphoric acid catalyst, high temperature and pressure

c (hydrogen) with platinum catalyst room temperature and pressure

d hydrogen with nickel catalyst 150 °C, 5 atm pressure

5 Shake the organic compound in a test-tube with bromine water (or 'aqueous bromine'). Colour change from brown/orange/yellow to colourless.

6 a Polarised hydrogen bromine approaches C=C bond. Curly arrow from C=C bond to H$^{\delta+}$ and curly arrow from H—Br bond to Br$^{\delta-}$. Curly arrow from electron pair on Br⁻ to C⁺ of carbocation. Bromo-propene.

b Bromine atom drawn on the other carbon atom of the C=C bond compared to answer to **a**.

7 a Polarised bromine approaches C=C bond. Curly arrow from C=C bond to Br$^{\delta+}$ and another curly arrow from Br—Br bond to Br$^{\delta-}$. Curly arrow from Cl⁻ to C⁺ of carbocation. $CH_3CHBrCH_2Cl$ formed.

b Bromine and chlorine atom positions swapped. Mechanism reflects this.

c The second bromine is introduced by Br⁻ reacting with the carbocation.

The carbocation has the positive charge on the 'other' carbon (without Br).

DF 7

1 a

—C—C—C—C—C—C—

b

—C—C—C—C—C—C—

2

—C—C—C—C—C—C—C—C—C—C—C—C—C—C—

3 but-1-ene and propene

4 poly(ethene), poly(propene), poly(chloroethene)

5

$$\left[\begin{array}{c} H\ \ H\ \ H\ \ H \\ C—C—C—C \\ H\ \ Cl\ \ H\ \ O \end{array} \right]_n$$

DF 8

1 The particles in a gas are much further apart than in a liquid or solid. In a gas, therefore, the volume of the particles is a very small part of the total volume and does not significantly affect it. In a liquid or solid the particles are close together and their volumes must be taken into account when deciding on the total volume.

2 a $H_2(g) + \frac{1}{2}O_2(g) \rightarrow H_2O(l)$

b $5\,cm^3$

3 a 3 **b** 3 **c** 4 **d** 2 **e** C_3H_4

4 $1.19\,dm^3$ $(1190\,cm^3)$

5 a 0.25 **b** 2 **c** $48\,dm^3$ **d** $229\,dm^3$
e $30\,dm^3$

6 1.28

7 a 0.11 moles **b** $1.58 \times 10^5\,Pa$

8 $13.2\,cm^3$

9 $1.4 \times 10^5\,Pa$

DF 9

1 a C_6H_{14}, C_6H_{12} not isomers

b both C_4H_9Cl isomers

c C_3H_8O, C_3H_6O not isomers

d both C_7H_8O isomers

e both C_3H_9N isomers

2 a

H—C—C—C—C—H   H—C—C—C—H

b

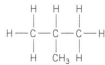

c

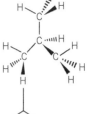

d butane   methylpropane

3

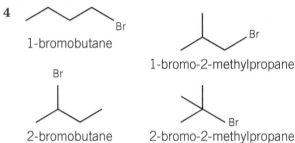

bond angle H—C—H is 109.5°

bond angle C—O—H is 104.5°

Four groups/pairs of electrons repel and get as far away as they can.

Lone pair–lone pair repulsion is greater than lone pair–bonding pair (or bonding pair–bonding pair) repulsion.

4

1-bromobutane

1-bromo-2-methylpropane

2-bromobutane

2-bromo-2-methylpropane

5

6

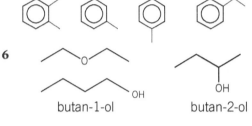

butan-1-ol

butan-2-ol

2-methylpropan-1-ol

2-methylpropan-2-ol

7

E-chloroethene

Z-chloroethene

E-chloroethene has the higher boiling point because of stronger intermolecular forces.

8 a E/Z isomers **b** 2

c citronellol is partially hydrogenated or citronellol has one fewer double bonds

d structural isomers

DF 10

1 a Primary pollutants are released directly into the atmosphere. Secondary pollutants are made by reactions of primary pollutants.

b Ozone made by the Sun or nitrogen oxides and hydrocarbon/SO_3 made by the oxidation of SO_2.

c Irritating, toxic gas/contributes to photochemical smog/weakens body's immune system/greenhouse gas/oxidises rubber.

2 a so they remove pollutants just after the car starts (when gases are cool)

b something that blocks the surface of a catalyst

c give maximum surface area

d they gradually get poisoned/degraded

e CO_2 is formed and this is a greenhouse gas

3 a catalyst

$$2CO + 2NO \rightarrow 2CO_2 + N_2$$

b excess oxygen in the exhaust so NO cannot be reduced

c lowering temperature by recycling exhaust gases through cylinder/using ammonia

$$4NO + 4NH_3 + O_2 \rightarrow 4N_2 + 6H_2O$$

4 a nitrogen and oxygen from the air react

$$N_2 + O_2 \rightarrow 2NO$$

b incomplete combustion of hydrocarbons

$$C_7H_{16} + 9O_2 \rightarrow 4CO_2 + 2CO + 8H_2O$$

c unburnt hydrocarbons are present in the exhaust gas (no equation)

d sulfur compounds in the fuel burn

$$S + O_2 \rightarrow SO_2$$

5 a For

- less CO_2 (per km)
- no (volatile) hydrocarbons
- less CO
- any other valid points

Against

- more particulates
- more NO_x
- any other valid points

b i $2CO + O_2 \rightarrow 2CO_2$

ii $C_{16}H_{34} + 24\frac{1}{2}O_2 \rightarrow 16CO_2 + 17H_2O$

DF 11

1 a $C_8H_{18} + 12\frac{1}{2}O_2 \rightarrow 8CO_2 + 9H_2O$

$C_{16}H_{34} + 24\frac{1}{2}O_2 \rightarrow 16CO_2 + 17H_2O$

$C_2H_5OH + 3O_2 \rightarrow 2CO_2 + 3H_2O$

b $0.035\,dm^3$

$0.036\,dm^3$

$0.035\,dm^3$

c very similar, so other factors must be important

2 $1.52 \times 10^6\,kJ$

3 a 10.6 kg

$1.27 \times 10^5 \, dm^3$

b i $1.06 \times 10^5 \, dm^3$

ii $8.5 \times 10^4 \, dm^3$

ES 1

1 a $Cl_2(aq) + 2NaI(aq) \rightarrow 2NaCl(aq) + I_2(aq)$

b $Br_2(aq) + 2KI(aq) \rightarrow 2KBr(aq) + I_2(aq)$

c no reaction

2 $2Br_2(aq) + 2e^- \rightarrow 2Br^-(aq)$

$2I^-(aq) \rightarrow I_2(aq) + 2e^-$ or $2I^-(aq) - 2e^- \rightarrow I_2(aq)$

3 a $Ag^+(aq) + I^-(aq) \rightarrow AgI(s)$

b $Ag^+(aq) + Br^-(aq) \rightarrow AgBr(s)$

c $Ag^+(aq) + Cl^-(aq) \rightarrow AgCl(s)$

4 a $0.065 \, mol \, dm^{-3}$ **b** $5.86 \, mol \, dm^{-3}$

c 1 : 90

5 a colourless solution **b** pale brown lower layer and violet upper layer

c pale brown lower layer and violet upper layer

ES 2

1 a i $K \rightarrow K^+ + e^-$ **ii** $H_2 \rightarrow 2H^+ + 2e^-$

iii $O + 2e^- \rightarrow O^{2-}$ **iv** $Cr^{3+} + e^- \rightarrow Cr^{2+}$

b i oxidation **ii** oxidation

iii reduction **iv** reduction

2 a silver = +1 **b** aluminium = +3, oxygen = −2

c sulfur = +6, oxygen = −2 **d** phosphorus = 0

e sulfur = +6, fluorine = −1

f phosphorus = +5, oxygen = −2

3 a +4 **b** +7 **c** −1 **d** +7 **e** −1 **f** +1

4 a i Bromine is not very soluble in water and so a lower and higher density bromine layer is formed. This can be run off from the water that floats on top.

ii Bromine is separated from chlorine in the distillation column because they have different boiling points.

b $Cl_2(aq) + 2Br^-(aq) \rightarrow 2Cl^-(aq) + Br_2(l)$

c 0.4 tonnes

d $0.75 \, dm^3$

5 a i hydrogen is oxidised from 0 to +1 chlorine is reduced from 0 to −1

ii Iron is oxidised from +2 to +3. Elemental chlorine is reduced from 0 to −1. Oxidation state of chlorine in $FeCl_2$ remains −1.

iii Oxygen is oxidised from −2 to 0. Fluorine is reduced from 0 to −1. Hydrogen remains +1.

b i Cl_2, Cl_2, F_2

ii H_2, Fe^{2+}, O^{2-}

6 a $2Br^- + 2H^+ + H_2SO_4 \rightarrow Br_2 + SO_2 + 2H_2O$

b $8I^- + 8H^+ + H_2SO_4 \rightarrow 4I_2 + H_2S + 4H_2O$

7 a tin(IV) oxide **b** iron(II) chloride

c nitrate(V) **d** lead(IV) chloride

e manganese(II) hydroxide **f** chromate(VI)

g vandate(V) **h** sulfate(IV)

8 a $KClO_2$ **c** $Fe(OH)_3$

b $NaClO_3$ **d** $Cu(NO_3)_2$

ES 3

Extracting iodine from seaweed

1 Breakdown the cells of the seaweed.

2 Chlorine and bromine are stronger oxidising agents than iodine so less readily oxidised.

3 $I^-(aq) + H_2O_2(aq) \rightarrow I_2(aq) + H_2O(l)$

4 violet

5 Cyclohexane is more volatile than iodine as it will evaporate at room temperature whilst iodine will not.

1 a lead at cathode and bromine (not bromide) at anode

b sodium at cathode and chlorine (not chloride) at anode

c zinc at cathode and iodine (not iodide) at anode

2 a 25 000 mol

b 0.5 mol

c 887 500 g

3 a hydrogen at the cathode and bromine at the anode

b hydrogen at the cathode and oxygen at the anode

c zinc at the cathode and bromine at the anode

4 a cathode: $Zn^{2+} + 2e^- \rightarrow Zn$ reduction

anode: $2Br^- \rightarrow Br_2 + 2e^-$ oxidation

b cathode: $2H^+ + 2e^- \rightarrow H_2$ reduction

anode: $2Br^- \rightarrow Br_2 + 2e^-$ oxidation

c cathode: $2H^+ + 2e^- \rightarrow H_2$ reduction

anode: $4OH^- \rightarrow O_2 + 2H_2O + 2e^-$ oxidation

d cathode: $2H^+ + 2e^- \rightarrow H_2$ reduction

anode: $2H_2O \rightarrow O_2 + 4H^+ + 4e^-$ oxidation

e cathode: $Cu^{2+} + 2e^- \rightarrow Cu$ reduction

anode: $Cu \rightarrow Cu^{2+} + 2e^-$ oxidation

ES 4

1 a $K_c = \dfrac{[NO_2]^2}{[NO]^2[O_2]}$ **b** $K_c = \dfrac{[C_2H_4][H_2]}{[C_2H_6]}$

c $K_c = \dfrac{[H_2][I_2]}{[HI]^2}$ **d** $K_c = \dfrac{[HCO_3^-][H^+]}{[CO_2][H_2O]}$

e $K_c = \dfrac{[CH_3COOC_3H_7][H_2O]}{[CH_3COOH][C_3H_7OH]}$

2 $2SO_2(g) + O_2(g) \rightleftharpoons 2SO_3(g)$

3 a $K_c = \dfrac{[NH_3]^2}{[N_2][H_2]^3}$

b 2.09

4 a $K_c = \dfrac{[PCl_3][Cl_2]}{[PCl_5]}$

b 0.196

5 Equilibrium constant is much greater than 1 so products favoured. Chloromethane concentration will be low.

6 (If acid was added) $[H^+(aq)]$ would increase. Some HCO_3^- and H^+ react making more CO_2 and H_2O. The equilibrium position moves to the left.

ES 5

1 a 2.0×10^{-2} moles

b 2.46×10^{-4} moles

2 $0.280\,mol\,dm^{-3}$

3 $Cl_2(g) + H_2O(l) \rightleftharpoons HCl(aq) + HClO(aq)$

4 $0.490\,mol\,dm^{-3}$

ES 6

1 a 60.0

b 42.0

c 70%

2 a $C_4H_9Br + H_2O \rightarrow C_4H_9OH + HBr$

b 47.8%

c $C_4H_9Br + NaOH \rightarrow C_4H_9OH + NaBr$

d Decreases the atom economy (41.8%)

3 a $NH_3(g) + HI(g) \rightarrow NH_4I(s)$

b $8HI(aq) + H_2SO_4(aq) \rightarrow$
$$H_2S(g) + 4H_2O(l) + 4I_2(s)$$

4 a 31.7%

b 3.168 tonnes

5 Sodium chloride reacts with concentrated acid to make pure hydrogen chloride gas.

$NaCl(s) + H_2SO_4(aq) \rightarrow NaHSO_4(aq) + HCl(g)$

Sodium bromide first of all reacts with concentrated sulfuric acid to make hydrogen bromide.

$NaBr(s) + H_2SO_4(aq) \rightarrow NaHSO_4(aq) + HBr(g)$

However, the bromide ions produced are strong enough reducing agents to reduce the sulfuric acid which is present to sulfur dioxide.

$2H^+(aq) + 2Br^-(aq) + H_2SO_4(aq) \rightarrow$
$$SO_2(g) + 2H_2O(l) + Br_2(l)$$

This means that adding concentrated sulfuric acid to sodium bromide would not be a good way to make hydrogen bromide gas because it won't be pure.

The gas made will be a mixture of:

- hydrogen bromide
- sulfur dioxide
- bromine vapour (since the reaction is exothermic).

ES 7

1 Chlorine is oxidised and oxygen is reduced.

2 B

3 a to the right   **d** to the right

b to the left   **e** to the right

c no effect

4 a $2H_2(g) + O_2(g) \rightleftharpoons 2H_2O(g)$

b $K_c = \dfrac{[H_2O]^2}{[H_2]^2[O_2]}$

c i Forward reaction is exothermic. The equilibrium position moves to the left.

ii The equilibrium position moves to the right as the products side has fewer molecules.

5 a The equilibrium position moves to the right to use up the extra oxygen.

b There are fewer molecules on the left hand side of the equation than on the right. The equilibrium position moves to the right.

c The forward reaction is exothermic. The equilibrium position moves to the right.

6 Adding acid removes NaOH so NaOH concentration decreases. The equilibrium position moves to the left producing more NaOH and Cl_2. Chlorine is a toxic gas.

OZ 1

The atmosphere past, present, and future

1

	^{52}Cr	^{53}Cr
proton	24	24
neutron	28	29
electron	24	24

2 0.0063%

3 Seasons where plants are experiencing lots of growth will cause lower concentrations of carbon dioxide as they will be absorbing more. Seasons where plants are not growing (such as in winter) concentrations of carbon dioxide will be higher as plants are not absorbing as much.

1 nitrogen, oxygen, argon

2 combustion of hydrocarbons, deforestation, cattle farming, landfill, changes in land use

3 a 10 000 ppm

b 0.001 82%

4 a $1.8 \times 10^{-6} dm^3$

b 0.000 18%

OZ 2

1 $5.5 \times 10^{13} Hz$

2 $1.15 \times 10^{-9} m$

3 a Specific frequencies corresponding to transitions between vibrational energy levels, making the bonds vibrate more.

b Molecules which have absorbed radiation have more kinetic energy. This energy is subsequently transferred to other molecules in the air by collisions.

4 a $5.43 \times 10^{-20} J$

b $8.20 \times 10^{-13} Hz$, infrared

c $3.66 \times 10^{-6} m$

5 a $1.88 \times 10^4 J$ **c** 0.978 J

b $1.62 \times 10^{-24} J$ **d** 19 200 moles

OZ 3

Other radical reactions

1 Radicals react with any species so hard to control where the monomer joins the polymer.

2 Under the right conditions (initiation), oxygen produces radicals allowing propagation reactions with the fuel to take place. Once all the fuel has been used up termination occurs.

1 a yes   **c** no   **e** yes

  b no   **d** yes   **f** yes

2 a homolytic

  b A is initiation, B and C are propagation.

  c i $2O_3 \rightleftharpoons 3O_2$

   ii catalyst

3 a oxidation of nitrogen in internal combustion engines

  b i $O_3 + O \rightarrow 2O_2$

   ii catalyst

   iii $(-100) + (-192) = -292 kJ mol^{-1}$

4 a initiation: reaction F
propagation: reactions G, H, and I
termination: reactions J and K

  b i endothermic: reaction F
exothermic: reaction K

   ii F is endothermic because C—C bond broken. K is exothermic because C—C bond formed.

  c $CH_3\cdot$ methyl radical, $C_2H_5\cdot$ ethyl radical, $H\cdot$ hydrogen radical.

5 a $Cl_2 \rightarrow 2Cl\cdot$ initiation
$CH_4 + Cl\cdot \rightarrow CH_3\cdot + HCl$ propagation
$CH_3\cdot + Cl_2 \rightarrow CH_3Cl + Cl\cdot$ propagation
$Cl\cdot + Cl\cdot \rightarrow Cl_2$ termination
$CH_3\cdot + CH_3\cdot \rightarrow C_2H_6$ termination

  b The chloromethane can react with $Cl\cdot$ radicals.
$CH_3Cl + Cl\cdot \rightarrow CH_2Cl\cdot + HCl$
$CH_2Cl\cdot + Cl_2 \rightarrow CH_2Cl_2 + Cl\cdot$
Further substitution produces $CHCl_3$ and CCl_4.

OZ 4

1 a Greater surface area, so more frequent collisions.

  b Insufficient energy to overcome the activation energy.

c Particles are in fixed positions and number of collisions is low.

d Very high surface area and a spark will easily overcome the activation energy.

2 a A b A c B d Mainly B with A to a minor extent.

3 a A, B, C, D b A, B c A, B, C

4 a The reaction has a high activation enthalpy that prevents it occurring at a significant rate at room temperature. However, the reaction is exothermic and once the spark has provided the energy needed to get it started, the reaction produces enough energy to sustain itself regardless of how much is present.

b The platinum catalyst offers an alternative pathway with a lower activation enthalpy close to the thermal energy of molecules at room temperature.

5 a Reaction profile shows reactants at higher enthalpy than products, and two lines showing activation enthalpy for uncatalysed reaction higher than activation enthalpy for catalysed reaction.

X-axis: progress of reaction, y-axis: enthalpy. E_a for catalysed reaction marked as $+36.4 \, \text{kJ} \, \text{mol}^{-1}$, and for uncatalysed marked as $+49.0 \, \text{kJ} \, \text{mol}^{-1}$.

b The rate will be faster for the catalysed reaction as the activation enthalpy is significantly lower, so more molecules have sufficient energy to react.

OZ 5

1 a B, C b A, D c D d B e B f D

2 a Increases in concentration mean more collisions so that there are more collisions with the minimum energy to react.

b As the activation energy is small, an increase in temperature will mean that many more collisions will have the minimum energy to react and so the rate of reaction increases rapidly.

3 a The shaded area in underneath the T_1 curve and to the right of E_a.

b The area shaded a different colour is underneath the T_2 curve and to the right of E_a, encompassing the first coloured area. The T_2 curve has a lower and broader maximum than the T_1 curve and the maximum value is shifted to the right. It tails off above the T_1 curve.

4 a Homogeneous means same physical state. Here the catalyst and reactants are gases.

b i It is a two step reaction with an activation energy for each step.

ii peaks – enough energy has been put in to start the reaction
troughs – intermediate

OZ 6

1 a trichloromethane

b 2-chloropropane

c 1,1,1-trichloro-2,2,2-trifluoroethane

d 2-chloro-1,1,1-trifluoropropane

e 2,2-dibromo-3-chlorobutane

2 In the solid or liquid state, noble gas atoms are held together by weak instantaneous dipole – induced dipole bonds. It takes very little energy to break these bonds and this results in very low melting and boiling points.

3

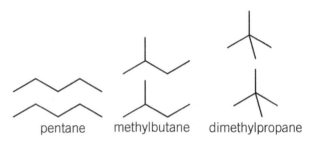

pentane   methylbutane   dimethylpropane

a Pentane is straight-chain and its molecules can approach each other closely so there are strong intermolecular forces.

Methylbutane and dimethylpropane have respectively more branching and cannot approach as closely, so the instantaneous dipole–induced dipole forces are weaker.

The stronger the intermolecular bonds, the more energy is required to break them.

b Pentane has the strongest intermolecular bonds and hence the highest boiling point. Dimethylpropane has the weakest intermolecular bonds and hence the lowest boiling point.

4 a CO_2 no dipole   d CH_3OH dipole

b $CHCl_3$ dipole   e $(CH_3)_2CO$ dipole

c C_6H_{12} (cyclohexane) no dipole

f benzene no dipole

5 a i 18 in SiH_4, 18 in H_2S.

ii The attractions will be similar because the number of electrons is the same.

iii H_2S has a permanent dipole – it is a bent molecule with two lone pairs. SiH_4 does not have an overall permanent dipole as it is a symmetrical molecule.

b Both compounds have similar instantaneous dipole–induced dipole bonds.
However H_2S also has permanent dipole–permanent dipole bonds.
So its boiling point is higher than that of SiH_4.

OZ 7

1 a Oxygen is more electronegative than hydrogen.

b **a** $C^{\delta+}-F^{\delta-}$ polar   **e** $H^{\delta+}-N^{\delta-}$ polar

   b C–H non polar   **f** S–Br non polar

   c C–S non polar   **g** $C^{\delta+}-O^{\delta-}$ polar

   d $H^{\delta+}-Cl^{\delta-}$ polar

2 a

b

3 Water has two O—H bonds and two lone pairs on the oxygen so more hydrogen bonding is possible.

OZ 8

1 a $CH_3CH_2CH_2Cl(l) + NaOH(aq) \rightarrow$
$CH_3CH_2CH_2OH(aq) + NaCl(aq)$

b The chlorine atom in 1-chloropropane has been replaced by a hydroxyl group, –OH.

c Curly arrow from lone pair of electrons on OH^- to carbon in C—Cl bond.
Curly arrow from C—Cl bond to chlorine.
Propanol and Cl^- formed.

$\delta+$ on carbon of C—Cl and $\delta-$ on chlorine of C—Cl bond.

2 a $C_2H_5I + OH^- \rightarrow C_2H_5OH + I^-$

b $C_2H_5Br + CN^- \rightarrow C_2H_5CN + Br^-$

c $C_5H_9Cl + OH^- \rightarrow C_5H_9OH + Cl^-$

d $CH_3C(CH_3)ClCH_3 + H_2O \rightarrow$
$CH_3C(CH_3)(OH)CH_3 + HCl$

e $CH_2BrCH_2Br + 2OH^- \rightarrow$
$CH_2OHCH_2OH + 2Br^-$

f $CH_3Br + C_2H_5O^- \rightarrow CH_3OC_2H_5 + Br^-$

g $CH_3CHClCH_3 + CH_3COO^- \rightarrow$
$CH_3CH(OOCCH_3)CH_3 + Cl^-$

3 a $CH_3CH_2Br + NH_3 \rightarrow CH_3CH_2NH_2 + H^+ + Br^-$

b

c Ammonia is the nucleophile because of its lone pair. It attacks the partial positive charge on the carbon atom and substitutes the bromine atom. There is a partial positive charge because bromine is more electronegative than carbon. After the NH_3 attacks the 1-bromoethane, a hydrogen ion is lost from the molecule to give the product ethylamine.

d Nucleophile: a species with a lone pair that can form a covalent bond.
Electronegativity: the power to attract electrons in a covalent bond.
Subsitution: where one atom or group replaces another atom or group.
Curly arrow: device to show the movement of a pair of electrons.

WM 1

1 a i primary alcohol   **v** diol

   ii secondary alcohol   **vi** primary alcohol

   iii secondary alcohol   **vii** ether

   iv tertiary alcohol

b i pentan-1-ol   **iv** 2-methylbutan-2-ol

   ii heptan-3-ol   **v** butane-2,3-diol

   iii cyclohexanol   **vi** decan-1-ol

2 Hydrogen bonding between ethanol and water molecules. As the hydrocarbon chain gets longer, the importance of the –OH group relative to that of the alkyl group becomes less and hexanol is unable to mix with water.

3 a A, B, D, E   **e** C

   b C   **f** A

   c A   **g** D

   d C, D   **h** E

4 a Ethanol has hydrogen bonds between molecules, ethane does not. Hydrogen bonds are the strongest intermolecular bond and so more energy is needed to break them to form gas. Therefore the boiling point is higher.

b Water forms more hydrogen bonds than ethanol so more energy is needed to break all the hydrogen bonds in water. Therefore the boiling point is higher.

c Both have an OH group and so form hydrogen bonds. Boiling point increases down a homologous series as M_r increases. Hence butan-1-ol has a higher boiling point than ethanol.

d Butan-1-ol forms hydrogen bonds, ethoxyethane does not. Hence more energy is needed to break the intermolecular bonds in butan-1-ol and so the boiling point is higher.

5 a butan-1-ol

b

CH₃CH₂CH₂CH₂OH ⟶ CH₃CH₂CH = CH₂ + H₂O

6 a i fizzes **iv** no reaction
ii no reaction **v** fizzes
iii no reaction

b i stays orange **iv** stays orange
ii orange to green **v** stays orange
iii orange to green

7 a i esterification **vi** oxidation
ii oxidation **vii** dehydration
iii dehydration **viii** nucleophilic substitution
iv no reaction **ix** oxidation
v esterification

b i ester **vi** aldehyde
ii carboxylic acid **vii** alkene
iii alkene **viii** haloalkane
iv no reaction **ix** ketone
v ester

WM 2

1 a i **b** iv **c** iii **d** v **e** vi
f iii **g** vii **h** ii and iv **i** vii

2 a

b

3 a $C_6H_5OH + NaOH \rightarrow C_6H_5O^-Na^+ + H_2O$

b $CH_3CH_2COOH + KOH \rightarrow$
$CH_3CH_2COO^-K^+ + H_2O$

c $2CH_3CH_2CH_2COOH + Na_2CO_3 \rightarrow$
$2CH_3CH_2CH_2COO^-Na^+ + CO_2 + H_2O$

4 a

b

WM 3

1 a 4.24

b $7.08 \times 10^{13}\,Hz$

2 O—H 3660 m^{-1}
C—H(arene) 3060 cm^{-1}

3 a $CH_3CH_2CH(OH)CH_3$
$CH_3CH_2COCH_3$

b A O—H 3660 B C—H 2990
  C—H 2970   C=O 1730

c A butan-2-ol B butanone

4 a C O—H 3580 D O—H 3670 E C—H 2990
  C—H 2990   C—H 2950 C=O 1770
  C=O 1775   (C—O 1050-1300)
       (C—O 1050—1300)

b C carboxylic acid
D alcohol
E ester

5 O—H 3600-3640 phenol
C—H 2850-2950 aliphatic
C=O 1735-1750 ester
C—H 3000-3100 aromatic

WM 4

1 a $C_4H_8O^+$

b isotope peak for ^{13}C

c the first four peaks are fragments from the molecular ion peak

2 a same molecular ion peak at 72 m/z

b the fragmentation pattern would differ

WM 5

1 Dissolve the solute in the minimum quantity of hot solvent.
Allow to cool and crystallise.
Filter off the crystals and wash with a small quantity of cold solvent.
Dry the crystals.

2 Better atom economy.
Prevention of waste products.
Reduce reagents and steps.
Use catalysts and more selective catalysts.

3 addition   atom economy = 100%
substitution   atom economy <100%
elimination   atom economy <100%

4

heated oil bath	cheap but messy, heated oil smells
heated metal block	reasonably cheap, robust, melting points easily repeated
electrically heated	expensive, robust, melting points very easily repeated

CI 1

1 Phosphorus does not form triple bonds like nitrogen. The triple bond in nitrogen is very strong (bond enthalpy = 945 kJ mol^{-1}). Large activation enthalpy needed before N_2 will react. P_4 only needs enough energy to break one P—P bond (bond enthalpy = 198 kJ mol^{-1}).

2 a +2 to +1

b $2NO(g) + 2H^+(aq) + 2e^- \rightarrow N_2O(g) + H_2O(l)$

3 a A: +3 → + 5, B: +5 → 0, C: −3 → −3, D: 0 → −3, E: +2 → +4

b A: oxidation, B: reduction, C: neither oxidation or reduction, D: reduction, E: oxidation

4 a $NO_2^-(aq) + H_2O(l) \rightarrow NO_3^-(aq) + 2H^+(aq) + 2e^-$

b $2NO_3^-(aq) + 12H^+(aq) + 10e^- \rightarrow N_2(g) + 6H_2O(l)$

c $NH_4^+(aq) + OH^-(aq) \rightarrow NH_3(g) + H_2O(l)$

d $N_2(g) + 6H^+(aq) + 6e^- \rightarrow 2NH_3(g)$

e $NO(g) + H_2O(l) \rightarrow NO_2(g) + 2H^+(aq) + 2e^-$

5 a $NO_3^- + 3Fe^{2+} + 4H^+ \rightarrow 3Fe^{3+} + NO + 2H_2O$
$[Fe(H_2O)_6]^{2+} + NO \rightarrow [Fe(H_2O)_5(NO)]^{2+} + H_2O$

b The presence of nitrite ions will interfere with this test by oxidizing Fe^{2+} ions. Hence, the test is not reliable if the solution has nitrite ions.

CI 2

1 a $K_c = \dfrac{[NO_2(g)]^2}{[NO(g)]^2[O_2(g)]}$ mol^{-1} dm^3

b $K_c = \dfrac{[C_2H_4(g)][H_2(g)]}{[C_2H_6(g)]}$ mol dm^3

c $K_c = \dfrac{[H_2(g)][(I_2(g)]}{[HI(g)]^2}$ (no units)

d $K_c = \dfrac{[HCO_3^-(aq)][H^+(aq)]}{[CO_2(aq)][H_2O(l)]}$ (no units)

e $K_c = \dfrac{[In_2(aq)]^3}{[In_6(aq)]}$ mol^2 dm^{-6}

f $K_c = \dfrac{[CH_3COOC_3H_7(l)][H_2O(l)]}{[CH_3COOH(l)][C_3H_7OH(l)]}$ (no units)

2 $2SO_2(g) + O_2(g) \rightleftharpoons 2SO_3(g)$

3 a $K_c = \dfrac{[NH_3(g)]^2}{[N_2(g)][H_2(g)]^3}$

b 2.09 mol^{-2} dm^6

4 a $2H_2(g) + O_2(g) \rightleftharpoons 2H_2O(g)$

b $K_c = \dfrac{[H_2O(g)]^2}{[H_2(g)]^2[O_2(g)]}$

c i Equilibrium moves towards reactants (reaction is exothermic).

  ii Equilibrium moves towards products (fewer gaseous molecules).

d i decreases   **ii** no effect

5 a $K_c = \dfrac{[PCl_3(g)][Cl_2(g)]}{[PCl_5(g)]}$

b 0.196 mol dm^{-3}

6 a $K_c = \dfrac{[H_2S(g)]^2}{[H_2(g)]^2[S_2(g)]}$

b $[S_2(g)] = \dfrac{0.442^2}{9.4 \times 10^5 \times 0.234^2}$
$= 3.80 \times 10^{-6}$ mol dm^{-3}

7 a $K_c = \dfrac{[NO(g)]^2[O_2(g)]}{[NO_2(g)]^2}$

b 0.083 mol dm^{-3}

8 a $K_c = \dfrac{[CH_3CH(OC_2H_5)_2(l)][H_2O(l)]}{[C_2H_5OH(l)]^2[CH_3CHO(l)]}$

b reactants

c low

d $0.074\,mol^{-1}\,dm^3$

9 very high K_c so chloromethane concentration is likely to be low.

10 a $x - 0.2096$

b $\dfrac{\dfrac{0.1048}{2}}{(x - 0.2096)^{2/4}} = 215.5$

$x = 0.24\,mol$

CI 3

1 Measure volume of CO_2 released.

2 Colorimetry as reaction mixture starts colourless and becomes orange/brown.

3 Add sodium hydrogencarbonate, which neutralises the acid catalyst produced.

CI 4

1 The reactant may not be involved in the rate-determining step.

2 a The reaction is first order with respect to bromoethane and zero order with respect to hydroxide ion.

b The reaction is first order with respect to methylmethanoate, zero order with respect to water and first order with respect to H^+.

c The reaction is first order with respect to urea, zero order with respect to water and first order with respect to urease.

d The reaction is a single step in the mechanism. It is first order with respect to the methyl radical and first order with respect to the chlorine molecule.

e The reaction is order ½ with respect to carbon monoxide and first order with respect to chlorine.

f The reaction is second order with respect to nitrogen dioxide.

3 a rate = $k[CH_3CH_2CH_2CH_2Cl]\,[OH^-]$
b rate = $k[C_{12}H_{22}O_{11}]\,[H^+]$

4 2.095×10^{-7} and 1.1967×10^{-6} so, 5.7 times (allow answers between 5.3 and 6.1)

5 410.28 times (allow answers between 405 and 415)

CI 5

1 Half-life of reaction is the time taken for half of a certain reactant to get used up.
Plot a graph of amount of reactant against time. Calculate three subsequent half-lives. If these half-lives are the same the reaction is first order. Zero order and second order reactions do not have constant half-lives.

2 a

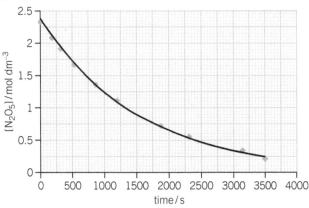

b All are approximately 1150 s (allow 1100 –1200 d). First order with respect to N_2O_5

c Five tangents should be drawn on the graph from **a**, and the gradient for each should be calculated to give the reaction rate.

d Answer dependent on graph.

e rate = $k[N_2O_5]$

f $6.2 \times 10^{-4}\,s^{-1}$ (allow for an answer between 5 and $7 \times 10^{-4}\,s^{-1}$)

3 a

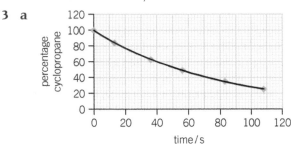

b half-lives are constant – reaction first order

4 a i first order **ii** second order

b rate = $k[H_2][NO]^2$

c $k = 0.384\,mol^{-2}\,dm^6\,s^{-1}$

5 a rate = $k[CH_3COCH_3]\,[H^+]$

b Yes. One of the steps has one mole of propanone reacting with one mole of hydrogen ions. Iodine is not involved in this step. This is consistent with the rate equation.

c Step 1

d Neutralise with an alkali to remove the acid catalyst from the system.

6 Mechanism 1, as it is the only mechanism that has a step involving two molecules of B only (step 1). This is consistent with the rate equation. Mechanism 2 has no such step.

Cl 6

1 a $2CH_4 + Cl_2 \rightarrow CH_3CH_3 + 2HCl$

$2CH_4 + 2Cl_2 \rightarrow CH_3CH_2Cl + 3HCl$

b co-product: CH_3CH_3 and HCl
by-product: CH_3CH_2Cl
Co-product is produced with the expected main product, by-product is the product of unwanted side reaction.

2 low temperature – forward reaction is exothermic (energy given out) so favoured by a low temperature
high pressure – fewer number of molecules in the forward reaction
use a catalyst – increases the rate of the forward reaction

PL 1

1 a propanoic acid

b butanedioic acid

c butanoic anhydride

d benzene-1,3-dicarboxylic acid

e methanol

f 3-methylbutan-2-one

2 a

b HCOOH   **c** $(HCOO)_2O$

d $HOCH_2CH_2CH_2OH$

e $HOOC(CH_2)_4COOH$

f $CH_3CH_2CH_2CH_2CHO$

g $CH_3COCH_2CH(CH_3)CH_3$

3 $HOCH(CH_3)COOH$

4 a methyl ethanoate

b ethyl methanoate

c methyl butanoate

d ethyl ethanoate

5 a $CH_3CH(OH)CH_3 + CH_3CH_2COOH \rightarrow$
$CH_3CH_2COOCH(CH_3)_2 + H_2O$

b $OHCH_2CH_2OH + 2CH_3COOH \rightarrow$
$CH_3COOCH_2CH_2OCOCH_3 + 2H_2O$

6 $—OCH_2CH_2OCOCH_2COOCH_2CH_2OCOCH_2COO—$
or
$—OCCH_2COOCH_2CH_2OCOCH_2COOCH_2CH_2—$

7 a $C_6H_5OH + NaOH \rightarrow C_6H_5ONa + H_2O$

b $2C_2H_5COOH + CaCO_3 \rightarrow (C_2H_5COO)_2Ca + CO_2 + H_2O$

c $2HCOOH + Mg \rightarrow (HCOO)_2Mg + H_2O$

d $(CH_3CO)_2O + C_2H_5OH \rightarrow CH_3COOC_2H_5 + CH_3COOH$

8 $CH_3CH_2COOCH_3$ methyl propanoate
$CH_3COOCH_2CH_3$ ethyl ethanoate
$HCOOCH_2CH_2CH_3$ propyl methanoate
$CH_3CH_2CH_2COOH$ butanoic acid
$CH_3CH(CH_3)COOH$ methylpropanoic acid
$HCOOCH(CH_3)_2$ methylethyl methanoate

9

$2\ CH_3CH(OH)COOH \rightarrow$

$+ 2H_2O$

10 a

b

$+ H_2O$

c acid anhydride

d because molecule has to rotate/rearrange

PL 2

1 a $—NH(CH_2)_6NHCO(CH_2)_4CO—$

b $—NH(CH_2)_6NHCO(CH_2)_8CO—$

c $—NH(CH_2)_5CO—$

2 a methanediamine

b 1-aminopropane OR propylamine

c 2-aminobutane

d butanoyl chloride

3 a nylon-6,8   **b** nylon-9,9

c nylon-4,4

4 a $—OC(CH_2)_5NH—$ nylon-6

b $—NH(CH_2)_5NHCO(CH_2)_5CO—$ nylon-5,7

5 a $H_2N(CH_2)_5NH_2$ $ClCO(CH_2)_8COCl$

b HCl

c $—NH(CH_2)_5NHCO(CH_2)_8CO—$

6 a $^+H_3NCH(CH_3)CH_3$

b $ClH_3NCH(CH_3)CH_3$

c $CH_3CH(CH_3)NHCOCH_3 + HCl$

7 a

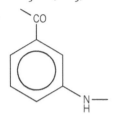

b Different arrangement of benzene rings means hydrogen bonds are not as regular.

8 a amide

b n

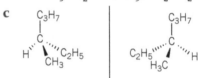

$\rightarrow +CO(CH_2)_5NH+_n$

c nylon-6

d addition, no loss of small molecule

e

O
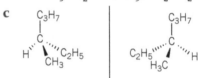
$+CO(CH_2)_5O+$

PL 3

1 a $C_2H_5OCOC_2H_5 + H_2O \rightarrow C_2H_5COOH + C_2H_5OH$

b $C_2H_5CONHCH_3 + OH^- \rightarrow C_2H_5COO^- + CH_3NH_2$

c $C_2H_5CONH_2 + H_2O + H^+ \rightarrow C_2H_5COOH + NH_4^+$

2 sodium ethanoate ethylamine

3 a $-OCH(CH_3)CO-$

b $-OCH_2CO-$

4 $CH_3CH(OH)COOH$ 2-hydroxypropanoic acid
$CH_2(OH)COOH$ (2-)hydroxyethanoic acid

5 a HCl/ reflux

b $CH_3CH_2COOH + (CH_3)_2NH_2^+Cl^-$

c

—NHCOCH₃

d $H_2N(CH_2)_3COO^-Na^+$

6 a

$HOOC$ —◯— $COOH$ 　　$HOCH_2CH_2OH$

b $HOOC(CH_2)_8COOH$ 　$^+H_3N(CH_2)_6NH_3^+$

c $H_2N(CH_2)_5COO^-$

d

^+H_3N—◯—NH_3^+ 　$HOOC$—◯—$COOH$

7 a The C—O bond next to the C=O

b The C=O bond is very polar with a δ+ charge on the carbon atom. Hydrolysis occurs by nucleophilic attack on this carbon atom by a lone pair on the oxygen-18 of a water molecule and the C—O bond breaks.

c A hydrogen ion attaches to the oxygen in the carbonyl group, making the carbon more susceptible to attack by a nucleophile.

PL 4

1 $^+H_3NCHRCOO^-$

2 $H_2NCH(CH_3)COOH$
$H_2NCH(CH(CH_3)_2)COOH$

3 a A carbon atom attached to four different groups.

b Only **iii** since it alone has four different groups on the same C atom (or vice-versa).

4 a/b $CH_3CH_2C^*H(CH_3)CH_2CH_2CH_3$

c

```
    C₃H₇                    C₃H₇
     |                       |
     C                       C
H  ⁞    C₂H₅      C₂H₅        ⁞   H
     CH₃                    H₃C
```

d $CH_3CHBrCH_2CH_3$

e

```
    Br                       Br
     |                        |
     C                        C
H  ⁞    C₂H₅      C₂H₅         ⁞   H
     CH₃                    H₃C
```

5 a

b

c no, ring is now symmetrical

6 a $H_2NCH(CH_3)COOH + HCl \rightarrow Cl^-{}^+H_3NCH(CH_3)COOH$

b $H_2NCH(CH_2OH)COOH + NaOH \rightarrow H_2NCH(CH_2OH)COO^-Na^+ + H_2O$

c $H_2NCH((CH_2)_4NH_2)COOH \rightarrow Cl^-{}^+H_3NCH((CH_2)_4NH_3^+Cl^-)COOH + 2HCl$

d $H_2NCH(CH_2COOH)COOH \rightarrow H_2NCH(CH_2COO^-Na^+)COO^-Na^+ + 2NaOH \qquad + 2H_2O$

7 a $H_2NCH(CH_3)CONHCH_2COOH \rightarrow Cl^- {}^+H_3NCH(CH_3)COOH$
    $+ H_2O + 2HCl \qquad + Cl^- {}^+H_3NCH_2COOH$

b $H_2NCH(CH_2OH)CONHCH_2COOH \rightarrow H_2NCH(CH_2OH)COO^-Na^+$
    $+ 2NaOH \qquad + H_2NCHCH_2COO^-Na^+$
    $\qquad\qquad\qquad\qquad + 2H_2O$

c Spot hydrolysate on to paper and place in solvent.
Run chromatogram and allow it to dry spray with ninhydrin and place in oven measure R_f values of spots.
Match against database to find amino acids

8 a HOOCC*H(OH)C*H(OH)COOH

b two are object and mirror image
one has internal symmetry so is not an entantiomer

c HOOCC(CH$_2$OH)(OH)COOH

PL 5

1 a many examples, eg glu and thr:

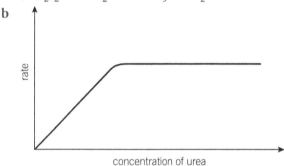

other possibilities: pro, trp, asp, glu, gln, tyr, his, lys, arg

b e.g., glu and arg:
$H_2NCH(CH_2COO^-)$ COOH
$H_2NCH((CH_2)_3 NHC(NH)NH_3^+)$ COOH

c Any two from: gly, ala, val, ile, instantaneous dipole-induced dipole

2 a peptide links between amino acids hydrogen bonds between C=O of peptide and NH of another peptide further up the helix

b (peptide links between amino acids) hydrogen bonds between C=O of peptide and NH of a peptide on another chain

c How the secondary structure/helices/sheets are folded.
Held together by bonds between the R groups of amino acids in different parts of the chain.

3 a washing can break hydrogen bonds but not covalent bonds

b -S—S- $\rightarrow$ -SH + -SH
proteins lose (part of) their tertiary structure

c -S—S- bonds reform and hair is re-shaped

4 a enzyme or hormone, structural proteins are fibrous

b the helical parts

c the whole structure

d helices of amino acid chains

e bonds between R-groups of amino acids

PL 6

1 a they are more rapidly denatured

b active site is destroyed hydrogen bonds broken

c Enzymes are rapidly denatured/active sites destroyed, so the rate at which vitamins are broken down is dramatically reduced.

2 a Enzymes are denatured by acid.

b active site is destroyed
ionic bonds broken

3 a butenedioic acid

b

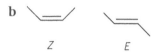

$Z \qquad\qquad E$

c $HOOC(CH_2)_3COOH$
product: $HOOCCH_2CH{=}CHCOOH$
EITHER it might/would catalyse because the COOH groups will still fit
OR it would not catalyses because the COOH groups would not fit

4 3-hydroxybenzoic acid fits into the active site and either
blocking it from the substrate
OR it is an inhibitor

5 a $(NH_2)_2CO + H_2O \rightarrow 2NH_3 + CO_2$

b

concentration of urea

c first part of graph: E + S $\rightarrow$ ES is rate determining
so first order with respect to urea
second part ES (or EP) $\rightarrow$ products is rate determining
so zero order with respect to urea

d solution becomes alkaline as reaction proceeds take samples and titrate against (standard) acid to follow reaction

PL 7

1 histidine and leucine

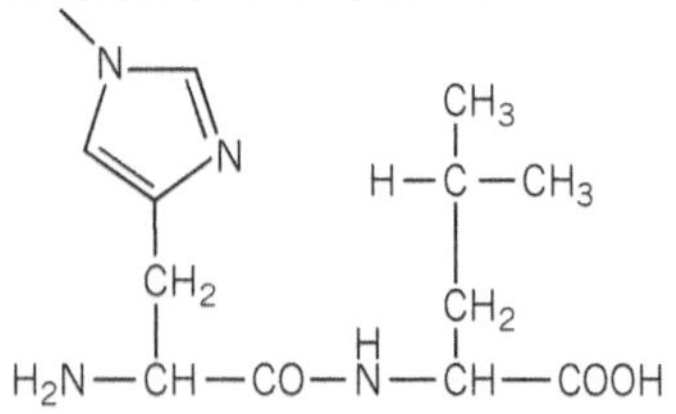

2 a it is the enantiomer at the upper carbon

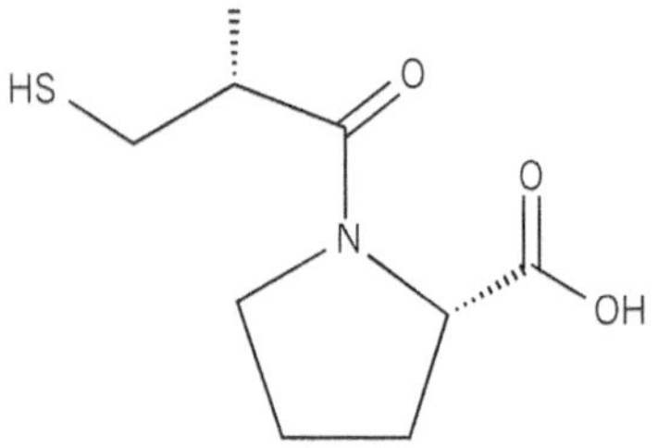

b it does not fit the active site so well

c —COOH group

3 a

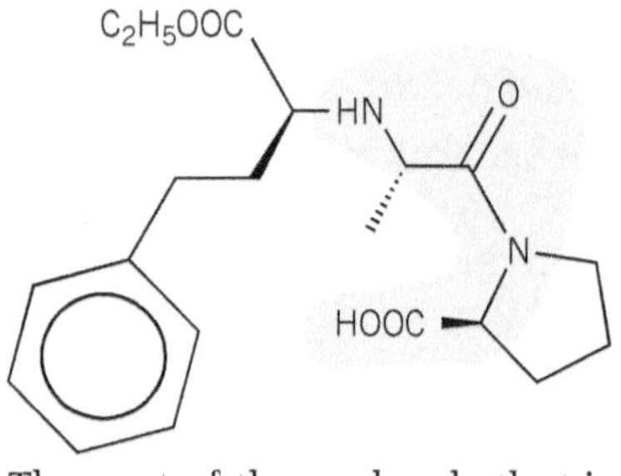

b The part of the molecule that is pharmacologically active.

PL 8

1 a i condensation   **ii** condensation

b

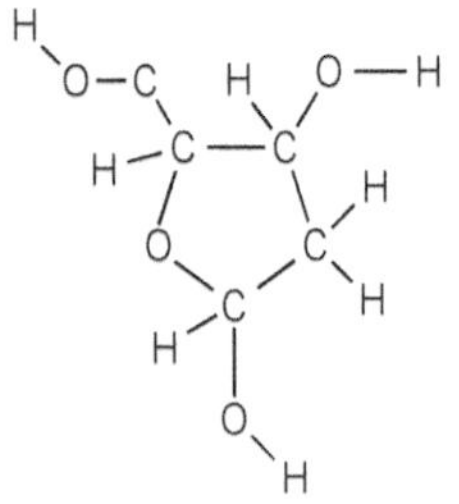

2 a third does not matter: ser, leu, pro, arg, thr, val, ala, gly
one codon: trp, met

3 a i lys lys

ii arg ala arg

iii tyr leu thr

3 b i ACC

ii CUA or CUG (both required)

4 a GUCA   **b** GTCA

5 a

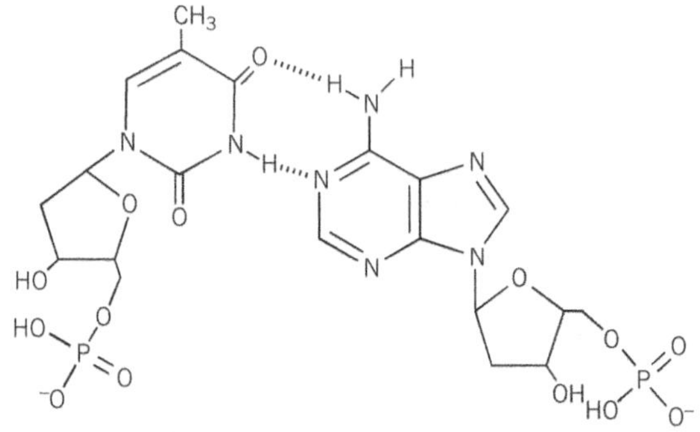

b

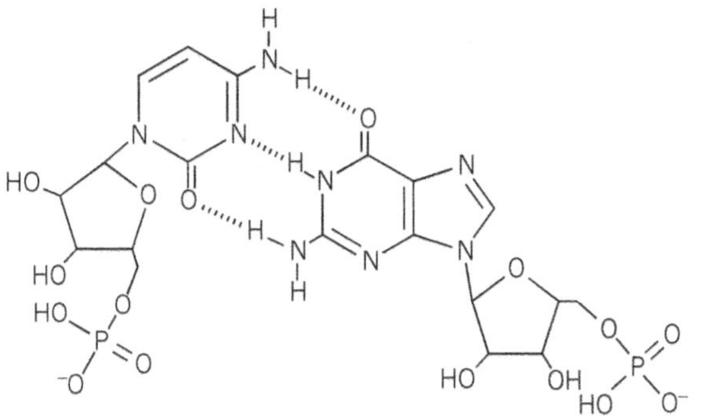

PL 9

Why does the $n + 1$ rule work?

The hydrogens of the CH_3 group are adjacent to one hydrogen. The single hydrogen could:

- all align with the external field
  N–S   N–S

- one with and one against the external field
  N–S   S–N

Hydrogens of the CH_3 group can absorb two different frequencies, so give two signals centred on the expected chemical shift.

1 a A ethanoic acid **B** ethanol

b A 43 CH_3CO^+; 45 $COOH^+$; 60 CH_3COOH^+
B 31 CH_2OH^+; 46 $C_2H_5OH^+$

2 a 1   **b** 2   **c** 3   **d** 2

3 a, b tartaric: 3 1:1:1
succinic: 2 2:1
reason: number of different proton environments

c tartaric: 2
succinic: 2
reason: number of different carbon environments

4 a 58 $C_4H_{10}^+$; 43 $C_3H_7^+$; 29 $C_2H_5^+$; 15 CH_3^+

b $CH_3CH_2CH_2CH_3$ and $CH_3CH(CH_3)CH_3$

c $CH_3CH_2CH_2CH_3$ since the alternative would not give rise to $C_2H_5^+$

5 **a** C_2H_5CHO and CH_3COCH_3

b $(58 - 43 = 15)$ CH_3

c H

d $C_2H_5^+$

e **D** is the aldehyde, since it forms $C_2H_5^+$ which C is unlikely to do/cannot.
C is the ketone that gives CH_3 which is much more likely to be high intensity

6 **a** $(CH_3)_3COH$

b Peak of intensity 9 would not be split since the adjacent carbon atom has no hydrogen atoms. Peak of intensity 1 would not be split as it has no adjacent C atoms/is adjacent to O

c two
0–50 and 50–90

7 **a** $CH_3COOCH_2CH_3$
Four marks for four reasons from: Ester since C=O at 1740; Not acid since no broad absorption around 3000 OR not alcohol since no peaks above 3000; three proton environments; Peak at 2.0 is CH_3CO, singlet since no H on adjacent C; Peak at 1.3 is CH_3 split into triplet by adjacent CH_2; Peak at 4.0 is CH_2-O split into a quartet by adjacent CH_3

b 4 – four different C environments

c $(CH_3OOCCH_2CH_3)$
triplet at 0.6 – 1.9, relative height 3
quartet at 2.0 – 3.0, relative height 2
singlet at 3.3 – 4.8, relative height 3

8 **a** CH-O 3.1 – 4.3
arene C-H 6.0 – 8.0

b Peak at 3.3 – 4.8 no split
two peaks in range 6.0 – 9.0. Allow any split

c 5 peaks
1 in range 160 – 220
3 in range 110 – 160
1 in range 50 – 90

9 **a** C_4H_8O

b $CH_3CH_2COCH_3$
since 57 is $CH_3CH_2CO^+$
OR 43 is CH_3CO^+

O 1

1 **a** $Ca^{2+}(aq) + 2OH^-(aq)$

b $Mg^{2+}(aq) + SO_4^{2-}(aq)$

c $K^+(aq) + OH^-(aq)$

d $Ag^+(aq) + NO_3^-(aq)$

e $2Al^{3+}(aq) + 3SO_4^{2-}(aq)$

2 **a** NaBr **b** $Mg(OH)_2$ **c** Na_2S

d BaO **e** $CaCO_3$ **f** $Ca(NO_3)_2$ **g** K_2CO_3

3 **a** When 1 mole of sodium fluoride is formed from 1 mole of $Na^+(g)$ and 1 mole of $F^-(g)$, $915\,kJ\,mol^{-1}$ of energy are released. This is the lattice enthalpy of sodium fluoride.

b The lattice enthalpy becomes more negative as the ionic radii decrease.

4 **a** LiF; Li^+ has a smaller radius than Na^+ and attracts F^- ions more strongly.

b Na_2O; Na^+ has a smaller radius than Rb^+ and attracts O^{2-} more strongly.

c MgO; Mg^{2+} is smaller and more highly charged than Na^+ and attracts O^{2-} more strongly.

d KF; F^- has a smaller radius than Cl^- and attracts K^+ more strongly.

5 **a** Li^+ attracts water molecules more strongly than Na^+ because of its smaller size.

b Mg^{2+} attracts water molecules more strongly than Ca^{2+} because of its smaller size.

c Ca^{2+} and Na^+ have similar sizes, but Ca^{2+} is more highly charged and so attracts water molecules more strongly.

6 **a** The ions in the lattice attract each other less strongly as the size of the anion increases from F^- to Cl^-.

b $\Delta_{hyd}H$ becomes less exothermic as the anion becomes bigger and attracts water molecules less strongly.

O 2

1 **a** $HNO_3 + H_2O \rightarrow H_3O^+ + NO_3^-$
 acid   base

b $NH_3 + H_2O \rightarrow NH_4^+ + OH^-$
 base   acid

c $NH_4^+ + OH^- \rightarrow NH_3 + H_2O$
 acid   base

d $SO_4^{2-} + H_3O^+ \rightarrow HSO_4^- + H_2O$
 base   acid

e $H_2O + H^- \rightarrow H_2 + OH^-$
 acid   base

f $H_3O^+ + OH^- \rightarrow 2\,H_2O$
 acid   base

g $NH_3 + HBr \rightarrow NH_4^+ + Br^-$
 base   acid

h $H_2SO_4 + HNO_3 \rightarrow HSO_4^- + H_2NO_3^+$
 acid   base

2 a acid–base   **b** acid–base

  c redox   **d** redox

3 a 2.0   **b** 0.7   **c** 0.4   **d** 0.4

4 a 2.9   **b** 3.0   **c** 3.6   **d** 1.7

5 a strong acid – hydrochloric acid
weak acid – nitrous acid (nitric(III) acid)

  b The position of the equilibrium for the reaction of the strong acid with water is completely to the right.

    The position of the equilibrium for the reaction of the weak acid with water is more to the left.

6 a 14   **b** 13.3

7 a In alkaline solution, the equilibrium shifts to the right as $H^+(aq)$ is removed by reaction with $OH^-(aq)$, so the indicator will be present as the pink In^- form.

  b $K_a = \dfrac{[H^+(aq)][In^-(aq)]}{[HIn(aq)]}$

  c pH at end point = 9.3

○ 3

1 A buffer solution is one that resists changes in pH when a small amount of acid or alkali is added.

2 a 3.8   **b** 4.7   **c** 3.8

3 a 3.6   **b** 4.3   **c** 5.1

4 a Ethanoate ions from the salt react with the extra $H^+(aq)$ ions to form ethanoic acid and water and so prevent a fall in the pH.

  b The addition of $OH^-(aq)$ ions removes $H^+(aq)$ but these are replaced by further dissociation of the ethanoic acid so the pH will remain constant.

  c Addition of a small amount of water will change the concentration of the acid and the salt by the same factor which means the ratio of [salt] to [acid] will remain constant and the pH will remain constant.

○ 4

1 a $AgI(s) \rightleftharpoons Ag^+(aq) + I^-(aq)$

  b $BaSO_4(s) \rightleftharpoons Ba^{2+}(aq) + SO_4^{2-}(aq)$

  c $PbI_2(s) \rightleftharpoons Pb^{2+}(aq) + 2I^-(aq)$

  d $Fe(OH)_3(s) \rightleftharpoons Fe^{3+}(aq) + 3OH^-(aq)$

2 a $K_{sp} = [Ag^+(aq)][I^-(aq)]\,mol^2\,dm^{-6}$

  b $K_{sp} = [Ba^{2+}(aq)][SO_4^{2-}(aq)]\,mol^2\,dm^{-6}$

  c $K_{sp} = [Pb^{2+}(aq)][I^-(aq)]^2\,mol^3\,dm^{-9}$

  d $K_{sp} = [Fe^{3+}(aq)][OH^-(aq)]^3\,mol^4\,dm^{-12}$

3 $[Ag^+(aq)][BrO_3^-(aq)] = 2.5 \times 10^{-7}\,mol^2\,dm^{-6}$ at 298 K
Less than K_{sp} so no precipitate would be observed.

○ 5

1 a increase   **b** decrease   **c** increase

  d increase   **e** decrease   **f** decrease

2 a molten wax (Liquids have higher entropies than solids.)

  b $Br_2(g)$ (Gases have higher entropies than liquids.)

  c brass (Mixtures have higher entropies than the pure substances.)

  d octane (Complex molecules have higher entropies than simpler molecules.)

3 The entropies increase for the first four alkanes as the molecules become heavier and composed of more atoms (the number of energy levels increases with the number of atoms). Pentane is a liquid and so has a lower entropy than butane.

4 a $\Delta S = +98.8\,JK^{-1}mol^{-1}$, entropy increase because 1 mole of liquid reactant is replaced by 1 mole of gaseous product

  b $\Delta S = +133.8\,JK^{-1}mol^{-1}$, entropy increase because the number of moles of gas doubles during the reaction and a solid has a much lower entropy than gas.

  c $\Delta S = -175.8\,JK^{-1}mol^{-1}$, entropy decrease because the number of moles of gas is reduced by half as the reaction proceeds.

  d $\Delta S = -198.8\,JK^{-1}mol^{-1}$, entropy decrease because the number of moles of gas is reduced by half as the reaction proceeds.

  e Entropy increase; 5 moles of gaseous product are formed.

5

	$\Delta_{sys}S/$ $JK^{-1}mol^{-1}$	$\Delta_{surr}S/$ $JK^{-1}mol^{-1}$	
a	+203	−44	spontaneous: total entropy change positive
b	+63	+329	spontaneous: total entropy change positive
c	+25	−604	not spontaneous: total entropy change negative
d	+209	+416	spontaneous: total entropy change positive
e	−4	−6.7	not spontaneous: total entropy change negative

DM 1

1 a $1s^22s^22p^63s^23p^6$

b $1s^22s^22p^63s^23p^63d^4$

c $1s^22s^22p^63s^23p^63d^7$

2 a $1s^22s^22p^63s^23p^63d^6$

b $1s^22s^22p^63s^23p^63d^2$

c $1s^22s^22p^63s^23p^63d^3$

3 Sc^{3+}: $1s^22s^22p^63s^23p^6$

Ni^{2+}: $1s^22s^22p^63s^23p^63d^7$

Zn^{3+}: $1s^22s^22p^63s^23p^63d^{10}$

Only Nickel has an ion with a partially filled d-orbital.

4 a $1s^22s^22p^63s^23p^63d^{10}4s^24p^64d^8$

b $1s^22s^22p^63s^23p^63d^{10}4s^24p^64d^{10}$

c $1s^22s^22p^63s^23p^63d^{10}4s^24p^64d^{10}$

5 Mn^{2+}: $[Ar]3s^23p^63d^5$

Mn^{3+}: $[Ar]3s^23p^63d^4$

Mn^{2+} has a half filled 3d orbital, which is more stable than the $3d^4$ electron configuration of Mn^{3+}.

6 a $0.02655 \times 0.0150 = 3.983 \times 10^{-4}$ moles

b $6 \times 3.983 \times 10^{-4} = 2.390 \times 10^{-3}$ moles

c $20 \times 2.390 \times 10^{-3} \times 56 = 2.2676\,g$

d $2.676 \div 11.05 \times 100 = 24.22\%$

DM 2

1 Heterogeneous catalysis. The hydrogen peroxide forms weak bonds with MnO_2, the catalytic step of the reaction occurs, and the weak bonds break.

2 a $S_2O_8^{2-} + 2I^- \rightarrow I_2 + 2SO_4^{2-}$

b Fe^{2+} is not used up in the reaction.

c Attempt the reaction without the presence of Fe^{2+} ions. If the reaction does not occur or occurs as a slower reaction rate, Fe^{2+} is acting as a catalyst.

DM 3

> ## Raman spectroscopy – it's more than skin deep
>
> 1 4.48×10^{-19}
> 2 blue

1 a Ti^{4+} [Ar] no 3d electron transitions possible

b Sc^{3+} [Ar] Zn^{2+} [Ar] $3d^{10}$ Cu^+ [Ar] $3d^{10}$ in all ions no 3d electron transitions possible, hence colourless/white

2 Cu^{2+} has a partially filled d-orbital ($[Ar]3d^9$) where as Zn^{2+} does not ($[Ar]3d^{10}$). Compounds of copper are therefore coloured, whilst compounds of zinc are not.

3 a monodentate b bidentate c monodentate

DM 4

1 a i 3.16 V   ii 1.10 V   iii 0.32 V

b i $Ag^+(aq)/Ag(s)$ ii $Cu^{2+}(aq)/Cu(s)$

iii $Fe^{2+}(aq)/Fe(s)$

c i $Ag^+(aq) + e^- \rightarrow Ag(s)$
$Mg^{2+}(aq) + 2e^- \rightarrow Mg(s)$
$2Ag^+(aq) + Mg(s) \rightarrow 2Ag(s) + Mg^{2+}(aq)$

ii $Cu^{2+}(aq) + 2e^- \rightarrow Cu(s)$
$Zn^{2+}(aq) + 2e^- \rightarrow Zn(s)$
$Cu^{2+}(aq) + Zn(s) \rightarrow Cu(s) + Zn^{2+}(aq)$

iii $Fe^{2+}(aq) + 2e^- \rightarrow Fe(s)$
$Zn^{2+}(aq) + 2e^- \rightarrow Zn(s)$
$Fe^{2+}(aq) + Zn(s) \rightarrow Fe(s) + Zn^{2+}(aq)$

2 a yes   b yes   c no   d yes

3 a $I_2 + 2S_2O_3^{2-} \rightarrow 2I^- + S_4O_6^{2-}$

b $Cl_2 + 2OH^- \rightarrow ClO^- + Cl^- + H_2O$

c $2H^+ + VO_2^+ \rightarrow VO^{2+} + H_2O$

d $Cr_2O_7^{2-} + 14H^+ + 6I^- \rightarrow 3I_2 + 2Cr^{3+} + 7H_2O$

e $2H^+ + H_2O_2 + Fe^{2+} \rightarrow 2H_2O + Fe^{3+}$

4 a all are feasible

b i $CH_4(g) + 2O_2(g) \rightarrow CO_2(g) + 2H_2O(l)$

ii $2CH_3OH(aq) + O_2(g) \rightarrow 2HCHO(aq) + 2H_2O(l)$

iii $2HCHO + O_2(g) \rightarrow 2HCOOH(aq)$

DM 5

1 a At the edge of the water surface where it is in contact with oxygen in the air and the steel. This is a cathode site.

b Under the middle of the drop. This is an anode site.

d

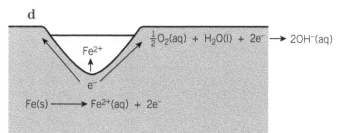

601

2 a This is the anode site where iron is oxidised and corrodes. Electrons flow away from this site to areas with higher oxygen concentration to where the reduction of oxygen occurs at the cathode site.

b It is not possible to see how bad the deterioration is without actually removing and checking the fastenings (coach screws).

c The head, unlike where the bolt was screwed into the wood, is always in contact with oxygen from the air (and regularly with water), therefore this is unlikely to be the anodic site and corrode.

3 Corrosion is electrochemical in nature and ions are necessary to act as a salt bridge between anode and cathode sites. Salt water contains a high concentration of ions.

4 Electrochemical cell set up between the two different metals in the presence of sea water.

Aluminium has a much greater tendency to be oxidised (lose electrons) than mercury and so aluminium becomes an anode site and corrodes.

DM 6

1 a 2 **b** 4 **c** 6 **d** 6

2 a +1 **b** +2 **c** +3 **d** +3

3 a $[Mn(H_2O)_6]^{2+}$ **b** $[Zn(NH_3)_4]^{2+}$
c $[FeF_6]^{3-}$ **d** $[Cr(H_2O)_5OH]^2$

4 a hexaaquavanadium(III) ion
b hexacyanoferrate(II) ion
c tetrachlorocobaltate(II) ion
d diamminesilver(I) ion
e tetraaquadichlorochromium(III) ion

5 a 6 **b** +4
c hexachlorotitanate(IV) ion
d

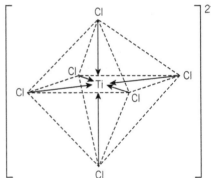

CD 1

1 Electrons are excited to higher energy levels/an excited state.

2 Yellow is reflected/transmitted, the complementary colour of yellow (violet) is absorbed.

3 a Electrons from π-bonds/p-oribitals are delocalised over several carbon atoms in a conjugated system.

b Reduces the energy gap between ground and excited state molecule, absorbs light in the visible region.

4 There is no conjugated system, so the energy gap between ground and excited state is very large and molecule absorbs in the ultraviolet part of the spectrum.

CD 2

1 Six p-electrons are not associated with a specific pair of carbon atoms but can spread out over all six carbon atoms, forming rings of charge above and below the plane of the benzene molecule.

2 a Bond angle is 120°/the molecule is planar – three groups of electrons repel each other and get as far apart as possible.

b Bond lengths are equal – Kekulé structure predicts C=C shorter than C—C.

3 −240

In cyclohexa-1,3-diene, there are two double bonds in a six-carbon ring. Double and single bonds do not form a delocalised benzene ring, so there is no extra stability. Enthalpy of hydrogenation will be two times the enthalpy of hydrogenation of cyclohexene.

4 a

b Three as the first two structures would be identical.

CD 3

Aromaticity

1 a $-120 \times 4 = -480\,kJ\,mol^{-1}$
b carbon–carbon single bonds would be longer than carbon–carbon double bonds
2 A, C, and D

1 a propylbenzene

  b 1,4-dimethylbenzene

  c 1-ethyl-3-methylbenzene

2 a   b

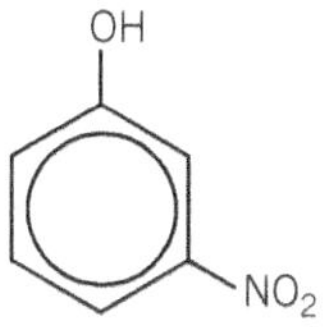

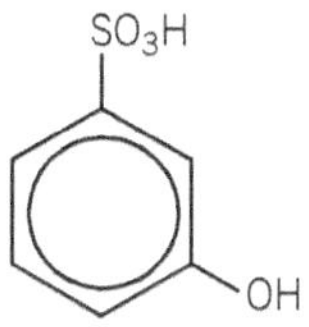

  c

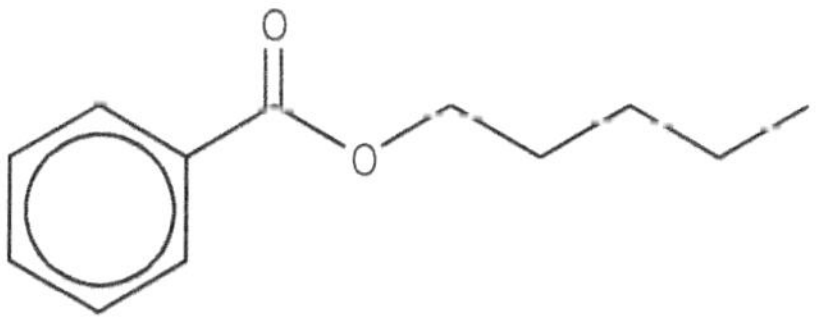

3 a $C_6H_6O_2$ b $C_{14}H_8O_5$

4 10 p-electrons are involved. The electrons are not associated with a specific pair of carbon atoms but can spread out over all 10 carbon atoms. Form rings of charge above and below the plane of the carbon atoms.

CD 4

1 a electrophilic substition

  b i presence of iron catalyst/FeBr$_3$/AlCl$_3$

    ii heat

2 a   b

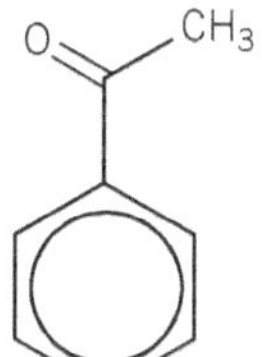

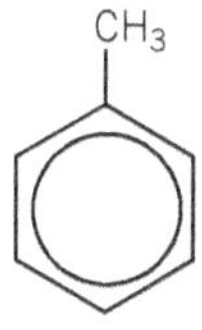

3 a 2-chloropropane   b butanoyl chloride

4

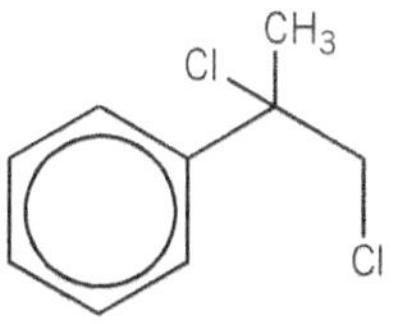

Benzene ring will not react at room temp or without a catalyst but the alkene group will.

CD 5

1 a concentrated nitric acid, concentrated sulfuric acid, temperature <55 °C

  b concentrated sulfuric acid and reflux

2 Nitro: modify the chromophore to produce a different colour

  Amine: modify the chromophore to produce a different colour

Sulfonate: increase solubility of the dye OR form bonds to (positively charged groups) on fibres

3 a

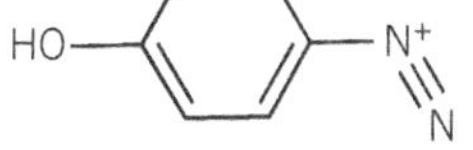

  b i

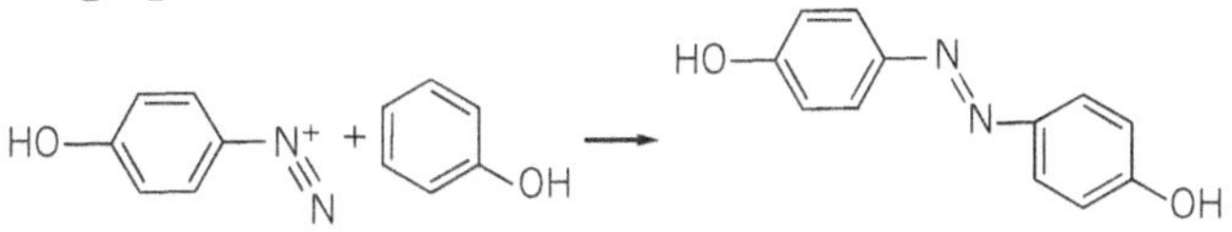

    ii

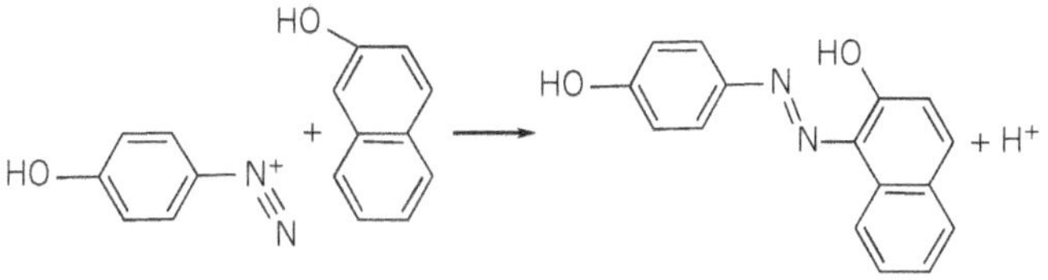

4

	Diazonium compound(s)	Coupling agent
chrysoidine	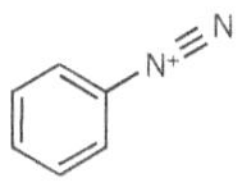	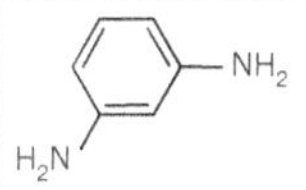
Acid Clack 1	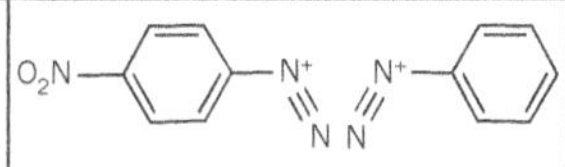	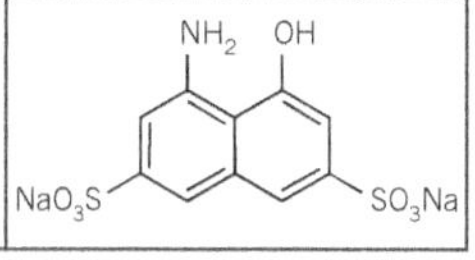

CD 6

1 sulfonate group   2 polyester

3 dye A and fabric E (polyester) – fabric is non-polar and dye is small and non-polar
dye B and fabric F (cotton/cellulose) – fabric has frequent OH groups and dye has several NH$_2$ groups and is linear
dye C and fabric D (polyamide) – fabric will have NH$_3^+$ groups at end of molecule (in acidic conditions) and dye has negatively charged SO$_3^-$ groups

4 Strongest forces between dye molecules will be instantaneous dipole–induced dipole. Strongest force between water molecules is hydrogen bonding. Strongest forces possible between dye and water is limited hydrogen bonding. Energy released by forming these bonds is insufficient to break bonds between solute molecules/water molecules.

CD 7

1 They are formed when three carboxylic acid molecules form ester links with a single molecule that contains three alcohol groups.

2

$$H_2C-O-\overset{\overset{\displaystyle O}{\|}}{C}-(CH_2)_7CH=CHCH_2\,CH=CH(CH_2)_4CH_3$$

$$HC-O-\overset{\overset{\displaystyle O}{\|}}{C}-(CH_2)_{14}CH_3$$

$$H_2C-O-\overset{\overset{\displaystyle O}{\|}}{C}-(CH_2)_{14}CH_3$$

3

$$H_2C-O-\overset{\overset{\displaystyle O}{\|}}{C}-(CH_2)_7\,CH=CHCH_2CH=CH(CH_2)_4CH_3$$

$$HC-O-\overset{\overset{\displaystyle O}{\|}}{C}-(CH_2)_{14}CH_3 \qquad + 3NaOH$$

$$H_2C-O-\overset{\overset{\displaystyle O}{\|}}{C}-(CH_2)_{14}CH_3$$

↓

$$H_2C-OH \qquad Na^+O^-\overset{\overset{\displaystyle O}{\diagup}}{\diagdown}C-(CH_2)_7CH=CH(CH_2)_7CH_3$$

$$HC-OH + \qquad \qquad \overset{\overset{\displaystyle O}{\diagup}}{\diagdown}C-(CH_2)_{10}CH_3$$
$$\qquad\qquad\quad Na^+O^-$$

$$H_2C-OH \qquad \overset{\overset{\displaystyle O}{\diagup}}{\diagdown}C-(CH_3)_{10}CH_3$$
$$\qquad\qquad Na^+O^-$$

4 Intermolecular bonds between oil/fat molecules are instantaneous dipole–induced dipole. Intermolecular bonds in water are hydrogen bonds. Only instantaneous dipole–induced dipole/permanent dipole–induced dipole are possible between water and oil/fat. Energy released when oil/fat dissolves is not sufficient to break bonds in solute and solvent.

CD 8

1 high boiling-point liquid on an inert solid support

2 Changes in temperature will affect the retention times temperature needs to be known in order to calibrate instrument using compounds at the same temperature.

3 a Z: 42%, E: 58%

b mass / m/z ratio of each peak will be the same

4 a

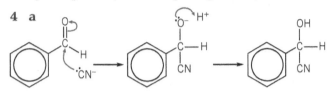

b Methyl esters have lower boiling points. This is necessary in order for molecules to vapourise in the chromatograph.

c E as it will have the highest boiling point and therefore will spend more time in the stationary (liquid) phase/most volatile components usually emerge first.

CD 9

1 a Mix Fehling's solutions A and B. Add mixture to the solution and warm.

b appearance of a brick-red precipitate

2 a $CH_3-CH_2-CH_2-CH(OH)CN$

b $CH_3-CH_2-CH(OH)(CN)-CH_2-CH_3$

3 cyanohydrins (allow 2-hydroxynitriles)

4 a

b CN^- has a lone pair of electrons which bonds to δ+ carbon atom. Two molecules combine to form a single product molecule.

5 CN^- reactions enable C—C bonds to form and this extends the length of the carbon chain. CN^- groups can be converted into other important functional groups.

CD 10

1 A: phenol and aldehyde, B: carboxylic acid and aldehyde

2 a paracetamol: phenol, amide
ibuprofen: carboxylic acid

b Paracetamol will react with iron chloride to give purple colour/can be hydrolysed under alkaline conditions to form ammonia/can be used as a coupling agent to form azo dyes. Ibuprofen does not react under these conditions.

Ibuprofen will react with sodium carbonate to form carbon dioxide, paracetamol does not.

3

4 Step 1: concentrated nitric acid, concentrated sulfuric acid, temperature $< 55\,^{\circ}C$

Step 2: tin, concentrated hydrochloric acid, reflux

CD 11

Nucleophilic addition reactions and optical isomers

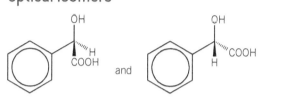

1 a substitution **b** elimination **c** condensation
 d oxidation **e** hydrolysis

2 a Route 1 = 25.2%, Route 2 = 30.1%, so Route 2

 b two from: availability and cost of reagents, atom economy of reactions, separation costs, safety issues, time taken/rate of reactions, value of any by-products or co-products

3 a A: $CH_3CHBrCH_3$ B: $CH_3CHCNCH_3$

 b Reaction 1: addition

 Reaction 2: substitution

 Reaction 3: hydrolysis

Index